Construction Materials

Fourth edition

The biggest thing university taught me was that, with perseverance and a book, you can do anything you want to.

It doesn't matter what the subject is; once you've learnt how to study, you can do anything you want.

George Laurer, inventor of the barcode

Construction Materials

Their nature and behaviour

Fourth edition

Edited by

Peter Domone and
John Illston

 Spon Press
an imprint of Taylor & Francis
LONDON AND NEW YORK

First published as *Concrete Timber and Metals* 1979
by Chapman and Hall

Second edition published 1994
by Chapman and Hall

Third edition published 2001
by Spon Press

This edition published 2010
by Spon Press
2 Park Square, Milton Park, Abingdon, Oxon OX14 4RN

Simultaneously published in the USA and Canada
by Spon Press
711 Third Avenue, New York, NY 10017, USA

Spon Press is an imprint of the Taylor & Francis Group, an informa business

© 2010 Spon Press

Typeset in Sabon by
Graphicraft Limited, Hong Kong
Printed and bound in Great Britain by
MPG Books Group, UK

This publication presents material of a broad scope and applicability.
Despite stringent efforts by all concerned in the publishing process, some
typographical or editorial errors may occur, and readers are encouraged to
bring these to our attention where they represent errors of substance. The
publisher and author disclaim any liability, in whole or in part, arising from
information contained in this publication. The reader is urged to consult
with an appropriate licensed professional prior to taking any action or
making any interpretation that is within the realm of a licensed
professional practice.

British Library Cataloguing in Publication Data
A catalogue record for this book is available from the British Library

Library of Congress Cataloging-in-Publication Data
Construction materials : their nature and behaviour / [edited by]
Peter Domone and J. M. Illston. – 4th ed.
p. cm.
Includes bibliographical references.
1. Building materials. I. Domone, P. L. J. II. Illston, J. M.
TA403.C636 2010
624.1'8–dc22   2009042708

ISBN10: 0-415-46515-X (hbk)
ISBN10: 0-415-46516-8 (pbk)
ISBN10: 0-203-92757-5 (ebk)

ISBN13: 978-0-415-46515-1 (hbk)
ISBN13: 978-0-415-46516-8 (pbk)
ISBN13: 978-0-203-92757-1 (ebk)

Contents

Contents

Contributors

Professor Gordon Airey
Nottingham Transportation Engineering Centre
Department of Civil Engineering
University of Nottingham
University Park
Nottingham
NG7 2RD
(Bituminous materials)

Professor John Dinwoodie
16 Stratton Road
Princes Risborough
Bucks
HP27 9BH
(Timber)

Graham Dodd
Arup Materials Consulting
13 Fitzroy Steet
London
W1T 4BQ
(Glass)

Dr Peter Domone
Dept of Civil, Environmental and Geomatic
  Engineering
University College London
Gower St
London
WC1E 6BT
(Editor and Fundamentals, Metals and
alloys, Concrete, and Selection, use and
sustainability)

Professor Len Hollaway
Faculty of Engineering and Physical Sciences
University of Surrey
Guildford
Surrey
GU2 7XH
(Polymers, Polymer composites)

Dr Phil Purnell
School of Civil Engineering
University of Leeds
Woodhouse Lane
Leeds
LS2 9JT
(Fibre-reinforced cements and composites)

Dr Bob de Vekey
215 Hempstead Road
Watford
Herts
WD17 3HH
(Masonry)

G D Airey
Gordon Airey is Professor of Pavement Engineering Materials at the University of Nottingham and Deputy Director of the Nottingham Transportation Engineering Centre (NTEC). After graduating from the University of Cape Town with a First-Class Honours Degree in Civil Engineering, he worked for a major research organisation in South Africa before taking a research associate position at the University of Nottingham. He obtained his Ph.D. from the University of Nottingham in 1997 before being appointed to the academic staff in the Department of Civil Engineering in 1998. Professor Airey's research is in the field of pavement engineering with particular emphasis on the rheological characterisation and durability of bituminous materials. He has over a hundred and fifty technical publications and has attracted financial support from research councils, government and industry.

J M Dinwoodie OBE
Professor John Dinwoodie graduated in Forestry from the University of Aberdeen, and was subsequently awarded his M.Tech. in Non-Metallic Materials from Brunel University, and both his Ph.D. and D.Sc.

in Wood Science subjects from the University of Aberdeen. He carried out research at the UK Building Research Establishment (BRE) for a period of 35 years on timber and wood-based panels with a special interest in the rheological behaviour of these materials. For this work he was awarded with a special merit promotion to Senior Principal Scientific Officer. Following his retirement from BRE in 1995, he was employed for ten years as a consultant to BRE to represent the UK in the preparation of European standards for wood-based panels. In 1985 he was awarded the Sir Stuart Mallinson Gold Medal for research on creep in particleboard and was for many years a Fellow of the Royal Microscopical Society and a Fellow of the Institute of Wood Science. In 1994 he was appointed an Honorary Professor in the Department of Forest Sciences, University of Wales, Bangor, and in the same year was awarded an O.B.E. He is author, or co-author, of over one-hundred-and-fifty technical papers and author of three textbooks on wood science and technology.

G S Dodd

Graham Dodd is a Chartered Mechanical Engineer with twenty years experience in the structural design of glass, in applications from high-rise façades to one-off suspended sculptures. Having worked in appliance manufacturing, glass processing, contracting and façade engineering, he is now responsible for design advice on a wide range of materials and production processes, particularly in relation to façades, within the Materials Consulting group of Arup.

P L J Domone

Dr Peter Domone graduated in civil engineering from University College London, where he subsequently completed a Ph.D. in concrete technology. After a period in industrial research with Taylor Woodrow Construction Ltd, he was appointed to the academic staff at UCL in 1979, first as lecturer and then as senior lecturer in concrete technology. He teaches all aspects of civil engineering materials to undergraduate students, and his principle research interests have included non-destructive testing, the rheology of fresh concrete, high-strength concrete and more recently, self-compacting concrete. As well editing the third edition and now the fourth edition of this book, he has over fifty technical publications including contributions to five books on concrete technology. He is also a course tutor on the Institute of Concrete Technology diploma in Advanced Concrete Technology distance learning course.

L C Hollaway

Professor Len Hollaway is Emeritus Professor of Composite Structures in the Faculty of Engineering and Physical Sciences – Civil Engineering, University of Surrey and in 1996 he was appointed as visiting Research Professor at the University of Southampton. He is a Fellow of the Institution of Civil Engineers, a Fellow of the International Institute for FRP in Construction, Member of the Institution of Structural Engineers and a Euro Engineer. He has considerable research experience in advanced polymer composite systems. He has had over 200 refereed technical papers published and is the author, co-author or editor of eight books on various aspects of polymer composites in the civil engineering industry.

P Purnell

Dr Phil Purnell was appointed to a Readership in the School of Civil Engineering at Leeds University in 2009, having previously been a Senior Lecturer at Warwick University. He took a Ph.D. from Aston University in 1998 and a B.Eng. in Engineering (Civil) from Exeter University in 1994. His research concerns composite resilience and durability, including durability of fibre-reinforced cements and polymers, non-destructive testing of concrete, novel applications for cementitious materials and life-cycle assessment of construction components. His teaching has included general engineering concepts, civil engineering materials and life-cycle assessment to all undergraduate and graduate levels. He has contributed to books on aspects of concrete technology such as durability and recycling and is Director of the Institute for Resilient Infrastructure.

R C de Vekey

Dr Bob de Vekey studied chemistry at Hatfield Polytechnic and graduated with the Royal Society of Chemistry. He continued his studies at Imperial College, London gaining a D.I.C. in materials science and an external Ph.D. for work on glass ceramics. At the Building Research Establishment (BRE) he worked on the manufacturing and performance aspects of masonry and its components and fibre reinforced cements. Between 1978 and 2000 he led a section concerned mainly with safe design, structural behaviour, durability, testing and standardisation of brick and block masonry buildings and their components. From September 2000 he has been an associate to the BRE. He has an extensive catalogue of written work including research papers, guidance documents, draft developmental standards/codes and contributions to several books.

Acknowledgements

I must first of all acknowledge the tremendous support given to me by my wife, Jenny, and my children, James and Sarah, during the production of this book; this would not have been completed without their encouragement and tolerance during the many hours that I have spent away from them in my study.

My thanks once again go to John Illston for his work and vision that resulted in the first two editions of this book and for his encouragement and advice during the preparation of both the third and this edition. My thanks must also go to all the contributors, both those who have revised and updated their previous contributions and those who have contributed for the first time. They have all done an excellent job and any shortfalls, errors or omissions in the book are entirely my fault.

Finally I must acknowledge the advice and inspiration of my colleagues at UCL and elsewhere, but particularly of the students at UCL from whom I have learnt so much.

Peter Domone

I wish to express my appreciation to the following individuals who most kindly read my draft on certain topics and who subsequently provided me with excellent advice: Chris Holland of BRE for his assistance on the structural use of timber and the history of the development of Greenweld; David Hunt, formerly of the University of the South Bank, for his detailed help on the challenging subject of modelling mechano-sorptive behaviour of timber under load and moisture change; and Peter Jackman, Technical Director of International Fire Consultants Limited, for much guidance on the confusing issue of fire resistance of timber in the UK and Europe.

Further thanks are due to the following people who answered my specific enquiries or who supplied me with relevant publications: John Brazier formerly of BRE; Vic Kearley of TRADA; Alastair Kerr of the WPPF; Nicolas Llewelin of the TTF; Tim Reynolds and Ed Suttie of BRE; and lastly John Wandsworth of Intermark Ltd.

I am indebted to the Building Research Establishment (BRE) for permission to use many plates and figures from the BRE collection and also to a number of publishers for permission to reproduce figures in journals.

Lastly to the many colleagues who have so willingly helped me in some form or other in the production of the first three editions of this text, I would like to record my very grateful thanks as those editions formed a very sound foundation on which to construct the fourth.

John Dinwoodie

Preface

Peter Domone

This book is an updated and extended version of the third edition, which was published in 2001. This has proved to be as popular and successful as the first two editions, but the continuing advances in construction materials technology and uses, not least in the many factors and issues relating to the sustainability of construction, have resulted in the need for this new fourth edition.

The first edition was published under the title *Concrete, Timber and Metals* in 1979. Its scope, content and form were significantly changed for the second edition, published in 1994, with the addition of three further materials – bituminous materials, masonry and fibre composites – with a separate part of the book devoted to each material, following a general introductory part on 'Fundamentals'.

This overall format was well received by both students and teachers, and was retained in the third edition, with a short section on polymers added. In this new edition, this format has again been retained; the principal modifications and extensions are:

- the 'Fundamentals' section has been reformatted into chapters which can more readily be studied independently if required
- a new section on glass has been added, reflecting its increasing use as a structural material;
- for each material the issues concerned with end-of-life and recycling, now major considerations, have been discussed
- a new section on 'Selection, use and sustainability' has been added, which compares the mechanical properties of all the materials and considers some of the factors relating to their selection for use and the consequences for society and the environment. This brings together much of the property data presented in the individual sections, and leads on to issues of sustainability that will increasingly dominate the life and careers of many who read this book.

Three of the contributors to the third edition, John Dinwoodie (timber), Len Hollaway (polymers and polymer composites) and Bob de Vekey (masonry) were able and willing to contribute again. Others were not due to changes in interests or retirement, but fortunately, Gordon Airey (bituminous materials) and Phil Purnell (fibre-reinforced cements and composites) have stepped in and taken over their respective sections. Graham Dodd has contributed the new section on glass.

The co-author of the first edition, editor of the second edition and inspiration for the third edition, John Illston, is still flourishing in his retirement and again provided encouragement for me to continue as editor.

Objectives and scope

As with the previous editions, the book is addressed primarily to students taking courses in civil or structural engineering, where there is a continuing need for a unified treatment of the kind that we have again attempted. We believe that the book provides most if not all of the information required by students for formal courses on materials throughout three- or four-year degree programmes, but more specialist project work in third or fourth years may require recourse to the more detailed texts that are listed in 'Further reading' at the end of each section. We also believe that our approach will continue to provide a valuable source of interest and stimulation to both undergraduates and graduates in engineering generally, materials science, building, architecture and related disciplines.

The objective of developing an understanding of the behaviour of materials from a knowledge of their structure remains paramount. Only in this way can information from mechanical testing, experience in processing, handling and placing, and materials science, i.e. empiricism, craft and science, be brought together to give the sound foundation for materials technology required by the practitioner.

The 'Fundamentals' section provides the necessary basis for this. Within each of the subsequent sections on individual materials, their structure and composition from the molecular level upwards is discussed, and then the topics of manufacture and processing, deformation, strength and failure, durability and recycling are considered. A completely unified treatment for each material is not possible owing to their different natures and the different requirements for manufacture, processing and handling, but a look at the contents list will show how each topic has been covered and how the materials can be compared and contrasted. Cross-references are given throughout the text to aid this, from which it will also be apparent that there are several cases of overlap between materials, for example concrete and bituminous composites use similar aggregates, and Portland cement is a component of masonry, some fibre composites and concrete. The final section enables comparison of mechanical properties of the materials, from which it is possible to get an idea of how each fits into the broad spectrum available to construction engineers, and then discusses some of the sustainability issues relating to all the materials.

It is impossible in a single book to cover the field of construction materials in a fully comprehensive manner. Not all the materials used in construction are included, and although some design considerations are included the book is in no way a design guide or manual – there are more than adequate texts on this available for all materials that we have included. Neither is this book a manual of good practice. Although some tables of the various properties discussed have been included, we have not attempted to provide a compendium of materials data – again this can be found elsewhere.

Nevertheless we hope that we have provided a firm foundation for the application and practice of materials technology.

Levels of information

The structure of materials can be described on dimensional scales varying from the smallest, atomic or molecular, through materials structural to the largest, engineering. *Figure 0.1* shows that there is considerable overlap between these for the different materials that we consider in this book.

THE MOLECULAR LEVEL

This considers the material at the smallest scale, in terms of atoms or molecules or aggregations of molecules. It is very much the realm of materials science, and a general introduction for all materials is given in Part 1 of the book. The sizes of the particles range from less than 10^{-10} to 10^{-2} m, clearly an enormous range. Examples occurring in this book include the crystal structure of metals, cellulose molecules in timber, calcium silicate hydrates in

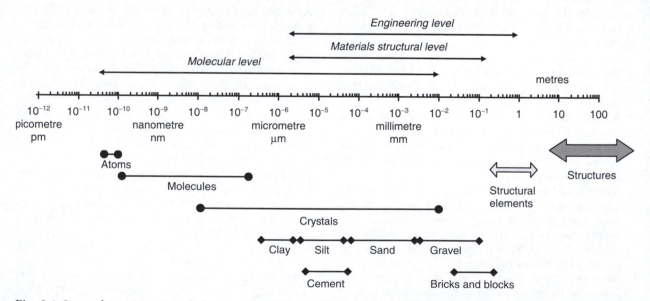

Fig. 0.1 *Sizes of constituents and components of structural materials and the levels considered in the discussions in this book.*

hardened cement paste and the variety of polymers, such as polyvinyl chloride, included in fibre composites.

As shown in Part 1 consideration of established atomic models leads to useful descriptions of the forms of physical structure, both regular and disordered, and of the ways in which materials are held together. Chemical composition is of fundamental importance in determining this structure. This may develop with time as chemical reactions continue; for example, the hydration of cement is a very slow process and the structure and properties show correspondingly significant changes with time. Chemical composition is of special significance for durability, which is often determined, as in the cases of timber and metals, by the rate at which external substances such as oxygen or acids react with the chemicals of which the material is made.

Chemical and physical factors also come together in determining whether or not the material is porous, and what degree of porosity is present. In materials such as bricks, timber and concrete, important properties such as strength and rigidity are inversely related to their porosities. Similarly, there is often a direct connection between permeability and porosity.

Some structural phenomena, such as dislocations in metals, are directly observable by microscopic and diffractometer techniques, but more often mathematical and geometrical models are employed to deduce both the structure of the material and the way in which it is likely to behave. Some engineering analyses, like fracture mechanics, come straight from molecular scale considerations, but they are the exception. Much more often the information from the molecular level serves to provide mental pictures that aid engineers' understanding so that they can deduce likely behaviour under anticipated conditions. In the hands of specialists knowledge of the chemical and physical structure may well offer a route to the development of better materials.

MATERIALS STRUCTURAL LEVEL

This level is a step up in size from the molecular level, and the material is considered as a composite of different phases, which interact to realise the behaviour of the total material. This may be a matter of separately identifiable entities within the material structure as in cells in timber or grains in metals; alternatively, it may result from the deliberate mixing of disparate parts, in a random manner in concrete or asphalt or some fibre composites, or in a regular way in masonry. Often the material consists of particles such as aggregates distributed in a matrix such as hydrated cement or bitumen. The dimensions of the particles differ considerably, from the wall thickness of a wood cell at 5×10^{-6} m to the length of a brick at 0.225 m. Size itself is not an issue; what matters is that the individual phases can be recognised independently.

The significance of the materials structural level lies in the possibility of developing a more general treatment of the materials than is provided from knowledge derived from examination of the total material. The behaviours of the individual phases can be combined in the form of multiphase models that allow the prediction of behaviour outside the range of normal experimental observation. In formulating the models consideration must be given to three aspects.

1. Geometry: the shape, size and concentration of the particles and their distribution in the matrix or continuous phase.
2. State and properties: the chemical and physical states and properties of the individual phases influence the structure and behaviour of the total material.
3. Interfacial effects. The information under (1) and (2) may not be sufficient because the interfaces between the phases may introduce additional modes of behaviour that cannot be related to the individual properties of the phases. This is especially true of strength, the breakdown of the material often being controlled by the bond strength at an interface.

To operate at the materials structural level requires a considerable knowledge of the three aspects described above. This must be derived from testing the phases themselves, and additionally from interface tests. While the use of the multiphase models is often confined to research in the interest of improving understanding, it is sometimes taken through into practice, albeit mostly in simplified form. Examples include the estimation of the elastic modulus of concrete, and the strength of fibre composites.

THE ENGINEERING LEVEL

At the engineering level the total material is considered; it is normally taken as continuous and homogeneous and average properties are assumed throughout the whole volume of the material body. The materials at this level are those traditionally recognised by construction practitioners, and it is the behaviour of these materials that is the endpoint of this book.

The minimum scale that must be considered is governed by the size of the representative cell, which is the minimum volume of the material that represents the entire material system, including its regions of disorder. The linear dimensions of this cell varies considerably, from say 10^{-6} m for metals to 0.1 m for concrete and 1 m for masonry. Properties measured over volumes greater than the unit cell can be taken to apply to the material at large. If the properties are the same in all directions then the material is isotropic and the representative cell is a cube, while if the properties can only be described with reference to orientation, the material is anisotropic, and the representative cell may be regarded as a parallelepiped.

Most of the technical information on materials used in practice comes from tests on specimens of the total material, which are prepared to represent the condition of the material in the engineering structure. The range of tests, which can be identified under the headings used throughout this book, includes strength and failure, deformation and durability. The test data are often presented either in graphical or mathematical form, but the graphs and equations may neither express the physical and chemical processes within the materials, nor provide a high order of accuracy of prediction. However, the graphs or equations usually give an indication of how the property values are affected by significant variables, such as the carbon content of steel, the moisture content of timber, the fibre content and orientation in composites or the temperature of asphalt. It is extremely important to recognise that the quality of information is satisfactory only within the ranges of the variables used in the tests. Extrapolation beyond those ranges is very risky and all too easy to do when using best-fit equations generated by tools contained within spreadsheet software. This is a common mistake made not only by students, but also by more experienced engineers and technologists who should know better.

A note on units

In common with all international publications, and with national practice in many countries, the SI system of units has been used throughout this text. Practice does however vary between different parts of the engineering profession and between individuals over whether to express quantities which have the dimensions of [force]/[length]^2 in the units of its constituent parts, e.g. N/m^2, or with the internationally recognised combined unit of the Pascal (Pa). In this book the latter is generally used, but you may find the following relationships useful whilst reading:

$$1 \text{ Pa} = 1 \text{ N/m}^2 \text{ (by definition)}$$
$$1 \text{ kPa} = 10^3 \text{ Pa} = 10^3 \text{ N/m}^2 = 1 \text{ kN/m}^2$$
$$1 \text{ MPa} = 10^6 \text{ Pa} = 10^6 \text{ N/m}^2 = 1 \text{ N/mm}^2$$
$$1 \text{ GPa} = 10^9 \text{ Pa} = 10^9 \text{ N/m}^2 = 1 \text{ kN/mm}^2$$

The magnitude of the unit for a particular property is normally chosen such that convenient numbers are obtained e.g. MPa (or N/mm^2) strength and GPa (or kN/mm^2) for the modulus of elasticity of structural materials.

PART 1

FUNDAMENTALS

Revised and updated by Peter Domone, with acknowledgements to the previous authors, Bill Biggs, Ian McColl and Bob Moon

Introduction

We conventionally think of a material as being either a solid or a fluid. These states of matter are conveniently based on the response of the material to an applied force. A solid will maintain its shape under its own weight, and resist applied forces with little deformation.[1] An unconfined fluid will flow under its own weight or applied force. Fluids can be divided into liquids and gases; liquids are essentially incompressible and maintain a fixed volume when placed in a container, whereas gases are greatly compressible and will also expand to fill the volume available. Although these divisions of materials are often convenient, we must recognise that they are not distinct, and some materials display mixed behaviour, such as gels, which can vary from near solids to near liquids.

In construction we are for the most part concerned with solids, since we use these to carry the applied or self-weight loads, but we do need to understand some aspects of fluid behaviour, for example when dealing with fresh concrete or the flow of water or gas into and through a material.

Intermediate viscoelastic behaviour is also important.

This first part of the book is aimed at both describing and explaining the behaviour of materials in general, without specifically concentrating on any one type or group of materials. That is the purpose of the later sections. This part therefore provides the basis for the later parts, and if you get to grips with the principles then much of what follows will be clearer.

In the first chapter we start with a description of the building blocks of all materials – atoms – and how they combine in single elements and in compounds to form gases, liquids and solids. We then introduce some of the principles of thermodynamics and the processes involved in changes of state, with an emphasis on the change from liquid to solid. In the next two chapters we describe the behaviour of solids when subjected to load and then consider the structure of the various types of solids used in construction, thereby giving an explanation for and an understanding of their behaviour.

This is followed in subsequent chapters by consideration of the process of fracture in more detail (including an introduction to the subject of fracture mechanics), and then by brief discussions of the behaviour of liquids, viscoelastic materials and gels, the nature and behaviour of surfaces and the electrical and thermal properties of materials.

---

[1] But note that the deformation may still be significant on an engineering scale, as we shall see extensively in this book.

Chapter 1

Atoms, bonding, energy and equilibrium

As engineers we are primarily concerned with the properties of materials at the macrostructural level, but in order to understand these properties (which we will introduce in Chapter 2) and to modify them to our advantage, we need an understanding of the structure of materials at the atomic level through bonding forces, molecules and molecular arrangement. Some knowledge of the processes involved in changes of state, particularly from liquids to solids, is also valuable.

The concept of 'atomistics' is not new. The ancient Greeks – and especially Democritus (*ca.* 460BC) – had the idea of a single elementary particle but their science did not extend to observation and experiment. For that we had to wait nearly 22 centuries until Dalton, Avogadro and Cannizzaro formulated atomic theory as we know it today. Even so, very many mysteries still remain unresolved. So in treating the subject in this way we are reaching a long way back into the development of thought about the universe and the way in which it is put together. This is covered in the first part of this chapter.

Concepts of changes of state are more recent. Engineering is much concerned with change – the change from the unloaded to the loaded state, the consequences of changing temperature, environment, etc. The first scientific studies of this can be attributed to Carnot (1824), later extended by such giants as Clausius, Joule and others to produce ideas such as the conservation of energy, momentum, etc. Since the early studies were carried out on heat engines it became known as the science of thermodynamics,[1]

but if we take a broader view it is really the art and science of managing, controlling and using the transfer of energy – whether the energy of the atom, the energy of the tides or the energy of, say, a lifting rig. The second part of this chapter therefore deals with the concepts of energy as applied to changes of state, from gases to liquid, briefly, and from liquid to solid, more extensively, including consideration of equilibrium and equilibrium diagrams. If these at first seem daunting, you may skip past these sections on first reading, but come back to them, as they are important.

1.1 Atomic structure

Atoms, the building block of elements, consist of a nucleus surrounded by a cloud of orbiting electrons. The nucleus consists of positively charged protons and neutral neutrons, and so has a net positive charge that holds the negatively charged electrons, which revolve around it, in position by an electrostatic attraction.[2] The charges on the proton and electron are equal and opposite (1.602×10^{-19} coulombs) and the number of electrons and protons are equal and so the atom overall is electrically neutral.

Protons and neutrons have approximately the same mass, 1.67×10^{-27} kg, whereas an electron has a mass of 9.11×10^{-31} kg, nearly 2000 times less. These relative densities mean that the size of the nucleus is very small compared to the size of the atom. Although the nature of the electron cloud makes it difficult to define the size of atoms precisely, helium has the smallest atom, with a radius of about

---

[1] In many engineering courses thermodynamics is treated as a separate topic, or not considered at all. But, because its applications set rules that no engineer can ignore, a brief discussion is included in this chapter. What are these rules? Succinctly, they are:

- You cannot win, i.e. you cannot get more out of a system than you put in.
- You cannot break even – in any change something will be lost or, to be more precise, it will be useless for the purpose you have in mind.

---

[2] Particle physicists have discovered or postulated a considerable number of other sub-atomic particles, such as quarks, muons, pions and neutrinos. It is however sufficient for our purposes in this book for us to consider only electrons, protons and neutrons.

Table 1.1 Available electron states in the first four shells and sub-shells of electrons in the Bohr atom (after Callister, 2007)

Principal quantum number (n)	Shell	Sub-shell (l)	Number of energy states (m₁)	Maximum number of electrons	
				Per sub-shell	Per shell
1	K	s	1	2	2
2	L	s	1	2	8
		p	3	6	
3	M	s	1	2	18
		p	3	6	
		d	5	10	
4	N	s	1	2	32
		p	3	6	
		d	5	10	
		f	7	14	

0.03 nanometers, while caesium has one of the largest, with a radius of about 0.3 nanometres.

An element is characterised by:

- the *atomic number*, which is the number of protons in the nucleus, and hence is also the number of electrons in orbit;
- the *mass number*, which is sum of the number of protons and neutrons. For many of the lighter elements these numbers are similar and so the mass number is approximately twice the atomic number, though this relationship breaks down with increasing atomic number. In some elements the number of neutrons can vary, leading to *isotopes*; the atomic weight is the weighted average of the atomic masses of an element's naturally occurring isotopes.

Another useful quantity when we come to consider compounds and chemical reactions is the mole, which is the amount of a substance that contains 6.023×10^{23} atoms of an element or molecules of a compound (*Avogadro's number*). This number has been chosen because it is the number of atoms that is contained in the atomic mass (or weight) expressed in grams. For example, carbon has an atomic weight of 12.011, and so 12.011 grams of carbon contain 6.023×10^{23} atoms.

The manner in which the orbits of the electrons are distributed around the nucleus controls the characteristics of the element and the way in which atoms bond with other atoms of the same element and with atoms from different elements.

For our purposes it will be sufficient to describe the structure of the so-called Bohr atom, which arose from developments in quantum mechanics in the early part of the 20th century. This overcame the problem of explaining why negatively charged electrons would not collapse into the positively charged nucleus by proposing that electrons revolve around the nucleus in one of a number of discrete orbitals or *shells*, each with a defined or quantised energy level. Any electron moving between energy levels or orbitals would make a quantum jump with either emission or absorption of a discrete amount or quantum of energy.

Each electron is characterised by four quantum numbers:

- the principal quantum number ($n = 1, 2, 3, 4 \ldots$), which is the quantum shell to which the electron belongs, also denoted by K, L, M, N . . ., corresponding to $n = 1, 2, 3, 4 \ldots$;
- the secondary quantum number ($l = 0, 1, 2 \ldots n - 1$), which is the sub-shell to which the electron belongs, denoted by s, p, d, f, g, h for $l = 1, 2, 3, 4, 5, 6$, according to its shape;
- the third quantum number (m_l), which is the number of energy states within each sub-shell, the total number of which is $2l + 1$;
- the fourth quantum number (m_s) which describes the electron's direction of spin and is either $+\frac{1}{2}$ or $-\frac{1}{2}$.

The number of sub-shells that occur within each shell therefore increases with an increase in the principal quantum number (n), and the number of energy states within each subshell (m_l) increases with an increase in the secondary quantum number (l). *Table 1.1* shows how this leads to the maximum number electrons in each shell for the first four shells.

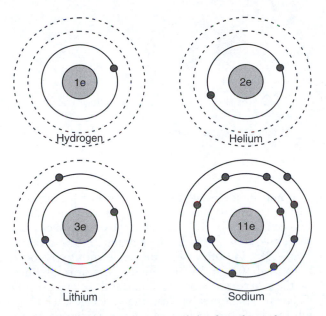

Fig. 1.1 *The atomic structure of the first three elements of the periodic table and sodium.*

filled by a total of eight electrons; such octets are found in neon, argon, krypton, xenon etc., and these 'noble gases' form very few chemical compounds for this reason. The exception to the octet rule for stability is helium; the outermost (K) shell only has room for its two electrons.

The listing of the elements in order of increasing atomic number and arranging them into groups of the same valence is the basis of the periodic table of the elements, which is an extremely convenient way of categorising the elements and predicting their likely properties and behaviour. As we will see in the next section, the number of valence electrons strongly influences the nature of the interatomic bonds.

1.2 Bonding of atoms

1.2.1 IONIC BONDING

If an atom (A) with one electron in the outermost shell reacts with an atom (B) with seven electrons in the outermost shell, then both can attain the octet structure if atom A donates its valence electron to atom B. However, the electrical neutrality of the atoms is disturbed and B, with an extra electron, becomes a negatively charged *ion* (an anion), whereas A becomes a positively charged ion (a cation). The two ions are then attracted to each other by the electrostatic force between them, and an ionic compound is formed.

The number of bonds that can be formed with other atoms in this way is determined by the valency. Sodium has one electron in its outer shell; it is able to give this up to form the cation whereas chlorine, which has seven electrons in its outer shell, can accept one to form the anion, thus sodium chloride has the chemical formula NaCl (*Fig. 1.2*). Oxygen, however, has six valence electrons and needs to

Each electron has a unique set of quantum numbers and with increasing atomic number, and hence increasing number of electrons, the shells and sub-shells fill up progressively, starting with the lowest energy state. The one electron of hydrogen is therefore in the only sub-shell in the K shell (denoted as $1s^1$), the two electrons of helium are both in this same shell (denoted as $1s^2$) and in lithium, which has three electrons, two are in the $1s^1$ shell and the third is in the $2s^1$ shell. By convention, the configuration of lithium is written as $1s^2 2s^1$. The configuration of subsequent elements follows logically (for example, sodium with 11 electrons is $1s^2 2s^2 2p^6 3s^1$). The structures of these elements are illustrated in *Fig. 1.1*.

An extremely important factor governing the properties of an element is the number of electrons in the outermost shell (known as the *valence electrons*), since it is these that are most readily available to form bonds with other atoms. Groups of elements with similar properties are obtained with varying atomic number but with the same number of outer shell electrons. For example, the 'alkali metals' lithium, sodium, potassium, rubidium and caesium all have one electron in their outermost shell, and all are capable of forming strong alkalis.

A further factor relating to this is that when the outermost electron shell is completely filled the electron configuration is stable. This normally corresponds to the s and p states in the outermost shell being

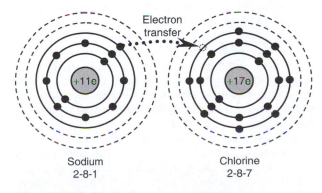

Fig. 1.2 *Ionic bonding.*

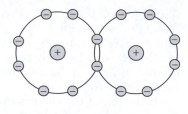 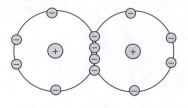

| (a) Between chlorine atoms | (b) Between oxygen atoms |

Fig. 1.3 Covalent bonding.

'borrow' or 'share' two; since sodium can only donate one electron, the chemical formula for sodium oxide is Na$_2$O. Magnesium has two valence electrons and so the chemical formula for magnesium chloride is MgCl$_2$ and for magnesium oxide MgO. Thus, the number of valence electrons determines the relative proportions of elements in compounds.

The strength of the ionic bond is proportional to $e_A e_B/r$ where e_A and e_B are the charges on the ions and r is the interatomic separation. The bond is strong, as shown by the high melting point of ionic compounds, and its strength increases, as might be expected, where two or more electrons are donated. Thus the melting point of sodium chloride, NaCl, is 801°C; that of magnesium oxide, MgO, where two electrons are involved, is 2640°C; and that of zirconium carbide, ZrC, where four electrons are involved, is 3500°C. Although ionic bonding involves the transfer of electrons between different atoms, the overall neutrality of the material is maintained.

The ionic bond is always non-directional; that is, when a crystal is built up of large numbers of ions, the electrostatic charges are arranged symmetrically around each ion, with the result that A ions surround themselves with B ions and vice versa, with a solid being formed. The pattern adopted depends on the charges on, and the relative sizes of, the A and B ions, i.e. how many B ions can be comfortably accommodated around A ions whilst preserving the correct ratio of A to B ions.

1.2.2 COVALENT BONDING

An obvious limitation of the ionic bond is that it can only occur between atoms of different elements, and therefore it cannot be responsible for the bonding of any of the solid elements. Where both atoms are of the electron-acceptor type, i.e. with close to 8 outermost electrons, octet structures can be built up by the sharing of two or more valence electrons between the atoms, forming a covalent bond.

For example, two chlorine atoms, which each have seven valence electrons, can achieve the octet struc-

ture and hence bond together by contributing one electron each to share with the other (*Fig. 1.3a*). Oxygen has six valence electrons and needs to share two of these with a neighbour to form a bond (*Fig. 1.3b*). In both cases a molecule with two atoms is formed (Cl$_2$ and O$_2$), which is the normal state of these two gaseous elements and a few others. There are no bonds between the molecules, which is why such elements are gases at normal temperature and pressure.

Covalent bonds are very strong and directional; they can lead to very strong two- and three-dimensional structures in elements where bonds can be formed by sharing electrons with more than one adjacent atom, i.e. which have four, five or six valence electrons. Carbon and silicon, both of which have four valence electrons, are two important examples. A structure can be built up with each atom forming bonds with four adjacent atoms, thus achieving the required electron octet. In practice, the atoms arrange themselves with equal angles between all the bonds, which produces a tetrahedral structure (*Fig. 1.4*). Carbon atoms are arranged in this way in diamond, which is one of the hardest materials known and also has a very high melting point (3500°C).

Covalent bonds are also formed between atoms from different elements to give compounds. Methane (CH$_4$) is a simple example; each hydrogen atom achieves a stable helium electron configuration by sharing one of the four atoms in carbon's outer shell and the carbon atom achieves a stable octet figuration by sharing the electron in each of the four hydrogen atoms (*Fig. 1.5*). It is also possible for carbon atoms to form long chains to which other atoms can bond along the length, as shown in *Fig. 1.6*. This is the basis of many polymers, which occur extensively in both natural and manufactured forms.

A large number of compounds have a mixture of covalent and ionic bonds, e.g. sulphates such as Na$_2$SO$_4$ in which the sulphur and oxygen are covalently bonded and form sulphate ions, which form an ionic bond with the sodium ions. In both the

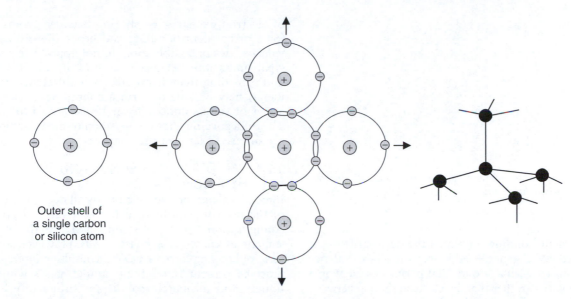

Fig. 1.4 *Covalent bonding in carbon or silica to form a continuous structure with four bonds orientated at equal spacing giving a tetrahedron-based structure.*

Outer shell of
a single carbon
or silicon atom

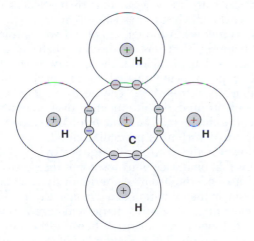

Fig. 1.5 *Covalent bonding in methane, CH₄.*

ionic and covalent bonds the electrons are held fairly strongly and are not free to move far, which accounts for the low electrical conductivity of materials containing such bonds.

1.2.3 METALLIC BONDS

Metallic atoms possess few valence electrons and thus cannot form covalent bonds between each other; instead they obey what is termed the *free-electron theory*. In a metallic crystal the valence electrons are detached from their atoms and can move freely between the positive metallic ions (*Fig. 1.7*). The positive ions are arranged regularly in a crystal lattice, and the electrostatic attraction between the positive ions and the free negative electrons provides the cohesive strength of the metal. The metallic bond may thus be regarded as a very special case

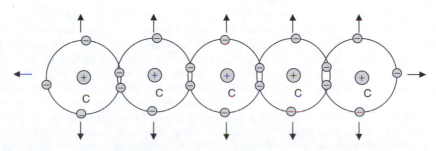

Fig. 1.6 *Covalent bonding in carbon chains.*

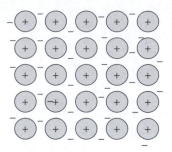

Fig. 1.7 The free electron system in the metallic bond in a monovalent metal.

of covalent bonding, in which the octet structure is satisfied by a generalised donation of the valence electrons to form a 'cloud' that permeates the whole crystal lattice, rather than by electron sharing between specific atoms (true covalent bonding) or by donation to another atom (ionic bonding).

Since the electrostatic attraction between ions and electrons is non-directional, i.e. the bonding is not localised between individual pairs or groups of atoms, metallic crystals can grow easily in three dimensions, and the ions can approach all neighbours equally to give maximum structural density. The resulting structures are geometrically simple by comparison with the structures of ionic compounds, and it is this simplicity that accounts in part for the ductility (ability to deform non-reversibly) of the metallic elements.

Metallic bonding also explains the high thermal and electrical conductivity of metals. Since the valence electrons are not bound to any particular atom, they can move through the lattice under the application of an electric potential, causing a current flow, and can also, by a series of collisions with neighbouring electrons, transmit thermal energy rapidly through the lattice. Optical properties can also be explained. For example, if a ray of light falls on a metal, the electrons (being free) can absorb the energy of the light beam, thus preventing it from passing through the crystal and rendering the metal opaque. The electrons that have absorbed the energy are excited to high energy levels and subsequently fall back to their original values with the emission of the light energy. In other words, the light is reflected back from the surface of the metal, which when polished is highly reflective.

The ability of metals to form alloys (of extreme importance to engineers) is also explained by the free-electron theory. Since the electrons are not bound, when two metals are alloyed there is no question of electron exchange or sharing between atoms in ionic or covalent bonding, and hence the ordinary valence laws of combination do not apply. The principal limitation then becomes one of atomic size, and providing there is no great size difference, two metals may be able to form a continuous series of alloys or solid solutions from 100% A to 100% B. The rules governing the composition of these solutions are discussed later in the chapter.

1.2.4 VAN DER WAALS BONDS AND THE HYDROGEN BOND

Ionic, covalent and metallic bonds all occur because of the need for atoms to achieve a stable electron configuration; they are strong and are therefore sometimes known as *primary bonds*. However, some form of bonding force between the resulting molecules must be present since, for example, gases will all liquefy and ultimately solidify at sufficiently low temperatures.

Such secondary bonds of forces are known as *Wan der Waals bonds* or *Wan der Waals forces* and are universal to all atoms and molecules; they are however sufficiently weak that their effect is often overwhelmed when primary bonds are present. They arise as follows. Although in *Fig. 1.1* we represented the orbiting electrons in discrete shells, the true picture is that of a cloud, the density of the cloud at any point being related to the probability of finding an electron there. The electron charge is thus 'spread' around the atom, and, over a period of time, the charge may be thought of as symmetrically distributed within its particular cloud.

However, the electronic charge is moving, and this means that on a scale of nanoseconds the electrostatic field around the atom is continuously fluctuating, resulting in the formation of a dynamic electric dipole, i.e. the centres of positive charge and negative charge are no longer coincident. When another atom is brought into proximity, the dipoles of the two atoms may interact co-operatively with one another (*Fig. 1.8*) and the result is a weak non-directional electrostatic bond.

As well as this fluctuating dipole, many molecules have permanent dipoles as a result of bonding between different types of atom. These can play a considerable part in the structure of polymers and organic compounds, where side-chains and radical groups of ions can lead to points of predominantly positive or negative charges. These will exert an electrostatic attraction on other oppositely charged groups.

The strongest and most important example of dipole interaction occurs in compounds between hydrogen and nitrogen, oxygen or fluorine. It occurs

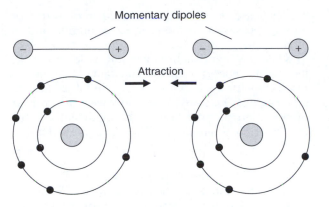

Fig. 1.8 *Weak Van der Waals linkage between atoms due to fluctuating electrons fields.*

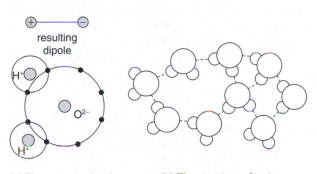

(a) The water molecule   (b) The structure of water

Fig. 1.9 *The hydrogen bond between water molecules.*

because of the small and simple structure of the hydrogen atom and is known as the *hydrogen bond*. When, for example, hydrogen links covalently with oxygen to form water, the electron contributed by the hydrogen atom spends the greater part of its time between the two atoms. The bond acquires a definite dipole with the hydrogen becoming virtually a positively charged ion (*Fig. 1.9a*).

Since the hydrogen nucleus is not screened by any other electron shells, it can attract to itself other negative ends of dipoles, and the result is the hydrogen bond. It is considerably stronger (about 10 times) than other Van der Waals linkages, but is much weaker (by 10 to 20 times) than any of the primary bonds. *Figure 1.9b* shows the resultant structure of water, where the hydrogen bond forms a secondary link between the water molecules, and acts as a bridge between two electronegative oxygen ions. Thus, this relatively insignificant bond is one of the most vital factors in the evolution and survival of life on Earth. It is responsible for the abnormally high melting and boiling points of water and for its high specific heat, which provides an essential global temperature control. In the absence of the hydrogen bond, water would be gaseous at ambient temperatures, like ammonia and hydrogen sulphide, and we would not be here.

The hydrogen bond is also responsible for the unique property of water of expansion during freezing i.e. a density decrease. In solid ice, the combination of covalent and strongish hydrogen bonds result in a three-dimensional rigid but relatively open structure, but on melting this structure is partially destroyed and the water molecules become more closely packed, i.e. the density increases.

1.3 Energy and entropy

The bonds that we have just described can occur between atoms in gases, liquids and solids and to a large extent are responsible for their many and varied properties. Although we hope construction materials do not change state whilst in service, we are very much concerned with such changes during their manufacture, e.g. in the cooling of metals from the molten to the solid state. Some knowledge of the processes and the rules governing them are therefore useful in understanding the structure and properties of the materials in their 'ready-to use' state.

As engineers, although we conventionally express our findings in terms of force, deflection, stress, strain and so on, these are simply a convention. Fundamentally, we are really dealing with energy. Any change, no matter how simple, involves an exchange of energy. The mere act of lifting a beam involves a change in the potential energy of the beam, a change in the strain energy held in the lifting cables and an input of mechanical energy from the lifting device, which is itself transforming electrical or other energy into kinetic energy. The harnessing and control of energy are at the heart of all engineering.

Thermodynamics teaches us about energy, and draws attention to the fact that every material possesses an *internal* energy associated with its structure. We begin this section by discussing some of the thermodynamic principles that are of importance to understanding the behaviour patterns.

1.3.1 STABLE AND METASTABLE EQUILIBRIUM
We should recognise that all systems are always seeking to minimise their energy, i.e. to become more stable. However, although thermodynamically correct, some changes toward a more stable condition proceed so slowly that the system appears to be stable

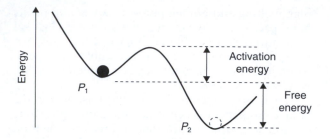

Fig. 1.10 *Illustration of activation and free energy.*

even though it is not. For example, a small ball sitting in a hollow at the top of a hill will remain there until it is lifted out and rolled down the hill. The ball is in a *metastable* state and requires a small input of energy to start it on its way down the main slope.

Figure 1.10 shows a ball sitting in a depression with a potential energy of P_1. It will roll to a lower energy state P_2, but only if it is first lifted to the top of the hump between the two hollows. Some energy has to be lent to the ball to do this, which the ball returns when it rolls down the hump to its new position. This borrowed energy is known as the *activation energy* for the process. Thereafter it possesses free energy as it rolls down to P_2. However, it is losing potential energy all the time and eventually (say, at sea level) it will achieve a stable equilibrium. However, note two things. At P_1, P_2, etc. it is apparently stable, but actually it is metastable, as there are other more stable states available to it, given the necessary activation energy. Where does the activation energy come from? In materials science it is extracted mostly (but not exclusively) from heat. As things are heated to higher temperatures the atomic particles react more rapidly and can break out of their metastable state into one where they can now lose energy.

1.3.2 MIXING

If whisky and water are placed in the same container, they mix spontaneously. The internal energy of the resulting solution is less than the sum of the two internal energies before they were mixed. There is no way that we can separate them except by distillation, i.e. by heating them up and collecting the vapours and separating these into alcohol and water. We must, in fact, put in energy to separate them. But, since energy can be neither be created nor destroyed, the fact that we must use energy, and quite a lot of it, to restore the status quo must surely pose the question 'Where does the energy come from initially?' The answer is by no means simple but, as

we shall see, every particle, whether of water or whisky, possesses kinetic energies of motion and of interaction.

When a system such as a liquid is left to itself, its internal energy remains constant, but when it interacts with another system it will either lose or gain energy. The transfer may involve work or heat or both and the first law of thermodynamics, 'the conservation of energy and heat', requires that:

$$dE = dQ - dW \qquad (1.1)$$

where E = internal energy, Q = heat and W = work done by the system on the surroundings. What this tells us is that if we raise a cupful of water from 20°C to 30°C it does not matter how we do it. We can heat it, stir it with paddles or even put in a whole army of gnomes each equipped with a hot water bottle, but the internal energy at 30°C will always be above that at 20°C by exactly the same amount. Note that the first law says nothing about the sequences of changes that are necessary to bring about a change in internal energy.

1.3.3 ENTROPY

Classical thermodynamics, as normally taught to engineers, regards entropy, S, as a capacity property of a system which increases in proportion to the heat absorbed (dQ) at a given temperature (T). Hence the well known relationship:

$$dS \geq dQ/T \qquad (1.2)$$

which is a perfectly good definition but does not give any sort of picture of the meaning of entropy and how it is defined. To a materials scientist entropy has a real physical meaning, it is a measure of the state of disorder or chaos in the system. Whisky and water combine; this simply says that, statistically, there are many ways that the atoms can get mixed up and only one possible way in which the whisky can stay on top of, or, depending on how you pour it, at the bottom of, the water. Boltzmann showed that the entropy of a system could be represented by:

$$S = k \ln N \qquad (1.3)$$

where N is the number of ways in which the particles can be distributed and k is a constant (Boltzmann's constant $k = 1.38 \times 10^{-23}$ *J/K*). The logarithmic relationship is important; if the molecules of water can adopt N_1 configurations and those of whisky N_2 the number of possible configurations open to the mixture is not $N_1 + N_2$ but $N_1 \times N_2$. It follows from this that the entropy of any closed system not in equilibrium will tend to a maximum since this represents the most probable array of configurations. This is the second law of thermodynamics, for which you should be very grateful. As you read these words, you are keeping alive by breathing a randomly distributed mixture of oxygen and nitrogen. Now it is statistically possible that at some instant all the oxygen atoms will collect in one corner of the room while you try to exist on pure nitrogen, but only statistically possible. There are so many other possible distributions involving a more random arrangement of the two gases that it is most likely that you will continue to breathe the normal random mixture.

1.3.4 FREE ENERGY

It must be clear that the fundamental tendency for entropy to increase, that is, for systems to become more randomised, must stop somewhere and somehow, i.e. the system must reach equilibrium. If not, the entire universe would break down into chaos. As we have seen in the first part of this chapter, the reason for the existence of liquids and solids is that their atoms and molecules are not totally indifferent to each other and, under certain conditions and with certain limitations, will associate or bond with each other in a non-random way.

As we stated above, from the first law of thermodynamics the change in internal energy is given by:

$$dE = dQ - dW$$

From the second law of thermodynamics the entropy change in a reversible process is:

$$TdS = dQ \tag{1.4}$$

Hence:

$$dE = TdS - dW \tag{1.5}$$

In discussing a system subject to change, it is convenient to use the concept of free energy. For irreversible changes, the change in free energy is always negative and is a measure of the driving force leading to equilibrium. Since a spontaneous change must lead to a more probable state (or else it would not happen) it follows that, at equilibrium, energy is minimised while entropy is maximised.

The Helmholtz free energy is defined as:

$$H = E - TS \tag{1.6}$$

and the Gibbs free energy as:

$$G = pV + E - TS \tag{1.7}$$

and, at equilibrium, both must be a minimum.

1.4 Equilibrium and equilibrium diagrams

Most of the materials that we use are not pure but consist of a mixture of one or more constituents. Each of the three material states of gases, liquids and solids may consist of a mixture of different components, e.g. in alloys of two metals. These components are called *phases*, with each phase being homogeneous. We need a scheme that allows us to summarise the influences of temperature and pressure on the relative stability of each state (and, where necessary its component phases) and on the transitions that can occur between these. The time-honoured approach to this is with *equilibrium diagrams*. Note the word equilibrium. Thermodynamics tells us that this is the condition in which the material has minimum internal energy. By definition, equilibrium diagrams tell us about this minimum energy state that a system is trying to reach, but when using these we should bear in mind that it will always take a finite time for a transition from one state to another to occur or for a chemical reaction to take place. Sometimes, this time is vanishingly small, as when dynamite explodes. At other times, it can be a few seconds, days or even centuries. Glass made in the Middle Ages is still glass and shows no sign of crystallising. So, not every substance or mixture that we use has reached thermodynamic equilibrium.

We only have space here to introduce some of the elements of the great wealth of fundamental theory underlying the forms of equilibrium diagrams.

1.4.1 SINGLE-COMPONENT DIAGRAMS

The temperature–pressure diagram for water (*Fig. 1.11*) is an important example of a single-component diagram, and we can use this to establish some ground rules and language for use later.

The diagram is in 'temperature–pressure space' and a number of lines are marked which represent boundary conditions between differing phases, i.e. states of H_2O. The line *AD* represents combinations of temperature and pressure at which liquid water and solid ice are in equilibrium, i.e. can coexist. A small heat input will alter the proportions of ice and water by melting some of the ice. However, it is absorbed as a change in internal energy of the mixture, the latent heat of melting. The temperature is not altered, but if we put in large amounts of heat, so that all the ice is melted and there is some heat left over, the temperature rises and we end up with slightly warmed water. Similarly, line *AB* represents the equilibrium between liquid water and

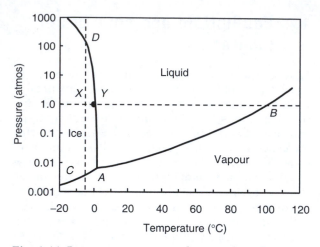

Fig. 1.11 *Pressure–temperature diagram for water (from Kingery et al., 1976).*

gaseous steam, and line *AC* the equilibrium between solid ice and rather cold water vapour.

It is helpful to consider what happens if we move around within the diagram. First, let us start at point *X*, representing −5°C at atmospheric pressure. We know we should have ice and, indeed, the point *X* lies within the phase field labelled ice. Adding heat at constant pressure takes the temperature along the broken line. This crosses the phase boundary, *AD*, at 0°C (point *Y*) and the ice begins to melt as further heat is added. Not until all the ice has melted does the temperature continue to rise. We now have liquid water until we reach 100°C (point *B*). Now, again, heat has to be added to boil the water but there is no temperature increase until all the liquid water has gone. We now have steam and its temperature can be increased by further heat input.

Next think of keeping temperature constant and increasing pressure, again starting at point *X*. If the pressure is raised enough, to about 100 atmospheres (≈10 MPa, point *D*) we reach the ice–water equilibrium and the ice can begin to melt. This accounts for the low friction between, for example, an ice skate and the ice itself: local pressures cause local melting. It is a factor that engineers need to consider when contemplating the use of locally refrigerated and frozen ground as coffer dams or as foundations for oil rigs in Alaska.

The Gibbs phase rule is a formal way of summarising the relationship between the number of phases (*P*) that can coexist at any given point in the diagram and the changes brought about by small changes in temperature or pressure. This states that:

$$P + F = C + 2 \tag{1.8}$$

Here, *C* is the number of components in the system; in this case we have only H_2O so $C = 1$. *F* is the number of degrees of freedom allowed to change. To illustrate, at point *X* in *Fig. 1.11* there is just one phase, ice, so $P = 1$ and $F = 2$. This means that both temperature and pressure can be changed independently without bringing about a significant change to the material. At *Y* both solid and liquid can coexist, so $P = 2$ and $F = 1$. To maintain the equilibrium, temperature and pressure must be changed in a co-ordinated way so that the point *Y* moves along the boundary *AD*. At *A*, all three phases can coexist so $P = 3$, therefore $F = 0$, i.e. any change at all will disturb the equilibrium.

1.4.2 TWO-COMPONENT DIAGRAMS

We now go on to look at two-component diagrams, such as we get with alloys between two metals or between iron and carbon. We now have a further variable, composition and, strictly, we should consider the joint influences of this variable in addition to temperature and pressure. We would therefore need three-dimensional diagrams, but to simplify things we usually take pressure to be constant. After all, most engineering materials are prepared and used at atmospheric pressure, unless you work for NASA! This leaves us with a composition–temperature diagram, the lifeblood of materials scientists.

The alloys formed between copper, Cu, and nickel, Ni (*Fig. 1.12*) produce an example of the simplest form of two-component diagram. This is drawn with composition as the horizontal axis, one end representing pure (100%) Cu, the other pure (100%) Ni. The vertical axis is temperature.

Let us think about an alloy that is 50%Cu:50%Ni by mass. At high temperatures, e.g. at *A*, the alloy is totally molten. On cooling, we move down the

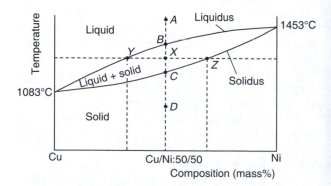

Fig. 1.12 *Equilibrium phase diagram for copper–nickel.*

composition line until we arrive at B. At this temperature, a tiny number of small crystals begin to form. Further reduction in temperature brings about an increase in the amount of solid in equilibrium with a diminishing amount of liquid. On arriving at C, all the liquid has gone and the material is totally solid. Further cooling brings no further changes. Note that there is an important difference between this alloy and the pure metals of which it is composed. Both Cu and Ni have well defined unique melting (or freezing) temperatures but the alloy solidifies over the temperature range BC; metallurgists often speak of the 'pasty range'.

We now need to examine several matters in more detail. First, the solid crystals that form are what is known as a 'solid solution'. Cu and Ni are chemically similar elements and both, when pure, form face-centred cubic crystals (see Chapter 3). In this case, a 50:50 alloy is also composed of face-centred cubic crystals but each lattice site has a 50:50 chance of being occupied by a Cu atom or a Ni atom.

If we apply Gibbs's phase rule at point A, $C = 2$ (two components, Cu & Ni) and $P = 1$ (one phase, liquid) and so $F = 3$ (i.e. 3 degrees of freedom). We can therefore independently alter composition, temperature and pressure and the structure remains liquid. But remember, we have taken pressure to be constant and so we are left with 2 practical degrees of freedom, composition and temperature. The same argument holds at point D, but, of course, the structure here is the crystalline solid solution of Cu and Ni.

At a point between B and C we have liquid and solid phases coexisting, so $P = 2$ and $F = 2$. As before, we must discount one degree of freedom because pressure is taken as constant. This leaves us with $F = 1$, which means that the status quo can be maintained only by a coupled change in both composition and temperature. Therefore, it is not only that the structure is two phase, but also that the proportions of liquid and solid phases remain unaltered.

We can find the proportions of liquid and solid corresponding to any point in the two phase field using the so-called *Lever rule*. The first step is to draw the constant temperature line through the point X, *Fig. 1.12*. This intersects the phase boundaries at Y and Z. The solid line containing Y represents the lower limit of 100% liquid, and is known as the *liquidus*. The solid line containing Z is the upper limit of 100% solid and known as the *solidus*.

Neither the liquid nor solid phases corresponding to point X have a composition identical with that of the alloy as a whole. The liquid contains more Cu and less Ni, the solid less Cu and more Ni. The

composition of each phase is given by the points Y and Z, respectively. The proportions of the phases balance so that the weighted average is the same as the overall composition of the alloy. It is easy to show that:

$$\text{(Weight of liquid of composition } Y) \times YX = \text{(Weight of solid of composition } Z) \times XZ \quad (1.9)$$

This is similar to what would be expected of a mechanical lever balanced about X, hence the name Lever rule.

One consequence of all this can be seen by re-examining the cooling of the 50:50 alloy from the liquid phase. Consider *Fig. 1.13*. At point X_1 on the liquidus, solidification is about to begin. At a temperature infinitesimally below X_1 there will be some crystals solidifying out of the liquid; their composition is given by Z_1. At a temperature about halfway between solidus and liquidus (X_2), we have a mixture of solid and liquid of compositions Z_2 and Y_2. In general, the proportion of liquid to solid halfway through the freezing range need not be $\approx 50:50$, but in this case it is. Finally, at a temperature infinitesimally above X_3, which is on the solidus, we have nearly 100% solid of composition Z_3 together with a vanishingly small amount of liquid of composition Y_3. When the temperature falls to just below X_3, the alloy is totally solid and Z_3 has become identical with X_3.

Note two important features. First, Z_3 is the same as the average composition we started with, X_1. Second, solidification takes place over a range of temperatures, and as it occurs the compositions of liquid and solid phases change continuously. For this to actually happen, substantial amounts of diffusion

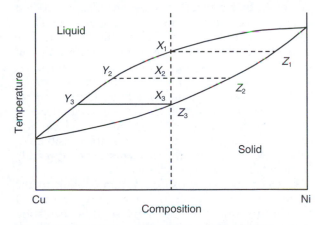

Fig. 1.13 Equilibrium phase diagram for Cu–Ni (Fig. 1.12 redrawn to show composition variations with temperature).

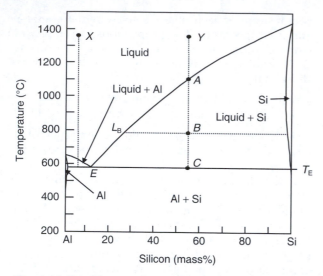

Fig. 1.14 Equilibrium phase diagram for aluminium–silicon.

must occur in both liquid and solid. Diffusion in solids is very much slower than that in liquids and is the source of some practical difficulty. Either solidification must occur slowly enough for diffusion in the solid to keep up or strict equilibrium conditions are not met. The kinetics of phase transformations is therefore of interest, but for the moment, we will continue to discuss very slowly formed, equilibrium or near equilibrium structures.

1.4.3 EUTECTIC SYSTEMS

Let us now examine another diagram, that for aluminium–silicon (Al–Si) alloys (*Fig. 1.14*). Pure Al forms face-centred crystals (see Chapter 3) but Si has the same crystal structure as diamond. These are incompatible and extensive solid solutions like those for Cu:Ni cannot be formed. Si crystals can dissolve only tiny amounts of Al. For our purposes, we can ignore this solubility, although we might recognise that the semiconductor industry makes great use of it, small as it is. Al crystals can dissolve a little Si, but again not very much, and we will ignore it. Thus, two solid phases are possible, Al and Si. When liquid, the elements dissolve readily in the melt in any proportions.

Consider the composition Y. On cooling to the liquidus line at A, pure (or nearly pure) crystals of Si begin to form. At B we have solid Si coexisting with liquid of composition L_B in proportions given by the Lever rule. At C we have solid Si in equilibrium with liquid of composition close to E.

Now consider alloy X. The sequence is much the same except the first solid to form is now Al. When the temperature has fallen to almost T_E we have solid Al in equilibrium with liquid of composition close to E. Note that both alloy X and alloy Y, when cooled to T_E, contain substantial amounts of liquid of composition E. An infinitesimal drop in temperature below T_E causes this liquid to solidify into a mixture of solid Al and solid Si. At E we have 3 phases which can coexist; liquid, solid Al and solid Si. The system has two components and thus the phase rule gives us no degrees of freedom once we have discounted pressure. E is an invariant point; any change in temperature or composition will disturb the equilibrium.

The point E is known as the eutectic point and we speak of the eutectic composition and the eutectic temperature, T_E. This is the lowest temperature at which liquid can exist and the eutectic alloy is that which remains liquid down to T_E. It solidifies at a unique temperature, quite unlike Cu–Ni or Al–Si alloys of other compositions. Alloys close to the eutectic composition ($\approx 13\%$ Si) are widely used because they can be easily cast into complex shapes, and the Si dispersed in the Al strengthens it. Eutectic alloys in other systems find similar uses (cast-iron is of near eutectic composition) as well as uses as brazing alloys etc.

1.4.4 INTERMEDIATE COMPOUNDS

Often, the basic components of a system can form compounds. In metals we have $CuAl_2$, Fe_3C and many more. Some other relevant examples are:

- SiO_2 and corundum, Al_2O_3, which form mullite, $3(Al_2O_3)2(SiO_2)$, an important constituent of fired clays, pottery and bricks. *Figure 1.15* shows the SiO_2–Al_2O_3 diagram. It can be thought of as two diagrams, one for 'SiO_2-mullite' and the other for 'mullite–Al_2O_3', joined together. Each part diagram is a simple eutectic system like Al–Si;
- lime, CaO, and silica, SiO_2, which form the compounds $2(CaO)SiO_2$, $3(CaO)SiO_2$ and others, which have great technological significance as active ingredients in Portland cement (to be discussed in detail in Chapter 13). In a similar way to mullite, the lime (CaO)–silica (SiO_2) diagram (*Fig. 1.16*) can be thought of as a series of joined together eutectic systems.

In many cases we do not have to think about the whole diagram. *Figure 1.17* shows the Al–$CuAl_2$ diagram, again a simple eutectic system. A notable feature is the so-called *solvus* line, AB, which represents the solubility of $CuAl_2$ in solid crystals

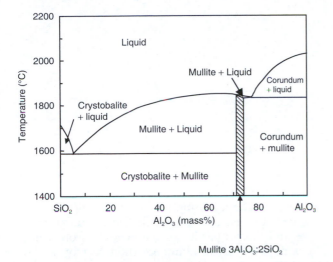

Fig. 1.15 Equilibrium phase diagram for silica (SiO_2)–alumina (Al_2O_3).

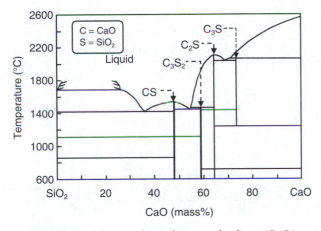

Fig. 1.16 Equilibrium phase diagram for lime (CaO)–silica (SiO_2).

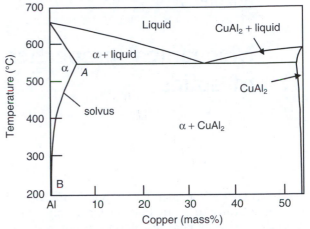

Fig. 1.17 Equilibrium phase diagram for Al–$CuAl_2$.

of Al. This curves sharply, so that very much less $CuAl_2$ will dissolve in Al at low temperatures than will at high temperatures. This is a fortunate fact that underlies our ability to alter the microstructures of some alloys by suitable heat-treatments, discussed in more detail later.

We have not yet considered the iron–carbon diagram, which is perhaps the most important diagram for nearly all engineers. This is of particular relevance in civil and structural engineering since steel in all its forms is used extensively. We will leave discussing this until Chapter 11.

References

Kingery WD, Bowen HK and Uhlmann DR (1976). *Introduction to Ceramics*, 2nd edition, John Wiley and Sons, New York.

Callister WD (2007). Materials science and engineering. An introduction 7th edn., John Wiley and Sons, New York.

Chapter 2

Mechanical properties of solids

We have seen in Chapter 1 how bonds are formed between atoms to form bulk elements and compounds, and how changes of state occur, with an emphasis on the formation of solids from molten liquids. The behaviour of solids is of particular interest to construction engineers for the obvious reason that these are used to produce load-bearing structures; in this chapter we define the properties and rules used to quantify the behaviour of solids when loaded. To understand this behaviour and therefore to be able to change it to our advantage we need to consider some other aspects of the structure and nature of the materials beyond those discussed in Chapter 1; we will do this in Chapter 3.

You will find it necessary to refer to the definitions etc. given in this chapter when reading the subsequent on individual materials. Although we will include here some examples of the behaviour of construction materials, all of the definitions and explanations are applicable to any materials being used by engineers of any discipline.

2.1 Stress, strain and stress–strain curves

Loading causes materials to deform and, if high enough, to break down and fail. All loading on materials can be considered as combinations of three basic types – tension, compression and shear. These are normally shown diagrammatically as in *Fig. 2.1*.

Clearly the deformation from loading on an element or test specimen will depend on both its size and the properties of the material from which it is made. We can eliminate the effect of size by converting:

- the load to *stress*, σ, defined as *load, P, divided by the area, A, to which is applied*, i.e.

$$\sigma = P/A \qquad (2.1)$$

and
- the deformation to *strain*, ε, defined as *change in length, Δl, divided by original length, l*, i.e.

$$\varepsilon = \Delta l/l \qquad (2.2)$$

These definitions are illustrated for simple tension in *Fig. 2.2a*. Compressive stress and strain are in

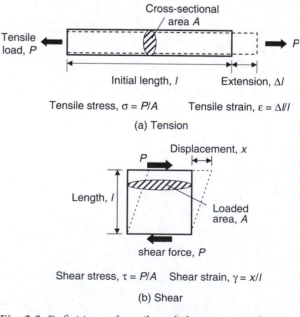

Tensile stress, $\sigma = P/A$ Tensile strain, $\varepsilon = \Delta l/l$

(a) Tension

Shear stress, $\tau = P/A$ Shear strain, $\gamma = x/l$

(b) Shear

Fig. 2.2 *Definitions of tensile and shear stress and strain.*

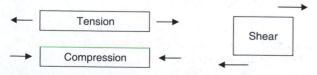

Fig. 2.1 *Basic types of load.*

the opposite directions. The equivalent definitions of shear stress, τ, and shear strain, γ, which are not quite so obvious, are shown in *Fig. 2.2b*.

As with all quantities, the dimensions and units must be considered:

- stress = load/area and therefore its dimensions are [Force]/[Length]^2. Typical units are N/mm^2 (or MPa in the SI system), lb/in^2 and tonf/ft^2 in the Imperial system.
- strain = change in length/original length and therefore its dimensions are [Length]/[Length], i.e. it is dimensionless.

However, strain values can be very small and it is often convenient to use either:

percentage strain (or % strain) = strain × 100
or   microstrain (μs) = strain × 10^6

As well as the linear strain, we can also similarly define

volumetric strain (ε_v)
  = change in volume(ΔV)/original volume(V) (2.3)

The relationship between stress and strain is an extremely important characteristic of a material. It varies with the rate of application of stress (or load); we will consider four cases:

a) *steadily increasing* – zero to failure in a few minutes, e.g. as in a laboratory test
b) *permanent or static* – constant with time, e.g. the self weight of the upper part of a structure acting on the lower part
c) *impact or dynamic* – very fast, lasting a few microseconds, e.g. the impact of a vehicle on a crash barrier, or an explosion
d) *cyclic* – variable with load reversals, e.g. earthquake loading – a few cycles in a few minutes, and wave loading on an offshore structure – many cycles over many years.

For the moment, we will confine ourselves to case (a): steady loading to failure in a few minutes. This is what is used in the most common types of laboratory tests that are used to measure or characterise a material's behaviour.

There are a wide variety of different forms of stress–strain behaviour for different materials; *Fig. 2.3* shows those for some common materials. Most have at least two distinct regions:

- An initial linear or near-linear region in which the strain is fully recovered when the load is removed, i.e. the material returns to its initial shape and dimensions. This is called *elastic deformation*,

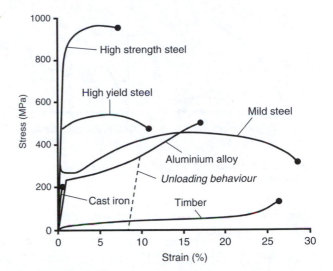

Fig. 2.3 *Typical tensile stress–strain curves for some structural materials.*

and this portion of the graph the *elastic region*; if the behaviour is also linear, we call this *linear elasticity*. The strains involved are usually small.
- An increasingly non-linear region in which the strains can increase significantly with progressively smaller increments of stress. Unloading at any point in this region will result in the strain reducing along a line parallel to the initial elastic region, and hence there is a permanent deformation at zero load (as shown on the graph for an aluminium alloy). This is known as the *plastic region*, and the permanent deformation as *plastic deformation*.

Eventually, of course the material breaks, which may occur at the end of either the elastic or the plastic region, sometimes after an apparent reduction of stress (as shown for mild steel).

We now take a close look at each of these regions in turn, starting with the elastic region, defining some of the *material constants* that are used to quantify the behaviour as we go.

2.2 Elastic behaviour and the elastic constants

In service most materials will be operating in the elastic region most of the time. Design engineers organise things so that, as far as it is possible to predict, this will always be the case (but sometimes things do go wrong). A number of *elastic constants*,

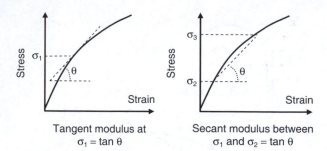

Fig. 2.4 *Definitions of tangent and secant moduli of elasticity.*

defined as follows, are used to calculate deflections and movement under load.

2.2.1 THE ELASTIC MODULI

For *linear elastic* materials, stress is proportional to strain (Hooke's law) and for uniaxial tension or compression we can define:

Young's modulus (E) = slope of stress–strain graph = stress/strain = σ/ε (2.4)
[E is also known as the modules of elasticity, the E-*modulus* or simply the *stiffness*.]

Since strain is dimensionless, the dimensions of E are the same as those of stress i.e. [Force]/[Length]^2. Convenient SI units to avoid large numbers are kN/mm^2 or GPa.

For materials that have non-linear elastic behaviour (quite a few, particularly non-metals) a modulus value is still useful and there are some alternative definitions, illustrated in *Fig. 2.4*:

- The *tangent modulus* is the slope of the tangent to the curve at any stress (which should be quoted). A special case is the *tangent modulus at the origin* i.e. at zero stress.
- The *secant modulus* is the slope of the straight line joining two points on the curve. Note that stress levels corresponding to the two points must be given. If only one stress is given, then it is reasonable to assume that the other is zero.

E-values for construction materials range from 0.007 GPa for rubber to 200 GPa for steel (diamond is stiffer still at 800 GPa, but this is hardly a construction material). Values therefore vary very widely, by more than 4 orders of magnitude from rubber to steel.[1]

---

[1] We discuss values for the major groups of materials in the relevant parts of the book, and then make comparisons of this and other key properties in Chapter 61.

For shear loading and deformation, the equivalent to E is the

shear modulus (G)
= shear stress(τ)/shear strain(γ) (2.5)

G, which is sometimes called the *modulus of rigidity*, is another elastic constant for the material, and it has a different numerical value to E.

The *bulk modulus* is used when estimating the change in volume of a material under load. In the case of uniform stress on a material in all directions i.e. a pressure (p) as might be found by submergence of the specimen to some depth in a liquid:

The *volumetric strain* (ε_v) = change (reduction) in volume/original volume   (2.6)

and

the *bulk modulus* $(K) = p/\varepsilon_v$ (2.7)

2.2.2 POISSON'S RATIO

When a material is loaded or stressed in one direction, it will deform (or strain) in the direction of the load, i.e. longitudinally, *and* perpendicular to the load i.e. laterally. The *Poisson's ratio* is the ratio of the strain in the direction to. Thus in *Fig. 2.5*:

$$\varepsilon_x = x/L \text{ (extension)}$$
$$\varepsilon_y = -y/a \text{ and } \varepsilon_z = -z/b$$
(The −ve sign indicates contraction.)

The y and z directions are both perpendicular to the direction of loading, x,

$$\therefore \quad \varepsilon_y = \varepsilon_z$$
$$\therefore \quad \textit{Poisson's ratio } (\upsilon) = -\varepsilon_y/\varepsilon_x = -\varepsilon_z/\varepsilon_x \quad (2.8)$$

The Poisson's ratio is another elastic constant for the material. The minus sign ensures that it is a positive number. Values for common materials vary from 0.15 to 0.49 (see *Table 61.1* in Chapter 61).

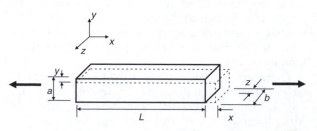

Longitudinal strain, $\varepsilon_x = x/L$ Lateral strain $\varepsilon_y = -y/a = \varepsilon_z = -z/b$

Poisson's ratio, $\upsilon = -\varepsilon_y/\varepsilon_x = -\varepsilon_z/\varepsilon_x$

Fig. 2.5 *Definition of Poisson's ratio.*

We should note that the above definitions of E, G and υ assume that the material has similar properties in all directions (i.e. it is *isotropic*) and therefore there is a single value of each elastic constant for any direction of loading. *Anisotropic* materials, i.e. those which have different properties in different directions, e.g. timber, will have different values of E, G and υ in each direction, and clearly the direction and well as the value itself must then be stated.

2.2.3 RELATIONSHIPS BETWEEN THE ELASTIC CONSTANTS

The four elastic constants that we have now defined, E, G, υ and K, might at first glance seem to describe different aspects of behaviour. It is possible to prove that they are not independent and that they are related by the simple expressions:

$$E = 2G(1 + \upsilon) \qquad (2.9)$$

and

$$K = E/3(1 - 2\upsilon) \qquad (2.10)$$

The proof of these expressions is not unduly difficult (see for example Case, Chilver and Ross, 1999) but what is more important is the consequence that if you know, or have measured, any two of the constants then you can calculate the value of the others. Many materials have a Poisson's ratio between 0.25 and 0.35, and so the shear modulus (G) is often about 40% of the elastic modulus (E).

Equation 2.10 tells us something about the limits to the value of Poisson's ratio. We have defined the bulk modulus, K, by considering the case of the change in volume of a specimen under pressure (equation 2.7). This change must always be a reduction, as it would be inconceivable for a material to expand under pressure – i.e. in the same direction as the pressure. Therefore K must always be positive and since E is also positive (by definition) then $(1 - 2\upsilon)$ must be positive, and so

$$\upsilon \le 0.5 \qquad \text{ALWAYS!} \qquad (2.11)$$

A material with $\upsilon > 0.5$ cannot exist; if you have carried out some tests or done some calculations that give such a value, then you must have made a mistake. It also follows that if $\upsilon = 0.5$ then K is zero and the material is incompressible.

2.2.4 WORK DONE IN DEFORMATION

The work done by a load when deforming a material, although not an elastic constant, is another useful value. Work is force x distance, and so

$$W = \int_0^e P \mathrm{d}e \qquad (2.12)$$

where W = work done by the load P in causing an extension e.

The work done on unit volume of the material of length l and cross-section A is:

$$W = \int_0^e P \mathrm{d}e/Al = \int_0^e \frac{P}{A} \cdot \frac{\mathrm{d}e}{l} = \int_0^\varepsilon \sigma \mathrm{d}\varepsilon \qquad (2.13)$$

which is the area under the stress–strain curve.

This work must go somewhere, and it is stored as internal *strain energy* within the material. With elastic deformation, it is available to return the material to its zero state on unloading; in plastic deformation, it permanently deforms the material and, eventually, it is sufficient to cause fracture. We will explore the relationship between this energy and fracture in more detail in Chapter 4.

2.3 Plastic deformation

As we have said, deformation is *plastic* if it results in *permanent deformation after load removal*.

In very broad terms, materials can be divided into those that are:

Ductile – which have large plastic deformation before failure (say strains > 1%)

and those that are:

Brittle – with little or no plastic deformation before failure (say strains < 0.1%)

Some examples of stress–strain curves of each type of material have been shown in *Fig. 2.3*. It follows from equation (2.13) that ductile materials require much greater amounts of work and have much greater amounts of internal strain energy before failure. There are clearly some intermediate materials, but engineers generally prefer to use ductile materials that give warning of distress before failure in the event of overload. Brittle materials fail suddenly without warning – often catastrophically.

Significant plastic deformation obviously occurs only in ductile materials. We can use the idealised stress–strain curve for mild steel shown in *Fig. 2.6* to illustrate common features of this behaviour.

- There is a sharp and distinct end to the linear elastic behaviour (point A), called the *limit of elasticity* or the *yield point*.
- There is a region of increasing strain with little or no increase in stress (AB), often very short.
- Unloading in the plastic region (say from a point x) produces behaviour parallel to the initial (linear) elastic behaviour. Reloading produces

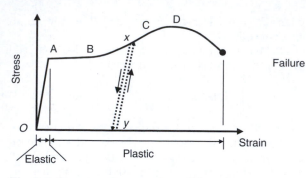

Fig. 2.6 *Stress–strain curve for mild steel.*

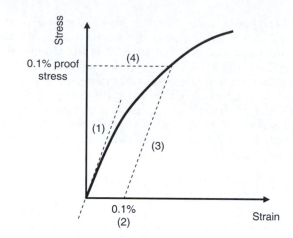

Fig. 2.7 *Determination of proof stress.*

similar elastic behaviour up to the unload point, and the deformation then continues as if the unload/reload had not occurred, i.e. the material 'remembers' where it was.

- Another feature of plastic behaviour, not apparent from *Fig. 2.6*, is that the deformation takes place at constant volume, i.e. the Poisson's ratio is 0.5 for deformation beyond the yield point.

Two important implications for engineers are:

1. The stress at the yield point A, called the *yield stress* (σ_y), is an important property for design purposes. Working stresses are kept safely below this.
2. If, before use, the material is loaded or strained to say a point *x* beyond the constant stress region, i.e. beyond B, and then unloaded, it ends up at point *y*. If it is then used in this state, the yield stress (i.e. at *x*) is greater than the original value (at A) i.e. the material is 'stronger'. This is known as *work hardening* or *strain hardening* (or sometimes *cold working*) to distinguish it from other methods of strengthening that involve heat treatment (which we will discuss in Chapter 8). The working stresses can therefore be increased. The drawback is that the failure strain of the work-hardened material (from *y* to failure) is less than that of the original material (from O to failure) and so therefore is the total work to fracture. The work-hardened material is therefore more brittle.

If there is no distinct end to the elastic behaviour i.e. the graph gradually becomes non-linear, then an alternative to the yield stress called the *proof stress* is used instead. This is defined and obtained as shown in *Fig. 2.7*:

1. A tangent is drawn to the stress–strain curve at the origin.

2. A low value of strain is selected – normally either 0.1% (as in the figure) or 0.2%.
3. A line is drawn through this point parallel to the tangent at the origin.
4. The stress value at the point where this intersects the stress–strain curve is the *0.1% proof stress*. (If a strain value of 0.2% is chosen, then the result is the 0.2% proof stress.)

2.4 Failure in tension

The form of failure in uniaxial tension depends on whether a material is brittle or ductile. As we have already said, brittle materials fail with little or no plastic deformation; failure occurs suddenly without warning, and the fracture surface is perpendicular to the direction of loading (*Fig. 2.8*).

Ductile materials not only undergo large strains before failure, but often have an apparent reduction of stress before failure (i.e. beyond point *D* in *Fig. 2.6*). Up to the maximum stress (D), the elongation is uniform, but after this, as the load starts to decrease, a localised narrowing or *necking* can be seen somewhere along the length (*Fig. 2.9a*).

As the stress continues to fall (but still at increasing strain) the diameter at the neck also decreases, until, with very ductile materials, it reaches almost zero before failure, which takes the form of a sharp

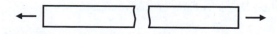

Fig. 2.8 *Brittle failure in tension.*

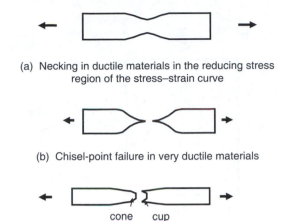

(a) Necking in ductile materials in the reducing stress region of the stress–strain curve

(b) Chisel-point failure in very ductile materials

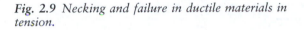

cone   cup

(c) 'Cup and cone' failure in medium-strength metals

Fig. 2.9 *Necking and failure in ductile materials in tension.*

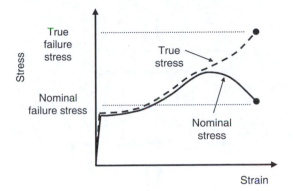

Fig. 2.10 *True and nominal stress/strain behaviour.*

point (*Fig. 2.9b*). This form of failure is extreme, and occurs only in very ductile materials such as pure metals, e.g. lead and gold, or chewing gum (try it for yourself). These materials tend to be weak, and most ductile structural materials fail at a stress and strain some way down the falling part of the curve but with the stress well above zero. Necking still occurs after the maximum stress and failure occurs at the narrowest section in the form of a '*cup and cone*' (*Fig. 2.9c*). The inner part of the failure surface is perpendicular to the applied load, as in a brittle failure, and the cracks first form here. The outer rim, at about 45° to this, is the final cause of the failure.

2.5 True stress and strain

The behaviour shown in *Fig. 2.6* shows the failure occurring at a lower stress than the maximum, i.e. the material seem to be getting weaker as it approaches failure. It fact, the opposite is occurring, and the reason why the stress appears to fall is because of the way we have calculated it. We have defined stress as load/area and *Fig. 2.6* has been obtained by dividing the load (P) by the original area before loading (A_0). The stress that we have obtained should strictly be called the *nominal stress* (σ_{nom}), i.e.

$$\sigma_{nom} = P/A_0 \qquad (2.14)$$

In fact, the cross-sectional area (A) is reducing throughout the loading i.e. $A < A_0$. At any load,

the *true stress* (σ_{true}) will therefore be higher than the nominal stress, i.e.

$$\sigma_{true} = P/A > \sigma_{nom} \qquad (2.15)$$

In the elastic and plastic regions the reduction is uniform along the length (the Poisson's ratio effect) but the magnitude of the strains involved are such that the difference between the nominal and the actual area is very small. However, once necking starts the area of the neck reduces at a rate such that the true stress continues to increase up to failure, as shown in *Fig. 2.10*.

In the case of strain, the relation between the increment of change in length (de), the increment of strain ($d\varepsilon$) and the length (l) is, by definition

$$d\varepsilon = de/l \qquad (2.16)$$

and so the *true strain* (ε) is

$$\varepsilon = \int_{l_0}^{l} \frac{dl}{l} = \ln\left(\frac{l}{l_0}\right) \qquad (2.17)$$

where l_0 is the initial length

True strain is not difficult to calculate, but measurement of the cross-sectional area throughout the loading, and hence calculation of the true stress, is more difficult, and therefore true stress–strain graphs are rarely obtained, except perhaps for research purposes. However, measurement of the size of the neck after fracture is easy, which enables the true fracture stress to be readily obtained.

2.6 Behaviour in compression

The elastic behaviour and constants discussed in section 2.2 apply equally to tensile and compressive loading. There are however differences in the observed behaviour during plastic deformation and failure.

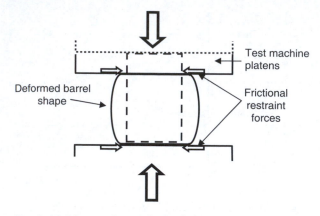

Fig. 2.11 *Non-uniform plastic deformation in a compression test.*

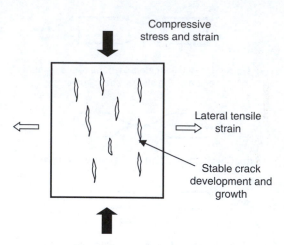

Fig. 2.12 *Multiple crack pattern in a brittle material in compression leading to higher strength than in tension.*

2.6.1 PLASTIC DEFORMATION OF DUCTILE MATERIALS

Values of yield stresses for ductile materials are similar to those in tension, but the subsequent behaviour in a laboratory test is influenced by the loading system. Test machines apply the load through large blocks of steel called *platens*, which bear on the specimen. These are stiffer than the specimen and therefore the lateral expansion of the specimen is opposed by friction at the platen/specimen interface. This causes a confining force or restraint at either end of the specimen. The effect of this force reduces with distance from the platen i.e. towards the centre of the specimen, with the result that a cylindrical specimen of a ductile material of say, mild steel will plastically deform into a barrel shape, and the sides will not stay straight, as in *Fig. 2.11*.

Continued loading of ductile materials to higher and higher stress will simply result in a flatter and flatter disc i.e. more and more plastic deformation, but no failure in the sense of cracking or breakdown of structure. In fact the area is increasing, and therefore very high loads are required to keep the true stress (see section 2.5) increasing. Tests can therefore easily reach the capacity of the test machine.

2.6.2 FAILURE OF BRITTLE MATERIALS

Failure stresses of brittle materials in compression are much higher than those in tension – up to twenty times higher for some materials, e.g. concrete. This results from a very different cracking and failure mechanism. Cracking is a pulling apart of two surfaces, and therefore occurs by the action of a tensile strain. In uni-axial compressive loading, the strains

in the direction of loading are obviously compressive and it is the lateral strains that are tensile (*Fig. 2.12*). The cracks are formed perpendicular to these strains, i.e. parallel to the load direction. A single small crack will not immediately grow to cause failure, and a whole network of cracks needs to be formed, grow and intersect before complete material breakdown occurs. This requires a much higher stress than that necessary to cause the single failure crack under tensile loading.

There is a further effect resulting from the friction restraint of the platens discussed above that causes the failure stress (i.e. the apparent compressive strength) to be dependent on the specimen geometry, specifically the height/width ratio. In the part of the specimen near the platen, this restraint opposes and reduces the lateral tensile strain. This increases the load required for complete breakdown, i.e. failure (in effect, this part of the specimen is under a tri-axial compressive stress system). The effect of the restraint reduces with distance from the platen (*Fig. 2.13*). Short fat specimens will have most of their volume experiencing high restraint, whereas the central part of longer, thinner specimens will be nearer to a uni-axial stress system, and will therefore fail at a lower average applied stress.

The typical effect of the height/width ratio is shown in *Fig. 2.14* from tests on concrete; the strength (i.e. the failure stress) expressed relative to that at a height/width ratio of 2. We will discuss measurement of the compressive strength of concrete in more detail in Chapter 21.

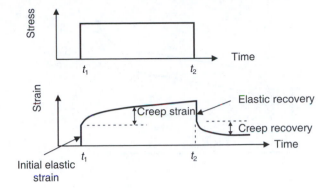

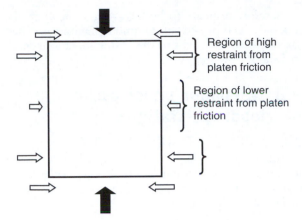

Fig. 2.13 *Variation of restraint from platen friction in a compression test leading to the size effect on compressive failure stress.*

Fig. 2.15 *Schematic of creep behaviour due to a constant applied stress.*

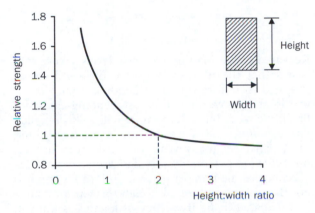

Fig. 2.14 *The effect of height/width ratio on the compressive strength of brittle materials.*

2.7 Behaviour under constant load – creep

Constant load or stress is a very common occurrence, e.g. the stress due to the self-weight of a structure. Materials respond to this stress by an immediate strain deformation, normally elastic, followed by an increase in strain with time, called *creep*. Typical behaviour is illustrated in *Fig. 2.15*. A stress applied at time t_1 and maintained at a constant level until removal at time t_2 results in:

- an initial elastic strain on stress application (related to the stress by the modulus of elasticity)
- an increase in this strain due to creep during the period of constant stress – fairly rapid at first but then at a decreasing rate

- an immediate elastic recovery on stress removal, often similar in magnitude to the initial elastic strain
- further recovery with time (called *creep recovery*) again at decreasing rate. This is normally less than the creep strain, so that the material does not return to zero position, i.e. there is some permanent deformation.

For calculation purposes, we define:

creep coefficient
= creep strain/initial elastic strain   (2.18)

effective elastic modulus
= stress/(total strain)
= stress/(elastic + creep strain)   (2.19)

Both of these will obviously vary with time.

Creep increases with time, with the applied stress (sometimes values of the *specific creep = creep/unit stress* are quoted) and with temperature. The magnitude of creep varies widely in different materials. For example, most metals and metallic alloys only start to creep at temperatures approaching half of their melting point (expressed in degrees Kelvin), whereas with concrete and many polymers the creep strain can be as great or greater than the initial elastic strain at room temperature.

Creep curves typically have three parts, illustrated in *Fig. 2.16* for two different levels of stress:

- *primary creep:* initially rapid, but at a reducing rate with time
- *secondary creep:* a steady rate of strain ($d\varepsilon/dt$) often expressed as

$$d\varepsilon/dt = C\sigma^n \quad (2.20)$$

where σ = applied stress, C is a constant and n is the creep

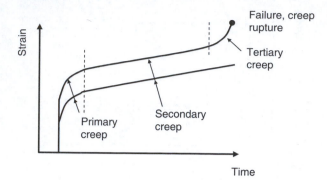

Fig. 2.16 Sub-divisions of creep curves.

exponent, which usually lies between 3 and 8

• *tertiary creep:* at high stress levels, after a period of time (which can be very lengthy) there may be an increasing rate with time leading to failure, a process known as *creep rupture*. This only occurs if the stress is high, typically more than 70–80% of the failure stress measured in a short-term test.

In some situations, the strain is constant e.g. a cable stretched between two fixed supports or a tensioned bolt clamping two metal plates together. The stress reduces with time, as shown in *Fig. 2.17*, a process known as **stress relaxation**. In extreme cases, the stress reduces to zero, i.e. the cable or bolt become slack.

During creep and the stress relaxation the material is, in effect, flowing, albeit at a very slow rate.

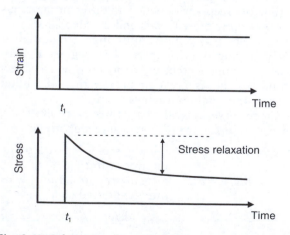

Fig. 2.17 Schematic of stress relaxation at constant strain.

It therefore appears to be behaving somewhat like a liquid. Such mixed solid/liquid behaviour is called viscoelasticity; we will be discussing this more detail in Chapter 5.

2.8 Behaviour under cyclic loading – fatigue

2.8.1 FATIGUE LIFE AND *S/N* CURVES

Cyclic loading is very common, e.g. wind and wave loading, vehicle loading on roads and bridges. We can define the characteristics of the loading as shown in *Fig. 2.18*, in which:

p = period of loading

∴ frequency = $1/p$ (e.g. in cycles/sec or Hertz)

σ_{max} = maximum applied stress

σ_{min} = minimum applied stress

σ_m = mean stress = $(\sigma_{max} + \sigma_{min})/2$

S = stress range = $(\sigma_{max} - \sigma_{min})$

Repeated cyclic loading to stress levels where σ_{max} is less that the ultimate (or even the yield stress) can lead to failure (think of bending a wire backwards and forwards – the first bending does not break the wire, but several more will). This is called *fatigue failure*.[2] This can be sudden and brittle, and can occur after many years of satisfactory service. It is therefore potentially very dangerous. *Fatigue life* is defined as the number of cycles (N) to failure. It is not the time to failure, although this can of course be calculated if the frequency of loading is known.

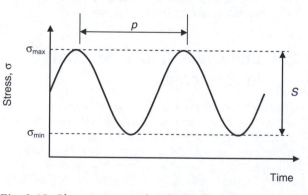

Fig. 2.18 Characteristics of cyclic loading.

---

[2] But do not make the common mistake of calling this 'metal fatigue'. All structural materials, not just metals, are subject to fatigue failure under appropriate combinations of stress range and time.

From testing, it has been found that:

- N is independent of frequency (except at very high frequencies, above 1 kHz)
- N is dependent on the stress range (S) rather than the individual values of σ_{max} or σ_{min}, provided σ_{max} or σ_{min} does not approach the yield strength. For example, the fatigue life under a stress cycling between −50 and +50 MPa is the same as that under stress cycling between +25 and +125 MPa
- a higher stress range results in a shorter fatigue life, i.e. N increases with decreasing S. The relationship is often of the form

$$S.N^a = C \qquad (2.21)$$

where a and C are constants (a is between 0.12 and 0.07 for most materials).

The fatigue performance of materials is normally given as S/N curves i.e. graphs of stress range (S) vs. fatigue life (N). Typical S/N curves for mild steel and a copper alloy are shown in *Fig. 2.19*. Fatigue lives are often very long (e.g. thousands, tens of thousands, or hundreds of thousands of cycles) so a log scale is normally used for N.

The individual data points shown for mild steel illustrate the considerable scatter that is obtained from test programmes. Apart from the obvious superior performance of steel, the best-fit line through the data shows a discontinuity at about 240 MPa/10^7 cycles where it becomes parallel to the x-axis for higher fatigue lives. This means that at values of S below 240 MPa the fatigue life is infinite, which is very useful for design purposes. This stress range is called the fatigue endurance limit, and is a typical characteristic of ferrous metals. Non-ferrous metals such as copper do not show such a limit i.e. there is no 'safe' stress range below which fatigue failure will not occur eventually.

2.8.2 CUMULATIVE FATIGUE DAMAGE: MINER'S RULE

S/N curves define the fatigue life at any given stress range S. However, in most situations in practice the material or structural component will not be subjected to cycles of a single stress range, but to differing numbers of cycles of different stress ranges. For example wind action will result in a few cycles at a high stress range from severe storms and a large number of cycles at lower stresses from lesser strength winds. To estimate the effect of this cumulative fatigue damage *Miner's rule* is used. This accounts for the partial effect of the number of cycles at each particular stress range by considering that, if the material is stressed for n_1 cycles at a stress range that will cause failure in a total of N_1 cycles, then a fraction n_1/N_1 of the fatigue life is used up; failure occurs when the sum of all the fractions, $\Sigma n_i/N_i$, reaches 1, irrespective of the sequence of application of the various cycles of loading.

Figure 2.20 illustrates the case of three stress ranges, S_1, S_2 and S_3, being applied for n_1, n_2 and n_3 cycles respectively, and for which the total fatigue lives are N_1, N_2 and N_3. The n_1 cycles at the stress range S_1 use up n_1/N_1 of the total fatigue life. The same applies for the n_2 cycles at S_2 and the n_3 cycles at S_3. Therefore the proportion of the total fatigue life used up by all three stress ranges is:

$$n_1/N_1 + n_2/N_2 + n_3/N_3$$

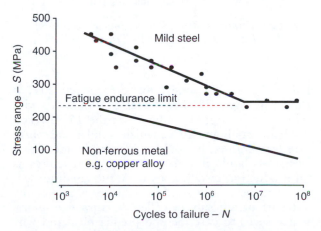

Fig. 2.19 Fatigue life data (S/N curves) for mild steel and a copper alloy.

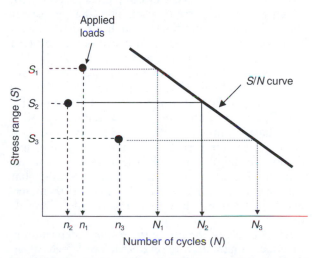

Fig. 2.20 Example of the application of Miner's rule of cumulative fatigue damage.

Failure will occur when this sum reaches 1. This will be achieved with continued cyclic loading, in which one or all of n_1, n_2 or n_3 may increase, depending on the nature of the loading. The general expression of Miner's rule is that for failure:

$$\sum n_1/N_1 = 1 \qquad (2.22)$$

2.9 Impact loading

Structures and components of structures can be subjected to very rapid rates of application of stress and strain in a number of circumstances, such as explosions, missile or vehicle impact, and wave slam. Materials can respond to such impact loading by:

- an apparent increase in elastic modulus, but this is a third or fourth order effect only – a 10^4 times increase in loading rate gives only a 10% increase in elastic modulus
- an increase in brittle behaviour, leading to fast brittle fracture in normally ductile materials. This can be very dangerous – we think we are using a ductile material that has a high work to fracture and gives warning of failure, but this reacts to impact loading like a brittle material. The effect is enhanced if the material contains a pre-existing defect such as a crack.

The latter effect cannot be predicted by extrapolating the results of laboratory tensile or compression tests such as those described earlier, and impact test procedures have been developed to assess the behaviour of specimens containing a machined notch, which acts a local stress raiser. The Charpy test for metals is a good example of such a test. In this, a heavy pendulum is released and strikes the standard specimen at the bottom of its swing (*Fig. 2.21*). The specimen breaks and the energy needed for the fracture is determined from the difference between the starting and follow-through positions of the pendulum. The energy absorbed in the fracture is called the Charpy impact value. As we have discussed earlier in the chapter, brittle materials require less energy for failure than ductile materials, and an impact value of 15J is normally used as a somewhat arbitrary division between the two, i.e. brittle materials have a value below this, and ductile materials above.

An example of the use of the test is in determining the effect of temperature on ductile/brittle behaviour. Many materials that are ductile at normal temperatures have a tendency to brittleness at reducing

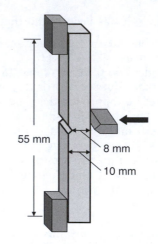

Fig. 2.21 Charpy impact test specimen (from dimensions specified in BS EN ISO 148-3:2008).

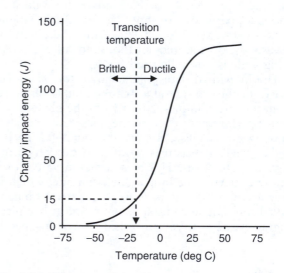

Fig. 2.22 Variation of the Charpy impact energy of a steel with temperature (after Rollason, 1961).

temperatures. This effect for a particular steel is shown in *Fig. 2.22*. The decrease in ductility with falling temperature is rapid, with the 15J division occurring at about −20°C, which is called the *transition temperature*. It would, for example, mean that this steel should not be used in such structures as oil production installations in Arctic conditions.

Impact behaviour and fast fracture are an important part of the subject called *fracture mechanics*, which seeks to describe and predict how and why cracking and fracture occur. We will consider this in more detail in Chapter 4.

2.10 Variability, characteristic strength and the Weibull distribution

Engineers are continually faced with uncertainty. This may be in the estimation of the loading on a structure (e.g. what is the design load due to a hurricane that has a small but finite chance of occurring sometime in the next 100 years?), analysis (e.g. what assumptions have been made in the computer modelling and are they valid?) or with the construction materials themselves. When dealing with uncertainty in materials, with natural materials such as timber we have to cope with nature's own variations, which can be large. With manufactured materials, no matter how well and carefully the production process is controlled, they all have some inherent variability and are therefore not uniform. Furthermore, when carrying out tests on a set of samples to assess this variability there will also be some unavoidable variation in the testing procedure itself, no matter how carefully the test is carried out or how skilful the operative. Clearly there must be procedures to deal with this uncertainty and to ensure a satisfactory balance between safety and economy. Structural failure can lead to loss of life, but the construction costs must be acceptable.

In this section, after some basic statistical considerations for describing variability, we will discuss two approaches to coping with variations of strength – characteristic strength and the Weibull distribution. We will take strength as being the ultimate or failure stress of a material as measured in, say, a tension, compression or bending test (although the arguments apply equally to other properties such as the yield or proof stress of a material).

2.10.1 DESCRIPTIONS OF VARIABILITY

A series of tests on nominally identical specimens from either the same or successive batches of material usually gives values of strength that are equally spread about the mean value with a normal or Gaussian distribution, as shown in *Fig. 2.23*.

The mean value, σ_m, is defined as the arithmetic average of all the results, i.e.:

$$\sigma_m = (\Sigma \sigma)/n \qquad (2.23)$$

where n is the number of results.

The degree of spread or variation about the mean is given by the *standard deviation, s*, where

$$s^2 = \Sigma(\sigma - \sigma_m)^2/(n - 1) \qquad (2.24)$$

s^2 is called the variance, and s has the same units as σ.

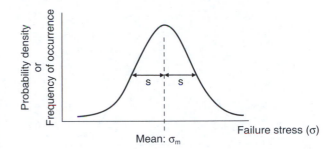

Fig. 2.23 *Typical normal distribution of failure stresses from successive tests on samples of a construction material.*

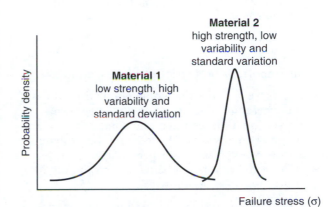

Fig. 2.24 *Two combinations of mean strength and variability.*

Materials can have any combination of mean strength and variability (or standard deviation) (*Fig. 2.24*).

For comparison between materials, the *coefficient of variation, c*, is used, where

$$c = s/\sigma_m \qquad (2.25)$$

c is non-dimensional, and is normally expressed as a percentage. Typical values of c are 2% for steel, which is produced under carefully controlled conditions, 10–15% for concrete, which is a combination of different components of different particle sizes, and 20–30% for timber, which has nature's own variations. Steel and timber are at the two ends of the variability scale of construction materials.

For structural use of a material, we need a 'safe' stress that takes into account of both the mean failure stress and the variability. This is done by considering the normal distribution curve (*Fig. 2.23*) in more detail. The equation of the curve is

$$y = \frac{1}{s\sqrt{2\pi}} \exp\left[-\frac{(\sigma - \sigma_m)^2}{2s^2}\right] \qquad (2.26)$$

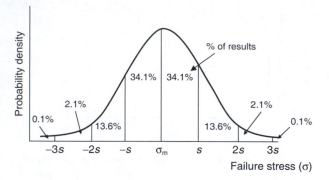

Fig. 2.25 *Proportion of results in the regions of the normal distribution curve.*

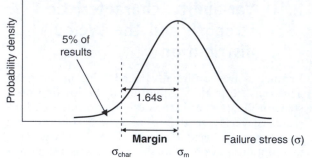

Fig. 2.26 *Definition of characteristic strength (σ_{char}) and margin for a 1 in 20 (5%) failure rate criterion.*

Some important properties of this equation are:

- The curve encloses the whole population of data, and therefore not surprisingly, integrating the above equation between the limits of $-\infty$ and $+\infty$ gives an answer of 1, or 100 if the probability density is expressed as a percentage.
- 50% of the results fall below the mean and 50% above, but also, as shown in *Fig. 2.25*:

  - 68.1% of results lie within one standard deviation of the mean
  - 95.5% of results lie within two standard deviations of the mean
  - 99.8% of results lie within three standard deviations of the mean.

2.10.2 CHARACTERISTIC STRENGTH

A guaranteed minimum value of stress below which no sample will ever fail is impossible to define – the nature of the normal distribution curve means that there will always be a chance, albeit very small, of a failure below any stress value. A value of stress called the *characteristic strength* is therefore used, which is defined as the stress below which an acceptably small number of results will fall. Engineering judgement is used to define 'acceptably small'. If this is very small, then there is a very low risk of failure, but the low stress will lead to increased cross-sectional area and hence greater cost. If it is higher, then the structure may be cheaper but there is an increased risk of failure.

Clearly a balance is therefore required between safety and economy. For many materials a stress below which 1 in 20 of the results occurs is considered acceptable, i.e. there is a 5% failure rate. Analysis of the normal distribution curve shows that this stress is 1.64 standard deviations below

the mean. This distance is called the *margin* and so, as shown in *Fig. 2.26*:

characteristic strength = mean strength – margin

$$\sigma_{char} = \sigma_m - ks \qquad (2.27)$$

where k, the standard deviation multiplication factor, is 1.64 in this case.

The value of k varies according to the chosen failure rate (*Table 2.1*), and, as we said above, judgement and consensus are used to arrive at an acceptable failure rate. In practice, this is not always the same in all circumstances; for example, 5% is typical for concrete (i.e. $k = 1.64$), and 2% for timber (i.e. $k = 1.96$).

There is a further step in determining an allowable stress for design purposes. The strength data used to determine the mean and standard deviation for the above analysis will normally have been obtained from laboratory tests on small specimens, which generally will have no apparent defects or damage. They therefore represent the best that can be expected from the material in ideal or near ideal circumstances. In practice, structural elements and members contain a large volume of material, which

Table 2.1 Values of k, the standard deviation multiplication factor, for various failure rates

Failure rate (%)	k
50	0
16	1
10	1.28
5	1.64
2	1.96
1	2.33

has a greater chance of containing manufacturing and handling defects. This size effect is taken into account by reducing the characteristic strength by a *partial materials' safety factor*, γ_m.

It is normal practice for γ_m to be given as a value greater than one, so the characteristic strength has to be *divided* by γ_m to give the allowable stress. Hence:

allowable design stress
= characteristic strength/γ_m
= (mean strength − margin)/γ_m (2.28)

As with the failure rate, the value of γ_m is based on knowledge and experience of the performance of the material in practice. For example, typical values recommended in the European standard for structural concrete design (Eurocode 2, BS EN 1992) are 1.15 for reinforcing steel and 1.6 for concrete.

2.10.3 THE WEIBULL DISTRIBUTION

An alternative statistical approach to the distribution of strength, particularly for brittle materials, was developed by the Swedish engineer Waloddi Weibull. As we have discussed in section 2.8 (and will consider further in Chapter 4) brittle fracture is initiated at flaws or defects, which are present to a greater or lesser extent in all materials. Therefore the variations of strength can be attributed to variations in the number and, more particularly, the maximum size of defect in a test specimen. Larger specimens have a higher probability of containing larger defects and therefore can be expected to have a lower mean strength (as just discussed in relation to the partial materials safety factor, γ_m).

Weibull defined the survival probability, $P_s(V_0)$, as the fraction of identical samples of volume V_0 that survive after application of a stress σ. He then proposed that:

$$P_s(V_0) = \exp\{-(\sigma/\sigma_0)^m\} \quad (2.29)$$

where σ_0 and m are constants. Plots of this equation for three values of m are shown in *Fig. 2.27*. In each case, when the stress is low, no specimens fail and so the survival probability is 1, but at increasing stress more and more samples fail until eventually, when they have all failed, the survival probability is zero. Putting $\sigma = \sigma_0$ in equation (2.29) gives $P_s(V_0) = 1/e = 0.37$, so σ_0 is the stress at which 37% of the samples survive. The value of m, which is called the Weibull modulus, is a measure of the behaviour on either side of σ_0, and therefore indi-

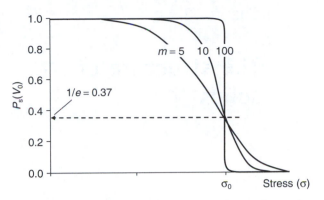

Fig. 2.27 *The Weibull distribution for three values of the Weibull modulus, m.*

cates the degree of variability of the results (and in this sense has a similar role to the coefficient of variation as defined in equation 2.25). Lower values indicate greater variability; m for concrete and bricks is typically about 10, whereas for steel it is about 100.

We can extend this analysis to give an estimate of the volume dependence of survival probability. $P_s(V_0)$ is the probability that one specimen of volume V_0 will survive a stress σ. If we test a batch of n such specimens, then the probability that they will all survive this stress is $\{P_s(V_0)\}^n$. If we then test a volume $V = nV_0$ of the material, which is the equivalent of combining all the smaller specimens into a single large specimen, then the survival probability, $P_s(V)$, is still $\{P_s(V_0)\}^n$.

$$P_s(V) = [P_s(V_0)]^n = [P_s(V_0)]^{v/v_0} \quad \text{(from eqn (2.29))}$$
$$= \{\exp\{-(\sigma/\sigma_0)\}^m\}^{v/v_0}$$

which gives

$$P_s(V) = \exp\{-(V/V_0)(\sigma/\sigma_0)^m\} \quad (2.30)$$

So, having determined σ_0 and m from tests on samples of volume V_0 and selected an acceptable value for $P_s(V)$, the design stress for structural element of volume V can be calculated.

References

Case J, Chilver H and Ross C (1999). *Strength of Materials and Structures*, Elsevier Science & Technology, London, p. 720.

Rollason EC (1961). *Metallurgy for Engineers*, Edward Arnold, London.

Chapter 3

The structure of solids

In Chapter 1 we discussed the various ways in which atoms bond together to form solids, liquids and gases, and some of the principles involved in the changes between these states. In Chapter 2 we described the behaviour of solids when subjected to load or stress and the various rules and constants used to characterise and quantify this. We now go on to consider the structure of solids in more detail, which will provide an explanation for much of the behaviour described in Chapter 2. Although the type of bonding between atoms goes some way towards explaining the properties of the resulting elements or compounds, it is equally important to understand the ways in which the atoms are arranged or packed together. We start by considering the relatively ordered structure of crystalline solids, and then discuss some aspects of the less ordered structures of ceramics and polymers.

3.1 Crystal structure

Many construction materials, particularly metals and some ceramics, consist of small *crystals* or *grains* within which the atoms are packed in regular, repeating, three-dimensional patterns giving a *long-range order*. The grains are 'glued' together at the grain boundaries; we will consider the importance of these later, but first we will discuss the possible arrangements of atoms within the grains. For this, we will assume that atoms are hard spheres – a considerable but convenient simplification.

It is also convenient to start with the atomic structure of elements (which of course consist of single-sized atoms) that have non-directional bonding (e.g. pure metals with metallic bonds). The simplest structure is one in which the atoms adopt a cubic pattern i.e. each atom is held at the corner of a cube. For obvious reasons, this is called the *simple cubic (SC) structure*. The atoms touch at

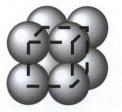 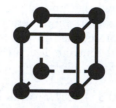

(a) Hard sphere model   (b) Reduced sphere model

Fig. 3.1 *The simple cubic structure of atoms in crystals and the unit cell.*

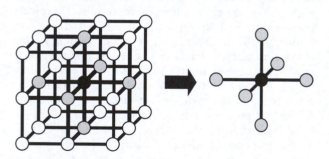

Fig. 3.2 *The coordination number = 6 for the simple cubic structure.*

the centre of each edge of the cube (*Fig. 3.1a*). The structure is sometimes more conveniently shown as in *Fig. 3.1b*. We can use this figure to define some properties of crystalline structures:

- the *unit cell*: the smallest repeating unit within the structure, in this case a cube (*Fig. 3.1*)
- the *coordination* number: the number of atoms touching a particular atom or the numbers of its nearest neighbours, in this case 6 (*Fig. 3.2*)
- the *closed-packed direction*: the direction along which atoms are in continuous contact, in this case any of the sides on the unit cell

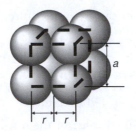

Fig. 3.3 *Unit cell dimensions and atomic radii in the simple cubic structure.*

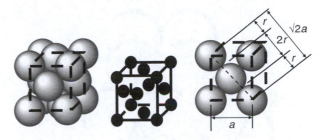

Fig. 3.5 *The face-centred cubic structure, unit cell and close-packed direction.*

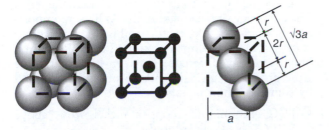

Fig. 3.4 *The body-centred cubic structure, unit cell and close-packed direction.*

- the *atomic packing factor (APF)*: the volume of atoms in the unit cell/volume of the unit cell, which therefore represents the efficiency of packing of the atoms.

The APF can be calculated from simple geometry. In this case:

- there are eight corner atoms, and each is shared between eight adjoining cells
    ∴ each unit cell contains $8 \times 1/8 = 1$ atom
- the atoms are touching along the sides of the cube (the close-packed direction)
    ∴ radius of each atom, $r = 0.5a$ (*Fig. 3.3*) when a = length of the side of the unit cube
- the volume of each atom = $4/3\pi r^3 = 4/3\pi(0.5a)^3$
    ∴ APF = [atoms/cell].[volume each atom]/ volume of unit cell
    = $[1] \times [(4/3\pi(0.5a)^3)/[a^3] = 0.52$.

There are two other crystal structures that have cubic structures with atoms located at the eight corners but which have additional atoms:

- the *body-centred cubic (BCC) structure*, which also has an atom at the centre of the cube (*Fig. 3.4*)
- the *face-centred cubic (FCC) structure*, which also an atom at the centre of each of the six faces (*Fig. 3.5*).

With the body-centred cubic structure, the coordination number is 8 (the atom in the cell centre touches the eight corner atoms) and the close-packed direction is the cell diagonal. It should be apparent from *Fig. 3.4* that:

- each unit cell contains $8/8 + 1 = 2$ atoms
- considering the close-packed direction gives:
    $4r = \sqrt{3}a$ or $r = \sqrt{3}a/4$.
    ∴ APF = $[2] \times [(4/3\pi(\sqrt{3}a/4)^3)/[a^3] = 0.68$.

With the face-centred cubic structure, a little thought should convince you that the coordination number is 12 and the close-packed direction is the face diagonal. From *Fig. 3.5*:

- each unit cell contains $8/8 + 6/2 = 4$ atoms
- considering the close-packed direction gives:
    $4r = \sqrt{2}a$ or $r = \sqrt{2}a/4$.
    ∴ APF = $[4] \times [(4/3\pi(\sqrt{2}a/4)^3)/[a^3] = 0.74$.

Moving from the SC to the BCC to the FCC structure therefore gives an increase in the coordination number (from 6 to 8 to 12) and in the efficiency of packing (from an APF of 0.52 to 0.68 to 0.74).

One further structure that might be expected to have efficient packing needs consideration – the *hexagonal close-packed (HCP)* structure. If we start with a single plane, then the most efficient packing is a hexagonal layout, i.e. as the atoms labelled A in *Fig. 3.6*. In adding a second layer the atoms (labelled B) place themselves in the hollows in the first layer. There are then two possible positions for the atoms in the third layer, either directly above the A atoms or in the positions labelled C. The first of these options gives the structure and unit cell shown in *Fig. 3.7*. Two dimensions, a and c, are required to define the unit cell, with $c/a = 1.633$. The coordination number is 12 and atomic packing factor 0.74, i.e. the same as for the face-centred cubic structure.

If we know the crystal structure and the atomic weight and size of an element then we can make an estimate of its density. For example, take copper, which has a face-centred cubic structure, an atomic

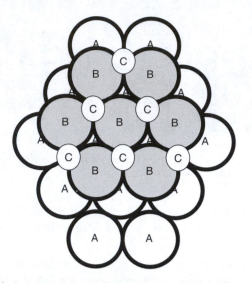

Fig. 3.6 *Arrangement of atoms in successive layers of the hexagonal close-packed structure.*

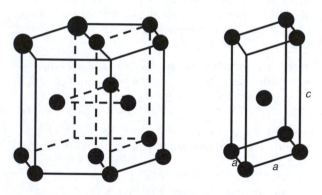

Fig. 3.7 *Atomic arrangement and unit cell of the hexagonal close-packed structure.*

weight of 63.5, and an atomic radius of 0.128 nm (atomic weights and sizes are readily available from tables of properties of elements).

- atomic weight = 63.5, therefore 63.5 g[1] of copper contain 6.023×10^{23} atoms[2]

  ∴ mass of one atom = 10.5×10^{-23} g
- In the FCC structure there are 4 atoms/unit cell

  ∴ mass of unit cell = 4.22×10^{-22} g

---

[1] The atomic weight in grams is normally called the molar mass, and is the weight of one mole.

[2] This is Avogadro's number, which we defined in Chapter 1 as the number of atoms in one mole. By definition it has the same value for all elements.

- As above, in the FCC structure, length of side of unit cell (a) = $4r/\sqrt{2}$

  ∴ $a = 4 \times 0.128/\sqrt{2} = 0.362$ nm
- ∴   unit cell volume = 4.74×10^{-2} nm³

  ∴   density = weight/volume

  = 4.22×10^{-22} g/4.74×10^{-2} nm³

  = 8900 kg/m³

A typical measured value of the density of copper is 8940 kg/m³, so our estimate is close.

We generally expect that elements that adopt one of the crystal structures described above will prefer to adopt the one that has the lowest internal energy. The efficiency of packing (i.e. the APF) is an important, but not the sole, factor in this. In practice, no common metals adopt the simple cubic structure, but the energy difference between the other three structures is often small, and the structures adopted by some common metals are:

- FCC – aluminium, copper, nickel, iron (above 910°C), lead, silver, gold
- HCP – magnesium, zinc, cadmium, cobalt, titanium
- BCC – iron (below 910°C), chromium, molybdenum, niobium, vanadium.

The two structures for iron show that the crystal structure can have different minimum energies at different temperatures. These various forms are called *allotropes*. Changes from one structure to another brought about by changes of temperature are of fundamental importance to metallurgical practice. For example, as we will see in Chapter 11, the change from FCC to BCC as the temperature of iron is reduced through 910°C forms the foundation of the metallurgy of steel.

3.2 Imperfections and impurities

In practice it is impossible for a perfect and uniform atomic structure to be formed throughout the material and there will always be a number of imperfections. *Point defects* occur at discrete sites in the atomic lattice and can be either missing or extra atoms, called *vacancies* and *interstitial atoms* respectively, as shown in *Fig. 3.8*. A linear dislocation is a one-dimensional defect; an example is when part of a plane of atoms is missing and causes an *edge dislocation*, as shown in *Fig. 3.9*. The result of all such defects is that the surrounding atomic structure is distorted and so is not in its preferred lowest energy state. This has important consequences during loading; when the internal strain energy is sufficient to locally rearrange the structure a dislocation is, in effect,

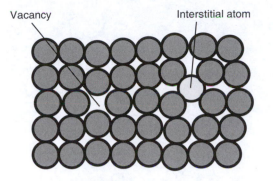

Fig. 3.8 *Vacancy and interstitial defects in a crystal lattice.*

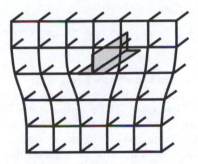

Fig. 3.9 *Edge dislocation.*

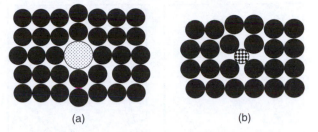

Fig. 3.10 *Distortions from (a) a substitutional impurity and (b) an interstitial impurity.*

moved. The dislocation does not move back to its original position on unloading, and so the resulting deformation is irreversible i.e. it is *plastic*, as shown in *Fig. 2.6*. If the required internal energy needed to trigger the dislocation movement is sharply defined then this gives rise to a distinct yield point (point A in *Fig. 2.6*). We will discuss this dislocation movement in metals in more detail in Chapter 8.

It is also impossible to produce a completely pure material, and some foreign atoms will also be present, thus producing a solid solution. A *substitutional impurity* occurs when the foreign atoms take the place of the parent atoms, resulting in a *substitutional solid solution* (*Fig. 3.10a*). If the atoms of the two materials are of a similar size then there will be little distortion to the atomic lattice, but if their size differs significantly then some distortion will occur. An *interstitial impurity*, as the name implies, occurs when the foreign atoms are forced between the parent atoms, resulting in an *interstitial solid solution*; again the degree of distortion depends on the relative sizes of the atoms involved (*Fig. 3.10b*).

The impurities may occur by chance during manufacture, but nearly all metals used in construction are in fact *alloys*, in which controlled quantities of carefully selected impurities have been deliberately added to enhance one or more properties. We will discuss some important examples of alloys in Part 2 of this book.

3.3 Crystal growth and grain structure

Crystals are formed in a cooling liquid. In the liquid the atoms are in a state of constant motion and change positions frequently. During cooling this motion becomes more sluggish until, sooner or later, the atoms arrange themselves locally into a pattern, often one of those described above, that forms the nucleus of the solid material. The kinetics of nucleation are quite complex, but it almost always begins from an impurity particle in the melt. The nucleus will have the form of the unit cell, which is often a cube. As the liquid solidifies it gives up its latent heat of solidification. The corners of the cube lose heat faster than the edges so that atoms from the melt attach themselves to the corners first, then to the edges and last of all to the faces of the elementary cube. Thus a branching or *dendritic* pattern is built up from each nucleation site (*Fig. 3.11a*) and dendrites will grow from each site until they are stopped by interference from other dendrites (*Fig. 3.11b* and *c*).

Eventually all the liquid is used up by infilling between the arms of the dendrites and the characteristic polycrystalline grain structure results (*Fig. 3.11d*). There are three important facts to note here:

1. Within each grain the atoms are arranged in a regular lattice, albeit containing some or all the defects and imperfections described above.
2. The orientation of the crystal lattice differs from grain to grain.
3. At each grain boundary there is a line of mismatch in the atomic arrangement.

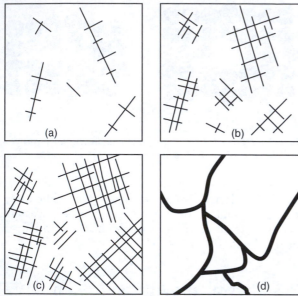

Fig. 3.11 Schematic of dendritic crystal growth and resulting grain boundaries.

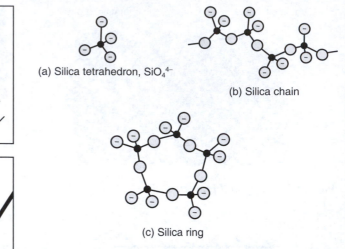

(a) Silica tetrahedron, SiO_4^{4-}

(b) Silica chain

(c) Silica ring

Fig. 3.12 Ionic and structural arrangements of silica, SiO_2.

The size of the individual grains depends on the type of material and more significantly, the cooling rate; larger gains are formed with a slower cooling rate. In many metals, the grains are large enough to be viewed with optical microscopy, which is extremely useful to metallurgists. As we shall see in Chapter 8, the grain size and the grain boundaries have important influences on the mechanical properties of the metal.

3.4 Ceramics

Most ceramics are compounds of metallic and non-metallic elements e.g. silica, which is silicon oxide, SiO_2, or alumina, aluminium oxide, Al_2O_3. The atomic bonding ranges from ionic to covalent and indeed, many ceramics have a combination of the two types (e.g. about half the bonds in silica are ionic and half covalent). Covalent bonds in particular are highly directional and therefore the structure of ceramics is more complex that of the single-element metallic solids described above.

Silicon and oxygen are the two most abundant elements in the earth's crust and so it is not surprising that silica and silicates are both important and widespread. Silica in various forms is a major component of many construction materials, including concrete, aggregates, bricks and glass, and so we will use it to illustrate the type of structures that ceramics can adopt.

Silicon is tetravalent and can form four equally spaced covalent bonds by sharing each of its valence electrons with one of the valence electrons of a divalent oxygen atom. In the resulting tetrahedron each of the four oxygen atoms requires an extra electron to achieve a stable configuration and therefore this is, in effect, an SiO_4^{4-} ion (Fig. 3.12a). This basic unit of silica has the ability to combine with other units and with other elements in a wide variety of ways of varying complexity, giving rise to an enormous number of silica-based materials with a wide range of properties.

For example, if two of the oxygen atoms are shared with other tetrahedra, then either a chain or ring structure can be formed, as shown in Fig. 3.12b and c, respectively, with the overall composition SiO_2. If the ring structure has long-range order then a regular, crystalline material is obtained, but if it has a more random, non-ordered structure then an amorphous, non-crystalline or glassy material results (Fig. 3.13). In general glassy structures are produced by rapid cooling from the molten liquid, as a result of which the basic units do not have time to align themselves in their preferred ordered state.

The oxygen atoms that are not part of the chain or ring bonds are available to form ionic or covalent bonds with other atoms or atomic groups. For example, there is a series of compounds between silica and varying ratios of the oxides of calcium, magnesium and sodium to give, among others, calcium silicates with overall compositions of $CaSiO_3$, Ca_3SiO_5, Ca_2SiO_4, and $Ca_3Si_2O_7$, magnesium silicates such as talc, $Mg_3Si_4O_{10}(OH)_2$, and asbestos, $Mg_3Si_2O_5(OH)_4$, and sodium silicate or water glass (Na_2SiO_3).

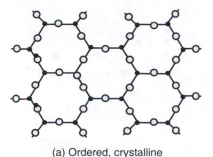 

(a) Ordered, crystalline   (b) Amorphous, glassy, non-crystalline

Fig. 3.13 *Two-dimensional views of the forms of silica, SiO$_2$.*

The strong, directional covalent bonds give rise to the brittle nature of most ceramics, with failure often initiated at a defect in the structure. We will discuss the mechanisms of such failure in some detail when considering the subject of fracture mechanics in the next chapter.

3.5 Polymers

Polymers contain long-chain or string-like molecules of very high molecular weight. They occur naturally in plants and animals or can be synthesised by *polymerisation* of the small molecules in a *monomer*. In construction timber and rubber have for many centuries been the most widely used naturally occurring polymers, but synthetic polymers such as plastics, polyester and epoxy resins, and many types of rubber are of increasing importance.

The backbone of the chain normally consists of covalently-bonded carbon atoms (although some silicon-based polymers, known collectively as *silicones*, are made). The monomer molecule typically contains a double bond between carbon atoms, which reduces to a single bond on polymerisation. The monomer therefore provides the repeat unit in the chain; two examples, polyethylene and polyvinyl chloride (PVC), are shown in *Fig. 3.14*. Polymerising a mixture of more that one type of monomer produces a *copolymer*.

In *linear polymers*, the repeat units are joined in single chains, which intertwine like a mass of string, as illustrated schematically in *Fig. 3.15a*. The covalent bonds in the chains are strong, but the bonding between the chains is due to secondary, Van der Waals bonds (see Chapter 1), which are weaker but in many cases sufficiently strong for the polymer to exist as a solid at normal temperatures.

If the chain has side-branches then a *branched polymer* is formed (*Fig. 3.15b*), often with a lower packing efficiency and hence a lower density than

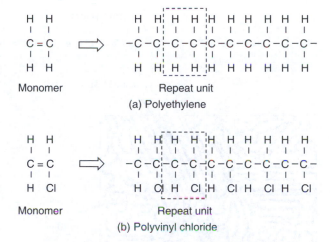

Fig. 3.14 *Monomer and polymer molecules for two common polymers.*

a linear polymer. It is also possible for the chains to be linked by other atoms or molecules that are covalently bonded to adjacent chains, thus forming a *cross-linked polymer* (*Fig. 3.15c*). With sufficient cross-linking then a *networked polymer* results (*Fig. 3.15d*). Cross-linked or network polymers have a more rigid structure than linear polymers, and are often therefore stronger but also more brittle. A polymer is one of two types depending on its behaviour with rising temperature. Thermoplastic polymers soften when heated and harden when cooled, both processes being totally reversible. Most linear polymers and some branched polymers with flexible chains fall into this group. Common examples are polyethylene, polystyrene and polyvinyl chloride. Thermosetting polymers, which harden during their formation, do not soften upon heating; these are mostly cross-linked and networked polymers, and they are generally harder and stronger than thermoplastics. Examples are vulcanised rubber and epoxy resins.

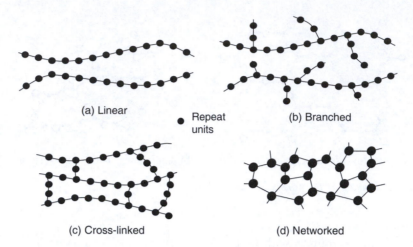

Fig. 3.15 *Schematics of molecular structures of polymers (after Callister, 2007).*

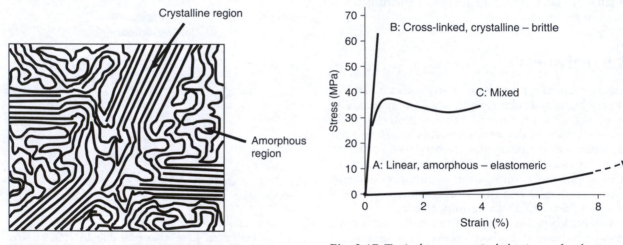

Fig. 3.16 *Schematic of molecular arrangement in polymers.*

Fig. 3.17 *Typical stress–strain behaviour of polymer types (after Callister, 2007).*

The polymer chains can pack together either in a 'random-walk' disordered manner, giving an amorphous structure, or in regular repeating patterns, giving a crystalline structure. Often both types of structure will occur at in different regions of the same polymer, as illustrated in *Fig. 3.16*.

The stress–strain behaviour of polymers is dependent on the extent of the crystallinity and cross-linking. *Figure 3.17* shows three possible forms of stress–strain curve. Curve A is typical of a linear polymer; there are large recoverable strains at low stresses while the intertwined long molecular chains are pulled straighter, followed by an increase in the stiffness as the chains become aligned. Materials exhibiting this type of behaviour are known as *elastomers*.

Polymers that are heavily cross-linked and crystalline can have a high elastic modulus and be very brittle with low failure strains, as in curve B. Many polymers with a mixture of crystalline and amorphous regions and an intermediate level of cross-linking behave as in curve C, i.e. with distinct elastic and plastic behaviour very similar in nature to that of mild steel, which we discussed in Chapter 2.

Reference

Callister WD (2007). *Materials science and engineering. An introduction*, 7th edition, John Wiley and Sons, New York.

Chapter 4

Fracture and toughness

An important consequence of the structures of solids described in Chapter 3 is the nature of the fracture and cracking processes that occur when they are subjected to sufficiently high stress. This is the subject of the branch of materials science called *fracture mechanics*; we will introduce some of the concepts of this in this chapter, including the important property of *toughness*.

The terms 'fracture' and 'failure' are often used synonymously but they are not necessarily describing the same process. In its broadest sense, failure means that a structure or component of a structure becomes unfit for further service; this can be due, for example, to excessive deformation or by reduction of area owing to corrosion or abrasion as well to local breakdown or fracture. Fracture is the separation of a component into two or more pieces under the action of an imposed load, at temperatures low compared to the melting temperature of the material. As we have seen in Chapter 2, this separation can occur under a gradually increasing load, a permanent or static load, leading to creep rupture, a fluctuating load, leading to fatigue, or an impact load. We have also seen that fracture can be of a brittle or ductile nature, depending on the amount of strain before fracture.

We need to be able to predict and analyse fracture, and so we will start by considering predictions of strength from a knowledge of the bonding forces between atoms. This is a logical place to start, but as we will see these estimates turn out to be wildly inaccurate and so we then need to turn to fracture mechanics.

4.1 Theoretical strength

To fracture a material, we need to break the bonds between the individual atoms and make sure that they do not reform. It is therefore instructive to start by considering the energies and forces within the bonds. As well as leading to an estimate of the tensile strength, we will establish the theoretical basis for some of the observed deformation behaviour along the way.

There are both attractive and repulsive forces between atoms, which balance one another when the atoms are in equilibrium. The causes of these forces are somewhat complex, but in simple terms they are mainly due to the gravitational attraction between the two masses (which is concentrated in the nucleus) and the repulsive force between the similarly (negatively) charged electron clouds as they start to overlap. However, whatever the cause, the energies tend to vary as the inverse of the distance between the atoms, raised to some power. So, if the distance between the atoms is r, and A, B, m and n are constants that vary with the material and its structure then:

- the atttractive energy is Ar^{-n}
- the repulsive energy is Br^{-m}
- the resultant energy is $U = Br^{-m} - Ar^{-n}$

Plotting these energies as a function of interatomic spacing gives the *Condon–Morse curves*, shown schematically in *Fig. 4.1a*. *Figure 4.1b* presents the same information, but in terms of the force between adjacent atoms, F, which is the differential of the energy with respect to distance. There are three things to note:

1. The bond energy U is a continuous function of r. Thus we can express the energy as a series:

$$U_{(r)} = U_{(r0)} + r(dU/dr)_{r0} + (r^2/2)(d^2U/dr^2)_{r0} + \ldots$$
$$(4.1)$$

where $U_{(r0)}$ is the energy at $r = r_0$, i.e. the interatomic separation at which the attractive and repulsive forces balance, and the differential is taken at $r = r_0$.

2. The minimum in the curve at r_0 allows the second term to be eliminated, since $(dU/dr) = 0$ at a minimum.

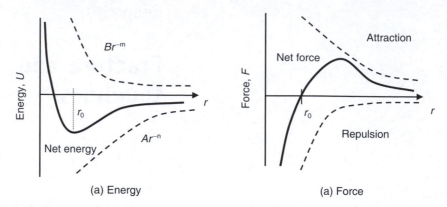

(a) Energy   (a) Force

Fig. 4.1 *Condon–Morse curves of variation of energy and interatomic force with* r, *the interatomic spacing.*

3. The displacement from r_0 is small, so ignoring terms higher than r^2 we obtain:

$$U(r) = U(r_0) + (r^2/2)(d^2U/dr^2)_{r0} \quad (4.2)$$

and hence

$$F = dU/dr = r(d^2U/dr^2)_{r0} \quad (4.3)$$

i.e. the force is proportional to displacement via a constant (d^2U/dr^2).

In other words, the constant of proportionality is the slope of the F–r graph at the equilibrium position where $r = r_0$.

We can use these mathematical facts about these graphs to predict some consequences and try and relate them to the real world. There are in fact a great many consequences but those most relevant to the subject of this book are as follows.

1. When a material is extended or compressed a little, the force is proportional to the extension (equation 2.4). This is Hooke's law. The slope of the F–r curve at $r = r_0$ is the fundamental origin of the elastic constant E (or stiffness).
2. Since the F–r curve is nearly symmetrical about the equilibrium position, the stiffness of a material will be nearly the same in tension and compression. This is, in fact, the case.
3. At large strains, greater than about 10%, the F–r curve can no longer be considered straight and so Hooke's law should break down. It does.
4. There should be no possibility of failure in compression since the repulsive force between the atoms increases *ad infinitum*. This is so.
5. There should be a limit to the tensile strength, since the attractive force between the atoms has a maximum value. This is so.

We can make a theoretical estimate of this tensile strength (σ_{ft}) by assuming that, on fracture, the internal strain energy due to the loading goes to creating new surface. We will discuss surfaces in more detail in Chapter 6, but for the moment we need to use the concept of *surface energy*. Atoms at the surface are bonded to other surface atoms and atoms further into the material; they therefore have asymmetric bonding forces leading to a higher energy state than that of atoms within the material, which have uniform bonding in all directions (*Fig. 6.1*). This excess energy is the surface energy (γ) of the material; it gives rise to the surface tension of liquids, but is perhaps not quite so obvious in solids.

Analysis equating estimates of the surface energy to the internal strain energy immediately before fracture gives, after making some simplifying assumptions about the Condon–Morse curves:

$$\sigma_{ft} = (E\gamma/r_0)^{0.5} \quad (4.4)$$

For many materials, this gives a value for σ_{ft} of about $E/10$. On this basis we would expect the strength of steel to be approximately 20,000 MPa, which is about 10 times higher than the strongest steel that we are capable of producing – a problem!

4.2 Fracture mechanics

Clearly some other explanation than that above is required to explain the values of tensile strength that we obtain in practice. This is provided by fracture mechanics, which arose from the studies of A. A. Griffith in the 1920s on the brittle fracture of glass. Griffith recognised that all materials, no matter how carefully made and how uniform in appearance, contain defects and flaws and it is the propagation or growth of these defects that leads to fracture. These may be microscopic e.g. as in the case of a metal that is made up of fine grains or crystals (as described in Chapter 3) or macroscopic as in the

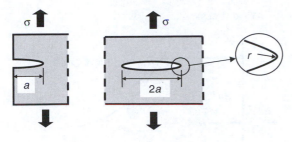

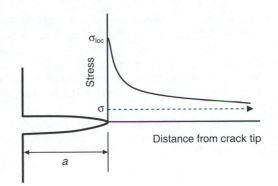

Fig. 4.2 Surface and internal crack geometry.

Fig. 4.3 Variation in local internal stress with distance from crack tip (from Ashby and Jones, 2005).

case of concrete, with large aggregate particles bonded imperfectly together by the surrounding hardened cement. So, in the above analysis we have therefore made some incorrect assumptions about both the stresses within a material and the nature of the fracture process.

First, consider the stresses within the material. We have assumed that the stress acts uniformly across a section, and is therefore simply the load divided by the cross-sectional area over which it is acting. We can think of defects and flaws as cracks, which may be either at the surface or contained within the material. Cracks are usually long and narrow with a sharp tip, and so we can draw them as shown in *Fig. 4.2*, with a length a for a surface crack and $2a$ for an internal crack, and a tip radius r in each case. The cracks act as local stress raisers, with the stress at the crack tip being many times greater than the average stress in the material. It is possible with stress analysis techniques to show that the local stress (σ_{loc}) is highest at the crack tip and is given by

$$\sigma_{loc} = \sigma[1 + 2(a/r)^{0.5}] \qquad (4.5)$$

$$= 2\sigma(a/r)^{0.5} \quad \text{for small } r \qquad (4.6)$$

We can also define the *stress concentration factor*, k_t, as

$$k_t = \sigma_{loc}/\sigma = [1 + 2(a/r)^{0.5}] \text{ or } 2(a/r)^{0.5} \text{ for } r << a \qquad (4.7)$$

For a circle, $a = r$ and so $k_t = 3$, but for a small sharp crack, say $a = 1$ mm and $r = 1$ μm then $k_t = 63$. It is therefore easy to see how the local stress can reach the theoretical strength that we estimated above, and hence the atoms will be pulled apart.

This is, however, only part of the picture. Further stress analysis shows that that the local stress σ_{loc} quickly diminishes with distance from the crack, as shown in *Fig. 4.3*. The atoms will be torn apart near the crack tip, but the crack will only grow (or propagate) through the material if there is also

sufficient energy in the system to keep driving it. This energy is, of course, the internal strain energy caused by the loading. In ductile materials, as well as creating new surface this energy is also consumed in plastic deformation of the material in the region where the local stress is higher than the yield stress, which may be some distance in advance of the crack tip. Even in brittle materials with no plastic deformation some localised microcracking may occur in this region.

Therefore when considering the balance between the energy consumed by the crack propagation that is available from the internal strain energy we need to consider more than just the energy of the new crack surface. We do this by defining the total energy consumed when a unit area of new crack is formed as the *toughness* (G_c) of the material. G_c is a material property, with units of energy/area e.g. J/m².

In brittle materials, such as glass, the surface energy is a significant part of the total energy, G_c is low and crack propagation occurs readily at low strains. In ductile materials, such as mild steel, the opposite holds; the energy required for plastic deformation as a crack advances is many orders of magnitude higher than the new surface energy, G_c is high and failure requires much higher strains.

Analysis of the energy balance gives the value of the stress to cause fracture (σ_f) as

$$\sigma_f = (G_c E/\pi a)^{0.5} \qquad (4.8)$$

for the condition of plane stress, which occurs when the material is relatively thin in the direction perpendicular to the applied load, and

$$\sigma_f = [G_c E/\pi(1 - \upsilon^2)a]^{0.5} \qquad (4.9)$$

(where υ = Poisson's ratio)

for the condition of plane strain in thicker sections.

We therefore now have equations for the failure stress, σ_f, in terms of three material properties (G_e, E and υ) and a defect size, a. A greater defect size would lead to a lower strength, which is logical, but this is a fundamental difference to the previous idea that the strength of a material is a constant when defined in terms of stress.

Rearranging equation (4.8) shows that failure will occur when the combination of applied stress and crack size satisfy

$$\sigma(a\pi)^{0.5} = (G_c E)^{0.5} \qquad (4.10)$$

For any combination of applied stress and crack length the term $\sigma(a\pi)^{0.5}$ is called the *stress intensity factor*, denoted by K. At the combination of stress and crack length to cause fracture the value of K is called the *critical stress intensity factor*, K_c. K_c is more commonly known as the *fracture toughness*, and it follows that:

$$K_c = (G_c E)^{0.5} \qquad (4.11)$$

This choice of name is a little confusing, but you must remember that toughness, G_c, and fracture toughness, K_c, are different properties with different values and different units. K_c has units of force/length$^{3/2}$, e.g. MN/m$^{3/2}$. The values of G_c and K_c can vary widely for different materials, and ranges of both properties for the most common construction materials are shown in *Table 61.1*.

Substitution of K_c into equation 4.10 and re-arranging shows that for any particular applied stress there is critical crack length, a_{crit}, that will result in fracture, given by

$$a_{crit} = K_c^2/\pi\sigma^2 \qquad (4.12)$$

When equation 4.12 is satisfied, the crack will propagate rapidly to failure i.e. there will be a *fast fracture*.

With brittle materials, without the capacity for plastic deformation and which consequently have low values of G_c and hence K_c, small defects, of the same order of size as might occur during manufacture, are sufficient for fast fracture. The result is a so-called *cleavage* type of facture. Conversely, ductile materials, which have the capacity for significant plastic flow and yielding, have higher G_c and K_c values, and therefore defects from manufacturing are not sufficient for fast fracture to occur before excessive yielding and a *tearing* form of facture occur. However, it follows that, with large cracks (e.g. from some previous damage) or stress concentrations from poor design detailing, fast fracture may occur before yielding, a potentially dangerous situation.

The crack size at which failure in a ductile material changes from yielding and tearing into fast fracture

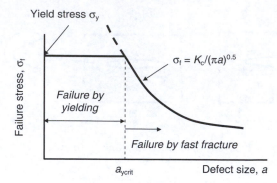

Fig. 4.4 *Variation of failure stress with defect size for a ductile material.*

and cleavage, the *critical yield crack length*, a_{ycrit}, is obtained by substituting the yield stress σ_y into equation 4.12, giving

$$a_{ycrit} = K_c^2/\pi\sigma_y^2 \qquad (4.13)$$

A plot of failure stress against defect size for ductile materials is of the form shown in *Fig. 4.4*. The value of a_{ycrit} is a significant property, since if the material contains cracks larger than this failure will not only occur at a stress lower than expected but also without warning; in other words a 'safe' ductile material behaves as a 'dangerous' brittle material. It is therefore useful to know a_{ycrit} not only for design but also when inspecting structures during their service life.

Toughness and fracture toughness are sometimes described as the ability of a material to tolerate cracks. Lack of toughness can lead to fast fracture, which is extremely dangerous as it can occur without warning and will often result in catastrophic failure and loss of life. Engineers therefore need to take great pains to avoid it occurring. Clearly one way of doing this is to avoid using materials with low G_c or K_c values. However, this is not always possible, e.g. construction without using any concrete is difficult to contemplate, and so we should then either:

- ensure that the loading will not cause high tensile stress and/or
- reinforce against fast fracture (or at least reduce its risk) by forming a composite, e.g. by adding steel reinforcement to concrete or fibres to resins.

Reference

Ashby MF and Jones DRH (2005). *Engineering Materials, Vol 1: An introduction to properties, applications and design*, 3rd edition, Elsevier Butterworth Heineman, London.

Chapter 5

Liquids, viscoelasticity and gels

5.1 Liquids

Liquids are effectively incompressible when subjected to direct stress (hydraulic power systems depend on this property), which implies that the elementary particles (atoms or molecules) are in direct contact. However, liquids obviously flow under the action of the shear stress, which shows that the particles are able to move relative to each other i.e. there are no primary bonds between them. In describing their behaviour we are therefore concerned with the relationship between the applied shear stress, τ, and the rate of shear strain, $d\gamma/dt$ (Fig. 5.1). Most simple liquids, such as water, white spirit, petrol, lubricating oil etc., and many true solutions, e.g. sugar in water, show 'ideal' or Newtonian behaviour where the two are directly proportional i.e.

$$\tau = \eta \, d\gamma/dt \qquad (5.1)$$

where η = coefficient of viscosity (or, strictly, the dynamic viscosity) which, as strain is dimensionless, has units of stress × time e.g. Pa.s.

This definition also applies to gases, but as might be expected, the viscosities of gases and liquids differ markedly. At 20°C, η for air is about 1.8×10^{-5} Pa.s and for water it is about 1×10^{-3} Pa.s. In both types of fluid at higher temperatures the particles possess more energy of their own and the stress required to move them is reduced, i.e. viscosity reduces rapidly as temperature is increased (think of asphalt road surfaces on very hot days).

We are often faced with liquids that contain dispersions of solid particles. These disturb and effectively increase the viscosity; for small volume fractions of particles the viscosity is given approximately by

$$\eta = \eta_0[1 + \alpha V_f] \qquad (5.2)$$

where η_0 = the viscosity of the pure fluid, V_f = the volume fraction of particles and α is a constant. The value of α varies with the shape of the particles; Einstein showed that $\alpha = 2.5$ for spheres, but it is higher for irregular particles.

This equation breaks down when the volume fraction of the particles increases to the point where the perturbed regions in the liquid begin to overlap, and terms in V_f^2 appear. Materials such as pastes, clays and freshly mixed mortar and concrete, which have solids contents in excess of, say 70%, can deform more or less elastically up to a certain *yield stress* and can preserve their shape against gravity. Above this stress, however, they behave like liquids and deform rapidly, e.g. toothpaste does not flow off your toothbrush but you can brush it around your teeth, clays can be moulded to the shape desired by the potter, concrete can be shovelled and vibrated into the formwork. The shear stress–strain rate curve (known as the *flow curve*) for these materials can take a variety of forms, as shown by the solid lines in *Fig. 5.2*. The general equation for the three types of behaviour that have a positive value of the yield stress is:

$$\tau = \tau_o + a(d\gamma/dt)^n \qquad (5.3)$$

where τ_o is the intercept on the shear stress axis and a and n are constants. The three curves have different values of n. In *shear thinning* behaviour, the curve is convex to the shear stress axis and $n < 1$; in *shear thickening*, the curve is concave to the shear stress axis and $n > 1$. The particular case of a

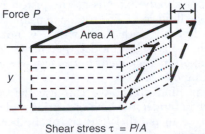

Force P

Area A

x

y

Shear stress $\tau = P/A$
Shear strain rate = $d\gamma/dt = d(x/y)dt$

Fig. 5.1 Shear stress and shear strain rate for fluid flow.

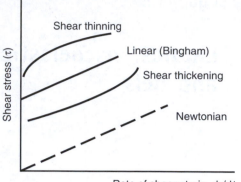

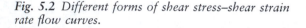

Fig. 5.2 *Different forms of shear stress–shear strain rate flow curves.*

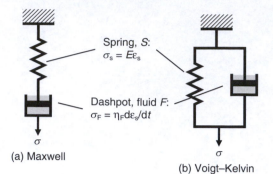

Fig. 5.3 *Viscoelastic models.*

straight-line relationship is called *Bingham* behaviour, for which $n = 1$. The equation for this is normally written as:

$$\tau = \tau_y + \mu.(d\gamma/dt) \qquad (5.4)$$

where τ_y is the *yield stress*, and μ is the *plastic viscosity*.

This is of particular interest for concrete technologists, as fresh concrete has been shown to conform reasonably well to this model. We will discuss this further in Chapter 18.

5.2 Viscoelastic behaviour

In many cases it is not possible to draw a sharp dividing line between the mechanical behaviour of liquids and solids; there is a large group of materials, known as *viscoelastics*, whose behaviour is part liquid and part solid. Many natural materials, e.g. tendons, plant fibres and wood, behave in this way. Of engineering materials, rubbers, many soft polymers and substances like tar and asphalt are examples.

We have already briefly discussed such behaviour in Chapter 2 when defining the two separate but allied cases of creep and stress relaxation. Under constant stress, a material responds by steadily increasing strain; under constant strain, stress relaxation occurs without dimensional change (*Figs 2.15* and *2.17*). Some of the microstructural mechanisms of this behaviour in different materials will be discussed in later parts of this book, but here we introduce how the behaviour can be modelled by using mechanical analogues consisting of arrays of springs that behave according to Hooke's law i.e. stress ∝ strain, and

viscous elements that behave as an ideal Newtonian liquid, i.e. stress ∝ rate of strain.

One such array, known as the Maxwell model, is shown in *Fig. 5.3a*. It consists of an elastic spring, S, of modulus E in series with a dashpot, i.e. a piston moving in a fluid, F, of viscosity η contained in a cylinder. Now think of suddenly applying a constant strain. At first all the strain is taken up by stretching the spring and the load required to do this is calculated from the strain in the spring. Later, the spring shortens by pulling the piston up through the fluid in the dashpot. Some of the total strain is now taken up by the movement of the piston and less by the stretch in the spring. The load required is now less than before, and thus the system is exhibiting stress relaxation. Mathematical analysis gives:

$$\sigma_t = \sigma_o \exp(-t/\tau) \qquad (5.5)$$

where σ_o is the initial applied stress, σ_t is the stress sustained at time t and $\tau = \eta/E$ is the so-called *relaxation time*. Under constant strain the stress decays exponentially, which is reasonably close to observed behaviour. In fact, τ is the time taken for the stress to decay to $1/e$ of its initial value.

Now take the case of applying a constant load or stress. The spring stretches and remains at that strain as long as the load remains. At the same time the dashpot slowly extends as the piston is pulled through the fluid in it. The total extension therefore increases linearly with time, which is not typical creep behaviour. This model therefore represents stress relaxation very well, but is less successful at representing creep.

For modelling creep, we can use the so-called Voigt–Kelvin model in which the spring and dashpot are arranged in parallel (*Fig. 5.3b*). Both elements must experience the same strain at any given time but load can be transferred over time from one element to the other. Analysis of the model gives:

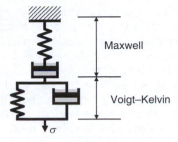

Fig. 5.4 *Four element viscoelastic model.*

$$\varepsilon_t = \sigma(1 - e^{-t/\tau})/E \qquad (5.6)$$

where ε_t = strain at time t, σ = applied stress and E and τ are as before.

This gives a good representation of creep behaviour, but not of relaxation. To get out of these difficulties, the two types of models are combined into what is known as the four-element model (*Fig. 5.4*). This gives a reasonable representation of both creep and relaxation in many cases, but where the viscoelastic material is a polymer consisting of many molecules and particles of varying size and properties, many elements with different relaxation times (i.e. a *relaxation spectrum*) need to be combined.

Nevertheless, the concept of relaxation times is important for two reasons. First, it helps us to distinguish between solids and liquids. A perfect solid will support the stress indefinitely, i.e. $\tau = \infty$, but for a liquid, relaxation is virtually instantaneous (for water $\tau \approx 10^{-11}$s). In between there is a grey area where stress relaxation may occur over a few seconds or centuries.

Second, we have the relationship between the relaxation time and the time scale of the loading t. If the load is applied so fast that relaxation cannot occur ($t \ll \tau$) the material will effectively behave elastically, but under slow loading ($t \gg \tau$) it will flow. This was one of the effects that we mentioned when considering impact loading in Chapter 2. An extreme case is the well-known 'potty putty', which bounces when dropped or thrown against a wall but collapses into a puddle under its own weight when left alone. Potty putty is a silicone-based inorganic polymer, and many other polymeric materials also show marked sensitivity to loading speed.

There are two important consequences of viscoelasticity. The first is that the stress–strain relationship is non-linear. We noted that in an elastic or Hookean solid the strain energy stored on loading is completely recovered when we unload. *Figure 5.5* shows that for a viscoelastic material the energy recovered on unloading is less than that stored during loading.

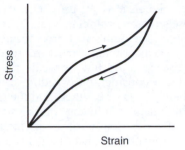

Fig. 5.5 *Loading/unloading behaviour for a viscoelastic material.*

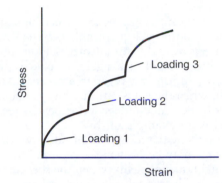

Fig. 5.6 *Boltzmann's superposition principle.*

This energy must go somewhere, and normally this is into heat, which explains why car tyres get hot after a few miles in which they are repeatedly loaded and unloaded.

The second consequence is known as Boltzmann's superposition theory. This states that each increment of load makes an independent and additive contribution to the total deformation. Thus, under the loading programme shown in *Fig. 5.6*, the creep response is additive and the total creep is the sum of all the units of incremental creep. This is useful in the analysis of varying load levels on the creep behaviour of concrete and soils.

5.3 Gels and thixotropy

There is a group of materials that show a mixture of solid and liquid behaviour because they are just that – a mixture of a solid and a liquid. One of the most familiar of these is the gel, known to most of us from childhood in the form of jellies and pastilles.

Gels are formed when a liquid contains a fairly concentrated suspension of very fine particles, usually of colloidal dimensions (<1 μm). The particles bond into a loose structure, trapping liquid in its interstices. Depending on the number of links formed, gels can vary from very nearly fluid structures to almost rigid solids. If the links are few or weak, the individual particles have considerable freedom of movement around their points of contact, and the gel deforms easily. A high degree of linkage gives a structure that is hard and rigid in spite of all its internal pores. The most important engineering gel is undoubtedly hardened cement, which develops a highly rigid but permeable structure of complex calcium silicates by the chemical reaction between the fresh, powdered cement and water. We discuss this in some detail in Chapter 13.

A feature of many gels is their very high specific surface area; if the gel is permeable as well as porous, the surface is available for adsorbing large amounts of water vapour, and such a gel is an effective drying agent. Adsorption is a reversible process (see section 6.4); when the gel is saturated it may be heated to drive off the water and its drying powers regained. Silica gel is an example of this.

If a gel sets by the formation of rather weak links, the linkages may be broken by vigorous stirring so that the gel liquefies again. When the stirring ceases, the bonds will gradually link up and the gel will thicken and return to its original set. Reversible behaviour of this sort, in which an increase in the applied shear rate causes the material to act in a more fluid manner, and vice-versa, is known as *thixotropy*. A familiar application of this is in non-drip paints, which liquefy when stirred and spread easily when being brushed on, but which set as a gel as soon as brushing is completed so that dripping and streaks on vertical surfaces are avoided. Clays can also exhibit thixotropy. This is turned to advantage on a potter's wheel and in the mixing of drilling muds for oil exploration. The thixotropic mud serves to line the shaft with an impermeable layer, whilst in the centre it is kept fluid by the movement of the drill and acts as a medium for removing the rock drillings. On the other hand, a thixotropic clay underlying major civil engineering works could be highly hazardous.

The reverse effect to thixotropy occurs when an increase in the applied shear rate causes a viscous material to behave more in the manner of a solid, and is known as *dilatancy*. It is a less familiar but rather more spectacular phenomenon. Cornflour–water mixtures demonstrate the effect over a rather narrow range of composition, when the viscous liquid will fracture if stirred vigorously. It is of short duration, however, since fracture relieves the stress, and the fractured surfaces immediately liquefy and run together again.

Chapter 6

Surfaces

All materials are bounded by surfaces, which are interfaces of varying nature. For the engineer the most important are the liquid–vapour, solid–vapour, solid–liquid and solid–solid interfaces. The last of these can be the boundary between two differing solid phases in a material, e.g. cement gel and aggregate in concrete, or between two similar crystals that differ only in orientation, e.g. the grain boundaries in a pure metal or, at the macroscopic level, as the interface between structural components, e.g. concrete and steel. Surfaces owe their interest and importance to two simple features:

- they are areas of abnormality in relation to the structure that they bound, and
- they are the only part of the material accessible to chemical change, i.e. all chemical changes and, for that matter, most temperature changes take place at or through surfaces.

The influence of surfaces on the bulk behaviour of materials depends on the ratio of surface area to the total mass. This in turn depends partly on the size and partly on the shape of the individual particles making up the bulk material. An extreme example is clay, which is composed of platelets typically 0.01 µm thick by 0.1 µm across; they are therefore very small with a high surface area/volume ratio. One gram of montmorillonite clay, rather smaller than a sugar cube, may contain a total surface area of over 800 square metres! Porous structures such as hardened cement and wood also contain enormous internal surface areas that exert a considerable effect on their engineering properties.

6.1 Surface energy

As we briefly discussed in Chapter 4, all surfaces have one thing in common: the atoms, molecules or ions at the surface are subjected to asymmetric or unsaturated bonding forces (*Fig. 6.1*). Since bond-

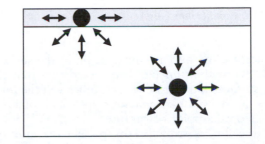

Fig. 6.1 *Asymmetric and symmetric forces in surface and internal atoms, molecules or ions.*

ing is taken to lower their energy (*Fig. 4.1*) the surface atoms or ions will be in a state of higher energy than interior ones. This excess energy is known as the *surface energy* of the material. In solids the presence of the surface energy is not immediately apparent, since the atoms in the surface are held firmly in position. However, with liquids the mobile structure permits the individual atoms to respond, and the result is the well-known surface-tension effect. Since surfaces are high-energy regions they will always act to minimise their area, and thus lower their energy, when possible. If a soap film is stretched across a frame with a movable wire as in *Fig. 6.2*, the force required to hold the wire in place is:

$$F = 2\gamma l \qquad (6.1)$$

where l is the length of the wire, γ is the surface tension of the soap film/air interface and the factor 2 is introduced because the film has two surfaces. The surface tension is in the plane of the soap–air interface and has units of force per unit length (N/m). It is important to note that surface tension differs from an elastic force acting between the surface atoms in that it remains constant whether the film is forced to expand or allowed to contract. This is because the work done in expanding the film is used to bring additional atoms to the surface rather than

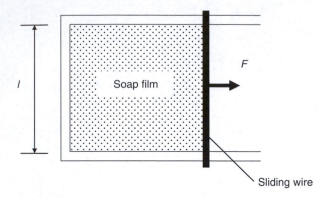

Fig. 6.2 Equilibrium between soap film and applied force.

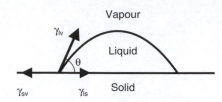

Fig. 6.3 Surface forces at the edge of a droplet.

to increase the interatomic spacing in the surface. Only when the film has become so thin that the two surfaces interact with each other will the force show partial elastic behaviour, by which time the film is on the point of rupture.

If the film is stretched by pulling the movable wire through a distance d, the work done on it, $2l\gamma d$, is stored as surface energy of the newly created surface of area $2ld$, therefore the surface energy per unit area is $2l\gamma d/2ld = \gamma$. Thus surface tension and surface energy are numerically equal with units of Jm^{-2} (= Nm^{-1}). It should be noted that γ tends to be used interchangeably for both surface tension and energy.

6.2 Wetting

We are all familiar with liquids wetting a solid surface. Clearly intermolecular forces are involved, and the behaviour is another example of a system seeking to minimise its total energy. The degree of wetting can be defined by the contact angle (θ) between the liquid–vapour interface and the solid–liquid interface at the edge of a droplet (*Fig. 6.3*). If the conditions for wetting are favourable, then the contact angle is low, and the liquid will spread over a large area. If the conditions are less favourable then the contact angle will be higher and the liquid will form droplets on the surface. If θ is greater than 90° then the surface is often said to be unwettable by the liquid.

The behaviour depends on the relative magnitudes of three surface tensions or energies: liquid–solid, γ_{ls}, liquid–vapour, γ_{lv}, and solid–vapour, γ_{sv}. At the edge of the droplet the three tensions will act as shown in *Fig. 6.3* and the equilibrium condition, resolved parallel to the solid surface is:

$$\gamma_{sv} = \gamma_{ls} + \gamma_{lv} \cos \theta \qquad (6.2)$$

Since the limits to $\cos \theta$ are ±1, a restriction of this equation is that it does not apply if either γ_{sv} or γ_{ls} is larger than the sum of the other two surface energies. If either of these is not the case, then:

- if $\gamma_{sv} \geq \gamma_{ls} + \gamma_{lv}$, then there is complete wetting (in effect, $\theta = 0°$)
- if $\gamma_{ls} \leq \gamma_{sv} + \gamma_{lv}$, then there is no wetting (in effect, $\theta = 180°$).

The quantity $\gamma_{sv} - (\gamma_{ls} + \gamma_{lv})$ is called the spreading force or spreading parameter and the behaviour depends on whether it is positive or negative:

- If it is positive then there is complete wetting (the first case above) of the solid surface; clearly the energy of such a system is lowered when the solid–vapour interface is replaced by a solid–liquid and a liquid–vapour interface.
- If it is negative then $\theta > 0°$ and partial or little wetting occurs.
  - if $\gamma_{sv} > \gamma_{ls}$ then $\gamma_{lv} \cos \theta$ (equation 6.2) is positive and $\theta < 90°$, giving partial wetting
  - if $\gamma_{sv} < \gamma_{ls}$ (which is comparatively rare, provided the surfaces are clean) then $\gamma_{lv} \cos \theta$ is negative and $\theta > 90°$, giving little or no tendency to wetting.

The rise of water in a capillary tube is a consequence of the ability of water to wet glass. If, in *Fig. 6.4*, θ is the angle of contact between water and glass, the water is drawn up the tube by a circumferential force $2\pi r \gamma_{ls} \cos \theta$, so that:

$$2\pi r \gamma_{lv} \cos \theta = \pi r^2 h \rho \qquad (6.3)$$

where $\pi r^2 h \rho$ is the weight of water in the capillary (ρ = unit weight of water), neglecting the weight of water contained in the curve of the meniscus. It follows that the height of the water in the capillary is:

$$h = 2\gamma_{lv} \cos \theta / \rho r \qquad (6.4)$$

If r is small, h will be large. This gives rise to the phenomenon with the general name of *absorption*,

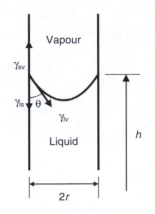

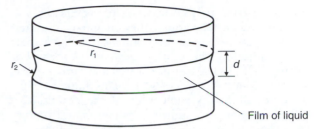

Fig. 6.5 *Adhesive effect of a thin film of liquid between two flat plates.*

Fig. 6.4 *Capillary rise of liquid up a tube.*

where water (or any other liquid) is sucked into the continuous capillaries within a porous material. Two examples of such materials are brick and concrete; in both of these the pores are small and if they were all continuous then h could reach 10 m – extreme rising damp! In practice, the pores are not continuous and evaporation keeps the level lower than this, but it is still a significant problem.

6.3 Adhesives

The ability of adhesives to spread and thoroughly wet surfaces is critical. The adhesion of a liquid to a solid surface is clearly relevant and the liquid may also have to penetrate a thin joint e.g. when repairing cracks with a resin of when soldering or brazing metals. The work needed to break away the adhesive (which may be considered as a viscous liquid) from the solid is the work required to create a liquid–vapour and a solid–vapour interface from an equivalent area of liquid–solid interface, i.e. it is the work to totally 'de-wet' the solid surface. Hence the work to cause breakage at the interface, per unit area, is given by:

$$W = \gamma_{lv} + \gamma_{sv} - \gamma_{ls} \qquad (6.5)$$

But from equation 6.2:

$$\gamma_{sv} - \gamma_{ls} = \gamma_{lv} \cos\theta \qquad (6.6)$$

and therefore:

$$W = \gamma_{lv}(1 + \cos\theta) \qquad (6.7)$$

Thus, the liquid–solid adhesion increases with the ability of the adhesive to wet the solid, reaching a maximum – when $\theta = 0°$ and wetting is complete – given by:

$$W = 2\gamma_{lv} \qquad (6.8)$$

For this to be the case $\gamma_{sv} > \gamma_{lv}$ (equation 6.6) and under these conditions fracture will occur within the adhesive, since the energy necessary to form two liquid–vapour interfaces is less than that needed to form a liquid–vapour and a solid–vapour interface.

Surface tension is also the cause of the adhesion between two flat surfaces separated by a thin film of liquid. Where the surface of the liquid is curved (as for example in *Fig. 6.5*) there will be a pressure difference p across it; if the curvature is spherical of radius r, then:

$$p = 2\gamma/r \qquad (6.9)$$

In the case of two circular discs, however, the surface of the film has two radii of curvature, as shown in *Fig. 6.5*; r_1 is approximately equal to the radius of the discs and presents a convex surface to the atmosphere whilst $r_2 \approx d/2$, where d is the thickness of the film between the plates, and presents a concave surface to the atmosphere. The pressure difference between the liquid film and its surroundings is now given by:

$$p = \gamma\left(\frac{1}{r_1} - \frac{1}{r_2}\right) = \gamma\left(\frac{1}{r_1} - \frac{2}{d}\right) \qquad (6.10)$$

If $d \ll r_1$, then:

$$p = -\frac{2\gamma}{d} \qquad (6.11)$$

the negative sign indicating that the pressure is lower within the liquid than outside it. Since the pressure acts over the whole surface of the discs, the force needed to overcome it and separate the discs is given by:

$$F = \frac{2\pi r_1^2 \gamma}{d} \qquad (6.12)$$

The magnitude of F thus depends on the factor r_1^2/d, and it is therefore important that surfaces to be joined should be as flat and closely spaced as possible. If you have tried to pull apart two wet glasses you will know how tenaciously they cling to each other; by contrast, however, they can easily be slid apart since liquid films have little resistance to shear. For example, If $d = 0.01$ mm, $r = 100$ mm and γ(water) $= 0.073$ Nm^{-1} then $F \approx 460$ N. This value of F for a liquid film gives some idea of the potential of adhesives that gain additional strength and rigidity by setting to highly viscous materials on polymerisation or solvent evaporation.

6.4 Adsorption

The ability of liquids to wet solids depends very much on the cleanliness of the solid, as anyone with any experience of soldering will appreciate. The presence of any dirt, such as oxide or grease films, will totally alter the balance of surface tensions discussed above and usually prevents wetting.

Clean surfaces, in fact, are so rare as to be virtually non-existent, since the broken surface bonds will readily attract to themselves any foreign atoms or molecules that have a slight affinity for the surface material. This effect is known as *adsorption* and by satisfying or partially satisfying the unsaturated surface bonds it serves to lower surface energy. Adsorption is a dynamic process, i.e. molecules are constantly alighting on and taking off from the surface. Different molecules adsorb with varying degrees of intensity, depending on the nature of the bond that they are able to form at the interface. The strength of the bond may be expressed in terms of ϕ_a, the *energy of adsorption*. As in the case of interatomic bonds, a negative value of ϕ_a is taken to indicate positive adsorption, i.e. the molecules are attracted to the interface, and the surface energy (tension) is thereby lowered. A positive value of ϕ_a indicates a repulsive interaction and the molecules avoid the surface. Typical plots of ϕ_a against the distance of the adsorbed layer from the surface are given in *Fig. 6.6*. They closely resemble the Condon–Morse curves (*Fig. 4.1*) and their shape is due to the same circumstance of equilibrium between attractive and repulsive forces, although the attraction is far weaker than that of the principal interatomic bonds.

If the molecule being adsorbed is non-polar and does not react chemically with the surface, absorption, if it occurs, will be by Van der Waals bonds, and the minimum value of ϕ_a is small (curve 2 in *Fig. 6.6*). If on the other hand the molecule is strongly polar, as is the case with water or ammonia, the electrostatic forces between the surface and the charged portion of the molecule give rise to stronger

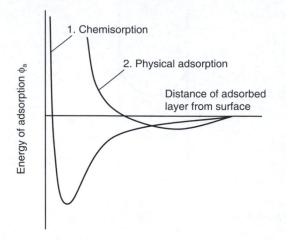

Fig. 6.6 Energies of adsorption for different adsorption mechanisms.

bonding. If a chemical reaction occurs as part of the bonding mechanism, e.g. when fatty acid in a lubricant forms an adsorbed layer of metallic soap on a metal surface, the bonding is still stronger (curve 1) and the effect is referred to as *chemisorption*.

The behaviour of water is of particular importance in this context. Because of its ability to form hydrogen bonds with neighbouring molecules, water adsorbs rapidly and strongly on most solid surfaces. Despite the tenacity with which such a layer is held (clay does not lose all its adsorbed water until heated to 300°C), the interaction cannot be thought of as chemisorption; rather it is, in a sense, a halfway stage to solution or alternatively to the taking up of water of hydration (see below). Bonding is strong enough to maintain a surface layer perhaps several molecules thick, but the affinity is not sufficient for the molecules to penetrate into the interstices of the structure.

The physical nature of such a film is difficult to visualise; it cannot be thought of as a fluid in the accepted sense of the term even when more than one molecule thick, as in the case of the clays and cements. Yet the molecules are mobile in this situation. They will not desorb readily, but they can diffuse along the surface under the impetus of pressure gradients. Such movements, occurring over the vast internal surface area of cement gels, are primarily responsible for the slow creep of concrete under stress (see section 20.6). The ability of water molecules to penetrate solid–solid interfaces in clays and build up thick adsorbed layers results in the swelling of clays, and has caused considerable structural damage to buildings erected on clays that are liable to behave in this manner. The readiness with which water will adsorb on surfaces is advantageous in the case of porous silica gel and molecular sieves being used as drying agents.

6.5 Water of crystallisation

As well as water being associated with a material by absorption into capillaries or adsorption by the surface, many ionic crystals contain water molecules locked up in their structure as water of crystallisation. Such crystals are known in general as *hydrates*, and their formation can be very important in the development of bulk strength. Both cement and 'plaster of Paris' owe their importance to their ability to take up water and form a rigid mass of interlocking crystals. However, in the case of cement the crystals are so small that it is difficult to decide whether to classify the structure as a crystalline hydrate or a hydrated gel.

Water and ammonia – both of which are small and strongly polar – are the only two molecules that can be taken up as a structural part of crystals. Their small size permits them to penetrate into the interstices of crystal structures where close packing of ions is not possible. This is particularly the case where the negative ion is large, such as SO_4 (sulphate) or SiO_4 (silicate). It must be emphasised that this process is not to be thought of as a capillary action, as in the take-up of large amounts of water by clays. The water molecules are bonded into definite sites within the crystal structure, and the crystal will only form a stable hydrate if ions of the appropriate signs are available and correctly placed to form bonds with the positively and negatively charged regions of the water molecule. We have already mentioned (section 1.2) the abnormal properties of water arising from the ability of the molecules to link up by means of hydrogen bonds; the formation of crystalline hydrates is an extension of the same behaviour.

Normally the water molecules cluster round the positive ions in the crystal, forming hydrated ions. This has the effect of making the small positive ion behave as if it were a good deal larger. As a result, the size difference between the positive and negative ions is effectively reduced, thus making possible simpler and more closely packed crystal structures. Water bonded in this manner is very firmly held in many instances, so that hydration becomes virtually irreversible. Cement, for example, retains its water of crystallisation up to temperatures of ~900°C.

Chapter 7

Electrical and thermal properties

Electrical conductivity is not normally a constraint in structural design, but thermal conductivity and thermal expansion are important, for example in the fabric of a building and when estimating diurnal or seasonal expansion and contraction. We shall nevertheless first briefly consider electrical conductivity since it provides a basis for the more complex ideas of thermal properties.

7.1 Electrical conductivity

Electrical conductivity is defined as the current per unit cross-sectional area of a conductor per unit voltage gradient along the conductor, and hence has units of reciprocal ohm.m or $(ohm.m)^{-1}$. Current flow involves the flow of electrons, and the conductivity depends on the ease of this flow. When an electrostatic field is applied, the electrons 'drift' preferentially and, being negatively charged, this drift is towards the positive pole of the field. The resistance to drift is provided by the 'stationary' ions. The force F tending to accelerate each electron is:

$$F = Ee \qquad (7.1)$$

where e is the charge on the electron and E is the electrostatic field. The bodily movement or flux of the electrons J, i.e. the current per unit cross-sectional area, can be expressed as:

$$J = neV_a \qquad (7.2)$$

where n is the concentration of free electrons and V_a is the *drift velocity*.

In metals the metallic bond (see section 1.2) results in a large concentration of free electrons and hence the flux and the conductivity are high, being in the range of 0.6×10^7 to 6×10^7 $(ohm.m)^{-1}$ (but note that this is still a range of an order of magnitude). In materials with strong ionic or covalent bonds, e.g. many polymers, the electrons are much less mobile and hence the conductivity is up to 27 orders

of magnitude lower; values range from 10^{-10} to 10^{-20} $(ohm.m)^{-1}$ e.g. for polyethelene it is 10^{-16} $(ohm.m)^{-1}$. For most applications these conductivities are effectively zero and hence such materials are used as insulators. There is an intermediate group of materials with weaker covalent bonds and therefore more mobile electrons, for example silicon and germanium. These are the semiconductors that are so important electronic applications, with conductivities ranging from 10^{-6} to 10^4 $(ohm.m)^{-1}$.

7.2 Thermal conductivity

Thermal conductivity is defined as the rate of heat transfer across a unit area of a material due a unit temperature gradient. It therefore has units of $Wm^{-1}K^{-1}$. In metals, heat transfer follows much the same general principle as electrical conductivity although it is not the bodily movement of electrons but rather the transference of energy between them by collision. The analysis is, however, much more complicated but it is intuitively obvious that the higher the temperature, the greater the excitation of the electrons and the larger the number of collisions. Typical values range from about 20 to 400 $Wm^{-1}K^{-1}$. Since both thermal and electrical conductivities have their origins in the same structural features they are roughly proportional.

The thermal conductivity of non-metals is more complex still, since it involves energy transfer between the atoms that make up the material. Heating results in increasing vibration of the nominally stationary atoms, increasing energy transfer between them. Values for polymers, in which the atoms are rigidly held, are typically in the range 0.2 to 0.5 $Wm^{-1}K^{-1}$. Ceramics also have thermal conductivities generally lower than metals, but the values, which vary over a wide range, are dependent on the microstructure. Where this is crystalline, with the atoms tightly packed and therefore readily able to transmit energy, the conductivity can be relatively high, for example 90 $Wm^{-1}K^{-1}$ for silicon carbide, but glasses, which have an amorphous, loosely packed structure, have values of the order of 1 $Wm^{-1}K^{-1}$. Not surprisingly, thermal conductivity is also related to density, with light materials with an open structure, such as cork and many timbers, being better insulators than heavy more compact materials.

Moisture also has a significant effect on the thermal conductivity of porous materials. If the pores are filled, the water acts as a bridge and since the conductivity of water is many times greater than that of air, the resulting conductivity is greater.

7.3 Coefficient of thermal expansion

Thermal expansion of a material results from increased vibration of the atoms when they gain thermal energy. Each atom behaves as though it has a larger atomic radius, causing an overall increase in the dimensions of the material. The linear coefficient of thermal expansion is defined as change in length per unit length per degree, and hence has units of K^{-1}. The volume coefficient of thermal expansion, defined as the change in volume per unit volume per degree, is sometimes used; for an isotopic material this is three times the linear coefficient.

As with the thermal conductivity, the coefficient of thermal expansion depends on the ease with which the atoms can move from their equilibrium position. Values for metals range from about 1×10^{-5} to 3×10^{-5} K^{-1} (but are significantly less for some alloys); most ceramics, which have strong ionic or covalent bonds, have lower coefficients, typically of the order of 10^{-7} K^{-1}. Although the bonding within the chains of polymers is covalent and therefore strong the secondary bonds between the chains are often weak, leading to higher coefficients, typically of the order of 10^{-5} K^{-1}.

Further reading for
Part 1 Fundamentals

Ashby MF and Jones DRH. *Engineering Materials,* Elsevier Butterworth Heineman, London
Vol 1 (3rd edition, 2005). *An introduction to properties, applications and design*
Vol 2 (3rd edition, 2005). *An introduction to microstructures, processing and design*
Vol 3 (1993), *Materials failure analysis: case studies and design implications*

> *Certainly books to be dipped into. Very thorough with much detail; as well as the basics and theory, some excellent case studies throughout illustrate the engineer's approach.*

Callister WD Jr (2007). *Materials Science and Engineering, An introduction,* 7th edition, John Wiley & Sons Inc, New York.

> *A comprehensive introduction to the science and engineering of materials.*

Gordon JE (1976). *The new science of strong materials – or Why you don't fall through the floor,* Penguin Books, Harmondsworth, Middlesex.

Gordon JE (1978). *Structures – or Why things don't fall down,* Penguin Books, Harmondsworth, Middlesex.

> *Excellent and very readable. Read them in bed, on the bus or on the train. Despite being more than thirty years old now, they will tell you more than many hours of library study.*

Petrowski H (1982). *To Engineer is Human,* Macmillan, London.

> *Petrowski considers what it is like to be an engineer in the twentieth century and lays some emphasis on the things that have gone wrong. Not a book for those lacking in self-confidence but good (and easy) reading.*

Cottrell AH (1964). *The Mechanical Properties of Matter,* John Wiley, New York.

> *First class, scientific and of much wider coverage than the title suggests. Essential reading for any student wishing to follow up the concepts herein and highly desirable reading for all students of all branches of engineering.*

PART 2

METALS AND ALLOYS

Revised and updated by Peter Domone, with acknowledgments to the previous authors, Bill Biggs, Ian McColl and Bob Moon

Introduction

Useful metals have been known to mankind for a long time and probably came into service very gradually. When metals became available they offered many advantages over stone and timber tools and weapons. They could be strong and hard, but their chief advantages were in their ductility. This enabled them to withstand a blow, and a range of shaping procedures and hence products became possible. The significance to us all is encapsulated in the ideas of the Stone Age, Bronze Age and Iron Age. Metals and their differences have even been the subject of poetry. Thus, from Rudyard Kipling:

> Gold is for the mistress – silver for the maid,
> Copper for the craftsman – cunning at his trade,
> "Good!" said the Baron, sitting in his hall,
> "But iron – cold iron – is master of them all."

Most metals are found in nature as ores – oxides, sulphides, carbonates, etc. The basic chemistry of extraction is generally fairly simple, but the industrial problem is to do the job on a big enough scale to make it economically worthwhile. The converse problem also exists. When a metal is exposed to a working environment it will tend to revert to the appropriate compound, i.e. it corrodes. Rust on steel is almost the same as the ore from which iron is extracted. The metallurgist therefore has two tasks, to extract the metal from its ore and then to keep it that way.

The origins of extraction are lost in prehistory and the early discoveries were probably made accidentally: a piece of rather special rock when heated in the reducing atmosphere of a fire gave up some metal. Copper, lead and tin were among the earliest to be produced this way, and alloys such as bronze (copper and tin) and brass (copper and zinc) followed. Bronze was much prized for its ability to be cast into shapes or mechanically formed as well as for its combination of strength, hardness and toughness.

Unaided fires can reach temperatures of about 1100–1200°C on a good day. This is sufficient to melt all the metals mentioned above, but is not hot enough for iron, which melts at about 1550°C. The early history of iron involved extraction processes that gave rise to solid lumps of very porous and friable metal, mixed with a glassy slag. This was formed into useful articles by hammering at temperatures high enough to melt the slag. The slag was partially squeezed out and the pores closed. The product is wrought iron, which has a specific meaning in metallurgy. Temperatures sufficient to melt the iron can be produced with a forced air blast and the resultant product after cooling is cast iron. This was probably first produced in China and then in modest and somewhat variable quantities in medieval Europe, being used for pots, cannon and shot. In the 18th century, early blast furnaces used coke as fuel and a forced air blast to raise the temperature sufficiently to enable production on larger and more controlled scales. In the furnace, the iron picked up about 4% by mass of carbon from the fuel, which had the advantage of also lowering the melting temperature of the alloy. The product, pig iron or cast iron, was brittle but easily cast into moulds and found a wide range of uses in engineering and as household articles. Abraham Darby's famous Ironbridge at Coalbrookdale in the West Midlands is the first example of its use on a significant scale for structural purposes. The structural use of cast iron and wrought iron on increasing scales was a feature, and perhaps the driving force, of the Industrial Revolution in the early 19th century. Victorian engineers used cast iron extensively in bridges and structural beams and columns.

The Bessemer converter, developed in the mid-19th century, blew cold air through a molten bath of pig iron to oxidise the impurities, including carbon, to leave relatively pure liquid iron. The oxidation process generated enough heat to raise the melt temperature sufficiently to keep the material molten as it became purer, resulting in the production of steel, which had an excellent combination of ductility and tensile strength. The whole process, from start to finish, took about an hour and produced about 10 to 20 tonnes of steel, not much by today's standards, but revolutionary at the time. Impressive structures such as the Forth Railway Bridge and the Eiffel Tower became possible.

The slower but more easily controlled Siemens–Martin process also came into use. Large pools of metal were melted in shallow open hearths and oxidation was achieved by reactions with a covering slag and controlled additions of iron oxide.

Both processes have been now superseded by more efficient methods. The basic oxygen converter is based on Bessemer's principle, but uses oxygen rather than air. No heat is wasted in heating the 80% of air that is nitrogen. The oxygen is blown in from the top through a lance. Several hundred tonnes per hour are possible from modern furnaces. Much steel is also produced from remelted scrap, for which purpose electric arc furnaces are often used: good control over material quality is possible using well-characterised scrap.

The development of freely available supplies of electricity in the late 19th century allowed aluminium to be extracted on a commercial scale. This has to be extracted electrolytically from molten salts, as are many other metals such as magnesium, copper and titanium as well as the non-metal silicon.

In the first three chapters of this part of the book we discuss some of the general principles of metallurgy that are relevant to all metals used in construction. In Chapter 8 we describe how they deform, referring to the descriptions of microstructure that we gave in Chapter 3, and then how knowledge of this is used to advantage in the improvement of strength, ductility and toughness. In Chapter 9 we briefly cover how metals are formed into the shapes required for our structures, and then in Chapter 10 we introduce the principles of corrosion and its prevention – essential subjects for ensuring long-lasting and satisfactory performance. In the last two chapters we concentrate on iron and steel, fairly extensively since this is the most important metal for construction (as well as for many other fields of use) and finally, more briefly, aluminium.

Chapter 8

Deformation and strengthening of metals

In Part 1 of the book we used many examples of metals when explaining and discussing features of materials' behaviour and structure. In this chapter, we will extend some of these aspects that are of particular importance and relevance to metals, notably, ductility, plasticity and grain structure, and then show how these lead to methods of strengthening and forming metals, subjects often collectively known as *physical metallurgy*.

To gain an understanding of the mechanisms and controlling factors of plastic deformation (or the lack of it) we need to consider the imperfections and impurities in the crystal lattice and, on a slightly larger scale, the grain structure; we introduced and discussed these briefly in section 3.2. These are inevitable results of production, but in many cases they are deliberately introduced to enhance some desirable properties, particularly strength.

8.1 Elasticity and plasticity

In Chapter 2 we defined elastic and plastic deformation with the help of typical stress–strain diagrams. In elastic deformation, which occurs at stresses below a defined level (the elastic limit or the yield stress) the strain is fully recoverable on unloading. Clearly this is a valuable property for a structural material, and structures are designed such that stresses in normal working conditions are kept below the elastic limit.

At the microstructural level, the bonds between the atoms are stretched or compressed and act like springs in storing strain energy, which is then released on unloading. This restores the material to its original shape, with no relative movement of the atoms occurring and each atom bonded to the same neighbours as before the loading.

Many metals (and other materials such as polymers) will, when loaded beyond the elastic limit, undergo plastic deformation, which is not recoverable on load removal. In ductile materials large plastic strains can occur before failure (*Fig. 2.3*). Although such strains are not desirable in normal working conditions, ductility and plasticity in metals are essential properties for both their production and use. They can be hammered, squeezed or rolled to shape at temperatures below their melting point and in service they can absorb overload by deforming and not fracturing.

8.2 Dislocation movement

Dislocations occur when the atoms do not arrange themselves in a perfect regular repeating pattern when the metal solidifies from the melt. An edge dislocation (*Fig. 3.9* simplified as *Fig. 8.1*) is a common example. This takes the form of an extra half plane of atoms inserted into the regular array. The bonds between the atoms in the region near the dislocation core are distorted, and with a sufficiently high applied stress these will break and re-form, resulting in an apparent movement or slip of the dislocation by an amount called the Burgers vector (*Fig. 8.1*), which is roughly the same as the interatomic distance. But note that it is the dislocation that has moved, with little movement of each individual atom. This can be thought of as similar to moving a carpet across a floor by pushing a ruck from one edge to the other (*Fig. 8.2*). This is much easier than trying to pull the whole carpet across the floor in one go. Clearly the movement is not reversed when the stress is removed and hence permanent (or plastic) deformation results.

The movement of the dislocation results in a shear displacement along a *slip plane* (*Fig. 8.3*) which must be coincident with one of the planes of atoms within the lattice, for preference a close-packed plane (see Chapter 3). The hexagonal close-packed structure has a fairly restricted number of slip planes; the body-centred cubic and face-centred

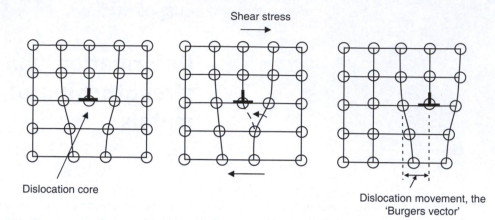

Shear stress

Dislocation core

Dislocation movement, the 'Burgers vector'

Fig. 8.1 Movement of an edge dislocation by breaking and reforming of bonds (adapted from Ashby and Jones, 2005).

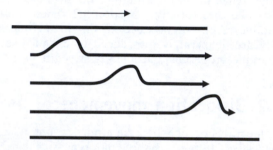

Fig. 8.2 How to move a heavy carpet.

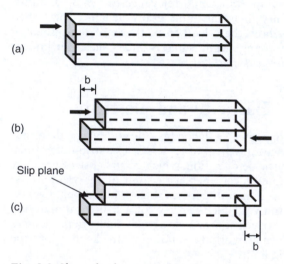

(a)

(b)

Slip plane

(c)

Fig. 8.3 Shear displacement during the slip process.

cubic structures have more. Metals contain an enormous number of dislocations within their atomic lattice, and if there are no obstacles to get in the way of their movement, then very large plastic strains will occur. This results in the high ductility of pure metals.

8.3 Dislocation energy

The atoms at the core of the dislocation are displaced from their proper positions. The strains are approximately 0.5 so that the corresponding stress is of the order of $G/2$, where G is the shear modulus, and so the strain energy per unit volume is about $G/8$. If we now assume that the radius of the core is about the size of the atom b, the cross-sectional area of the core is πb^2 and its total volume is $\pi b^2 l$, where l is the length of the dislocation line. Hence, the total dislocation energy per unit length of line is:

$$U/l = \pi Gb^2/8 \approx Gb^2/2 \qquad (8.1)$$

which has units of J/m or N.

In order to minimise the energy, the dislocation line tries to be as short as possible. It behaves as if it were an elastic band under a tension T, the value of which is identical with U/l. $T \approx Gb^2/2$ is very small indeed, but it is large in relation to the size of the dislocation and it plays an important role in determining the way in which obstacles of one sort or another can obstruct the movement of dislocations.

8.4 Strengthening of metals

Pure metals in the 'as-cast' condition after slow cooling are generally soft, have low yield stresses and are very ductile; this is a consequence of the

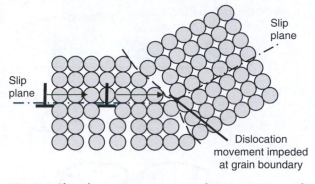

Fig. 8.4 Slip-plane orientations in adjoining grains and dislocation 'pile-up' (adapted from Callister, 2007).

ease of dislocation movement or slip. They are therefore unsuitable for use as a structural material, with perhaps the most critical property for design being the low yield stress. Clearly if we can increase the yield stress then the safe working stress can be correspondingly increased. In using the term 'strengthening', we are concerned with ways by which we can make the start of slip more difficult. We now consider some of these ways and their consequences.

8.4.1 GRAIN SIZE

In a single crystal of a pure metal the shear stress required to move a dislocation is small, in some cases maybe only ~1 MPa. However, most materials are polycrystalline, and the grain boundaries are discontinuities in the atomic lattice, which will have differing orientations on either side of the boundary, as illustrated in *Fig. 8.4* (but note that the numerous atomic bonds across the grain boundary are usually strong enough not to weaken the material). A dislocation that reaches a grain boundary cannot produce a slip step there unless the neighbouring grain also deforms to accommodate the shape change. A dislocation in the second grain cannot move until the shear stress, resolved on to the new slip plane and in the new slip direction, reaches the value needed to continue movement. Back in the first grain, the dislocation is stuck and other dislocations will pile up behind, like a traffic jam, exerting a force on it, until, ultimately, the push is too great and it is forced through the grain boundary. The stress on the leading dislocation is a simple function of the number of dislocations in the pile-up. In a coarse-grained structure many dislocations can pile up and the critical stress is reached early, whereas in a fine-grained structure the length of the pile-up is smaller and more stress must be applied from external forces, i.e.

the yield point is raised. The outcome is summarised in the famous Hall–Petch equation:

$$\sigma_y = \sigma_0 + kd^{-1/2} \qquad (8.2)$$

where σ_y is the yield strength of our polycrystalline material, σ_0 is the yield strength of one crystal on its own, k is a proportionality constant and d is the grain size of the material. Mild steel with a grain size of 250 μm has $\sigma_y \approx 100$ MPa, but when $d = 2.5$ μm, $\sigma_y \approx 500$ MPa. The incentive for making fine-grained steels is clear.

Control of grain size in castings is generally achieved by 'inoculating' the liquid metal with substances that can react with ingredients in the metal to form small solid particles that act as nucleation sites for crystal growth. In wrought products, the thermal and mechanical history of the working process can be controlled to give fine grains, as discussed below. Rolling and forging are therefore used not only to shape materials but also, perhaps more importantly, to control their microstructures and hence their properties.

8.4.2 STRAIN HARDENING

We discussed the process of increasing the yield stress of a material by work or strain hardening in Chapter 2, section 2.3. This involves loading into the region of plastic deformation with a positive slope of stress–strain behaviour (BC in *Fig. 2.6*). In a tension test, a reasonably ductile metal becomes unstable and begins to form a neck at strains of only about 30% or so. But when we roll the same metal or form it into wire by drawing, the deformation in the local area being worked is essentially compressive. This allows us, for example, to draw a wire to many times its original length with relative ease. The work hardening is extended well beyond what can be achieved in a tension test; for example with some steels, the yield strength can be increased by 4 or 5 times by drawing it to a thin wire.

Metals, especially those with the face-centred and body-centred cubic systems, have many different planes on which dislocations can move to produce slip. But none of these are markedly different from the others and, under increasing stress, all dislocations try to move at once. If the slip planes intersect each other, as indeed they do, the dislocations on one slip plane act as a barrier to dislocations trying to move across them. With any significant amount of plastic deformation, many millions of dislocations are on the move, the traffic pile-up is considerable and the dislocations get jammed. Very much more stress needs to be applied to get things moving again and so strain hardening is the result. It is one of the most effective ways of raising the yield strength of a metal, though if carried too far it results in fracture, as we have seen.

8.4.3 ANNEALING

An undesirable effect of strain hardening at room temperatures (or cold working) is that it can cause local internal stresses and hence non-uniformity of the metal. Since each dislocation is a region of high strain in the lattice, they are not thermodynamically stable and comparatively little energy is required to cause a redistribution and cancellation of the trapped dislocation arrays. The energy is most conveniently supplied in the form of heat, which gives the atoms enough energy to move spontaneously and to form small areas that are relatively free of dislocations. This is called *recovery* but, since the dislocation density is only slightly reduced, the yield strength and ductility remain almost unchanged. The major change involves *recrystallisation*. New grains nucleate and grow, the material is restored to its original dislocation density and the yield point returns to its original value. This process is known as *annealing*, and the annealing temperature is normally kept fairly low (say at most to around 0.6 T_m, where T_m is the melting point in degrees K) so that the increase in strength due to cold working is not affected. Annealing is also a useful way of controlling grain size; we will discuss its importance in the preparation of metals and alloys for commercial use later in the chapter.

8.4.4 ALLOYING

One of the most powerful ways of impeding dislocation movement, and hence of increasing the yield strength, is to add another element or elements to the metal in order to distort the atomic lattice. We have seen in Chapter 3, section 3.2 how foreign atoms can be located as either interstitial or substitutional impurities. Deliberate introduction of an appropriate type and quantity of the foreign element(s) produces alloys whose properties are significantly enhanced over that of the parent metal. Nearly all metals used in construction are alloys, the most notable being steel, an alloy of iron and carbon (and normally other elements as well).

Dispersion hardening is a particular form of alloying in which the alloying element or impurity combines with the parent metal. The impurity is added to the molten metal at high temperature and then, as the alloy cools and solidifies, the impurity–metal compound precipitates as small, hard, often brittle, particles dispersed throughout the structure. Examples of such particles are $CuAl_2$ formed after

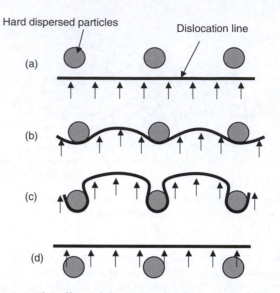

Fig. 8.5 *The effect of dispersion hardening on dislocation movement (adapted from Ashby and Jones, 2005).*

adding small quantities of copper to aluminium or iron carbides formed after adding small quantities of carbon to iron.

Figure 8.5 illustrates how such particles obstruct the movement of a dislocation line. An increased stress is required to push the line between the particles ((a) and (b)), but eventually it is forced through ((c) and (d)); it will of course, soon encounter, more obstacles. Clearly the greatest hardening is produced by strong, closely spaced precipitates or dispersions.

8.4.5 QUENCHING AND TEMPERING

Many of us know that if you take a piece of steel containing, say, 0.5% carbon, heat it to glowing red (~900°C) and then quench it by placing it in a bath of water, the outcome is a very hard but brittle substance. Indeed, it could be used to cut a piece of steel that had not been so treated. The quenching of steel is an example in which an unstable microstructure is generated when there is no time for diffusion to keep up with the requirements of thermodynamic equilibrium. The procedure generates a new and unexpected structure (called *martensite*) in which there are large internal locked-in stresses. In the as-quenched condition this is too brittle to be useful, but if it is heated to just a few hundred degrees C a number of subtle changes come about. The steel is softened a little, not much, but a useful degree of toughness is restored. This second heating is called *tempering*, and gives us tempered steel. In this state it finds many uses as components in

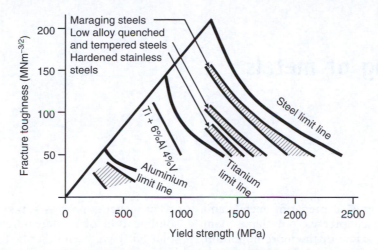

Fig. 8.6 Relationship between fracture toughness and yield strength for a range of alloys (after Dieter, 1986).

machinery, gears, cranks, etc. A difficulty with this method of obtaining strength is that it works well only for certain types of steel.

8.5 Strengthening, ductility and toughness

As in most things in nature, you do not get something for nothing, and there is a cost to be paid for increasing the yield strength of metals by the above processes. This is that as yield and tensile strength increase ductility, toughness and fracture toughness (see Chapter 4) are reduced. *Figure 8.6* shows the relationship between yield strength and fracture toughness for a range of alloys. We must, however, take care that the reduction in toughness is not too excessive. For example, continued cold working will raise the yield strength ever closer to the tensile strength but at the same time the reserve of ductility is progressively diminished and, in the limit, the material will snap under heavy cold working. This is familiar to anyone who has broken a piece of wire by continually bending and rebending it.

References

Ashby MF and Jones DRH (2005). *Engineering Materials Vol 1. An introduction to properties, applications and design*, 3rd edition, Elsevier Butterworth Heineman, London.

Callister WD (2007). *Materials science and engineering. An introduction*, 7th edition, John Wiley and Sons, New York.

Dieter EG (1986). *Mechanical Metallurgy*, McGraw-Hill Science/Engineering/Math, New York.

Chapter 9

Forming of metals

There are many methods of preparing metals and alloys for use; in this chapter we only have room for briefly describing a few of the more important ones that are used in the production of both metals and alloys. Before starting, we must recognise that metallurgists look on these not only as ways of shaping materials but also as ways of controlling their microstructure and, consequently, their properties. *Figure 9.1* outlines the processing routes for most of the more common metals and alloys used in structural engineering.

9.1 Castings

Most common metals can be produced by melting and casting into moulds. The cast may be of the shape and dimensions required for the component, or a prism of material may be produced for further processing.

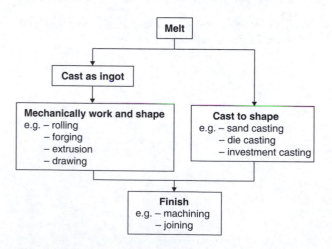

Fig. 9.1 Processing procedures for the more common metals.

The general processes taking place during the solidification of molten pure metals and metallic solutions have been described in Chapter 1. Solidification of alloys often gives rise to compositional variations from place to place in a casting and on a microstructural scale within the dendrites. When intended for further processing, little attempt is made to control grain size, and the metal often solidifies to a rather coarse grain structure containing a number of casting defects, such as porosity, compositional variations and shrinkage. These are not disastrous because further processing will rectify them. Shaped castings need more care. To ensure that the desired mechanical properties are achieved the castings are normally degassed, the grain size is carefully controlled by one or more of the means described in Chapter 8, and compositional variations minimised by attention to solidification patterns within the mould.

9.2 Hot working

The working of metals and alloys by rolling, forging, extrusion etc. (*Fig. 9.2*) depends upon plasticity, which is usually much greater at high temperatures, i.e. temperatures above the metals' recrystallisation temperature. This allows all the common metals to be heavily deformed, especially in compression, without breaking. For structural steel members, the most usual method is by hot rolling between simple cylindrical or shaped rolls at temperatures around 1000°C or higher. After rolling, the members are left to cool naturally and end up with annealed microstructures and grain sizes that depend on the extent of the deformation, the maximum temperature used and the cooling rate. It follows that all these process variables need to be controlled to give products with consistent properties. Another feature of hot-working processes is that exposure to air at high temperatures causes a heavy film of oxide to form on the surface. Thus steel sections delivered 'as rolled'

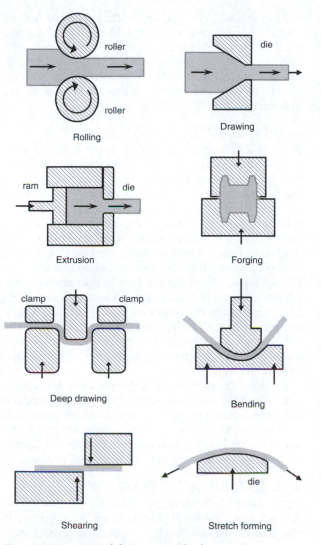

Fig. 9.2 *Some metal-forming methods.*

can be produced. Aluminium glazing bars are a familiar example.

One disadvantage of hot forming arises from the contraction of the article on cooling and from such problems as oxidation. These and other factors conspire to limit the precision of the product. In some cases the tolerances are acceptable, but to meet more demanding tolerances further cold forming or machining is required.

9.3 Cold working

Because of their ductility at room temperature many metals and alloys can be cold worked, that is to say, shaped at temperatures below their recrystallisation temperature. As we discussed in Chapter 8, this creates an immense number of dislocations and, as a consequence, the metal work-hardens and its yield point is raised. Indeed, for pure metals and some alloys it is the only way of increasing the yield strength.

There are many cold-working processes. Rolling is extensively used to produce sheet material, while high-strength wire, as used for pre-stressing strands and cables, is cold drawn by pulling through a tapered die. Metal sheets can be shaped into cups, bowls or motor-car body panels by deep drawing or stretch forming (*Fig. 9.2*). Some metals can be cold extruded.

Clearly there comes a limit beyond which the ductility is exhausted and the metal will fracture. If further cold work is required the metal must be annealed by heating it to a temperature where recrystallisation occurs, when the original ductility is restored and further working is possible.

Cold working using well designed tools and careful control is capable of delivering to demanding tolerances. From the metallurgist's point of view, control over rolling and annealing schedules is a very effective way of controlling the grain size of the product.

9.4 Joining

The design and fabrication of joints between metallic structural components are obviously crucial factors in ensuring the success of the structure. Design engineers have Codes of Practice etc. to help them in their task, but some understating of the processes involved and relevant materials' behaviour is also important. Although adhesive bonding is being used increasingly for joining metal parts, the commonest methods are still welding, brazing, soldering or by mechanical fasteners, such as rivets and bolts.

are covered with iron oxide (mill scale) and need to be shot-blasted or sand-blasted before receiving any protective coating. For many structural steels further heat treatment is required, as we shall see in Chapter 11.

Many familiar articles, e.g. engine crankshafts, are forged into shape. This involves placing a hot blank into one half of a shaped mould and then impressing the other half of the mould onto the blank (*Fig. 9.2*). This can be done under impact using such methods as drop forging and die stamping or more slowly using large hydraulic presses.

Many metals can be extruded. This has the advantage that very long lengths with complex sections

9.4.1 WELDING

It is beyond the scope of this book to list the various welding processes that are available (see the further reading list at the end of Part 2 for more comprehensive texts). However, all welding involves essentially the same sequence of operations at the joint. The material is heated locally to its melting temperature, additional metal may or may not be added and the joint is then allowed to cool naturally. Some protection to the weld to avoid oxidation of the metal when molten and during cooling is often provided by a slag layer (which is knocked-off when the weld has cooled) or by working in an atmosphere of an inert gas such as argon. Whatever the material or process all welds should comply with the two following ideal requirements:

1. There should be complete continuity between the parts to be joined, and every part of the joint should be indistinguishable from the parent metal. In practice this is not always achieved, although welds giving satisfactory performance can be made.
2. Any additional joining metal should have metallurgical properties that are no worse than those of the parent metal. This is largely the concern of the supplier of welding consumables, though poor welding practice can significantly affect the final product.

The weld itself is a small and rapidly formed casting. However, during welding a temperature gradient is created in the parent material which results in a heat affected zone (HAZ) surrounding the weld. This gradient ranges from the melting temperature at the fusion zone to ambient temperature at some distant point. In the regions that have been exposed to high temperature and fast cooling rates, metallurgical changes can occur. The quality of the joint is therefore affected by both the structure and properties of the weld metal and the structure and properties of that part of the parent material that has undergone a significant thermal cycle (the HAZ).

Both of these are significantly affected by the rate of cooling after welding – the slower the rate of cooling the closer the structure is to equilibrium. Cooling occurs principally by conduction in the parent metal and, since the thermal conductivity is a constant, the controlling factor is the thermal mass, i.e. the thickness and size of the material to be welded. The greater the thermal mass the faster the cooling rate. Responses to rapid cooling differ markedly from metal to metal, not only from say aluminium to steel but also from one steel to another. Structural steels are designed to be weldable, i.e. they are capable of being welded without serious loss of performance in the weld and HAZ. Nevertheless, the job must be carried out with thought, care and skill, with due allowance made if the welding is being carried out in difficult conditions such as on a construction site in poor weather. Most jobs are best carried out by welding specialists.

9.4.2 BRAZING, SOLDERING AND GLUING

Brazing and soldering, and in some cases gluing, involve joining by means of a thin film of a material that has a melting temperature lower than that of the parent material and which, when melted, flows into the joint, often by capillary action, to form a thin film which subsequently solidifies. A sound well-brazed or soldered joint should have a strength that is not too different from that of the parent material. Quite high forces are needed to break a film of liquid provided the film is thin enough (see Chapter 6) and the same applies to thin solid films. This is not quite the whole explanation but is a very significant part of it. The rest is associated with the behaviour of materials under complex stress conditions, biaxial and triaxial, and is beyond the scope of this chapter.

Although it may seem strange to say so, gluing works in a very similar way. Thin layers of modern adhesives bond well to the substrate material and are strong in shear. Design of joints to be made by gluing, soldering or brazing should avoid potential failure by peeling and aim to use the adhesive in shear.

9.4.3 BOLTING AND RIVETING

Some materials (such as cast iron) do not lend themselves to joining by welding. Even with materials that can be welded (such as structural steel) it may not possible to weld prefabricated elements on a construction site owing to difficulties of access and working conditions for both the welder and the welding equipment. Gluing and brazing may be valid options but require thought about the joint design. Bolting and riveting are by far the most common ways of making joints in such circumstances. Both rely on friction. A tightened bolt forces the two members together and the friction between nut and bolt at the threads holds it in place. In riveting, the hot rivet is hammered into prepared holes and the end hammered flat as a 'head' on the surface of the sheet; as it cools it contracts and develops a tensile stress that effectively locks the members together. High-strength friction grip bolts used in structural steelwork combine both aspects; the nut is tightened to place the bolt into tension and this tensile pre-stress acts in the same way as the tensile stress in a rivet.

Chapter 10

Oxidation and corrosion

Having produced and formed a metal, it is necessary to ensure that it performs well during service. A major consideration in this is corrosion. This involves loss of material from the metal's surface and can be divided into two processes: dry oxidation and wet corrosion.

10.1 Dry oxidation

The earth's atmosphere is oxidising. Nearly all of the earth's crust consists of oxides, which indicates that this is the preferred minimum energy state for most materials. Gold and silver are the only two metals that are found in their native, unoxidised state. The general oxidising reaction can be written as:

$$M + O \rightarrow MO \qquad (10.1)$$

where M is the metal and O is oxygen. For all metals, except gold and silver, this reaction is accompanied by release of energy, indicating the unstable nature of the metals. In fact this characteristic is shared with many other materials, which is why, for example, burning hydrocarbons is a useful source of heat.

Normally the oxidation takes place in two steps. First, the metal forms an ion, releasing electrons, and the electrons are accepted by oxygen to form an ion:

$$M \rightarrow M^{2+} + 2e \quad \text{and} \quad O + 2e \rightarrow O^{2-} \qquad (10.2)$$

Secondly, the ions attract one another to form the oxide compound:

$$M^{2+} + O^{2-} \rightarrow MO \qquad (10.3)$$

At the metal surface the oxygen ions attach themselves to the metal to form a thin layer of oxide. Thereafter, for the oxidation to continue, the metal M^{2+} ions and the electrons must diffuse outwards through this layer to form and meet more oxygen O^{2-} ions at the outer surface, or the oxygen ions must diffuse inwards. The rate of oxidation is determined by whichever reaction can proceed the faster and, largely, this is controlled by the thickness and structure of the oxide skin.

On some metals the oxide occupies a lower volume than the metal from which it was formed. If it is brittle (and oxides usually are), it will crack and split, exposing fresh metal to more corrosion. On other metals the oxide occupies a higher volume and it will tend to wrinkle and spring away, again exposing fresh metal. Even in these circumstances, the rates of reaction are generally low. In some other cases, however, the oxide volume matches the metal volume and thin adherent films form that act as near total barriers to further oxidation. This is true of aluminium, which is why it does not need protection against corrosion when used, for example in window frames, and chromium and nickel, which are therefore the essential components of so-called 'stainless steel'.

10.2 Wet corrosion

In the presence of moisture the situation changes drastically and the loss of metal by corrosion becomes much more significant. Indeed, in several countries, including Japan, the UK and the USA, estimated losses to the national economy due to corrosion could be as high as up to 5% of their gross domestic product (GDP). The explanation for the high corrosion rates is that the metal ions formed in equation (10.2) are soluble in the corroding medium (water); the electrons produced are then conducted through the metal to a nearby place where they are consumed in the reaction with oxygen and water to produce hydroxyl ions, which in turn link up with the metal ions to give a hydroxide.

The corrosion of iron, illustrated in *Fig. 10.1*, is a useful example. The reaction in which the iron atoms pass into solution as Fe^{2+} ions leaving behind

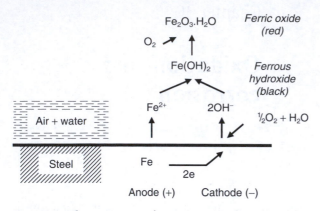

Fig. 10.1 *The corrosion of iron in aerated water.*

two electrons is called the *anodic reaction* and takes place at the *anode* of a resulting *corrosion cell*. The hydroxyl ions are produced in the *cathodic reaction*, which takes place at the *cathode* of the cell, and the hydrated iron oxides ($Fe(OH)_2$ or $Fe(OH)_3$) are deposited either loosely on the metal surface or away from it, thus giving little or no protection. This, coupled with the ready conduction of the electrons through the iron, results in high rate of attack – many millions of times faster than that in air.

10.3 The electromotive series

Because wet oxidation involves electron flow in conductors, the application of an external voltage will either slow down or increase the rate of reaction depending on whether the applied voltage is negative or positive. The values of the voltage that cause the anodic reaction to stop in different metals form the so-called *electromotive series*. By convention, the voltage is given in relation to that for the ionisation of hydrogen:

$$H \rightarrow H^+ + e \qquad (10.4)$$

This therefore appears as zero volts in the series. Some selected values from this are given in *Table 10.1*. Those metals which are more positive than the reference value are anodic and will corrode, metals which are relatively cathodic will not.

The series also tells us what happens when two different metals are connected while in an electrolyte. The more anodic metal will form the anode of the corrosion cell and hence will corrode, while the less anodic one forms the cathode and will not corrode. The resulting voltage difference between the metals can be useful, and was used in one of the first

Table 10.1 The electromotive series: some standard electrode potentials

Electrode	Voltage
Na^+	+2.71*
Al^{+++}	+1.66
Zn^{++}	+0.76
Fe^{++}	+0.46
Ni^{++}	+0.25
Sn^{++}	+0.14
H^+	0.0†
Cu^{++}	−0.34
Ag^+	−0.80
Pt^{++}	−1.20
Au^{+++}	−1.50‡

Note: Conventions differ as to which are negative and which are positive in the series. This does not matter too much since it is the relative position that is important. *, base, anodic, corrodes; †, reference; ‡, noble, cathodic.

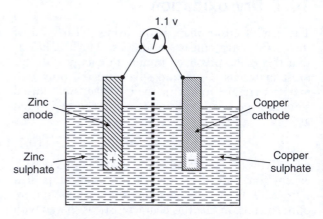

Fig. 10.2 *The Daniell cell.*

electrical batteries – the Daniell cell (*Fig. 10.2*), developed in 1836. This consists of a zinc anode (which corrodes) in a zinc sulphate solution and a copper cathode (which does not corrode) in a copper sulphate solution; the two solutions are separated by a semi-permeable membrane, which prevents the copper ions in the copper sulphate solution from reaching the zinc anode and undergoing reduction. The voltage produced is the difference in standard electrode potential of the two elements i.e. 1.1v.

The voltages in *Table 10.1* must, however, be used with some caution. They apply in controlled laboratory conditions with the ions in solution having specific concentrations. In other environments e.g. seawater, and at other temperatures, the voltages will

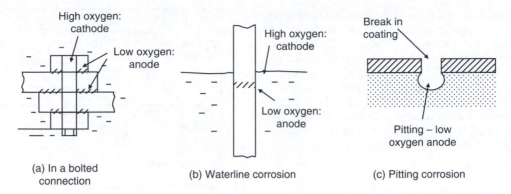

Fig. 10.3 *Localised corrosion due to differential oxygen concentration.*

vary in magnitude, and even in order. Despite this, they provide a very useful guide to performance.

A further limitation of the electromotive series is that although it tells us about where corrosion is likely to occur, it tells us little about the likely corrosion rate. For example, aluminium, which is high in the anodic voltage range, corrodes extremely slowly in moist atmospheres because a thin dry film of Al_2O_3 forms on the metal surface. However, in seawater the chloride ions tend to breakdown this protective film and aluminium corrodes very rapidly. Corrosion rates are also temperature dependent; for example mild steel in aerated water corrodes six times faster at 100°C than at 0°C.

10.4 Localised corrosion

There are some circumstances in which wet corrosion can occur at selected localised sites. In these circumstances the corrosion can be intense and lead to premature failure of a component rather than generalised loss of material.

10.4.1 INTERGRANULAR ATTACK
The grain boundaries in metals have different corrosion properties from the rest of the grain and can become the anodic region of the corrosion cell at which the corrosion is concentrated.

10.4.2 CONCENTRATION CELL CORROSION
The localised corrosion is a consequence of a difference in the constitution of the electrolyte itself. For example, consider water containing dissolved oxygen with differing concentrations in different regions. The reaction:

$$2H_2O + O_2 + 4e^- \rightarrow 4(OH)^- \qquad (10.5)$$

removes electrons, and these must be supplied from adjacent areas, which then become deficient in oxygen. These act as the anode and hence corrode. Thus, in a bolted connection, corrosion will occur in the inaccessible (i.e. oxygen-poor) areas, e.g. under the bolt head (*Fig. 10.3a*). A classic case is the 'waterline' corrosion of steel piling in stagnant water. Here the surface layers of the water are richer in oxygen and become the cathode. The lower, oxygen-deficient layers are anodic and corrosion occurs locally (*Fig. 10.3b*). Much the same mechanism applies to pitting corrosion, which typically occurs where the metal is exposed at a break in a protective coating. The oxygen-poor region at the bottom of the pit is anodic and the pit therefore tends to deepen, often rapidly (*Fig. 10.3c*).

Note that waterline corrosion can be confused with a different phenomenon. We are all familiar with the enhanced corrosion that is seen on steel supports of seaside piers etc. Here we have a region that is washed by wave and tidal action. The region is alternately wetted and dried and it is this that accelerates corrosion.

10.4.3 STRESS CORROSION CRACKING
With some materials, when loaded or stressed in a corrosive environment, cracks can grow steadily under a stress intensity factor (K) that is much less than the critical stress intensity factor K_c (see Chapter 4). This is clearly dangerous and can lead to brittle failure, even in a ductile material, after many years of apparently normal structural behaviour. Examples are stainless steels in chloride solutions and brass in ammonia.

10.4.4 CORROSION FATIGUE
The combination of cycling loading and a corrosive environment can lead to significant reductions in the

fatigue life of a material (see Chapter 2), far greater than would be expected from the sum of the cyclic loading and the corrosion acting independently. With ferrous metals, the fatigue endurance limit also disappears (see *Fig. 2.19*), which makes safe design more difficult. A classic situation is steel in sea water, which necessitates great care being taken when designing offshore structures, e.g. for oil and gas production, which will be subjected to wave action and seawater throughout their working life.

10.5 Corrosion prevention

10.5.1 DESIGN

At ambient temperatures significant corrosion occurs only if moisture is present. Thus, surfaces should be exposed as little as possible to moisture and arranged so that they dry out quickly after wetting. In practice all surfaces are at risk, vertical surfaces suffer 'run off', flat surfaces retain moisture on their top side and can attract dew and condensation on the under side. Water retention by 'V', 'H' and other channel sections is obvious and drain holes should be provided, if mechanically acceptable. Over-laps and joints should be arranged to avoid the formation of water channels. Porous materials that can retain moisture should not be in contact with metals. The design should also make provision for inspection and maintenance during the service life of the structure.

10.5.2 COATINGS

Application of one or more coatings to a suitably prepared surface will isolate the metal from a cor-roding environment. Organic coatings, such as paints, pitch, tar, resins etc., form a protective barrier and are commonly used, often in conjunction with a metallic primer. There is a wide range of products available, often for specialist purposes. Some metallic coatings will form a simple protective barrier, e.g. nickel or chromium on steel. However, with chrome-plated steel, the chromium is more cathodic than iron so that if a small pit appears in the chromium, the steel underneath rusts away quite rapidly, as owners of old cars will testify.

All paint coatings, even of the highest quality and meticulous application, are only as good as the qual-ity of the preceding surface preparation. Application, whether by brushing or spraying, should always be carried out on dry surfaces and in conditions of low humidity. Steel that has been allowed to rust on site can be a problem, as the methods available for cleaning steel are often less than adequate, and some

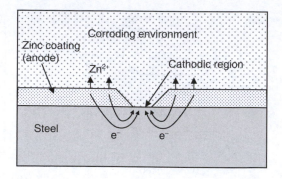

Fig. 10.4 *Cathodic protection of steel by galvanising (adapted from Callister, 2007).*

rust will inevitably remain at the bottom of the pits formed during rusting. These will contain sufficient active material for rusting to continue below any paint film. The only real remedy is not to let rusting start by protecting the steel by priming coats as an integral part of the manufacturing process and, if these are damaged during erecting, to repair the damage as soon as possible.

We will discuss the particular problem of the corrosion of steel in concrete, which is the cause of much deterioration and hence cost, in Chapter 24.

10.5.3 CATHODIC PROTECTION

If a metal in a corroding environment is connected to another metal that is more reactive, i.e. higher up the electromotive series, then the second metal will form the anode of the corrosion cell and hence will preferentially oxidise, thus protecting the first metal from corrosion. This can be used for example to protect buried steel pipelines by connecting them to zinc slabs buried nearby. The slabs corrode, and are therefore called *sacrificial anodes*; they are per-iodically replaced. The same principle is applied in *galvanising*, in which a layer of zinc is deposited on the surface of steel by hot dipping. In moist air and in most other aqueous environments zinc is anodic to the steel and will thus protect it if there is any surface damage. Furthermore the corrosion rate of the zinc is very slow because of the high ratio of anode to cathode surface area (*Fig. 10.4*).

In addition to the use of sacrificial anodes, cathodic protection can be achieved by the use of an external power source to make the metal cathodic to its surroundings. Inert anodes are used, commonly carbon, titanium, lead or platinum. The procedure is not without its problems. For example in many cases the cost of replacement anodes is greater than the cost of the impressed power supply. This method of cathodic protection has been quite widely used in marine environments, especially on offshore oilrigs. However in buried structures secondary reactions with other nearby buried structures may enhance, rather than control, corrosion and there is the possibility of hydrogen evolution at the cathode. This can diffuse into the metal and embrittle it.

10.6 Corrosion control

The management and control of corrosion comprise one of the most difficult problems facing the design engineer. It is critical to recognise that the problems start at, and must be tackled at, the design stage. There are three requirements, all easy in theory but difficult in practice.

1. Understand the environment in which the metal must work, whether polluted or not, whether facing or away from pervading sources of corrosion, whether wet and/or humid or dry, whether these conditions are stable or variable.

2. Consider the 'design life'. How long before the first major maintenance? Are you designing a 'throwaway' structure like a modern motor-car or are you designing a bridge for a century of service? If the component is not expected to outlive the structure as a whole, how easy is it to inspect and replace?

3. Select the most appropriate method of control from those outlined above. You may be excused for imagining that the 'most appropriate' method is that one which involves the longest life, but you will, of course, be wrong. In the commercial world the most appropriate means of control is that one which produces the longest life at the least annual cost. So, on the whole, you would be best to master such matters as payback, rate of return, and discounted cash flow before deciding upon an appropriate technology.

Reference

Callister WD (2007). *Materials science and engineering. An introduction*, 7th edition, John Wiley and Sons, New York.

Chapter 11

Iron and steel

Ferrous (iron-based) metals have widespread use throughout all branches of engineering but are particularly important in construction. Iron and carbon-based alloys, i.e. cast iron and steel, are the principal forms of interest and this chapter is primarily concerned with these, either for structural use in their own right or in the form of reinforcing or pre-stressing steel for concrete construction. We will also more briefly consider products formed by alloying with other elements, for example stainless steel.

11.1 Extraction of iron

As with all metals iron is extracted from naturally occurring ores. Iron ores consist of a wide range of mixtures of complex chemical compounds, but they have a common feature of being rich in iron oxides, often in the form of magnetite (Fe_3O_4) or haematite (Fe_2O_3). The iron oxide is reduced to iron in a blast furnace, which is a large steel vessel up to 30 m high lined with refractory brick. A mixture of raw materials – the iron ore, carbon in the form of coke and limestone – is fed into the top of the furnace and hot air at a temperature of 900–1300°C is blasted through this from the bottom of the furnace. The materials take from 6 to 8 hours to descend to the bottom of the furnace, during which time they are transformed into molten iron and molten slag. The primary reducing agent is carbon monoxide, which is produced by the reaction of the coke and the hot air

$$2C + O_2 \rightarrow 2CO + \text{heat} \qquad (11.1)$$

The heat from this reaction raises the temperature, and the main chemical reaction producing the molten iron is then:

$$Fe_2O_3 + 3CO \rightarrow 2Fe + 3CO_2 \qquad (11.2)$$

At intermediate temperatures in the middle zone of the furnace the limestone decomposes to calcium oxide and carbon dioxide.

$$CaCO_3 \rightarrow CaO + CO_2 \qquad (11.3)$$

The calcium oxide formed by decomposition reacts with some of the impurities in the iron ore (particularly silica), to form a slag, which is essentially calcium silicate, $CaSiO_3$.

The molten iron, covered by a layer of molten slag, collects at the bottom the furnace. Both the iron and the slag are then tapped off at intervals and allowed to cool – the iron into ingots of pig iron. This has a relatively high carbon content of around 4–5% (and other impurities such as silica) making it very brittle and of little use. It therefore has to be further refined in a secondary process to convert it into usable cast iron or steel. We described the principles of two of these processes that were developed in the mid-19th century, the Bessemer converter and the Siemens–Martin open hearth method, in the Introduction to this part of the book. A third method, the *basic oxygen (BOS) process*, is now the current primary method of steel production; it is more efficient and is capable of producing steel in larger quantities than either of the above processes.

In this, the furnace converter is a large vessel that has a top opening that can be rotated to either receive the charge or discharge the final products. It is first charged with scrap steel, which acts as a coolant to control the high temperatures produced by the subsequent exothermic reactions. About three or four times as much molten metal from the blast furnace is then poured in using a ladle. The furnace is then 'blown' by blasting oxygen through a lance that is lowered into the molten metal. No heating is required because the reaction of the oxygen with the impurities of carbon, silicon, manganese and phosphorus is exothermic. Carbon monoxide is given off and the other acidic oxides are separated from the metal by adding calcium oxide to the furnace, thus producing a slag. After a blow of about 20 minutes the metal is sampled, and tapped if a suitable composition has been achieved. The slag becomes a solid waste.

Table 11.1 *Carbon contents and typical uses of the main iron-based metals (after Ashby and Jones, 2005)*

Type	Carbon content (% by weight)	Typical uses
Cast iron	1.8–4	Low-stress uses – machine bases, heavy equipment, tunnel linings
High-carbon steel	0.7–1.7	High-stress uses – springs, cutting tools, dies
Medium-carbon steel	0.3–0.7	Medium-stress uses – machine parts, nuts and bolts, gears, drive shafts
Low-carbon or mild steel	0.04–0.3	Low-stress uses – construction steel (suitable for welding)

Table 11.2 *The main solid phases in the iron–carbon equilibrium diagram (after Ashby and Jones, 2005)*

Phase	Atomic packing structure	Description	Stability of pure form
Austenite (also known as 'γ')	Face-centred cubic	Random interstitial solid solution of C in Fe. Maximum solubility of 1.7% C at 1130°C	Stable between 914 and 1391°C
Ferrite (also known as 'α')	Body-centred cubic	Random interstitial solid solution of C in Fe. Maximum solubility of 0.035% C at 723°C	Stable below 914°C
Fe_3C – iron carbide (also known as 'cementite')	Complex	Hard and brittle compound of Fe and C; 6.7% C by weight	

In the alternative *electric arc process* the charge of metal and lime is melted by heat from an electric arc between graphite electrodes that are lowered into the furnace. As in the basic oxygen process, oxygen is then blown in to convert the impurities into oxides. The method is particularly suitable for the reprocessing of scrap steel, which can form 100% of the charge.

11.2 Iron–Carbon equilibrium diagram

Reducing the carbon content of the pig iron gives rise to either cast iron or a range of steel types, generally with increasing ductility; the broad types are listed in *Table 11.1*, together with some typical uses. As we will see, many of these will in practice also contain some other alloying elements (for example, most contain about 0.8% manganese), but as the behaviour of the iron–carbon system is fundamental to the properties of cast iron and steel,

it is first appropriate discuss its equilibrium phase diagram. We need only be concerned with the part of this up to about 5% carbon, as shown in *Fig. 11.1*. At first glance this looks somewhat complex, but it is apparent that it consists of a combination of the features of such diagrams that we described in Chapter 1.

The liquid is a solution of carbon in iron; the solid parts of the diagram consist of mixtures of the phases austenite, ferrite and iron carbide (Fe_3C). The main features of these are given in *Table 11.2*.

The diagram has a eutectic at ~4.3% C and 1150°C. This is in the cast iron region, and this low melting temperature allows cast irons to be melted with relative ease and cast into complex shapes. Sometimes the products of solidification are Fe and Fe_3C, sometimes Fe and graphite (pure carbon), and sometimes Fe, Fe_3C and graphite.

At lower carbon contents, say less than ~2%, the liquidus climbs to temperatures that make melting more difficult. But by far the most important feature of the diagram arises from the different allotropic

69

Table 11.3   Ranges of properties of cast irons

Type	Elastic modulus (Gpa)	Yield strength (MPa)	Tensile strength (MPa)	Compressive strength (MPa)	Elongation at failure (%)
White	170	–	275	–	0
Malleable		220–310	350–450		10–5
Grey	100–145		150–400	600–1200	0.7–0.2
Spheroidal graphite	165–170	240–530	400–700		7–2

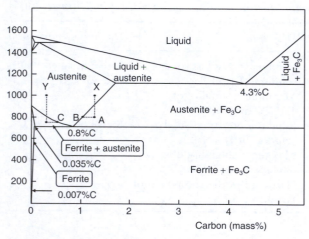

Fig. 11.1 The iron–carbon equilibrium phase diagram.

forms of iron. At temperatures below 910°C, pure Fe forms into body-centred cubic crystals known as ferrite. At higher temperatures, the crystals have face-centred cubic structures and are known as austenite. Up to 1.7% carbon can dissolve in austenite at 1130°C, but this rapidly reduces and Fe_3C is precipitated as the temperature falls to 723°C, at which a maximum of 0.8% carbon can dissolve. Almost no carbon will dissolve in ferrite, but that which does has very profound effects. Transitions from austenite to lower temperature forms of the alloys give rise to a part of the diagram that is reminiscent of a eutectic diagram.

Consider the alloy X at 1000°C (*Fig. 11.1*). It is fully austenitic; on cooling to the point A some Fe_3C is precipitated and the composition of the austenite in equilibrium with the Fe_3C is given by point B. Similarly, on cooling alloy Y from 1000°C to say 750°C, ferrite is precipitated and the austenite in

equilibrium with it has composition C. At 723°C, we have the eutectoid point and the austenite contains 0.8%C. Further cooling causes the austenite to decompose into a mixture of ferrite and Fe_3C. It consists of alternating lamellae (lathes) of Fe and Fe_3C, arranged in 'colonies' within which the lamellae are nearly parallel. The scale is such that the structure acts as a diffraction grating to light and gives the microstructures an iridescent and pearly appearance. Consequently, the mixture is known as pearlite. We can now discuss the consequences of all this to the metallurgy and properties of cast irons and steels.

11.3 Cast irons

Cast iron was one of the dominant structural materials of the nineteenth century and can be found in beams, columns and arches in many rehabilitation and refurbishment projects. It has also been used extensively in pipes and fittings for services, although current practice favours plastic pipes wherever possible. An important use in construction is for tunnel segments and mine-shaft tubing. As we shall see it should be treated with respect.

Most cast irons have carbon contents between 3 and 4.5% by weight and their relatively low melting temperatures mean that they are easily melted and therefore suitable for casting. The properties of the four principal types are summarised in *Table 11.3*.

In white cast iron, the phases are ferrite and Fe_3C, and the large proportion of Fe_3C gives a hard but very brittle substance. This makes it unsuitable for structural uses. It is very difficult to machine, but its high resistance to wear and abrasion makes it suitable for, for example, the facings of earth-moving machines. If, however, it is heated to temperatures between 800 and 900°C for a lengthy time then the Fe_3C can decompose and produce graphite in the form of clusters that are surrounded by a ferrite matrix. Graphite is soft, and the result is the more ductile but still reasonably strong malleable iron.

Grey cast iron also contains about 2% silicon. In this system, iron–graphite is more stable than iron–Fe_3C and so there are significant quantities of graphite, which can increase ductility and toughness (and lead to the grey colour when the metal is cut). The properties are however dependent on the shape of the graphite particles. If these are in the form of flakes then toughness can be low because the flakes are planes of weakness. However, if a small proportion of magnesium (< 0.1%) is added to form an alloy then, with correct casting procedures, the graphite is induced to form spherical particles and the resulting spheroidal graphite (SG) iron (also known as nodular or ductile iron) has good strength, ductility and toughnesss. A typical use is for modern cast-iron tunnel linings.

11.4 Steel

The thermochemistry of steel making is very complex but, as we have described above, the most important reaction is simply that of reducing the carbon content by a process of controlled oxidation:

$$2Fe[C] + O_2 \leftrightarrow 2Fe + 2CO \qquad (11.4)$$

A considerable amount of oxygen is required, some of which dissolves in the liquid steel. If not removed this would form hard, brittle iron oxide FeO and we would be back at square one. So, when the required carbon content is reached the residual oxygen is 'fixed' as an oxide which, after a period of resting, rises to the surface and is removed as slag. The substances commonly used to fix the oxygen in this way are manganese and silicon, and steels treated in this way are known as *killed steels*. Manganese also reacts with sulphur, a very persistent impurity, to form MnS. This avoids the formation of iron sulphide FeS which, if present in even small quantities, can cause a defect known as *hot shortness*, in which the steel cracks disastrously if it is stressed when hot. Even the simplest steels therefore contain silicon and manganese.

As shown in *Fig. 11.1* and discussed above most steels at normal temperature consist of two phases, ferrite and Fe_3C, which combine in laminar regions of alternating layers about 0.5 μm thick to give the uniform appearance of pearlite. The overall composition of pearlite is about 0.8% carbon, and steels containing less than this are mixtures of ferrite and regions of pearlite. Ferrite contains very little carbon

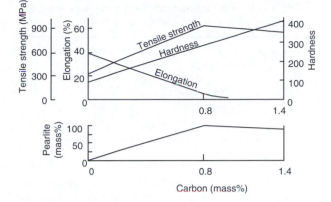

Fig. 11.2 *The influence of carbon content on the pearlite content, strength and ductility of steels (adapted from Rollason, 1968).*

and so the relative proportions of ferrite and pearlite depend linearly on the carbon content between 100% ferrite at 0% carbon to 100% pearlite at 0.8% carbon. It follows that at carbon contents approaching 0.8% the properties of the steel are dominated by those of pearlite – high hardness, high strength and poor ductility and toughness. The properties of steel with low carbon contents, say 0.15%, are dominated by the converse properties of the ferrite.

It might therefore be expected that the properties of steels are strongly affected by their pearlite content, or to put it another way, by their carbon content; *Fig. 11.2* confirms this. The tensile strength increases approximately linearly from about 300 MPa at 0% carbon to about 900 MPa at 100% pearlite (0.8% carbon). Over the same range, the elongation to fracture decreases from about 40% to nearly zero. This is a bit of an over-simplification, because the properties of low-carbon steels are also affected by the grain size of the ferrite that occupies the greater part of the microstructure, with smaller grain increasing the yield stress and decreasing the ductility (as discussed in principle in Chapter 8).

We must also note another very important feature of the behaviour of steels, which we described in Chapter 2. Structural steels in particular can go through a ductile-to-brittle transition as their temperature changes over ranges that are typical of those due to variations in weather, season and climate. This phenomenon is usually shown up most clearly by impact tests, with typical results obtained from a Charpy test shown in *Fig. 2.22*. For this steel the ductile-to-brittle transition temperature is about –20°C. Even though this phenomenon appears to have been first noted by Isambard Kingdom Brunel in 1847, it still brings about its fair share of failures.

Table 11.4 The importance of grain size for the properties of a typical low-carbon steel.

Grain size (µm)	σ_y (MPa)	Ductile-to-brittle transition temperature (°C)
25	255	0
10	400	−40
5	560	−60

The trick is to formulate steels for which the ductile-to-brittle transition temperature is low but which can also be joined successfully by welding. To do this, we want the carbon content to be low, a high ratio of manganese to carbon and a small grain size of the ferrite. The importance of grain size in controlling both the strength and transition temperature is shown *Table 11.4*. We therefore have a powerful argument in favour of fine-grained steels. To produce and maintain fine grain sizes, careful control must be exercised over the temperatures of hot rolling, the amounts of deformation imposed and cooling rates. As might be expected, so-called controlled-rolled steels are more expensive than less carefully controlled products.

Although all steels are strictly alloys the term *non-alloy* or *plain-carbon* steel is used when carbon is the primary alloying element, even though there may be other minor constituents such as manganese and silicon (for the reasons given above), and *alloy steels* for those which contain appreciable concentrations of other elements.

11.4.1 HOT-ROLLED STRUCTURAL STEELS

Most structural steels are classified as low carbon, with carbon contents of up to about 0.3%. The steel production process results in semi-finished products called blooms, billets and slabs, which have to be transformed by hot rolling into the wide variety of shapes and sizes required for construction – I-sections, angles, hollow tubes, plates etc. These are produced by re-heating the steel up to about 1200°C and then passing it through rollers which squeeze it into the required shape (see Chapter 9, section 9.2). Several passes may be required before the required dimensions are achieved.

The mechanical working of the steel during rolling increases its strength (but reduces its ductility) and so the yield strength will reduce with increasing material thickness. The steel cools during rolling to about 750°C. After leaving to cool to room temperature it is termed *as-rolled steel*, and usually requires more heat treatment to achieve the required mechanical properties.

The main designations of steel produced reflect the different heat treatment processes:

- Normalised steel. The as-rolled element is re-heated to about 900°C, held there for a specific time and then allowed to cool naturally. This refines the grain size and improves the mechanical properties, particularly toughness. The properties are more uniform, and residual rolling stresses are removed.
- Normalised-rolled steel. The rolling finish temperature is above 900°C, and the steel is allowed to cool naturally. The re-heating for normalised steel is therefore not required, but the resulting properties are similar. Normalised and normalised rolled steels are denoted 'N'.
- Thermomechanically rolled steel. This has a different chemistry, which allows a lower rolling finish temperature of 700°C, followed by natural cooling. A greater rolling force is required, and the properties are retained unless reheated above 650°C. These steels are denoted 'M'.
- Quenched and tempered steel. Normalised steel at 900°C is rapidly cooled or 'quenched' to produce a steel with high strength and hardness, but low toughness. The fast cooling produces a hard brittle microstructure, known as *martensite*, in which all the carbon is trapped. The structure is, however, metastable and 'tempering' or reheating causes the carbon to be precipitated as tiny particles of carbides throughout the matrix. The loss of carbon from the martensite in this way allows it to become softer and more ductile. Control of the reheating temperature (typically about 600°C) and time, thus varying the amount of carbon left in the martensite, gives great control over the properties that can be achieved. These steels are denoted 'Q'.

The steel properties of most interest to structural designers are yield strength, ductility, toughness, modulus of elasticity, coefficient of thermal expansion, weldability and corrosion resistance. Most structural steels used in the UK and in Europe comply with BS EN 10025.[1] Part 2 of this standard specifies five grades for hot-rolled non-alloy structural steel: S185, S235, S275, S355 and S450, where 'S' denotes 'structural' and the numbers are minimum yield strengths (in MPa) for sections with a thickness of

[1] The British and European standards referred to in the text are listed in 'Further reading' at the end of this part of the book.

Table 11.5 Composition limits and properties of grades of hot-rolled structural steel (extracted from BS EN 10025 Part 2 – Non-alloy structural steels)

Grade	Min yield strength (MPa)	Tensile strength (MPa)	Min elongation at fracture (%)	Composition limits (max %)							
				C	Si	Mn	P	S	N	Cu	CEV
S235	235	360–510	26	0.19		1.5	0.04	0.04	0.014	0.6	0.35
S275	275	410–560	23	0.21		1.6	0.04	0.04	0.014	0.6	0.40
S355	355	470–630	22	0.23	0.6	1.7	0.04	0.04	0.014	0.6	0.47
S450	450	550–720	17	0.23	0.6	1.8	0.04	0.04	0.027	0.6	0.49

Notes:
1. Data for flat and long products, up to 16 mm thick, J0 toughness grade (see below), normalised steel.
2. The yield strength reduces for increased section thickness.
3. CEV = carbon equivalent content.
4. The toughness grade relates to the Charpy impact energy at a specified temperature:
   - J0: > 27 J impact energy at 0°C
   - J2: > 27 J impact energy at −20°C
   - K2: > 40 J impact energy at −20°C.

less than 16 mm. They are also given the designation 'N' or 'AR' depending on whether they are delivered in the normalised or as-rolled state. The yield strength reduces for increased section thickness because thick sections cool more slowly than thin ones and, consequently, the grain size is larger and the Fe_3C ends up differently distributed. The most commonly used grades are S275 and S355. The required properties and the composition limits for the four highest grades are given in *Table 11.5*. The ratio of the minimum tensile strength to yield strength reduces from 1.5 to 1.25 with increasing strength. The ductility (% elongation at fracture in a tensile test) reduces with increasing strength, but not prohibitively so. The toughness is specified as one of three Charpy impact values (J0, J2 or K2) at a specified temperature.

Other mechanical properties, not given in *Table 11.5*, but which can be considered as near constant for almost all steel types, are:

- modulus of elasticity, $E = 205$ GPa
- shear modulus, $G = 80$ GPa
- Poisson's ratio, $v = 0.3$
- coefficient of thermal expansion = 12×10^{-6}/C degree.

Weldability relates mainly to the susceptibility of the steel to embrittlement during the (often rapid) heating and cooling when welded (Chapter 9). This depends on the composition, mainly the carbon content, but also to some extent on the other alloying elements. The *Carbon Equivalent Value* (CEV) is used as a measure of this; it is defined as:

$$CEV = C + Mn/6 + (Cr + Mo + V)/5 + (Ni + Cu)/15$$

where C, Mn etc. are the percentage of each of the alloying elements.

The importance of grain size is recognised by separate parts of BS EN 10025 for normalised/normalised-rolled weldable fine-grained structural steels and for thermochemical rolled weldable fine-grain structural steels. These include alloy steels with significant quantities of chromium and nickel. The standard covers grades from S275 to S460.

The higher strengths that can result from quenching and tempering without unacceptable reduction in other properties are recognised in Part 6 of EN 10025, which includes seven strength grades from S460Q to S960Q. The properties and composition limits for the main alloying elements are given in *Table 11.6*; they are all classed as alloy-special steels. As might be expected from our previous discussion, the ductility (expressed as elongation at failure) decreases with increasing strength, but it is possible to produce steels with different toughness limits (expressed as minimum impact energy). The composition limits for each of the grades are similar, but the increasing total alloy contents with increasing strength grades are apparent from the increasing CEV limits. All grades are weldable in principle, but do not have unlimited suitability for all welding processes, so specialist advice is recommended. These steels are not yet in widespread use in construction, but they may become increasingly used for specific applications in future.

Table 11.6 Composition limits and properties of grades of hot-rolled structural steel (extracted from BS EN 10025 Part 6: Quenched and tempered condition)

Grade	Min yield strength (MPa)	Tensile strength (MPa)	Min elongation at fracture (%)	Composition limits of major elements (max %)									
				C	Si	Mn	Cr	Cu	Mo	Ni	V	CEV
S460Q	460	550–720	17									0.47
S500Q	500	590–770	17									0.47
S550Q	550	640–820	16									0.65
S620Q	620	700–890	15	0.22	0.86	1.8	1.6	0.55	0.74	2.1	0.14	0.65
S690Q	690	770–940	14									0.65
S890Q	890	940–1100	11									0.72
S960Q	960	980–1150	10									0.82

Notes:
1. Data for flat and long products ≤ 50 mm thick, yield and tensile strength reduce with increasing thickness.
2. CEV = carbon equivalent content.
3. Toughness designation within each grade:

	Minimum impact energy (J)			
	0°C	−20°C	−40°C	−60°C
Q	40	30		
QL	50	40	30	
QL1	60	50	40	30

Most structural steels have a similar low resistance to corrosion, and in exposed conditions they need to be protected by one of the protective systems described in Chapter 10. There are no special requirements of the steel material for ordinary coating systems, including both aluminium and zinc metal spray. However, if the steel is to be galvanised, then there is a need to control the alloy content (notably the silicon).

The exception is 'weather-resistant steels' or, more simply, 'weathering steel', also marketed as Corten steels. These fall into the class of alloy steels and have their own part of BS EN 10025 (Part 5). They contain a higher than normal copper content, the most significant alloying elements other than those shown in *Table 11.5* being chromium and nickel. Their composition is such that, when they are exposed to the atmosphere over a period of time, a tightly adhering oxidised steel coating or 'patina' is formed on the surface that inhibits further corrosion. This is often an attractive brown colour. Thus when used appropriately they do not require any protective coating. The layer protecting the surface develops and regenerates continuously when subjected to the influence of the weather. However, when designing a structural element an allowance must be made for the oxidised surface layers by subtracting a specified amount from all exposed surfaces.

11.4.2 COLD-ROLLED STEELS
Many lightweight sections are produced from cold-rolled steel of low carbon content. Strength is derived from work or strain hardening of the ferrite (see Chapters 2 and 8) and good control over section sizes and shapes is possible. Examples of applications include light steel framing, lightweight lintels, angle sections, and roadside crash barriers. Hollow sections can be made by welding two angles together, but welding will locally anneal the material, with consequent changes to properties in the heat-affected zone (section 9.4).

The relevant BS EN standard for sections formed in this way (BS EN 10219) includes grades of S235 to S355 for non-alloy steels and S275 to S460 for fine-grained steels. As with other steels, the ductility reduces with increasing strength.

11.4.3 STAINLESS STEEL
The term 'stainless steel' covers a wide range of ferrous alloys, all of which contain at least 10.5% chromium, which produces a stable passive oxide

Table 11.7   Ranges of properties of the main types of stainless steels (extracted from EN 10088-2 2005)

Type	0.2% proof stress (MPa)	Tensile strength (MPa)	Elongation at fracture (%)
Martensitic	400–800	550–1100	20–11
Ferritic	210–300	380–640	23–17
Austenitic	190–430	500–1000	45–40

film when exposed to air. Other alloying elements, notably nickel and molybdenum, may also be present. There are three basic types, grouped according to their metallurgical structure:

1. Martensitic (410 series) are low-carbon steels containing 12–14% chromium. They are heat treatable and can be made very hard. Since they retain a keen cutting edge they are particularly useful for cutlery, but are not as corrosion resistant as the other two types.
2. Ferritic (430 series) contain between 10.5 and 27% chromium with very low carbon and little, if any, nickel. They are not heat treatable but are reasonably ductile, middle-strength steels.
3. Austenitic (300 series) contain a maximum of 0.15% carbon and have a basic composition of 18% chromium and 8 or 10% nickel though other additions may be made. Like ferritic steels, they are not heat treatable, are reasonably ductile and have good strength.

Typical ranges of properties of each type as included in BS EN 10088 are shown *Table 11.7*. The higher strengths of the martensitic types are a consequence of their ability to be heat treated.

A further group, Duplex stainless steels, is so-called because they have a two-phase microstructure consisting of grains of ferritic and austenitic stainless steel. When solidifying from the liquid phase a completely ferritic structure is first formed, but on cooling to room temperature, about half of the ferritic grains transform to austenitic grains, appearing as 'islands' surrounded by the ferritic phase. The result is a microstructure of roughly 50% austenite and 50% ferrite.

All these steels offer good resistance to corrosion as long as the passive film can be maintained. All will corrode in solutions low in oxygen and this has been the cause of some embarrassing disasters. The austenitic steels are the most resistant to pitting corrosion, though they may suffer from stress corrosion cracking in chloride solutions at slightly elevated temperatures.

Type 316 (18% Cr, 10% Ni%, 3% Mo) is recommended for all external applications. Ferritic steels should be limited to internal use.

A similar range of section types and sizes to structural steel is available. For all practical purposes, martensitic and ferritic stainless steels should be regarded as unweldable, since both undergo significant changes in structure and properties as a result of the thermal cycle. Ordinarily, austenitic stainless steels can be welded, but they can suffer from a form of intergranular attack (weld decay), and grades recommended for welding, i.e. stabilised by the use of titanium, should be specified.

11.4.4 STEEL REINFORCEMENT FOR CONCRETE

All structural concrete contains steel reinforcement in the form of bars or welded mesh to compensate for the low tensile strength of the concrete. Bars with nominal diameters from 4 to 50 mm diameter are available. The steel is produced in either the basic oxygen process, in which up 30% scrap steel can be added to the pig iron from the converter, or in the electric arc furnace process, in which 100% scrap steel can be used for the charge. Billets are produced from continuous casting, which are then reheated to 1100–1200°C and hot rolled to the required bar diameter, which increases strength and closes any defects in the billets. A pattern of ribs is rolled onto the steel in the last part of the rolling process to improve the bond between the steel and the concrete in service.

The steel is low carbon, with typical levels of 0.2% carbon, 0.8% manganese and 0.15% silicon. If the steel is obtained from electric arc furnaces then the larger quantities of scrap steel used for the charge can lead to significant proportions of other alloying elements from the scrap.

Nearly all reinforcement in current use has a yield stress of 500 MPa. The strength is achieved by one of four processes:

- Micro-alloying, in which smaller quantities of specific alloying metals that have a strong effect on the strength are added, the most common being vanadium at 0.05–0.1%.

- Quenching and self tempering (QST), in which water is sprayed onto the bar for a short time as it comes out of the rolling mill; this transforms the bar surface region into hard martensite, allowing the core to cool to a softer, tougher mixture of ferrite and pearlite. Heat diffusing from the core during cooling also tempers the martensite and the result is a bar with a relatively soft ductile core and stronger harder surface layer.
- Cold rolling, in which a hot-rolled round section bar is squeezed by a series of rollers, thus cold-working the steel.
- Cold stretching or drawing, in which the hot-rolled steel is drawn through a series of dies, thus reducing the cross-sectional area and producing wire with a plain round section.

These processes produce steel with somewhat different ductilities. BS 4449 specifies three grades: B500A, B500B and B500C. The first B in each case is for 'bar', 500 is the yield strength in MPa, and the final letter, A, B or C, is the ductility class. The minimum elongations at maximum force for classes A, B and C are 2.5, 5.0 and 7.5%, respectively (with the tensile:yield strength ratios being 1.05, 1.08 and 1.15–1.35, respectively). Micro-alloying and QST can produce higher-ductility grades B and C, cold rolling the lower-ductility grade A and cold stretching grade B. The grades can be identified by differing rib patterns, defined in BS 4449. Other important properties are:

- *Bendability.* The bars are made from relatively high-strength steels and because the surface ribs acts as stress concentrators, may fracture on bending to the required shape for construction if the bend radius is too tight. BS 4449 specifies that bars with diameters ≤ 16 mm should be capable of being bent around a former with a minimum diameter of 4 times the bar diameter, and bars with higher diameters around a former of 7 times the bar diameter.
- *Fatigue properties.* Fatigue cracking under cycling load will initiate at the root of the ribs and therefore a sharply changing cross-section at this point should be avoided in the rolling process.
- *Bond to concrete.* This is a function of the surface and rib geometry, and is independent of the steel properties. BS 4449 gives examples and limits to the dimensions of suitable geometries and bond test methods.
- *Weldability.* Welding of bars is required when forming mesh or prefabricated cages of reinforce-ment, which are increasingly important for the reduction of labour-intensive bar-fixing operations on site. As with other steels this depends mainly on the CEV. Typical values are 0.3–0.35 for QST bar, 0.4–0.5 for micro-alloy and stretched bar and 0.2–0.3 for cold-rolled bar. The differences in values should therefore be taken into account when selecting a welding procedure.
- *Corrosion resistance.* Although concrete normally provides an excellent protective medium for steel, there are circumstances in which this protection can break down and the steel can corrode. Stainless steel reinforcing is produced for use in such situations. We will discuss the corrosion of steel in concrete in some detail in Chapter 24.

11.4.5 PRE-STRESSING STEEL

Pre-stressed concrete, in which a compressive stress is applied to the concrete before the service loads by means of tensioned steel running through the concrete, became feasible for large-scale construction when high-strength steel and pre-stressing systems were developed in the 1940s. The tension can be applied by either single wires, strands consisting of a straight core wire around which six helical wires are spun in a single outer layer, or bars.

Wires are produced by cold drawing hot-rolled rods; they are subsequently stress-relieved by heating to about 350°C for a short time. If the stress-relieving is carried out when the steel is longitudinally strained then low-relaxation steel can be produced, which reduces the loss of stress with time during service. Indentations or crimps to improve bond to the concrete may be mechanically rolled into the surface after the cold drawing. Strand is produced from smooth surface wires. Properties of wires and strands as included in BS 5896 are given in *Table 11.8*. The very high strengths that are achieved by the drawing process are apparent. The available diameters of the strands depend on the diameters of the wires from which they are formed – nominal diameters of 8, 9.6, 11, 12.5 and 15.2 mm are listed in BS 5986.

Bars for pre-stressing are available in diameters from 20 to 75 mm. They are made from a carbon–chrome alloy steel, and all sizes have an ultimate tensile strength of 1030 MPa and a 0.1% proof stress of 835 MPa, i.e. lower than for the smaller diameter wire and strand, but still much higher than for reinforcing steel. All the bars are produced by hot-rolling, with the strength being achieved by subsequent cold working for diameters of 25 to 40 mm and quenching and tempering for diameters from 50 to 75 mm. The smaller diameters have an elastic (secant) modulus of 170 GPa, and the larger 205 MPa. BS 4486 specifies

Table 11.8 Available sizes and properties of pre-stressing wire and strands (from BS 5896)

Cold drawn wire					
Diameter (mm)	7	6	5	4.5	4
Tensile strength (MPa)	1570–1670	1670–1770	1670–1770	1620	1670–1770
0.1% proof stress (MPa)	1300–1390	1390–1470	1390–1470	1350	1390–1470
Relaxation after 1000 hrs at 80% ultimate load					
Class 1			12%		
Class 2			4.5%		
Elastic modulus (GPa)			205 ± 10		

7-wire standard strand				
Diameter (mm)	15.2	12.5	11	9.6
Tensile strength (MPa)	1670–1860	1770–1860	1770	1770–1860
Relaxation after 1000 hrs at 80% ultimate load				
Class 1			12%	
Class 2			4.5%	
Elastic modulus (GPa)			195 ± 10	

a maximum relaxation of 3.5% when a bar is loaded to 70% of its failure load.

Maximum available lengths are typically limited to 6 m, but threads can be rolled on to the bars, which can then be joined with couplers. Corrosion-resistant bars with the same mechanical properties are available; these are made from a martensitic nickel–chrome alloy steel, and have corrosion resistant properties similar to those of austenitic stainless steels.

No pre-stressing steel should be welded, as this would cause a potentially catastrophic local reduction in strength.

11.5 Recycling of steel

Re-use of steel components of structures after de-molition is feasible since, in the majority of cases, the properties of the steel will not have changed since it was first produced. However section sizes may have reduced owing to corrosion, which would preclude their use, and also the design of the new structures may not be able to accommodate the component sizes available. Scrap steel, however, is readily recycled and forms a large part of the feedstock for converters, particularly electric arc furnaces, in which the scrap can form up to 100% of the charge. Some care may have to be taken with the composition since the scrapped steel will contain alloying elements that may have undesirable effects in the steel being produced.

References

Ashby MF and Jones DRH (2005). *Engineering Materials, Vol 2: An introduction to microstructures, processing and design*, 3rd edition, Elsevier Butterworth Heinemann, London.

Rollason EC (1968). *Metallurgy for Engineers*, Edward Arnold, London.

Chapter 12

Aluminium

The use of aluminium in construction is second only to that of steel. In comparison with structural steel, aluminium alloys are lightweight, resistant to weathering and have a lower elastic modulus, but can be produced with similar strength grades. They are easily formed into appropriate sections and can have a variety of finishes. They are however generally more expensive than steels.

12.1 Extraction

Aluminium is strongly reactive and forms a strong bond with oxygen, and it therefore requires more energy than other metals to produce it from its naturally occurring oxide, Al_2O_3 (alumina). The most important ore is bauxite; this contains only 35–40% alumina, which must first be extracted by the Bayer process. In this, the bauxite is washed with a solution of sodium hydroxide, NaOH, at 175°C, which converts the alumina to aluminium hydroxide, $Al(OH)_3$, which dissolves in the hydroxide solution, leaving behind the other constituents of the ore, mainly a mixture of silica, iron oxides and titanium dioxide. The solution is filtered and cooled, and the aluminium hydroxide precipitates as a white, fluffy solid. When heated to 1050°C this decomposes to alumina.

Direct reduction of alumina with carbon, as is used in the analogous process to produce iron, is not possible since aluminium is a stronger reducing agent than carbon, and a process involving electrolysis must be used in the second stage of aluminium production. In the Hall–Héroult process, the alumina is dissolved in a carbon-lined bath of molten cryolite, Na_3AlF_6, operating at a temperature of about 1000°C. At this temperature some of the alumina (which has a melting point of over 2,000°C) dissolves; a low-voltage high-amp current is passed through the mixture via carbon anodes, causing liquid aluminium to be deposited at the cathode and the anodes

to be oxidised to carbon dioxide. The liquid aluminium sinks to the bottom of the bath, where it is periodically collected and either cast into its final form after adding any required alloying materials or cast into ingots for subsequent remelting.

12.2 Aluminium alloys

The term 'aluminium' is normally used to include aluminium alloys. These can be formulated and processed to have a wide variety of properties which are used for wide variety of products – drinks cans, kitchen utensils, automobiles, aircraft frames etc. etc. as well as structural elements for construction. Either cast or wrought aluminium products can be produced.

Casting alloys are generally based on a eutectic alloy system, aluminium combined with up to 13% silicon being widely used. Solidification is over a narrow temperature range (see *Fig. 1.14*), which makes such alloys very suitable for casting into moulds that allow rapid solidification. Other alloying elements that are added in various combinations include iron, copper, manganese, magnesium, nickel, chromium, zinc, lead, tin and titanium. Property ranges listed in BS EN 1706 include 0.2% proof stress, 70–240 MPa; tensile strength, 135–290 MPa; and elongation at failure, 1–8%.

Wrought aluminium alloys are also produced with a wide range of compositions. In the classification scheme adopted by many countries and described in BS EN 573, these are divided into eight series depending on the principal alloying element:

- 1000 series: ≥99% pure aluminium
- 2000 series: aluminium–copper alloys
- 3000 series: aluminium–manganese alloys
- 4000 series: aluminium–silicon alloys
- 5000 series: aluminium–magnesium alloys
- 6000 series: aluminium–magnesium–silicon alloys

Table 12.1 Ranges of properties of aluminium and aluminium alloys (from AluSelect http://aluminium.matter.org.uk/aluselect/ (accessed 5/2/09))

Alloy and treatment	Yield or proof stress (MPa)	Tensile strength (MPa)	Ductility (% elongation to fracture)
Pure aluminium (1000 series)			
Annealed	30	70	43
Strain-hardened	125	130	6
Cast alloys			
As-cast	80–140	150–240	2–1
Heat-treated	180–240	220–280	2–< 1
Wrought alloys			
2000 series, heat-treated	270–425	350–485	18–11
3000 series, strain-hardened	90–270	140–275	11–4
5000 series, strain-hardened	145–370	140–420	15–5
6000 series, heat-treated	100–310	150–340	25–11
7000 series, heat-treated	315–505	525–570	14–10

Constant properties of all types: Density, 2700 kg/m³; modulus of elasticity, 70; GPa, coefficient of thermal expansion, 23.6×10^{-6} per C degree.

- 7000 series: aluminium–zinc–magnesium alloys
- 8000 series: miscellaneous alloys.

Within each series designation, the second, third and fourth digits are used to indicate the proportions of all the alloying elements. These fall into two groups:

- work- or strain-hardened alloys (e.g. the 3000 and 5000 series), where strength is achieved by cold working
- heat-treatable or precipitation hardening alloys, such as the 2000 and 6000 series, where the strength and other required properties are achieved by heat-treatment processes, often complex.

Table 12.1 gives the range of properties that can be obtained for cast and the different series of wrought aluminium alloys. The ranges are relatively large in each group, reflecting both the alloy composition and the degree of treatment imposed. Specific alloys from all series can be used in construction, either in the form of extruded sections or sheets for cladding panels etc.

The alloys that are suitable for heat-treatment are also known as age-hardenable alloys. In Chapter 8 we discussed how the dislocation movement in plastic flow can be impeded by suitable barriers. A classic example is provided by the original aluminium alloy, first developed in 1906, Duralumin, which contains 4% Cu, and upon which the whole of the aircraft industry has depended for many years. When heated to around 550°C the copper dissolves into solid solution in the aluminium and then remains in solution when the alloy is rapidly cooled. Thereafter, even at room temperature, a fine dispersion of a hard intermediate compound, $CuAl_2$, forms slowly. Because the particles are small, actually submicroscopic, and evenly dispersed throughout the matrix, they offer maximum resistance to dislocation movement, and the yield stress is consequently considerably higher than that of pure aluminium. This process, known as ageing, can be speeded up by reheating to temperatures of about 150°C. But, if reheated to too high a temperature (~250°C) the minute particles of $CuAl_2$ coalesce and clump together. They are then more widely separated, the dislocations can pass easily through the matrix and the yield strength is correspondingly reduced. This is known as over-ageing. This phenomenon places a restriction on the operating temperatures of the alloy if the properties are not to deteriorate during use. Modern alloys are more sophisticated and capable of use at higher temperatures, but the same principles apply.

The durability of aluminium alloys is, generally, greater than that of steel. Aluminium has a high affinity for oxygen, and an inert oxide film forms on its surface when it is exposed to air. Although this is very thin, between 0.5 and 1 µm, it forms an effective barrier to water and a variety of other chemical agents. The film is stable within a pH range of about 4 to 8, with acid and alkaline dissolution occurring in environments below and above this range, respectively. This corrosion resistance in neutral or near-neutral environments is an extremely useful property, but it does mean that aluminium is not suitable for use as reinforcement of concrete which has a pH of between 12 and 13, and care must be taken if it comes into contact with concrete, when, for example, aluminium window frames are used in a concrete frame structure. The corrosion resistance does, however, depend on an alloy's composition and heat treatment.

Welding of casting alloys and non-heat treatable alloys is possible with the usual care. However, the welding of heat-treated aluminium alloys can be problematic, since the thermal cycle will, inevitably, produce an over-aged structure in the parent metal.

Although techniques for welding are now well established, bolting, and to a lesser extent riveting, are often preferred. If the bolts or rivets are non-aluminium (e.g. steel) then arrangements must be made to keep them electrically isolated from the aluminium, or else bimetallic corrosion can lead to rapid attack of the aluminium.

12.3 Recycling of aluminium

Aluminium can in principle be recycled indefinitely, as remelting and refining results in no loss of its properties. Aluminium is also a cost-effective material to recycle, since it requires only 5% of the energy and produces only 5% of the CO_2 emissions compared with primary production. Think of the demand for recycled drinks cans!

Further reading for Part 2 Metals and Alloys

BOOKS

Some more extensive texts that expand on the subject matter of this part of the book are:

Callister WD Jr (2007). *Materials Science and Engineering, An introduction*, 7th edition, John Wiley & Sons Inc, New York.

> *A comprehensive introduction to the science and engineering of materials.*

Brandon D and Kaplan WD (1997). *Joining processes – An Introduction*, John Wiley and Sons, Chichester.

> *Aimed at students, this includes more detailed sections on surface science and welding than we have room for in this book.*

Pascoe KJ (1978). *An Introduction to the Properties of Engineering Materials*, Van Nostrand Reinhold, New York.

> *Very much a beginner's text.*

Evans UR (1960). *The Corrosion and Oxidation of Metals*, Edward Arnold, London.

> *Size and cost also make this a reference book, but it is the classic textbook on the subject.*

British Constructional Steelwork Association (1986) *Guides for Protection against Corrosion in Steelwork*.

Llewellyn DT (1994). *Steels, Metallurgy and Applications*, Butterworth – Heinemann, Oxford.

ON-LINE INFORMATION ON STEEL

The Steel Construction Institute

The SCI has produced a comprehensive set of publications on all aspects of the use of steel in construction. A list can be found on their website: http://www.steel-sci.org/Information/Publications.htm

They are not free, but your library may have access to these.

Corus

The UK steel producers, Corus, have a series publications of the manufacture, uses and properties of steel for use in construction that are available to download from their web-site: http://www.corusconstruction.com/en/

These include brochures on design guidance, corrosion protection and weathering steel.

The CARES Guide to Reinforcing Steels

CARES is a product certification scheme for reinforcing steel in the UK, and has produced a series of guides covering the production, properties, methods of use, welding and specification of reinforcing steel, aimed at designers and users. These can be found on: http://www.ukcares.co.uk/

BRITISH AND EUROPEAN STANDARDS

The list below is of those standards, specifications and design codes published by the British Standards Institution that are mentioned in the text. It is not intended to be an exhaustive list of all those concerned with steel and its use in structures. These can be found by looking at the BSI website: www.bsi-global.com/

Structural steel

BS EN 10025 2004 Hot-rolled products of structural steels.

BS EN 10219-1:2006 Cold-formed welded structural hollow sections of non-alloy and fine grained steels Part 1: Technical delivery conditions.

EN 10088-2 2005 Stainless steels – Part 2: Technical delivery conditions for sheet/plate and strip of corrosion-resisting steels for general purposes.

Reinforcement and pre-stressing steel for concrete

BS EN 10080:2005 Steel for the reinforcement of concrete – Weldable reinforcing steel – General.

BS 4449:2005 Steel for the reinforcement of concrete – Weldable reinforcing steel – Bar, coil and decoiled product – Specification.

BS 5896: 1980, amended 2007, High-tensile steel wire and strand for the pre-stressing of concrete.

BS 4486: 1980 Hot-rolled and hot-rolled and processed high-tensile alloy steel bars for the prestressing of concrete.

Aluminium

BS EN 573 (1995, 2044 and 2007) Aluminium and aluminium alloys: Chemical composition and form of wrought products.

PART 3

CONCRETE

Peter Domone

Introduction

Concrete is a ubiquitous material and its versatility and ready availability have ensured that it has been and will continue to be of great and increasing importance for all types of construction throughout the world. In volume terms it is the most widely used manufactured material, with nearly 2 tonnes produced annually for each living person. It can be found above ground, in housing, industrial and commercial buildings, bridges etc., on the ground in roads, airport runways etc., under the ground in foundations, tunnels, drainage systems, sewers etc., and in water in river and harbour works and off-shore structures. Many structures have concrete as their principal structural material, either in a plain, mass form, as for example in gravity dams, but more often as a composite with steel, which is used to compensate for concrete's low tensile strength thus giving either reinforced or pre-stressed concrete. However, even in those structures where other materials such as steel or timber form the principal structural elements, concrete will normally still have an important role, for example in the foundations. Not surprisingly, concrete has been described as the essential construction material.

Historical background

Even though our knowledge and understanding of the material are still far from complete, and research continues apace, concrete has been successfully used in many cultures and in many civilisations. It is not just a modern material; various forms have been used for several millennia. The oldest concrete discovered so far is in southern Israel, and dates from about 7000 BC. It was used for flooring, and consists of quicklime – made by burning limestone – mixed with water and stone, which set into a hard material. Mortars and concretes made from lime, sand and gravels dating from about 5000 BC have been found in Eastern Europe, and similar mixtures were used by the ancient Egyptians and Greeks some three to four thousand years later. Early concretes produced by the Romans were also of this type, but during the second century BC it was the Romans who first made concrete with a *hydraulic cement*, i.e. one that reacts chemically with the mix water, and is therefore capable of hardening under water and is subsequently insoluble. The cement was a mixture of lime and volcanic ash from a source near Pozzuoli. This ash contained silica and alumina in a chemically active form that combined with the lime to give calcium silicates and aluminates; the term *pozzolana* is still used to describe such materials, and as we will see in Chapter 15, various types of these are in common use in concrete today. Concretes produced by combining this cement with aggregates were used in many of the great Roman structures, for example in the foundations and columns of aqueducts and, in combination with pumice, a lightweight aggregate, in the arches of the Colosseum and in the dome of the Pantheon in Rome.

Lime concretes were used in some structures in the Middle Ages and after, particularly in thick walls of castles and other fortifications, but it was not until the early stages of the Industrial Revolution in the second half of the eighteenth century that a revival of interest in calcium silicate-based cements led to any significant developments. In 1756, John Smeaton required a mortar for use in the foundations and masonry of the Eddystone Lighthouse and, after many experiments, he found that a mixture of burnt Aberthaw blue lias, a clay-bearing limestone from South Wales, and an Italian pozzolana produced a suitable hydraulic cement.

In the 1790s, James Parker developed and patented *Roman cement* (a confusing name since it bore little resemblance to the cement of Roman times). This was made from nodules of a calcareous clay from North Kent, which were broken up, burnt in a kiln or furnace, and then ground to a powder to produce the cement. Alternative sources of suitable clay were soon identified, and production of significant quantities continued until the 1860s. The cement was used in many of the pioneering civil engineering structures of the period, such as Brunel's Thames Tunnel and the foundations of Stephenson's Britannia Bridge over the Menai Straits.

Roman cement, and some others of a similar type developed at about the same time, relied on using a raw material that was a natural mixture of clay (silica-rich) and calcareous (calcium-rich) minerals. Methods of producing an 'artificial' cement from separate clay- and lime-bearing materials were therefore sought, resulting in the patenting by Joseph Aspdin in 1824 of *Portland cement*. A mixture of clay and calcined (or burnt) limestone was further calcined until carbon dioxide was expelled, and the product was then ground to give the fine cement powder. This had hydraulic cementitious properties when mixed with water; it was called Portland cement because Aspdin considered the hardened product to have a resemblance to Portland stone – an attractive and popular building material. In 1828, Brunel found the hardened mortar to be three times stronger than that made from Roman cement, and he used it for repairs in his Thames Tunnel. However, Portland cement was relatively expensive, and it did not come into widespread use until larger-scale production processes with higher burning temperatures, which gave enhanced properties and more careful control over the composition and uniformity of supply, had been developed. In particular, the replacement of single-shaft kilns by continuous-process rotary kilns in the 1880s was critical. Increasingly larger-capacity kilns have met the enormous worldwide demand of the twentieth century. A measure of the importance of Portland cement is that it was the subject of one of the first British Standards (BS 12) in 1904, subsequently revised several times before being subsumed in the recent European standard (EN 197). Although the constituent materials have remained essentially the same, refinements in the production processes, in particular higher burning temperatures and finer grinding, and a greater knowledge of cement chemistry and physics have led to steadily increasing quality and uniformity of the cement. From the closing years of the nineteenth century, the vast majority of concrete has been made with Portland cement. However, as we will see in the next chapter, this is not a single material, and there are a considerable number of varieties and types, with an ever increasing number of international standards.

Over the last sixty years or so, there has also been increasing use of other materials incorporated either in small quantities to enhance the fresh and/or hardened properties (termed *admixtures*) or to replace some the Portland cement (currently termed *additions*). These have been developed and exploited to give concrete with an increasingly wide range of fresh and hardened properties, making it possible to produce structures of increasing complexity, size and durability in severe environments effectively and efficiently.

Concrete Technology

In this part of the book we will be considering the constituents, composition, production, structure and properties of concrete itself, i.e. the topics that form the subject of *concrete technology*. Most students of civil engineering will also study the behaviour, design and production of reinforced and pre-stressed concrete, but that is not our function here.

A simple definition of concrete is that it is a mixture of cement, water and aggregates in which the cement and water combine to bind the aggregate particles together to form a monolithic whole. This may sound straightforward but concrete technology has many complexities for a number of reasons, including:

- The series of chemical reactions of hydration between Portland cement and water are complex, and produce a hardened cement that has an equally complex composition and microstructure. Furthermore, Portland cement is not a single uniform material but, as mentioned above, has a range of compositions and hence properties when obtained from different sources or even from the same source over a period of time.
- The use of admixtures and additions, although advantageous, adds to the complexity.
- The aggregates are normally obtained from local sources and, although they are carefully selected for size, strength etc., a range of types, including both natural and artificial (mainly lightweight) are used, each of which will affect the concrete's properties to a greater or lesser extent.
- Although the hardened properties are obviously of paramount importance, the properties in the

newly mixed, fresh (or fluid) state must be such that the concrete can be transported from the mixer, handled, placed in the moulds or formwork and compacted satisfactorily. This requirement can be demanding, for example with *in-situ* concrete being placed in extreme weather conditions in parts of a structure with difficult access. Although this gives rise to one of the great advantages of concrete – its ability to be placed in complex shapes and forms – the responsibility for ensuring that these operations are carried out satisfactorily rests with the engineers in charge of the construction operations. In this respect concrete is different to most other structural materials, which are supplied in a ready-to-use state, with the exception of factory-produced pre-cast elements.

- Even when hardened, the concrete's structure and properties are not static, but continue to change with time. For example, about 50–60% of the ultimate strength can be developed in 7 days, 80–85% in 28 days, and small but measurable increases in strength have been found in 30-year-old concrete.
- Long term movements due to both load (i.e. creep), and changes in moisture (i.e. swelling and shrinkage) can be significant.
- The concrete and any steel contained within in it can deteriorate for a variety of reasons, and so ensuring adequate durability as well as mechanical properties such as strength and stiffness is a major consideration.

At first glance this may therefore seem daunting, but it is the intention of this part of the book to consider all of these, and some other, aspects of concrete technology in sufficient detail for you to take forward into structural design and production, and to be able to access the many and varied more advanced publications on the subject.

A look at the contents list will show you how this will be achieved. We start by describing the constituent materials of concrete: Portland cement (in some detail), additions, admixtures, alternatives to Portland cement (briefly) and aggregates. We then discuss the fresh and early age properties before going on to consider the hardened properties of deformation and strength. The principles of mix design, the process of selecting the relative proportions of the constituents to give the required properties, are then presented. We then consider some methods of non-destructive testing before we discuss various aspects of durability in some detail. We then come right up-to-date by describing a number of 'special concretes' that are produced for specific purposes, such as lightweight and sprayed concrete, and some recent developments in high-performance concrete that are extending the properties and uses of the material in exciting ways. Finally we discuss the recycling of concrete, an increasingly important factor in the sustainability of construction.

This is a logical sequence of presentation, but not all courses in concrete technology follow this order, and the chapters and sections within them are written so that they need not be read consecutively.

SOME DEFINITIONS

When reading this part of the book, there are some key terms and definitions that are worth having at your finger tips, or at least not too far from them.

From above:

- *Concrete* is a mixture of *cement*, water, *fine aggregate* (sand) and *coarse aggregate* (gravel or crushed rocks) in which the cement and water have hardened by a chemical reaction – *hydration* – to bind the nearly (non-reacting) aggregate.
- Other materials in addition to the above are often incorporated, such as fine powders that can substitute some of the cement, known as *additions*, and small quantities of chemicals, known as *admixtures*, which can alter and improve some properties.
- The use of additions, most of which are fine powders like the cement and which participate in the hydration reactions, requires the definition of the *binder* as the mixture of cement and addition(s).

Also:

- *Grout* or *cement paste* is a mixture of cement and water only; it will hydrate and gain strength, but it is rarely used for structural purposes since it is subject to much higher dimensional changes than concrete under loading or in different environments, and it is more expensive.
- *Mortar*, a mixture of cement, water and fine aggregate (sand), is more commonly used for small volume applications, for example in brickwork.
- The aggregates form the bulk of the concrete volume, typically 70–80%. Most of the remainder of the hardened concrete is the hydrated cement (or binder) and water, often called the *hardened cement paste* (HCP). There is also a small quantity of air voids (typically 1–3% of the concrete volume) due to the presence of air that was not expelled when the concrete was placed.

- Aggregate is divided at a particle size of 4 mm, all particles with a diameter smaller than this being referred to as fine aggregate and all larger particles being coarse aggregate. The maximum particle size of coarse aggregate can be 10, 20 or 40 mm. In most concrete, the fine aggregate is somewhere between 30 and 45% of the total aggregate. On mixing, the volume of water is normally in the range of 50–75% of the cement paste and therefore, ignoring any air, most freshly mixed concrete comprises, by volume:

  6–16% cement or binder
  12–20% water
  20–30% fine aggregate
  40–55% coarse aggregate

  So, although cement (or binder) is the key component of concrete, it occupies the smallest proportion by volume.

- The *mix proportions* are the amounts of each of the constituents that are mixed together to form a unit quantity of concrete. These are most commonly expressed as the weight of each material in a unit volume of concrete e.g. if the proportions are:

  cement or binder: 350 kg/m^3
  water: 200 kg/m^3
  coarse aggregate: 1100 kg/m^3
  fine aggregate: 750 kg/m^3

  then a cubic metre of the fresh concrete will comprise 350 kg of cement or binder, 200 kg of water, 1100 kg of coarse aggregate and 750 kg of fine aggregate. This often causes some confusion to those new to concrete, since the units for each material are the same as those of density – take care not to make this mistake.

- We will see that nearly all of the properties of concrete are affected by the relative or absolute amounts of the constituents. Therefore, to ensure that satisfactory properties are achieved, the mix proportions must be carefully chosen and controlled. Measuring exact volumes of the materials is difficult, so the weights required are normally specified and used for concrete production; thus the mix proportions are most conveniently expressed as the weight of each material required for unit volume of the concrete, as above.

- The relative particle density of Portland cement is about 3.15, most binders have values in the range 2.2 to 2.9 and most aggregates used for concrete have values of 2.55 to 2.65 (the exceptions being lightweight and high-density aggregate used for more specialised concrete). A few calculations using these figures and the volumes given above show that the ranges of the mix proportions by weight for most concrete are:

  cement (or binder) 150–600 kg/m^3
  water 110–250 kg/m^3
  aggregates (coarse + fine) 1600–2000 kg/m^3.

The total of these for any particular mix gives, of course, the concrete density, which can vary from 2200 to 2450 kg/m^3 with normal density aggregates. As we shall see, the ratio of the weight of water to that of cement or binder, normally referred to just as the *water/cement ratio* or the *water/binder ratio*, is an important factor influencing many of the concrete's properties. Values are typically in the range 0.3 to 1.0.

Chapter 13

Portland cements

Cement is the essential component of concrete which, when hydrated, binds the aggregates together to form the hard, strong and monolithic whole that is so useful. Well over 95% of the cement used in concrete throughout the world is Portland cement in its various forms. It is by no means a simple material, and its complexities have an impact on the properties and behaviour of concrete from mixing right through to the end of its life. It is therefore important to have some understanding of its manufacture, its composition, the processes involved in its hydration and of its final hardened structure if it is to be used effectively.

13.1 Manufacture

The crucial components of Portland cement are calcium silicates, which in the manufacturing process are formed by heating a mixture of calcium oxide (CaO) and silicon dioxide (or silica, SiO_2) to high temperatures. Both of these occur in the earth's crust in large quantities, the former in various forms of calcium carbonate ($CaCO_3$), e.g. chalk and limestone, and the latter in a variety of mineral forms in sand, clay or shale. Cement production is a large-scale operation requiring huge quantities of the raw materials, and the production plants are therefore normally sited close to a suitable source of one or both of these, which occasionally even occur in a single source such as marl. The raw materials all contain some other components, and in particular clays contain oxides of aluminium, iron, magnesium, sodium and potassium. These cannot be avoided; the first two have a significant effect on the manufacture and composition of the resulting cement, and as we will see when discussing durability, some of the others can have significant effects even though they are present only in small quantities.

The manufacturing process is relatively simple in principle, although the high temperatures and large quantities involved required sophisticated monitoring and control systems to ensure that a uniform high-quality product is obtained. The stages are:

1. Initially the limestone or chalk and clay or shale are blended in carefully controlled proportions (normally about 80/20) and interground in ball or roller mills until most or of all the particles are smaller than 90 μm. The composition of the mixture is critical, and it may be necessary to add small quantities of other materials such as ground sand or iron oxide.

2. The heart of the manufacturing process consists of heating this mixture (known as the raw meal) to about 1400–1500°C. In modern cement plants this takes place in two stages. First the raw meal is fed into the top of a pre-heater tower that includes a *pre-calcining vessel* (whose use improves the overall energy efficiency of the whole process). As it falls through this it is flash-heated to about 900°C for a few seconds, during which about 90% of the carbonate component decomposes into calcium oxide and carbon dioxide (the *calcining* reaction). The mixture then passes into a heated rotary kiln that takes the form of an inclined steel cylinder lined with refractory bricks; it can be up to tens of metres long and several metres in diameter (depending on the capacity of the plant) and it is rotated about its longitudinal axis, which is set at a slope of about 3 degrees (*Fig. 13.1*).

3. The kiln is heated at its lower end to about 1500°C by the combustion of a fuel–air mixture. The most common fuel is powdered coal, but oil and natural gas are also used; waste organic materials such as ground tyres are often added to the main fuels. The pre-heated meal from the pre-calciner is fed into the higher end of the kiln, and it takes between 20 and 30 minutes to reach and pass out of the lower heated end as a granular material called *clinker*. As the temperature

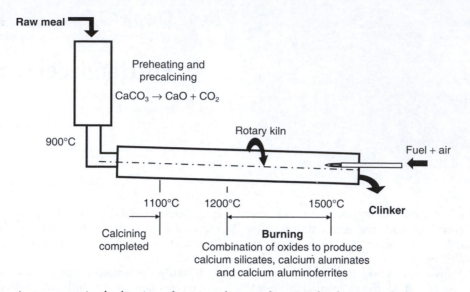

Fig. 13.1 *The main processes in the heating of raw meal to produce Portland cement clinker.*

of the feed increases as it moves through the kiln, decarbonation becomes complete at about 1100°C and then, in the so-called burning zone, the oxides start to combine to form a mixture consisting mainly of calcium silicates, calcium aluminates and calcium aluminoferrites. The chemistry involved is fairly complex, with compound formation at 1400–1500°C being greatly helped by the small quantities of alumina and iron oxide that are present (typically 5% and 3% respectively) and that act as a molten flux.

4. The clinker emerges from the kiln at about 1200°C and is then cooled to about 60°C before being mixed with a small quantity (3–5%) of gypsum (calcium sulphate dihydrate, $CaSO_4.2H_2O$), and sometimes a small quantity (up to 5%) of a filler such as limestone powder, and then ground, usually in a ball mill, to give the Portland cement. The grinding process also increases the temperature of the clinker/gypsum mixture so cooling by water sprayed onto the outside of the grinding mill is required. The increased temperature causes some dehydration of the gypsum.

13.2 Physical properties

Portland cements are fine grey powders. The particles have a relative density of about 3.14, and most have a size of between 2 and 80 μm. The particle size is, of course, dependent on the clinker grinding process, and it can be and is varied depending on

the requirements of the cement, as will be discussed in section 13.7. The particles are too small for their distribution to be measured by sieve analysis (as used for aggregates, see Chapter 17), and instead the specific surface area (SSA), the surface area per unit weight, is normally used as an alternative measurement. This increases as the particle size reduces i.e. a higher value means smaller average particle size. There are a number of ways of measuring this, but unfortunately they all give somewhat different values. It is therefore necessary to define the method of measurement when specifying, quoting or using a value. The Blaine method, which is the most commonly used, is based on measuring the rate of flow of air under a constant pressure through a small compacted sample of the cement. Values of SSA measured with this method range from about 300 to 500 m²/kg for most cements in common use.

13.3 Chemical composition

We have seen that Portland cement consists of a mixture of compounds formed from a number of oxides at the high temperatures in the burning zone of the kiln. For convenience, a shorthand notation for the principal oxides present is often used:

CaO (lime) = C; SiO_2 (silica) = S;
Al_2O_3 (alumina) = A; Fe_2O_3 (iron oxide) = F.

The four main compounds, sometimes called *phases*, in the cement are:

Tricalcium silicate	$3CaO.SiO_2$
	in short C_3S
Dicalcium silicate	$2CaO.SiO_2$
	in short C_2S
Tricalcium aluminate	$3CaO.Al_2O_3$
	in short C_3A
Tetracalcium aluminoferrite	$4CaO.Al_2O_3.Fe_2O_3$
	in short C_4AF

Strictly, C_4AF is not a true compound, but represents the average composition of a solid solution.

These compounds start to form at somewhat different temperatures as the clinker heats up when passing down the kiln. C_2S (often known as *belite*) starts to form at about 700°C, C_3S (known as *alite*) starts to form at about 1300°C, and as the temperature increases to the maximum of about 1450°C most of the belite formed at lower temperatures is transformed into alite. C_3A and C_4AF both start to form at about 900°C.

Each grain of cement consists of an intimate mixture of these compounds, but it is difficult to determine the amounts of each by direct analysis; instead the oxide proportions are determined, and the compound composition then calculated from these using a set of equations developed by Bogue (1955). These assume:

- all the Fe_2O_3 is combined as C_4AF
- the remaining Al_2O_3, after deducting that combined in the C_4AF, is combined as C_3A.

The equations in shorthand form are:

$$(C_3S) = 4.07(C) - 7.60(S) - 6.72(A) - 1.43(F) - 2.85(Š) \qquad (13.1)$$

$$(C_2S) = 2.87(S) - 0.754(C_3S) \qquad (13.2)$$

$$(C_3A) = 2.65(A) - 1.69(F) \qquad (13.3)$$

$$(C_4AF) = 3.04(F) \qquad (13.4)$$

Where $Š = SO_3$, (C_3S), (C_2S) etc. are the percentages by weight of the various compounds, and (C), (S) etc. are the percentages by weight of the oxides from the oxide analysis. The value of (C) should be the total from the oxide analysis less the free lime, i.e. that not compounded.

The Bogue equations do not give exact values of the compound composition, mainly because these do not occur in a chemically pure form, but contain some of the minor oxides in solid solution (strictly alite and belite are slightly impure forms of C_3S and C_2S, respectively). For this reason, the calculated composition is often called the *potential compound composition*. However, the values obtained are suf-

ficiently accurate for many purposes, including consideration of the variations in the composition for different types of Portland cement, and their effect on its behaviour.

The approximate range of oxide proportions that can be expected in Portland cements is given in the first column of figures in *Table 13.1*. As might be expected from our description of the raw materials and the manufacturing process, CaO and SiO_2 are the principal oxides, with the ratio of CaO:SiO_2 normally being about 3:1 by weight. The two calcium silicates (C_3S and C_2S) therefore form the majority of the cement. However the composition of any one cement will depend on the composition, quality and proportions of the raw materials, and will therefore vary from one cement plant to another and even with time from a single plant. *Table 13.1* illustrates the effects of this on the compound composition by considering four individual cements, A, B, C and D, whose oxide proportions vary slightly (by at most 3%), but which are all well within the overall ranges. The compound compositions calculated with the Bogue formulae show that:

- The principal compounds, C_3S and C_2S, together amount to 71–76% of the cement.

Table 13.1 Ranges of oxide proportions and compound composition of four typical Portland cements (all proportions percent by weight)

		Cement			
		A	B	C	D
Oxide	Range	Proportion			
CaO	60–67	66	67	64	64
SiO_2	17–25	21	21	22	23
Al_2O_3	3–8	7	5	7	4
Fe_2O_3	0.5–6	3	3	4	5
$Na_2O + K_2O$	0.2–1.3	1	1	1	1
MgO	0.1–4	2	2	2	2
Free CaO	0–2				
SO_3	1–3				
Compound composition					
C_3S		48	65	31	42
C_2S		24	11	40	34
C_3A		13	8	12	2
C_4AF		9	9	12	15

- The relative proportions of each compound vary considerably, by at least two orders of magnitude more than the small variations in the oxide composition. For example, the four ratios of C_3S/C_2S are 2, 5.9, 0.8 and 1.2, and the C_3A content of cement D is 4 to 6 times less than that of the other cements.

As we shall see, such variations have considerable effects on the hydration process and properties of the hardened cement, and therefore careful control of the raw materials and manufacturing processes is vital if cement of uniform quality is to be produced. Cement A can be considered to have a 'typical' or 'average' composition for Portland cement (most modern cements have a C_3S content in the range 45–65% and a C_2S content in the range 10–30%). Cements B, C and D are common and useful variations of this, i.e. they have higher early strength, low heat and sulphate-resisting properties respectively, all of which are discussed in more detail in section 13.7. (Note: the compound compositions in *Table 13.1* do not add up to 100% – the remainder comprises the minor compounds, which include the gypsum added to the clinker before grinding.)

13.4 Hydration

For an initial period after mixing, the fluidity or consistence of a paste of cement and water appears to remain relatively constant. In fact, a small but gradual loss of fluidity occurs, which can be partially recovered on remixing. At a time called the *initial set*, normally between two and four hours after mixing at normal temperatures, the mix starts to stiffen at a much faster rate. However, it still has little or no strength, and hardening, or strength gain, does not start until after the *final set*, which occurs some hours later. The rate of gain of strength is rapid for the next few days, and continues, but at a steadily decreasing rate, for at least a few months.

Setting times are measured by somewhat arbitrary but standardised methods that involve measuring the depth of penetration of needles or plungers into the setting paste.[1] They do not mark a sudden change in the physical or chemical nature of the cement paste, but the initial set defines the time

limit for handling and placing the concrete (and thus cement standards set a minimum time for this) and the final set indicates the start of the development of mechanical strength (and so standards set a maximum time for this).

The cement paste also gets noticeably warm, particularly during the setting and early hardening periods. In other words, the hydration reactions are exothermic. The amount of heat released is sufficient to raise the temperature to 100°C or more in a day or so if the paste is kept in adiabatic (zero heat loss) conditions. However, measurement of the rate of heat output at constant temperature is a more useful direct indication of the rate of reaction, and *Fig. 13.2* shows a typical plot of rate of heat output with time after mixing. Immediately on mixing, there is a high but very short peak (A), lasting only a few minutes or less. This quickly declines to a low constant value for the so-called *dormant period*, when the cement is relatively inactive; this may last for up to two or three hours. The rate then starts to increase rapidly, at a time corresponding roughly to the initial set, and reaches a broad peak (B), some time after the final set. The reactions then gradually slow down, with sometimes a short spurt after one or two days giving a further narrow peak (C).

The hydration reactions causing this behaviour involve all four main compounds simultaneously. The physical and chemical processes that result in the formation of the solid products of the hardened cement paste are complex, but the following simplified description, starting by considering the chemical reactions of each of the compounds individually, is nevertheless valuable.

The main contribution to the short intense first peak (A) is rehydration of calcium sulphate hemihydrate,

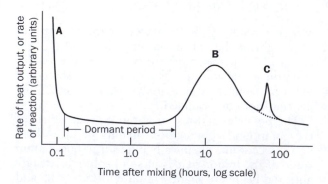

Fig. 13.2 Typical rate of reaction of hydrating cement paste at constant temperature (after Forester, 1970).

---

[1] As in the other parts of the book, a list of relevant standards is included in 'Further reading' at the end of the section.

which arises from the decomposition of the gypsum in the grinding process. Gypsum is reformed:

$$2C\check{S}(0.5H) + 3H \rightarrow 2C\check{S}.2H$$
$$[H = H_2O \text{ in shorthand form}] \quad (13.5)$$

Additional contributions to this peak come from the hydration of the free lime, the heat of wetting, heat of solution and the initial reactions of the aluminate phases. The behaviour of the aluminates is particularly important in the early stages of hydration. In a pure form, C_3A reacts very violently with water, resulting in an immediate stiffening of the paste or a *flash set*. This must be prevented, which is why gypsum is added to the clinker. The initial reaction of the gypsum and C_3A is

$$C_3A + 3C\check{S}.2H + 26H \rightarrow C_3A.3C\check{S}.32H \quad (13.6)$$

The product, calcium sulphoaluminate, is also known as *ettringite*. This is insoluble and forms a protective layer on the C_3A, thus preventing rapid reaction. Usually about 5–6% of gypsum by weight of the cement is added and, as this is consumed, the ettringite reacts with the remaining C_3A to give to calcium monosulphoaluminate, which has a lower sulphate content:

$$C_3A.3C\check{S}.32H + 2C_3A + 4H \rightarrow 3(C_3A.C\check{S}.12H) \quad (13.7)$$

Eventually, if all the gypsum is consumed before all the C_3A, the direct hydrate, $C_3A.6H$, is formed. This causes the short third peak C, which can occur some 2 or 3 days after hydration starts. Whether this peak occurs at all depends on the relative amounts of gypsum and C_3A in the unhydrated cement, and it follows that it tends to be a feature of high C_3A content cements.

The C_4AF phase reaction is similar to that of the C_3A, also involving gypsum, but it is somewhat slower. The products have an imprecise and variable composition, but include high- and low-sulphate forms approximating to $C_3(A.F).3C\check{S}.32H$ and $C_3(A.F).C\check{S}.16H$, respectively, i.e. similar to the C_3A products. The reactions or products contribute little of significance to the overall behaviour of the cement.

As we have seen, the two calcium silicates C_3S and C_2S form the bulk of unhydrated cement, and it is their hydration products that give hardened cement most of its significant engineering properties such as strength and stiffness; their reactions and reaction rates therefore dominate the properties of the hardened cement paste (HCP) (and concrete) and are extremely important. The C_3S (or, more accurately, the alite) is the faster to react, producing

a calcium silicate hydrate with a Ca:Si ratio of between 1.5 and 2 and calcium hydroxide (deposited in a crystalline form often referred to by its mineral name *portlandite*). A somewhat simplified but convenient form of the reaction is:

$$2C_3S + 6H \rightarrow C_3S_2.3H + 3CH \quad (13.8)$$

Most of the main peak B in the heat evolution curve (*Fig. 13.2*) results from this reaction, and it is the calcium silicate hydrate (often simply referred to as C-S-H) that is responsible for the strength of the HCP.

The C_2S (or, strictly, the belite) reacts much more slowly, but produces identical products, the reaction in its simplified form being:

$$2C_2S + 4H \rightarrow C_3S_2.3H + CH \quad (13.9)$$

This reaction contributes little heat in the timescales of *Fig. 13.2*, but it does make an important contribution to the long-term strength of HCP.

The cumulative amounts of individual products formed over timescales a few days longer than those of *Fig. 13.2* are shown in *Fig. 13.3*. The dominance of the C-S-H after a day or so is readily apparent; this is accompanied by an increase in the amount of calcium hydroxide, which, together with some of the minor oxides, results in the HCP being highly alkaline, with a pH between of 12.5 and 13. As we shall see in Chapter 24, this alkalinity has a significant influence on some aspects of the durability of concrete construction.

The timescales and contributions of the reactions of the individual compounds to the development of the cement's strength are shown in *Fig. 13.4*. This further emphasises the long-term nature of the strength-giving reactions of the calcium silicates,

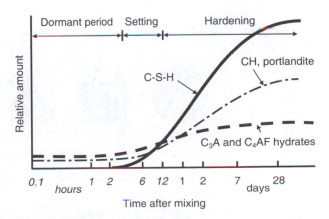

Fig. 13.3 *Typical development of hydration products of Portland cement (after Soroka, 1979).*

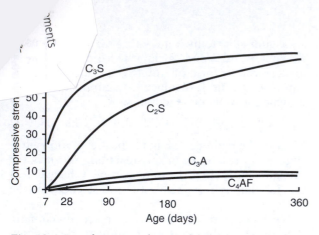

Fig. 13.4 *Development of strength of compounds in Portland cement on hydration (after Bogue, 1955).*

particularly of the C_2S (or, more correctly, the belite). In fact the reactions can never be regarded as complete, and the extent of their completeness is called *the degree of hydration*.

In common with most chemical processes, increasing temperature accelerates all of the above reactions. With decreasing temperature, hydration will continue even below 0°C, but stops completely at about −10°C. We will be discussing the effect of temperature in relation to the development of the strength of cement and concrete in Chapter 19.

The physical processes occurring during hydration and the resulting microstructure of the hardened cement paste are equally, if not more, important than the chemical reactions, and numerous studies have been made of these by scanning, transmission and analytical electron microscopy. *Fig. 13.5* illustrates schematically the hydration of a single grain of cement in a large volume of water. The important features are:

- The processes take place at the solid–liquid interface, with solid products being deposited in the region around the diminishing core of unhydrated cement in each cement grain.
- The very early products form a surface layer on the cement grain, which acts a barrier to further reactions during the dormant period.
- The dormant period ends when this layer is broken down by either a build-up of internal pressure by osmosis, or by portlandite ($Ca(OH)_2$), or both, enabling hydration to proceed more rapidly.
- The hydration products (known as the *gel*) consist of:

  - needle-like crystals of ettringite, deposited early in the hydration
  - an amorphous mass, mainly C-S-H, of small, irregular fibrous particles, some solid, some hollow and some flattened, typically 0.5–2 μm long and less than 0.2 μm diameter, with very high surface area estimated to be of the order of 200,000 m²/kg, i.e. approaching a thousand times greater than the fresh cement grains from which it has been formed
  - large hexagonal crystals of portlandite interspersed in the fibrous matrix.

- The gel contains many small *gel pores*, typically between 0.5 and 5 nm wide, in between the fibrous particles, and as hydration continues, new product is deposited within the existing matrix, decreasing the gel porosity.
- There is some difference in density and structure between the hydrates deposited within the original surface of the cement grain, known as *inner product*, and the less dense hydrates deposited in the original water-filled space, which contain more

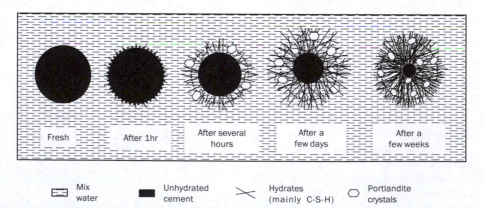

| Mix water | Unhydrated cement | Hydrates (mainly C-S-H) | Portlandite crystals |

Fig. 13.5 *Illustration of the hydration of a single grain of Portland cement.*

crystals of portlandite and aluminoferrite and are known as *outer product*.

- The rate of hydration reduces over a long period after peak B owing to the increased difficulty of diffusion of water through the hydration products to the unhydrated cement. It has been estimated that, for this reason, complete hydration is not possible for cement grains of more than 50 μm in diameter – even after many years there is a residual core of unhydrated cement.
- At complete hydration:
  - the gel porosity reaches a lower limit of about 28%
  - the volume of the products of hydration is little more than twice that of the unhydrated cement, but about two-thirds of the combined initial volume of the unhydrated cement and the water which it consumes.

In reality of course, hydration is occurring simultaneously in a mass of cement grains in the mix water, and so the hydration productions interact and compete for the same space. An important and vital feature of hydration is that it occurs at a (nearly) constant overall volume, i.e. the mixture does not swell or contract and the HCP or concrete is the same size and shape when hardened as the mould in which was placed after mixing. Using this fact, and the measured properties of the fresh and hydrated materials[2] it can be shown that:

- At a water:cement ratio of about 0.43, there is just sufficient mix water to hydrate all the cement and fill all of the resulting gel pores. Therefore at water:cement ratios lower than this, full hydration can never occur unless there is an available external source of water, for example if the cement or concrete is immersed in water. This is the *condition of insufficient water*, and the paste is subject to *self-desiccation*. In practice, in a sealed specimen the hydration will cease somewhat before all of the available water is consumed, and an initial water:cement ratio of about 0.5 is required for full hydration. As we will see in Chapter 19, self-desiccation can also have other effects.
- At a water:cement ratio of about 0.38, the volume of hydration products, i.e. the gel, exactly matches

that of the fresh cement and water. At values lower than this, hydration will be stopped before completion, even if an external source of water is available. This is called the *condition of insufficient volume*. At water:cement ratios higher than this there is an increasing amount of unfilled space between the original grains in the form of *capillary pores*, between about 5 nm and 10 μm wide, and so on average they are about a hundred times larger than the gel pores within the gel itself. Calculations give the relative volumes of unhydrated cement, gel and capillary pores at complete hydration shown in *Fig. 13.6*. In reality, for the reasons discussed above, hydration is never complete and therefore the volumes in *Fig. 13.6* are never achieved, but they may be approached. However, at any stage of hydration, the volume of capillary pores will increase with the water:cement ratio.

The diagrams in *Fig. 13.7* provide a visual illustration of this. These show idealised diagrams of the structure of two cement pastes with high and low water:cement ratios, say of the order of 0.8 and 0.4 respectively, on mixing and when mature, say after several months. In the high water:cement ratio paste the grains are initially fairly widely dispersed in the mix water and, when mature, there is still a significant capillary pore volume. On the other hand, in the low water:cement ratio paste, the grains are initially much more closely packed, and the hydrates occupy a greater volume of the mature paste, which

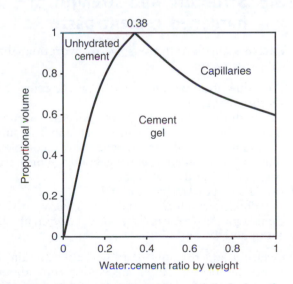

Fig. 13.6 Volumetric composition of fully hydrated cement paste after storage in water (after Hansen, 1970).

[2] Relative densities: unhydrated cement, 3.15; gel solids, 2.61; saturated gel, 2.16; unsaturated gel, 1.88. Gel porosity 28%. 1 g of cement chemically combines with 0.23 g of water during hydration. The analysis derives from the work of Powers in the 1950s; a full summary can be found in Neville (1995).

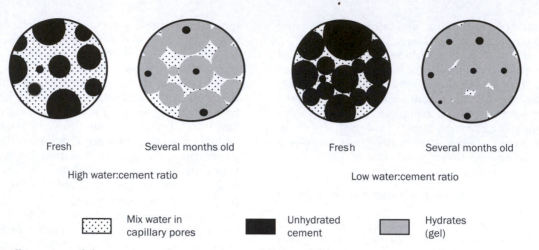

Fresh   Several months old   Fresh   Several months old

High water:cement ratio   Low water:cement ratio

Mix water in capillary pores   Unhydrated cement   Hydrates (gel)

Fig. 13.7 Illustration of the structure of cement pastes of high and low water:cement ratios.

therefore has a greater volume of capillary pores (but which, if the water:cement ratio is low enough, may eventually disappear altogether).

Although it is important to distinguish between capillary and gel pores, in practice there is a near continuous distribution of pore sizes. *Figure 13.8* shows typical measurements that illustrate this, and also provides direct evidence of the substantial reduction in both overall pore volume and pore size with reducing water:cement ratio for pastes of similar age, in this case 28 days.

13.5 Structure and strength of hardened cement paste

We have seen that, at any stage of hydration, the HCP consists of:

- a residue of unhydrated cement, at the centre of the original grains
- the hydrates (the gel), chiefly calcium silicates (C-S-H) but also some calcium aluminates, sulphoaluminates and ferrites, which have a complex fibrous form and contain the gel pores, which are between 0.5 and 5 nm wide
- crystals of portlandite ($Ca(OH)_2$)
- the unfilled residues of the spaces between the cement grains – the capillary pores, between about 5 nm and 10 μm wide.

We should add that the paste will also contain a varying number of larger air voids, from about 5 μm upwards, which have become entrapped in the paste during mixing and have not subsequently been expelled during placing and compaction.

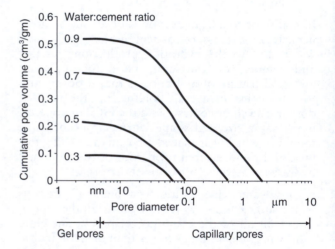

Fig. 13.8 Pore size distribution in 28-day-old hydrated cement paste (adapted from Mehta, 1986).

The significant strength of HCP derives from van der Waals type bonds between the hydrate fibres (see Chapter 1). Although each individual bond is relatively weak, the integrated effect over the enormous surface area is considerable. The unhydrated cement is in itself strong and its presence is not detrimental to overall strength, and it can even be beneficial since it is exposed if the paste or concrete is subsequently cracked or fractured and can therefore form new hydrates to seal the crack and restore some structural integrity provided, of course, some water is present. No other common structural materials have this self-healing property.

For any particular cement, the compressive strength of specimens stored at constant temperature and humidity increases with age and decreasing

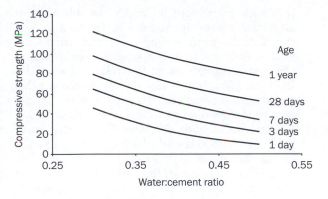

Fig. 13.9 *Compressive strength development of Portland cement paste stored in water at 20°C (after Domone and Thurairatnam, 1986).*

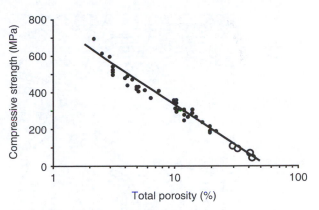

Fig. 13.10 *The dependence of the strength of hardened cement paste on porosity (after Roy and Gouda, 1975).*

water:cement ratio; *Fig. 13.9* shows typical behaviour. The change with age reflects the progress in hydration reactions, i.e. the degree of hydration. At 28 days (a typical testing age when comparing cements) the reactions are about 90% complete for a typical Portland cement. We should also note that the strength continues to increase at water:cement ratios below 0.38, even though *Fig. 13.6* shows that there is an increasing volume of unhydrated cement in the 'end state'. This is direct evidence that unhydrated cement is not detrimental to strength – it is the quality of the hydrates that is the governing factor (there are, however, lower practical limits to the water:cement ratio, which we will discuss in Chapter 20).

We have seen that both the size and volume of the capillary pores are also influenced by age and water:cement ratio (*Figs. 13.6, 13.7* and *13.8*) and it is therefore not surprising that the strength and porosity are closely linked. In simple terms: less porosity (due to either increasing age or lower water:cement ratio or both), means higher strength. The relationship between the two was shown by Powers (1958) to be of the form

$$\sigma = k(1 - P)^3 \qquad (13.10)$$

where k is a constant, σ = compressive strength and P = porosity = pore volume/total paste volume.

Note that in this expression the porosity is raised to power three, showing its great significance. Powers' experiments were on 'normally' cured pastes, i.e. kept in water at ambient temperature and pressure, with variations in porosity obtained by varying the water:cement ratio. This resulted in total (capillary plus gel) porosities ranging from about 25 to 50%. Porosities down to about 2% were obtained by Roy and Gouda (1975) by curing pastes with water:cement ratios down to 0.093 at higher temperatures (up to

250°C) and pressures (up to 350 MPa). *Figure 13.10* shows that at these very low porosities they achieved compressive strengths of more than 600 MPa, Powers' results being consistent with their overall relationship of the form

$$\sigma = A \log(P/P_{crit}) \qquad (13.11)$$

where A is a constant and P_{crit} is a critical porosity giving zero strength, shown by *Fig. 13.10* to be about 55%.

The size of the pores has also been shown to be an important factor. Birchall *et al.* (1981) reduced the volume of the larger pores (greater than about 15 μm diameter) by incorporating a polymer in pastes of water:cement ratios of about 0.2, and curing initially under pressure. The resulting 'macrodefect free' (MDF) cement had compressive strengths of 200 MPa and above, with flexural strengths of 70 MPa, a much higher fraction of compressive strength than in 'normal' pastes or concrete.

Clearly, the extremes of low porosity and high strength cannot be achieved in concretes produced on a large scale by conventional civil engineering practice, but results such as those shown in *Fig. 13.10* are useful *per se* in helping to understand the behaviour of HCP. We will discuss concrete strength in detail in Chapter 21, and in Chapter 24 we will see that porosity is also a significant factor influencing the durability of concrete.

13.6 Water in hardened cement paste and drying shrinkage

The large surface areas in the gel give the HCP a considerable affinity for water, and make its overall

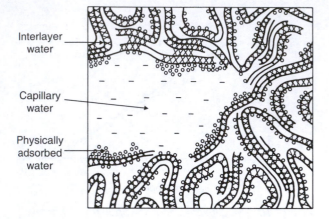

Interlayer water

Capillary water

Physically adsorbed water

Fig. 13.11 *Schematic of types of water within calcium silicate hydrate (after Feldman and Sereda, 1970).*

dimensions water-sensitive, i.e. loss of water results in shrinkage, which is largely recoverable on regain of water. We will discuss the magnitude of these effects and their consequences in Chapter 20, but for the moment we will consider the various ways in which the water is contained in the paste and how its loss can lead to shrinkage. The possible sites of the water are illustrated in the diagram of the gel structure shown in *Fig. 13.11*, and given in the following list:

1. *Water vapour*. The larger voids may be only partially filled with water, and the remaining space will contain water vapour at a pressure in equilibrium with the relative humidity and temperature of the surrounding environment.
2. *Capillary water*. This is located in the capillary and larger gel pores (wider than about 5 nm). Water in the voids larger than about 50 nm can be considered as free water, as it is beyond the reach of any surface forces (see Chapter 6), and its removal does not result in any overall shrinkage; however, the water in pores smaller than about 50 nm is subject to capillary tension forces, and its removal at normal temperatures and humidities may result in some shrinkage.
3. *Adsorbed water*. This is the water that is close to the solid surfaces, and under the influence of surface attractive forces. Up to five molecular layers of water can be held, giving a maximum total thickness of about 1.3 nm. A large proportion of this water can be lost on drying to 30% relative humidity, and this loss is the main contributing factor to drying shrinkage.
4. *Interlayer water*. This is the water in gel pores narrower than about 2.6 nm; it follows from (3)

that such water will be under the influence of attractive forces from two surfaces, and will therefore be more strongly held. It can be removed only by strong drying, for example, at elevated temperatures and/or relative humidities less than 10%, but its loss results in considerable shrinkage, the van der Waals forces being able to pull the solid surfaces closer together.
5. *Chemically combined water*. This is the water that has combined with the fresh cement in the hydration reactions discussed in section 13.4. This is not lost on drying, but is only evolved when the paste is decomposed by heating to high temperatures (in excess of 1000°C).

The above divisions should not be thought of as having distinct boundaries, but the removal of the water does become progressively more difficult as one proceeds down the list. An arbitrary but often useful division is sometimes made between *evaporable* and *non-evaporable* water. There are a number of way of defining this, the simplest being that evaporable water is that lost on drying at 105°C. This encompasses all the water in (1) to (3) above, and some of (4). The non-evaporable water includes the rest of (4) and all of (5); its amount expressed as a proportion of the total water content increases as hydration proceeds, and this can be used to assess the progress of the hydration reactions.

13.7 Modifications of Portland cement

When discussing the properties and compositions of cements in sections 13.2 and 13.3 we pointed out that these can be altered either by variations in the composition of the raw material or by changes in the manufacturing process. In this section we will discuss ways in which the cement can be altered from 'average' or 'normal' to obtain properties that are more useful for specific purposes.

13.7.1 SETTING, STRENGTH GAIN AND HEAT OUTPUT

The relative timescales of the dormant, setting and strength-gain periods govern some of the critical operations in concrete practice, for example the transport and placing of the concrete, and the time at which formwork can safely be removed. One way of modifying these properties is to alter the compound composition by varying the type and relative proportions of the raw materials used in the cement manufacture. For example, increased proportions of C_3S and C_3A can reduce the setting time, and if a cement with a higher C_3S and lower C_2S content is produced, as in cement B in *Table 13.1*, this will have a higher rate of strength gain than cement A (but it is important to understand the difference between rapid setting and rapid strength gain – the two do not necessarily go together). Rapid hardening properties can also be achieved by finer grinding of the cement, which gives an increased surface area exposed to the mix water, and therefore faster hydration reactions.

Since the hydration reactions are exothermic, a consequence of rapid hardening is a higher rate of heat output in the early stages of hydration, which will increase the risk of thermal cracking in large concrete pours from substantial temperature differentials at early ages, i.e. during the first few days after casting. To reduce the rate of heat of hydration output a 'low-heat' cement with a lower C_3S and higher C_2S content may be used, i.e. as in cement C in *Table 13.1*, or by coarser grinding. The disadvantage is a lower rate of gain of strength.

13.7.2 SULPHATE RESISTANCE

If sulphates from external sources, such as groundwater, come into contact with the HCP, reactions can take place with the hydration products of the calcium aluminate phases, forming calcium sulphoaluminate – *etttringite* – or, strictly, reforming it, since it was also formed very early in the hydration process (as described in section 13.4). Crucially the reaction is expansive and can therefore lead to disruption, cracking and loss of strength in the relatively brittle, low-tensile-strength HCP. (Its earlier formation would not have had this effect, as the paste would have still been fluid, or at least plastic.) The solution is a low-C_3A-content cement such as cement D in *Table 13.1*, which is therefore an example of a sulphate-resisting cement. We will return to this when discussing sulphate attack in more detail in Chapter 24.

13.7.3 WHITE CEMENT

The grey colour of most Portland cements is largely due to ferrite in the C_4AF phase, which derives from the ferrite compounds in the clay or shale used in the cement manufacture. The use of non-ferrite-containing material, such as china clay, results in a near-zero C_4AF-content cement, which is almost pure white, and therefore attractive to architects for exposed finishes. White cement is significantly more expensive than normal Portland cements owing to the increased cost of the raw materials, and the greater care needed during manufacture to avoid discoloration. As we shall see in the next two chapters, it is also possible to modify the properties of concrete by other means, involving the use of admixtures and/or cement replacement materials.

13.8 Cement standards and nomenclature

The first edition of the UK standard for Ordinary Portland Cement was issued in 1904, since when there have been a further 14 editions with increasingly complex and rigorous requirements. The last of these was in 1996, and a unified European standard, BS-EN 197-1:2000, has now replaced this. This covers five types of cement – CEM I, CEM II, CEM III, CEM IV and CEM V. The last four these are mixtures or blends of Portland cement with other materials of similar or smaller particle size, and we will leave discussion of these until Chapter 15. The cement and variations described in this chapter are type CEM I. The standard states that at least 95% of this should be ground clinker and gypsum – the remaining maximum 5% can be a 'minor additional constituent' (such as limestone powder).

There are sub-divisions within the main type that reflect the performance of the cement as altered by composition and/or fineness. The strength characteristics are determined by measuring the compressive strength of prisms made of a standard mortar with a sand:cement:water ratio of 3:1:0.5 by weight, which has been mixed, cast and stored under defined and carefully controlled conditions. The cement is then given a number – 32.5, 42.5 or 52.5 – depending on the strength in MPa achieved at 28 days, and a letter, either N or R (N for normal and R for rapid), depending on the strength at either 2 or 7 days. The requirements of the strength classes are set out in *Table 13.2*. Limits to initial setting time are also included in the standard. The previous Ordinary Portland Cement from BS 12 roughly corresponds to a CEM I 42.5N, and although it is strictly not now correct to use the term 'OPC', it will take a long time before it dies out.

Sulphate-resisting Portland Cement has a separate standard (BS 4027), for which there is no European equivalent; the most significant difference to other Portland cements is the requirement for a C_3A content of less than 3.5%. There is no separate standard for white cement.

Many other countries have their own standards. For example, in the USA the American Society for Testing and Materials (ASTM) classifies Portland cement in their specification C-150-94 by type number:

Table 13.2 BS EN-197 strength classes for Portland cement

Class	Compressive strength from mortar prisms (MPa)			
	2 days	7 days	28 days	
32.5 N		≥16		
32.5 R	≥10		≥32.5	<52.5
42.5 N	≥10			
42.5 R	≥20		≥42.5	<62.5
52.5 N	≥20			
52.5 R	≥30		≥52.5	<72.5

- type I is ordinary Portland cement
- type II is moderate sulphate-resistant or moderate heat cement
- type III is a rapid-hardening cement
- type IV is low heat cement
- type V is sulphate-resistant cement.

It is beyond the scope of this book, and potentially very boring for most readers, to go into further details about these standards. They can be found in most libraries and on-line when necessary.

References

Birchall JD, Howard AJ and Kendall K (1981). Flexural strength and porosity of cements. *Nature*, **289** (No. 5796), 388–390.

Bogue RH (1955). *Chemistry of Portland Cement*, Van Nostrand Reinhold, New York.

Domone PL and Thurairatnam H (1986). Development of mechanical properties of ordinary Portland and Oilwell B cement grouts. *Magazine of Concrete Research*, 38 (No. 136), 129–138.

Feldman RF and Sereda PJ (1970). A new model for hydrated Portland cement paste and its practical implications. *Eng J (Canada)*, 53 (No. 8/9), 53–59.

Forester J (1970). A conduction calorimeter for the study of cement hydration, *Cement Technology*, 1 (No. 3), 95–99.

Hansen TC (1970). Physical composition of hardened Portland cement paste. *Proceedings of the American Concrete Institute*, 67 (No. 5), 404–407.

Mehta PK (1986). *Concrete: Structure, Properties and Materials*, Prentice-Hall, New Jersey, p. 450.

Neville AM (1995). *Properties of concrete*, 4th edition, Pearson Education, Harlow, UK, p. 844.

Powers TC (1958). Structure and physical properties of hardened cement paste. *Journal of the American Ceramic Society*, 41 (No. 1), 1–6.

Roy DM and Gouda GR (1975). Optimization of strength in cement pastes. *Cement and Concrete Research*, 5 (No. 2), 153–162.

Soroka I (1979). *Portland Cement Paste and Concrete*, Macmillan, London.

Chapter 14

Admixtures

Admixtures are chemicals that are added to concrete during mixing and significantly change its fresh, early age or hardened state to economic or physical advantage. They are usually defined as being added at rates of less than 5% by weight of the cement, but the typical range for most types is only 0.3–1.5%. They are normally supplied as aqueous solutions of the chemical for convenience of dispensing and dispersion through the concrete during mixing. Their popularity and use have increased considerably in recent years; estimates for the UK are that about 12% of all concrete produced in 1975 contained an admixture, and that this increased to 50% by 1991 and is now well over 75%. In some places, notably parts of Europe, North America, Australia and Japan, the proportion is even higher.

14.1 Action and classification of admixtures

An extremely large number of commercial products are available, which work by one or more of the following mechanisms:

- interference with the hydration reactions to accelerate or retard the rate of hydration of one or more of the cement phases
- physical absorption onto the surface of cement particles causing increased particle dispersion
- altering the surface tension of the mix water causing air entrainment
- increasing the viscosity of the mix water resulting in an increased plastic viscosity or cohesion of the fresh concrete
- incorporating chemicals into the hardened cement paste to enhance particular properties such as increased protection to embedded steel or water repellence.

These result in admixtures usually being classified or grouped according to their mode of action rather than by their chemical constituents. For example the European standard (BS EN 934) includes requirements for:

- water-reducing/plasticising admixtures
- high-range water-reducing/superplasticising admixtures
- set and hardening accelerating admixtures
- set retarding admixtures
- air-entraining admixtures
- water-resisting admixtures
- water-retaining admixtures
- set-retarding/water-reducing/plasticising admixtures
- set-retarding/high-range water-reducing/ superplasticising admixtures
- set-accelerating/water-reducing/plasticising admixtures.

Clearly the last three are admixtures with a combination of actions.

We shall consider the five distinct types which together make up more than 80% of the total quantities used in concrete – plasticisers, superplasticisers, accelerators, retarders and air-entraining agents – and briefly mention others.

14.2 Plasticisers

Plasticisers, also called *workability aids*, increase the fluidity or workability of a cement paste or concrete. They are long-chain polymers, the main types being based on either lignosulphonates, which are derived in the processing of wood for paper pulp, or polyycarboxylate ether. They are relatively inexpensive but lignosulphonates in particular can contain significant levels of impurities depending on the amount of processing.

Their plasticising action is due to the surface-active nature of the component polymer molecules, which are adsorbed on to the surface of the cement

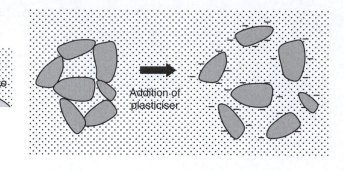

(a) Adsorption on to cement
particle surface

(b) Dispersion of particle flocs and realease of
entrapped water to give greater fluidity

Fig. 14.1 Mode of action of plasticisers.

grains. In their normal state the surfaces of cement particles carry a mixture of positive and negative residual charges (a property of all surfaces), which means that when mixed with water the particles coalesce into flocs, thus trapping a considerable amount of the mix water and leaving less available to provide fluidity. In solution the plasticiser molecules have negative ionic groups that form an overall negative charge of the order of a few millivolts on the cement particles after they are absorbed onto the cement particle surface. The particles therefore now repel each other and become more dispersed, thus releasing the trapped water and increasing the fluidity, as illustrated in *Fig. 14.1*. The particles also become surrounded by a sheath of oriented water molecules, which prevent close approach of the cement grains, a phenomenon known as *steric hindrance* or *steric repulsion*. The overall effect is one of greater lubrication and hence increased fluidity of the paste or concrete.

If a constant consistence or fluidity is required then the water content can now be reduced, thus leading to a lower water:cement ratio and increased strength; this is why plasticisers are often known as *water-reducers*. BS EN 934 requires that the water reduction for constant consistence should be greater than 5%. Values are normally between 5 and 12%. The use of plasticisers has been increasingly widespread since their first appearance in the 1930s; the quantitative benefits that can be obtained will be discussed when considering mix design in Chapter 22.

Significant, and sometimes undesirable, secondary effects with some plasticisers are that they act as retarders, delaying the set and decreasing the early strength gain, and/or that they entrain air in the form of small bubbles. Depending on the amount of processing in manufacture they may also contain impurities that have other undesirable side-effects at increasing doses, and therefore the magnitude of the primary effects that can be satisfactorily achieved with plasticisers is relatively modest, though nevertheless useful and cost effective.

14.3 Superplasticisers

As the name implies superplasticisers are more powerful than plasticisers and they are used to achieve increases in fluidity and workability of a much greater magnitude than those obtainable with plasticisers. They are also known as *high-range water-reducers*. They were first marketed in the 1960s, since when they have been continually developed and increasingly widely used. They have higher molecular weights and are manufactured to higher standards of purity than plasticisers, and can therefore be used to achieve substantially greater primary effects without significant undesirable side-effects. They are a crucial ingredient of many of the special or so-called 'high-performance' concretes, which we will discuss in Chapter 25.

BS EN 934 requires that the water reduction for constant consistence should be greater than 12%. Values vary between 12 and about 30%, depending on the types and efficiency of the constituent chemicals. Currently three main chemical types are used (Dransfield 2003):

1. Sulphonated melamine formaldehyde condensates (SMFs), normally the sodium salt.
2. Sulphonated naphthalene formaldehyde condensates (SNFs), again normally the sodium salt.

3. Polycarboxylate ethers (PCLs). These have been the most recently developed, and are sometimes referred to as 'new generation' superplasticisers.

These basic chemicals can be used alone or blended with each other or lignosulphonates to give products with a wide range of properties and effects. A particular feature is that polycarboxylates in particular can be chemically modified or tailored to meet specific requirements, and much development work has been carried out to this end by admixture suppliers in recent years. This has undoubtedly led to improvements in construction practice, but a consequence is that the websites of the major suppliers contain a confusing plethora of available products, often with semi-scientific sounding names.

The mode of action of superplasticisers is similar to that of plasticizers, i.e. they cause a combination of mutual repulsion and steric hindrance between the cement particles. Opinions differ about the relative magnitude and importance of these two effects with different superplasticisers, but a consensus (Collepardi, 1998; Edmeades and Hewlett, 1998) is that:

- with SMFs and SNFs, electrostatic repulsion is the dominant mechanism
- with PCLs, steric hindrance is equally if not more important. This is due to a high density of polymer side-chains on the polymer backbone, which protrude from the cement particle surface (*Fig. 14.2*). This leads to greater efficiency, i.e. similar increases in fluidity require lower admixture

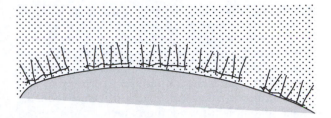

Fig. 14.2 *'Comb-type' molecules of polycarboxylic superplasticisers on the surface of a cement grain, leading to steric hindrance between grains.*

dosages. The term 'comb polymer' has been used to describe this molecular structure.

Some typical fluidity effects of admixtures of different types, measured by spread tests on a mortar, are shown in *Fig. 14.3*. The limited range and effectiveness of a lignosulphonate-based plasticiser and the greater efficiency of a PCL superplasticiser (in this case a polyacrylate) compared to an SNF-based material are apparent.

Some of the more important features of the behaviour of superplasticisers, which directly effect their use in concrete, can be summarised as follows.

- The behaviour of any particular combination of superplasticiser and binder will depend on several factors other than the admixture type, including the binder constituents, the cement composition, the cement fineness and the water:binder ratio (Aitcin *et al.*, 1994).

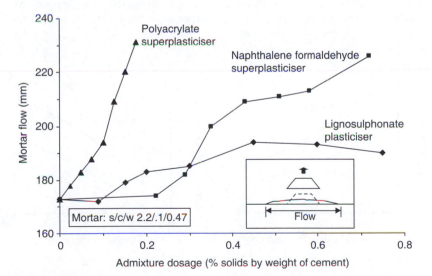

Fig. 14.3 *Typical effects of plasticising and superplasticising admixtures on flow of mortars (after Jeknavorian et al., 1997).*

- Substantially increased performance can be obtained if the superplasticiser is added a short time (1–2 minutes) after the first contact of the mix water with the cement. It appears that if the superplasticiser is added at the same time as the mix water, a significant amount is incorporated into the rapid C_3A/gypsum reaction, hence reducing that available for workability increase. This effect has been clearly demonstrated for lignosulphonate, SMF and SNF based admixtures, but has been reported as being less significant for at least some PCLs, which are therefore more tolerant of mixing procedures.

- The superplasticising action occurs for only a limited time, which may be less than that required if, for example, the concrete has to be transported by road from mixing plant to site. Methods of overcoming this include:

  - blending a retarder with the superplasticiser
  - addition of the superplasticiser on site just before discharge from the mixer truck
  - repeated additions of small extra doses of the admixture.

The losses with some PCLs have been shown to be lower than with other types, at least over the critical first hour after mixing.

- For any particular binder/superplasticiser combination there is a 'saturation point' or optimum dosage beyond which no further increases in fluidity occur (*Fig. 14.4*). At dosages higher than this, not only is there no increase in fluidity, but detrimental effects such as segregation, excessive retardation or entrapment of air during mixing – which is suddenly released – can occur.

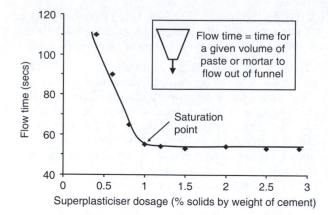

Fig. 14.4 *The saturation point for a cement/ superplasticiser combination (after Aitcin et al., 1994).*

We will discuss the quantitative benefits that can be obtained when describing concrete mix design in Chapter 22.

14.4 Accelerators

An accelerator is used to increase the rate of hardening of the cement paste, thus enhancing the early strength, particularly in the period of 24–48 hours after placing, perhaps thereby allowing early removal of formwork, or reducing the curing time for concrete placed in cold weather. They may also reduce the setting time. Calcium chloride ($CaCl_2$) was historically very popular as it is readily available and very effective. *Figure 14.5a* shows that

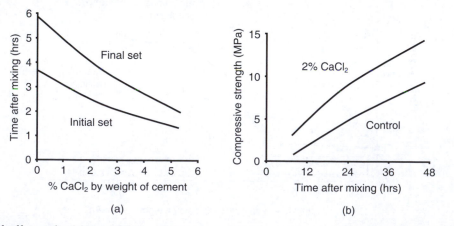

(a) (b)

Fig. 14.5 *Typical effects of calcium chloride admixture on (a) setting times and (b) early strength of concrete (after Dransfield and Egan, 1988).*

it accelerates both the initial and final set, and *Fig. 14.5b* shows that a 2% addition by weight of cement can result in very significant early strength increases. This effect diminishes with time, and the long-term strength is similar to that of non-accelerated concrete.

The calcium chloride becomes involved in the hydration reactions involving C_3A, gypsum and C_4AF, but the acceleration is caused by its acting as a catalyst in the C_3S and C_2S reactions (Edmeades and Hewlett, 1998). There is also some modification to the structure of the C-S-H produced.

Of great significance is the increased vulnerability of embedded steel to corrosion owing to the presence of the chloride ions. This has led to the use of calcium chloride being prohibited in reinforced and pre-stressed concrete, and to the development of a number of alternative chloride–free accelerators, most commonly based on either calcium formate, sodium aluminate or triethanolamine. However, as with plasticisers and superplasticisers the magnitude of the effects of these depends on the binder constituents and composition and cannot be predicted with certainty, and so should be established by testing. We shall discuss the corrosion of steel in concrete in some detail when considering durability in Chapter 24.

14.5 Retarders

Retarders delay the setting time of a mix, and examples of their use include:

- counteracting the accelerating effect of hot weather, particularly if the concrete has to be transported over a long distance
- controlling the set in large pours, where concreting may take several hours, to achieve concurrent setting of all the concrete, hence avoiding cold joints and discontinuities, and achieving uniform strength development.

The retardations resulting from varying doses of three different retarding chemicals are shown in *Fig. 14.6*. Sucrose and citric acid are very effective retarders, but it is difficult to control their effects, and lignosulphonates, often with a significant sugar content, are preferred. The retarding action of normal plasticisers such as some lignosulphonates and carboxylic acids has already been mentioned; most commercial retarders are based on these compounds, and therefore have some plasticising action as well.

The mode of action of retarders involves modification of the formation of the early hydration products,

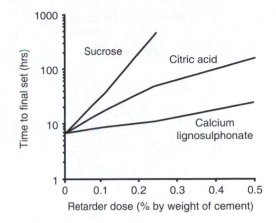

Fig. 14.6 *Influence of retarders on the setting time of cement paste (after Ramachandran et al., 1981).*

including the portlandite crystals. As with other admixtures, temperature, mix proportions, fineness and composition of the cement and time of addition of the admixture all affect the degree of retardation, and it is therefore difficult to generalise.

14.6 Air-entraining agents

Air-entraining agents (AEAs) are organic materials which, when added to the mix water, entrain a controlled quantity of air in the form of microscopic bubbles in the cement paste component of the concrete. The bubble diameters are generally in the range 0.02–1 mm, with an average distance between them of about 0.2 mm. They are sufficiently stable to be unchanged during the placing, compaction, setting and hardening of the concrete. Entrained air should not be confused with entrapped air, which is normally present as the result of incomplete compaction of the concrete, and usually occurs in the form of larger irregular cavities.

AEAs are powerful surfactants, which change the surface tension of the mix water and act at the air–water interface within the cement paste. Their molecules have a hydrocarbon chain or backbone terminated by a hydrophilic polar group, typically of a carboxylic or sulphonic acid. This becomes orientated into the aqueous phase, with the hydrocarbon backbone pointing inwards towards the air, thus forming stable, negatively charged bubbles that become uniformly dispersed (*Fig. 14.7*). Only a limited number of materials are suitable, including vinsol resins extracted from pinewood and synthetic alkylsulphonates and alkylsulphates.

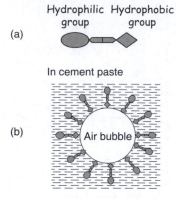

(a) Hydrophilic group   Hydrophobic group

In cement paste

(b) Air bubble

Stabilised air bubble

Fig. 14.7 Schematic of air entrainment by surface-active molecules. (a) Molecular structure; (b) stable air bubble (adapted from Mindess et al., 2003).

The major reason for entraining air is to provide freeze–thaw resistance to the concrete. Moist concrete contains free water in entrapped and capillary voids, which expands on freezing, setting up disruptive internal bursting stresses. Successive freeze–thaw cycles, say, over a winter, may lead to progressive deterioration. Entrained air voids, uniformly dispersed throughout the HCP, provide a reservoir for the water to expand into when it freezes, thus reducing the disruptive stresses. Entrained-air volumes of only about 4–7% by volume of the concrete are required to provide effective protection, but the bubble diameter and spacing are important factors. We will consider freeze–thaw damage in more detail when discussing the durability of concrete in Chapter 24.

Air entrainment has two important secondary effects:

1. There is a general increase in the consistence of the mix, with the bubbles seeming to act like small ball-bearings. The bubbles' size means that they can compensate for the lack of fine material in a coarse sand, which would otherwise produce a concrete with poor cohesion. (Aggregate grading will be discussed in Chapter 17.)
2. The increase in porosity results in a drop in strength, by a factor of about 6% for each 1% of air. This must therefore be taken into account in mix design, but the improvement in workability means that the loss can at least be partly offset by reducing the water content and hence the water:cement ratio.

AEAs have little influence on the hydration reactions, at least at normal dosages, and therefore have no effect on the resulting concrete properties other than those resulting from the physical presence of the voids, as described above.

14.7 Other types of admixture

Other admixtures include pumping aids, water-resisting of waterproofing admixtures, anti-bacterial agents, bonding agents, viscosity agents or thickeners, anti-washout admixtures for underwater concrete, shrinkage-reducing admixtures, foaming agents, corrosion inhibitors, wash-water systems and pigments for producing coloured concrete. Some selected texts that contain information on these, and give a more detailed treatment of the admixtures we have described, are included in 'Further reading' at the end of this part of the book. Admixtures collectively contribute to the great versatility of concrete and its suitability for an ever-increasing range of applications, some of which we shall discuss when considering 'Special Concretes' in Chapter 25.

References

Aitcin P-C, Jolicoeur C and MacGregor JG (1994). Superplasticizers: how they work and why they occasionally don't. *Concrete International*, **16** (No. 5), 45–52.

Collepardi M (1998). Admixtures used to enhance placing characteristics of concrete. *Cement and Concrete Composites*, **20**, 103–112.

Dransfield JM and Egan P (1988). *Accelerators* in Cement Admixtures: Use and Applications, 2nd edition (ed. Hewlett PC), Longman, Essex, pp. 102–129.

Dransfield JM (2003). Admixtures for concrete, mortar and grout. Chapter 4 of *Advanced Concrete Technology, Vol I: Constituent Materials* (eds Newman JB and Choo BS), Butterworth Heinemann, Oxford, pp. 4/3–4/36.

Edmeades RM and Hewlett PC (1998). Cement admixtures. In *Lea's Chemistry of Cement and Concrete* (ed. Hewlett PC), Arnold, London, pp. 837–901.

Jeknavorian AA, Roberts LR, Jardine L, Koyata H and Darwin DC (1997). *Condensed polyacrylic acid–aminated polyether polymers as superplasticizers for concrete*. ACI SP-173. Proceedings of Fifth CANMET/ACI International Conference on Superplasticizers and Other Chemical Admixtures in Concrete, Rome, Italy (ed. Malhotra VM). American Concrete Institute, Detroit, USA, pp. 55–81.

Mindess S, Young JF and Darwin D (2003). *Concrete*, 2nd edition, Prentice Hall, New Jersey.

Ramachandran VS, Feldman RF and Beaudoin JJ (1981). *Concrete Science*, Heyden and Sons, London.

Chapter 15

Additions

Additions are defined as 'finely divided materials used in concrete in order to improve certain properties or to achieve special properties' (BS EN 206, 2000). Somewhat confusingly, there are a number of alternative names favoured in different countries and at different times: cement replacement materials, fillers, mineral additives, mineral admixtures, supplementary cementing materials, cement substitutes, cement extenders, latent hydraulic materials or, simply, cementitious materials. They are nearly always inorganic materials with a particle size similar to or smaller than that of the Portland cement, and they are normally used to replace some of the cement in the concrete mix (or sometimes supplement it) for property and/or cost and/or environmental benefits. Several types of materials are in common use, some of which are by-products from other industrial processes, hence their potential for economic advantages and environmental and sustainability benefits (we will discuss the latter in more detail in Chapter 62). However the principal reason for their use is that they can give a variety of useful enhancements of or modifications to the properties of concrete.

They can be supplied either as separate materials that are added to the concrete at mixing, or as pre-blended mixtures with the Portland cement. The former case allows choice of the rate of addition, but means that an extra material must be handled at the batching plant; a pre-blended mixture overcomes the handling problem but means that the rate of addition is fixed. Pre-blended mixtures have the alternative names of extended cements, Portland composite cements or blended Portland cements. Generally, only one material is used in conjunction with the Portland cement, but there are an increasing number of examples of the combined use of two or even three materials for particular applications.

The incorporation of additions leads to a rethink about the definition of cement content and water:cement ratio. Logically these should still mean what they say, with cement being the Portland cement component. Since, as we will see, additions contribute to the hydration reactions, the Portland cement and additions together are generally known as the *binder*, and hence we can refer to the *binder content* and *water:binder ratio* when discussing mix proportions. (To add to the confusion, the alternative terms *powder*, *powder content* and *water:powder ratio* are sometimes used when additions that make little contribution to the hydration reactions are incorporated.)

BS-EN 206 recognises two broad divisions of additions:

- Type 1: nearly inert additions
- Type 2: pozzolanic or latent hydraulic additions.

This reflects the extent to which the additions are chemically active during the hydration process and therefore the extent to which they contribute to or modify the structure and properties of the hardened paste. It will be useful at this stage to explain what is meant by *pozzolanic behaviour* before going on to consider some of the most commonly used additions.

15.1 Pozzolanic behaviour

Type 2 additions exhibit pozzolanic behaviour to a greater or lesser extent. A pozzolanic material is one that contains active silica (SiO_2, or S in shorthand form) and is not cementitious in itself but will, in a finely divided form and in the presence of moisture, chemically react with calcium hydroxide at ordinary temperatures to form cementitious compounds. The key to the pozzolanic behaviour is the structure of the silica; this must be in a glassy or amorphous form with a disordered structure, which is formed by rapid cooling from a molten state. Many of the inter-molecular bonds in the structure are then not at their preferred low-energy orientation and so can readily be broken and link with the oxygen component of the calcium hydroxide. A uniform crystalline structure that is formed in slower cooling, such as is found in silica sand, is not chemically active.

Naturally-occurring pozzolanic materials were used in early concretes, as mentioned in the Introduction to this part of the book, but when a pozzolanic material is used in conjunction with a Portland cement, the calcium hydroxide (CH in shorthand) that takes part in the pozzolanic reaction is the portlandite produced from hydration of the cement (see equations 13.8 and 13.9). Further quantities of calcium silicate hydrate are produced:

$$S + CH + H \rightarrow C\text{-}S\text{-}H \qquad (15.1)$$

The reaction is clearly secondary to the hydration of the Portland cement, which has led to the name 'latent hydraulic material' in the list of alternatives above. The C-S-H produced is very similar to that from the primary cement hydration (the molar ratios of C/S and H/S may differ slightly with different pozzolanic materials) and therefore make their own contribution to the strength and other properties of the hardened cement paste and concrete.

15.2 Common additions

The most commonly used type 1 addition is ground limestone, normally known as limestone powder. As we mentioned in Chapter 13, addition of up to 5% of this to Portland cement is permitted in many countries without declaration. For example, BS EN 197-1:2000 allows this amount to be included in a CEM I cement as a 'minor additional constituent'. Higher additions are also used in some types of concrete, most notably self-compacting concrete in which high powder contents are required for stability and fluidity of the fresh concrete (see Chapter 25). The main enhancement of properties is physical – the fine powder particles can improve the consistence and cohesiveness of the fresh paste or concrete. However, although there is no pozzolanic reaction, there is some enhancement to the rate of strength gain due to the 'filler effect' of improved particle packing and the powder particles acting as nucleation sites for the cement hydration products, and there is some reaction between the calcium carbonate in the limestone with the aluminate phases in the cement.

The main Type 2 additions in use worldwide are:

- *fly ash*, also known as pulverised fuel ash (pfa) – the ash from pulverised coal used to fire power

stations, collected from the exhaust gases before discharge to the atmosphere; not all ashes have a suitable composition and particle size range for use in concrete
- *ground granulated blast furnace slag* (ggbs) – slag from the 'scum' formed in iron smelting in a blast furnace, which is rapidly cooled in water and ground to a similar fineness to Portland cement
- *condensed silica fume* (csf), often called microsilica – extremely fine particles of silica condensed from the waste gases given off in the production of silicon metal
- *calcined clay or shale* – a clay or shale heated, rapidly cooled and ground
- *rice husk ash* – ash from the controlled burning of rice husks after the rice grains have been separated
- *natural pozzolans* – some volcanic ashes and diatomaceous earth.

We will now discuss the first four of the materials in the above list in more detail, using metakaolin (also known as HRM – high reactivity metakaolin) as an example of a calcined clay. All these four are somewhat different in their composition and mode of action, and therefore in their uses in concrete. Rice husk ash has similarities with microsilica, and natural pozzolans are not extensively used.

15.3 Chemical composition and physical properties

Typical chemical compositions and physical properties of these four materials are given in *Table 15.1*, together with typical equivalent properties of Portland cement for comparison. Two types of fly ash are included, high- and low-lime, which result from burning different types of coal. High-lime fly ash is not available in many countries, and the low-lime form is most commonly available. It is normally safe to assume that when fly ash is referred to in textbooks, papers etc. it is the low-lime version unless specifically stated otherwise.

The following features can be deduced from the table:

- All the materials contain substantially greater quantities of silica than does Portland cement, but crucially, most of this is in the active amorphous or glassy form required for the pozzolanic action.
- Microsilica is almost entirely active silica.
- The alumina in the fly ash, ggbs and metakaolin are also in an active form, and becomes involved

Table 15.1 Typical composition ranges and properties of additions

Addition	Fly ash		ggbs	Microsilica	Metakaolin	Portland cement
	Low lime (class F)	High lime (class C)				
Oxides						
SiO$_2$	44–58	27–52	30–37	94–98	50–55	17–25
CaO	1.5–6	8–40	34–45	<1	<1	60–67
Al$_2$O$_3$	20–38	9–25	9–17	<1	40–45	3–8
Fe$_2$O$_3$	4–18	4–9	0.2–2	<1	5	0.5–6
MgO	0.5–2	2–8	4–13	<1	<1	0.1–4
Particle size range (microns)	1–80		3–100	0.03–0.3	0.2–15	0.5–100
Specific surface area (m^2/kg)	350		400	20000	12000	350
Relative particle density	2.3		2.9	2.2	2.5	3.15
Particle shape	Spherical		Irregular	Spherical	Irregular	Angular

ggbs, ground granulated blast furnace slag.

Notes:
- Most but not all specific materials will have a composition within the above ranges. However, there are exceptions.
- All materials for use in concrete have to comply with the relevant standards. A list of these is included in 'Further reading' at the end of this part of the book.

in the pozzolanic reactions, forming complex products. The metakaolin comprises nearly all active silica and alumina.

- Two of the materials, high-lime fly ash and ggbs, also contain significant quantities of CaO. This also takes part in the hydration reactions, and therefore neither material is a true pozzolan, and both are to a certain extent self-cementing. The reactions are very slow in the neat material, but they are much quicker in the presence of the cement hydration, which seems to act as a form of catalyst for the production of C-S-H.
- The above considerations lead to maximum effective Portland cement replacement levels of about 90% for high-lime fly ash and ggbs, 40% for low-lime fly ash and metakaolin and 25% for csf. At higher levels than these, there is insufficient Portland cement to produce the required quantities of calcium hydroxide for the secondary reactions to be completed. However, high-volume fly ash (HVFA) concrete, with up to 70% fly ash and low water:binder ratios, has been of increasing interest in recent years.
- Fly ash and ggbs have particle sizes similar to those of Portland cement, whereas the metakaolin particles are on average nearly ten times smaller and the microsilica particles 100 times smaller (although the ggbs and metakaolin are both ground specifically for use in concrete, and so their

fineness can be varied). The consequences of the associated differences in surface area are:

- the rate of reaction of the metakaolin is higher than that of fly ash and ggbs, and that of microsilica highest of all (but remember that all are still secondary to that of the Portland cement)
- both metakaolin and microsilica result in a loss of fluidity of the cement paste and concrete if no other changes are made to the mix, with again the effect of csf being greater that that of metakaolin. To maintain fluidity, either the water content must be increased, or a plasticiser or superplasticiser added (see Chapter 14). The latter is the preferred option, since other properties such as strength are not compromised. With a sufficient dosage of superplasticiser to disperse the fine particles, a combination of excellent consistence with good cohesion and low bleed can be obtained.
- The spherical shape of the fly-ash particles leads to an increase in fluidity if no other changes are made to the mix. Some increase is also obtained with ggbs.
- All the materials have lower relative densities than Portland cement, and therefore substitution of the cement on a weight-for-weight basis will result in a greater volume of paste.

We should also note that variability of fly ash due to changes in the coal supply and power station demands can be a significant problem. Some processing of the ash is therefore often carried out to ensure a more uniform, high-quality material for use in concrete. This includes screening to remove large particles, and the removal of particles of unburnt carbon, which are very porous and can reduce the consistence of the fresh concrete.

All the above considerations have led to an ever-increasing use of the various additions in all types of concrete in the last few decades. We will discuss their use and their effect on the properties of fresh and hardened concrete at appropriate places in subsequent chapters, from which their advantages and disadvantages will become even more apparent.

15.4 Supply and specification

As we said earlier, additions can be supplied as separate materials or pre-blended with Portland cement. For many years blends with ggbs have been known as Portland Blast Furnace cements and blends with fly ash Portland Pozzolanic cements. There is an array of relevant standards throughout the world, covering the materials individually but also as blends. It is worthwhile briefly considering the designations for the latter in the current European standard BS EN 197-1:2000. As we discussed at the end of Chapter 13, this includes five main types of cement, with CEM I being Portland cement containing at least 95% ground clinker and gypsum with up to 5% minor additional constituents. The other four types are:

- *CEM II Portland composite cement*: Portland cement with up to 35% of another single constituent, which can be ggbs, microsilica, a natural or calcined pozzolan, fly ash, burnt shale or limestone powder, or with up 35% of a mixture of these additions
- *CEM III Blast furnace cement*: Portland cement with 35–95% ggbs
- *CEM IV Pozzolanic cement*: Portland cement with 11–35% of any combination of microsilica, natural or calcined pozzolan or fly ash
- *CEM V Composite cement*: Portland cement with 35–80% of a mixture of blast furnace slag with natural or calcined pozzolan or fly ash.

Within each main type there are a number of sub-types for the different types and quantities of additions, which results in 27 products within the whole family of cements. This may seem unnecessarily complex, but the standard covers all of the cements produced throughout Europe, and in any country or region only a few of the 27 will be available. Each of the products has its own unique letter-code designation which, together with the strength-class designations described at the end of Chapter 13, leads to a lengthy overall designation for any one cement. You should consult the standard itself for a complete list and full details if and when you need these.

Chapter 16

Other types of cement

In the last three chapters we have discussed Portland cement, including variations in its physical and chemical properties, and how additions and admixtures can be used to modify and improve the properties of concrete containing Portland cement. In this chapter we will relatively briefly discuss some alternatives to Portland cement. These include calcium aluminate cement, which has been in use for a hundred years or so, and some cements that have had more limited use or are currently being developed and are not yet in widespread use. This latter group is of particular interest because of their potential of being produced at lower temperatures than Portland cement, thereby requiring less energy for production; as we shall see in Chapter 62 this is a major concern for Portland cement in relation to sustainability issues, particularly carbon emissions.

16.1 Calcium aluminate cement

Calcium aluminate cement (CAC) is, as can be deduced from the name, based on calcium aluminate rather than the calcium silicate of Portland cement. It is also known as high-alumina cement (HAC); the two names are synonymous. It was first developed in France in the early years of the 20th century to overcome the problem of sulphate attack that was being experienced by Portland cement concrete. Its French name is Ciment Fondu, and supplies in Europe are covered by BS EN 14647.

16.1.1 MANUFACTURE AND COMPOSITION
The manufacture of CAC has some parallels with that of Portland cement. The raw materials are usually limestone and bauxite, which contains alumina (aluminium oxide), iron and titanium oxides and some silica. After crushing and blending these are fed into a furnace and heated to about 1600°C, where they fuse into a molten material. This is drawn off, cooled and ground to the required particle size, which is

normally of the same order as that of Portland cement, i.e. a specific service area of 290 to 350 m^2/kg. The relative particle density is 3.20, marginally higher than that of Portland cement (3.15), and the powder is very dark grey owing mainly to the significant amounts of iron oxide usually present in the bauxite, but if a white bauxite with little or no iron oxide is used then the cement is light grey to white.

The oxide proportions are typically 35–40% each of alumina (Al_2O_3 or A in shorthand form) and lime (CaO or C), about 15% iron oxides (Fe_2O_3 or F), about 5% of silica (SiO_2 or S) and some other minor compound present in the raw materials. The principal cementitious compound is CA, with some $C_{12}A_7$, C_2S and C_2AS and C_6A_4FS.

16.1.2 HYDRATION AND CONVERSION
The setting time of CAC is about 30 minutes longer than that of a typical Portland cement. At temperatures up to about 35°C the first hydration reaction is simply:

$$CA + 10H \rightarrow CA.10H \qquad (16.1)$$

This reaction is relatively rapid and gives rise to an initial rate of strength gain much greater than that of Portland cements (*Fig. 16.1*), but as with Portland cement the hydration reaction is exothermic and consequently there is a more rapid rate of heat evolution.

Between 35 and 65°C the dominant hydration reaction is:

$$2CA + 11H \rightarrow C_2A.8H + A.3H \qquad (16.2)$$

and above 65°C it is:

$$3CA + 12H \rightarrow C_3A.6H + 2A.3H \qquad (16.3)$$

The products of this last reaction are in fact stable at all temperatures, whereas those of the first two reactions are metastable, and with time will transform or convert to the stable phases:

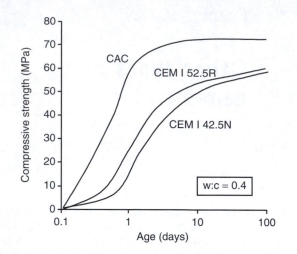

Fig. 16.1 *Typical strength gains of concrete containing Portland and calcium aluminate cements (adapted from Neville and Wainwright, 1975).*

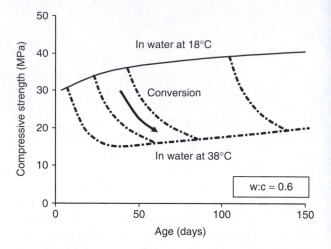

Fig. 16.3 *Strength changes in CAC concrete due to conversion (adapted from Neville and Wainwright, 1975).*

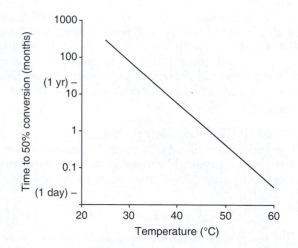

Fig. 16.2 *Effect of storage temperature on rate of conversion of CAC concrete (after Neville, 1995).*

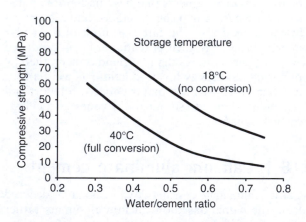

Fig. 16.4 *The effect of water:cement ratio on the strength of CAC concrete stored for 100 days at two temperatures (after Neville, 1995).*

$$2CA.10H \rightarrow C_2A.8H + A.3H + 9H \quad (16.4)$$

$$3C_2A.8H \rightarrow 2C_3A.6H + A.3H + 9H \quad (16.5)$$

The rate of conversion is temperature dependent, and takes years to complete at 20°C but only days at 60°C (*Fig. 16.2*). A major consequence of conversion arises from the solid density of the final stable product ($C_3A.6H$) being greater than that of the metastable products of reactions (16.1) and (16.2). Hardened cement that has been formed at low temperatures will therefore become more porous

as its hydrates convert, with a consequent loss of strength. This loss can occur at any time in the life of the concrete in the event of an increase in temperature, as shown in *Fig. 16.3*.

An essential feature of the conversion is that it does not result in disintegration of the concrete and the products have a significant and stable strength, albeit lower than that of the initial hydrates. The strength – both before and after conversion – depends on the initial water:cement ratio, as shown in *Fig. 16.4*.

16.1.3 USES

As mentioned above the initial uses of CAC concrete were to provide sulphate resistance. However the rapid rate of strength gain made it particularly suitable for the production of pre-cast pre-stressed concrete beams that could be demoulded a few hours after casting and placed in service within a few days. These beams were a feature of the new and high-rise building construction in the period following the Second World War. Conversion was known about at this time, but it was not thought to be significant if section sizes were limited to avoid high temperatures from the exothermic hydration reactions, the water:cement ratio was limited to a maximum of 0.5 and warm moist service environments were avoided.

However, a number of failures of ceilings and roofs constructed with CAC beams occurred in the UK in the 1970s. Fortunately none of these involved loss of life but subsequent investigations found significant conversion of the CAC. However the primary causes of failures were identified as design and tolerance problems, particularly relating to the bearing area and structural link provided between the beams and their support walls. Subsequent inspection of more than a thousand buildings containing CAC concrete found only one case of a problem attributed to strength loss during conversion. Despite this CAC was omitted from the list of cements permitted in design standards for reinforced and pre-stressed concrete, and so it was effectively banned from use for this purpose in the UK and in many other countries. Further problems involving similar CAC beams occurred in some Spanish apartment blocks in the 1990s, including one collapse; the concrete was found to have disintegrated owing to its high porosity and the use of poor quality aggregates, indicating the importance of understanding the complexity of the materials issues involved.

Two recent papers (Neville, 2009) have discussed the background to the original failures, the investigation that followed and its consequences; they make interesting reading. A Concrete Society report (Concrete Society, 1997) concluded that in some circumstances CAC concrete could be safely used if its long-term strength is taken into account for design purposes, but there has been no inclusion of this provision in UK or European codes of practice.

There are, however, several current important uses of CAC cement and concrete that take advantage of its superior properties compared to other cements:

- Its excellent resistance to acids, particularly those derived from bacteria, means that it is ideal for use for tunnel linings and pipes in sewage networks.
- Its rapid strength gain lends itself to uses in applications where rapid service use is required, such as temporary tunnel linings; blends of CAC and ggbs have been used for this purpose.
- Its strength and hard-wearing characteristics make it useful in non-structural finishing operations such as floor levelling, and in mortars for fixing and rapid repairs. In these materials mixed binders of CAC, Portland cement and calcium sulphate are often used.
- CAC concrete has excellent resistance to high temperatures and thermal shock and so it is used in foundry floors and in refractory bricks for furnace linings.

16.2 Alkali–activated cements

These are cements in which materials that are not cementitious in themselves, or are only weakly so, such as the Type 2 additions described in Chapter 15, are activated by alkalis to form cementitious compounds. These are generally calcium silicate hydrates as in hardened Portland cement. A number of strong alkalis or salts derived from them can be used, including caustic soda (sodium hydroxide, NaOH), soda ash (sodium carbonate, Na_2CO_3), sodium silicates (a range of compounds with the general formula $Na_2O, nSiO_2$) and sodium sulphate (Na_2SO_4). The cementing components include:

- slags such as ground granulated blast furnace slag, granulated phosphorus slag, steel slag (from the basic oxygen or electric arc process, see Chapter 11), all of which should be rapidly cooled to give an unstable or active microstructure
- pozzolans such as volcanic ash, fly ash, metakaolin and condensed silica fume; these are often mixed with lime as well as the activator.

There is thus a considerable number of possible combinations of activator and cementing compound, with slag activated by sodium hydroxide, sodium carbonate or sodium silicates having been widely studied and used for concrete, primarily in Eastern Europe and China. Although the resulting concretes obey the same or similar rules to Portland cement concrete, such as the influence of water:cement ratio on strength, and can achieve similar mechanical properties, they are not without their problems, such as high drying shrinkage, variability of the raw materials and inadequate understanding of long-term properties. Also, some of the activators, particularly sodium hydroxide, are corrosive and require very careful handling. We do not have room to describe these in any detail here, but if you are feeling confident, a comprehensive treatment has been produced by Shi *et al.* (2006).

16.3 Geopolymer cements

This is the name for a group of cements that are produced from alumina- and silica-containing raw materials. These are transformed into silico-aluminates by heating to relatively low temperatures (about 750°C). The silico-aluminates have a ring polymer structure, hence the name 'geo-polymer'; the more specific name polysialates has also been suggested (Davidovits, 2002). These compounds have some cementitious properties but if they are blended with an alkali-silicate activator and ground blast furnace slag the resulting cement has accelerated setting time and a high rate of strength gain (up to 20 MPa after four hours at 20°C) as well as high longer-term strength (more than 70–100 MPa at 28 days). There are therefore some similarities with the alkali-activated cements discussed above. These properties are very useful for repair concrete, for example for roads and runways. Blends with Portland cement are also used.

16.4 Magnesium oxide–based cements

These can be derived from:

- *Magnesium carbonate.* On heating to about 650°C magnesium carbonate dissociates to reactive magnesium oxide (magnesia). This is mixed with Portland cement and other industrial by-products such as slag or fly ash to form the binder (Teccement, 2009). During hydration the magnesia forms magnesium hydroxide (brucite), which has some cementitious properties in itself, but which may also enhance the hydration of the Portland cement.
- *Magnesium silicate.* This decomposes at about 650°C yielding a cementitious product. The process is being developed and the resulting cement investigated (Novacem, 2009), but so far little information and few results have been published.

16.5 Waste-derived cements

The search for uses of industrial waste as an alternative to sending it to landfill has resulted in the development of some potential processes for conversion to cementitious materials. In Japan an 'Eco-cement' has been produced in which up to 50% of the raw materials for Portland cement production is substituted by municipal solid waste in the form of incinerator ash and sewage sludge (Shimoda and Yokoyama, 1999). The required clinkering temperature (1350°C) is a little lower than that for Portland cement (1450°C); the resulting cement contains similar compounds to Portland cement albeit with a generally higher C_3A content. Its properties are, not surprisingly, claimed to be similar to those of Portland cement. However care has to be taken during the production process to remove and recover chlorides and toxic heavy metals. A rapid-hardening version in which the chlorides are not recovered is also available.

In the UK a process has been developed that involves blending a variety of industrial waste products and treating these in a low-temperature, low-emission process (Celtic Cement Technology, 2009). The result is a cement-substitute material which, it is claimed, will outperform blast furnace slag when used as an addition to Portland cement. Altering the combination of waste and the particle size distribution gives products with specific compositions for particular applications, for example for concrete with high early or high long-term strength or fast or slow setting.

References

Celtic Cement Technology (2009). *Engineering ultra-low carbon cement replacements*, http://www.celticcement.com (accessed 25-5-09).

Concrete Society (1997). *Calcium aluminate cement in construction: A reassessment*, Technical Report No 46, Concrete Society, Slough, UK.

Davidovits J (2002). *30 years of successes and failures in geopolymer applications: market trends and potential breakthroughs*. Proceedings of Geopolymer 2002 Conference, Melbourne, Australia.

Neville AM (1995). *Properties of concrete*, 4th edition, Pearson Education, Harlow, p. 844.

Neville AM (2009). A History of high-alumina cement. Part 1: Problems and the Stone report. Part 2: Background to issues. *Proceedings of Institution of Civil Engineers – Engineering History and Heritage*, May Issue, EH2, pp. 81–91 and pp. 93–101.

Neville AM and Wainwright PL (1975). *High Alumina Cement concrete*, The Construction Press, Lancaster.

Novacem (2009). http://www.novacem.com/ (accessed 23-5-09).

Shi C, Krivenko PV and Roy D (2006). *Alkali-activated cements and concrete*, Taylor and Francis, Abingdon, 2006.

Shimoda T and Yokoyama S (1999). *Eco-Cement: a new Portland cement to solve municipal and industrial waste problems*. Proceedings of the International Conference on Modern Concrete Materials: Binders, Additions and Admixtures, Dundee, 1999 (ed. Dhir RK and Dyer TD). Thomas Telford, London, pp. 17–30.

Tec-cement (2009). *Tec-cement and Eco-cement* http://www.tececo.com/ (accessed 23-5-09).

Chapter 17

Aggregates for concrete

In the preceding three chapters we have seen that hardened cement paste (HCP) formed from the hydration of mixtures of Portland cement, admixtures and additions has strength and other properties that could make it suitable for use as a construction material in its own right. However, it suffers from two main drawbacks – high dimensional changes, in particular low modulus, high creep and shrinkage, and cost. Both of these disadvantages are overcome, or at least modified, by adding aggregates to the cement paste, thus producing concrete. The objective is to use as much aggregate as possible, binding the particles together with the HCP. This means that:

- the largest possible aggregate size consistent with the mixing, handling and placing requirements of fresh concrete should be used
- a continuous range of particle sizes from fine sand up to coarse stones is desirable; this minimises the void content of the aggregate mixture and therefore the amount of HCP required, and helps the fresh concrete to flow more easily. Normally aggregates occupy about 65–80% of the total concrete volume.

With one or two notable exceptions, aggregates can be thought of as being inert fillers; for example, they do not hydrate, and they do not swell or shrink. They are distributed throughout the HCP, and it is sometimes useful to regard concrete as a two-phase material of either coarse aggregate dispersed in a mortar matrix, or coarse and fine aggregate dispersed in an HCP matrix. Models based on this two-phase material are of value in describing deformation behaviour, as discussed in Chapter 20 but, when cracking and strength are being considered, a three-phase model of aggregate, HCP and the transition or interfacial zone between the two (about 30–50 μm wide) is required, since the transition zone can have a significantly different microstructure from the rest of the HCP, and is often the weakest phase and the source of cracks as applied stress increases. We will

discuss this in more detail when considering the strength of concrete in Chapter 21.

There are three general types or groups of aggregate depending on their source:

- *primary*, which are specifically produced for use in concrete
- *secondary*, which are by-products of other industrial processes not previously used in construction
- *recycled*, from previously used construction materials e.g. from demolition.

Primary aggregates form by far the greatest proportion of those used and so we will concentrate on discussing the sources, properties and classification of these. We will also make some brief comments about secondary aggregates but leave discussing recycled aggregates until considering recycled concrete in Chapter 26.

17.1 Types of primary aggregate

These can either be obtained from natural sources, such as gravel deposits and crushed rocks, or be specifically manufactured for use in concrete. It is convenient to group them in terms of their relative density.

17.1.1 NORMAL-DENSITY AGGREGATES

Many different natural materials are used for making concrete, including gravels, igneous rocks such as basalt and granite and the stronger sedimentary rocks such as limestone and sandstone. They should be selected to have sufficient integrity to maintain their shape during concrete mixing and to be sufficiently strong to withstand the stresses imposed on the concrete. Stress concentration effects within the concrete result in local stresses at aggregate edges about three times greater than the average stress on the concrete, and so the aggregates should have an inherent compressive strength about three times greater than the required concrete strength if they are not to crack before the HCP. This becomes a particular consideration with high-strength concrete (Chapter 25). Provided that the mechanical properties are acceptable the mineral constituents are not generally of great importance, notable exceptions being those that can participate in alkali–silica reactions and in the thaumasite form of sulphate attack, both of which will be discussed in Chapter 24.

All of the above rock types have relative densities within a limited range of approximately 2.55–2.75, and therefore all produce concretes with similar densities, normally in the range 2250–2450 kg/m³, depending on the mix proportions.

Gravels from suitable deposits in river valleys or shallow coastal waters have particles that for the most part are of a suitable size for direct use in concrete, and therefore only require washing and grading, i.e. subdividing into various sizes, before use. Bulk rock from quarries, e.g. granites and limestones, require crushing to produce suitably sized material. The particles are therefore sharp and angular and distinctly different from the naturally more rounded particles in a gravel; we will see in later chapters that particle shape has a significant effect on fresh and hardened concrete properties.

17.1.2 LIGHTWEIGHT AGGREGATE

Lightweight aggregates are used to produce lower-density concretes, which are advantageous in reducing the self-weight of structures and also have better thermal insulation than normal-weight concrete. The reduced relative density is obtained from air voids within the aggregate particles. We will leave discussion of lightweight aggregates and lightweight aggregate concrete to Chapter 25 – 'Special concretes'.

17.1.3 HEAVYWEIGHT AGGREGATES

Where concrete of high density is required, for example in radiation shielding, heavyweight aggregates can be used. These may be made with high-density ores such as barytres and haematite, or manufactured, such as steel shot. Again, we will discuss these further in Chapter 25 when we consider high-density concrete.

17.2 Aggregate classification – shape and size

Within each of the types described above, aggregates are classified principally by shape and particle size. Normal-density aggregates in particular may contain a range of particle shapes, from well rounded to

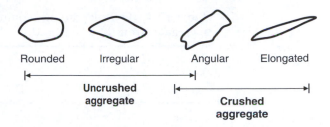

Fig. 17.1 *Aggregate particle shapes.*

angular, but it is usually considered sufficient to classify the aggregate as uncrushed, i.e. coming from a natural gravel deposit, with most particles rounded or irregular, or crushed, i.e. coming from a bulk source, with all particles sharp and angular (*Fig. 17.1*).

The principal size division is that between fine and coarse aggregate at a particle size of 4 mm (although some countries divide at 5, 6 or 8 mm). Coarse aggregate can have a maximum size of 10, 16, 20, 32 or 40 mm (although, again, some countries use different values). In Europe, the size is described by designation d/D, where d is the smallest nominal particle size and D the nominal largest. We say 'nominal' because in practice a few particles may be a smaller than d and a few a little larger than D. Thus:

- 0/4 is a fine aggregate with a maximum particle of 4 mm (with the '0' indicating a near zero lower size limit)
- 4/20 is a coarse aggregate with a minimum particle size of 4 mm and a maximum particle size of 20 mm
- 10/20 is a coarse aggregate with a minimum particle size of 10 mm and a maximum particle size of 20 mm.

The distribution of particle sizes within each of these major divisions is also important both for classification and for determining the optimum combination for a particular mix (a part of the mix design process to be discussed in Chapter 22). To determine this, a *sieve analysis* is carried out using a series of standard sieves with, in European practice, apertures ranging from 0.063 to 63 mm, each sieve having approximately twice the aperture size of the previous one, i.e. in the geometric progression 0.063, 0.125, 0.25, 0.5, 1, 2, 4, 8, 16, 32 and 63 mm. Some countries also use supplementary sizes in the coarse aggregate range, e.g. 10, 20 and 37.5 (40) mm in the UK.

The analysis starts with drying and weighing a representative sample of the aggregate, and then

Table 17.1 Overall grading requirements for coarse and fine aggregate (from BS EN 12620)

Aggregate	Percent passing by weight				
	2D	1.4D	D	d	d/2
Coarse					
$D/d \leq 2$ or $D \leq 11.2$ mm	100	98–100	85–99	0–20	0–5
$D/d > 2$ and $D > 11.2$ mm	100	98–100	90–99	0–15	0–5
Fine					
$D \leq 4$ mm and $d = 0$	100	95–100	85–99	–	–

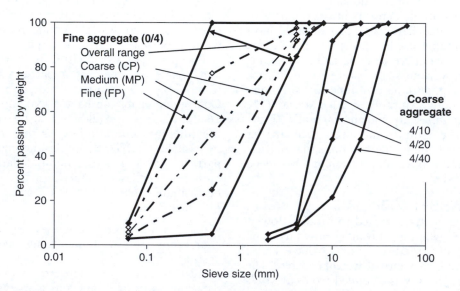

Fig. 17.2 Gradings of aggregates at mid-range of BS EN 12620 limits.

passing this through a stack or nest of the sieves, starting with that with the largest aperture. The weights of aggregate retained on each sieve are then measured. These are converted first to percentage retained and then to cumulative, i.e. total, percent passing, which are then plotted against the sieve size to give a *grading curve* or *particle-size distribution*.

Standards for aggregate for use in concrete contain limits inside which the grading curves for coarse and fine aggregate must fall. In the European standard (BS EN 12620:2002[1]), these are given in terms of the required percentage passing sieves with various ratios of D and d; *Table 17.1* gives examples of the values for coarse and fine aggregate from this standard.

For fine aggregate the definition of some intermediate values gives a useful addition to these overall limits when considering their use in concrete. The European standard suggests using the percentage passing the 0.5-mm sieve (called the P value), and give ranges of:

- for coarse graded fine aggregate, CP = 5–45%
- for medium graded fine aggregate, MP = 30–70%
- for fine graded fine aggregate, FP = 55–100%.

The overlap of the limits means that it is possible for an aggregate to fall into two classes – which can cause confusion. The grading curves for the mid-points of the ranges for the most commonly used aggregates grades are plotted in *Fig. 17.2*.

A single number, the *fineness modulus*, is sometimes calculated from the results of the sieve analysis. The cumulative percent passing figures are converted to

[1] A list of all standards referred to in the text is included in 'Further reading' at the end of this part of the book.

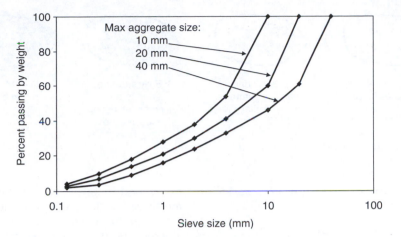

Fig. 17.3 *Examples of preferred overall aggregate gradings for use in concrete.*

Table 17.2 Fineness modulus values for aggregates with the grading shown in *Fig. 17.2*

Aggregate size	Fineness modulus
Coarse aggregate	
4/10	5.95
4/20	6.5
4/40	7.25
Fine aggregate	
0/4 FP	1.7
0/4 MP	2.65
0/4 CP	3.4

cumulative percent *retained*, and the fineness modulus is defined as the sum of all of these starting with that for the 125 µm sieve and increasing in size by factors of two, divided by 100. A higher fineness modulus indicates a coarser material; the values for the grading curves in *Fig. 17.2* are given in *Table 17.2*. It is important to remember that the calculation is carried out only with those sieves in the geometric progression, not intermediate sizes, and that for coarse aggregate with all particles larger than, say, 4 mm the cumulative percent retained on all sieves smaller than 4 mm should be entered as 100.

During the process of mix design, the individual subdivisions or *fractions* of aggregates are combined in proportions to give a suitable overall grading for good concrete consistence and stability. This should be continuous and uniform. Examples for maximum coarse aggregates sizes of 10, 20 and 40 mm that are produced by the mix design process to be discussed

in Chapter 22 are shown in *Fig. 17.3*. These result from using aggregates with ideal gradings; in practice it is normally not possible to achieve these exactly, but they are good targets.

Sieve analysis and grading curves take no account of particle shape, but this does influence the voids content of the aggregate sample – more-rounded particles will pack more efficiently and will therefore have a lower voids content. According to Dewar (1999) it is sufficient to use only three numbers to characterise an aggregate for mix design purposes – specific gravity (or particle relative density); mean particle size; and voids content in the loosely compacted state.

We should also mention here the *bulk density*. This is the weight of aggregate occupying a unit overall volume of both the particles and the air voids between them. It is measured by weighing a container of known volume filled with aggregate. The value will clearly depend on the grading, which will govern how well the particles fit together, and also on how well the aggregate is compacted. Unlike the relative particle density, which is more useful, it is not therefore a constant for any particular aggregate type.

17.3 Other properties of aggregates

It is important that aggregates are clean and free from impurities such as clay particles or contaminants that would affect the fresh or hardened properties of the concrete. Other properties that influence their suitability for use in concrete include porosity

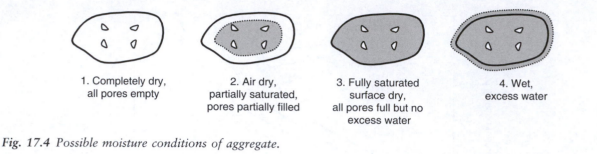

1. Completely dry, all pores empty

2. Air dry, partially saturated, pores partially filled

3. Fully saturated surface dry, all pores full but no excess water

4. Wet, excess water

Fig. 17.4 Possible moisture conditions of aggregate.

and absorption, elasticity, strength and surface characteristics.

17.3.1 POROSITY AND ABSORPTION

All aggregates contain pores, which can absorb and hold water. Depending on the storage conditions before concrete mixing, the aggregate can therefore be in one of the four moisture conditions shown in *Fig. 17.4*. In the freshly mixed concrete, aggregate that is in either of conditions (1) or (2) will absorb some of the mix water, and aggregate in condition (4) will contribute water to it. Condition (3), *saturated surface dry*, is perhaps most desirable, but is difficult to achieve except in the laboratory. It also leads to the definition of the *absorption* of an aggregate:

$$\text{Absorption (\% by weight)} = 100(w_2 - w_1)/w_1 \tag{17.1}$$

where w_1 is weight of a sample of aggregate in the completely (oven) dry state and w_2 is the weight in the saturated surface dry state. Clearly, the absorption is related to the porosity of the aggregate particles. Most normal weight aggregates have low but nevertheless significant absorptions in the range 1–3%.

Of prime importance to the subsequent concrete properties is the amount of water available for cement hydration, i.e. the amount that is non-absorbed or 'free'; therefore, to ensure that the required free water:cement ratio is obtained, it is necessary to allow for the aggregate moisture condition when calculating the amount of water to be added during concrete mixing. If the aggregate is drier than saturated surface dry, extra water must be added; if it is wetter, then less mix water is required.

17.3.2 ELASTIC PROPERTIES AND STRENGTH

Since the aggregate occupies most of the concrete volume, its elastic properties have a major influence on the elastic properties of the concrete, as we shall discuss in Chapter 20. Normal-weight aggregates

are generally considerably stronger than the HCP and therefore do not have a major influence on the strength of most concretes. However, in high-strength concrete (with compressive strengths in excess of, say, 80 MPa – see Chapter 25) careful aggregate selection is important. There are a number of tests used to characterise the strength and other related properties of aggregates – such as abrasion resistance – that may be important for particular uses of the concrete. A look at any typical aggregate standard will lead you to these.

17.3.3 SURFACE CHARACTERISTICS

The surface texture of the aggregate depends on the mineral constituents and the degree to which the particles have been polished or abraded. It seems to have a greater influence on the flexural strength than on the compressive strength of the concrete, probably because a rougher texture results in better adhesion to the HCP. This adhesion is also greatly affected by the cleanliness of the surface – which must therefore not be contaminated by mud, clay or other similar materials. The interface or transition zone between the aggregate surface and the HCP has a major influence on the properties of concrete, particularly its strength, and is discussed in some detail in Chapter 21.

17.4 Secondary aggregates

In principle, any by-product from other processes or waste material that is inert and has properties that conform to the requirements for primary aggregates – strength, particle size etc. – are suitable for use in concrete. Examples that have been used include power station ash, ferro-silicate slag from zinc production, shredded rubber from vehicle tyres and crushed glass. With materials such as ferro-silicate slag, a problem may be the variability of supply (particularly particle-size distribution) since this was not an issue for the producers. Clearly with crushed glass and shredded tyres some processing of the waste is first required. Crushed glass is not suitable for high-strength concrete, and there may be some issues with long-term durability owing to alkali–silica reaction between glass and the cement (see Chapter 24). Shredded rubber will result in a concrete with a low elastic modulus but this may not be a problem if, for example, shock absorbent properties are required. Some case studies involving several of these aggregate types can be found on the Aggregain web-site (WRAP-AggRegain, 2009).

References

Dewar JD (1999). *Computer modelling of concrete mixtures*, E & FN Spon, London, p. 272.

WRAP-AggRegain (2009). *Case studies* http://www.aggregain.org.uk/case_studies/index.html (accessed 20/8/09).

Chapter 18

Properties of fresh concrete

Civil engineers are responsible for the production, transport, placing, compacting and curing of fresh concrete. Without adequate attention to all of these the potential hardened properties of the concrete, such as strength and durability, will not be achieved in the finished structural element. It is important to recognise that it is not sufficient simply to ensure that the concrete is mixed and placed correctly; the behaviour and treatment of the concrete during the period before setting, typically some six to ten hours after casting, and during the first few days of hardening have a significant effect on its long-term performance.

It is beyond the scope of this book to discuss the operations and equipment used to batch, mix, handle, compact and finish concrete, but there are many options available, particularly for site operations for *in-situ* concrete. These include transport of the concrete from the point of delivery by pump, skip, conveyor belt or wheelbarrow and compaction by internal poker vibrators or external vibratrors clamped on to the formwork. (Some publications describing these are included in 'Further reading' at the end of this part of the book.)

The aim of all of these practices is to produce a homogeneous structure with minimum air voids as efficiently as possible; it is also necessary to ensure that the concrete is then stable and achieves its full, mature properties. We therefore need to consider the properties when freshly mixed, between placing and setting, and during the early stages of hydration. We will discuss the former in this chapter, and the latter two in the next chapter.

18.1 General behaviour

Experience in mixing, handling and placing fresh concrete quickly gives concrete workers (and students) a subjective understanding of its behaviour and an ability to recognise 'good' and 'bad' concrete. A major problem is that a wide variety of subjective terms are used to describe the concrete, e.g. harsh, cohesive, lean, stiff, rich, which can mean different things to different people and do not quantify the behaviour in any way. However, the main properties of interest can be grouped as follows:

1. *Fluidity*. The concrete must be capable of being handled and of flowing into the formwork and around any reinforcement, with the assistance of whatever equipment is available. For example, concrete for a lightly reinforced shallow floor slab need not be as fluid as that for a tall narrow column with congested reinforcement.
2. *Compactability*. All, or nearly all, of the air entrapped during mixing and handling should be capable of being removed by the compacting system being used, such as poker vibrators.
3. *Stability or cohesiveness*. The concrete should remain as a homogeneous uniform mass throughout. For example, the mortar should not be so fluid that it flows out of or segregates from the coarse aggregate.

The first two of these properties, fluidity and compactability, have traditionally been combined into the general property called *workability*, but this has now been replaced by the term *consistence* in some current standards, including those in Europe. We will use the latter term in this book, although the two can be considered synonymous.

Although consistence (or workability) might seem a fairly obvious property, engineers and concrete technologists have struggled since concrete construction became popular early in the last century to produce an adequate definition. Two examples illustrate the difficulty:

- 'that property of freshly mixed concrete or mortar which determines the ease and homogeneity with which it can be mixed, placed, consolidated and finished' (ACI, 1990)

• 'that property determining the effort required to manipulate a freshly mixed quantity of concrete with minimum loss of homogeneity' (ASTM, 1993).

These both relate to the requirements in very general terms only, but the biggest problem is that neither makes any reference to a quantitative measurable property, which engineers need and have for most other properties, e.g. elastic modulus, fracture toughness, etc., etc. As we will see, the measurement of consistence is by no means straightforward.

In general, higher-consistence concretes (however defined or measured) are easier to place and handle, but if higher consistence is obtained, for example, by an increased water content, then a lower strength and/or durability will result if no other changes to the mix are made. The more widespread use of plasticisers and superplasticisers (Chapter 14) has therefore been a key factor in the trend towards the use of higher-consistence concrete in recent years in many countries. It is clear that a proper understanding of the fresh properties and the factors that affect them is important. Achieving a balance between consistence and strength is part of the mix design process, which we will be discussing in Chapter 22.

As mentioned in the Introduction to this part of the book, for most concrete about 65–80% of the volume consists of fine and coarse aggregate. The remainder is cement paste, which in turn consists of 30–50% by volume of cement, the rest being water. Cement paste, mortar and concrete are all therefore concentrated suspensions of particles of varying sizes, but all considerably denser than the mix water. Surface attractive forces are significant in relation to gravitational forces for the cement particles, but less so for the aggregate particles, where the main resistance to flow comes from interference and friction between them. The behaviour is therefore far from simple.

18.2 Measurement of consistence

18.2.1 FUNDAMENTAL PROPERTIES

Rigorous measurement of the flow behaviour of any fluid is normally carried out in a rheometer or viscometer of some sort. We do not have space to describe these, but they apply a shear stress to the fluid and measure its consequent rate of shear, for example in a concentric cylinder viscometer an inner cylinder or bob is rotated in an outer cylinder or cup of the fluid. Any respectable undergraduate fluid mechanics textbook will describe such instruments,

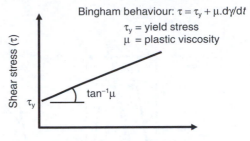

Fig. 18.1 *Flow curve of fresh concrete and the definitions of yield stress and plastic viscosity.*

and a test will result in a *flow curve* of shear stress vs. shear rate (we discussed the nature of this relationship in Chapter 5). Several such tests have been developed for concrete, involving either a mixing or a shearing action, and for which the apparatus is of sufficient size to cope with coarse aggregate particles of up to 20 mm (RILEM, 2000). There is general agreement that the behaviour of fresh paste, mortar and concrete all approximate reasonably closely to the Bingham model illustrated in *Fig. 18.1*. Flow only starts when the applied shear stress reaches a yield stress (τ_y) sufficient to overcome the inter-particle interference effects, and at higher stresses the shear rate varies approximately linearly with shear stress, the slope defining the plastic viscosity (μ). Thus two constants, τ_y and μ, are required to define the behaviour, unlike the simpler and very common case of a Newtonian fluid that does not have a yield stress, and which therefore requires only a single constant, *viscosity* (Chapter 5). Because at least two data points are required to define the flow curve, the first satisfactory test that was devised to measure this on concrete was called the *two-point workability test* (Tattersall and Banfill, 1983; Tattersall, 1991; Domone *at al.*, 1999).

18.2.2 SINGLE-POINT TESTS

A large number of simple but arbitrary tests for consistence or workability have been devised over many years, some only being used by their inventors. These all measure only one value, and can therefore be called *single-point tests*. Four are included in European standards, and have also been adopted elsewhere, and are therefore worth considering in some detail.

The simplest, and crudest, is the *slump test* (BS EN12350-2, *Fig. 18.2*). The concrete is placed in the frustum of a steel cone and hand compacted in three successive layers. The cone is lifted off, and

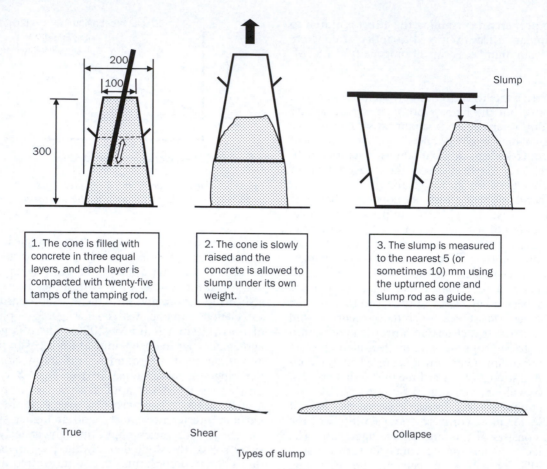

1. The cone is filled with concrete in three equal layers, and each layer is compacted with twenty-five tamps of the tamping rod.

2. The cone is slowly raised and the concrete is allowed to slump under its own weight.

3. The slump is measured to the nearest 5 (or sometimes 10) mm using the upturned cone and slump rod as a guide.

True   Shear   Collapse

Types of slump

Fig. 18.2 The slump test.

slump is defined as the downward movement of the concrete. A true slump, in which the concrete retains the overall shape of the cone and does not collapse, is preferred, which gives a limit to the slump measurement of about 180 mm. A shear slump invalidates the test, and may indicate a mix prone to segregation owing to lack of cement or paste. A collapsed slump is not ideal, but the trend mentioned above of the increasing use of high-consistence mixes, which produce collapsed slumps with little or no segregation, means that slump values up to, and even above, 250 mm are considered valid in many standards. For such very high consistence mixes an alternative is to measure the final diameter or 'flow' of the concrete, which is more sensitive to changes in the mix than the change in height. Indeed, for self-compacting mixes (see Chapter 25) the *slump-flow test* is carried out without any initial compaction when filling the cone.

As a general guide, mixes with slumps ranging from about 10 mm upwards can be handled with conventional site equipment, with higher slumps (100 mm and above) being more generally preferred and essential to ensure full compaction of the concrete in areas with limited access or congested reinforcement. However, some zero-slump mixes have sufficient consistence for some applications.

The *degree of compactability test* (BS EN 12350-4, Fig. 18.3), which has replaced the *compacting factor test* in many standards, is able to distinguish between low-slump mixes. A rectangular steel container is filled with concrete by allowing it to drop from a trowel under its own weight from the top of the container. It is therefore only partially compacted. The concrete is then compacted, e.g. by vibration, and its final height measured. The difference between the initial and final heights is a measure of the amount of compaction the concrete undergoes when loaded into the container, and will be lower with high-consistence concrete. The degree of compactability is defined as the ratio of the initial height to the final height. Values over 1.4 indicate a very low

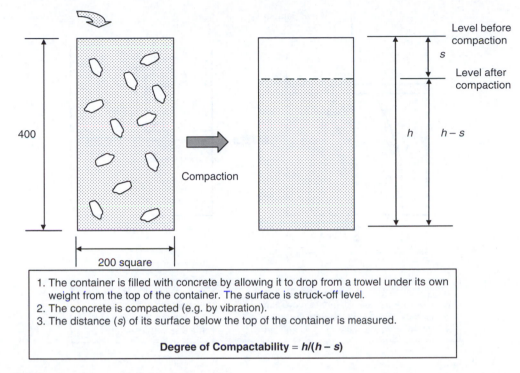

- Level before compaction
- s
- Level after compaction
- h
- $h - s$
- 400
- Compaction
- 200 square

1. The container is filled with concrete by allowing it to drop from a trowel under its own weight from the top of the container. The surface is struck-off level.
2. The concrete is compacted (e.g. by vibration).
3. The distance (s) of its surface below the top of the container is measured.

Degree of Compactability = $h/(h - s)$

Fig. 18.3 The degree of compactability test.

consistence, and as the consistence increases the value gets closer and closer to 1.

In the *Vebe test* (BS EN 12350-3, *Fig. 18.4*), the response of the concrete to vibration is determined. The Vebe time is defined as the time taken to completely remould a sample of concrete from a slump test carried out in a cylindrical container. Standard vibration is applied, and remoulding times from 1 to about 25 seconds are obtained, with higher values indicating lower consistence. It is often difficult to define the end-point of complete remoulding with a sufficient degree of accuracy.

The *flow table test* (BS EN 12350-5, *Fig. 18.5*) was devised to differentiate between high consistence mixes. It is essentially a slump test with a lower volume of concrete in which, after lifting the cone, some extra work is done on the concrete by lifting and dropping one edge of the board (or table) on which the test is carried out. A flow or spread of 400 mm indicates medium consistence, and 500 mm or more high consistence.

Apart from only giving a single test value, these four tests (or five if we consider the slump-flow test to be distinct from the slump test) all measure the response of the concrete to specific, but arbitrary and different, test conditions. The slump, slump-flow and flow table tests provide a measure of the fluidity or mobility of the concrete; the slump test after a standard amount of compaction work has been done on the concrete, the slump-flow test after the minimal amount of work of pouring into the cone, and the flow table test with a combination of compaction work and energy input. The degree of compactability test assesses the response of the concrete to applied work, but the amount of work done in falling from the top of the container is much less than the energy input from practical compaction equipment such as a poker vibrator. The Vebe test comes closest to assessing the response to realistic energy levels, but it is the most difficult test to carry out and it is not able to distinguish between the types of high-consistence mixes that are becoming increasingly popular.

There is some degree of correlation between the results of these tests, as illustrated in *Fig. 18.6*, but as each of the tests measures the response to different conditions the correlation is quite broad. It is even possible for the results to be conflicting – e.g. for say the slump test to show that Mix *A* has a higher consistence than Mix *B*, and for the degree of compactability test to give the opposite ranking. The result therefore depends on the choice of test, which is far from satisfactory.

123

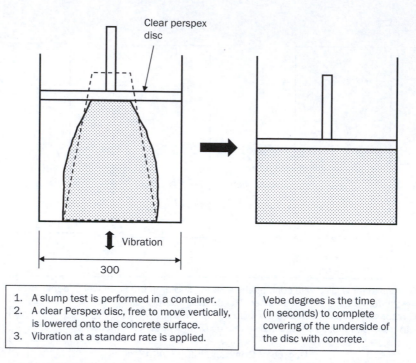

1. A slump test is performed in a container.	Vebe degrees is the time
2. A clear Perspex disc, free to move vertically, is lowered onto the concrete surface.	(in seconds) to complete covering of the underside of
3. Vibration at a standard rate is applied.	the disc with concrete.

Fig. 18.4 The Vebe test.

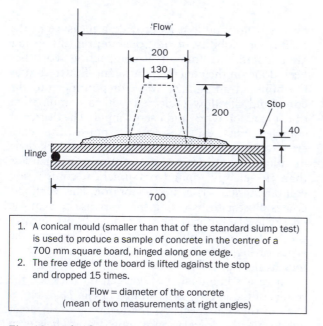

1. A conical mould (smaller than that of the standard slump test) is used to produce a sample of concrete in the centre of a 700 mm square board, hinged along one edge.
2. The free edge of the board is lifted against the stop and dropped 15 times.

Flow = diameter of the concrete
(mean of two measurements at right angles)

Fig. 18.5 The flow table test.

The slump and slump-flow tests clearly involve very low shear rates, and therefore, not surprisingly, reasonable correlations are obtained with yield stress (e.g. *Fig. 18.7*). No correlation is obtained with plastic viscosity. The test therefore indicates the ease with which the concrete starts to flow, but not its behaviour thereafter.

Despite their limitations, single-point tests, particularly the slump test and, to a somewhat lesser extent, the flow table and slump-flow tests, are popular and in regular use, both for specification and for compliance testing of the concrete after production. Perhaps the main reason for this is their simplicity and ease of use both in the laboratory and on site, but specifiers and users must be aware of the potential pitfalls of over-reliance on the results.

18.3 Factors affecting consistence

Lower values of yield stress (τ_y) and plastic viscosity (μ) indicate a more fluid mix; in particular, reducing τ_y lowers the resistance to flow at low shear stresses, e.g. under self-weight when being poured, and reducing μ results in less cohesive or 'sticky' mixes and increased response during compaction by vibration, when the localised shear rates can be high. Some of the more important effects of variation of mix proportions and constituents on τ_y and μ, shown schematically in *Fig. 18.8*, are as follows:

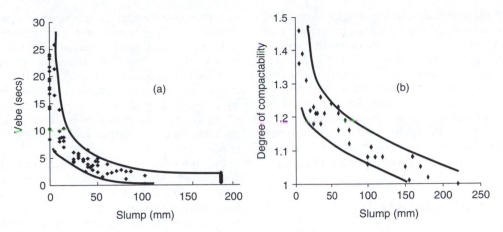

Fig. 18.6 *Typical relationships between results from single-point workability tests (data from (a) Ellis, 1977; (b) UCL tests).*

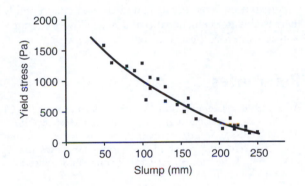

Fig. 18.7 *The relationship between yield stress and slump of fresh concrete (from Domone et al., 1999).*

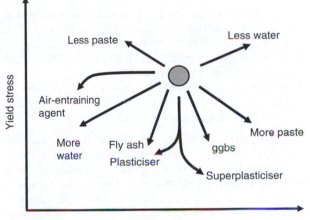

Fig. 18.8 *Summary of the effect of varying the proportions of concrete constituents on the Bingham constants.*

- Increasing the water content while keeping the proportions of the other constituents constant decreases τ_y and μ in approximately similar proportions.
- Adding a plasticiser or superplasticiser decreases τ_y but leaves μ relatively constant. In essence, the admixtures allow the particles to flow more easily but in the same volume of water. The effect is more marked with superplasticisers, which can even increase μ, and can therefore be used to give greatly increased flow properties under self-weight, while maintaining the cohesion of the mix. This is the basis for a whole range of high-consistence or *flowing concretes*, which will be discussed further in Chapter 25.
- Increasing the paste content will normally increase μ and decrease τ_y i.e. the mix may start to flow more easily but will be more cohesive or 'stickier', and vice versa.

- Replacing some of the cement with fly ash or ggbs will generally decrease τ_y, but may either increase or decrease μ, depending on the nature of the addition and its interaction with the cement.
- The small bubbles of air produced by air-entraining agents provide lubrication to reduce the plastic viscosity, but at relatively constant yield stress.

An important consequence of these considerations is that yield stress and plastic viscosity are independent properties, and different combinations can be obtained by varying the mix constituents and their relative proportions.

There is a great deal of information available on the effect of mix constituents and proportions on

125

consistence measurements using single-point tests, particularly slump. Many mix design methods (see Chapter 22) take as their first assumption that, for a given aggregate type and size, slump is a direct function of the water content. This is very useful and reasonably accurate – other factors such as cement content and aggregate grading are of secondary importance for slump, but are of greater importance for cohesiveness and stability. The effectiveness of admixtures, particularly plasticisers and super-plasticisers, is also often given in terms of slump.

18.4 Loss of consistence

Although concrete remains sufficiently workable for handling and placing for some time after it has been mixed, its consistence continually decreases. This is due to:

- mix water being absorbed by the aggregate if this is not in a saturated state before mixing
- evaporation of the mix water
- early hydration reactions (but this should not be confused with cement setting)
- interactions between admixtures (particularly plasticisers and superplasticisers) and the cementitious constituents of the mix.

Absorption of water by the aggregate can be avoided by ensuring that saturated aggregate is used, for example by spraying aggregate stockpiles with water and keeping them covered in hot/dry weather, although this may be difficult in some regions. Evaporation of mix water can be reduced by keeping the concrete covered during transport and handling as far as possible.

Most available data relate to loss of slump, which increases with higher temperatures, higher initial slump, higher cement content and higher alkali and lower sulphate content of the cement. At an ambient temperature of 20°C, slump may reduce to about half its initial value in two hours, but the loss is more acute, and can have a significant effect on concrete operations, at ambient temperatures in excess of 30°C. The rate of loss of consistence can be reduced by continued agitation of the concrete, e.g. in a ready-mix truck, or modified by admixtures, particularly retarders (Chapter 14). In hot weather, the initial concrete temperature can be reduced by cooling the constituents before mixing (adding ice to the mix water is a common practice) and the concrete can be transported in cooled or insulated trucks.

In principle, re-tempering, i.e. adding water to compensate for slump loss, should not have a significant effect on strength if only that water that has been lost by evaporation is replaced. Also, studies have shown that water can be added during retempering to increase the initial water:cement ratio by up to 5% without any loss in 28-day strength (Cheong and Lee 1993). However, except in very controlled circumstances, retempering can lead to an unacceptably increased water:cement ratio and hence lower strength, and is therefore best avoided.

References

ACI (1990). *Cement and concrete terminology*, ACI 116R-90, American Concrete Institute, Detroit, USA.

ASTM (1993). *Standard definitions and terms relating to concrete and concrete aggregates*, Specification C 125-93, American Society for Testing and Materials, West Conshohocken, USA.

Cheong HK and Lee SC (1993). Strength of retempered concrete. *ACI Materials Journal*, **90** (No. 3), 203–206.

Domone PL, Xu Y and Banfill PFG (1999). Developments of the two-point workability test for high-performance concrete. *Magazine of Concrete Research*, **51** (No. 3), 171–179.

Ellis C (1977). *Some aspects of pfa in concrete*, MPhil thesis, Sheffield City Polytechnic.

RILEM (2000). *Compendium of concrete workability tests*, TC 145-WSM, RILEM, Paris.

Tattersall GH (1991). *Workability and quality control of concrete*, E & FN Spon, London.

Tattersall GH and Banfill PFG (1983). *The rheology of fresh concrete*, Pitman, London.

Chapter 19

Early age properties of concrete

Successful placing of concrete is not enough. It is necessary to ensure that it comes through the first few days of its life without mishap, so that it goes on to have the required mature properties. Immediately after placing, before the cement's initial set (see Chapter 13, section 13.4) the concrete is still in a plastic and at least semi-fluid state, and the component materials are relatively free to move. Between the initial and final set, it changes into a material which is stiff and unable to flow, but which has no strength. Clearly it must not be disturbed during this period. After the final set hardening starts and the concrete develops strength, initially quite rapidly.

In this chapter we will discuss the behaviour of concrete during each of these stages and how they affect construction practice. The hydration processes and the timescales involved have been described in some detail in Chapter 13, and their modification by admixtures and cement replacement materials in Chapters 14 and 15. In particular, we discussed the exothermic nature of the hydration reactions, and we will see that this has some important consequences.

19.1 Behaviour after placing

The constituent materials of the concrete are of differing relative particle density (cement 3.15, normal aggregates approx. 2.6 etc.) and therefore while the concrete is in its semi-fluid, plastic state the aggregate and cement particles tend to settle and the mix water has a tendency to migrate upwards. This may continue for several hours, until the time of final set and the onset of strength gain. Inter-particle interference reduces the movement, but the effects can be significant. There are four interrelated phenomena – bleeding, segregation, plastic settlement and plastic shrinkage.

19.1.1 SEGREGATION AND BLEEDING
Segregation involves the larger aggregate particles falling towards the lower parts of the pour, and

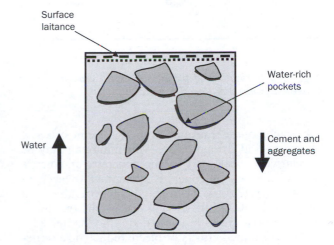

Fig. 19.1 Segregation and bleeding in freshly placed concrete.

bleeding is the process of the upward migration or upward displacement of water. They often occur simultaneously (*Fig. 19.1*).

The most obvious manifestation of bleeding is the appearance of a layer of water on the top surface of concrete shortly after it has been placed; in extreme cases this can amount to 2% or more of the total depth of the concrete. In time this water either evaporates or is re-absorbed into the concrete with continuing hydration, thus resulting in a net reduction of the concrete's original volume. This in itself may not be of concern, but there are two other effects of bleeding that can give greater problems, illustrated in *Fig. 19.1*. Firstly, the cement paste at or just below the top surface of the concrete becomes water rich and therefore hydrates to a weak structure, a phenomenon known as surface laitance. This is a problem in, for example, floor slabs, which are required to have a hard-wearing surface. Secondly, the upward migrating water can be trapped under aggregate particles, causing a local enhanced

127

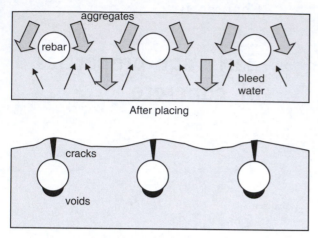

After placing

After several hours

Fig. 19.2 Formation of plastic settlement cracks (adapted from Day and Clarke, 2003).

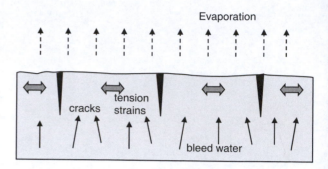

Fig. 19.3 Formation of plastic shrinkage cracks (adapted from Day and Clarke, 2003).

weakening of the transition or interface zone between the paste and the aggregate, which may already be a relatively weak part of the concrete, and hence an overall loss of concrete strength. However, in most concrete some bleed may be unavoidable, and may not be harmful. We will discuss the transition zone and its effects in some detail in Chapter 21.

The combined effects of bleed and particle settlement are that after hardening the concrete in the lower part of a pour of any significant depth is stronger than that in the upper part, possibly by 10% or more, even with a cohesive and well produced concrete.

19.1.2 PLASTIC SETTLEMENT
Overall settlement of the concrete will result in greater movement in the fresh concrete near the top surface of a deep pour. If there is any local restraint to this movement from, say, horizontal reinforcing bars, then plastic settlement cracking can occur, in which vertical cracks form along the line of the bars, penetrating from the surface to the bars (*Fig. 19.2*).

19.1.3 PLASTIC SHRINKAGE
Bleed water arriving at an unprotected concrete surface will be subject to evaporation; if the rate of evaporation is greater than the rate of arrival of water at the surface, then there will be a net reduction in water content of the surface concrete, and plastic shrinkage, i.e. drying shrinkage while the concrete is still plastic, will occur. The restraint of the mass of concrete will cause tensile strains to be set up in the near-surface region, and as the concrete

has near-zero tensile strength, plastic shrinkage cracking may result (*Fig. 19.3*). The cracking pattern is a fairly regular 'crazing' and is therefore distinctly different from the oriented cracks resulting from plastic settlement.

Any tendency to plastic shrinkage cracking will be encouraged by greater evaporation rates of the surface water, which occurs, for example, with higher concrete temperature or ambient temperature, or if the concrete is exposed to wind.

19.1.4 METHODS OF REDUCING SEGREGATION AND BLEED AND THEIR EFFECTS
A major cause of excessive bleed is the use of a poorly graded aggregate, a lack of fine material below a particle size of 300 μm being most critical. This can be remedied by increasing the sand content, but if this is not feasible for some reason, or if a particularly coarse sand has to be used, then air entrainment (see Chapter 14) can be an effective substitute for the fine particles.

Higher bleeds may also occur with higher consistence mixes, and if very high consistence is required it is preferable to use superplasticisers rather than high water contents, as discussed in Chapter 14. Microsilica, with its very high surface area, is also an effective bleed-control agent.

Bleed, however, cannot be entirely eliminated, and so measures must be taken in practice to reduce its effects if these are critical. Plastic settlement and plastic shrinkage cracks that occur soon after placing the concrete can be overcome by re-vibrating the surface region, particularly in large flat slabs.

19.2 Curing

All concretes, no matter how great or small their tendency to bleed, must be protected from moisture loss from as soon after placing as possible, and for the first few days of hardening. This will not only reduce or eliminate plastic shrinkage cracking, but also ensure that there is an adequate supply of water for continued hydration and strength gain. This protection is called *curing*, and is an essential part of any successful concreting operation, although often overlooked. Curing methods include:

- spraying or ponding the surface of the concrete with water
- protecting exposed surfaces from wind and sun by windbreaks and sunshades
- covering surfaces with wet hessian and/or polythene sheets
- applying a curing membrane, usually a spray-applied resin seal, to the exposed surface; this prevents moisture loss, and weathers away in a few weeks.

Extended periods of curing are required for mixes that gain strength slowly, such as those containing additions, particularly fly ash and ggbs, and in conditions of low ambient temperature.

19.3 Strength gain and temperature effects

19.3.1 EFFECT OF TEMPERATURE

We mentioned in Chapter 13 that the hydration reactions between cement and water are temperature-dependent and their rate increases with curing or storage temperature. The magnitude of the effect on the development of strength for concrete continuously stored at various temperatures at ages of up 28 days is apparent from *Fig. 19.4*. There is, however, evidence that at later ages higher strengths are obtained from concrete cured at lower temperatures, perhaps by as much as 20% for concrete stored at 5°C compared to that at 20°C (Klieger, 1958). Explanations for this behaviour have been conflicting, but it would seem that, as similar behaviour is obtained with cement paste, the C-S-H gel more rapidly produced at higher temperatures is less uniform and hence weaker than that produced at lower temperatures. There also appears to be an optimum temperature for maximum long-term strength of between 10 and 15°C, although this varies with the type of concrete.

The hydration reactions do still proceed at temperatures below the freezing point of water, 0°C. In fact they only cease completely at about −10°C. However, the concrete must only be exposed to such temperatures after a significant amount of the mix

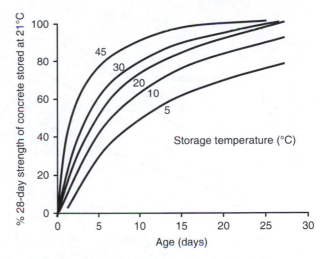

Fig. 19.4 Effect of storage temperature on strength development of concrete (adapted from Mehta and Monteiro, 2006).

water has been incorporated in the hydration reactions, since the expansion of free water on freezing will disrupt the immature, weak concrete. A degree of hydration equivalent to a strength of 3.5 MPa is considered sufficient to give protection against this effect.

19.3.2 MATURITY

Temperature effects such as those shown in *Fig. 19.4* have led to the concept of the *maturity* of concrete, defined as the product of time and curing temperature, and its relationship to strength. For the reasons given above, −10°C is taken as the datum point for temperature, and hence:

$$\text{maturity} = \sum t(T + 10) \qquad (19.1)$$

where t and T are the time (normally in hours or days) and curing temperature (in °C), respectively. *Figure 19.5* shows the relationship between strength and maturity for concrete with three water:cement ratios. These results were obtained with each mix being cured at 4, 13 and 21°C for periods of up to 1 year; the results for each mix fall on or very near to the single lines shown, thus demonstrating the usefulness of the maturity approach. If the temperature history of a concrete is known, then its strength can be estimated from the strength–age relationship at a standard curing temperature (e.g. 20°C).

Figure 19.5 shows that over much of the maturity range:

$$\text{strength} = a + b \log_{10}(\text{maturity}) \qquad (19.2)$$

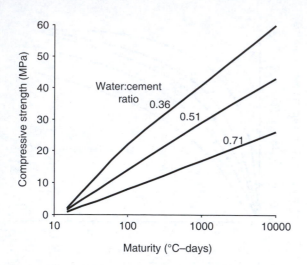

Fig. 19.5 *Strength–maturity relationship for concrete with three water:cement ratios (adapted from Neville, 1995).*

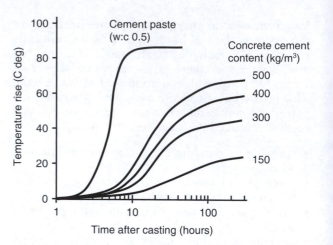

Fig. 19.6 *Typical temperature rise during curing under adiabatic conditions for a neat cement paste and concrete with varying cement content (after Bamforth, 1988).*

which is a convenient relationship for estimating strength. The constants *a* and *b* will be different for different mixes and will generally need to be established experimentally.

A slightly different approach is to express the maturity as being equivalent to a certain number of days at the standard curing temperature of control cubes (normally 20°C). On this basis, for example, a maturity of 1440°C hrs has an equivalent age of 3 days at 20°C. Equation 19.1 then becomes

$$\text{equivalent age at } 20°C = \sum kt \qquad (19.3)$$

where *k* is the *maturity function*. Various forms for this function have been proposed, as summarised by Harrison (2003).

19.3.3 HEAT OF HYDRATION EFFECTS

As well as being temperature dependent, the hydration of cement is exothermic, and in Chapter 13 we discussed in some detail the rate of heat output at constant temperature (i.e. isothermal) conditions in relation to the various hydration reactions. The opposite extreme to the isothermal condition is adiabatic (i.e. perfect insulation or no heat loss), and in this condition the exothermic reactions result in heating of a cement paste, mortar or concrete. This leads to progressively faster hydration, rate of heat output and temperature rise, the result being substantial temperature rises in relatively short times (*Fig. 19.6*). The temperature rise in concrete is less than that in cement paste as the aggregate acts as a heat sink and there is less cement to react. An

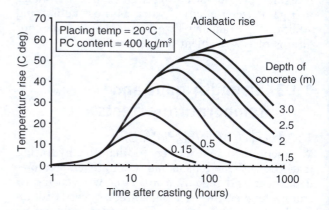

Fig. 19.7 *Temperature rise at mid-depth of a concrete pour during hydration (after Browne and Blundell, 1973).*

average rise of 13°C per 100 kg of cement per m³ of concrete has been suggested for typical structural concretes.

When placed in a structure, concrete will lose heat to its surrounding environment either directly or through formwork, and it will therefore not be under truly adiabatic or isothermal conditions, but in some intermediate state. This results in some rise in temperature within the pour followed by cooling to ambient. Typical temperature–time profiles for the centre of pours of varying depths are shown in *Fig. 19.7*; it can be seen that the central regions of a pour with an overall thickness in excess of about 1.5–2 m will behave adiabatically for the first few days after casting.

Such behaviour has two important effects. First, the peak temperature occurs after the concrete has hardened and gained some strength and so the cool down will result in thermal contraction of the concrete, which if restrained will result in tensile stresses that may be sufficiently large to crack the concrete. Restraint can result from the structure surrounding the concrete, e.g. the soil underneath a foundation, or from the outer regions of the concrete pour itself, which will have been subject to greater heat losses, and therefore will not have reached the same peak temperatures, or from reinforcement within the concrete. The amount of restraint will obviously vary in different structural situations.

As an example, a typical coefficient of thermal expansion for concrete is 10×10^{-6} per C degree, and therefore a thermal shrinkage strain of 300×10^{-6} would result from a cool down of 30°C. Taking a typical elastic modulus for the concrete of 30 GPa, and assuming complete restraint with no relaxation of the stresses due to creep, the resulting tensile stress would be 9 MPa, well in excess of the tensile strength of the concrete, which would therefore have cracked.

Rigorous analysis of the thermal strains and the consequent stresses is complex but in structural concrete, control of the likelihood and consequences of any cracking can be obtained by design of the reinforcement system and in pours of any substantial size to limit the temperature differentials. Insulation by way of increased formwork thickness or thermal blankets will have some beneficial effect but, more commonly, or in addition, low heat mixes are used. If strength or durability criteria mean that a sufficiently low cement content cannot be used, then either a low-heat Portland cement (discussed in Chapter 13) can be used or, more conveniently, the use of additions; fly ash or ground granulated blast furnace slag (ggbs) are effective solutions, as shown in *Fig. 19.8* (these materials were described in Chapter 15). As alternative or additional measures, the temperature of the fresh concrete can be reduced by pre-cooling the mix water or the aggregates, or by injecting liquid nitrogen.

Second, much of the concrete will have hydrated for at least a few days after casting at temperatures higher than ambient, and the long-term strength may therefore be reduced, owing to the effects described above. Typical effects of this on the development of strength are shown in *Fig. 19.9*. By comparing *Fig. 19.9a* and *Fig. 19.9b* it can be seen that fly ash and ggbs mixes do not suffer the same strength losses as 100% Portland cement mixes. Measurement of the concrete's properties after being subjected

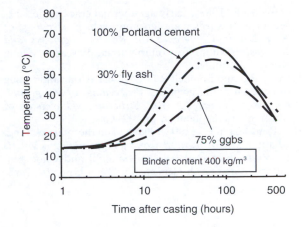

Fig. 19.8 The effect of additions on the temperature variation at mid-height of 2.5-m deep concrete pour during hydration (after Bamforth, 1980).

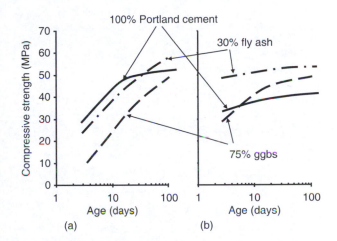

Fig. 19.9 The effect of additions on strength development of concrete (a) with standard curing at 20°C; and (b) when subjected to the temperature cycles of Fig. 19.8 (after Bamforth, 1980).

to such 'temperature-matched curing' is therefore extremely important if a full picture of the *in-situ* behaviour is to be achieved.

References

Bamforth PB (1980). In-situ measurement of the effect of partial Portland cement replacement using either fly ash or ground granulated blast furnace slag on the performance of mass concrete. *Proceedings of the Institution of Civil Engineers*, Part 2, **69**, 777–800.

Concrete

Bamforth PB (1988). Early age thermal cracking in large sections: Towards a design approach. *Proceedings of Asia Pacific Conference on Roads, Highways and Bridges*, Institute for International Research, Hong Kong, September.

Browne RD and Blundell R (1973). Behaviour and testing of concrete for large pours. *Proceedings of Symposium on Large Pours for RC Structures*, University of Birmingham, September, pp. 42–65.

Day R and Clarke J (2003). Plastic and thermal cracking. Chapter 2 of *Advanced Concrete Technology, Vol II: Concrete Properties* (eds Newman JB and Choo BS), Elsevier/Butterworth Heinemann, Oxford, pp. 2/1–2/17.

Harrison T (2003). Concrete properties: setting and hardening. Chapter 4 of *Advanced Concrete Technology, Vol II: Concrete properties* (eds Newman JB and Choo BS), Elsevier/Butterworth Heinemann, Oxford, pp. 4/1–4/33.

Klieger P (1958). Effect of mixing and curing temperature on concrete strength. *Journal of the American Concrete Institute*, 54 (No. 12), 1063–1081.

Mehta PK and Monteiro PJM (2006). *Concrete: Microstructure, Properties and Materials*, 3rd edition, McGraw Hill, New York, p. 64.

Neville AM (1995). *Properties of concrete*, 4th edition, Pearson Education, Harlow, p. 307.

Chapter 20

Deformation of concrete

Deformation of concrete results both from environmental effects, such as moisture gain or loss and heat, and from applied stress, both short- and long-term. A general view of the nature of the behaviour is given in *Fig. 20.1*, which shows the strain arising from a uniaxial compressive stress applied to concrete in a drying environment. The load or stress is applied at a time t_1, and held constant until removal at time t_2.

- Before applying the stress, there is a net contraction in volume of the concrete, or shrinkage, associated with the drying. The dotted extension in this curve beyond time t_1 would be the subsequent behaviour without stress, and the effects of the stress are therefore the differences between this curve and the solid curves.
- Immediately on loading there is an instantaneous strain response, which for low levels of stress is approximately proportional to the stress, and hence an elastic modulus can be defined.
- With time, the strain continues to increase at a decreasing rate. This increase, after allowing for shrinkage, represents the creep strain. Although reducing in rate with time, the creep does not tend to a limiting value.
- On unloading, at time t_2, there is an immediate (elastic) strain recovery, which is often less than the initial strain on loading. This is followed by a time-dependent creep recovery, which is less than the preceding creep, i.e. there is permanent deformation but, unlike creep, this reaches completion in due course.

In this chapter we discuss the mechanisms and factors influencing the magnitude of all the components of this behaviour, i.e. shrinkage, elastic response and creep, and also consider thermally

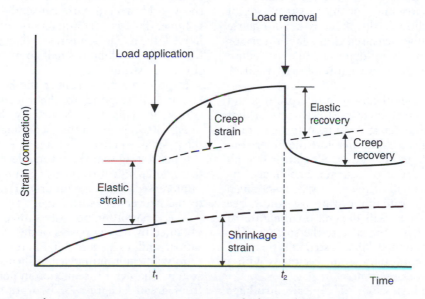

Fig. 20.1 The response of concrete to a compressive stress applied in a drying environment.

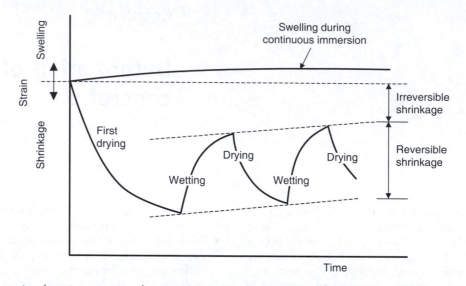

Fig. 20.2 Schematic of strain response of cement paste or concrete to alternate cycles of drying and wetting.

induced strains. We will for the most part be concerned with the behaviour of HCP and concrete when mature, but some mention of age effects will be made.

20.1 Drying shrinkage

20.1.1 DRYING SHRINKAGE OF HARDENED CEMENT PASTE

In section 13.6 we described the broad divisions of water in HCP and how their removal leads to a net volumetric contraction, or drying shrinkage, of the paste. Even though shrinkage is a volumetric effect, it is normally measured in the laboratory or on structural elements by determination of length change, and it is therefore expressed as a linear strain.

A considerable complication in interpreting and comparing drying shrinkage measurements is that specimen size will affect the result. Water can only be lost from the surface and therefore the inner core of a specimen will act as a restraint against overall movement; the amount of restraint and hence the measured shrinkage will therefore vary with specimen size. In addition, the rate of moisture loss, and hence the rate of shrinkage, will depend on the rate of transfer of water from the core to the surface. The behaviour of HCP discussed in this section is therefore based on experimental data from specimens with a relatively small cross-section.

A schematic illustration of typical shrinkage behaviour is shown in *Fig. 20.2*. Maximum shrinkage

occurs on the first drying and a considerable part of this is irreversible, i.e. is not recovered on subsequent rewetting. Further drying and wetting cycles result in more or less completely reversible shrinkage; hence there is an important distinction between reversible and irreversible shrinkage.

Also shown in *Fig. 20.2* is a continuous, but relatively small, swelling of the HCP on continuous immersion in water. The water content first increases to make up for the self-desiccation during hydration (see Chapter 13, section 13.4), and to keep the paste saturated. Secondly, additional water is drawn into the C-S-H structure to cause the net increase in volume. This is a characteristic of many gels, but in HCP the expansion is resisted by the skeletal structure, so the swelling is small compared to the drying shrinkage strains.

In principle, the stronger the HCP structure, the less it will respond to the forces of swelling or shrinkage. This is confirmed by the results shown in *Fig. 20.3*, in which the increasing total porosity of the paste is, in effect, causing a decrease in strength. It is interesting that the reversible shrinkage appears to be independent of porosity, and the overall trend of increased shrinkage on first drying is entirely due to the irreversible shrinkage.

The variations in porosity shown in *Fig. 20.3* were obtained by testing pastes of different water:cement ratios, and in general an increased water:cement ratio will result in increasing shrinkage. As we have seen in Chapter 13, reduction in porosity also results from greater degrees of hydration of pastes with the same water:cement ratio, but the effect of the

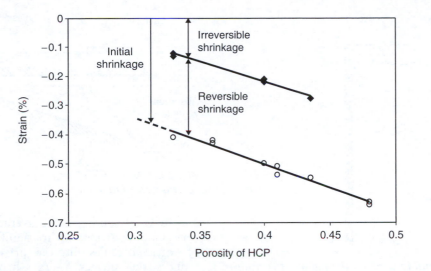

Fig. 20.3 *Reversible and irreversible shrinkage of HCP after drying at 47% relative humidity (after Helmuth and Turk, 1967).*

degree of hydration on shrinkage is not so simple. The obvious effect should be that of reduced shrinkage with age of the paste if properly cured; however, the unhydrated cement grains provide some restraint to the shrinkage, and as their volume decreases with hydration, an increase in shrinkage would result. Another argument is that a more mature paste contains more water of the type whose loss causes greater shrinkage, e.g. less capillary water, and so loss of the same amount of water from such a paste would cause more shrinkage. It is thus difficult to predict the net effect of age on the shrinkage of any particular paste.

Since shrinkage results from water loss, the relationship between the two is of interest. Typical data are given in *Fig. 20.4*, which shows that there is a distinct change of slope with increased moisture losses, in this case above about 17% loss. This implies that there is more than one mechanism of shrinkage; as other tests have shown two or even three changes of slope, it is likely that in fact several mechanisms are involved.

20.1.2 MECHANISMS OF SHRINKAGE AND SWELLING

Four principal mechanisms have been proposed for shrinkage and swelling in cement pastes, which are now summarised.

Capillary tension
Free water surfaces in the capillary and larger gel pores (section 13.4) will be in surface tension, and

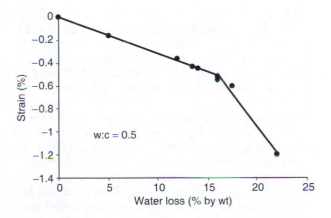

Fig. 20.4 *The effect of water loss on the drying shrinkage of hardened cement paste (after Verbeck and Helmuth, 1968).*

when water starts to evaporate owing to a lowering of the ambient vapour pressure the free surface becomes more concave and the surface tension increases (*Fig. 20.5*). The relationship between the radius of curvature, r, of the meniscus and the corresponding vapour pressure, p, is given by Kelvin's equation:

$$\ln(p/p_0) = 2T/R\theta\rho r \qquad (20.1)$$

where p_0 is the vapour pressure over a plane surface, T is the surface tension of the liquid, R is the gas constant, θ is the absolute temperature and ρ the density of the liquid.

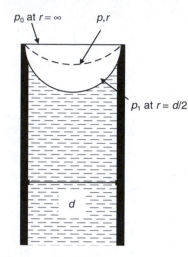

Fig. 20.5 *Relationship between the radius of curvature and vapour pressure of water in a capillary (after Soroka, 1979).*

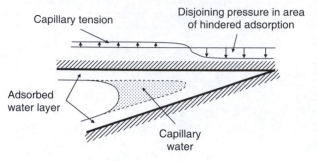

Fig. 20.6 *Water forces in a gel pore in hardened cement paste (after Bazant, 1972).*

The tension within the water near the meniscus can be shown to be $2T/r$, and this tensile stress must be balanced by compressive stresses in the surrounding solid. Hence evaporation, which causes an increase in the tensile stress, will subject the HCP solid to increased compressive stress, which will result in a decrease in volume, i.e. shrinkage. The diameter of the meniscus cannot be smaller than the diameter of the capillary, and the pore therefore empties at the corresponding vapour pressure, p_1. Hence on exposing a cement paste to a steadily decreasing vapour pressure, the pores gradually empty according to their size, the widest first. Pastes with higher water:cement ratios and higher porosities will therefore shrink more, thus explaining the general form of *Fig. 20.3*. As a pore empties, the imposed stresses on the surrounding solid reduce to zero and so full recovery of shrinkage would be expected on complete drying. Since this does not occur, it is generally accepted that other mechanisms become operative at low humidity, and that this mechanism only applies at a relative humidity above about 50%.

Surface tension or surface energy

The surface of both solid and liquid materials will be in a state of tension owing to the net attractive forces of the molecules within the material. Work therefore has to be done against this force to increase the surface area, and the surface energy is defined as the work required to increase the surface by unit area. Surface tension forces induce compressive stresses in the material of value $2T/r$ (see above)

and in HCP solids, whose average particle size is very small, these stresses are significant. Adsorption of water molecules onto the surface of the particles reduces the surface energy, hence reducing the balancing internal compressive stresses, leading to an overall volume increase, i.e. swelling. This process is also reversible.

Disjoining pressure

Figure 20.6 shows a typical gel pore, narrowing from a wider section containing free water in contact with vapour to a much narrower space between the solid, in which all the water is under the influence of surface forces. The two layers are prevented from moving apart by an inter-particle van der Waals type bond force (see Chapter 1). The adsorbed water forms a layer about five molecules or 1.3 nm thick on the solid surface at saturation, which is under pressure from the surface attractive forces. In regions narrower than twice this thickness, i.e. about 2.6 nm, the interlayer water will be in an area of hindered adsorption. This results in the development of a swelling or disjoining pressure, which is balanced by tension in the inter-particle bonds. On drying, the thickness of the adsorbed water layer reduces, as does the area of hindered adsorption, hence reducing the disjoining pressure. This results in an overall shrinkage.

Movement of interlayer water

The mechanisms described above concern the free and adsorbed water. The third type of evaporable water, the interlayer water, may also have a role. Its intimate contact with the solid surfaces and the tortuosity of its path to the open air suggest that a steep hygrometric energy gradient is needed to move it, but also that such movement is likely to result in significantly higher shrinkage than the movement of an equal amount of free or adsorbed water. It is likely that this mechanism is associated with the

Table 20.1 Summary of suggested shrinkage mechanisms (after Soroka, 1979)

Mechanism	Author(s)	Range of relative humidity (%)
Capillary tension	Powers (1965)	60–100
	Ishai (1965)	40–100
	Feldman and Sereda (1970)	30–100
Surface energy	Ishai (1965)	0–40
	Feldman and Sereda (1970)	30–100
	Wittman (1968)	0–40
Disjoining pressure	Powers (1965)	0–100
	Wittman (1968)	40–100
Interlayer water	Feldman and Sereda (1970)	0–40

steeper slope of the graph in *Fig. 20.4* at the higher values of water loss.

The above discussion applies to the reversible shrinkage only, but the reversibility depends on the assumption that there is no change in structure during the humidity cycle. This is highly unlikely, at least during the first cycle, because:

- the first cycle opens up interconnections between previously unconnected capillaries, thereby reducing the area for action of subsequent capillary tension effects
- some new inter-particle bonds will form between surfaces that move closer together as a result of movement of adsorbed or interlayer water, resulting in a more consolidated structure and a decreased total system energy.

Opinion is divided on the relative importance of the above mechanisms and their relative contribution to the total shrinkage. These differences of opinion are clear from *Table 20.1*, which shows the mechanisms proposed by five authors and the suggested humidity levels over which they act.

20.1.3 DRYING SHRINKAGE OF CONCRETE

Effect of mix constituents and proportions

The drying shrinkage of concrete is less than that of neat cement paste because of the restraining influence of the aggregate which, apart from a few exceptions, is dimensionally stable under changing moisture states.

The effect of aggregate content is shown in *Fig. 20.7*. It is apparent that normal concretes have a shrinkage of some 5 to 20% of that of neat paste. Aggregate stiffness will also have an effect. Normal-density aggregates are stiffer and therefore give more restraint than lightweight aggregates, and therefore

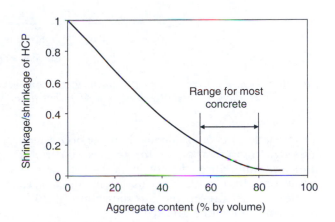

Fig. 20.7 Effect of aggregate content of concrete on the shrinkage of concrete compared to that of cement paste (after Pickett, 1956).

lightweight aggregate concretes will tend to have a higher shrinkage than normal-density concretes of similar volumetric mix proportions.

The combined effect of aggregate content and stiffness is contained in the empirical equation:

$$\varepsilon_c/\varepsilon_p = (1 - g)^n \qquad (20.2)$$

where ε_c and ε_p are the shrinkage strains of the concrete and paste, respectively, g is the aggregate volume content, and n is a constant that depends on the aggregate stiffness, and has been found to vary between 1.2 and 1.7.

The overall pattern of the effect of mix proportions on the shrinkage of concrete is shown in *Fig. 20.8*; the separate effects of increased shrinkage with increasing water content and increasing water:cement ratio can be identified.

The properties and composition of the cement and the incorporation of fly ash, ggbs and microsilica

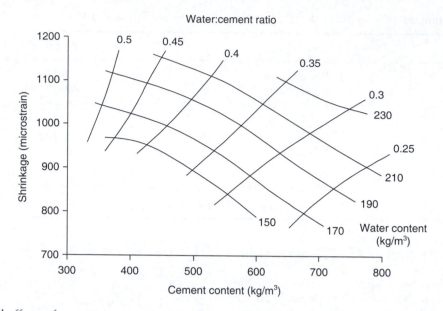

Fig. 20.8 *Typical effects of cement content, water content and water:cement ratio on shrinkage of concrete – moist curing for 28 days followed by drying for 450 days (after Shoya, 1979).*

all have little effect on the drying shrinkage of concrete, although interpretation of the data is sometimes difficult. Admixtures do not in themselves have a significant effect, but if their use results in changes in the mix proportions then, as shown in *Fig. 20.8*, the shrinkage will be affected.

Effect of specimen geometry

The size and shape of a concrete specimen will influence the rate of moisture loss and the degree of overall restraint provided by the central core, which will have a higher moisture content than the surface region. The rate and amount of shrinkage and the tendency for the surface zones to crack are therefore affected.

In particular, longer moisture diffusion paths lead to lower shrinkage rates. For example, a member with a large surface area to volume ratio, e.g. a T-beam, will dry and therefore shrink more rapidly than, say, a beam with a square cross-section of the same area. In all cases, however, the shrinkage process is protracted. In a study lasting 20 years, Troxell *et al.* (1958) found that in tests on 300 × 150 mm diameter cylinders made from a wide range of concrete mixes and stored at relative humidities of 50 and 70%, an average of only 25% of the 20-year shrinkage occurred in the first two weeks, 60% in three months, and 80% in one year.

Non-uniform drying and shrinkage in a structural member will result in differential strains and hence shrinkage-induced stresses – tensile near the surface and compressive in the centre. The tensile stresses may be sufficient to cause cracking, which is the most serious consequence for structural behaviour and integrity. However, as discussed above, the effects in practice occur over protracted timescales, and the stresses are relieved by creep before cracking occurs. The structural behaviour is therefore complex and difficult to analyse with any degree of rigour.

20.1.4 PREDICTION OF SHRINKAGE

It is clear from the above discussion that although much is known about shrinkage and the factors that influence its magnitude, it is difficult to estimate its value in a structural situation with any degree of certainty. It has been shown that it is possible to obtain reasonable estimates of long-term shrinkage from short-term tests (Neville *et al.*, 1983), but designers often require estimates long before results from even short-term tests can be obtained. There are a number of methods of varying degrees of complexity for this, often included in design codes, e.g. Eurocode 2 (BS EN 1992) or ACI (2000), all of which are based on the analysis and interpretation of extensive experimental data.

20.2 Autogenous shrinkage

Continued hydration with an adequate supply of water leads to slight swelling of cement paste, as shown in *Fig. 20.2*. Conversely, with no moisture movement to or from the cement paste, self-desiccation leads to removal of water from the capillary pores (as described in section 13.4) and *autogenous shrinkage*. Most of this shrinkage occurs when the hydration reactions are proceeding most rapidly, i.e. in the first few days after casting. Its magnitude is normally at least an order of magnitude less than that of drying shrinkage, but it is higher and more significant in higher-strength concrete with very low water:cement ratios. It may be the only form of shrinkage occurring in the centre of a large mass of concrete, and can lead to internal cracking if the outer regions have an adequate supply of external water, e.g. from curing.

20.3 Carbonation shrinkage

Carbonation shrinkage differs from drying shrinkage in that its cause is chemical and it does not result from loss of water from the HCP or concrete. Carbon dioxide, when combined with water as carbonic acid, reacts with many of the components of the HCP, and even the very dilute carbonic acid resulting from the low concentrations of carbon dioxide in the atmosphere can have significant effects. The most important reaction is that with the portlandite (calcium hydroxide):

$$CO_2 + Ca(OH)_2 \rightarrow CaCO_3 + H_2O \quad (20.3)$$

Thus water is released and there is an increase in weight of the paste. There is an accompanying shrinkage, and the paste also increases in strength and decreases in permeability. The most likely mechanism to explain this behaviour is that the calcium hydroxide is dissolved from more highly stressed regions, resulting in shrinkage, and the calcium carbonate crystallises out in the pores, thus reducing the permeability and increasing the strength.

The rate and amount of carbonation depend in part on the relative humidity of the surrounding air and within the concrete. If the pores are saturated, then the carbonic acid will not penetrate the concrete, and no carbonation will occur; if the concrete is dry, then no carbonic acid is available. Maximum carbonation shrinkage occurs at a humidity of about 50% and it can be of the same order of magnitude as drying shrinkage (*Fig. 20.9*). The porosity of the concrete is also an important controlling factor. With average-strength concrete, provided it is well compacted and cured, the carbonation front will only penetrate a few centimetres in many years, and with high-strength concrete even less. However, much greater penetration can occur with poor quality

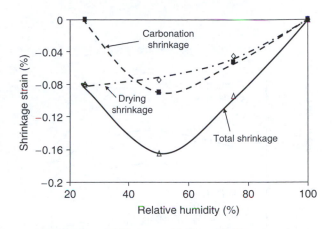

Fig. 20.9 *The effect of surrounding relative humidity on drying and carbonation shrinkage of mortar (after Verbeck, 1958).*

concrete or in regions of poor compaction, and this can lead to substantial problems if the concrete is reinforced, as we shall see in Chapter 24.

20.4 Thermal expansion

In common with most other materials, cement paste and concrete expand on heating. Knowledge of the coefficient of thermal expansion is needed in two main situations: firstly to calculate stresses due to thermal gradients arising from heat of hydration effects or continuously varying diurnal temperatures, and secondly to calculate overall dimensional changes in structures such as bridge decks due to variations in ambient temperature.

The measurement of thermal expansions on laboratory specimens is relatively straightforward, provided sufficient time is allowed for thermal equilibrium to be reached (at most a few hours). However, the *in-situ* behaviour is complicated by differential movement from non-uniform temperature changes in large members resulting in time-dependent thermal stresses; as with shrinkage, it is therefore difficult to extrapolate movement in structural elements from that on laboratory specimens.

20.4.1 THERMAL EXPANSION OF HARDENED CEMENT PASTE

The coefficient of thermal expansion of HCP varies between about 10 and 20×10^{-6} per °C, depending mainly on the moisture content. *Figure 20.10* shows typical behaviour, with the coefficient reaching a maximum at about 70% relative humidity. The value at 100% relative humidity, i.e. about 10×10^{-6}

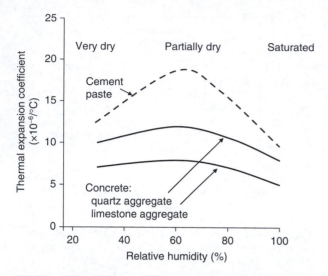

Fig. 20.10 The effect of dryness on the thermal expansion coefficient of hardened cement paste and concrete (after Meyers, 1950).

per °C, probably represents the 'true' inherent value for the paste itself. The behaviour does, however, show some time dependence, with the initial expansion on an increase in temperature showing some reduction over a few hours if the temperature is held constant.

Explanations for this behaviour have all involved the role of water, and relate to the disturbance of the equilibrium between the water vapour, the free water, the freely adsorbed water, the water in areas of hindered adsorption and the forces between the layers of gel solids (section 13.6). Any disturbances will have a greater effect at intermediate humidities, when there is a substantial amount of water present with space in which to move. On an increase in temperature, the surface tension of the capillary water will decrease and hence its internal tension and the corresponding compression in the solid phases will decrease, causing extra swelling, as observed. However, changes in internal energy with increased or decreased temperature will stimulate internal flow of water, causing the time-dependent volume change in the opposite sense to the initial thermal movement mentioned above.

20.4.2 THERMAL EXPANSION OF CONCRETE

The thermal expansion coefficients of the most common types of rock used for concrete aggregates vary between about 6 and 10×10^{-6} per °C, i.e. lower than either the 'true' or 'apparent' values for cement paste. The thermal expansion coefficient of concrete is therefore lower than that of cement paste, as shown in *Fig. 20.10*. Furthermore, since the aggregate occupies 70–80% of the total concrete volume, there is a considerable reduction of the effects of humidity that are observed in the paste alone, to the extent that a constant coefficient of thermal expansion over all humidities is a reasonable approximation. The value depends on the concrete mix proportions, chiefly the cement paste content, and the aggregate type; for normal mixes the latter tends to dominate. The curves for quartz and limestone aggregate concrete shown in *Fig. 20.10* represent the two extremes of values for most normal aggregate concrete. Such values apply over a temperature range of about 0 to 60°C. At higher temperatures, the differential stresses set up by the different thermal expansion coefficients of the paste and aggregate can lead to internal microcracking and hence non-linear behaviour. We shall discuss this further when considering fire damage in Chapter 24.

20.5 Stress–strain behaviour

20.5.1 ELASTICITY OF THE HARDENED CEMENT PASTE

HCP has a near linear compressive stress–strain relationship for most of its range and therefore a modulus of elasticity can readily be determined from stress–strain data. Water-saturated pastes generally have a slightly higher modulus than dried pastes, indicating that some of the load is carried by the water in the pores. Nevertheless, the skeletal lattice of the paste carries most of the load, and the elastic response is governed by the lattice properties. As might therefore be expected, the elastic modulus (E_p) is highly dependent on the capillary porosity (p_c); the relationship has been found to be of the form:

$$E_p = E_g(1 - p_c)^3 \qquad (20.4)$$

where E_g is the modulus when $p_c = 0$, i.e. it represents the modulus of elasticity of the gel itself. This is a similar expression to equation (13.10) for the strength of the paste, and therefore it is to be expected that the same factors will influence both strength and modulus. This is indeed the case; for example, *Fig. 20.11* shows that a decreasing water:cement ratio and increasing age both increase the elastic modulus, an effect directly comparable to that on strength shown in *Fig. 13.9*.

20.5.2 MODELS FOR CONCRETE BEHAVIOUR

Concrete is, of course, a composite multiphase material, and its elastic behaviour will depend on

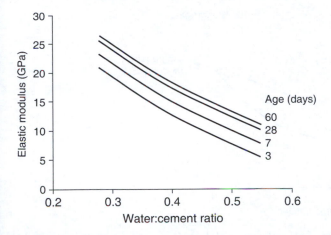

Fig. 20.11 *Effect of water:cement ratio and age on the elastic modulus of hardened cement paste (after Hirsch, 1962).*

the elastic properties of the individual phases – unhydrated cement, cement gel, water, coarse and fine aggregate and their relative proportions and geometrical arrangements. The real material is too complex for rigorous analysis, but if it is considered as a two-phase composite consisting of HCP and aggregate, then analysis becomes possible and instructive.

Models for the behaviour of concrete require the following:

1. The property values for the phases; in this simple analysis, three are sufficient:

   * the elastic modulus of the aggregate, E_a
   * the elastic modulus of the HCP, E_p
   * the volume concentration of the aggregate, g.

2. A suitable geometrical arrangement of the phases; three possibilities are shown in *Fig. 20.12*. All

the models consist of unit cubes. Models A and B have the phases arranged as adjacent layers, the difference being that in A the two phases are in parallel, and therefore undergo the same strain, whereas in B the phases are in series and are therefore subjected to the same stress. Model C has the aggregate set within the paste such that its height and base area are both equal to √g, thus complying with the volume requirements. This intuitively is more satisfactory in that it bears a greater resemblance to concrete.

Analysis of the models is not intended to give any detail of the actual distribution of stresses and strains within concrete, but to predict average or overall behaviour. Three further assumptions are necessary:

1. The applied stress remains uniaxial and compressive throughout the model.
2. The effects of lateral continuity between the layers can be ignored.
3. Any local bond failure or crushing does not contribute to the deformation.

Model A – phases in parallel

Strain compatibility. The strain in the concrete, ε_c is equal to the strain in the aggregate, ε_a, and the paste, ε_p, i.e.

$$\varepsilon_c = \varepsilon_a = \varepsilon_p \qquad (20.5)$$

Equilibrium. The total force is the sum of the forces on each of the phases. Expressed in terms of stresses and areas this gives:

$$\sigma_c.1 = \sigma_a.g + \sigma_p.(1 - g) \qquad (20.6)$$

Constitutive relations. Both of the phases and the concrete are elastic, hence:

$$\sigma_c = \varepsilon_c.E_c \quad \sigma_a = \varepsilon_a.E_a \quad \text{and} \quad \sigma_p = \varepsilon_p.E_p \qquad (20.7)$$

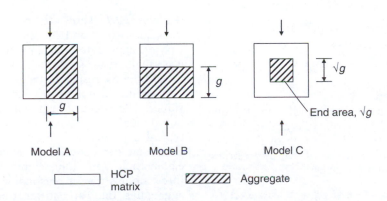

Fig. 20.12 *Simple two-phase models for concrete (after Hansen, 1960; Counto, 1964).*

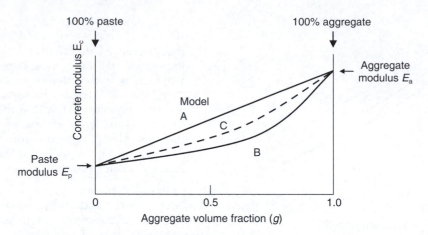

Fig. 20.13 *The effect of volume concentration of aggregate on the elastic modulus of concrete calculated from the simple two-phase models of* Fig. 20.12.

where E_c, E_a and E_p are the elastic moduli of the concrete, aggregate and paste, respectively.

Substituting into equation (20.6) from equation (20.7) gives:

$$\varepsilon_c.E_c = \varepsilon_a.E_a.g + \varepsilon_p.E_p.(1 - g)$$

and hence, from equation (20.5)

$$E_c = E_a.g + E_p.(1 - g). \qquad (20.8)$$

Model B – phases in series

Equilibrium. The forces and hence the stresses (since the forces act on equal areas) in both phases and the composite are equal, i.e.

$$\sigma_c = \sigma_a = \sigma_p \qquad (20.9)$$

Strains. The total displacement is the sum of the displacements in each of the phases; expressed in terms of strain this gives:

$$\varepsilon_c = \varepsilon_a.g + \varepsilon_p.(1 - g) \qquad (20.10)$$

Substituting from equations (20.7) and (20.9) into (20.10) and rearranging gives:

$$1/E_c = g/E_a + (1 - g)/E_p \qquad (20.11)$$

Model C – combined

This is a combination of two layers of HCP alone in series with a third layer of HCP and aggregate in parallel, as in model A. Repetition of the above two analyses with substitution of the appropriate geometry and combination gives:

$$1/E_c = (1 - \sqrt{g})/E_p + \sqrt{g}/(E_a.g + E_p[1 - \sqrt{g}]) \qquad (20.12)$$

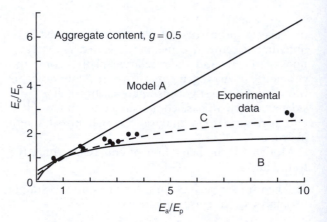

Fig. 20.14 *Prediction of the elastic modulus of concrete* (E_c) *from the moduli of the cement paste* (E_p) *and the aggregate* (E_a) *for 50% volume concentration of the aggregate.*

Figure 20.13 shows the predicted results of equations 20.8, 20.11 and 20.12 from the three models for varying aggregate concentrations, with $E_p < E_a$ as is normally the case. Models A and B give upper and lower bounds, respectively, to the concrete modulus, with model C, not surprisingly, giving intermediate values. The effect of aggregate stiffness is shown in non-dimensional form in *Fig. 20.14*, on which some typical experimental results are also plotted. It is clear that for concrete in which E_a/E_p is near 1, e.g. with low-modulus lightweight aggregates, all three models give a reasonable fit, but for normal aggregates that are stiffer than the paste – i.e. $E_a/E_p > 1$ – model C is preferable.

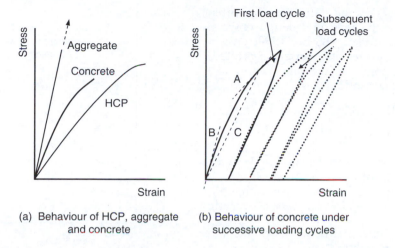

(a) Behaviour of HCP, aggregate and concrete

(b) Behaviour of concrete under successive loading cycles

Fig. 20.15 *Stress–strain behaviour of cement paste, aggregate and concrete.*

20.5.3 MEASURED STRESS–STRAIN BEHAVIOUR OF CONCRETE

The stress–strain behaviour of both HCP and aggregate is substantially linear over most of the range up to a maximum. However, that of the composite concrete, although showing intermediate stiffness as predicted from the above analysis, is markedly non-linear over much of its length, as shown in *Fig. 20.15a*. Furthermore, successive unloading/loading cycles to stress levels below ultimate show substantial, but diminishing, hysteresis loops, and residual strains at zero load, as in *Fig. 20.15b*.

The explanation for this behaviour lies in the contribution of microcracking to the overall concrete strain, i.e. assumption (3) in the above analysis is invalid. As we will see in Chapter 21, the transition zone between the aggregate and the HCP or mortar is a region of relative weakness, and in fact some microcracks will be present in this zone even before loading. The number and width of these will depend on such factors as the bleeding characteristics of the concrete immediately after placing and the amount of drying or thermal shrinkage. As the stress level increases, these cracks will increase in length, width and number, thereby making a progressively increasing contribution to the overall strain, resulting in non-linear behaviour. Cracking eventually leads to complete breakdown and failure, therefore we will postpone more detailed discussion of cracking until the next chapter.

Subsequent cycles of loading will not tend to produce or propagate as many cracks as the initial loading, provided the stress levels of the first or previous cycles are not exceeded. This explains the diminishing size of the hysteresis loops shown in *Fig. 20.15b*.

20.5.4 ELASTIC MODULUS OF CONCRETE

The non-linear stress–strain curve of concrete means that a number of different elastic modulus values can be defined. These include the slope of the tangent to the curve at any point (giving the *tangent modulus*, A or B in *Fig. 20.15b*) or the slope of the line between the origin and a point on the curve (giving the *secant modulus*, C in *Fig. 20.15b*).

A typical test involves loading to a working stress, say 40% of ultimate, and measuring the corresponding strain. Cylindrical or prism specimens are usually used, loaded longitudinally, and with a length at least twice the lateral dimension. Strain measurements are usually taken over the central section of the specimen to avoid end effects. To minimise hysteresis effects, the specimens are normally subjected to a few cycles of loading before the strain readings are taken over a load cycle lasting about five minutes. It is usual to calculate the secant modulus from these readings. This test is often called the static test – and the resulting modulus the static modulus – to distinguish it from the dynamic modulus test, which we will describe in Chapter 23.

The elastic modulus increases with age and decreasing water:cement ratio of the concrete, for the reasons outlined above and, as with paste, these two factors combine to give an increase of modulus with compressive strength, but with progressively smaller increases at higher strength. However, there is no simple relationship between strength and modulus since, as we have seen, the aggregate modulus and its volumetric concentration, which can vary at constant concrete strength, also have an effect. The modulus should therefore be determined experimentally if its value is required with any certainty. This is not always possible, and estimates are often needed, e.g. early in the structural design process; for example the Eurocode 2 (BS EN 1992) gives a table of values for concrete with quartzite aggregates and compressive strengths in the range 20 to 110 MPa, which have the relationships:

$$E_c = 10 f_{cube}^{0.31} \qquad (20.13)$$

and

$$E_c = 11 f_{cyl}^{0.3} \qquad (20.14)$$

where E_c is the secant modulus of elasticity between zero and 40% of the ultimate stress;
f_{cube} is the mean cube compressive strength;
and f_{cyl} is the mean cylinder compressive strength.

[We will discuss the difference between cylinder and cube compressive strength in the next chapter.] For concrete with limestone and sandstone aggregates the values derived from these expressions should be reduced by 10 and 30%, respectively, and for basalt aggregates they should be increased by 20%.

20.5.5 POISSON'S RATIO

The Poisson's ratio of water-saturated cement paste varies between 0.25 and 0.3; on drying it reduces to about 0.2. It seems to be largely independent of the water:cement ratio, age and strength. For concrete, the addition of aggregate again modifies the behaviour, lower values being obtained with increasing aggregate content. For most concrete, values lie within the range 0.17–0.2.

20.6 Creep

The general nature of the creep behaviour of concrete was illustrated in *Fig. 20.1*. The magnitude of the creep strains can be higher than the elastic strains on loading, and they therefore often have a highly significant influence on structural behaviour. Also, creep does not appear to tend to a limit, as shown in *Fig. 20.16* for tests of more than 20 years duration. This figure also shows that creep is substantially increased when the concrete is simultaneously drying, i.e. creep and shrinkage are interdependent. This leads to the definitions of creep strains shown in *Fig. 20.17*. Free shrinkage (ε_{sh}) is defined as the shrinkage of the unloaded concrete in the drying condition, and basic creep (ε_{bc}) as the creep of a similar specimen under load but not drying, i.e. sealed so that there is no moisture movement to or from the surrounding environment. The total strain (ε_{tot}) is that measured on the concrete while simultaneously shrinking and creeping and, as shown in *Fig. 20.17*, it is found that:

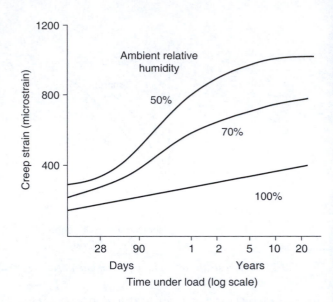

Fig. 20.16 *Creep of concrete moist-cured for 28 days, then loaded and stored at different relative humidities (after Troxell et al., 1958).*

$$\varepsilon_{tot} > \varepsilon_{sh} + \varepsilon_{bc} \qquad (20.15)$$

The difference, i.e. $\varepsilon_{tot} - (\varepsilon_{sh} + \varepsilon_{bc})$, is called the drying creep (ε_{dc}). It follows that the total creep strain (ε_{cr}) is given by:

$$\varepsilon_{cr} = \varepsilon_{dc} + \varepsilon_{bc} \qquad (20.16)$$

It also follows that the total creep of a specimen or structural member will be dependent on its size, since this will affect the rate and uniformity of drying.

20.6.1 FACTORS INFLUENCING CREEP

Apart from the increase in creep with simultaneous shrinkage just described, the following factors have a significant effect on creep.

- A reduced moisture content before loading, which reduces creep. In fact, completely dried concrete has very small, perhaps zero, creep.
- The level of applied stress; for any given concrete and loading conditions, the creep is found to increase approximately linearly with the applied stress up to stress:strength ratios of about 0.4–0.6 (different studies have indicated different limits). It is therefore often useful to define the *specific creep* as the creep strain per unit stress in this region. At higher stress levels increased creep is observed, which can ultimately result in failure, as will be discussed in the next chapter.
- Increasing concrete strength, which decreases the creep.

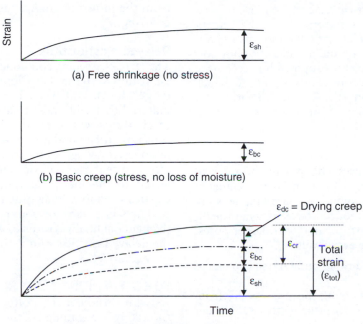

Fig. 20.17 *Definitions of strains due to shrinkage, creep and combined shrinkage and creep of hardened cement paste and concrete.*

- Increasing temperature, which increases the creep significantly for temperatures up to about 70°C. Above this, moisture migration effects lead to lower creep.
- The aggregate volume concentration, illustrated in *Fig. 20.18*, which shows that the aggregate is inert as regards creep, and hence the creep of concrete is less than that of cement paste. This is therefore directly comparable to the shrinkage behaviour shown in *Fig. 20.7*.

Neville (1964) suggested a relationship between the creep of concrete (C_c) and that of neat cement paste (C_p) of the form:

$$C_c/C_p = (1 - g - u)^n \qquad (20.17)$$

where g and u are the volume fractions of aggregate and unhydrated cement, respectively, and n is a constant that depends on the modulus of elasticity and Poisson's ratio of the aggregate and the concrete. This therefore shows that:

- the properties of the aggregate are important, and they can have a substantial effect of the magnitude of the creep
- the effect of the water:cement ratio and age of the concrete need not be considered separately, since they both affect the elastic modulus

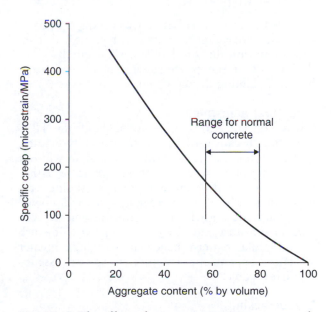

Fig. 20.18 *The effect of aggregate content on creep of concrete (after Concrete Society, 1973).*

- the effect of other materials that also affect the rate of gain of strength, such as admixtures and cement replacement materials, can be treated similarly.

145

20.6.2 MECHANISMS OF CREEP

Since the creep process occurs within the cement paste, and the moisture content and movement have a significant effect on its magnitude, it is not surprising that the mechanisms proposed for creep have similarities with those proposed for shrinkage, which we discussed in section 20.1.2. As with shrinkage, it is likely that a combination of the mechanisms now outlined is responsible.

Moisture diffusion

The applied stress causes changes in the internal stresses and strain energy within the HCP, resulting in an upset to the thermodynamic equilibrium; moisture then moves down the induced free-energy gradient, implying a movement from smaller to larger pores, which can occur at several levels:

- in capillary water as a rapid and reversible pressure drop
- in adsorbed water moving more gradually from zones of hindered adsorption – this movement should be reversible
- in interlayer water diffusing very slowly out of the gel pores. Some extra bonding may then develop between the solid layers, so this process may not be completely reversible.

In sealed concrete there are always enough voids to allow the movement of moisture, hence basic creep can occur with this mechanism. With simultaneous drying, all of the processes are much enhanced, hence explaining drying creep.

Structural adjustment

Stress concentrations arise throughout the HCP structure because of its heterogeneous nature, and consolidation to a more stable state without loss of strength occurs at these points by either viscous flow, with adjacent particles sliding past each other, or local bond breakage, closely followed by reconnection nearby after some movement. Concurrent moisture movement is assumed to disturb the molecular pattern, hence encouraging a greater structural adjustment. The mechanisms are essentially irreversible.

Microcracking

We have seen that HCP and concrete contain defects and cracks before loading, and propagation of these and the formation of new cracks will contribute to the creep strain, particularly at higher levels of stress. This is the most likely explanation of the non-linearity of creep strain with stress at high stress levels. In a drying concrete, the stress gradient arising from the moisture gradient is likely to enhance the cracking.

Delayed elastic strain

The 'active' creeping component of HCP or concrete, i.e. mainly water in its various forms in the capillary or gel pores, will be acting in parallel with inert material that will undergo an elastic response only. In HCP this will be solid gel particles, unhydrated cement particles and portlandite crystals, augmented in concrete by aggregate particles. The stress in the creeping material will decline as the load is transferred to the inert material, which then deforms elastically as its stress gradually increases. The process acts in reverse on removal of the load, so that the material finally returns to its unstressed state; thus the delayed elastic strain would be fully recoverable in this model.

20.6.3 PREDICTION OF CREEP

As with shrinkage, it is often necessary to estimate the likely magnitude of the creep of a structural element at the design stage but, again, because of the number of factors involved, prediction of creep with a degree of certainty is problematic. Brooks and Neville (1978) have suggested that a satisfactory method is to carry out short-term (28-day) tests, and then estimate creep at a later age by extrapolation using the expressions:

$$\text{basic creep} \quad c_t = c_{28} \times 0.5 t^{0.21} \qquad (20.18)$$

$$\text{total creep} \quad c_t = c_{28} \times (-6.19 + 2.15 \log_e t)^{0.38} \qquad (20.19)$$

where t = age at which creep is required (days, > 28)

c_{28} = measured specific creep at 28 days

c_t = specific creep at t days in microstrain per MPa.

If short-term tests are not feasible, there are, as for shrinkage, a number of empirical methods of varying degrees of complexity for estimating creep, often included in design codes, e.g. Eurocode 2 (BS EN 1992) and ACI (2000).

References

ACI (2000). ACI 209R-92. Prediction of creep, shrinkage and temperature effects in concrete structures, *ACI Manual of Concrete Practice, Part 1: Materials and General Properties of Concrete*, American Concrete Institute, Michigan, USA.

Bazant ZP (1972). Thermodynamics of hindered adsorption and its implications for hardened cement paste and concrete. *Cement and Concrete Research*, **2** (No. 1), 1–16.

Brooks JJ and Neville AM (1978). Predicting the long-term creep and shrinkage from short-term tests. *Magazine of Concrete Research*, **30** (No. 103), 51–61.

Concrete Society (1973). *The Creep of Structural Concrete*, Technical Report No. 101, London.

Counto UJ (1964). The effect of the elastic modulus of aggregate on the elastic modulus, creep and creep recovery of concrete. *Magazine of Concrete Research*, **16** (No. 48), 129–138.

Feldman RF and Sereda PJ (1970). A new model for hydrated Portland cement and its practical implications. *Engineering Journal*, **53** (No. 8/9), 53–59.

Hansen TC (1960). Creep and stress relaxation of concrete. *Swedish Cement and Concrete Research Institute*, Report No. 31, p. 112.

Helmuth RA and Turk DM (1967). The reversible and irreversible drying shrinkage of hardened Portland cement and tricalcium silicate pastes. *Journal of the Portland Cement Association, Research and Development Laboratories*, **9** (No. 2), 8–21.

Hirsch TJ (1962). Modulus of elasticity of concrete as affected by elastic moduli of cement paste matrix and aggregate. *Proceedings of the American Concrete Institute*, **59**, (No. 3), 427–452.

Ishai O (1965). *The time-dependent deformational behaviour of cement paste, mortar and concrete*. Proceedings of the Conference on Structure of Concrete and Its Behaviour Under Load, Cement and Concrete Association, London, September, pp. 345–364.

Meyers SL (1950). Thermal expansion characteristics of hardened cement paste and concrete. *Proceedings of the Highway Research Board*, **30**, 193–200.

Neville AM (1964). Creep of concrete as a function of its cement paste content. *Magazine of Concrete Research*, **16** (No. 46), 21–30.

Neville AM, Dilger WH and Brooks JJ (1983). *Creep of plain and structural concrete*, Construction Press, London.

Pickett G (1956). Effect of aggregate on shrinkage of concrete and hypothesis concerning shrinkage. *Journal of the American Concrete Institute*, **52**, 581–590.

Powers TC (1965). *Mechanisms of shrinkage and reversible creep of hardened cement paste*. Proceedings of the Conference on Structure of Concrete and Its Behaviour Under Load, Cement and Concrete Association, London, September, pp. 319–344.

Shoya M (1979). Drying shrinkage and moisture loss of superplasticizer admixed concrete of low water/cement ratio. *Transactions of the Japan Concrete Institute*, II-5, 103–110.

Soroka l (1979). *Portland Cement Paste and Concrete*, Macmillan, London.

Troxell GE, Raphael JM and Davis RE (1958). Long-time creep and shrinkage tests of plain and reinforced concrete. *Proceedings of the American Society for Testing and Materials*, **58**, 1101–1120.

Verbeck GJ (1958). Carbonation of hydrated Portland cement, *American Society for Testing and Materials Special Publication*, No. 205, 17–36.

Verbeck GJ and Helmuth RA (1968). *Structure and physical properties of cement paste*. Proceedings of the Symposium on the Chemistry of Cement, Tokyo, **3**, pp. 1–37.

Wittman FH (1968). Surface tension, shrinkage and strength of hardened cement paste. *Materials and Structures*, **1** (No. 6), 547–552.

Chapter 21

Strength and failure of concrete

Strength is probably the most important single property of concrete, since the first consideration in structural design is that the structural elements must be capable of carrying the imposed loads. The maximum value of stress in a loading test is usually taken as the strength, even though under compressive loading the test piece is still whole (but with substantial internal cracking) at this stress, and complete breakdown subsequently occurs at higher strains and lower stresses. Strength is also important because it is related to several other important properties that are more difficult to measure directly, and a simple strength test can give an indication of these properties. For example, we have already seen the relation of strength to elastic modulus; we shall discuss durability in Chapter 24, but in many cases a low-permeability, low-porosity concrete is the most durable and, as discussed when we considered the strength of cement paste in Chapter 13, this also means that it has high strength.

We are primarily concerned with compressive strength since the tensile strength is very low, and in concrete structural elements reinforcement is used to carry the tensile stresses. However, in many structural situations concrete may be subject to one of a variety of types of loading, resulting in different stress conditions and different potential modes of failure, and so knowledge of the relevant strength is therefore important. For example, in columns or reinforced concrete beams, compressive strength is required; for cracking of a concrete slab the tensile strength is important. Other situations may require torsional strength, fatigue or impact strength or strength under multiaxial loading. As we shall see, most strength testing involves the use of a few, relatively simple tests, generally not related to a particular structural situation. Procedures enabling data from the tests described in this chapter to be used in design have been obtained from empirical test programmes at an engineering scale on large specimens. You should refer to texts on

structural design for a description of these design procedures.

In this chapter we shall describe the most common test methods used to assess the strength of concrete and then discuss the factors influencing the results obtained from them. We follow this with a more detailed consideration of the cracking and fracture processes taking place within concrete. Finally, we shall briefly discuss strength under multiaxial loading conditions.

21.1 Strength tests

21.1.1 COMPRESSIVE STRENGTH

The simplest compressive strength test uses a concrete cube, and this is the standard test in the UK and many other countries. The cube must be sufficiently large to ensure that an individual aggregate particle does not unduly influence the result; 100 mm is recommended for maximum aggregate sizes of 20 mm or less, 150 mm for maximum sizes up to 40 mm. The cubes are usually cast in lubricated steel moulds, accurately machined to ensure that opposite faces are smooth and parallel. The concrete is fully compacted by external vibration or hand tamping, and the top surface trowelled smooth. After demoulding when set, the cube is normally cured under water at constant temperature until testing.

The cube-testing machine has two heavy platens through which the load is applied to the concrete. The bottom one is fixed and the upper one has a ball-seating that allows rotation to match the top face of the cube at the start of loading. This then locks in this position during the test. The load is applied to a pair of faces that were cast against the mould, i.e. with the trowelled face to one side. This ensures that there are no local stress concentrations, which would result in a falsely low average failure stress. A very fast rate of loading gives strengths

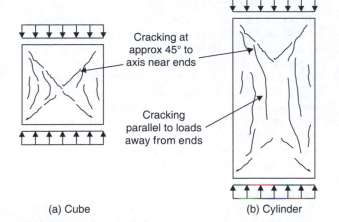

(a) Cube   (b) Cylinder

Fig. 21.1 Cracking patterns during testing of concrete specimens in compression.

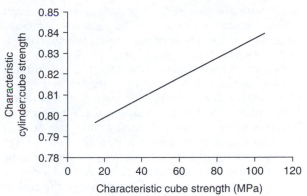

Fig. 21.2 Variation of cylinder/cube strength ratio with strength (from Eurocode 2 strength classes (BS EN 1992)).

that are too high, and a rate to reach ultimate in a few minutes is recommended. It is vital that the cube is properly made and stored; only then will the test give a true indication of the properties of the concrete, unaffected by such factors as poor compaction, drying shrinkage cracking, etc.

The cracking pattern within the cube (*Fig. 21.1a*) produces a double pyramid shape after failure. From this it is immediately apparent that the stress within the cube is far from uniaxial. The compressive load induces lateral tensile strains in both the steel platens and the concrete owing to the Poisson effect. The mismatch between the elastic modulus of the steel and the concrete and the friction between the two results in lateral restraint forces in the concrete near the platen, partially restraining it against outward expansion. This concrete is therefore in a triaxial stress state, with consequent higher failure stress than the true, unrestrained strength. This is the major objection to the cube test. The test is, however, relatively simple and capable of comparing different concretes. (We shall consider triaxial stress states in more detail later in the chapter.)

An alternative test, which at least partly overcomes the restraint problem, uses cylinders; this is popular in North America, most of Europe and in many other parts of the world. Cylinders with a height:diameter ratio of 2, most commonly 300 mm high and 150 mm in diameter, are tested vertically; the effects of end restraint are much reduced over the central section of the cylinder, which fails with near uniaxial cracking (*Fig. 21.1b*), indicating that the failure stress is much closer to the unconfined compressive strength. As a rule of thumb, it is often assumed that the cylinder strength is about 20%

lower than the cube strength, but the ratio has been found to depend on several factors, and in particular, increases with increasing strength. The relationship derived from values given in the Eurocode 2 (BS EN 1992) is:

$$f_{cyl} = 0.85 f_{cube} - 1.6 \qquad (21.1)$$

where f_{cyl} = characteristic cylinder strength, and f_{cube} = characteristic cube strength (values in MPa). *Figure 21.2* shows how the ratio of the two strengths varies with strength.

A general relationship between the height:diameter ratio (*h/d*) and the strength of cylinders for low- and medium-strength concrete is shown in *Fig. 21.3*. This is useful in, for example, interpreting the results from testing cores cut from a structure, where *h/d* often cannot be controlled. It is preferable to avoid an *h/d* ratio of less than 1, where sharp increases

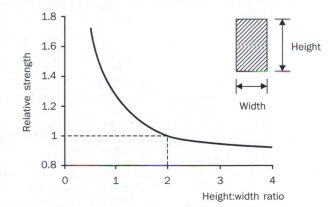

Fig. 21.3 The relationship between height:width (or diameter) ratio and strength of concrete in compression.

in strength are obtained, while high values, although giving closer estimates of the uniaxial strength, result in excessively long specimens which can fail due to slenderness ratio effects.

Testing cylinders have one major disadvantage; the top surface is finished by a trowel and is not plane and smooth enough for testing, and it therefore requires further preparation. It can be ground, but this is very time consuming, and the normal procedure is to cap it with a thin (2–3 mm) layer of high-strength gypsum plaster, molten sulphur or high-early-strength cement paste, applied a day or two in advance of the test. Alternatively, the end of the cylinder can be set in a steel cap with a bearing pad of an elastomeric material or fine dry sand between the cap and the concrete surface. Apart from the inconvenience of having to carry this out, the failure load is sensitive to the capping method, particularly in high-strength concrete.

21.1.2 TENSILE STRENGTH

Direct testing of concrete in uniaxial tension, as shown in *Fig. 21.4a*, is more difficult than for, say, steel or timber. Relatively large cross-sections are required to be representative of the concrete and, because concrete is brittle, it is difficult to grip and align. Eccentric loading and failure at or in the grips are then difficult to avoid. A number of gripping systems have been developed, but these are somewhat complex, and their use is confined to research laboratories. For more routine purposes, one of the following two indirect tests is preferred.

Splitting test

A concrete cylinder, of the type used for compression testing, is placed on its side in a compression-testing machine and loaded across its vertical diameter (*Fig. 21.4b*). The size of cylinder used is normally either 300 or 200 mm long (l) by 150 or

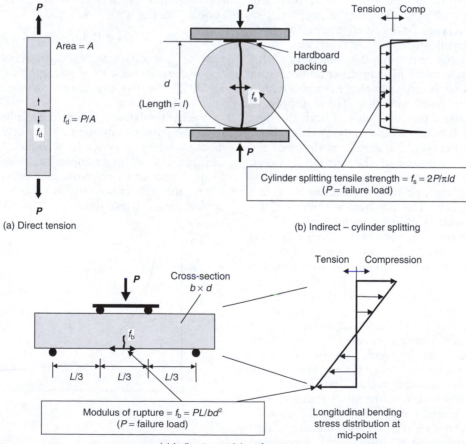

Fig. 21.4 *Tensile testing methods for concrete.*

100 mm diameter (d). The theoretical distribution of horizontal stress on the plane of the vertical diameter, also shown in *Fig. 21.4b*, is a near uniform tension (f_s), with local high compression stresses at the extremities. Hardboard or plywood strips are inserted between the cylinder and both top and bottom platens to reduce the effect of these and ensure even loading over the full length. Failure occurs by a split or crack along the vertical plane, the specimen falling into two neat halves. The cylinder splitting strength is defined as the magnitude of the near-uniform tensile stress on this plane, which is given by:

$$f_s = 2P/\pi l d \tag{21.2}$$

where P is the failure load.

The state of stress in the cylinder is biaxial rather than uniaxial (on the failure plane the vertical compressive stress is about three times higher than the horizontal tensile stress) and this, together with the local zones of compressive stress at the extremes, results in the value of f_s being higher than the uniaxial tensile strength. However, the test is very easy to perform with standard equipment used for compressive strength testing, and gives consistent results; it is therefore very useful.

Flexural test

A rectangular prism, of cross-section $b \times d$ (usually 100 or 150 mm square) is simply supported over a span L (usually 400 or 600 mm). The load is applied at the third points (*Fig. 21.4c*), and since the tensile strength of concrete is much less than the compressive strength, failure occurs when a flexural tensile crack at the bottom of the beam, normally within the constant bending moment zone between the loading points, propagates upwards through the beam. If the total load at failure is P, then analysis based on simple beam-bending theory and linear elastic stress–strain behaviour up to failure gives the stress distribution shown in *Fig. 21.4c*, with a maximum tensile stress in the concrete, f_b, as:

$$f_b = PL/bd^2 \tag{21.3}$$

f_b is known (somewhat confusingly) as the *modulus of rupture*.

However, as we have seen in the preceding chapter, concrete is a non-linear material and the assumption of linear stress distribution is not valid. The stress calculated from equation (21.3) is therefore higher than that actually developed in the concrete. The strain gradient in the specimen may also inhibit crack growth. For both these reasons the

modulus of rupture is also greater than the direct tensile strength.

21.1.3 RELATIONSHIP BETWEEN STRENGTH MEASUREMENTS

We have already discussed the relationship between cube and cylinder compressive strength measurements. The tensile strength, however measured, is roughly one order of magnitude lower that the compressive strength. The relationship between the two is non-linear, with a good fit being an expression of the form:

$$f_t = a(f_c)^b \tag{21.4}$$

where f_t = tensile strength, f_c = compressive strength, and a and b are constants. Eurocode 2 (BS EN 1992) gives $a = 0.30$ and $b = 0.67$ when f_c is the characteristic cylinder strength and f_t is the mean tensile strength. This relationship, converted to cube compressive and tensile strengths, is plotted in *Fig. 21.5* together with equivalent data from cylinder splitting and modulus of rupture tests obtained over a number of years by UCL undergraduate students.

It is clear from this figure that, as we have already said, both the modulus of rupture and the cylinder splitting tests give higher values than the direct tensile test. The modulus of rupture is the higher value, varying between about 8 and 17% of the cube strength (the higher value applies to lower strengths). The cylinder splitting strength is between about 7 and 11% of the cube strength, and the direct tensile strength between about 5 and 8% of the cube strength. *Figure 21.5* also shows that, as with all such relationships, there is a considerable scatter of individual data points about the best-fit line (although in this case some of this may be due to the inexperience of the testers).

21.2 Factors influencing the strength of Portland cement concrete

In this section we will consider the strength of concrete with Portland cement as the sole binder. The effect of additions will be discussed in the next section.

21.2.1 TRANSITION/INTERFACE ZONE

Before looking at the relationships between the strength of concrete and the many factors that influence it, we need to introduce an extremely important aspect of concrete's structure. In Chapter 13 we

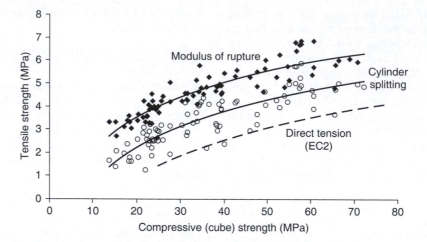

Fig. 21.5 *The relationship between direct and indirect tensile strength measurements and compressive strength of concrete (from Eurocode 2 BS EN 1992, EC2, 2004; UCL data).*

described the microstructure of the hardened cement paste that is formed during hydration. Concrete is, of course, a mixture of paste and aggregate, and it is the interface between these that is of great significance. The paste close to the aggregate surface is substantially different to that of the bulk paste, and crucially this *transition* or *interface zone* is significantly more porous and therefore weaker than the rest of the paste. As the load on the concrete increases, cracking will start in this zone, and subsequently propagate into the HCP until crack paths are formed through the concrete, as shown in *Fig. 21.6*, which when sufficiently extensive and continuous will result in complete breakdown, i.e. failure. The overall effect is that the strength of a sample of concrete is nearly always less than that of the bulk cement paste within it.

The formation, structure and consequences of the transition zone have been the focus of much research since the mid-1980s. Suggested mechanisms for its formation include an increased water:cement ratio at the paste aggregate/interface due to:

- the 'wall effect', whereby the cement grains cannot pack as efficiently next to the aggregate surface as they can in the bulk paste
- mix water separation at the interface due to the relative movement of the aggregate particles and cement paste during mixing, leading to a higher local water:cement ratio.

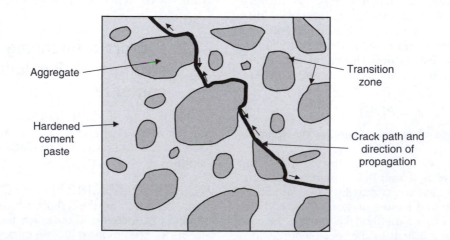

Fig. 21.6 *Cracking pattern in normal-strength concrete.*

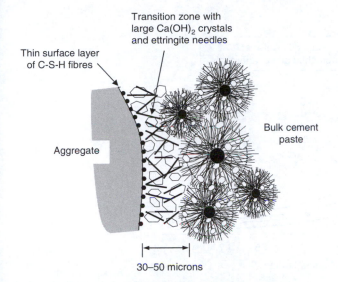

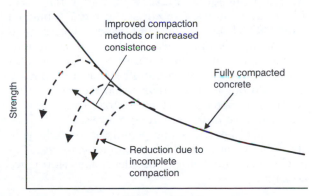

Transition zone with large Ca(OH)$_2$ crystals and ettringite needles

Thin surface layer of C-S-H fibres

Aggregate

Bulk cement paste

30–50 microns

Fig. 21.7 Features of the transition zone at the paste–aggregate interface (adapted from De Rooij et al., 1998).

Although there are some differences of opinion, there is a general consensus that the zone is between 30 and 50 microns wide and that its structure – in a much simplified form – is as shown in *Fig. 21.7*. This shows two main features:

- a very thin surface layer of calcium silicate hydrate fibres on the aggregate, also containing some small calcium hydroxide (portlandite) crystals
- a greater concentration of larger calcium hydroxide crystals and fine needles of calcium sulphoaluminate (ettringite) than in the bulk paste and hence a greater porosity.

Although the zone's porosity will reduce with time with the continuing deposition of hydration products (chiefly C-S-H), we should think of concrete as a three-phase material – HCP, aggregate and the transition zone. It will be useful to bear this model in mind during the discussion of the more important factors that affect concrete strength that now follows. We will discuss some further aspects of the cracking and failure process later in the chapter.

21.2.2 WATER:CEMENT RATIO

In Chapter 13 we saw that the strength of cement paste is governed by its porosity, which in turn depends on the water:cement ratio and degree of hydration. The overall dependence of the strength of concrete on the amount of cement, water and

air voids within it was recognised in 1896 by Feret, who suggested a rule of the form:

$$f_c = K(c/\{c + w + a\})^2 \qquad (21.5)$$

where f_c = strength, c, w and a are the absolute volumetric proportions of cement, water and air, respectively, and K is a constant. Working independently, Abrams, in 1918, demonstrated an inverse relationship with concrete strength of the form:

$$f_c = k_1/(k_2^{w/c}) \qquad (21.6)$$

This has become known as Abrams' Law, although strictly, as it is based on empirical observations, it is a rule. The constants K, k_1 and k_2 are empirical and depend on age, curing regime, type of cement, amount of air entrainment, test method and, to a limited extent, aggregate type and size.

Feret's rule and Abrams' law both give an inverse relationship between strength and water:cement ratio for a fully compacted concrete of the form shown in *Fig. 21.8*. It is important to recognise the limitations of such a relationship. First, at low water:cement ratios, the concrete's consistence decreases and it becomes increasingly more difficult to compact. Feret's rule recognises that increasing air content will reduce the strength, and in general the strength will decrease by 6% for each 1% of included air by volume. This leads to the steep reduction in strength shown by the dashed lines in *Fig. 21.8*. The point of divergence from the fully compacted line can be moved further up and to the left by the use of more efficient compaction and/or by improvements in consistence without increasing

Improved compaction methods or increased consistence

Fully compacted concrete

Strength

Reduction due to incomplete compaction

Water:cement ratio

Fig. 21.8 The general relationship between strength and water:cement ratio of concrete (adapted from Neville, 1995).

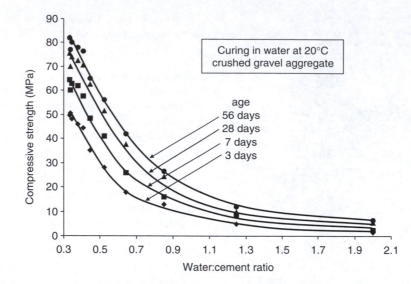

Fig. 21.9 *Compressive strength vs. water:cement ratio for concrete made with a CEM I 42.5N Portland cement (after Balmer, 2000).*

the water:cement ratio, for example by using plasticisers or superplasticisers, which were discussed in Chapter 14. Without such admixtures it is difficult to achieve adequate consistence for most normal compaction methods at water:cement ratios much below 0.4; with admixtures this limit can be reduced to 0.25 or even less.

Abrams himself showed that, at the other end of the scale, his rule was valid for water:cement ratios of up to 2 or more. However, at these high values the paste itself is extremely fluid, and it is very difficult to achieve a homogeneous, cohesive concrete without significant segregation. In practice water:cement ratios in excess of 1 are rarely used. *Figure 21.9* shows a recent set of results obtained with CEM I 42.5N Portland cement, where good compaction was achieved in laboratory conditions at water:cement ratios down to 0.33. This gives a good idea of typical concrete performance, but we must add our normal proviso that the use of other constituent materials (cement source, aggregate type etc.) will give different strength levels.

We will be discussing examples of achieving strengths significantly higher than those in *Fig. 21.9* in Chapter 25, when we consider high-performance concrete.

21.2.3 AGE
The degree of hydration increases with age, leading to the effect of age on strength apparent from *Fig. 21.9*. As discussed in Chapter 13, the rate of

hydration depends on the cement composition and fineness and so both of these will affect the rate of gain of strength. This is taken into account in the classification of cement described in Chapter 13. The strength at 28 days is often used to characterise the concrete for design, specification and compliance purposes, probably because it was originally thought to be a reasonable indication of the long-term strength without having to wait too long for test results. However strengths at other ages will often be important, for example during construction and when assessing long-term performance. Eurocode 2 (BS EN 1992) gives the following relationships for estimating the strength at any age from the 28-day strength for concrete kept at 20°C and high humidity:

$$f_c(t) = \beta(t).f_c(28) \qquad (21.7)$$

where $f_c(t)$ is the strength at age t days, $f_c(28)$ is the 28-day strength and:

$$\beta(t) = \exp\{s[1 - (28/t)^{0.5}]\} \qquad (21.8)$$

where s is a coefficient depending on the cement strength class:

$s = 0.2$ for CEM 42.5R, 52.5N and 52.5R
$s = 0.25$ for CEM 32.5R and 42.5N
$s = 0.38$ for CEM 32.5N

These equations have been used to produce the relationships shown in *Fig. 21.10*. Depending on the strength class the 3-day strength is between 40 and 65% of the 28-day strength and the 7-day

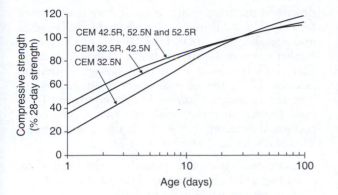

Fig. 21.10 *Effect of cement class on rate of gain of strength of concrete (from Eurocode 2 (BS EN 1992) relationships).*

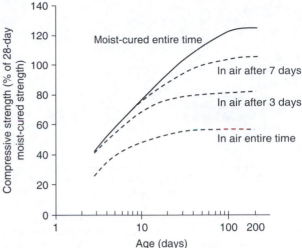

Fig. 21.11 *The influence of curing conditions on the development of concrete strength (from Portland Cement Association, 1968).*

strength between 60 and 80% of the 28-day strength. This figure also shows that:

- the strength gain continues well beyond 28 days. As discussed in Chapter 13, the hydration reactions are never complete and, in the presence of moisture, concrete will in fact continue to gain strength for many years, although, of course, the rate of increase after such times will be very small
- the long-term strength beyond 28 days is higher with cements that give lower short-term strength. This is because the microstructure is more efficiently formed at slower rates of hydration; in other words, if you can afford to wait long enough, then the final result will be better.

21.2.4 TEMPERATURE

As we discussed in Chapter 19, a higher temperature maintained throughout the life of a concrete will result in higher short-term strengths but lower long-term strengths, a similar effect to that just described for cement class. As also discussed, an early age heating–cooling cycle from heat of hydration effects can lead to lower long-term strength, but the effect can be reduced or even eliminated by the incorporation of fly ash or ground granulated blast furnace slag. We shall discuss the effect of transient high temperatures when considering the durability of concrete in fire in Chapter 24.

21.2.5 HUMIDITY

The necessity of a humid environment for adequate curing has already been discussed; for this reason concrete stored in water will achieve a higher strength than if cured in air for some or all of its life, as shown in *Fig. 21.11*. Also, specimens cured

in water will show a significant increase in strength (5% or more) if allowed to dry out for a few hours before testing.

21.2.6 AGGREGATE PROPERTIES, SIZE AND VOLUME CONCENTRATION

As discussed above, for normal aggregate it is the strength of the paste–aggregate bond or transition zone that has a dominant effect on concrete strength; the aggregate strength itself is generally significant only in very-high-strength concrete or with the relatively weaker lightweight aggregates. Tests have shown that with some carbonate and siliceous aggregates there is evidence that the structure and chemistry of the transition zone are influenced by the aggregate mineralogy and surface texture; for example limestone aggregates give excellent bond (Struble *et al.*, 1980). Crushed rocks tend to have rougher surfaces which, together with the increased mechanical interlocking of the angular aggregate particles, means that concretes made with crushed rocks are typically some 15–20% stronger than those made with uncrushed gravels, provided all other mix proportions are the same. *Figure 21.12* shows the range of strengths that are obtained from some cements and aggregates used in the UK, from which the effect of aggregate type is apparent.

The use of a larger maximum aggregate size reduces the concrete's strength, again provided all other mix proportions are the same. The reduction is relatively small – about 5% – with an increase in aggregate size from 5 to 20 mm at normal

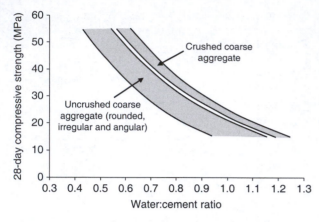

Fig. 21.12 *Strength ranges for concrete from UK materials (data for 7 cement and 16 aggregate sources, from Foulger, 2008).*

concrete strengths, but greater reductions (up to 20%) are obtained at higher strength and with larger aggregate particle sizes. Larger aggregates have a lower overall surface area with a weaker transition zone, and this has a more critical effect on the concrete's strength at lower water:cement ratios. In fresh concrete, the decreased surface area with increased aggregate size leads to increased consistence for the same mix proportions, and therefore for mix design at constant consistence the water content can be reduced and a compensating increase in strength obtained.

Increasing the volumetric proportion of aggregate in the mix will, at a constant water:cement ratio, produce a relatively small increase in concrete strength (typically a 50% increase in aggregate content may result in a 10% increase in strength). This has been attributed, at least in part, to the increase in aggregate concentration producing a greater number of secondary cracks prior to failure, which require greater energy, i.e. higher stress, to reach fracture. This effect is only valid if the paste content remains high enough to at least fill the voids in the coarse/fine aggregate system, thereby allowing complete consolidation of the concrete. This therefore imposes a maximum limit to the aggregate content for practical concretes.

21.3 Strength of concrete containing additions

We discussed the nature, composition and behaviour of additions in Chapter 15, and in particular we

described the pozzolanic or secondary reactions of four Type 2 additions – fly ash, ggbs, microsilica and metakaolin – that lead to the formation of further calcium silicate hydrates. When each addition is used within its overall dosage limitation (section 15.3), the general effect is an increase in long-term strength compared to the equivalent Portland cement mix (i.e. with the two mixes being compared differing only in the binder composition). This is owing to a combination of:

- better packing of the particles in the fresh state, leading to an overall reduced porosity of the hardened cement paste after hydration
- preferential enhancement of the transition zone which, as we have seen, is of higher porosity and is rich in portlandite and is therefore a prime target for the secondary reactions.

Not surprisingly, the strength of mixes containing additions does take some time to reach and overtake that of the equivalent Portland cement mix. As an example, *Fig. 21.13* shows the strength gain of mixes with binders of up to 60% fly ash compared to that of a mix with 100% Portland cement. With 20% fly ash the strength exceeds that of the 100% PC mix after about 3 months; with 40% fly ash the 'cross-over' is some months later; and with 60% fly ash it appears never to reach the strength of the PC mix. In this last case there is insufficient calcium hydroxide produced by the cement, even after complete hydration, to react with all the silica in the fly ash.

Figure 21.14 is a schematic that illustrates the general strength-gain behaviour of mixes with all the four Type 2 additions previously considered when each is used within its normal dosage limitation. The 'cross-over' point for microsilica mixes is very early, sometimes within a day; metakaolin mixes

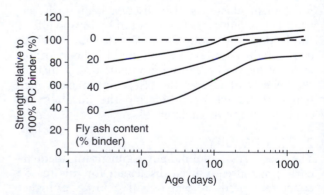

Fig. 21.13 *Strength gain of concrete containing fly ash (from data in Neville, 1995).*

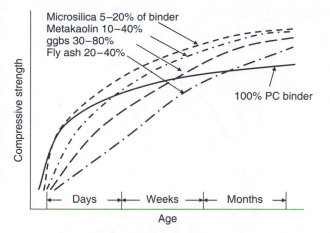

Fig. 21.14 *Schematic of typical strength gain characteristics of concrete containing Type 2 additions.*

take a few days, ggbs mixes days or weeks, and pfa mixes weeks or months. The reasons for the different rates of strength gain with the four different materials lie in their characteristics (as given in Chapter 15), which can be summarised as:

- the extremely fine particles, which act as nucleation sites for hydrate deposition, and very high active silica content of the microsilica result in the strength quickly overtaking that of the equivalent Portland cement mix
- metakaolin is somewhat coarser so it is a little slower to react, but it contains a high active silica content and is therefore not far behind the microsilica mixes
- ggbs and fly ash have similar particle sizes to Portland cement, but ggbs also contains its own calcium oxide, which contributes to the secondary reactions and so it is more than just a pozzolanic material.

If the slower rate of gain of strength is a problem during construction, mixes can of course be modified accordingly, e.g. by using plasticisers to maintain workability at a reduced water:cement ratio.

It is however difficult to do more than generalise on the timescales and magnitude of the strength characteristics, for two reasons:

- The vast amount of published information on the properties of concrete containing additions shows that with each one there is a wide range of performance (not just of strength, but also of all other properties), owing mainly to the differences in physical and chemical composition of both the addition and the Portland cement in the various test programmes.

- Each set of tests will have been designed for a different purpose and therefore will have a different set of variables, such as changing the water content to obtain equal workability or equal 28-day strength, and therefore it is often difficult to compare like with like (indeed, this a problem facing students in nearly all areas of concrete technology).

The contribution of the addition to strength is often expressed in terms of an *activity coefficient* or *cementing efficiency factor (k)*, which is a measure of the relative contribution of the addition to strength compared to an equivalent weight of Portland cement. This means that if the amount of the addition is x kg/m^3, then this is equivalent to kx kg/m^3 of cement, and the concrete strength is that which would be achieved with a cement content of $c + kx$, where c is the amount of cement.

If k is greater than 1, then the addition is more active than the cement, and if less than 1, it is less active. Its value will clearly increase with the age of the concrete, and will also vary with the amount of addition and other mix proportions. For 28-day-old concrete and proportions of the addition within the overall limits of *Fig. 21.14*, values of 3 for microsilica, 1 for ggbs and 0.4 for fly ash have been proposed (Sedran *et al.*, 1996), although again, a considerable range of values has been suggested by different authors. The cementing efficiency factor approach is useful in mix design, as will be discussed in Chapter 22.

21.4 Cracking and fracture in concrete

21.4.1 DEVELOPMENT OF MICROCRACKING

As we discussed in Chapter 20, the non-linear stress–strain behaviour of concrete in compression is largely due to the increasing contribution of microcracking to the strain with increasing load. Four stages of cracking behaviour have been identified (Glucklich, 1965):

- Stage 1, up to about 30% maximum stress. The pre-existing transition-zone cracks remain stable, and the stress–strain curve remains approximately linear.
- Stage 2, about 30–50% maximum stress. The cracks begin to increase in length, width and number, causing non-linearity, but are still stable and confined to the transition zone.
- Stage 3, about 50–75% maximum stress. The cracks start to spread into the matrix and become unstable, resulting in further deviation from linearity.

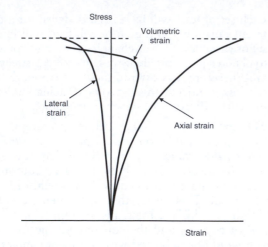

Fig. 21.15 Stress–strain behaviour of concrete under compressive loading (after Newman, 1966).

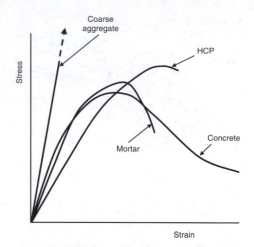

Fig. 21.16 Typical stress–strain characteristics of aggregate, hardened cement paste, mortar and concrete under compressive loading (after Swamy and Kameswara Rao, 1973).

• Stage 4, above about 75% ultimate stress. Spontaneous and unstable crack growth becomes increasingly frequent, leading to very high strains. Also at this stage the excessive cracking results in the lateral strains increasing at a faster rate than the axial strains, resulting in an overall increase in volume (*Fig. 21.15*).

Complete breakdown, however, does not occur until strains significantly higher than those at maximum load are reached. *Fig. 21.16* shows stress–strain curves from strain-controlled tests on paste, mortar and concrete. The curve for HCP has a small descending branch after maximum stress; with the mortar it is more distinct, but with the concrete it is very lengthy. During the descending region, excess cracking and slip at the paste–aggregate interface occur before the cracking through the HCP is sufficiently well developed to cause complete failure.

21.4.2 CREEP RUPTURE

We discussed in Chapter 20 the contribution of microcracking to creep. This increases with stress level to the extent that if a stress sufficiently close to the short-term ultimate is maintained then failure will eventually occur, a process known as *creep rupture* (see Chapter 2, section 2.7). There is often an acceleration in creep rate shortly before rupture. The behaviour can be shown by stress–strain relationships plotted at successive times after loading, giving an ultimate strain envelope, as shown for

compressive and tensile loading in *Fig. 21.17a* and *Fig. 21.17b*, respectively. The limiting stress below which creep rupture will not occur is about 70% of the short-term maximum for both compression and tension.

21.4.3 THE FRACTURE MECHANICS APPROACH

Griffith's theory for the fracture of materials and its consequent development into fracture mechanics were described in general terms in Chapter 4. Not surprisingly, there have been a number of studies attempting to apply linear fracture mechanics to concrete, with variable results; some of the difficulties encountered have been:

1. Failure in compression, and to a lesser extent in tension, is controlled by the interaction of many cracks, rather than by the propagation of a single crack.
2. Cracks in cement paste or concrete do not propagate in straight lines, but follow tortuous paths around cement grains, aggregate particles etc., which both distort and blunt the cracks (*Fig. 21.6*).
3. The measured values of fracture toughness are heavily dependent on the size of the test specimen, and so could not strictly be considered as a fundamental material property.
4. Concrete is a composite made up of cement paste, the transition zone and the aggregate, and each

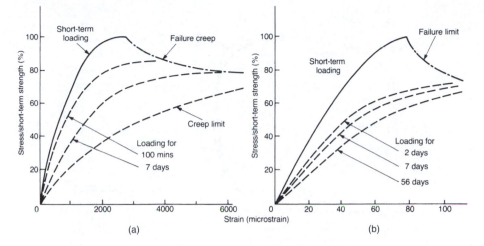

Fig. 21.17 *The effect of sustained compressive and tensile loading on the stress–strain relationship for concrete: (a) compressive loading; (b) tensile loading (after Rusch, 1960; Domone, 1974).*

has its own fracture toughness (K_c), each of which is difficult to measure.

Despite these difficulties, K_c values for cement paste have been estimated as lying in the range 0.1 to 0.5 MN/m$^{3/2}$, and for concrete between about 0.45 and 1.40 MN/m$^{3/2}$ (Mindess and Young, 1981). K_c for the transition zone seems to be smaller, about 0.1 MN/m$^{3/2}$, confirming the critical nature of this zone. Comparison of these values with those for other materials given in *Table 61.1* shows the brittle nature of concrete.

21.5 Strength under multiaxial loading

So far in this chapter our discussions on compressive strength have been concerned with the effects of uniaxial loading, i.e. where σ_1 (or σ_x) is finite, and the orthogonal stresses σ_2 (or σ_y) and σ_3 (or σ_z) are both zero. In many, perhaps most, structural situations concrete will be subject to a multiaxial stress state (i.e. σ_2 and/or σ_3 as well as σ_1 are finite). This can result in considerable modifications to the failure stresses, primarily by influencing the cracking pattern.

A typical failure envelope under biaxial stress (i.e. $\sigma_3 = 0$) is shown in *Fig. 21.18*, in which the applied stresses, σ_1 and σ_2, are plotted non-dimensionally as proportions of the uniaxial compressive strength, σ_c. Firstly, it can be seen that concretes of different

strengths behave very similarly when plotted on this basis. Not surprisingly, the lowest strengths in each case are obtained in the tension–tension quadrant. The effect of combined tension and compression is to reduce considerably the compressive stress needed for failure even if the tensile stress is significantly less than the uniaxial tensile strength. The cracking pattern over most of this region (Type 1 in *Fig. 21.18*) is a single tensile crack, indicating that the failure criterion is one of maximum tensile strain, with the tensile stress enhancing the lateral tensile strain from the compressive stress. In the region of near uniaxial compressive stress, i.e. close to the compressive stress axes, the cracking pattern (Type 2) is essentially the same as that in the central region of the cylinder shown in *Fig. 21.1b*, i.e. the cracks form all around the specimen approximately parallel to the compressive load. In the compression–compression quadrant, the cracking pattern (Type 3) becomes more regular, with the cracks forming in the plane of the applied loads, splitting the specimen into slabs. Under equal biaxial compressive stresses, the failure stress is somewhat larger than the uniaxial strength. Both Type 2 and Type 3 crack patterns also indicate a limiting tensile strain failure criterion, in the direction perpendicular to the compressive stress(es).

With triaxial stresses, if all three stresses are compressive then the lateral stresses (σ_2 and σ_3) act in opposition to the lateral tensile strain produced by σ_1. This in effect confines the specimen, and results in increased values of σ_1 being required for failure, as illustrated in *Fig. 21.19* for the case of

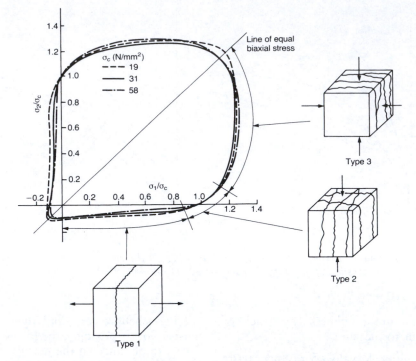

Fig. 21.18 *Failure envelopes and typical fracture patterns for concrete under biaxial stress* σ_1 *and* σ_2 *relative to uniaxial stress* σ_c *(after Kupfer et al., 1969; Vile, 1965).*

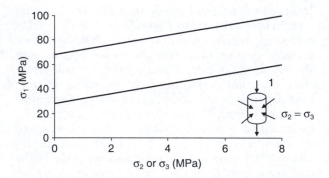

Fig. 21.19 *The effect of lateral confining stress* (σ_2, σ_3) *on the axial compressive strength* (σ_1) *of concretes of two different strengths (from FIP/CEB, 1990).*

uniform confining stress (i.e. $\sigma_2 = \sigma_3$); the axial strength (σ_{1ult}) can be related to the lateral stress by the expression:

$$\sigma_{1ult} = \sigma_c + K\sigma_2 \text{ (or } \sigma_3) \qquad (21.9)$$

where K has been found to vary between about 2 and 4.5.

In describing strength tests in Section 21.1.1, we said that when a compressive stress is applied to a specimen by the steel platen of a test machine, the lateral (Poisson effect) strains induce restraint forces in the concrete near the platen owing to the mismatch in elastic modulus between the concrete and the steel. This is therefore a particular case of triaxial stress, and the cause of the higher strength of cubes compared to longer specimens such as cylinders.

References

Balmer T (2000). *Investigation into the effects on the main concrete relationship using class 42.5N Portland cement at varying compliance levels.* Diploma in Advanced Concrete Technology project report, Institute of Concrete Technology, Camberley, UK.

De Rooij MR, Bijen JMJM and Frens G (1998). *Introduction of syneresis in cement paste.* Proceedings of International Rilem Conference on the Interfacial Transition Zone in Cementitious Composites, Israel (eds Katz A, Bentur A, Alexander M and Arliguie G). E&FN Spon, London, pp. 59–67.

Domone PL (1974). Uniaxial tensile creep and failure of concrete. *Magazine of Concrete Research*, **26** (No. 88), 144–152.

FIP/CEB (1990). *State-of-the-art report on high strength concrete*, Thomas Telford, London.

Foulger D (2008). *Simplification of Technical Systems.* Diploma in Advanced Concrete Technology project report, Institute of Concrete Technology, Camberley, UK.

Glucklich J (1965). Proceedings of the International Conference on Structure of Concrete and Its Behaviour Under Load, Cement and Concrete Association, London, September, pp. 176–189.

Kupfer H, Hilsdorf HK and Rusch H (1969). Behaviour of concrete under biaxial stress. *Proceedings of the American Concrete Institute*, **66** (No. 88), 656–666.

Maso JC (ed.) (1992). Proceedings of International RILEM Conference on Interfaces in Cementitious Composites, Toulouse, October. E&FN Spon, London, p. 315.

Mindess S and Young JF (1981). *Concrete*, PrenticeHall, New Jersey.

Neville AM (1995). *Properties of concrete*, 4th edition, Pearson Education, Harlow, p. 844.

Newman K (1966). Concrete systems. In *Composite Materials* (ed. L. Hollaway), Elsevier, London.

Portland Cement Association (1968). *Design and Control of Concrete Mixes*, 11th edition, Stokie, Illinois, USA.

Rusch H (1960). Researches toward a general flexural theory for structural concrete. *Proceedings of the American Concrete Institute*, **57** (No. 7), 1–28.

Sedran T, de Larrard F, Hourst F and Contamines C (1996). *Mix design of self-compacting concrete (SCC)*. Proceedings of the International RILEM Conference on Production Methods and Workability of Concrete (eds Bartos PJM, Marrs DL and Cleland DJ). E&FN Spon, London, pp. 439–450.

Struble L, Skalny J and Mindess S (1980). A review of the cement–aggregate bond. *Cement and Concrete Research*, **10** (No. 2), 277–286.

Swamy RN and Kameswara Rao CBS (1973). Fracture mechanism in concrete systems under uniaxial loading. *Cement and Concrete Research*, **3** (No. 4), 413–428.

Vile GWD (1965). Proceedings of the International Conference on Structure of Concrete and Its Behaviour Under Load, Cement and Concrete Association, London, September, pp. 275–288.

Chapter 22

Concrete mix design

Mix design is the process of selecting the proportions of cement, water, fine and coarse aggregates and, if they are to be used, additions and admixtures to produce an economical concrete mix with the required fresh and hardened properties. It is often, perhaps justifiably, referred to as 'mix proportioning' rather than 'mix design'. The cement and other binder constituents are usually the most expensive component(s), and 'economical' usually means keeping its/their content as low as possible, without, of course, compromising the resulting properties. There may be other advantages, such as reduced heat of hydration (Chapter 19), drying shrinkage or creep (Chapter 20).

22.1 The mix design process

Figure 22.1 shows the stages in the complete mix design process; we will discuss each of these in turn.

22.1.1 SPECIFIED CONCRETE PROPERTIES

The required hardened properties of the concrete result from the structural design process, and are therefore provided to the mix designer. Strength is normally specified in terms of a characteristic strength (see section 2.9) at a given age. In Europe there are a discrete number of *strength classes* that

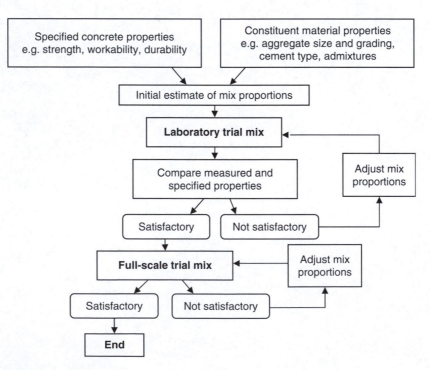

Fig. 22.1 The mix design process.

can be specified (BS EN 206); for normal-weight concrete these are:

C8/10, C12/15, C16/20, C20/25, C25/30, C30/37, C35/45, C40/50, C45/55, C50/60, C55/67, C60/75, C70/85, C80/95, C90/105 and C100/115

In each case the first number of the pair is the required minimum characteristic cylinder strength and the second number the required minimum characteristic cube strength. This reflects the different methods of measuring the compressive strength in different countries within Europe with, as described in the preceding chapter, the latter giving a higher value than the former for the same concrete.

Durability requirements, to be discussed in Chapter 24, may impose an additional limit on some mix proportions e.g. a minimum cement content or maximum water:cement ratio, or demand the use of an air-entraining agent or a particular aggregate type.

The choice of consistence will depend on the methods selected for transporting, handling and placing the concrete (e.g. pump, skip etc.), the size of the section to be filled and the congestion of the reinforcement. *Table 22.1* shows the consistence classes in the European standard for each of the single-point test methods described in Chapter 18. The consistence must clearly be sufficient at the point of placing, which in the case of ready-mixed concrete transported by road to site, may be some time after mixing.

22.1.2 CONSTITUENT MATERIAL PROPERTIES

As a minimum, the fine and coarse aggregate size, type and grading and the cement type must be known. The relative density of the aggregates, the cement composition, and details of any additions and admixtures that are to be used or considered may also be needed.

22.1.3 INITIAL ESTIMATE OF MIX PROPORTIONS

An initial best estimate of the mix proportions that will give concrete with the required properties is then made. In this, as much use as possible is made of previous results from concrete made with the same or similar constituent materials. In some cases, for example in producing a new mix from an established concrete production facility, the behaviour of the materials will be well known. In other circumstances there will be no such knowledge, and typical behaviour such as that given in the preceding few chapters will have to be used.

There are a considerable number of step-by-step methods of varying complexity that can be used to produce this 'best estimate'. Many countries have their own preferred method or methods and, as an example, we will describe a current UK method below. Whichever method is used, it is important to recognise that the result is only a best estimate, perhaps even only a good guess; because the constituent materials will not be exactly as assumed and their interaction cannot be predicted with any great certainty, the concrete is unlikely to meet the requirements precisely, and some testing will be required.

22.1.4 LABORATORY TRIAL MIXES

The first stage of the testing to verify the mix properties is normally a trial mix on a small scale in a laboratory. The test results will often show that the required properties have not been obtained with sufficient accuracy and so some adjustment to the mix portions will be necessary e.g. a decrease in the water:cement ratio if the strength is too low. A second trial mix with the revised mix proportions is then carried out, and the process is repeated until a mix satisfactory in all respects is obtained.

22.1.5 FULL-SCALE TRIAL MIXES

Laboratory trials do not provide the complete answer. The full-scale production procedures will not

Table 22.1 Consistence classes for fresh concrete from BS EN 206

Class	Slump (mm)	Class	Vebe time (secs)	Class	Degree of compactability	Class	Flow diameter (mm)
S1	10–40	V0	≥31	C0	1.46	F1	340
S2	50–90	V1	30–21	C1	1.45–1.26	F2	350–410
S3	100–150	V2	20–11	C2	1.25–1.11	F3	420–480
S4	160–210	V3	10–6	C3	1.10–1.04	F4	490–550
S5	≥220	V4	5–3	C4	<1.04	F5	560–620
						F6	≥630

be exactly the same as those in the laboratory, and this may cause differences in the properties of the concrete. Complete confidence in the mix can therefore only be obtained with further trials at full scale, again with adjustments to the mix proportions and re-testing if necessary.

22.2 The UK method of 'Design of normal concrete mixes' (BRE 1997)

This method of mix design provides a good example of the process of making an initial estimate of the mix proportions. It has the advantage of being relatively straightforward and producing reasonable results with the materials most commonly available in the UK. It should be emphasised that it is not necessarily the 'best' method available worldwide, and that it may not give such good results with other materials.

The main part of the method is concerned with the design of mixes incorporating Portland cement, water and normal-density coarse and fine aggregates only, and with characteristic cube strengths of up to about 70 MPa (since it is a UK method, all the strengths referred to are cube strengths). It encompasses both crushed and uncrushed coarse aggregate. The steps involved can be summarised as follows.

22.2.1 TARGET MEAN STRENGTH

As described in Chapter 2, the specified *characteristic strength* is a lower limit of strength to be used in structural design. As with all materials,

concrete has an inherent variability in strength, and an average cube compressive strength (or *target mean strength*) somewhat above the characteristic strength is therefore required. The difference between the characteristic and target mean strength is called the *margin*; a 5% failure rate is normally chosen for concrete, and the margin should therefore be 1.64 times the standard deviation of the strength test results (*Table 2.1*).

This means that a knowledge of the standard deviation is required. For an existing concrete production facility this will be known from previous tests. Where limited or no data are available, this should be taken as 8 MPa for characteristic strengths above 20 MPa, and *pro rata* for strengths below this. When production is under way, this can be reduced if justified by sufficient test results (20 or more), but not to below 4 MPa for characteristic strengths above 20 MPa, and *pro rata* for strengths below this. The advantage of reducing the variability by good practice is clear.

22.2.2 FREE WATER:CEMENT RATIO

For a particular cement and aggregate type, the concrete strength at a given age is assumed to be governed by the free water:cement ratio only. The first step is to obtain a value of strength at a water:cement ratio of 0.5 from *Fig. 22.2* for the relevant age/aggregate type/cement type combination (note: this figure has been produced from tabulated data in the method document). This value is then plotted on the vertical line in *Fig. 22.3* to give a starting point for a line that is constructed parallel to the curves shown. The point of intersection of this line with the horizontal line of the required

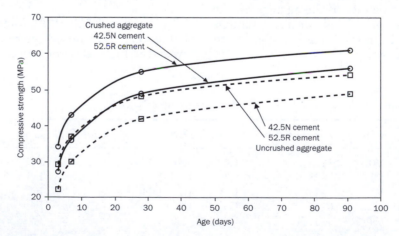

Fig. 22.2 Compressive strength vs. age for concrete with a water:cement ratio of 0.5 (after BRE, 1997).

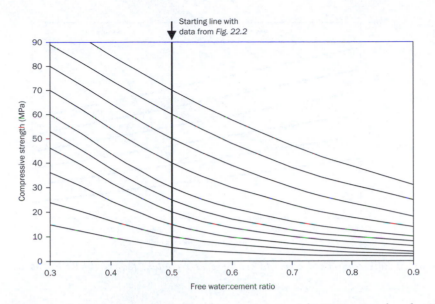

Fig. 22.3 *Compressive strength vs. water:cement ratio of concrete (copyright BRE, reproduced with permission).*

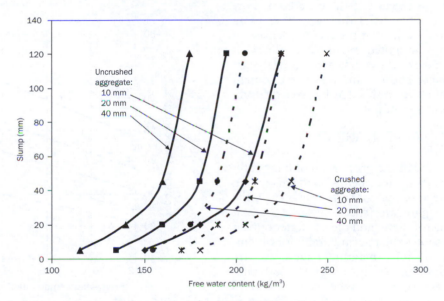

Fig. 22.4 *Slump vs. free water content of concrete (after BRE, 1997).*

target mean strength then gives the required free water:cement ratio. The ranges of the axes in *Fig. 22.3* indicate the limits of validity of the method.

22.2.3 FREE WATER CONTENT

It is now assumed that, for a given coarse aggregate type and maximum size, the concrete consistence is governed by the free water content only. The consistence can be specified in terms of either slump or Vebe time (see Chapter 18), although slump is by far the most commonly used. *Figure 22.4* is a graph of data for slump, again produced from tabulated data in the method document, from which the free water content for the appropriate aggregate can be obtained.

22.2.4 CEMENT CONTENT

This is a simple calculation from the values of the free water:cement ratio and free water content just calculated.

165

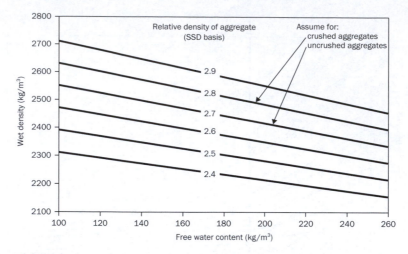

Fig. 22.5 *Wet density of fully compacted concrete vs. free water content (copyright BRE, reproduced with permission).*

22.2.5 TOTAL AGGREGATE CONTENT

An estimate of the density of the concrete is now required. This is obtained from *Fig. 22.5*, using known or assumed values of the relative density of the aggregates. A weighted mean value is used if the specific gravities of the coarse and fine aggregate are different. Subtraction of the free water content and cement content from this density gives the total aggregate content per m³.

22.2.6 FINE AND COARSE AGGREGATE CONTENT

The estimated value of the proportion of fine aggregate in the total aggregate depends on the maximum size of the aggregate, the concrete consistence, the grading of fine aggregate (specifically the amount passing a 600-micron sieve) and the free water:cement ratio. *Fig. 22.6* shows the relevant graph for obtaining this proportion for a maximum aggregate size of 20 mm and slump in the range 60–180 mm. Sufficient fine aggregate must be incorporated to produce a cohesive mix that is not prone to segregation, and *Fig. 22.6* shows that increasing quantities are required with increasing water:cement ratio and if the aggregate itself is coarser. The mix design document also gives equivalent graphs for lower slump ranges and 10 and 40 mm coarse aggregate; less fine aggregate is required for lower slumps, between 5 and 15% more fine aggregate is required with 10 mm aggregate, and between 5 and 10% less with 40 mm aggregate.

The fine and coarse aggregate content is now calculated by simple arithmetic, and the amounts (in kg/m³) of free water, cement, coarse and fine

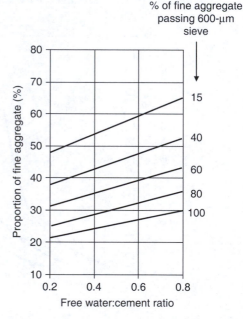

Fig. 22.6 *Proportions of fine aggregate according to percentage passing 600-μm sieve (for 60–180 mm slump and 20 mm max coarse aggregate size) (copyright BRE, reproduced with permission).*

aggregates for the laboratory trial mix have now all been obtained.

It is important to note the simplifying assumptions used in the various stages. These make the method somewhat simpler than some other alternatives, but highlight the importance of trial mixes and subsequent refinements.

22.3 Mix design with additions

As we have seen, additions affect both the fresh and hardened properties of concrete, and it is often difficult to predict their interaction with the Portland cement with any confidence. The mix design process for concretes including additions is therefore more complex and, again, trial mixes are essential.

The mix design method described above (BRE 1997) includes modifications for mixes containing good-quality low-lime fly ash or ggbs. With fly ash:

- the amount, expressed as a proportion of the total binder, first needs to be selected, for example for heat output, durability or economic reasons, subject to a maximum of 40%
- the increase in workability is such that the water content obtained from *Fig. 22.4* can be reduced by 3% for each 10% fly ash substitution of the cement
- the effect upon the strength is allowed for by the use of a *cementing efficiency factor*, k, which we discussed in Chapter 21. This converts the amount of fly ash to an equivalent amount of cement. The total equivalent cement content is then $C + kF$, where C = Portland cement content and F = fly ash content. The value of k varies with the type of ash and Portland cement and with the age of the concrete, but a value of 0.30 is taken for 28-day strength with a class 42.5 Portland cement. Thus, if W = water content, a value of the equivalent water:cement ratio – $W/(C + kF)$ – is obtained from *Fig. 22.3*
- subsequent calculations follow using $C + F$ when the total binder content is required.

With ggbs:

- Again the amount as a proportion of the binder is first chosen, with values of up to 90% being suitable for some purposes.

- The improvements in workability are such that the water content derived from *Fig. 22.4* can be reduced by about 5 kg/m^3.
- The cementing efficiency factor approach used for fly ash is more difficult to apply as the value of k is dependent on more factors, including the water:equivalent cement ratio, and for 28-day strengths it can vary from about 0.4 to over 1.0. It is assumed that for ggbs contents of up to 40% there is no change in the strength, i.e. $k = 1$, but for higher proportions information should be obtained from the cement manufacturer or the ggbs supplier.

22.4 Design of mixes containing admixtures

22.4.1 MIXES WITH PLASTICISERS

As we have seen in Chapter 15, plasticisers increase the fluidity or consistence of concrete. This leads to three methods of use:

1. To provide an increase in consistence, by direct addition of the plasticiser with no other changes to the mix proportions.
2. To give an increase in strength at the same consistence, by allowing the water content to be reduced, with consequent reduction in the water:cement ratio.
3. To give a reduction in cement content for the same strength and consistence, by coupling the reduction in water content with a corresponding reduction in cement content to maintain the water:cement ratio.

Methods (1) and (2) change the properties of the concrete, and method (3) will normally result in a cost saving, as the admixture costs much less than the amount of cement saved.

Table 22.2 gives typical figures for these effects on an average-strength concrete mix with a typical

Table 22.2 Methods of using a plasticiser in average-quality concrete (using typical data from admixture suppliers)

Mix	Cement (kg/m^3)	Water	Water:cement ratio	Plasticiser dose (% by weight of cement)	Slump (mm)	28-day strength (MPa)
Control	325	179	0.55	0	75	39
1	325	179	0.55	0.3*	135	39.5
2	325	163	0.5	0.3	75	45
3	295	163	0.55	0.3	75	39

*'standard' dose.

lignosulphonate-based plasticiser. The figures have been obtained using data provided by an admixture supplier. The admixture amount is a 'standard' dose. The important changes are, respectively:

- mix 1: an increase in slump from 75 to 135 mm
- mix 2: an increase in strength from 39 to 45 MPa
- mix 3: a reduction in cement content of 30 kg/m^3.

Plasticisers can have some effect on setting times, but mechanical properties and durability at later ages appear largely unaffected, and are similar to those expected for a plain concrete of the same water:cement ratio, with two relatively minor exceptions:

1. There is some evidence of a slight increase in 28-day strength, attributed to the dispersion of the particles causing an increased surface area of cement being exposed to the mix water (Hewlett, 1988).
2. Some plasticisers entrain about 1–2% air because they lower the surface tension of the mix water. This will reduce the density and strength of the concrete.

22.4.2 MIXES WITH SUPERPLASTICISERS

For the reasons explained in Chapter 14, it is very difficult to generalise about the effects and uses of superplasticisers other than to say that they can produce greater increases in consistence and/or strength and/or greater reductions in cement content than plasticisers. They are more expensive than plasticisers, and therefore the economic advantages of cement reduction may not be as great. Suppliers will provide information on each specific product or formulation, but a mix designer must ensure compatibility with the proposed binder. This can often be judged by tests on paste or mortar in advance of trial mixes on concrete (Aitcin *et al.*, 1994). Superplasticisers enable a much greater range of concrete types to be produced than with plasticisers e.g. high workability flowing concrete, self-compacting mixes and high-strength mixes with low water:cement ratios. These will be discussed in Chapter 25.

22.4.3 MIXES WITH AIR-ENTRAINING AGENTS

As discussed in Chapters 15 and 24, air entrainment is used to increase the resistance of concrete to freeze–thaw damage, but the entrained air increases the consistence and reduces the subsequent strength. The method of mix design described above (BRE 1997) includes the following modifications to allow for these effects if the specified air content is within the normal range of 3–7% by volume:

- It is assumed that the strength is reduced by 5.5% for each 1% of air; the target mean strength is therefore increased by the appropriate amount.
- The slump is reduced by a factor of about two for the selection of water content from *Fig. 22.4*.
- The concrete density obtained from *Fig. 22.5* is reduced by the appropriate amount.

22.5 Other mix design methods

The BRE mix design method described in this chapter is probably the most commonly used simple method in the UK. Methods used in other countries depend on similar principles but differ in their step-wise progression. The American Concrete Institute method is a good example (ACI, 2009). A number of more sophisticated computer-based methods have also been developed. Three of these have been described in Day (2006), de Larrard (1999) and Dewar (1999).

References

ACI 211.1-91 (2009). *Standard Practice for Selecting Proportions for Normal, Heavyweight, and Mass Concrete*. American Concrete Institute, Farmington Mills, Michigan, USA.

Aitcin P-C, Jolicoeur C, MacGregor JG (1994). Superplasticizers: how they work and why they occasionally don't. *Concrete International*, 16 (No. 5), 45–52.

BRE (1997). *Design of normal concrete mixes*, 2nd edition, Building Research Establishment, Watford.

Day KW (2006). *Concrete mix design, quality control and specification*, 2nd edition, E&FN Spon, London, p. 350.

de Larrard F (1999). *Concrete mixture proportioning: a scientific approach*. E&FN Spon, London, p. 350.

Dewar JD (1999). Computer modelling of concrete mixtures, E&FN Spon, London p. 272.

Hewlett PC (ed.) (1988). *Cement Admixtures: Use and Applications*, 2nd edition, Longman, Essex.

Chapter 23

Non–destructive testing of hardened concrete

There are a wide variety of methods and techniques available for the non-destructive testing of structural concrete, which can be broadly divided into those that assess the concrete itself, and those which are concerned with locating and determining the condition of the steel embedded in it. We are going to describe three well-established tests for concrete that are strictly non-destructive, more briefly discuss others that involve some minor damage to the concrete – the so-called partially destructive tests – and then list and briefly comment on some other methods. We do not have space to consider tests to assess the location and condition of reinforcing and prestressing steel, important though these are. Some texts describing these and other tests on concrete are included in 'Further reading' at the end of this part of the book.

Non-destructive testing of concrete is used for two main purposes:

1. In laboratory studies, where it is particularly useful for repeated testing of the same specimen to determine the change of properties with time, for example to provide information on degradation in different environments.

2. In *in-situ* concrete, to assess:

   - strength development, where this is critical for the construction sequence
   - compliance with specifications, particularly where the concrete has underperformed or been deemed 'unfit for purpose'
   - the residual strength after damage, e.g. by fire or overload
   - the cause of degradation or deterioration, often long term, associated with the durability issues that will be discussed in Chapter 24
   - strength when a change of use is proposed.

Two of the tests that we will describe, the rebound hammer and ultrasonic pulse velocity, are commonly used for both these purposes; the third, the resonant frequency test, can only be used on prepared specimens in the laboratory.

An estimation of the strength of concrete is often required, and therefore the degree of correlation of the non-destructive test measurement(s) with strength is important, and will be discussed in each case. It will be apparent that a single non-destructive test rarely gives a single definitive answer, and engineering judgement is required in interpreting the results. Nevertheless, the usefulness of such tests will become apparent.

23.1 Surface hardness – rebound (or Schmidt) hammer test

This is perhaps the simplest of the commonly available tests, and can be used on laboratory specimens or on *in-situ* concrete. Its use in Europe is covered by BS EN 12504-2. The apparatus is contained in a hand-held cylindrical tube, and consists of a spring-loaded mass that is fired with a constant energy against a plunger held against the surface of the concrete (*Fig. 23.1*). The amount of rebound of the mass expressed as the percentage of the initial extension of the spring is shown by the position of a rider on a graduated scale, and recorded as the *rebound number*. Less energy is absorbed by a harder surface, and so the rebound number is higher. A smooth concrete surface is required, but even then there is considerable local variation due to the presence of coarse aggregate particles (giving an abnormally high rebound number) or a void (giving a low number) just below the surface, and therefore a number of readings must be taken and averaged. Typical recommendations are for at least nine readings over an area of 300 mm^2, no two readings being taken within 25 mm of each other

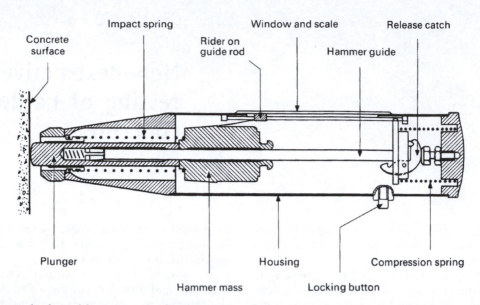

Fig. 23.1 A typical rebound hammer (after Bungey et al., 2006).

or from an edge. Also, the concrete being tested must be part of an unyielding mass; laboratory specimens such as cubes should therefore be held under a stress of about 7 MPa in a compression-testing machine.

The test clearly only measures the properties of the surface zone of the concrete, to a depth of about 25–30 mm. Although the hardness of the concrete cannot in principle be directly related to any other single property, calibrations tests produce empirical correlations with strength that depend on:

- the aggregate type
- the moisture condition of the surface
- the angle of the hammer with the vertical, which will vary since the test must be carried out with the plunger normal to the surface of the concrete.

There is therefore no single universal correlation. *Figure 23.2* shows the relationship between rebound number and strength obtained by students at UCL in laboratory classes over several years. The degree of scatter is somewhat higher than that reported by other workers, the most likely explanation being the inexperience of the operatives. Even with more skilful operatives, strength cannot be predicted with great certainty, but the test is very simple and convenient, and so is often used as a first step in an investigation of *in-situ* concrete, for example to assess uniformity or to compare areas of known good quality and suspect concrete.

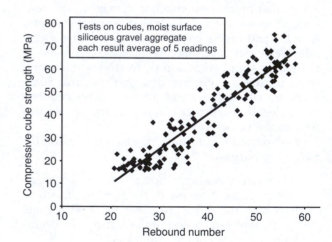

Fig. 23.2 *Relation between strength of concrete and rebound test results (UCL data).*

23.2 Ultrasonic pulse velocity (upv) test

This is an extremely versatile and popular test for both *in-situ* and laboratory use. Its use in Europe is covered by BS EN 12504-4. The test procedure involves measuring the time taken for an ultrasonic pulse to travel through a known distance in the concrete, from which the velocity is calculated. The ultrasonic signal is generated by a piezo-electric

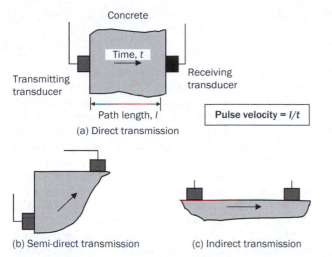

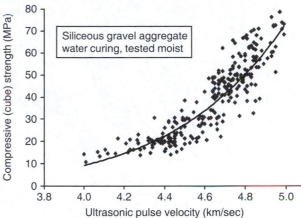

Fig. 23.3 *Measurement of ultrasonic pulse velocity in concrete.*

Fig. 23.4 *Relation between strength and ultrasonic pulse velocity of concrete (UCL data).*

crystal housed in a transducer, which transforms an electric pulse into a mechanical wave. The pulse is detected by a second similar transducer, which converts it back to an electrical pulse, and the time taken to travel between the two transducers is measured and displayed by the instrumentation. Various test arrangements, illustrated in *Fig. 23.3*, are possible. Efficient acoustic coupling between the transducers and the concrete is essential, and is usually obtained by a thin layer of grease. The pulse velocity is independent of the pulse frequency, but for concrete fairly low frequencies in the range 20–150 kHz (most commonly 54 kHz) are used to give a strong signal that is capable of passing through several metres of concrete. Transducers that produce longitudinal waves are normally used, although shear wave transducers are available.

The velocity (*V*) of the longitudinal ultrasonic pulse depends on the material's dynamic elastic modulus (E_d), Poisson's ratio (v) and density (ρ):

$$V = \sqrt{\frac{E_d(1 - v)}{\rho(1 + v)(1 - 2v)}} \qquad (23.1)$$

Hence the upv is related to the elastic properties of the concrete. As with E_d, it can be correlated empirically with strength, but with similar limitations of dependence on constituent materials and – as with the rebound hammer – moisture conditions, pulse velocity being up to 5% higher for the same concrete in a dry compared with a saturated state. *Figure 23.4*

shows UCL students' data obtained on cubes tested in a moist condition. The relation is clearly non-linear, which is to be expected since upv is related to the dynamic modulus, but shows a greater degree of scatter than the strength/E_d relationship in *Fig. 23.7*. Two factors contribute to this; first the upv test requires a little more skill than the resonant frequency test, e.g. in ensuring good acoustic coupling between the transducer and the concrete. Second, the results were obtained on 100 mm cubes, and therefore a smaller and inherently more variable volume of concrete was being tested. Both these factors should be borne in mind when interpreting any non-destructive test data.

The ultrasonic pulse travels through both the hardened cement paste and the aggregate, hence the pulse velocity will depend on the velocity through each and their relative proportions. The velocity through normal-density aggregate is higher than that through paste, which leads to the broad relationships between between upv and strength for paste, mortar and concrete shown in *Fig. 23.5*.

The upv test has the great advantage of being able to assess concrete throughout the signal path, i.e. in the interior of the concrete. Direct transmission is preferred, but for *in-situ* measurements, semi-direct or indirect transmission can be used if access to opposite faces is limited (*Fig. 23.3*). With *in-situ* testing, it is also very important to ensure that measurements are taken where they are not influenced by the presence of reinforcing steel, through which the pulse travels faster (upv = 5.9 km/sec), and which can therefore result in a falsely low transit time.

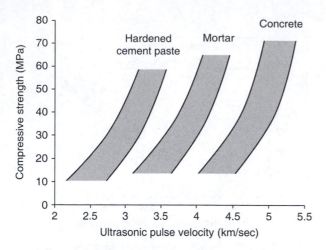

Fig. 23.5 Envelope of strength vs. ultrasonic pulse velocity for hardened cement paste, mortar and concrete (based on Sturrup et al., 1984, and UCL data).

23.3 Resonant frequency test

This is a laboratory test on prepared specimens, and can be used to assess progressive changes in the specimen due, for example, to freeze–thaw damage or chemical attack, so it is therefore particularly useful for generating data in durability testing. It is covered by BS 1881-209.

The specimen is in the form of beam, typically $500 \times 100 \times 100$ mm; the test normally consists of measuring the beam's fundamental longitudinal resonant frequency when it is supported at its mid-point. A value of elastic modulus called the *dynamic elastic modulus* can be obtained from this frequency

(n), the length of the beam (l) and its density (ρ) using the relationship:

$$E_d = 4.n^2.l^2.\rho \qquad (23.2)$$

The resonant frequency is measured with the test arrangement shown in *Fig. 23.6*. The vibration is produced by a small oscillating driver in contact with one end of the beam, and the response of the beam is picked up by a similar device at the other end (*Fig. 23.6a*). The amplitude of vibration varies along the beam as in *Fig. 23.6b*. The frequency of the driver is altered until the maximum amplitude of vibration is detected by the pick-up, indicating resonance (*Fig. 23.6c*). The frequency is normally displayed digitally and manually recorded.

The test involves very small strains but, as we have seen in Chapter 20, concrete is a non-linear material. The dynamic modulus, E_d, is therefore in effect the tangent modulus at the origin of the stress–strain curve, i.e. the slope of line B in *Fig. 20.15b*, and it is higher than the static or secant modulus (E_s) measured in a conventional stress–strain test i.e. the slope of line C in *Fig. 20.15b*. The ratio of E_s to E_d depends on several factors, including the compressive strength, but is normally between 0.8 and 0.85.

As with the static modulus, for a particular set of constituent materials the dynamic modulus and strength can be related; *Fig. 23.7* shows data obtained by students at UCL. The amount of scatter is less than that for rebound hammer vs. strength (*Fig. 23.2*), mainly because the dynamic modulus gives an average picture of the concrete throughout the beam, not just at a localised point. The relationship is clearly nonlinear, as with those for static modulus and strength given in equations 20.13 and

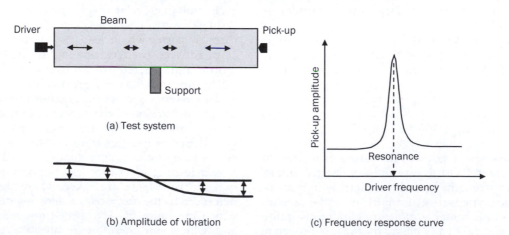

Fig. 23.6 Measurement of the longitudinal resonant frequency of a concrete beam.

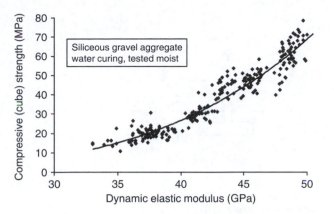

Fig. 23.7 Relation between strength and dynamic elastic modulus of concrete (UCL data).

20.14. The applicability of relationships such as those in *Fig. 23.7* to only a restricted range of parameters (aggregate type, curing conditions etc.) must be emphasised.

It is also possible to set up the support and driver system to give torsional or flexural vibration of the beam, so that the dynamic shear or flexural modulus can be obtained.

23.4 Near-to-surface tests

The specific need to assess the strength of *in-situ* concrete has led to the development of a range of tests in which the surface zone is penetrated or fractured. Either the amount of penetration or the force required for the fracture is measured, and the strength estimated from previous calibrations. The limited amount of damage incurred does not significantly affect the structural performance of the concrete elements or members, but it does normally require making good after the test for aesthetic or durability requirements. There are five main types of test, illustrated in *Fig. 23.8*:

- In penetration resistance tests (*Fig. 23.8a*) a high-strength steel bolt or pin is fired into the concrete and the depth of penetration measured.
- In pull-out tests, the force needed to pull out a bolt or similar device from the concrete is measured. The device can either be cast into the concrete, as in *Fig. 23.8b*, which involves preplanning (although a version that fits into an under-reamed drilled hole has been developed) or be inserted into a drilled hole, as in *Fig. 23.8c*, with fracture being caused by the expansion of the wedge

anchor. Their use in Europe is covered by BS EN 12504-3.

- In pull-off tests (*Fig. 23.8d*) a metal disk is resin-bonded to the concrete surface and is pulled off; the failure at rupture is essentially tensile.
- Break-off tests (*Fig. 23.8e*) involve partial drilling of a core, and then applying a transverse force to cause fracture. The results have been shown to have a reasonable correlation with modulus of rupture strength (see section 21.1.2).

There are a number of commercial versions of each test (Bungey *et al.*, 2006). In each case, to give an estimate of compressive strength prior collaboration in the laboratory is necessary and, as with the truly non-destructive tests already described, considerable scatter is obtained, which must be taken into account when interpreting the results. Also all of the correlations – particularly for the aggregate type – are dependent on a number of factors.

23.5 Other tests

Developments in instrumentation and increasingly sophisticated methods of analysis of the results have led to of a number of significant and useful methods of non-destructive testing. Some examples are:

- Maturity meters, in which a thermocouple is embedded in the concrete at casting and the temperature–time history recorded. This is particularly useful for estimating the early strength development of concrete, as discussed in section 18.3.2.
- Radiography and radiometry. Gamma-ray imaging can show the location of reinforcing and pre-stressing rods and voids within the concrete, and the absorption of gamma rays can give an estimation of density.
- Impact-echo and pulse-echo techniques, in which the response of the concrete to an impact on its surface is measured, by for example a geophone or an accelerometer. Voids beneath slabs or behind walls can be detected, and the pulse-echo technique in particular is useful for integrity testing of concrete piles.
- Acoustic emission, which can detect the sounds produced by concrete cracking. This is mainly suitable for laboratory use.
- Radar systems, in which the reflections and refractions of radar waves generated by a transmitter on the surface of the concrete can be interpreted to give an evaluation of the properties and geometry of subsurface features.

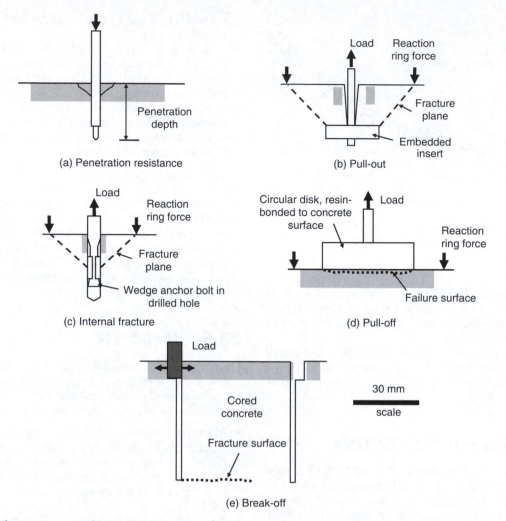

Fig. 23.8 *The main types of partially destructive tests for concrete (all to the same approximate scale).*

With the exception of maturity meters, most of these require considerable expertise in carrying out the tests and interpreting the results, which is best left to specialists. Bungey *et al.* (2006) provide a useful account and comparison of these and other methods.

Finally, in many cases of structural investigation, a direct measurement of compressive strength is often required, and so cores are drilled which are tested after appropriate preparation. This is costly and time-consuming, and is best carried only after as much information as possible has been obtained from non-destructive tests; a combination of tests is often used to give better estimates of properties than are possible from a single test.

References

Bungey JH, Millard SG and Grantham MG (2006). *Testing of concrete in structures*, 4th edition, Taylor and Francis, Abingdon, 2006.

Sturrup VR, Vecchio FJ and Caratin H (1984). Pulse velocity as a measure of concrete compressive strength. In *In-situ/non-destructive testing of concrete* (ed. Malhotra V), ACI Sp-82, American Concrete Institute, Detroit, USA, pp. 201–207.

Chapter 24

Durability of concrete

Durability can be defined as the ability of a material to remain serviceable for at least the required lifetime of the structure of which it forms a part. Standards and specifications increasingly include requirements for a design life, which can typically be 50 or 100 years, but for many structures this is not well defined, so the durability should then be such that the structure remains serviceable more or less indefinitely, given reasonable maintenance. For many years, concrete was regarded as having an inherently high durability, but experience in recent decades has shown that this is not necessarily the case. Degradation can result either from the environment to which the concrete is exposed, for example freeze–thaw damage, or from internal causes within the concrete, as in alkali–aggregate reaction. It is also necessary to distinguish between degradation of the concrete itself and loss of protection and subsequent corrosion of the reinforcing or pre-stressing steel contained within it.

The rate of many of the degradation processes is controlled by the rate at which moisture, air or other aggressive agents can penetrate the concrete. This *penetrability* is a unifying theme when considering durability, and for this reason we shall first consider the various transport mechanisms through concrete – pressure-induced flow, diffusion and absorption – their measurement and the factors that influence their rate. We shall then discuss the main degradation processes, firstly of concrete – chemical attack by sulphates, seawater, acids and the alkali–silica reaction, and physical attack by frost and fire – and then the corrosion of embedded steel. In each case a discussion of the mechanisms involved and the factors that influence these will show how potential problems can be eliminated, or at least minimised, by due consideration of durability criteria in the design and specification of new structures. By way of illustration, some typical recommendations from current European specifications and guidance documents are included. Ignorance of, or lack of attention to, such criteria in the past has led to a thriving and ever expanding repair industry in recent years; it is to be hoped that today's practitioners will be able to learn from these lessons and reduce the need for such activities in the future. It is beyond the scope of this book to discuss repair methods and processes.

24.1 Transport mechanisms through concrete

As we have seen in Chapter 13, hardened cement paste and concrete contain pores of varying types and sizes, and therefore the transport of materials through concrete can be considered as a particular case of the more general phenomenon of flow through a porous medium. The rate of flow will not depend simply on the porosity, but on the degree of continuity of the pores and their size – flow will not take place in pores with a diameter of less than about 150 nm. The term *permeability* is often loosely used to describe this general property (although we shall see that it also has a more specific meaning); *Fig. 24.1* illustrates the difference between permeability and porosity.

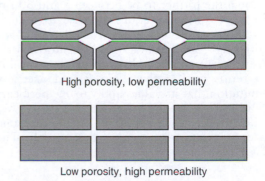

High porosity, low permeability

Low porosity, high permeability

Fig. 24.1 Illustration of the difference between porosity and permeability.

Flow can occur by one of three distinct processes:

- *permeation* – i.e. movement of a fluid under a pressure differential
- *diffusion* – i.e. movement of ions, atoms or molecules under a concentration gradient
- *sorption* – i.e capillary attraction of a liquid into empty or partially empty pores.

Each of these has an associated 'flow constant', defined as follows:

1. In the flow or movement of a fluid under a pressure differential, flow rates through concrete pores are sufficiently low for the flow of either a liquid or gas to be laminar, and hence it can be described by Darcy's law:

$$u_x = -K\partial h/\partial x \qquad (24.1)$$

where, for flow in the x-direction, u_x = mean flow velocity, $\partial h/\partial x$ = rate of increase in pressure head in the x-direction, and K is a constant called the *coefficient of permeability*, the dimensions of which are [length]/[time], e.g. m/sec. The value of K depends on both the pore structure within the concrete and the properties of the permeating fluid. The latter can, in theory, be eliminated by using the *intrinsic permeability* (k) given by:

$$k = K\eta/\rho \qquad (24.2)$$

where η = coefficient of viscosity of the fluid and ρ = unit weight of the fluid. k has dimensions of [length]^2 and should be a property of the porous medium alone and therefore applicable to all permeating fluids. However, for liquids it depends on the viscosity being independent of the pore structure, and for HCP with its very narrow flow channels, in which a significant amount of the water will be subject to surface forces, this may not be the case. Furthermore, comparison of k values from gas and liquid permeability tests has shown the former to be between 5 and 60 times higher than the latter, a difference attributed to the flow pattern of a gas in a narrow channel differing from that of a liquid (Bamforth, 1987). It is therefore preferable to consider permeability in terms of K rather than k, and accept the limitation that its values apply to one permeating fluid only, normally water.

2. The movement of ions, atoms or molecules under a concentration gradient is described by Fick's law:

$$J = -D\partial C/\partial x \qquad (24.3)$$

where, for the x-direction, J = transfer rate of the substance per unit area normal to the x-direction,

$\partial C/\partial x$ = concentration gradient and D is a constant called the *diffusivity*, which has the dimensions of [length]^2/[time], e.g. m^2/sec. Defining diffusivity in this way treats the porous solid as a continuum, but the complex and confining pore structure within concrete means that D is an effective, rather than a true, diffusion coefficient. We are also interested in more than one type of diffusion process, for example moisture movement during drying shrinkage, or de-icing salt diffusion through saturated concrete road decks. Furthermore, in the case of moisture diffusion (in, say, drying shrinkage) the moisture content within the pores will be changing throughout the diffusion process. There is, however, sufficient justification to consider D as a constant for any one particular diffusion process, but it should be remembered that, as with the permeability coefficient K, it is dependent on both the pore structure of the concrete and the properties of the diffusing substance.

3. Adsorption and absorption of a liquid into empty or partially empty pores occur by capillary attraction. Experimental observation shows that the relationship between the depth of penetration (x) and the square root of the time (t) is bi- or tri-linear (*Fig. 24.2*), with a period of rapid absorption in which the larger pores are filled being followed by more gradual absorption (Buenfeld and Okundi, 1998). A constant called the *sorptivity (S)* can be defined as the slope of the relationship (normally over the initial period), i.e.:

$$x = S.t^{0.5} \qquad (24.4)$$

As before, S relates to a specific liquid, often water. It has the dimensions of [length]/[time]$^{0.5}$, e.g. mm/sec$^{0.5}$.

Different mechanisms will apply in different exposure conditions. For example, permeation of seawater will occur in the underwater regions of concrete

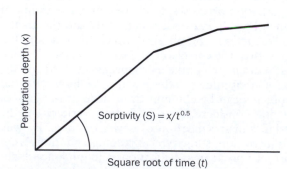

Fig. 24.2 Typical form of results from sorptivity tests.

offshore structures, diffusion of chloride ions will occur when de-icing salts build up on concrete bridge decks and rain water falling on dry concrete will penetrate by absorption.

24.2 Measurement of flow constants for cement paste and concrete

24.2.1 PERMEABILITY

Permeability is commonly measured by subjecting the fluid on one side of a concrete specimen to a pressure head, and measuring the steady-state flow rate that eventually occurs through the specimen, as illustrated in *Fig. 24.3*. The specimen is normally a circular disc, the sides of which are sealed to ensure uniaxial flow. If the fluid is incompressible, i.e. it is a liquid such as water, the pressure head gradient through the specimen is linear, and Darcy's equation reduces to:

$$\Delta Q/\Delta A = -K.\Delta P/l \qquad (24.5)$$

where ΔQ = volumetric flow rate, ΔA = total cross-sectional area of flow perpendicular to the z-direction, ΔP = pressure head and l = length of flow path.

Much of the fundamental work on the permeability of cement paste to water was carried out by Powers and colleagues (Powers *et al.*, 1954; Powers, 1958). As the cement hydrates, the hydration products infill the skeletal structures, blocking the flow channels and hence reducing the permeability. As might be expected from our earlier description of cement hydration in Chapter 13, the reduction of permeability is high at early ages, when hydration is proceeding rapidly. In fact, as shown in *Fig. 24.4*, it reduces by several orders of magnitude in the first 2–3 weeks after casting.

Although, as discussed above, permeability and porosity are not necessarily related (*Fig. 24.1*) there is a general non-linear correlation between the two for cement paste, as shown in *Fig. 24.5*. The greatest reduction in permeability occurs for porosities reducing from about 40 to 25%, where increased hydration product reduces both the pore sizes and the sizes of the flow channels between them. Further hydration product, although still reducing porosity significantly, results in much lower changes in permeability. This explains the general form of *Fig. 24.5*, and also accounts for the effect of water:cement ratio on permeability shown in *Fig. 24.6* for a constant degree of hydration. At water:cement ratios above about 0.5 the capillary pores form

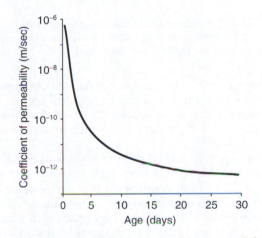

Fig. 24.4 *The effect of hydration on the permeability of cement paste (w:c = 0.7) (after Powers et al., 1954).*

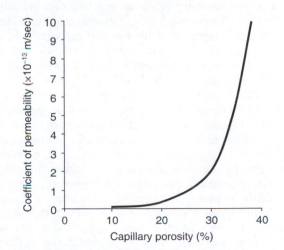

Fig. 24.5 *The relationship between permeability and capillary porosity of hardened cement paste (after Powers, 1958).*

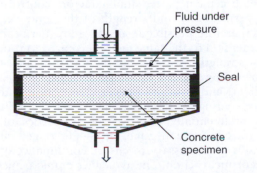

Fig. 24.3 *A simple test system for measuring concrete permeability under steady-state flow.*

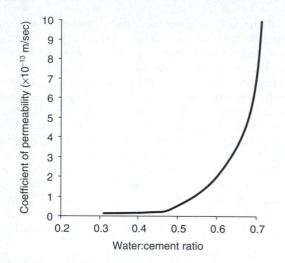

Fig. 24.6 *The relationship between permeability and water:cement ratio of mature cement paste (93% hydrated) (after Powers* et al., *1954).*

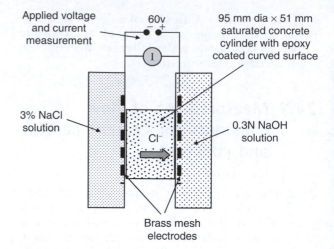

Fig. 24.7 *Rapid Chloride Permeability test (after ASTM C1202).*

an increasingly continuous system, with consequent large increases in permeability. We shall see later in the chapter that many recommendations for durable concrete limit the water:cement ratio to a maximum value below this.

It is apparent from the above arguments and from those in Chapter 13 that high strength and low permeability both result from low porosity, and in particular a reduction in the volume of the larger capillary pores. In general, higher strength implies lower permeability, although the relationship is not linear, and may be different for different curing histories and cement types.

The permeability of a concrete will also be influenced by the permeability of the aggregate. Many of the rock types used for natural aggregates have permeabilities of the same order as that of cement paste, despite having relatively low porosities. Lightweight aggregates, which are highly porous, can have much higher permeabilities. However, values for the permeability of the composite concrete, despite considerable variation in reported values from different sources, are normally in the range 10^{-8} to 10^{-12} m/sec (Lawrence, 1985), i.e. between two and four orders of magnitude higher than that of either the cement paste or aggregate. This is primarily owing to the presence of defects or cracks, particularly in the weaker interface or transition zone between the HCP and aggregate, which we saw in preceding chapters are present in the concrete before any load is applied.

Permeability testing of concrete by fluid penetration under pressure (as in *Fig. 24.3*) can have considerable

experimental difficulties, such as avoiding leaks around the specimen and the protracted timescales necessary for measuring flow rates through low-permeability concrete. An alternative indirect method of measuring permeability more rapidly that has become increasingly popular in recent years is the rapid chloride permeability test (ASTM C1202). The test, illustrated in *Fig. 24.7*, involves the application of a voltage between two sides of a concrete specimen with solutions of sodium hydroxide and sodium chloride on opposite sides. The chloride ions are driven through the concrete, and as they penetrate it the conductivity of the pore water and the current readings increase. The test is continued for six hours and the total charge passed (current × time) determined.

Some results from this test that show the effect of water:cement ratio and the incorporation of additions are shown in *Fig. 24.8*. These show that, as with cement paste, similar factors control both the permeability and strength of the concrete, and it is therefore possible to produce low permeability by attention to the same factors required to produce high strength. More generally, these include using a low water:cement ratio and an adequate cement content, and ensuring proper compaction and adequate curing. As discussed in Chapter 21, additions can preferentially improve the properties of the interface transition zone, although longer curing periods are necessary to ensure continuance of the pozzolanic reaction. The avoidance of microcracking from thermal or drying shrinkage strains and premature or excessive loading is also important.

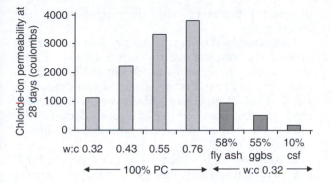

Fig. 24.8 Results from the Rapid Chloride Ion Permeability test (afer Zhang et al., 1999).

Table 24.1 Chloride ion diffusivities of paste and concrete

Binder	Water:cement ratio	Diffusivity (m^2/sec)
Paste*		
100% PC	0.4	2.6×10^{-12}
100% PC	0.5	4.4×10^{-12}
100% PC	0.6	12.4×10^{-12}
70% PC + 30% pfa	0.5	1.47×10^{-12}
30% PC + 70% ggbs	0.5	0.41×10^{-12}
Concrete†		
100% PC	0.4	18×10^{-12}
100% PC	0.5	60×10^{-12}
60% PC + 40% pfa	0.4	2×10^{-12}
25% PC + 75% ggbs	0.4	2×10^{-12}

*, from Page *et al.*, 1981; †, from Buenfeld *et al.*, 1998.
PC, Portland cement.

24.2.2 DIFFUSIVITY

The principle of diffusivity testing is relatively simple. A high concentration of the diffusant is placed on one side of a suitable specimen (normally a disc) of HCP, mortar or concrete, and the diffusion rate calculated from the increase of concentration on the other side. In the case of gas diffusion, the high-concentration side may be an atmosphere of the pure gas; in the case of salts, a high-concentration aqueous solution would be used. The test is therefore similar to the fluid permeability test without the complication of high pressure. It is generally found that, after an initial period for the diffusant to penetrate through the specimen, the concentration on the 'downstream' side increases linearly with time. The diffusivity will change if the moisture content of the concrete changes during the test, and so the specimens must be carefully conditioned before testing.

Control of test conditions is therefore important, and diffusivity measurements from different test programmes are not entirely consistent. *Table 24.1* shows values of chloride-ion diffusivity that were obtained on mature saturated pastes and concrete. As with permeability the values are higher for concrete than for paste, but in both cases the beneficial effects of low water:cement ratios and the use of additions are clear.

24.2.3 SORPTIVITY

Sorptivity can be calculated from measurements of penetration depth, and tests are carried out on samples in which penetration is restricted to one direction only, such as cylinders with the curved surface sealed with a suitable bitumen or resin coating. The penetration depth at a particular time can be measured by splitting a sample open, but this requires a considerable number of samples to obtain a significant number of results. It is often more convenient to measure weight gain, in which case the sorptivity is expressed as the amount of water absorbed per unit exposed surface/square root of time, e.g. with units of $kg/m^2/hr^{0.5}$ or similar. Penetration calculations can be made if the concrete's porosity is known (which can be conveniently found by drying the specimen after the test), and the results can be expressed in the normal way.

Values of sorptivity at various distances from the surface of a concrete slab are shown in *Fig. 24.9*. These were obtained on slices of cores cut from concrete slabs 28 days old, which had been moist-cured for 4 days and then air-cured for 24 days. The sorptivity decreases with depth, attributed to the air drying causing imperfect curing of the surface zone. However, although the similar strength mixes containing additions had similar sorptivities in the 15-mm thick surface zone, they generally had lower values than the plain Portland cement concrete at greater depth, again indicating the advantages to be gained from these materials with sufficient curing.

A number of tests have been developed to measure the absorption and permeability characteristics of *in-situ* concrete while still in place, i.e. avoiding the need to cut cores. These all measure the penetration rate of a fluid (normally air or water) into the concrete, either through the concrete surface or outwards from a hole drilled into the concrete.

One popular test of this type is the Initial Surface Absorption Test (ISAT), shown in *Fig. 24.10*. It is

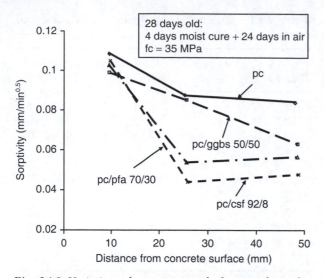

Fig. 24.9 *Variation of sorptivity with distance from the cast surface of concrete made with Porland cement and additions (after Bamforth and Pocock, 1990).*

covered by BS 1881-5. A cap is clamped to the concrete surface and a reservoir of water is set up with a constant head of 200 mm. The reservoir is connected through the cap to a capillary tube set level with the water surface. At the start of the test, water is allowed to run through the cap (thus coming into contact with the concrete surface) and to fill the capillary tube. The rate of absorption is then found by closing off the reservoir and observing the rate of movement of the meniscus in the capillary tube. Readings are taken at standard times after the start of the test (typically 10 mins, 20 mins, 30 mins, 1 hour and 2 hours) and expressed as flow rate per surface area of the concrete, e.g. in units of ml/m^2/sec. The rate drops off with time and in general increases with the sorptivity of the concrete.

Typical results showing the effect of the water:cement ratio of the concrete and the duration of the initial water curing period on the 10-min ISAT value for tests carried out on 28-day-old concrete are shown in *Fig. 24.11*. Not surprisingly, decreasing water:cement ratio and increased curing time both decrease the ISAT values; the results clearly reinforce the importance of curing.

In common with other tests of this type, the ISAT has two main disadvantages. Firstly, the results depend on the moisture state of the concrete at the start of the test, which is particularly difficult to control if the test is carried out *in situ*. Secondly, the flow-path of the fluid through the concrete is not unidirectional but diverges; a fundamental property of the concrete is therefore not measured and it is difficult to compare results from different test systems. However, the tests all measure some property of the surface layers of the concrete and, as we shall see, this is all important in ensuring good durability.

24.3 Degradation of concrete

The degradation agencies that affect concrete can be divided into two broad groups:

• those whose action is initially chemical, before subsequently leading to loss of physical integrity; these include sulphates, seawater, acids and alkali–silica reactions
• those which directly lead to physical effects, such as freeze–thaw and fire.

We will now consider each of these in turn.

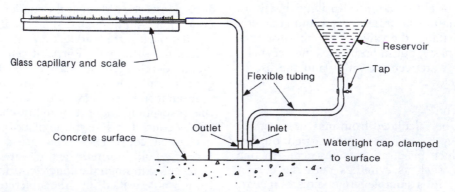

Fig. 24.10 *Initial Surface Absorption Test (after Bungey* et al., *2006).*

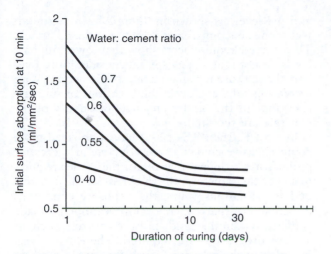

Fig. 24.11 *Effect of water:cement ratio and initial curing on surface absorption of concrete as measured by the ISAT test (after Dhir* et al., *1987).*

24.3.1 ATTACK BY SULPHATES

We have seen in Chapter 13 that a controlled amount of calcium sulphate, in the form of gypsum, is added to Portland cement during its manufacture to control the setting process. Further sulphates in fresh concrete can arise from contaminated aggregates, a particular problem in some Middle Eastern countries. Sources of sulphates that can penetrate hardened concrete include ground-water from some clay soils, fertilisers and industrial effluent, and so we can see that any problems mainly occur in concrete in contact with the ground e.g. in foundations, floor slabs and retaining walls. Sodium, potassium, calcium and magnesium sulphates are all common, and when in solution these will all react with components of the hardened cement paste.

We briefly described the nature of the problem when discussing sulphate-resisting Portland cement (SRPC) in Chapter 13; specifically, the sulphates and the hydrated aluminate phases in the hardened cement paste react to form ettringite. Using the cement chemists' shorthand notation that we described in Chapter 13, the reaction of calcium sulphate with the monosulphate hydrate is:

$$2C\check{S} + C_3A.C\check{S}.12H + 20H \rightarrow C_3A.3C\check{S}.32H \tag{24.6}$$

and with the direct hydrate:

$$3C\check{S} + C_3A.6H + 26H \rightarrow C_3A.3C\check{S}.32H \tag{24.7}$$

Both reactions are expansive, with the solid phases increasing significantly in volume, causing expansive forces and, possibly, disruption. Sodium sulphate (N$\check{S}$.10H where N = Na$_2$O) also forms ettringite by reacting with the hydrated aluminate:

$$3N\check{S}.10H + 2C_3A.6H \rightarrow$$
$$C_3A.3C\check{S}.32H + 2AH + 3NH + 5H \tag{24.8}$$

but in addition it reacts with the calcium hydroxide in the HCP:

$$N\check{S}.10H + CH \rightarrow C\check{S}.2H + NH + 8H$$
$$[NH = 2NaOH] \tag{24.9}$$

This is analogous to acid attack, and in flowing water it is possible for the calcium hydroxide to be completely leached out.

With magnesium sulphate (M$\check{S}$.7H where M = MgO), a similar reaction to (24.9) takes place, but the magnesium hydroxide formed is relatively insoluble and poorly alkaline; this reduces the stability of the calcium silicate hydrate, which is also attacked:

$$C_3S_2H_3 + 3M\check{S}.7H \rightarrow 3C\check{S}.2H + 3MH + 2S.aq \tag{24.10}$$

Since it is the calcium silicate hydrate that gives the hardened cement its strength, attack by magnesium sulphate can be more severe than that by other sulphates. In each case, attack occurs only when the amount of sulphate present exceeds a certain threshold; the rate of attack then increases with increasing concentration of sulphate, but at a reducing rate of increase above about 1% concentration. Also, the rate of attack will be faster if the sulphate is replenished, for example if the concrete is exposed to flowing groundwater.

Concrete that has been attacked has a whitish appearance; damage usually starts at edges and corners, followed by progressive cracking and spalling, eventually leading to complete breakdown. Although this stage can be reached in a few months in laboratory tests, it normally takes several years in the field.

For any given concentration and type of sulphate, the rate and amount of deterioration decrease with:

- the C$_3$A content of the cement, hence the low C$_3$A content of sulphate-resisting Portland cement
- a higher cement content and lower water:cement ratio of the concrete; higher quality concrete is less vulnerable owing to its lower permeability. *Figure 24.12* shows some results illustrating the combined effect of C$_3$A content and concrete composition

- the incorporation of additions, which as we have seen can decrease the permeability and reduce the amount of free lime in the HCP, but also 'dilute' the C_3A.

Recommendations for suitable concrete for use in sulphate-containing environments follow from a detailed knowledge of the above factors. For example, the relevant European specification, BS EN 206, classifies groundwater and soil as slightly, moderately or highly aggressive (denoted as exposure classes XA1, XA2 and XA3, respectively) according to the sulphate (SO_4^{2-}), carbon dioxide (CO_2), ammonium (NH_4^+) and magnesium (Mg^{2+}) contents

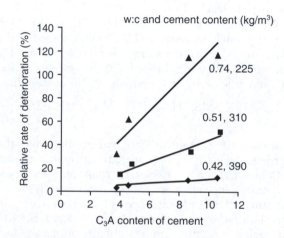

Fig. 24.12 *The effect of C_3A content of the cement and concrete mix proportions on the deterioration in a soil containing 10% Na_2SO_4 (after Verbeck, 1968).*

and pH level, as shown in *Table 24.2* for groundwater, thus encompassing sulphate and acid attack. The concrete requirements show that lower sulphate levels do not require any special considerations over and above those for all concrete construction. With increasing sulphate levels combinations of sulphate resistance of the binder and overall quality of the concrete are required.

In the UK, the BRE (2005) have published more comprehensive guidelines for concrete in aggressive ground that include provision for the service requirements of the structure, the thickness of the concrete and for situations where the exposure conditions are so severe that concrete alone cannot be made sufficiently durable and extra measures, such as surface protection, are required. These have been included in the relevant British Standard specification (BS 8500 part 1, 2006).

We should also mention here *delayed ettringite formation* (DEF). The ettringite that is formed from the aluminates during cement hydration at normal temperatures (described in Chapter 13) breaks down at temperatures higher than about 70°C, and both the sulphates and aluminates appear to be reabsorbed by the C-S-H. After cooling, the sulphate becomes available and ettringite is re-formed. By this time the cement paste has hardened, so the formation of expansive ettringite can lead to disruption. This is therefore a form of 'internal sulphate attack'. High temperatures may be deliberately applied to pre-cast concrete to increase strength gain, or more commonly may be the result of heat of hydration effects in large pours, which we described in Chapter 19. The damage may not become

Table 24.2 BS EN 206 exposure classes and concrete requirements for attack by ground water

Exposure class	XA1	XA2	XA3
Groundwater composition			
SO_4^{2-} (mg/l)	200–600	600–3000	3000–6000
pH	5.5–6.5	4.5–5.5	4.0–4.5
CO_2 (mg/l)	15–40	40–100	>100
NH_4^+ (mg/l)	15–30	30–60	60–100
Mg^{2+} (mg/l)	300–1000	1000–3000	>3000
Concrete requirements			
Minimum strength class (see section 22.1.1)	C30/37	C30/37	C35/45
Maximum water:cement ratio	0.55	0.50	0.45
Minimum cement content	300	320	360
Cement type	any	SRPC or CEM I + addition*	

*, CEM I + an addition with the equivalent performance.

apparent until some considerable time after casting. DEF can be avoided by reducing the temperature rise by any of the methods outlined in Chapter 19, the use of fly ash or ggbs being particularly useful as their effect on the chemistry and microstructure of HCP also seems to be beneficial.

24.3.2 THE THAUMASITE FORM OF SULPHATE ATTACK

This form of attack, known as TSA for short, also involves sulphates, but has distinct differences from the sulphate attack described above, and can have more serious consequences.

Thaumasite is a rare mineral that occurs naturally in some basic rocks and limestones. It is a compound of calcium silicate, carbonate and sulphate with the formula $CaSiO_3.CaCO_3.CaSO_4.15H_2O$. To be formed in concrete and mortar it requires:

- a source of calcium silicate, clearly available from the hydrated or unhydrated Portland cement
- a source of sulphate ions, for example from soil or groundwater
- a source of carbonate, most commonly limestone aggregate or fillers or formed from the bicarbonates arising from the atmospheric carbon dioxide dissolved in pore water
- a wet, cold (below 15°C) environment.

Clearly all these requirements do not often occur together, the most common case where they do being in concrete made with limestone aggregate used for the foundations of structures in sulphate-bearing soils in temperate or cold climates. Sulphides (e.g. pyrites, FeS_2) can also be a problem, as when they are exposed to the atmosphere when the soil is disturbed during excavation they will oxidise to form more sulphates. Because the attack involves the calcium silicate hydrates, it can lead to complete disintegration of the cement paste, which turns into a soft, white, mushy mass. The features of a concrete member that has been affected by TSA are a surface layer of the mushy mass, below which are regions that contain progressively decreasing amounts of thaumasite in cracks and voids, particularly around the aggregate particles. As well as reducing the strength of the concrete, the loss of effective cover to any reinforcing steel makes this more vulnerable to corrosion induced by, say, chloride penetration. (We will discuss this subject in some detail later in the chapter.)

Incidents of attack are significant but not widespread, the most notable involving buried concrete in house and bridge foundations in the west of England, column building supports in Canada, tunnel linings, sewage pipes and road sub-bases. The rate of attack is generally slow and the incidents that came to light in England during the 1990s were in structures that were several years old. These led to a major investigation, which included surveys of potentially vulnerable structures and laboratory studies. The reports of the investigation (Thaumasite Expert Group, 1999 and Clarke and BRE, 2002) make interesting reading, not only on the subject itself but also as an illustration of how government, industry and universities react together to a problem of this nature.

Following these extensive investigations, the most recent guidelines for concrete in aggressive ground published by the BRE (2005) do not include TSA as a separate consideration from other forms of sulphate attack. These differ from the previous recommendations because it has become apparent that the carbonates required for TSA can arise from atmospheric carbon dioxide (Collett et al., 2004), and hence protection against TSA is obtained by stricter requirements for all classes of exposure, rather than by treating it as a special case when limestone aggregates are used.

24.3.3 SEAWATER ATTACK

Concrete in seawater is exposed to a number of possible degradation processes simultaneously, including the chemical action of the sea salts, wetting and drying in the tidal zones and just above, abrasion from waves and water-borne sediment and, in some climates, freezing and thawing.

The total soluble salt content of seawater is typically about 3.5% by weight, the principal ionic contributors and their typical amounts being Cl^-, 2.0%; Na^+, 1.1%; SO_4^{2-}, 0.27%; Mg^{2+}, 0.12%; and Ca^{2+}, 0.05%. The action of the sulphate is similar to that of pure sulphate solutions described above, with the addition of some interactive effects. Importantly, the severity of the attack is not as great as for a similar concentration of sulphate acting alone, and there is little accompanying expansion. This is due to the presence of chloride ions; gypsum and ettringite are more soluble in a chloride solution than in pure water, and therefore tend to be leached out of concrete by seawater, and hence their formation does not result in expansive disruption. Magnesium sulphate also participates in the reactions as in equation 24.10, and a feature of concrete damaged by seawater is the presence of white deposits of $Mg(OH)_2$, often called *brucite*. In experiments on concrete permanently saturated with seawater, a form of calcium carbonate called *aragonite* has also been found, arising from the reaction of dissolved carbon dioxide with calcium hydroxide. The brucite and aragonite can have a pore-blocking effect, effectively reducing the permeability of the concrete near the surface (Buenfeld and Newman, 1984).

In areas subject to wetting and drying cycles, salts will crystallise as the water evaporates, which can lead to a high salt concentration on or in the concrete's surface and to potential disruption from the pressure exerted by the crystals as they rehydrate and grow during subsequent wetting–drying cycles – a process known as *salt weathering*. This can be compounded by damage from freeze–thaw cycles or wave action, depending on the environment. These areas therefore tend to be the most vulnerable. The key to the elimination or at least reduction of all these problems is, not surprisingly, the use of a low-permeability concrete, perhaps combined with some limits on the C_3A content of the cement, or the use of additions. However, for the reasons given above, the degradation processes in many climates do not cause rapid deterioration, which explains why concrete of even relatively modest quality has a long and distinguished history of use in marine structures, both coastal and offshore.

The salts in seawater can contribute to two other, potentially much more critical, degradation processes, namely alkali–aggregate reaction and corrosion of embedded steel. Both are discussed later.

24.3.4 ACID ATTACK

We have seen that the hardened cement paste binder in concrete is alkaline, and therefore no Portland cement concrete can be considered acid resistant. However, it is possible to produce a concrete that is adequately durable for many common circumstances by giving attention to low permeability and good curing. In these circumstances, attack is only considered significant if the pH of the aggressive medium is less than about 6.

Examples of acids that commonly come into contact with concrete are dilute solutions of carbon dioxide (CO_2) and sulphur dioxide (SO_2) in rain water in industrial regions, and carbon dioxide (CO_2) and hydrogen sulphide (H_2S)-bearing groundwater from moorlands. The acids attack the calcium hydroxide within the cement paste, converting it, in the case of CO_2, into calcium carbonate and bicarbonate. The latter is relatively soluble, and leaches out of the concrete, destabilising it. The process is thus diffusion controlled, and progresses at a rate approximately proportional to the square root of time. The C-S-H may also be attacked, as can calcareous aggregates such as limestone. The rate of attack increases with reducing pH.

As mentioned above, the quality of the concrete is the most important factor in achieving acid resistance, but well-cured concretes containing additions also have greater resistance owing to the lower calcium hydroxide content as a result of the pozzolanic reaction. In cases where some extra acid resistance is required, such as in floors of chemical factories, the surface can be treated with diluted water glass (sodium silicate), which reacts with the calcium hydroxide forming calcium silicates, blocking the pores. In more aggressive conditions, the only option is to separate the acid and the concrete by, for example, applying a coating of epoxy resin or other suitable paint system to the concrete.

24.3.5 ALKALI–AGGREGATE AND ALKALI–SILICA REACTION

We described the general nature and composition of natural aggregates in Chapter 17. Among many other constituents, they may contain silica, silicates and carbonates, which in certain mineral forms can react with the alkaline hydroxides in the pore water derived from the sodium and potassium oxides in the cement. The general term for this is *alkali–aggregate reaction* (AAR), but the most common and important reaction involves active silica, and is known as the *alkali–silica reaction* (ASR). The product is a gel, which can destroy the bond between the aggregate and the HCP, and which absorbs water and swells to a sufficient extent to cause cracking and disruption of the concrete. Compared to most other forms of degradation it is particularly insidious, as it starts within the concrete from reactions between the initial constituent materials.

For the reaction to occur, clearly both active silica and alkalis must be present. In its reactive form, silica occurs as the minerals opal, chalcedony, crystobalite and tridymite and as volcanic glasses. These can be found in some flints, limestones, cherts and tuffs. The sources of such aggregates include parts of the USA, Canada, South Africa, Scandinavia, Iceland, Australia, New Zealand and the midlands and southwest of England. Only a small proportion of reactive material in the aggregate (as low as 0.5%) may be necessary to cause disruption to the concrete.

In unhydrated cement, sodium and potassium oxides (Na_2O and K_2O) are present in small but significant quantities (see *Table 13.1*), either as soluble sulphates (Na_2SO_4 and K_2SO_4) or as a mixed salt ($Na,K)SO_4$. There is also a small amount of free CaO, which is subsequently supplemented by $Ca(OH)_2$ (portlandite) from the hydration reactions of C_3S and C_2S. During hydration, these sulphates take part in a reaction with the aluminate phases in a similar

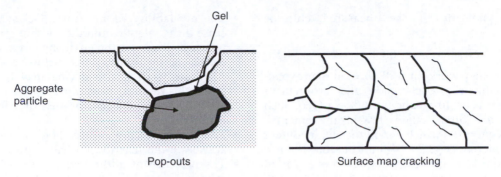

Fig. 24.13 Typical cracking patterns resulting from the alkai–silica reaction.

way to gypsum (see Section 13.1.3), the product again being ettringite. With sodium, potassium and hydroxyl ions going into solution:

$$3(Na,K)SO_4 + 3CaO.SiO_2 + 3Ca(OH)_2 + 32H_2O \rightarrow$$
$$(C_3A)$$

$$3CaO.SiO_2.3CaSO_4.32H_2O + 3Na^+ + 3K^+ + 6OH^-$$
$$\text{(ettringite)} \qquad \qquad \text{(in solution)}$$
$$(24.11)$$

The resulting pH of the pore water is 13–14, higher than that of saturated calcium hydroxide solution alone. Alkalis may also be contributed by some admixtures, fly ash and ggbs, and by external sources such as aggregate impurities, seawater or road de-icing salts.

The reactions between the reactive silica and the alkalis in the pore solution within the concrete to form the alkali–silicate gel occur first at the aggregate–cement paste interface. The nature of the gel is complex, but it is clear that it is a mixture of sodium, potassium and calcium alkali–silicates. It is soft, but imbibes a large quantity of water, possibly by osmosis, and the sodium and potassium silicates swell considerably (the calcium silicates are non-swelling). The hydraulic pressure that is developed leads to overall expansion of the concrete and can be sufficient to cause cracking of the aggregate particles, the HCP and the transition zone between the two.

Continued availability of water causes enlargement and extension of the cracks, which eventually reach the outer surface of the concrete, forming either 'pop-outs' if the affected aggregate is close to the surface, or more extensive crazing, or 'map cracking', on the concrete surface, as illustrated in *Fig. 24.13*. These surface cracks are often highlighted by staining from the soft gel oozing out of them. In general, cracking adversely affects the appearance

and serviceability of a structure before reducing its load-carrying capacity.

The whole process is often very slow, and cracking can take years to develop in structural concrete. A description was first published in the USA by Stanton in 1940, since when numerous examples have been reported in many countries. Over 100 cases were identified in the UK between 1976 and 1987, triggering much research aimed at understanding and quantifying the mechanisms involved, determining its effects on structural performance and providing guidance for minimising the risk in new concrete. The latter can be considered successful, as there have been very few, if any, confirmed cases of ASR in the UK since 1987.

Laboratory tests on ASR often take the form of measuring the expansion of concrete or mortar specimens stored over water at 38°C to accelerate the reactions. The mortar is made with crushed aggregates, thereby speeding up any reaction and expansion, but there are sometimes conflicting results from the two methods. Even though such tests have not always satisfactorily explained all field observations, the most important factors influencing the amount and rate of reaction can be summarised as follows:

• The amount of alkali available. Since sodium and potassium ions react in a similar way, it is normal practice to convert the amount of potassium oxide in a cement to an equivalent amount of sodium oxide (by the ratio of the molecular weights) and to express the alkali content as the weight of sodium oxide equivalent (Na_2Oeq), calculated as:

$$Na_2Oeq = Na_2O + 0.658.K_2O \qquad (24.12)$$

If the aggregates contain any salts, then their contribution can be obtained from the sodium oxide equivalent of the measured chloride content:

$$Na_2Oeq = 0.76.Cl^- \qquad (24.13)$$

Stanton's early work on ASR showed that expansion is unlikely to occur if the alkali content of the cement is less than about 0.6% Na_2Oeq. Such cements are often called 'low-alkali cements'. More recent tests, which varied the alkali content of the concrete (*Fig. 24.14*) indicate that there is a threshold level (typically about 3.5–4 kg/m^3 of concrete) below which no disruption will occur, even with reactive aggregates.

- The amount of active silica. Some results from tests on concrete are shown in *Fig. 24.15*. The expansion increases with active silica content in the aggregate, but in two of the three sets of results only up to a certain content, beyond which the expansion reduces. There is thus a pessimum ratio of silica:alkali for maximum expansion, which varies for different combinations of materials, but is probably the point at which the amount of reactive silica is just sufficient to react with all the alkali present. The ratio usually lies in the range 3.5 to 5.5. One explanation is that at high silica contents, above the pessimum, greater proportions of the sodium and potassium are tied up, reducing the pH and increasing the amount of the non-expansive calcium alkali silicate produced.
- The aggregate particle size, which affects the amount of reactive silica exposed to the alkali; fine particles (20–30 μm) can lead to expansion within a few weeks, larger ones only after many years.

- The availability of moisture. Gel swelling will cease if the relative humidity within the concrete, which depends on the environment and the concrete's permeability, falls below about 85%. Alternate wetting and drying may be the most harmful situation, possibly because it can lead to local high concentrations of the reacting materials.
- The ambient temperature. Higher temperatures accelerate the reaction, at least up to 40°C.
- The presence of additions, e.g. fly ash, ggbs or microsilica, and lithium salts. Microsilica is particularly effective (*Fig. 24.16*). Even though fly

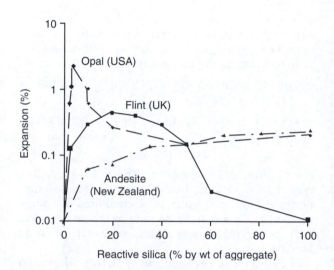

Fig. 24.15 The effect of active silica content of aggregates on the expansion of concrete from the alkali–silica reaction (adapted from Hewlett, 1998).

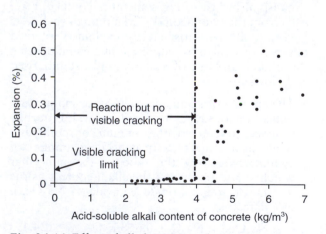

Fig. 24.14 Effect of alkali content of concrete on expansion and cracking after 200 days in alkali–silica reaction tests (after Hobbs, 1986).

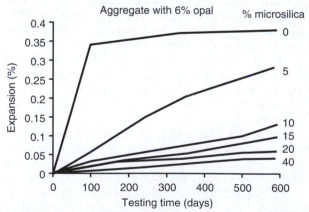

Fig. 24.16 The effect of the microsilica content of the binder on the expansion of concrete from the alkali–silica reaction (adapted from Sims and Poole, 2003).

ash and ggbs themselves contain quantities of alkalis, if the rate of addition is sufficiently high then these do not contribute to the formation of the gel and so need not be taken into account. It seems that the hydrates formed in the secondary pozzolanic reaction bind the alkalis, either from the cement or contained in the additions, which are therefore not available to react with the aggregate. However, the exact mechanisms and quantitative nature of the role of additions are complex and still unclear.

Once started, the only effective way of stopping ASR is by eliminating water and ensuring that the concrete remains dry throughout its life; this is clearly impractical in many structural situations. It follows that it is important to reduce or eliminate the risk of ASR occurring by careful selection of materials and concrete mix design. Consideration of the factors influencing the occurrence and rate of ASR described above leads to the following possibilities:

- Avoiding the use of reactive aggregates. This can be more difficult than it sounds, particularly with mixed mineral aggregates.
- Testing aggregates for their potential reactivity. There are a number of tests for aggregate reactivity, several involving accelerated expansion of mortar or concrete. A concrete prism test that was developed in the UK (BS 812-123) has been used by many organisations in the UK, and the results obtained show good correlation with the field performance of aggregates.
- Limiting the amount of alkali in the cement, for example by using a low-alkali cement, i.e. one with an alkali content of less than 0.6% by weight, as discussed above.
- Combining the Portland cement with an addition. However, to be wholly effective significant quantities must be added and minimum cement replacement levels of 25–40% fly ash, 40–50% ggbs and 8% microsilica have been suggested.
- Limiting the total alkali content of the concrete. Alkalis from all sources – cement, additions (but taking account of the factors discussed above), de-icing salts, etc. – should be safely below a threshold value such as that shown in *Fig. 24.14*.

BRE Digest 330 (Building Research Establishment, 2004) contains much more detailed explanations and comprehensive recommendations, and is a good example of the type of document that is of direct value to concrete practitioners.

24.3.6 FROST ATTACK – FREEZE–THAW DAMAGE

In cold climates frost attack is a major cause of damage to concrete unless adequate precautions are taken. We discussed this briefly when considering air-entraining agents in Chapter 14. When free water in the larger pores within the HCP freezes it expands by about 9% and, if there is insufficient space within the concrete to accommodate this, then potentially disruptive internal pressures will result. Successive cycles of freezing and thawing can cause progressive and cumulative damage, which takes the form of cracking and spalling, initially of the concrete surface.

It is the water in the larger capillary pores and entrapped air voids that has the critical effect; the water in the much smaller gel pores (see Chapter 13) is adsorbed onto the C-S-H surfaces, and does not freeze until the temperature falls to about $-78°C$. However, after the capillary water has frozen it has a lower thermodynamic energy than the still-liquid gel water, which therefore tends to migrate to supplement the capillary water, thus increasing the disruption. The disruptive pressure is also enhanced by osmotic pressure. The water in the pores is not pure, but is a solution of calcium hydroxide and other alkalis, and perhaps chlorides from road de-icing salts or seawater; pure water separates out on freezing, leading to salt concentration gradients and osmotic pressures, which increase the diffusion of water to the freezing front.

The magnitude of the disruptive pressure depends on the capillary porosity, the degree of saturation of the concrete (dry concrete will clearly be unaffected) and the pressure relief provided by a nearby free surface or escape boundary. The extent of this pressure relief will depend on:

- the permeability of the material
- the rate at which ice is formed
- the distance from the point of ice formation to the escape boundary. In saturated cement paste, the disruptive pressures will only be relieved if the point of ice formation is within about 0.1 mm of an escape boundary. A convenient way of achieving this is by the use of an air-entraining agent (see Chapter 14), which entrains air in the form of small discrete bubbles, and an average spacing of about 0.2 mm is required.

As we saw in Chapter 13, the capillary porosity of a cement paste or concrete, and hence its susceptibility to frost attack, can be reduced by lowering the water:cement ratio and ensuring that by proper

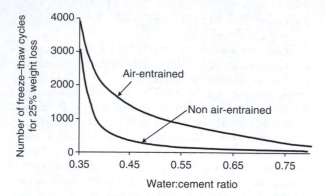

Fig. 24.17 The effect of air-entrainment and water:cement ratio on the freeze–thaw resistance of concrete moist-cured for 28 days (from US Bureau of Reclamation, 1955).

curing hydration is as complete as possible. Bleeding, which results in local high-porosity zones, should also be minimised. The combined effects of air entrainment and water:cement ratio are illustrated in *Fig. 24.17*.

Certain aggregates are themselves susceptible to freeze–thaw damage, and their use must be avoided if a durable concrete is to be achieved. The first sign of damage caused by aggregate disruption is normally pop-outs on the concrete surface. Vulnerable aggregates include some limestones and porous sand-

stones; these generally have high water absorption, but other rocks with high absorption are not vulnerable. Similar consideration of pore size and distribution as for cement paste apply to aggregates; for example, it has been found that pores of about 4–5 μm are critical, since these are large enough to permit water to enter but not large enough to allow dissipation of disruptive pressure. Aggregate size is also a factor, with smaller particles causing less disruption, presumably because the average distance to an escape boundary on the aggregate surface is lower. The only satisfactory way of assessing an aggregate is by its performance when incorporated in concrete, using field experience or laboratory testing.

Some of the recommendations for concrete exposed to freeze–thaw action in the UK (which has a relatively mild climate in this respect compared to many countries) are shown in *Table 24.3*. For each exposure class this gives the option of air entrainment or a higher-quality concrete, and attention to the properties of the aggregate is required for the two most severe cases. The use of a de-icing agent, for example on exposed road surfaces, is included in the definition of the exposure class as it can result in distress from thermal shock when applied.

24.3.7 FIRE RESISTANCE
Concrete is incombustible and does not emit any toxic fumes when exposed to high temperatures. It

Table 24.3   Recommendations for 100-year design life for concrete exposed to freeze–thaw attack (from BS 8500, 2006)

Exposure class and description		Min strength class (see section 22.1)	Max w:c	Min air content (%)	Min cement content (kg/m³)	Aggregate	Cement types (see sections 13.8 and 15.4)
XF1	Moderate water saturation, without de-icing agent	C25/35	0.6	3.5	280		
		C28/35	0.6	–	280		
XF2	Moderate water saturation, with de-icing agent	C25/35	0.6	3.5	280		CEM I, CEM II, CEM III (with max 80% ggbs) SRPC
		C32/40	0.55	–	300		
XF3	High water saturation, without de-icing agent	C25/35	0.6	3.5	280		
		C40/50	0.45	–	340		
XF4	High water saturation, with de-icing agent or seawater	C28/35	0.55	3.5	300	Freeze–thaw resisting aggregate	
		C40/50	0.45	–	340		

w:c, water:cement ratio.

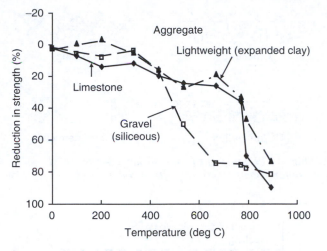

Fig. 24.18 *The effect of temperature and aggregate type on the compressive strength of concrete tested hot (average initial strength, 28 MPa) (after Abrams, 1971).*

is thus a favoured material, both in its own right and as protection for steelwork, when structural safety is being considered. However, although it can retain some strength for a reasonable time at high temperatures, it will eventually degrade; the rate and amount of degradation depend on the maximum temperature reached, the period of exposure, the induced temperature gradients, the concrete's constituents and moisture content and the size of the sample, and will therefore vary considerably.

Figure 24.18 shows typical results of testing small samples by holding them at elevated temperatures for a reasonable period of time. For temperatures up to about 500°C the reduction in strength is relatively gradual, but thereafter the decline is more rapid, giving almost total loss as the temperature approaches 1000°C. There are three main contributions to the degradation:

1. Evaporation of water within the concrete, which starts at 100°C, and continues with progressively more tightly held water being driven off. If the concrete is initially saturated and also of low permeability, then the water vapour cannot disperse quickly, and build-up of pressure can lead to cracking and spalling. This is therefore a particular problem with high-strength, low-porosity concrete. Even though the total volume of water in the concrete is low, the induced pressures are very high, and progressive explosive spalling of the surface layers can occur within a few minutes of exposure to the fire. There have been some notable examples of damage of this type, most

for example from two fires in heavy goods vehicles being carried on trains in the Channel Tunnel linking England and France, one in 1996 and one in 2008. Both of these resulted in extensive damage to the high-strength concrete tunnel requiring costly and time-consuming repairs. The inclusion of polypropylene fibres in the concrete during mixing is one way of overcoming this effect; these rapidly melt and provide pressure-relief channels.

2. Differential expansion between the HCP and aggregate, resulting in thermal stresses and cracking, initiated in the transition zone. This is mainly responsible for the more rapid loss of strength above about 500°C, and also explains the superior performance of limestone and lightweight aggregate concrete apparent in *Fig. 24.18*; the former has a coefficient of thermal expansion closer to that of the HCP (see section 20.4) and the latter is less stiff and hence the thermal stresses are lower. Lightweight aggregates have the additional advantage of decreasing the thermal conductivity of the concrete, thus delaying the temperature rise in the interior of a structural member.

3. Breakdown of the hydrates in the HCP, which is not complete until the temperature approaches 1000°C, but results in a total loss of strength at this point.

24.4 Durability of steel in concrete

Nearly all structural concrete contains steel, either in the form of reinforcement to compensate primarily for the low tensile and shear strength of the concrete, or as prestressing tendons that induce stresses in the concrete to oppose those due to the subsequent loading. Sound concrete provides an excellent protective medium for the steel, but this protection can be broken down in some circumstances, leaving the steel vulnerable to corrosion. Crucially, the corrosion products – rust in its various forms – occupy a considerably greater volume than the original steel. Rusting within concrete therefore causes internal expansive or bursting stresses, which eventually will result in cracking and spalling of the concrete covering the steel. Although unsightly, this will not immediately result in structural failure, but the remaining steel is then fully exposed, and undetected or unchecked the more rapid corrosion that results can lead, and has led, to collapse.

Although the processes involved are less complex than those of the various degradation mechanisms of the concrete itself, described above, they are much more difficult to avoid and control. Indeed corrosion of steel in concrete is the greatest threat to the durability and integrity of concrete structures in many regions. In the last few decades the concrete repair industry has benefited considerably and is thriving.

In this section we shall first describe the general nature of the phenomenon, and then consider the factors that control its onset and subsequent rate.

24.4.1 GENERAL PRINCIPLES OF THE CORROSION OF STEEL IN CONCRETE

The electrochemical nature of the corrosion of iron and steel was described in Chapter 10, and the processes involved in the corrosion of iron in an air/water environment were illustrated in *Fig. 10.1*. In the corrosion cell shown in this figure the anode and cathode are close together, e.g. across a single crystal or grain. The oxide is formed and deposited near but not directly on the metal surface, allowing the corrosion to be continuous. In concrete different conditions prevail. The electrolyte is the pore water in contact with the steel and, as we have seen, this is normally highly alkaline (pH = 12.5–13) owing to the presence of $Ca(OH)_2$ from the cement hydration and the small amounts of Na_2O and K_2O in the cement. In such a solution the primary anodic product is not Fe^{2+} as in *Fig. 10.1* but is a mixed oxide (Fe_3O_4), which is deposited at the metal surface as a tightly adherent thin film only a few nanometres thick. This stifles any further corrosion, and the steel is said to be *passive*. Thus sound concrete provides an excellent protective medium. However the passivity can be destroyed by either loss of alkalinity by carbonation of the concrete, in which the calcium and other hydroxides are neutralised by carbon dioxide from the air, producing calcium and other carbonates, or chloride ions, e.g. from road de-icing salts or seawater, which are able to breakdown or disrupt the passive film (a process known as *pitting*).

Either of these can therefore create conditions for the corrosion reactions in *Fig. 10.1*. The corrosion can be localised, for example in load-induced cracks in the concrete, or the corrosion cells can be quite large ('macrocells'), for example if anodic areas have been created by penetration of chloride ions into a locally poorly compacted area of concrete. However, it is important to remember that oxygen and water must still be available at the cathode to ensure that the corrosion continues.

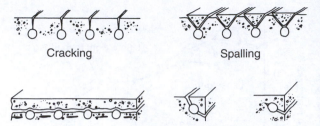

Fig. 24.19 *Different forms of damage from steel reinforcement corrosion (after Browne, 1985).*

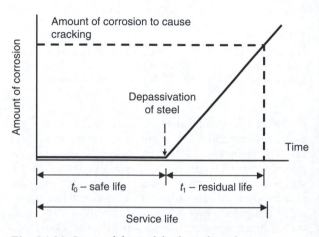

Fig. 24.20 *Service-life model of reinforced concrete exposed to a corrosive environment (after Tuutti, 1982).*

As mentioned above, the corrosion products (ferric and ferrous hydroxide) have a much larger volume than the original steel, by about 2–3 times, and can eventually lead to cracking, spalling or delamination of the concrete cover. This damage can take various forms, as illustrated in *Fig. 24.19*.

Since carbon dioxide or chloride ions will normally have to penetrate the concrete cover before corrosion can be initiated, the total time to concrete cracking (the service life of the structural element) will consist of two stages, illustrated in *Fig. 24.20*:

1. The time (t_0) for the depassivating agents (the carbon dioxide or chloride) to reach the steel in sufficient quantities to initiate corrosion; t_0 can be considered a 'safe life'.
2. The time (t_1) for the corrosion to then reach critical or limit-state levels, i.e. sufficient to crack the concrete; this is the 'residual life', and depends on the subsequent rate of corrosion.

As we shall see, in many situations estimation of t_1 is difficult and so design guidance and rules are normally framed so that t_0 is a large proportion or even all of the intended service life.

We shall now discuss the processes of carbonation-induced corrosion and chloride-induced corrosion separately.

24.4.2 CARBONATION-INDUCED CORROSION

We discussed carbonation and its associated shrinkage in Chapter 20. Neutralising the hydroxides in the HCP by atmospheric carbon dioxide in solution in the pore water reduces the pH from 12 or more to about 8. There are also some reactions between the carbon dioxide and the other hydrates, but these are not significant in this context.

The carbonation reaction occurs first at the surface of the concrete and then progresses inwards, further supplies of carbon dioxide diffusing through the carbonated layer. Extensive analysis by Richardson (1988) showed that the carbonation depth (x) and time (t) are related by the simple expression:

$$x = k.t^{0.5} \qquad (24.14)$$

where k is a constant closely related to the diffusion characteristics of the concrete. The form of this equation is the same as that of equation 24.4, which indicates that carbonation may be considered as a sorption process. The value of k depends on several factors, chiefly:

1. The degree of saturation of the concrete. It is necessary for the carbon dioxide to be dissolved in the pore water, and so concrete that has been dried at low relative humidities will not carbonate. At the other extreme, diffusion will be slow in concrete completely saturated with water, and so the fastest advance of the carbonation front occurs in partially saturated concrete at relative humidities of between 50 and 70%. Thus concrete surfaces that are sheltered will carbonate faster than those exposed to direct rainfall (Fig. 24.21).

2. The pore structure of the concrete. Parrott (1987) suggested that relating carbonation depth to concrete strength, as in Fig. 24.21, is a useful way of combining the effects of water:cement ratio, cement content and incorporation of additions. Adequate curing at early ages is also an important factor. Although additions can result in lower overall porosity with full curing, the pozzolanic reaction can also reduce the calcium hydroxide content before carbonation, and so additions do not necessarily have the same benefits as they do with other degradation processes.

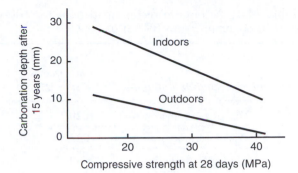

Fig. 24.21 *The relationship between carbonation depth and concrete strength (after Nagataki et al., 1986).*

3. The carbon dioxide content of the environment.

Observed rates of carbonation, such as those shown in Fig. 24.21, are such that with high-quality, well-cured concrete the carbonated region, even after many years' exposure to normal atmospheric conditions, is restricted to less than 20–30 mm of the surface of the concrete. It is difficult to estimate or predict the rate of corrosion once the steel has been depassivated, and therefore design recommendations are aimed at ensuring that the depth and quality of concrete cover are sufficient to achieve a sufficiently long initiation period (t_0). BS 8500 (2006) gives combinations of required concrete quality and cover to steel for various exposure classes or conditions; the minimum values, summarised in *Table 24.4*, clearly show how the factors discussed above have been take into account.

It should also be noted that carbonation is not entirely detrimental. The calcium carbonate formed occupies a greater volume that the calcium hydroxide, and so the porosity of the carbonated zone is reduced, increasing the surface hardness and strength, and reducing the surface permeability.

24.4.3 CHLORIDE-INDUCED CORROSION

There are four common sources of chloride ions:

- calcium chloride, a cheap and effective accelerator (see Chapter 14)
- contamination in aggregates
- seawater, for coastal or marine structures
- road de-icing salts, a particular problem on bridge decks.

Calcium chloride, or any chloride-containing admixture, is normally no longer permitted in concrete containing steel, and aggregates, particularly from marine sources, should be washed before use to remove chlorides and other contaminants.

Table 24.4 Minimum recommendations for 100-year design life for carbonation-induced corrosion of steel in concrete (from BS 8500, 2006)

Exposure class and description		Min strength class (see section 22.1)	Max w:c	Min cement content (kg/m³)	Minimum cover to steel (mm)	Cement type (see sections 13.8 and 15.4)
XC1	Dry or permanently wet	C20/25	0.70	240	15	CEM I, CEM II, CEM III (with max 80% ggbs) CEM IV, SRPC
XC2	Wet, rarely dry	C25/30	0.65	260	25	
XC3	Moderate humidity	C40/50	0.45	340	30	
XC4	Cyclic wet and dry					

w:c, water: cement ratio.

There has been considerable interest in the amount of chloride required to initiate corrosion, i.e. a threshold level that is required to depassivate the steel. In practice, corrosion in structures has been found to occur at a very wide range of total chloride content, but with increasing frequency with increasing chloride content. For example in a survey of UK concrete highway bridges, Vassie (1984) found that only 2% of the bridges showed corrosion-induced cracking when the chloride content level was less than 0.2% by weight of the cement, but the proportion rose progressively to 76% showing cracking at chloride levels greater than 1.5% by weight of the cement. It may, therefore, be better to think of the chloride content as giving a risk of corrosion, rather than there being an absolute threshold level below which no corrosion can ever occur.

The reasons for such variations in behaviour are not entirely clear, despite much research effort. One significant factor is that the C_3A component of cement binds some of the chloride ions as chloro-aluminates, thus reducing the amount available to depassivate the steel. However, in a recent review, Page and Page (2007) concluded that many other factors are also involved, and there is no straightforward answer as to the effect of, for instance, variations in cement composition or blends of cement and various additions in this respect. Despite this, standards and design recommendations have, since the 1970s, included allowable chloride levels. These have varied, but have generally been reduced as new or revised standards are published. For example the current European Standard specification (BS EN 206) has chloride content limits of 0.2% by weight of cement for concrete containing steel reinforcement and 0.1% for concrete containing pre-stressing steel.

If the chloride is included in the concrete on mixing, then the steel may never be passivated, and the initiation period, t_0, will be zero. Chlorides from

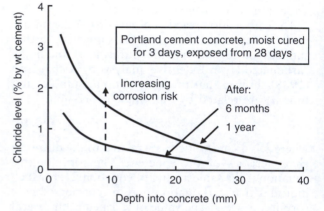

Fig. 24.22 *Chloride penetration profiles in concrete after exposure in marine tidal/splash zone (after Bamforth and Pocock, 1990).*

external sources (seawater or de-icing salts) have to penetrate the concrete cover in sufficient quantities, however defined, to depassivate the steel before the corrosion is initiated: t_0 is therefore finite in these circumstances. The transport mechanisms may be governed by: permeability in the case of, say, concrete permanently submerged in seawater; diffusivity, where salts are deposited onto saturated concrete; or sorptivity, where salts are deposited on to partially saturated concrete. The corrosion risk in situations in which the salts are water-borne and deposited onto the surface by evaporation, such as in the splash zone of marine structures or on run-off from bridge decks, is particularly high as the reservoir of salts is constantly replenished. An absorption mechanism may dominate in the early stages of such contamination, with diffusion being more important at later stages (Bamforth and Pocock, 1990).

These processes result in chloride profiles such as those shown in *Fig. 24.22*. A large number of such profiles showing the effect of a large number of variables have been generated both experimentally and analytically. In general, concrete with lower permeability, diffusivity or sorptivity will have lower rates of chloride penetration, and we have seen that these are achieved by lower water:cement ratios, adequate cement content, the use of additions and attention to good practice during and after placing the concrete. The amount of cover will also clearly affect the time needed for the chloride to reach the steel.

Although many recommendations for concrete cover and quality are aimed at extending the period t_0 as far as possible, there are circumstances in which it is impossible to prevent corrosion being initiated. Much research has therefore been carried out to determine the factors that control the rate of corrosion during the residual-life period. These have been found to include the following:

1. The spacing and relative size of the anode and cathode in the corrosion cell. Relatively porous areas of a concrete member, such as a poorly compacted underside of a beam, will allow rapid penetration of chloride, depassivating a small area of steel to form the anode. The reinforcement throughout the structure is normally electrically continuous, and so the remainder forms a large-area cathode, resulting in a concentration of corrosion current, and hence a high corrosion rate, at the anode.
2. The availability of oxygen and moisture, particularly to sustain the cathodic reaction. If the supply of either is reduced, then the corrosion rate is reduced. Hence little corrosion occurs in completely dry concrete, and only very low rates in completely and permanently saturated concrete through which diffusion of oxygen is difficult, although localised depletion of oxygen at the anode can increase corrosion rates.
3. The electrical resistivity of the electrolyte of the corrosion cell, i.e. the concrete. High resistivities reduce the corrosion current and hence the rate of corrosion, but increasing moisture content, chloride content and porosity all reduce the resistivity.

Analysis of the extensive and increasing amount of data on this subject has been used to produce guidelines to ensure adequate durability in all countries or regions where reinforced concrete is used. BS 8500 (2006) gives numerous combinations of required concrete quality and cover to steel for various exposure classes or conditions, which gives design engineers some flexibility of choice for the combination for each exposure condition; some of the minimum values are summarised in *Tables 24.5* and *24.6*. Different exposure classes apply when the corrosion is induced by chlorides from road de-icing salts (*Table 24.5*) to those that apply when the chlorides come from seawater (*Table 24.6*), but the requirements for the concrete quality and cover are largely similar for the equivalent exposure class.

There are, however, circumstances in which protection against corrosion cannot be guaranteed by selection of the materials and proportions of the concrete, depth of cover and attention to sound construction practice. These include, for example, marine exposure in extreme climatic conditions, and regions in which aggregates containing excess chloride must be used. One or more of the following extra protective measures may then be taken:

Table 24.5 Some minimum recommendations for 100-year design life for corrosion of steel in concrete induced by chlorides from road de-icing salts (from BS 8500, 2006)

Exposure class and description	Min strength class (see section 22.1)	Max w:c	Min cement content (kg/m³)	Min cover to steel (mm)	Cement type (see sections 13.8 and 15.4)
XD1, Moderate humidity	C45/55	0.40	380	30	CEM I, CEM II, CEM III (with max 80% ggbs) CEM IV, SRPC
XD2, Wet, rarely dry	C35/45	0.45	360	40	CEM I, CEM II with up 20% fly ash or 35% ggbs, SRPC
XD3, Cyclic wet and dry	C45/55	0.35	380	55	
	C40/50	0.35	380	45	CEM II with 21–35% fly ash CEM III with 36–65% ggbs

w:c, water: cement ratio.

Table 24.6 Some minimum recommendations for 100-year design life for corrosion of steel in concrete induced by chlorides from seawater (from BS 8500, 2006)

Exposure class and description	Min strength class (see section 22.1)	Max w:c	Min cement content (kg/m³)	Min cover to steel (mm)	Cement type (see sections 13.8 and 15.4)
XS1, Exposed to airborne salt but not in direct contact with seawater	C45/55	0.35	380	45	CEM I, CEM II with up 20% fly ash or 35% ggbs, SRPC
XS2, Permanently submerged	C35/45	0.45	360	40	
XS3, Tidal, splash and spray zones	C45/55	0.35	380	60	
	C40/50	0.35	380	45	CEM II with 21–35% fly ash CEM III with 36–65% ggbs

w:c, water: cement ratio.

- the addition of a corrosion-inhibiting admixture such as calcium nitrite to the fresh concrete
- the use of corrosion-resistant stainless steel reinforcement bars, or epoxy-coated conventional bars
- applying a protective coating to the concrete, to reduce chloride and/or oxygen ingress
- cathodic protection of the reinforcement, i.e. applying a voltage from an external source sufficient to ensure that all the steel remains permanently cathodic (see section 10.5.3).

References

Abrams MS (1971). *Temperature and Concrete*, American Concrete Institute Special Publication No. 25, pp. 33–58.

ASTM C1202 – 09. *Standard Test Method for Electrical Indication of Concrete's Ability to Resist Chloride Ion Penetration*, American Society for Testing and Materials, West Conshohocken, USA.

Bamforth PB (1987). The relationship between permeability coefficients for concrete obtained using liquid and gas. *Magazine of Concrete Research*, **39** (No. 138), 3–11.

Bamforth PB and Pocock DC (1990). Proceedings of Third International Symposium on Corrosion of Reinforcement in Concrete Construction, Elsevier Applied Science, pp. 11931.

Browne RD (1985). *Practical considerations in placing durable concrete*. Proceedings of Seminar on Improvements in Concrete Durability, Institute of Civil Engineers, London, pp. 97–130.

Buenfeld N and Newman JB (1984). *Magazine of Concrete Research*, **36** (No. 127), 67–80.

Buenfeld N and Okundi E (1998). Effect of cement content of transport in concrete. *Magazine of Concrete Research*, **50** (No. 4), 339–351.

Building Research Establishment (2004). *Digest 330: Alkali–silica reaction in concrete*, BRE, Watford.

Building Research Establishment (2005). *Special Digest 1 'Concrete in aggressive ground'*, 3rd edition, BRE, Watford.

Bungey JH Millard SG and Grantham MG (2006). *Testing of concrete in structures*, 4th edition, Taylor and Francis, Abingdon.

Clark LA and BRE (2002). *Thaumasite Expert Group Report: Review after three years experience* www.planningportal.gov.uk (accessed 20/7/09).

Collett G, Crammond NJ, Swamy RN and Sharp JH (2004). The role of carbon dioxide in the formation of thaumasite. *Cement and Concrete Research*, **34** (No. 9), 1599–1612.

Dhir RK, Hewlett PC and Chan YN (1987). Near-surface characteristics of concrete: assessment and development of in situ test methods. *Magazine of Concrete Research*, **39** (No. 141), 183–195.

Hewlett P (ed.) (1998). *Lea's Chemistry of Cement and Concrete*, 4th edition, ed Arnold, London, p. 963.

Hobbs DW (1986). Deleterious expansion of concrete due to alkali–silica reaction: influence of PFA and slag. *Magazine of Concrete Research*, **38** (No. 137), 191–205.

Lawrence CD (1985). *Permeability testing of site concrete*, Concrete Society Materials Research Seminar on Serviceability of Concrete, Slough, July.

Nagataki S, Ohga H and Kim EK (1986). *Proceedings of the 2nd International Conference on Fly Ash, Silica Fume, Slag and Natural Pozzolans in Concrete*. American Concrete Institute Special Publication SP-91, pp. 521–540.

Page CL and Page MM (2007). *Durability of concrete and cement composites*, Woodhead Publishing, Cambridge, pp. 157–158.

Powers TC (1958). Structure and physical properties of Portland cement paste. *Journal of the American Ceramic Society*, **41**, 1–6.

Powers TC, Copeland LE, Hayes JC and Mann HM (1954). Permeability of Portland cement paste. *Journal of the American Concrete Institute*, **51**, 285–298.

Sims I and Poole A (2003). Alkali–aggregate reactivity. In *Advanced Concrete Technology, Vol 2: Concrete Properties* (eds Newman J and Choo BS), Elsevier, Oxford, pp. 13/31.

Stanton TE (1940). The expansion of concrete through reaction between cement and aggregate. *Proceedings of the American Society of Civil Engineers*, **66**, 1781–1811.

Thaumasite Expert Group (1999). *The thaumasite form of sulphate attack: Risks, diagnosis, remedial works and guidance on new construction*. Dept of Environment, Transport and Regions, London p. 180.

Tuutti K (1982). *Corrosion of Steel in Concrete*, Report No. 482, Swedish Cement and Concrete Institute, Stockholm.

US Bureau of Reclamation (1955). *Concrete Laboratory Report No. C-810*, Denver, Colorado.

Vassie PRW (1984). Reinforcement corrosion and the durability of concrete bridges. *Proceedings of the Institution of Civil Engineers*, **76** (No. 1), 713–723.

Verbeck GJ (1968). In *Performance of Concrete* (ed Swenson EG), University of Toronto Press.

Zhang MH, Bilodeau A, Malhotra VM, Kim KS and Kim J-C (1999). Concrete incorporating supplementary cementing materials: effect of curing on compressive strength and resistance to chloride-ion penetration. *ACI Materials Journal*, **96** (No. 2).

Chapter 25

Special concretes

In this chapter we will describe some types of concrete that have been developed to extend the range of properties that we have described in the preceding chapters. These have been chosen to illustrate the tremendous versatility of concrete, which has led to its use in an ever-increasing range of applications and structural situations. Some of these properties have been obtained by the use of alternative materials, such as lightweight and high-density aggregate concrete. Others have involved modifications to mix proportions, as in no-fines concrete, and the more extensive use of admixtures, as in sprayed concrete, high-strength concrete, flowing concrete, underwater concrete, self-compacting concrete and foamed concrete. We will also mention aerated concrete, which is factory produced. Space limitations mean that we are not able to describe any of these in great detail, but references are provided to more extensive information in each case.

25.1 Lightweight aggregate concrete

Lightweight aggregates, which contain air voids within the aggregate particles, produce concretes with lower densities than those made with normal-density aggregates (see Chapter 17). The aggregate particles are generally weaker than those of normal-density aggregates, resulting in lower limits to concrete strengths; structural lightweight aggregate concrete is usually defined as having a strength greater than 15 MPa and a density of less than 2000 kg/m³. The main advantage is in reducing the weight of structures, leading to easier handling of pre-cast elements and lower loads on foundations, but the lower thermal conductivity can also be an advantage. Both fine and coarse aggregates can be lightweight, but for higher strengths and densities lightweight coarse aggregate and natural fine aggregate are often preferred.

Pumice, a naturally occurring volcanic rock of low density, has been used since Roman times, but it is only available at a few locations, and artificial lightweight aggregates are now widely available. They are of three main types:

1. Sintered fly ash, formed by heating pelletised ash from pulverised coal used in power stations until partial fusion and hence binding occur.
2. Expanded clay or shale, formed by heating suitable sources of clay or shale until gas is given off and trapped in the semi-molten mass.
3. Foamed slag, formed by directing jets of water, or steam, on to or through the molten slag from blast furnaces.

Many different products are available, particularly in industrialised countries. The overall range of strengths and densities that can be produced is shown in *Fig. 25.1*. The quality and properties of different aggregates vary considerably, and therefore

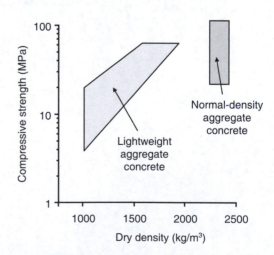

Fig. 25.1 Strength/density ranges for normal and lightweight aggregate concrete (compiled from aggregate manufacturers' information).

produce different strength/density relationships within this range. Sintered fly ash aggregates generally produce concrete in the upper part of the range shown.

The aggregates comply with the same requirements for size, shape and grading as described for normal-density aggregates in Chapter 17, but maximum particle sizes for the coarse aggregates are often limited to between 10 and 20 mm, depending on the production process.

The same general rules and procedures can be used for the design of lightweight and normal-density aggregates mixes, but as well as being generally weaker, lightweight aggregates are not as rigid as normal-weight aggregates, and therefore produce concrete with a lower elastic modulus and higher creep and shrinkage for the same strength. As with strength, the properties depend on the lightweight aggregate type and source, and also whether lightweight fines or natural sands are used. The porosity can also cause problems in the fresh concrete, as consistence can be lost with absorption of the mix water by the aggregate particles, so pre-soaking of the aggregate before mixing may be required. However, the internal reservoir of water that is created is subsequently available for continuing cement hydration, resulting in a degree of self-curing. Comprehensive information can be found in Newman and Owens (2003) and Clarke (1993).

25.2 High-density aggregate concrete

High-density aggregates can be used to produce high-density concrete for a number of specialised applications, such as radiation shielding, counter-weights in construction plant (and even in domestic washing machines) and ballasting of submerged structures. Aggregates that have been used include:

- those from natural sources, such as barytes (barium sulphate), which has a relative particle density of 4.2, and a range of iron ores such as magnetite and haematite, with relative particle densities of 4.9
- those which have been manufactured, including iron and lead shot, with relative particle densities of 7.6 and 11.3, respectively.

The resulting density of the concrete will obviously depend on the aggregate type and mix proportions, but can range from 3500 kg/m^3 with barytes up to 8600 kg/m^3 with lead shot. All except lead shot can be used to produce structural grade concrete, and strengths of more than 80 MPa are possible with some iron ores. Not surprisingly, freshly produced

mixes can have a tendency to segregate, and therefore low water:cement ratios and superplasticisers are normally recommended. Normal transporting and placing procedures can be used, but particularly with the high-density mixes only small volumes can be handled. For further information see Miller (2003).

25.3 No-fines concrete

No-fines mixes comprise cement, water and coarse aggregate with the fine aggregate omitted. During mixing, each coarse aggregate particle becomes coated with cement paste, which binds adjacent particles at their points of contact during hydration, thus giving large interconnecting inter-particle voids. Densities vary with mix proportions, but with normal-density aggregates can vary from about 1500 to 1900 kg/m^3 (somewhat less with lightweight aggregates). Not surprisingly in view of the large voids, the strengths are low, varying from 15 MPa down to less than 5 MPa, depending on the density.

No-fines concrete was traditionally used for *in-situ* internal wall construction in low-rise housing, providing good insulation when covered on each face by plasterboard. In recent years it has become increasingly popular for hardstanding areas such as car parks, where the high permeability enables surface water to drain through the concrete to the substrate and then replenish the groundwater rather than run off into storm water drains. When used for this purpose it is known as 'pervious concrete' and is a valuable material for use in so-called 'sustainable urban drainage systems'.

25.4 Sprayed concrete

Sprayed concrete, also known as gunite or shotcrete, has been in use for over 100 years. Concrete is projected from a nozzle at high velocity by compressed air on to a hard sloping, vertical or overhead surface; with suitable mix proportioning thicknesses of up to 150 mm can be built up with successive passes of the spray gun. Applications include tunnel linings, swimming pools, reservoirs, canal and watercourse linings and seawalls as well as freestanding structures. It is particularly useful for strengthening and repair of existing structures. There are two distinct processes, depending on the method of mixing of the concrete before it emerges from the nozzle:

- In the dry process, the cement additions and aggregate are dry mixed and then fed under pressure down a flexible hose, which can be many metres long, to the spray gun. A fine water spray is then also fed into the gun before the mixture is projected from the nozzle.

- In the wet process the entire mixture is batched and pumped to the gun, where compressed air is fed in to project it from the nozzle.

In the wet process, plasticisers and retarders may be required, depending on the pumping requirements and distance, but there is greater control of the final mix proportions than in the dry process. Also, in the dry process there tends to be more rebound of the aggregate particles as the concrete hits the surface.

Maximum aggregate size can range from 4 to 20 mm depending on the pump, hose and nozzle sizes. Water contents are generally low to ensure that the concrete stays in place while setting. Binder contents are generally in the range 350–450 kg/m^3, and strengths of up to 60 MPa with the wet process and 50 MPa with the dry process are possible. Mixes often include the additions discussed in Chapter 15, and short steel or synthetic fibres can readily be incorporated, either during pre-mixing or fed into the gun. A useful introduction to this subject has been produced by the Sprayed Concrete Association (1999) along with some more detailed publications.

25.5 High-strength concrete

The quantitative definition of the strength of high-strength concrete has continually increased as concrete technology has advanced. It is generally taken to be a strength significantly higher than that used in prevailing normal practice; an accepted current value is a characteristic strength in excess of about 80 MPa, but this may vary from country to country. Over the last twenty years or so there have been many examples of strengths of up to 130 MPa being successfully produced and placed with conventional mixing, handling and compaction methods. Although the cost increases with increasing strength (for reasons that will become apparent below), its use may lead to overall structural economies, for example in reduced section sizes and hence lower weights – a major factor in, e.g., a long span bridge – and in reduced column cross-sections in high-rise buildings, giving higher usable space, particularly in the lower storeys.

It will be clear from Chapter 20 that the use of a low water:cement or water:binder ratio is a prime requirement, i.e. using values less than the lower limit of 0.3 in *Fig. 22.3*, which is for normal-strength concrete. This by itself would result in impractically low consistence for conventional placing methods and so superplasticisers are an essential ingredient. Water:binder ratios as low as 0.2 have been used, but this exceptional. These low ratios are, however, not sufficient and the other main considerations arising from the many research studies and development programmes can be summarised as follows:

- Microsilica at levels of up to about 10% of the binder is important, particularly for achieving strengths in excess of about 100 MPa. The main benefit is improvement of the interface transition zone (see section 20.3).
- All materials must be very carefully selected for optimum properties:
  - high-quality aggregates are required with strong crushed rocks normally preferred, some limestones giving a particularly good performance. Limiting the maximum aggregate size to 10 mm is normally suggested
  - the binder (particularly the cement) and superplasticiser must be compatible to avoid problems such as rapid loss of workability.
- Even at slumps in excess of the 175 mm normally used the mixes can be cohesive and difficult to handle, i.e. they have a high plastic viscosity. Attention to aggregate grading and particle size can reduce this problem; the lubrication provided by the very fine spherical microsilica particles is beneficial.
- Problems that are of minor significance for normal-strength concrete, such as loss of workability and heat of hydration effects, will be exaggerated in high-strength mixes, and therefore may become critical without due consideration. The use of ternary cement blends – i.e. Portland cement plus microsilica and either fly ash or ggbs – can be helpful in many cases.
- Mix design is considerably more complex than for normal-strength concrete because of the larger number of variables involved and their interactive effects. A more extensive set of trial mixes at both laboratory and full scale are therefore often required.
- All production and quality-control issues need much greater attention than for normal-strength concrete, since the consequences of variations and fluctuations will be much more serious.
- The elastic modulus continues to increase non-linearly with strength (equation 20.13) but the concrete when under stress becomes distinctly more brittle, and it fails at increasingly lower strains as strength increases. This has consequences for reinforcement design.

The upper strength level of about 130 MPa mentioned above is about the limit that can be achieved with 'conventional' concrete materials and practice, but does not represent a ceiling or a limit if alternative production methods are considered. For example, DSP (densified with small particles) cement, in which the action of a superplasticiser and microsilica is used to produce low porosity, when combined with a strong aggregate of 4 mm maximum size produces compressive strengths of up to 260 MPa (Bache, 1994).

A further example is reactive powder concrete (RPC). This combines cement, microsilica and aggregate (quartz sand or steel shot) with a maximum aggregate size of 600 microns, in proportions to give maximum density. Water:powder ratios of 0.15–0.19 and superplasticisers give sufficient fluidity for placing in moulds. Curing under pressure immediately after placing at temperatures of 90°C for 3 days gives compressive strengths of up to 200 MPa, and at up to 400°C gives strengths of up to 800 MPa. The inclusion of short steel fibres provides ductility and flexural strengths of up to 25% of the compressive strength.

Although such materials may only find use in specialist applications, they do demonstrate that by understanding and applying the principles of materials science to cement composites a wide and continuous spectrum of performance can be achieved. Caldarone (2008) gives an up-to-date coverage of many aspects of high-strength concrete. Detailed information about RPC can be found in Richard and Cheyrezy (1995) and Bonneau *et al.* (1996).

25.6 Flowing concrete

The term 'flowing concrete' appeared in the 1970s to describe the high-consistence concretes with little bleeding or segregation that became feasible with the use of the newly developed superplasticisers. Slumps are generally in excess of 200 mm and flow table values (see section 18.2) in excess of 500 mm, roughly corresponding to the S5 slump class and F4, F5 and F6 flow classes, respectively, in *Table 22.1*. The concrete can be handled and placed with much less effort than lower-slump mixes; it is particularly useful for rapid placing in large flat slabs, and is readily pumped. As a rule of thumb, mixes can be obtained by proportioning as for a 75 mm slump without admixtures (e.g. by a method such as that outlined in Chapter 22), adding sufficient superplasticiser to give the required slump and, to ensure stability, increasing the fine aggregate content by

5%, with a corresponding reduction in the coarse aggregate. An alternative approach is to ensure that sufficient sand is added to give a total content of material smaller than 300 µm of at least 450 kg/m³. Further details can be found in Neville (1995).

25.7 Self-compacting concrete

Self-compacting concrete (SCC) can achieve full and uniform compaction without the need for any help from vibration. This in itself distinguishes it from other high-consistence concrete, such as flowing concrete, which needs some compaction, but also, and crucially, it is able to flow through and around heavily congested reinforcement while retaining its integrity and homogeneity. It was developed in Japan in the late 1980s in response to a lack of skilled construction workers; it was quickly adopted into Japanese construction practice and its use is now widespread throughout the world. Major advantages are that fewer workers are required for concrete placing, construction sites and pre-cast works are much less noisy, the health risks associated with hand-held vibrators are eliminated and the resulting quality of the concrete is high. In several countries, including the UK, SCC has been more widely used for pre-cast concrete manufacture than for *in-situ* concrete.

SCC requires a combination of:

- high fluidity and stability, which, in rheological terms, means a very low yield stress and a moderate to high plastic viscosity (but not so high that flow times are excessive). This is achieved by a combination of low water:binder ratios and superplasticisers, often supplemented by viscosity-enhancing agents or 'thickeners'. These two properties are described as *filling ability* and *segregation resistance*, respectively
- the avoidance of aggregate particles bridging between reinforcing bars and blocking the flow, achieved by an increase in the volume of paste or mortar, and a consequent reduction in the coarse aggregate volume (*Fig. 25.2*). This property is called *passing ability*.

This combination of properties is typically achieved with the following key mix proportions:

- coarse aggregate volumes in the range of 28–34% of the concrete volume (compared to 40–55% in normal concrete)
- water:binder ratios in the range 0.3–0.4
- binder contents in the range 450–600 kg/m³.

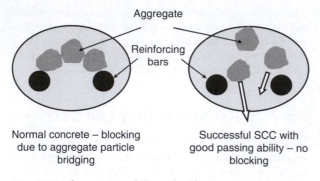

Aggregate

Reinforcing bars

Normal concrete – blocking due to aggregate particle bridging

Successful SCC with good passing ability – no blocking

Fig. 25.2 The passing ability of self-compacting concrete.

It is possible to produce most of the ranges of strengths and other properties of concrete described hitherto in this book. However the low water:binder ratios and high binder contents can lead to high strength and heat of hydration effects but, both of these can be controlled by the use of significant quantities of Type 1 and Type 2 additions (see Chapter 15), limestone powder and fly ash being particularly popular.

The combination of properties required has led to the introduction of numerous test methods of different forms. Five of these have now been incorporated into European guidelines (EFNARC, 2005), and are being incorporated into standards in Europe and elsewhere:

- The slump flow test, which uses the slump test apparatus described in Chapter 18, but the concrete is not compacted by rodding, the test is carried out on a large flat board and the final diameter of the spread is measured. Spread values are in excess of 550 mm and an indication of the viscosity of the concrete can be obtained from the time taken for the concrete to reach a diameter of 500 mm after lifting the cone. This is normally in the rage of 1 to 6 seconds.
- The V-funnel test, in which the time taken for 9.5 litres of concrete to flow out of a V-shaped funnel is measured, which gives an indication of the viscosity and any tendency to aggregate bridging and concrete blocking at the outlet.
- The J-ring and L-box tests, which assess the passing ability of the concrete by measuring the amount that flows under its own weight through a grid of reinforcing bars.
- The sieve segregation test, which measures the coarse aggregate content of the upper layer of concrete from a container that has been filled and allowed to stand for a period of time.

Classes of values for each of these tests have been published for use in specifications (EFNARC, 2005), i.e. the SCC equivalent to the consistence classes for normal vibrated concrete given in *Table 22.1*. De Schutter *et al.* (2008) give a full and up-to-date treatment of SCC.

25.8 Underwater concrete

Underwater concrete, as the name implies, is capable of being placed underwater and thus avoids the need to isolate the area to be concreted from the surrounding water, for example with a coffer dam. Its main application has been for the foundations of harbour and shallow-water structures, but off-shore deep-water placing has been carried out. It is possible to simply drop the concrete through the water into formwork that has been placed on the sea or river bed, but the preferred and more controlled method of placing is by the so-called *tremie* method. In this the concrete is fed by gravity from the surface through a vertical pipe (the tremie pipe), the open lower end of which is kept immersed in the fresh concrete. The concrete flows out of the pipe by self-weight, and mixes are designed to be sufficiently cohesive to not disperse into the surrounding water. As the concreting proceeds the tremie pipe is progressively raised while keeping its lower end within the fresh concrete. The particular requirements for the concrete are:

- high consistence, for flow and reasonable self-compacting properties since no compaction by vibration is possible
- sufficient viscosity to ensure minimum wash-out of the cement at the concrete–water interface.

These are achieved by a combination of:

- a low water:binder ratio, typically in the range 0.35–0.45
- a high binder or powder content, typically 350–450 kg/m³. This often consists of Portland cement and an addition (fly ash, ggbs or limestone powder), to reduce heat of hydration temperature rise effects. Up to 5% microsilica can also aid cohesion
- a moderately high sand content, typically 45–50% of the total aggregate
- a combination of a superplasticiser and an anti-wash-out viscosity-enhancing admixture and, for lengthy concreting operations, a retarder.

There are thus some clear parallels with the self-compacting concrete mixes described above. Consistence is normally in the range of slump flows of

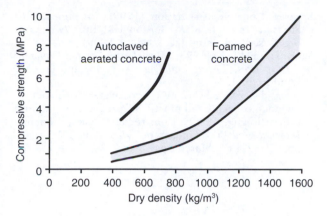

Fig. 25.3 Strength/density relationships for typical foamed and autoclaved aerated concrete (after Newman and Owens, 2003).

300–700 mm. Yao and Gerwick (2004) have produced a useful summary of all the main issues involved in underwater concreting.

25.9 Foamed concrete

Foamed concrete is a misleading title as it does not contain coarse aggregate, therefore strictly speaking it should be termed foamed mortar or foamed grout. It is produced by adding a preformed foam to a base mix of water, cement, sand or fly ash. The density can be controlled by the base mix composition and the amount of foam added. Air contents range upwards from 20% by volume, giving densities from 1700 down to 300 kg/m^3. Strengths are relatively low but, as with all concrete, depend on the density, as shown in *Fig. 25.3*.

When freshly mixed, foamed concrete is lightweight, free-flowing, easy to pump and does not require compaction. Its principal applications have therefore been where a relatively low-strength fill material is required, such as in trench reinstatement, filling of disused mine workings and subways, and road and floor foundations. It is particularly suitable were large volumes are required and access is limited. In may also be useful for the production of pre-cast building blocks and panels as an alternative to autoclaved aerated concrete, which is described below.

The foam is produced from a surfactant, which is mixed with water and passed through a foam generator. This is then blended with the base mix in either a mixing unit or in a ready-mixed concrete truck. The former system is more controlled, the latter requires the foam to be injected into the mixer

drum with some form of lance, but smaller quantities can be produced. The foam is sufficiently stable to withstand mixing and to maintain its void structure during cement hydration. When hardened it has insulating and frost-resistant properties. The Concrete Society (2009) has published a concise guide to the production and uses of foamed concrete.

25.10 Aerated concrete

Aerated concrete which, as is the case with foamed concrete, is strictly a mortar, is a factory produced product. A Portland cement paste or mortar, often with fly ash as an addition, is mixed with a small amount of finely divided aluminium powder (typically 0.2% by weight cement), which in the early stages of hydration reacts with the calcium hydroxide and other alkalis in the cement to produce hydrogen bubbles and hence expansion while the mortar is still plastic. When the required density is reached and some hardening has occurred the concrete is then cut into blocks of the required size and cured either in steam at atmospheric pressure or at about 180°C in an autoclave oven. The air content may be as high as 80% and the density may therefore be as low as 400 kg/m^3. As with foamed concrete, the density is related to strength (*Fig. 25.3*) and the thermal conductivity is low. The principal use of aerated concrete is for lightweight building blocks. Autoclaved aerated concrete is further discussed in Chapter 33, section 33.8 and a full treatment, including design and practical applications as well as properties, can be found in Wittmann (1993).

References

Bache HH (1994). Design for ductility. In *Concrete Technology: New trends, industrial applications* (eds Gettu R, Aguado A and Shah S), E & FN Spon, London, pp. 113–125.

Bonneau O, Poulin C, Dugat J, Richard P and Aitcin P-C (1996). Reactive powder concrete from theory to practice. *Concrete International*, **18** (No. 4), 47–49.

Caldarone MA (2008). *High-Strength Concrete – A Practical Guide*, Taylor and Francis, Abingdon, p. 272.

Clarke JL (ed.) (1993). *Structural Lightweight Aggregate Concrete*, 1st edition, Blackie Academic and Professional.

Concrete Society (2009). *Foamed Concrete – Application and specification*. Good Practice Guide No 7, Concrete Society, Camberley, Surrey.

De Schutter G, Bartos P, Domone P and Gibbs J (2008). *Self-Compacting Concrete*, Whittles Publishing, Caithness, Scotland.

EFNARC (2005). *The European Guidelines for Self-Compacting Concrete* http://www.efnarc.org/publications.html (accessed 2–3–09).

Miller E (2003). High density and radiation-shielding concrete and grout. Chapter 5 of *Advanced Concrete Technology – Processes* (eds Newman J and Choo BS), Butterworth Heinemann, Oxford pp. 5/1–5/15.

Newman J and Owens P (2003). Properties of lightweight concrete. Chapter 2 of *Advanced Concrete Technology – Processes* (eds Newman J and Choo BS), Butterworth Heinemann, Oxford pp. 2/11–2/25.

Richard P and Cheyrezy M (1995). Composition of reactive powder concrete. *Cement and Concrete Research* **25**, 7.

Sprayed Concrete Association (1999). *Introduction to Sprayed Concrete* SCA, Bordon UK. http://www.sca.org.uk/sca_pubs.html (accessed 14–8–09).

Wittmann FH (ed.) (1993). *Autoclaved Aerated Concrete – Properties, Testing and Design*, Taylor and Francis, London, p. 424.

Yao S and Gerwick BC (2004). Underwater Concrete Part 1 Design concepts and practices, Part 2 Proper mixture proportioning, Part 3 Construction issues, *Concrete International* Pt 1 January pp. 79980, Pt 2 February pp. 77–82, Pt 3 March pp. 60–64.

Chapter 26

Recycling of concrete

Recycling of elements of concrete structures at the end of their working life as components of new structures is difficult, with the exception perhaps of some pre-cast elements. We will therefore confine our discussions in this chapter to the recycling of unused fresh concrete and the recycling of structural concrete after it has been crushed and processed into aggregate-sized particles to produce recycled aggregates (as defined in Chapter 17).

26.1 Recycling of fresh concrete

Concrete is produced to order for specific applications and there is considerable economic incentive to avoid waste, e.g. by not over-ordering from ready-mixed concrete suppliers. However, as well as the fresh concrete, which may be returned to the read-mix plant unused, waste arises from the washing-out of truck mixing drums and from wash-down of the plant and equipment at the end of the working day. Freshly mixed cement and concrete is alkaline and the waste is therefore classed as hazardous.

Although the quantities may not be large in relation to the total amount of concrete produced, the avoidance of waste that requires disposal is important (Sealy *et al.*, undated). Most concrete plants now incorporate a reclaiming system whereby the wash-out (and unused fresh concrete) is passed through sieves that separate the aggregates, which are then returned to their stockpiles for reuse. Not surprisingly, these are often termed *reclaimed aggregates*. The wash water, which contains some fine particles of hydrated cement, is reused in new concrete. An alternative with unused concrete is to add a set-controlling admixture to the concrete so that it can be incorporated into a new batch of concrete on the following day.

26.2 Recycling of concrete after demolition

After demolition, the large lumps of concrete are fed through a crushing machine, the steel reinforcement and timber etc. are removed and the resulting particles then passed through screens to divide them into size fractions, as for primary aggregates (Chapter 17). The majority are then used as hardcore for foundations, sub-base for roads pavements and other fill applications, but an increasing amount is used in new concrete.

The crushed product can contain a mixture of materials – concrete, masonry, plaster etc. – depending on the structure being demolished. Masonry and plaster can cause significant problems when used in new concrete owing to their lower strength and high absorption, which leads to the classification of the material as either recycled concrete aggregate (RCA) – which is predominantly crushed concrete – and recycled aggregate (RA), which has a mixture materials. The BS 8500 composition requirements for these are shown in *Table 26.1*.

Recycled aggregate has been shown to be suitable for low-strength concrete blocks (Jones *et al.*, 2005), but recycled concrete aggregate can be used in structural strength concrete. Only the coarse aggregate sizes are normally used since the finer material often has a high absorption, which can lead to an excessive water demand in the fresh concrete. RCA particles will comprise the original aggregate with quantities of mortar and paste bonded to the surface, which can lead to:

- higher absorption than with typical primary aggregates, so pre-wetting of the aggregate before mixing is often recommended
- reduced concrete strength
- higher drying shrinkage and creep.

Table 26.1 BS 8500 requirements for maximum amounts of constituents of recycled concrete aggregate and recycled aggregate for use in new concrete (% by weight)

	Masonry	Fines	Lightweight material	Asphalt	Other foreign material	Sulphate
Recycled concrete aggregate (RCA)	5	5	0.5	5	1	1
Recycled aggregate (RA)	100	3	1	10	1	1

For these reasons, a mixture of recycled and primary aggregate is normally used. The reduction in properties with RCA content and concrete quality restricts the use of RCA (some data for strength are shown in *Fig. 26.1*). For example, the UK standard (BS 8500) limits its use to up to 20% replacement of the primary aggregate, to concrete with maximum strength class C40/50 and which will be subjected to the least severe of the exposure classes discussed in Chapter 24, unless satisfactory performance with the particular RCA can be demonstrated.

References

Limbachiya MC, Koulouris A, Roberts JJ and Fried AN (2004). *Performance of Recycled Aggregate in Concrete, Environment-Conscious Materials and Systems for Sustainable Development*. Proceedings of RILEM International Symposium RILEM, September, pp. 127–135.

Sealey BJ, Hill GJ and Phillips PS (undated). Review of Strategy for Recycling and Reuse of Waste Materials www.computing.northampton.ac.uk/~gary/cv/ReviewOfWasteStrategy.pdf (accessed 16–8–09).

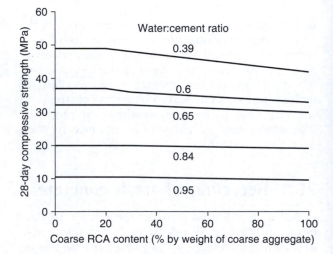

Fig. 26.1 Effect of recycled aggregate content on the compressive strength of concrete (after Limbachiya et al., 2004).

Soutsos MN, Millard SG, Bungey JH, Jones N, Tickell RG and Gradwell J (2004). Using recycled construction and demolition waste in the manufacture of precast concrete building blocks. *Proceedings of the Institution of Civil Engineers: Engineering Sustainability*, **157** (ES3), 139–148.

Further reading for Part 3 Concrete

BOOKS

General

Neville AM (1995). *Properties of concrete* 4th edition, Pearson Education, Harlow, 844 pages.

> *Since its first edition in 1963, this has been the definitive reference book on all aspects of concrete technology. Updated for this fourth edition, it is a valuable source of information for all those with an interest in concrete.*

Newman JB and Choo BS (2003). *Advanced Concrete Technology*, Butterworth Heinemann, Oxford *Vol I: Constituent Materials, Vol II: Concrete Properties, Vol. III: Processes, Vol IV: Testing and Quality*, 1920 pages.

> *This comprehensive four-volume set covers all aspects of concrete technology, from fundamentals through to practice and specialist applications. It covers all the subjects included in this part of the book, and many others. It is multi-authored and is an invaluable source for reference and further reading.*

Mehta PK and Monteiro PJ (2005). *Concrete: Microstructure, Properties, and Materials* 3rd edition, McGraw-Hill Education, UK, 659 pages.

> *A substantial and comprehensive text aimed at undergraduate students and professionals. Much fundamental content, with a bias towards North American practice.*

Forde M (ed.) (2009). *ICE Manual of Construction Materials*, Thomas Telford, London.

> *A comprehensive text (two volumes) covering concrete and all of the other materials discussed in this book. It is aimed at all professionals in civil engineering and the wider construction market, and so is not a teaching text, but could be useful to consult in the library during project work.*

Cements and additions

Some of the major cement producers and manufacturers have more information on cement, including animated illustrations of production, on their websites e.g. http://www.heidelbergcement.com/uk/en/hanson/products/cements/education.htm (accessed 10–5–09)

Hewlett PC (ed.) (1998). *Lea's Chemistry of Cement and Concrete*, Arnold, London, 1052 pages.

> *An update of a book first published in 1935. Multi-authored, and the authoritave text on the subject, with much detail at an advanced level. Not for the faint hearted, but worth consulting for project work etc.*

Bensted J and Barnes P (2002). *Structure and Performance of Cements*, 2nd edition, Spon Press, London, 565 pages.

> *An alternative multi-authored detailed text. Heavy on chemistry and microstructure.*

Aïtcin P-C (2007). *Binders for Durable and Sustainable Concrete*, Taylor and Francis, London, 528 pages.

> *Goes from fundamentals right through to practice. Contains much detail, and is suitable as a reference for advanced study.*

Winter N (2009). *Understanding Cement: An introduction to cement production, cement hydration and deleterious processes in concrete*, WHD Microanalysis Consultants Ltd, ebook available on http://www.understanding-cement.com/uceb.html

> *This is described as an informal introduction to cement, and is aimed at students and professionals new or newish to the subject. It covers much of the relevant content in this section of the book more extensively, but starts from the same point, i.e. no previous knowledge. Recommended.*

Admixtures

Rixom R and Mailvaganam N (1999). *Chemical admixtures for concrete*, 3rd edition, E&FN Spon, London, 456 pages.

> *Does not include recent developments, but useful.*

Fresh concrete

Tattersall GH and Banfill PFG (1983). *The rheology of fresh concrete*, Pitman, London.

> *Contains all the relevant theory, background and details of application of rheology to fresh cement and concrete, and summarises all the pioneering studies of the 1960 and 70s. An excellent reference text.*

Tattersall GH (1991). *Workability and quality control of concrete*, E&FN Spon, London.

Discusses the nature of workability and workability testing, summarises the background to rheological testing, and considers quality control issues in some detail.

Non-destructive testing

Bungey JH, Millard SG and Grantham MG (2006). *Testing of Concrete in Structures*, Taylor and Francis, London, 352 pages.

Fairly easy to read, and covers most aspects of NDT of concrete, including partially destructive tests.

Malhotra NJ and Carino VM (2003). *Handbook on non-destructive testing of concrete*, 2nd edition, CRC Press.

More detailed, a good reference source.

Durability

Page CM and Page MM (2007). *Durability of Concrete and Cement Composites*, CRC Press, London.

A detailed and valuable consideration of all aspects of concrete durability. Multi-authored.

Broomfield JP (2006). *Corrosion of Steel in Concrete: Understanding, Investigation and Repair*, Taylor and Francis, Abingdon.

Considers the corrosion of steel in concrete in much more detail than we have described in this book, and also covers investigation techniques and repair methods. Comprehensive.

Soutsos M (ed.) (2010). *Concrete durability: a practical guide to the design of durable concrete structures*, Thomas Telford, London.

This specialist text has contributions from many experts in their field. As the title implies it is aimed at designers, but includes descriptions of the processes involved in concrete degradation, and therefore it is a useful reference text if you want to go beyond the coverage given in this book.

Special concretes

Clarke J (ed.) (1993). *Structural Lightweight Aggregate Concrete*, Blackie Academic and Professional.

A useful source for all aspects of this subject.

Aitcin P-C (1998). *High-performance concrete*, E&FN Spon, London, 591 pages.

A comprehensive text on many aspects of high-performance concrete, takes the subject from the background science to many case studies of applications.

Price WF (2001). *The use of high-performance concrete*, E&FN Spon, London.

Consider the uses and practical applications of the various types of high-performance concrete, including high strength, controlled density, high durability, high workability and self-compacting concrete.

De Schutter G, Bartos P, Domone P and Gibbs J (2008). *Self-Compacting Concrete*, Whittles Publishing, Caithness, Scotland.

Gives a comprehensive coverage of all aspect of SCC including development, testing, mix design, properties and applications. Primarily aimed at advanced students and practitioners, but useful for undergraduate project work.

PUBLICATIONS AVAILABLE ON-LINE (BUT NOT ALL ARE FREE)

Concrete Society

The UK Concrete Society (www.concrete.org.uk) produce a whole range of publications written by and for professionals working in the concrete materials, supply, production, use, design and repair sectors, but nevertheless which are valuable for students who wish to know more about practical aspects of concrete. Some of the more recent ones of most relevance to the content of this book are listed below.

Technical Reports *(comprehensive and authoritative documents)*

Report No	Title	
TR18	A Guide to the Selection of Admixtures for Concrete (2nd edition)	2002
TR22	Non-Structural Cracks in Concrete	1992
TR30	Alkali–Silica Reaction – Minimizing the risk of damage to concrete	1999
TR31	Permeability testing of site concrete	2008
TR35	Underwater Concreting	1990
TR36	Cathodic Protection of Reinforced Concrete	1989
TR40	The use of GGBS and PFA in Concrete	1991
TR41	Microsilica in Concrete	1993
TR44	The Relevance of Cracking in Concrete to Corrosion of Reinforcement	1995
TR46	Calcium Aluminate Cements in Construction – A re-assessment	1997
TR48	Guidance on Radar Testing of Concrete Structures	1997
TR49	Design Guidance for High Strength Concrete	1998
TR51	Guidance on the use of Stainless Steel Reinforcement	1998

TR54	Diagnosis of Deterioration in Concrete Structures – Identification of defects, evaluation and development of remedial action	2000
TR56	Construction and Repair with Wet-process Sprayed Concrete and Mortar	2002
TR60	Electrochemical Tests for Reinforcement Corrosion	2004
TR61	Enhancing Reinforced Concrete Durability	2004
TR62	Self-Compacting Concrete – A Review	2005
TR65	Guidance on the use of Macro-synthetic Fibre-Reinforced Concrete	2007
TR67	Movement, restraint and cracking in concrete structures	2008
TR68	Assessment, Design and Repair of Fire-damaged Concrete Structures	2008
TR69	Repair of concrete structures with reference to BS EN 1504	2009

Current Practice Sheets *(short 2–3 page articles)*

CP No.	Title	Date
120	Half cell potential surveys of reinforced concrete structures	07/2000
123	Self-compacting concrete, part I. The material and its properties	07/2001
127	Bridge durability	01/2002
128	Measuring concrete resistivity to assess corrosion rate	02/2002
129	Cold-weather concreting	09/2002
131	Measuring depth of carbonation	01/2003
132	Measuring the corrosion rate of RC using LPR	03/2003
133	Measurement of chloride ion concentration of RC	09/2003
136	Portland-limestone cement – the UK situation	03/2004
139	Corrosion inhibitors	06/2004
140	Factory-produced cements	06/2004
141	Strengthening concrete bridges with fibre composites	06/2004
144	Controlled Permeability Formwork	10/2005
145	Self-compacting concrete	10/2005
146	Fly ash	03/2006
148	Cement Combinations	05/2006
149	Admixture current practice – parts 1 and 2	9 and 11/2006

Good Concrete Guides *(concise guidance on 'best practice')*

GCG1	Concrete for Industrial Floors – Guidance on specification and mix design	2007
GCG2	Pumping Concrete	2002
GCG6	Slipforming of vertical structures	2008
GCG7	Foamed Concrete	2007
GCG8	Concrete Practice – Guidance on the practical aspects of concreting	2008
GCG9	Designed and Detailed	2009

British Cement Association

The BCA (www.cementindustry.co.uk) have produced a wide range of publications with an emphasis on cements, but also covering concrete and its uses. Some of those of most relevance to the contents of this book are listed below.

Concrete-on-site
A series of guides (published in 1993) each a few pages long, on how to carry out concrete operations. Essential reading if you find yourself in that situation – either during or after your studies.

1	Ready-mixed concrete
2	Reinforcement
3	Formwork
4	Moving concrete
5	Placing and compacting
6	Curing
7	Construction joints
8	Making good and finishing
9	Sampling and testing fresh concrete
10	Making test cubes
11	Winter working

Fact sheets

Short (one or two page) documents containing 'essential' information.

Fact Sheet 1	Fire resistance of concrete	2006
Fact Sheet 2	Thaumasite form of sulphate attack (TSA)	2006
Fact Sheet 3	Delayed ettringite formation (DEF)	2006
Fact Sheet 4	Alkali–silica reaction (ASR)	2006
Fact Sheet 5	Self-compacting concrete (SCC)	2006
Fact Sheet 6	Use of recycled aggregate in concrete	2006
Fact Sheet 7	Using wastes as fuel and raw materials in cement kilns: cement quality and concrete performance	2006
Fact Sheet 8	Factory-made Portland limestone cement (PLC)	2006
Fact Sheet 10 (7 parts)	Chromium (VI) legislation for cement. I	2006
Fact Sheet 12	Novel cements: low-energy, low-carbon cements.	2006
Fact Sheet 14	Factory-made composite cements	2007
Fact Sheet 17	Cement, cement clinker and REACH	2009

British and European Standards referred to in the text

The list below is of those standards, specifications and design codes published by the British Standards Institution that are mentioned in the text. It is not intended to be an exhaustive list of all those concerned with concrete and its constituents. These can be found be looking at the BSI website: www.bsi-global.com/

Cement

BS EN 196-3:2005 Methods of testing cement. Determination of setting time and soundness

BS EN 197-1:2000 Cement. Composition, specifications and conformity criteria for common cements

BS 4027:1996 Specification for sulphate-resisting Portland cement

BS EN 14647:2005 Calcium aluminate cement. Composition, specifications and conformity criteria

Admixtures

BS EN 934-2:2009 Admixtures for concrete, mortar and grout. Concrete admixtures: Definitions, requirements, conformity, marking and labelling

Additions

BS EN 450-1:2005+A1:2007 Fly ash for concrete. Definition, specifications and conformity criteria

BS EN 15167-1:2006 Ground granulated blast furnace slag for use in concrete, mortar and grout. Definitions, specifications and conformity criteria

BS EN 13263-1:2005 Silica fume for concrete. Definitions, requirements and conformity criteria

Aggregates

BS EN 12620:2002+A1:2008 Aggregates for concrete

Concrete

BS EN 1992-1-1:2004 Eurocode 2: Design of concrete structures. General rules and rules for buildings

BS EN 206-1:2000 Concrete: Specification, performance, production and conformity

BS 8500-2:2006 Concrete. Complementary British Standard to BS EN 206-1. Specification for constituent materials and concrete

Test methods: Fresh concrete

BS EN 12350-1:2009 Testing fresh concrete. Sampling

BS EN 12350-2:2009 Testing fresh concrete. Slump-test

BS EN 12350-3:2009 Testing fresh concrete. Vebe test

BS EN 12350-4:2009 Testing fresh concrete. Degree of compactability

BS EN 12350-5:2009 Testing fresh concrete. Flow table test

BS EN 12350-6:2009 Testing fresh concrete. Density

BS EN 12350-7:2009 Testing fresh concrete. Air content. Pressure methods

Test methods: Hardened concrete

BS EN 12390-3:2009 Testing hardened concrete. Compressive strength of test specimens

BS EN 12390-5:2009 Testing hardened concrete. Flexural strength of test specimens

BS EN 12390-6:2000 Testing hardened concrete. Tensile splitting strength of test specimens

BS EN 12390-7:2009 Testing hardened concrete. Density of hardened concrete

BS EN 13791:2007 Assessment of in-situ compressive strength in structures and pre-cast concrete components

BS EN 12504-1. Testing concrete in structures. Part 1. Cored specimens. Taking, examining and testing in compression

BS EN 12504-2:2001 Testing concrete in structures. Non-destructive testing. Determination of rebound number

BS EN 12504-3:2005 Testing concrete in structures. Determination of pull-out force

BS EN 12504-4:2004 Testing concrete. Determination of ultrasonic pulse velocity

BS 1881-5:1970 Testing concrete. Methods of testing hardened concrete for other than strength. Part 6 ISAT test

BS 1881-209:1990 Testing concrete. Recommendations for the measurement of dynamic modulus of elasticity

Test methods: Aggregates

BS 812-123:1999 Testing aggregates. Method for determination of alkali–silica reactivity. Concrete prism method

PART 4

BITUMINOUS MATERIALS

Gordon Airey

Introduction

The term 'bituminous materials' is generally taken to include all materials consisting of aggregate bound with either bitumen (or previously) tar. Materials of this kind are used almost exclusively in road construction. However, bitumen is also used for industrial applications and other forms of construction, for example in roofing materials, paints, carpet tiles and as a protective/waterproof coating. This part of the book will concentrate on the use of bituminous materials in road construction.

Prior to the use of bitumen, tar was the binder used in bituminous materials. The use of tar in road-building materials in the UK began to grow significantly just after the turn of the 20th century following the advent of motor vehicles with pneumatic tyres. Up to that time, roads were constructed following the principles developed by Macadam using water-bound, graded aggregate. Under the action of pneumatic tyres and the higher speeds of motor vehicles, a great deal of dust was generated on macadam roads, which led to the use of tar as a dressing to bind the surface. Tar was eminently suitable for this purpose since it could be made sufficiently fluid by the use of heat to be sprayed, but stiffened on cooling. Furthermore, it protected the road from the detrimental effects of water. The benefits of using tar were quickly realised and a range of 'coated stone' materials, or 'tarmacadams', were developed.

References to natural sources of bitumen date back to biblical times. However, the first use of natural rock asphalt for paving roads was in the middle of the 19th century. The first refinery bitumens to be used in the UK came from the Mexican oilfields around 1913. But it was the opening of the Shell Haven refinery in 1920 that gave rise to the rapid development of bitumen for road construction.

Bitumen was found to be less temperature-susceptible than tar. Thus it was harder or stiffer than an equivalent grade of tar at high temperatures, making it more resistant to deformation, and softer than tar at low temperatures, making it less brittle and more resistant to cracking. As the quantity and weight of traffic increased, the performance required of bituminous materials increased and bitumen became more widely used than tar. Although tar-bound materials predominated during the Second World War owing to difficulties in importing crude oil, the introduction of North Sea gas in the late 1960s dramatically reduced the production of crude coal tar in the UK. In addition, the carcinogenic nature of tar made it an unacceptable binder in bituminous materials. Therefore the following chapters will deal only with bitumen.

A very wide range of bituminous mixtures have been developed to suit the wide variety of circumstances in which they are used. They vary according to their bitumen content and grade as well as their aggregate grading and size. Bituminous mixtures (also known as 'asphalt mixtures' or simply 'asphalt') have traditionally been classified in the UK into two groups, namely 'hot rolled asphalts' and 'coated macadams' or more simply 'asphalts' and 'macadams'. (It should be noted that the term 'asphalt' when used in North America means bitumen.) *Figure IV.1* illustrates the fundamentally different characteristics of asphalts and macadams. Asphalts rely on their dense, stiff mortar for strength and stiffness, whereas macadams rely on the stability of the aggregate through its grading. Thus the role of bitumen is quite different in each case, and the properties of asphalts are more strongly dependent on the nature of the bitumen than the properties of macadams. However, although *Fig. IV.1* portrays two very different types of material, there is in fact almost a continuous spectrum of materials between these

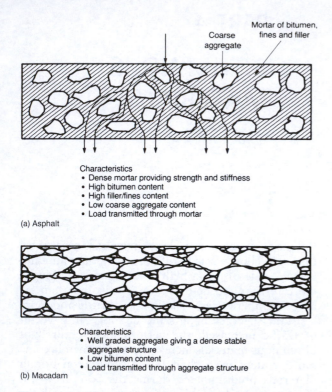

Coarse aggregate

Mortar of bitumen, fines and filler

Characteristics
- Dense mortar providing strength and stiffness
- High bitumen content
- High filler/fines content
- Low coarse aggregate content
- Load transmitted through mortar

(a) Asphalt

Characteristics
- Well graded aggregate giving a dense stable aggregate structure
- Low bitumen content
- Load transmitted through aggregate structure

(b) Macadam

Fig. IV.1 *The essential features of asphalts and macadams.*

Surface course (or wearing course) – bituminous materials
Binder course (or base course) – bituminous materials
Base – bituminous materials, concrete or granular
Sub-base – hydraulically-bound or granular
Capping – hydraulically-bound or granular
Subgrade (or substrate) – soil

Fig. IV.2 *Pavement layers of a flexible road.*

two extremes. Thus there are asphalts that have a larger coarse aggregate content than suggested in *Fig. IV.1*, such that the overall aggregate grading is more continuous and the materials resemble macadams in that respect. Similarly there are macadam mixtures that are dense and contain more bitumen, and tend towards the model for asphalts. With the implementation of European Specifications for asphalt, its constituents, and methods of testing, the British Standards BS 594 for Hot Rolled Asphalt and BS 4987 for Coated Macadam have now been withdrawn in favour of the BS EN 13108 series for

asphalt mixtures. This will be discussed in more detail in Chapter 31.

Bituminous materials are used in so-called 'flexible' pavement construction. The alternative is 'rigid' construction where the road consists essentially of a concrete slab. In flexible pavements there are a number of layers to the road structure, each having a specific function. *Figure IV.2* illustrates those layers and indicates where bituminous materials may be used. The nature of the materials will vary according to their position and function in the structure. Thus the surface course, binder course and base may be asphalts but the properties required of the surface course at the road surface are different from those required just below the surface in the binder course and base. Therefore a surface-course asphalt will differ from a binder-course and base-course asphalt. Particular types of material are selected according to their suitability. This will be discussed more fully in Chapter 29.

Chapter 27

Components of bituminous materials

27.1 Constituents of bituminous materials

Bituminous materials consist of a graded aggregate bound together with bitumen. In addition, the mixture contains a small proportion of air. Thus bituminous materials are three-phase materials and their properties depend upon the properties of the individual phases as well as the mixture proportions. The two solid phases are quite different in nature. While the aggregate is stiff and hard, the bitumen is flexible and soft and is particularly susceptible to temperature change. Therefore the proportion of bitumen in the asphalt mixture has a great influence on the mixture's properties and is crucial in determining the performance of the material.

Bitumen may be supplied in a number of forms either to facilitate the mixing and laying process or to provide a particular performance. Aggregates may come from a wide range of rock types or from artificial sources such as slag. The grading of the aggregate is important and ranges from continuous grading for mixture types known as asphalt concretes (previously known in the UK as 'macadams') through to gap grading for mixtures known as hot rolled asphalts or stone mastic asphalts (previously known in the UK as 'asphalts'). The very fine component of the aggregate (passing 63 microns) is called filler. Although the graded aggregate will normally contain some material of this size it is usually necessary to provide additional filler in the form of limestone dust, pulverised fuel ash, hydrated lime or ordinary Portland cement.

27.2 Bitumens

27.2.1 SOURCES
There are two sources of bitumen: natural deposits and refinery bitumen.

Natural asphalts
Bitumen occurs naturally, formed from petroleum by geological forces, and always in intimate association with mineral aggregate. Types of deposit range from almost pure bitumen to bitumen-impregnated rocks and bituminous sands with only a few per cent bitumen.

Rock asphalt consists of porous limestone or sandstone impregnated with bitumen with a typical bitumen content of 10%. Notable deposits are in the Val de Travers region of Switzerland and the 'tar sands' of North America.

Lake asphalt consists of a bitumen 'lake' with finely divided mineral matter dispersed throughout the bitumen. The most important deposit of this type, and the only one used as a source of road bitumen in the UK, is the Trinidad Lake. The lake consists of an area of some 35 ha and extends to a depth of 100 m. Asphalt is dug from the lake, partially refined by heating to 160°C in open stills to drive off water, then filtered, barrelled and shipped. The material consists of 55% bitumen, 35% mineral matter and 10% organic matter. It is too hard in this form to use directly on roads and is usually blended with refinery bitumen.

Refinery bitumen
This is the major source of bitumen in the UK. In essence, bitumen is the residual material left after the fractional distillation of crude oil. Crudes vary in their bitumen content. The lighter paraffinic crudes, such as those from the Middle East and North Sea, have a low bitumen content, which must be obtained by further processes after distillation. Heavier crudes, known as asphaltic crudes, such as those from the countries around the Caribbean, contain more bitumen, which is more easily extracted.

27.2.2 MANUFACTURE
The process of refining crude oil yields a range of products, as shown in *Fig. 27.1*. These products are

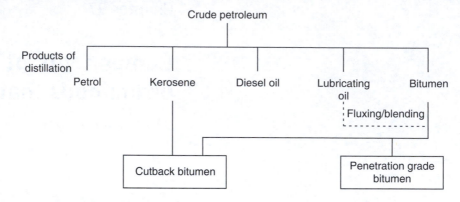

Fig. 27.1 Preparation of refinery bitumen.

released at different temperatures, with the volatility decreasing and viscosity increasing as the temperature rises. Bitumen is the residual material but its nature will depend on the distillation process and, in particular, on the extent to which the heavier oils have been removed. If the residual material contains significant amounts of heavy oils, it will be softer than if the heavy oils had been more thoroughly extracted. Modern refinery plant is capable of very precise control, which enables bitumen to be produced consistently at a required viscosity.

27.2.3 CHEMISTRY AND MOLECULAR STRUCTURE

Bitumen is a complex colloidal system of hydrocarbons and their derivatives which is soluble in trichloroethylene. The usual approach to the determination of the constituents of a bitumen is through the use of solvents. It may be subdivided into the following main fractions:

- *Asphaltenes* – fraction insoluble in light aliphatic hydrocarbon solvent, e.g. *n*-heptane.
- *Maltenes* – fraction soluble in *n*-heptane.

The maltenes may be further subdivided into resins (highly polar hydrocarbons) and oils (subdivided into aromatics and saturates). The asphaltenes have the highest molecular weight but their exact nature is dependent on the type of solvent and the volume ratio of solvent to bitumen. If small amounts of solvent are used, resins, which form part of the maltenes fraction, may be adsorbed on to the asphaltene surfaces, yielding a higher percentage of asphaltenes. Although they may vary according to the method of extraction, the appearance of asphaltenes is always of a dark brown to black solid that is brittle at room temperature. They have a

complex chemical composition but consist chiefly of condensed aromatic hydrocarbons and include complexes with nitrogen, oxygen, sulphur and metals such as nickel and vanadium. The structure of asphaltenes is not known with certainty. One suggestion is of two-dimensional condensed aromatic rings, short aliphatic chains and naphthenic rings combined in a three-dimensional network (Dickie and Yen, 1967). Another suggestion is that there are two different molecular types, one being a simple condensed aromatic unit and the other consisting of collections of these simple units (Speight and Moschopedis, 1979). It is likely that all of these may exist in bitumens from different sources since the nature of the molecules present in a crude oil will vary according to the organic material from which the crude was formed, and to the type of surrounding geology.

Maltenes contain lower molecular weight versions of asphaltenes, called resins, and a range of hydrocarbon compounds known as 'oils' including olefins, naphthenes and paraffins. The aromatic oils are oily and dark brown in appearance and include naphthenoaromatic type rings. The saturated oils are made up mainly of long straight saturated chains and appear as highly viscous whitish oil.

Bitumen is normally described as a colloidal system in which the asphaltenes are solid particles in the form of a cluster of molecules or micelles in a continuum of maltenes (Girdler, 1965). Depending on the degree of dispersion of the micelles in the continuous phase, the bitumen may be either a sol, where there is complete dispersal, or a gel, where the micelles are flocculated into flakes. Bitumens with more saturated oils of low molecular weight have a predominantly gel character. Those with more aromatic oils, which are more like asphaltenes, have a predominantly sol character.

Table 27.1 Specification for paving grade bitumens (from BS EN 12591)

Property	Unit	Test method	Grade designation		
			20/30	40/60	70/100
Penetration @ 25°C	0.1 mm	BS EN 1426	20–30	40–60	70–100
Softening point	°C	BS EN 1427	55–63	48–56	43–51
Resistance to hardening @ 163°C		BS EN 12607-1 or			
Change in mass, maximum, ±	%	BS EN 12607-3	0.5	0.5	0.8
Retained penetration, minimum	%		55	50	46
Softening point after hardening, minimum	°C	BS EN 1427	57	49	45
Flash point, minimum	°C	BS EN 22592	240	230	230
Solubility, minimum	% (m/m)	BS EN 12592	99.0	99.0	99.0

In terms of their influence on the properties of bitumen, asphaltenes constitute the body of the material, the resins provide the adhesive and ductile properties, and the oils determine the viscosity and rheology.

Although bitumens are largely complex mixtures of hydrocarbons, there are other elements present. The high-molecular-weight fraction contains significant amounts of sulphur, the amount of which influences the stiffness of the material. Oxygen is also present, and some complexes with oxygen determine the acidity of the bitumen. This is important in determining the ability of the bitumen to adhere to aggregate particles.

27.2.4 PHYSICAL AND RHEOLOGICAL PROPERTIES

Bitumen is a thermoplastic, viscoelastic material and as such its physical and rheological (flow) properties are a function of temperature, load (stress) level and load duration (time of loading). Under extreme conditions, such as low temperatures (< 0°C) and short loading durations (< 0.1 seconds) or high temperatures (> 60°C) and long loading times (> 1 second) it shows elastic and brittle behaviour or viscous and fluid-like behaviour, respectively. At intermediate temperatures and loading times, bitumen possesses both elastic and viscous properties (viscoelastic response), the relative proportions of these two responses depending on temperature, loading rate and applied stress level.

27.3 Types of bitumen

27.3.1 PENETRATION GRADE BITUMENS

Refinery bitumens are produced with a range of viscosities and are known as penetration grade bitumens. The term derives from the test that is used to characterise them according to hardness. The range of penetration grades for road bitumens is from 15 to 450, although the most commonly used are in the range 25 to 200. The range is produced partly through careful control of the distillation process and partly by fluxing 'residual' bitumen with oils to the required degree of hardness. The specification for penetration grade bitumens is contained in BS EN 12591.[1] *Table 27.1* indicates a range of tests with which penetration grade bitumens for road purposes must comply. These bitumens are specified by their penetration value (BS EN 1426) and softening point (BS EN 1427), which indicate hardness and equiviscosity temperature, respectively. However, they are designated only by their penetration, e.g. 40/60 pen bitumen has a penetration of 50 ± 10. In addition there are limits for loss on heating (BS EN 13303), which ensures that there are no volatile components present whose loss during preparation and laying would cause hardening of the bitumen, and solubility (BS EN 12592), which ensures that there are only negligible amounts of impurity.

27.3.2 OXIDISED BITUMENS

Refinery bitumen may be further processed by air blowing. This consists of introducing air under pressure into a soft bitumen under controlled temperature conditions. The oxygen in the air reacts with certain compounds in the bitumen, resulting in the formation of compounds of higher molecular weight. Thus the asphaltenes content increases at the expense of the maltenes content, resulting in harder bitumens that are also less ductile and less temperature susceptible.

---

[1] A list of all standards referred to in the text is included in 'Further reading' at the end of the section.

Table 27.2   Specification for cutback bitumens (from BS 3690: Part 1 and BS EN 12591)

Property	Test method	Grade of cutback		
		50 sec	100 sec	200 sec
Viscosity (STV) at 40°C, 10 mm cup, secs	BS 2000: Part 72	50 ± 10	100 ± 20	200 ± 40
Distillation	BS EN 13358			
(a) Distillate to 225°C, % by volume, max		1	1	1
360°C, % by volume,		55	50	46
(b) Penetration at 25°C of residue from distillation to 360°C, mm	BS EN 1426	100–350	100–350	100–350
Solubility in trichloroethylene, % by mass, min	BS EN 13303	99.5	99.5	99.5

Although these bitumens are mostly used for industrial applications such as roofing and pipe coatings, low penetration grade road bitumens (paving grade bitumens) can also be produced by this process.

27.3.3 CUTBACKS

Penetration grade bitumen is thermoplastic, thus its viscosity varies with temperature. At ambient temperature it can be more or less solid and to enable it to be used for road construction it must be temporarily changed into a fluid state. This may simply be achieved by raising the temperature. However, for surface dressing and some types of bituminous mixture it is necessary to have a fluid binder that can be applied and mixed at relatively low temperatures, but have an adequate hardness after laying. Cutback bitumens are penetration grade bitumens that have their viscosity temporarily reduced by dilution in a volatile oil. After application the volatile oil evaporates and the bitumen reverts to its former viscosity.

The curing time and viscosity of cutbacks can be varied according to the volatility of the diluting oil and the amount of diluent used. In the UK, cutbacks are manufactured using 70/100 or 160/220 pen bitumen diluted with kerosene. Three grades are produced to comply with a viscosity specification based on the standard tar viscometer (STV). *Table 27.2* from BS 3690 shows that cutbacks also must comply with solubility, distillation and recovered penetration requirements. The last two are to ensure that the diluent will evaporate at the required rate and that the residual bitumen will have an appropriate hardness for the performance requirements.

27.3.4 EMULSIONS

An emulsion is a two-phase system consisting of two immiscible liquids, one being dispersed as fine globules within the other. A bitumen emulsion consists of discrete globules of bitumen dispersed within a continuous phase of water, and is a means of enabling penetration grade bitumens to be mixed and laid.

Dispersal of the bitumen globules must be maintained electrochemically using an emulsifier which consists of a long hydrocarbon chain terminating with either a cationic or an anionic functional group. The hydrocarbon chain has an affinity for bitumen, whereas the ionic part has an affinity for water. Thus the emulsifier molecules are attracted to the bitumen globules with the hydrocarbon chain binding strongly to the bitumen, leaving the ionic part on the surface of the globule, as shown in *Fig. 27.2*. Each droplet then carries a like surface charge depending on the charge of the ionic part of the emulsifier. Cationic emulsions are positively charged and anionic emulsions are negatively charged. The globules therefore repel each other, making the emulsion stable. Cationic emulsions are preferred because they also aid adhesion, the positively charged bitumen globules being strongly attracted to the negatively charged aggregate surface.

Emulsions must satisfy two conflicting requirements in that they need stability for storage but also may need to break quickly when put into use. The stability of an emulsion depends on a number of factors, as follows:

1. The quantity and type of emulsifier present. Anionic emulsions require substantial water loss before they break, whereas cationic emulsions break by physicochemical action before much evaporation has taken place. The more free emulsifier ions there are in the continuous phase, the easier it is for the negatively charged aggregate surface to be satisfied without attracting the bitumen globules.

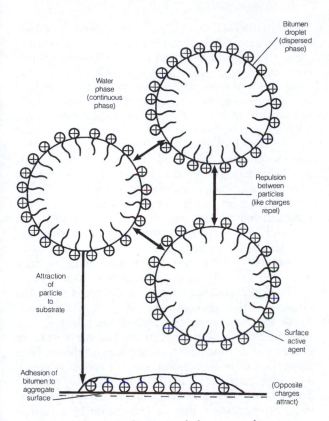

Fig. 27.2 Schematic diagram of charges on bitumen droplets (from The Shell Bitumen Handbook, *2003).*

2. Rate of water loss by evaporation. This in turn depends on ambient temperature, humidity and wind speed as well as rate and method of application.
3. The quantity of bitumen. Increasing the bitumen content will increase the breaking rate.
4. Size of bitumen globules. The smaller their size, the slower will be the breaking rate.
5. Mechanical forces. The greater the mixing friction or, in the case of surface dressing, the rolling and traffic action, the quicker the emulsion will break.

The viscosity of emulsions is important because a large proportion of emulsions are applied by spray. The viscosity increases with bitumen content and is very sensitive for values greater than about 60%. The chemistry of the aqueous phase is also important, viscosity being increased by decreasing the acid content or increasing the emulsifier content. The viscosity for road emulsions is specified in BS 434 Part 1 and BS EN 13808.

27.3.5 POLYMER-MODIFIED BITUMENS

Polymer-modified bitumens (PMBs) are penetration grade bitumens that have been modified in terms of their strength and rheological properties by the addition of small amounts (usually 2–8% by mass) of polymer. The polymers tend to be either plastics or rubbers and alter the strength and viscoelastic properties of the penetration grade bitumen by increasing its elastic response, improving its cohesive and fracture strength and providing greater ductility. Typical examples of rubbers (thermoplastic elastomers) used to modify bitumen include styrenic block copolymers such as styrene–butadiene–styrene (SBS), synthetic rubbers such as styrene–butadiene rubber (SBR), and both natural and recycled (crumb tyre) rubbers. Plastics (thermoplastic polymers) tend to include polyethylene (PE) and polypropylene (PP) as well as semi-crystalline polymers such as ethylene–vinyl acetate (EVA).

27.4 Aggregates

Aggregates make up the bulk of bituminous materials; the percentage by weight ranges from about 92% for a surface course asphalt to about 96% for a dense, heavy-duty and high-modulus base asphalt concrete. The aggregate has important effects on the strength and stiffness of bituminous mixtures. In asphalt concrete mixtures, the strength and resistance to deformation are largely determined by the aggregate grading, with the bitumen acting principally as an adhesive. Here the grading is continuous and provides a dense packing of particles, leading to a stable aggregate structure. In hot rolled asphalt and stone mastic asphalt mixtures, aggregate grading is again important, but the properties are largely determined by the matrix of fines and bitumen.

The majority of aggregates used in bituminous mixtures are obtained from natural sources, either sands and gravels or crushed rock. The main non-natural aggregate source is slag, with blast furnace slag being the most commonly used. As with concrete, aggregates in bituminous mixtures are regarded as inert fillers. However, whereas in concrete both the aggregate and the hardened cement paste are relatively stiff, in a bituminous mixture, the bitumen is very soft compared to the aggregate. Therefore the role of the aggregate in determining mixture stiffness and strength is more important in bituminous mixtures.

Three size ranges are recognised in aggregates for bituminous mixtures. These are coarse, fine and filler. Coarse material is that retained on a 2-mm sieve, fine material passes 2 mm but is retained on

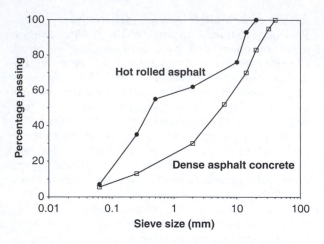

Fig. 27.3 Aggregate grading curves.

the 63-micron sieve, and filler is material passing 63 microns.

27.4.1 PROPERTIES

The importance of grading has already been mentioned. In addition, aggregates suitable for use in bituminous mixtures must have sufficient strength to resist the action of rolling during construction. For surfacing materials, they must also be resistant to abrasion and polishing in order to provide a skid-resistant surface. Here the shape and surface texture of aggregate particles are important.

Figure 27.3 gives typical grading curves for a dense asphalt concrete (previously known as 'macadam') and a hot rolled asphalt (previously known as 'asphalt'). The curve for the asphalt concrete clearly shows the continuous nature of the grading, whereas that for the hot rolled asphalt shows a gap grading with a lack of material in the range 500 microns to 10 mm. This is typical of a hot rolled asphalt, where the 'mortar' of fines and filler bound with bitumen characterises the material, the coarse aggregate providing the bulk.

The strength of aggregate is assessed in two ways. For resistance to crushing, the aggregate crushing value test is used. This test (BS 812: Part 110 and

BS EN 1097-2) determines the extent to which an aggregate crushes when subjected to a steadily increasing load of up to 40 tonnes. The test sample consists of single-sized particles, 10–14 mm, and the amount of fines produced in the test is expressed as a percentage by weight of the original sample. A variation of this test for weaker aggregates is the 10% fines test. Here the maximum crushing load that will produce 10% fines from the original single-sized sample is determined. The importance of this test is that it allows assessment of the extent of crushing that may occur during compaction.

Resistance to impact loads is also required for road aggregates. The impact value test (BS 812: Part 112 and BS EN 1097-2) determines the response of aggregate to blows from a heavy hammer. Once again the outcome of the test is the percentage of fines produced from the original single-sized sample.

The skid resistance of a road surface is provided largely by the aggregate exposed at the surface. There are two components, which are referred to as macrotexture and microtexture. The macrotexture is the overall road surface roughness, which arises from the extent to which there are spaces between aggregate particles. This is a function of mixture proportions. For example, a hot rolled asphalt provides an extremely low macrotexture because the coarse aggregate content is low and coarse particles are immersed in the fines/filler/bitumen mortar. Consequently, a layer of single-sized aggregate particles is rolled into the surface to provide the required macrotexture. These are precoated with a film of bitumen to improve adhesion to the asphalt surface and are known as coated chippings. Asphalt concretes, on the other hand, have a lower proportion of fines and filler and provide a rough surface.

Macrotexture is measured in terms of texture depth using the sand patch test (BS EN 13036-1). The test involves spreading a known volume of sand in a circular patch over the road surface until it can be spread no further. The sand fills the spaces between aggregate particles as shown in *Fig. 27.4*. The diameter of the patch is measured and, knowing the volume of sand, the average depth can be calculated.

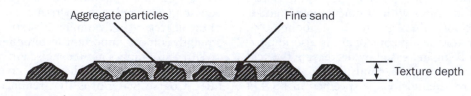

Fig. 27.4 Measurement of macrotexture using the sand patch test.

Microtexture refers to the surface texture of individual particles and varies according to the type of aggregate. Here it is important to use an aggregate which not only has a rough surface texture, but also will retain that texture under the action of traffic. This is assessed using the polished stone value test (BS 812: Part 114). Here samples of aggregate are subjected to a simulated wear test, where a pneumatic tyre runs continuously over the aggregate particles under the abrasive action of emery powder. The skidding resistance of the samples is determined after the test using the pendulum skid tester.

References

Dickie TP and Yen TF (1967). Macrostructure of asphaltic fractions by various instrumental methods. *Analytical Chemistry*, 39, 13–16.

Girdler RB (1965). Constitution of asphaltenes and related studies. *Proceedings of the Association of Asphalt Paving Technologists*, 34, 45.

Shell Bitumen UK (2003). *The Shell Bitumen Handbook*, Shell Bitumen, UK.

Speight JG and Moschopedis SE (1979). Some observations on the molecular 'nature' of petroleum asphaltenes. *American Chemical Society*, Division of Petroleum Chemistry, 24, 22–25.

Chapter 28

Viscosity, stiffness and deformation of bituminous materials

28.1 Viscosity and rheology of binders

The viscosity of a liquid is the property that retards flow, so when a force is applied to a liquid, the higher the viscosity, the slower will be its movement. The viscosity of bitumen is dependent upon both its chemical make-up and its structure. In sol-type bitumens, the asphaltene micelles are well dispersed within the maltenes continuum. The viscosity depends on the relative amounts of asphaltenes and maltenes, decreasing as the asphaltene content reduces. In gel-type bitumens, where the asphaltene micelles have aggregated, the viscosity is higher and dependent upon the extent of aggregation. The degree of dispersion of the asphaltenes is controlled by the relative amounts of resins, aromatics and saturated oils. If there are sufficient aromatics they form a stabilising layer around the asphaltene micelles, promoting the dispersion. However, if they are not present in sufficient quantity the micelles will tend to join together. A schematic representation of the two states is shown in *Fig. 28.1* (*The Shell Bitumen Handbook*, 1990). In practice most bitumens are somewhere between these two states. The maltenes continuum is influenced by the saturated oils, which have low molecular weight and consequently a low viscosity. These saturates have little solvent power in relation to the asphaltenes, so that as the saturate fraction increases, there is a greater tendency for the asphaltenes to aggregate to form a gel structure. Thus a high proportion of saturates on the one hand tends to reduce viscosity because of their low molecular weight, but on the other hand encourages aggregation of the asphaltene micelles, which increases viscosity. The relative importance of these two opposing effects depends on the stabilising influence on the asphaltenes of the aromatics.

The asphaltenes exert a strong influence on viscosity in three ways. Firstly, the viscosity increases as the asphaltene content increases. Secondly, the shape of the asphaltene particles governs the extent of the change in viscosity. The asphaltene particles are thought to be formed from stacks of plate-like sheets

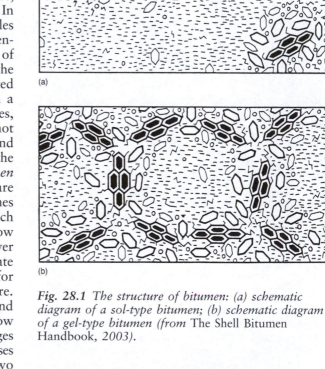

Fig. 28.1 *The structure of bitumen: (a) schematic diagram of a sol-type bitumen; (b) schematic diagram of a gel-type bitumen (from* The Shell Bitumen Handbook, *2003).*

of aromatic/naphthenic ring structures. These sheets are held together by hydrogen bonds. However, the asphaltenes can also form into extended sheets and combine with aromatics and resins so that the particle shape varies. Thirdly, the asphaltenes may tend to aggregate, and the greater the degree of aggregation the higher is the viscosity.

28.2 Empirical measurement of viscosity

The physical behaviour of bitumen is complex and to describe its properties over a wide range of operating conditions (temperature, loading rate, stress and strain) would require a large number of tests. To avoid this and simplify the situation, the mechanical behaviour and rheological properties of bitumen have traditionally been described using empirical tests and equations. The two consistency tests required in the European Standard BS EN 12591 to characterise different bitumen paving grades are the needle penetration test (BS EN 1426) and the ring and ball softening point test (BS EN 1427). These tests provide an indication of the consistency (hardness) of the bitumen without completely characterising the viscoelastic response, and form the basis of the bitumen specification.

The softening point is the temperature at which a bitumen reaches a specified level of viscosity. This viscosity is defined by the ring and ball test apparatus as the consistency at which a thin disc of bitumen flows under the weight of a 10 mm diameter steel ball by a distance of 25 mm. *Figure 28.2* shows a diagrammatic representation of the test. The more viscous the bitumen, the higher the temperature at which this level of viscosity is reached.

Another test that is commonly applied to bitumens, and is the basis for their characterisation, is the penetration test. The test measures hardness, but this is related to viscosity. The test consists of measuring the depth to which a needle penetrates a sample of bitumen under a load of 100 g over a period of 5 seconds at a temperature of 25°C. Thus the test differs from the softening point test in that, rather than determining an equiviscous temperature, the viscosity is determined at a particular temperature. However, because bitumen is viscoelastic, the penetration will depend on the elastic deformation as well as the viscosity. Therefore, since viscosity changes with temperature, different bitumens may have the same hardness at 25°C but different hardnesses at other temperatures. It is the varying elasticity of

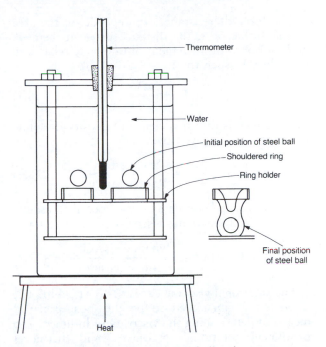

Fig. 28.2 *Apparatus for ring and ball softening point test.*

bitumens which prevents correlation between these empirical tests.

28.3 Measurement of viscosity

Viscosity is the measure of the resistance to flow of a liquid and, as discussed in section 5.1, is defined as the ratio between the applied shear stress and the rate of shear strain measured in units of Pascal seconds (Pa.s). In addition to this absolute or dynamic viscosity, viscosity can also be measured as kinematic viscosity in units of m^2/s or, more commonly, mm^2/s with 1 mm^2/s being equivalent to 1 centistoke (cSt).

The viscosity of bitumen can be measured with a variety of devices in terms of its absolute and kinematic viscosities. Specifications are generally based on a measure of absolute viscosity at 60°C and a minimum kinematic viscosity at 135°C, using vacuum and atmospheric capillary tube viscometers, respectively. Absolute viscosity can also be measured using a fundamental method known as the sliding plate viscometer. The sliding plate test monitors force and displacement on a thin layer of bitumen contained between parallel metal plates at varying combinations of temperature and loading time.

The force of resistance, F, depends on the area of the surfaces, A, the distance between them, d, and the speed of movement of one plate relative to the other, V, such that:

$$F = \eta \frac{AV}{d} \qquad (28.1)$$

The factor η is the coefficient of viscosity (absolute viscosity), and is given by:

$$\eta = \frac{Fd}{AV} = \frac{\text{Shear stress}}{\text{Rate of strain}} \qquad (28.2)$$

The relationship between dynamic viscosity (absolute viscosity) and kinematic viscosity is expressed as:

$$\text{Kinematic viscosity} = \frac{\text{Dynamic viscosity}}{\text{Mass density}} \qquad (28.3)$$

The rotational viscometer test (ASTM D4402-02) is currently considered to be the most practical means of determining the viscosity of bitumen. The Brookfield rotational viscometer and thermocel system allow the testing of bitumen over a wide range of temperatures (more so than most other viscosity measurement systems). The rotational viscometer consists of one cylinder rotating coaxially inside a second (static) cylinder containing the bitumen sample, all contained in a thermostatically controlled environment. The material between the inner cylinder and the outer cylinder (chamber) is therefore analogous to the thin bitumen film found in the sliding plate viscometer. The torque on the rotating cylinder or spindle is used to measure the relative resistance to rotation of the bitumen at a particular temperature and shear rate. The torque value is then altered by means of calibration factors to yield the viscosity of the bitumen.

28.4 Influence of temperature on viscosity

Bitumens are thermoplastic materials so that they soften as the temperature rises but become hard again when the temperature falls. The extent of the change in viscosity with temperature varies between different bitumens. It is clearly important, in terms of the performance of a bitumen in service, to know the extent of the change in viscosity with temperature. This is referred to as temperature susceptibility and, for bitumens, is determined from the penetration value, P, and softening point temperature, T. These are related empirically by the expression:

$$\log P = AT + k \qquad (28.4)$$

where A is the temperature susceptibility of the logarithm of penetration and k is a constant. From this relationship, an expression has been developed (Pfeiffer and Van Doormaal, 1936) that relates A to an index, known as the penetration index, PI, such that for road bitumens the value of PI is about zero.

$$A = \frac{d(\log P)}{dT} = \frac{20 - PI}{50(10 + PI)} \qquad (28.5)$$

It has been determined that, for most bitumens, the penetration at their softening point (SP) temperature is about 800. Thus if the penetration at 25°C and the softening point temperature are known, the PI can be evaluated from:

$$\frac{d(\log P)}{dT} = \frac{\log 800 - \log P}{SP - 25} = \frac{20 - PI}{50(10 + PI)} \qquad (28.6)$$

For example, for a 40/60 pen bitumen with a softening point of 48°C:

$$\frac{d(\log P)}{dT} = \frac{\log 800 - \log 50}{48 - 25} = \frac{1.204}{23} = 0.0523$$

$$(28.7)$$

Therefore:

$$0.0523 = \frac{20 - PI}{50(10 + PI)}$$

giving $PI = -1.7$.

Pfeiffer and Van Dormaal produced a nomograph (*Fig. 28.3*) to evaluate the above expression, and it can be seen that for the above example a similar result is obtained.

Bitumens for road use normally have a PI in the range -2 to $+2$. If the PI is low, bitumens are more Newtonian in their behaviour and become very brittle at low temperatures. High-PI bitumens have marked time-dependent elastic properties and give improved resistance to permanent deformation. The influence of chemical composition on temperature susceptibility is illustrated in *Fig. 28.4*. In general the PI increases as the asphaltene content increases at the expense of the aromatics. This change can be achieved by controlled air blowing.

28.5 Resistance of bitumen to deformation

Since bitumen is a viscoelastic material, the response to an applied load depends on the size of the load,

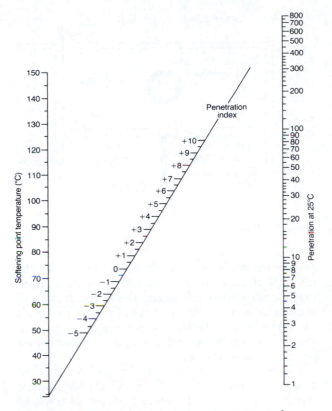

Fig. 28.3 *Nomograph to evaluate penetration index from softening point and penetration (after Pfeiffer and Doormaal, 1936).*

the temperature, and the duration of its application. In other words there is no simple relationship between stress and strain and it is therefore difficult to predict the elastic modulus (or equivalent Young's modulus) of bitumen. To take account of the viscoelastic nature of bitumen, Van der Poel (1954) introduced the concept of stiffness modulus. This modulus is dependent on both temperature and time of loading, and is given by:

$$S_{t,T} = \frac{\sigma}{\varepsilon_{t,T}} \qquad (28.8)$$

where σ is the tensile stress and $\varepsilon_{t,T}$ is the resultant strain after loading for time t at temperature T. *Figures 28.5* and *28.6* illustrate the effect of loading time and temperature for bitumens of different *PI*. For low-PI bitumens (*Fig. 28.5*) the stiffness is constant for very short loading times and virtually independent of temperature. This represents elastic behaviour. For longer loading times the curves have a consistent slope of 45° and have a significant variation with temperature, indicating viscous behaviour. The effect of increasing *PI* can be seen by comparing *Figs 28.5* and *28.6*. High-PI bitumens are much stiffer at high temperatures and longer loading times. Thus under conditions that are more likely to give rise to deformation, namely slow moving or stationary traffic and high temperatures, a high-PI bitumen offers greater resistance to deformation by virtue of its higher stiffness and more elastic response.

When considering a bituminous mixture consisting of a graded aggregate bound with bitumen, the

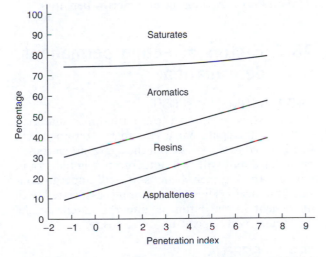

Fig. 28.4 *Relationship between chemical composition and penetration index (after Lubbers, 1985).*

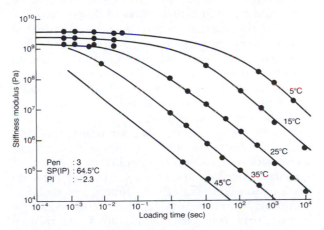

Fig. 28.5 *The effect of temperature and loading time on stiffness of a low-PI bitumen (from* The Shell Bitumen Handbook, *2003).*

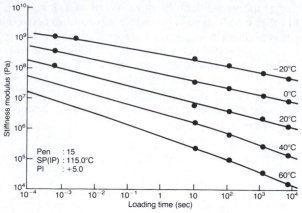

Fig. 28.6 *The effect of temperature and loading time on stiffness of a high-PI bitumen (from* The Shell Bitumen Handbook, *2003).*

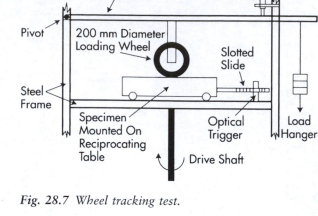

Fig. 28.7 *Wheel tracking test.*

stiffness of the mixture is dependent on the stiffness of the bitumen and the quantity and packing of aggregate in the mixture (Van der Poel, 1955). The quantity and packing of aggregate particles depend on grading, particle shape and texture, and method of compaction.

28.6 Determination of permanent deformation

Rutting of bituminous pavements is the most common type of failure in the UK. It is therefore important to be able to predict the permanent deformation for a bituminous mixture, and this depends on the low stiffness response, that is the stiffness at long loading times and high temperatures, as well as the balance between the viscous (non-recoverable) and elastic (recoverable) components of the mixture's deformation. Two tests that have been commonly used to determine the permanent deformation properties of bituminous mixtures are the creep test (usually under repeated loading) and the wheel tracking test.

In the creep test (known as the repeated load axial test (RLAT) in the UK), a repeating uniaxial load of 0.1 MPa, with a loading time of 1 second and a rest period of 1 second, is applied to a cylindrical specimen for 2 hours at 40°C. During the test, deformation is measured as a function of time.

Although simple, the repeated load axial test is extremely convenient and allows the relative performance of different bituminous mixtures to be easily determined. It is often criticised as being too severe as the test does not employ a confining stress.

In-situ materials will clearly be confined and the effect of the confining stress on the vertical strain may be important. However, the severe nature of the test does mean that intrinsically poor materials can easily be identified, and there is good correlation between creep tests and permanent deformation performance in the road.

The wheel tracking test can be considered to be a simulative test. *Figure 28.7* shows a diagrammatic representation of a laboratory-scale wheel tracking test. In the UK, the wheel tracking test is usually carried out at either 45°C or 60°C with an applied wheel load of 520 N. The performance of the bituminous mixture is assessed by measuring the resultant rut depth after a given number of passes or the rate of tracking in millimetres per hour.

28.7 Factors affecting permanent deformation

28.7.1 BITUMEN VISCOSITY
When a stress is applied to a bituminous material, both the aggregate particles and the bitumen will be subjected to the stress. But the aggregate particles, being hard and stiff, will undergo negligible strain, whereas the bitumen, being soft, will undergo considerable strain. Thus deformation is associated with movement in the bitumen, and the extent of the movement will depend on its viscosity.

28.7.2 AGGREGATE
Bituminous mixtures that utilise a continuously graded aggregate, such as asphalt concretes, rely mainly on aggregate particle interlock for their resistance to deformation. Thus the grading and particle shape

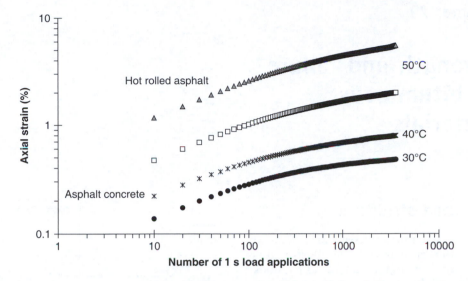

Fig. 28.8 *Comparison of permanent strain for asphalt concrete and hot rolled asphalt mixtures.*

of the aggregate are major factors governing deformation. The characteristics of the fine aggregate are particularly important in gap-graded materials, which rely on a dense bitumen and fine mortar for their strength. These are the hot rolled asphalt and stone mastic asphalt mixtures. Sand particles can vary considerably from spherical glassy grains in dune sands, to angular and relatively rough grains from some pits. Mixtures made with a range of sands all at the same bitumen content have been shown to give deformations, when tested in the laboratory wheel tracking test, that varied by a factor of 4 from the best to the worst sand (Knight *et al.*, 1979).

28.7.3 TEMPERATURE

Figure 28.8 shows permanent strain against number of test cycles in a repeated load axial test. It can be seen that permanent strain increases with temperature. This is due to the reduction in viscosity of bitumen, the consequent reduction in bitumen stiffness and the accumulation of repeated, non-recoverable viscous deformations. The figure also indicates the effect of the aggregate grading. At low temperatures, the permanent strain in continuously graded asphalt concretes and gap graded hot rolled asphalts will be very similar. Here the high degree of aggregate particle interlock in the asphalt concrete and the high viscosity bitumen in the hot rolled asphalt provide a similar resistance to deformation. How-

ever, at higher temperatures, the hot rolled asphalt deforms more due to the reduced bitumen viscosity, which is not compensated by the aggregate interlock effect. In the asphalt concrete, although the bitumen will also be less stiff and viscous, the aggregate grading continues to provide a compensating resistance to deformation.

References

Hills JF, Brien D and Van de Loo PJ (1974). *The correlation of rutting and creep tests on asphalt mixes*, Institute of Petroleum, IP-74-001.

Knight VA, Dowdeswell DA and Brien D (1979). Designing rolled asphalt wearing courses to resist deformation. In *Rolled Asphalt Road Surfacings*, ICE, London.

Lubbers HE (1985). *Bitumen in de weg- en waterbouw.* Nederlands Adviesbureau voor bitumentopassingen.

Pfeiffer JPh and Van Doormaal PM (1936). The rheological properties of asphaltic bitumens. *Journal of Institute of Petroleum*, **22**.

Shell Bitumen (2003). *The Shell Bitumen Handbook*, Shell Bitumen, UK.

Van der Poel C (1954). A general system describing the viscoelastic properties of bitumen and its relation to routine test data. *Journal of Applied Chemistry*, **4**.

Van der Poel C (1955). Time and temperature effects on the deformation of bitumens and bitumen mineral mixtures. *Journal of the Society of Plastics Engineers*, **11**.

Chapter 29

Strength and failure of bituminous materials

29.1 The road structure

A flexible road structure consists of a number of layers of different materials, as illustrated in *Fig. IV.2* in the Introduction to this part of the book. In structural terms, the purpose of the road is to distribute the applied load from the traffic to a level that the underlying subgrade can bear. The stresses induced by the loads are high at the surface but reduce with depth. Thus, the surfacing material must be of high quality, but at greater depths below the surface, economies can be achieved by using materials of lower strength.

29.2 Modes of failure in a bituminous structure

Roads deteriorate in a number of ways, but broadly there are two forms of failure – functional and structural. Functional failure (distress) is usually associated with the road surface through deterioration with time. This may be through breakdown of the surface material, for example through fretting or stone loss (ravelling), or alternatively, the surface texture of the surface course may be reduced, through polishing or abrasion, so that the skidding resistance drops below an acceptable level.

Structural deterioration (failure) develops gradually with the continued application of wheel loads. In the early stages, the rate of deterioration is very small and the structural changes are not perceptible and are difficult to measure. But with continued service, signs of structural change become clearer and the rate of deterioration accelerates. There are two modes of breakdown, which are illustrated in *Fig. 29.1*. Firstly, permanent deformation occurs in the wheel tracks. This 'rutting' is associated with deformation of all the pavement layers and is linked to a loss of support from the structural layers of the road and

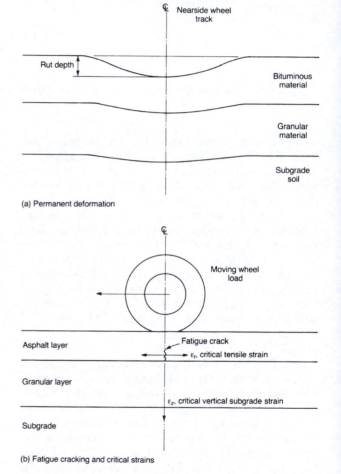

(a) Permanent deformation

(b) Fatigue cracking and critical strains

Fig. 29.1 Modes of failure and critical strains in a flexible pavement (after Brown, 1980).

the underlying subgrade. This structural rutting differs from the 'non-structural' permanent deformation found within the bituminous layers, which is an accumulation of the small irrecoverable part of the deformation caused by each wheel load due to

224

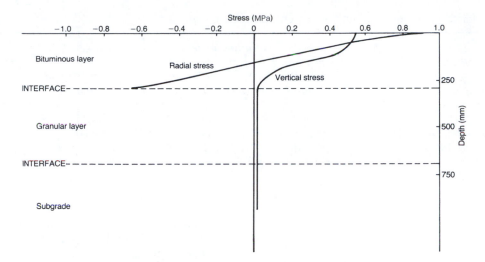

Fig. 29.2 *Variation of vertical and radial stresses below the centre-line of a 40 kN wheel load acting over a circular area of radius 160 mm, with a contact pressure of 0.5 MN/m² (after Peattie, 1978).*

the viscoelastic nature of the bitumen together with the mechanical support offered by the grading of the aggregate.

The second mode of failure is cracking, which appears along the wheel tracks. Cracking is caused by the tensile strain developed in the bound layers as each wheel load passes. It is therefore a function of both the size of tensile strain, and the repetitive nature of the loading; that is a fatigue failure. It is important to note, as *Fig. 29.1* shows, that the cracking initiates at the base of the bound layer. This is where the tensile stresses are highest, as shown in *Fig. 29.2*. It follows that, by the time the cracking is visible at the surface, the damage has been present for some time.

In both modes of failure, the breakdown is caused by (a) the repetitive nature of the loading, and (b) the development of excessive strains in the structure. This leads to the notion that, if failure can be defined, the life of a road can be determined provided that the loading can be assessed and the performance of the materials evaluated. Alternatively, the structural design of the road together with the make-up of the materials necessary to give the required properties may be determined, in order to provide a given life. In either case, it is necessary to define failure.

In most civil engineering structures, structural failure renders the structure unusable and is often associated with collapse. However, while roads may become less comfortable to drive on and less safe, they do not, except in very extreme cases, become unusable. Therefore failure for roads must be identified in

terms of serviceability and/or repairability; that is, the extent of cracking and deformation which is just acceptable to drivers and/or which represents a condition that may be economically restored by repair, must be defined. A definition is provided, in terms of both modes of failure, which identifies three conditions: sound, critical, and failed. *Table 29.1* shows that if any cracking is visible at the surface, then the road is regarded as being at a critical condition or as having failed. Here the term 'critical' means that failure is imminent but the road still has sufficient structural capability to support strengthening and to provide an extended life from the strengthened road. If there are no signs of cracking, then the condition is defined in terms of the extent of permanent deformation. Thus, if the rut depth reaches 20 mm, the road is regarded as having failed.

29.3 Fatigue characteristics

Fatigue cracking arises from the fact that under repeated applications of tensile stress/strain, a bituminous material will eventually fracture. The higher the level of stress and strain applied, the smaller the number of load applications before cracking occurs. For a particular level of stress and strain, the mixture proportions and nature of the bitumen dictate the number of cycles before cracking occurs.

A number of laboratory tests have been developed to assess the fatigue characteristics of bituminous materials. The tests, illustrated in *Fig. 29.3*, are

Table 29.1 Criteria for determining pavement condition

	Wheel-track rutting under a 2 m straight edge			
Wheel-track cracking	Less than 5 mm	From 5 mm to < 10 mm	From 10 mm to < 20 mm	20 mm or greater
None	Sound	Sound	Critical	Failed
Less than half width or single crack	Critical	Critical	Critical	Failed
More than half width	Failed	Failed	Failed	Failed

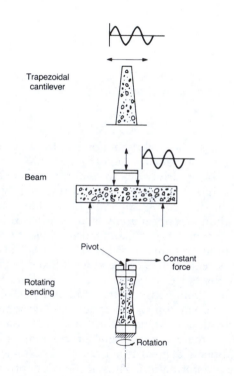

Fig. 29.3 *Methods of fatigue testing of bituminous materials (after Brown, 1978).*

flexure tests and simulate the repeated bending action in the stiff bound layer of a pavement caused by the passage of each wheel load. The number of load cycles that a particular specimen can endure before failure depends on a number of factors, discussed below.

29.3.1 STRESS AND STRAIN CONDITIONS

Fatigue tests may be conducted in two ways. They may be constant-stress tests, where each load application is to the same stress level regardless of the amount of strain developed, or they may be constant-strain tests, where each load application is to the same strain level regardless of the amount of stress required.

These two alternatives produce quite different results. *Figure 29.4a* shows the general pattern of results from constant-stress tests. Each line represents a different test temperature, i.e. a different stiffness, and it can be seen that mixtures with higher stiffness have longer lives. *Figure 29.4c* shows the general pattern of results from constant-strain tests. Again each line represents a different temperature or stiffness and it can be seen that the outcome is reversed, with the mixtures of higher stiffness having the shortest lives. This contrast may be explained in terms of the failure mechanism. Cracks initiate at points of stress concentration and propagate through the material until fracture occurs. If the stress level is kept constant, the stress level at the tip of the crack continues to be high so that propagation is rapid. However, in a constant-strain test, the development of a crack causes a steady reduction in the applied stress level because the cracks contribute more and more to the strain as they propagate. Thus the stress at the crack tips reduces and the rate of propagation is slow. Thus it is important to establish which test condition is most relevant to actual pavement behaviour. It has been shown (Monismith and Deacon, 1969) that strain control is appropriate to thin layers (for example surfacing layers), whereas stress control is appropriate to thicker structural layers. This is because pavements are subject to a stress-controlled loading system, so that the main (and normally thick) structural layers are stress controlled. However, the thin surface layer must move with the lower structural layers and so is effectively subject to strain control. Nevertheless, under low-temperature conditions giving high stiffness, crack propagation is relatively quick even under strain-controlled loading, so the difference between the two loading conditions is small.

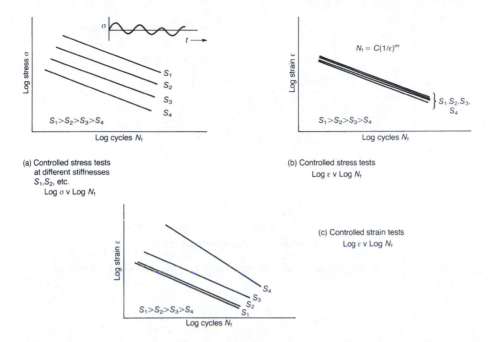

Fig. 29.4 *Fatigue lines representing number of cycles to failure, N_f, under different test conditions (after Brown, 1980); S, stiffness.*

29.3.2 THE STRAIN CRITERIA

If the results of a controlled-stress test are expressed in terms of an equivalent strain then the log–log plot of strain against number of load cycles produces a single linear relationship for all test conditions for a particular mixture, as shown in *Fig. 29.4b*. In other words the relationship is independent of mixture stiffness. This suggests that strain is the principal criterion governing fatigue failure, and it has been demonstrated (Cooper and Pell, 1974) that flexure tests on a wide range of mixtures produce unique fatigue lines for each mixture. The general relationship defining the fatigue line is:

$$N_f = C\left(\frac{1}{\varepsilon}\right)^m \qquad (29.1)$$

where N_f is the number of load cycles to initiate a fatigue crack, ε is the maximum applied tensile strain, and C and m are constants depending on the composition and properties of the asphalt mixture. The fatigue lines for a range of asphalt mixtures are shown in *Fig. 29.5*.

29.3.3 EFFECT OF MIXTURE VARIABLES

A large number of variables associated with the asphalt mixture affect the fatigue line. However, it

has been shown (Cooper and Pell, 1974) that two variables are of prime importance. These are:

1. The volume of bitumen in the mixture.
2. The viscosity of the bitumen as measured by the softening point.

As the volume of bitumen increases up to 15%, the fatigue life increases, and as the bitumen becomes more viscous, with softening point increasing up to 60°C, the fatigue life also increases.

Other factors are important insofar as they affect the two main variables. The void content of the mixture has an effect on the volume of bitumen. The total void content is in turn affected by the particle shape and grading of the aggregate, the compactive effort, and the temperature. In other words there is a link between workability, compactive effort and void content, which is controlled by the bitumen content. However, while higher bitumen content improves fatigue life, it also reduces stiffness, which leads to increased strain. *Figure 29.6* illustrates the double influence of void content. *Figure 29.6a* shows the effect that increased bitumen content has on stiffness. The stiffness is reduced, which increases the strain under constant-stress conditions, causing a shift to the left along the fatigue line. Thus the fatigue life is reduced. *Figure 29.6b* shows the influence

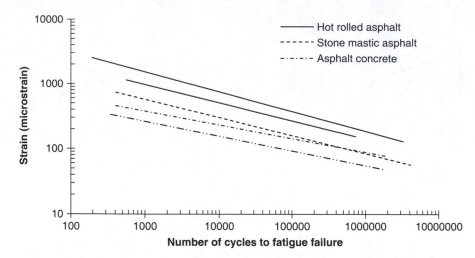

Fig. 29.5 *Fatigue lines under controlled stress loading conditions for typical asphalt mixtures.*

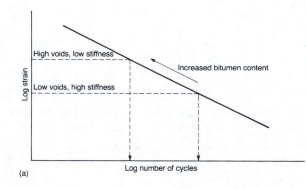
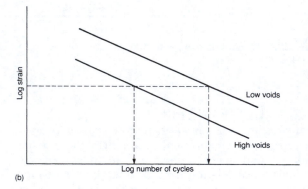

Fig. 29.6 *The influence of voids on fatigue life.*

that void content has on the fatigue line, so that for the same strain the fatigue life is reduced as void content increases. The change in position of the fatigue line corresponds to a change in material type, as was seen in *Fig. 29.5*, whereas the shift along a fatigue line due to a stiffness change corresponds to a change in degree of compaction. However, in practice, both effects occur if the change in void content is associated with a change in bitumen content.

References

Brown SF (1978). Material characteristics for analytical pavement design. In *Developments in Highway Pavement Engineering – 1*, Applied Science Publishers, London.

Brown SF (1980). *An introduction to the analytical design of bituminous pavements*, Department of Civil Engineering, University of Nottingham.

Cooper KE and Pell PS (1974). *The effect of mix variables on the fatigue strength of bituminous materials*, Transport and Road Research Laboratory Laboratory Report 633.

Monismith CL and Deacon JA (1969). Fatigue of asphalt paving mixtures. *Journal of Transport Engineering Division, ASCE*, **95**, 154–161.

Peattie KR (1978). Flexible pavement design. In *Developments in Highway Pavement Engineering*, Applied Science Publishers, London.

Chapter 30

Durability of bituminous structures

Durability is the ability to survive and continue to give an acceptable performance. In the case of roads, it is necessary that the structure should survive for the specified design life, although it is accepted that not all aspects of performance can be sustained for this duration without some restorative maintenance. The design guide for UK roads suggests a design life of 40 years (Highways Agency, 2006), although this can usually only be achieved in stages, as shown in *Fig. 30.1*. The durability of a flexible road structure depends on the durability of the materials from which it is constructed, in particular the bituminous materials. Bituminous materials may deteriorate in a number of ways. The bitumen itself will harden with exposure to oxygen and through temperature effects, the aggregate may not be of sufficient quality so that some individual particles may break down, or there may be loss of adhesion between the bitumen and aggregate particles. These forms of deterioration are caused by weathering (moisture) and the action of traffic. These agents act at the road surface, which is particularly vulnerable. However, deterioration can also occur in the body of the material and this is controlled by the void content and permeability of the material.

30.1 Ageing of bitumen

The ageing or hardening of bitumen is an inevitable result of exposure of bitumen to the atmosphere. The rate of hardening will depend on the conditions and the nature of the bitumen. There are two main processes that occur: oxidation and loss of volatiles.

30.1.1 OXIDATION
In the oxidation process, oxygen molecules from the air combine with the aromatics and resins to form asphaltenes. Thus there is an increase in the polar, high molecular weight fraction at the expense of the lower molecular weight components. This results in

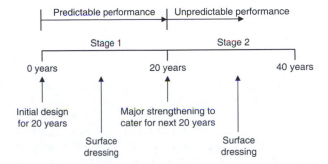

Fig. 30.1 The life of a flexible road.

an increase in the viscosity of the bitumen. Also, the bitumen becomes unstable owing to the discontinuity that develops between the saturates and the rest of the components. This instability causes a lack of cohesion within the bitumen, which may lead to cracking. The rate of oxidation is highly dependent on temperature, and is rapid at the high temperatures used for mixing and laying bituminous materials.

30.1.2 LOSS OF VOLATILES
Loss of volatiles will occur if there is a substantial proportion of low molecular weight components in the bitumen and if the bitumen is subjected to high temperatures. However, for penetration grade bitumens the loss of volatiles once the material has been laid is relatively small.

30.1.3 AGEING INDEX
The hardening of bitumen results in a lowering of penetration, an increase in softening point and an increase in penetration index. Therefore an excessive amount of hardening will cause the material to become brittle at low temperatures and vulnerable to cracking. A convenient way of representing the ageing of bitumen is by means of an ageing index, calculated as the ratio of the viscosity of the aged bitumen to that of the original bitumen.

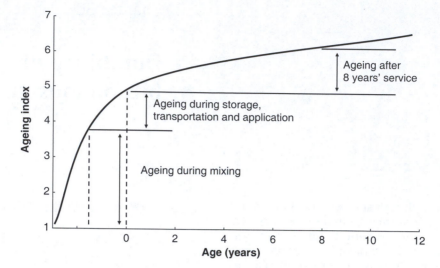

Fig. 30.2 *Ageing of bitumen during mixing, storage, transportation, application and service (from* The Shell Bitumen Handbook, 2003).

In practice, the ageing of bitumen is most marked during the mixing process because of the high temperatures involved in a batch or drum asphalt production plant. For example, the penetration value of a 50 pen-bitumen will fall to between 30 and 40 depending on the duration of the mixing and the temperature used; subsequent high-temperature storage will cause further ageing. Thus the penetration value could be reduced by as much as a half. Ageing of bitumen on the road is generally a much slower process. This is because the temperatures are much lower and the availability of oxygen is restricted by the void content and permeability of the mixture. In more open-textured (porous) asphalt mixtures with a large volume of interconnected voids, air can readily permeate the material, allowing oxidation to occur. However, in dense mixtures with high binder contents, such as hot rolled asphalts and stone mastic asphalts, the permeability is low and there will be very little movement of air through the material. In both cases, ageing will be more rapid at the surface than in the bulk of the material because there is a continual availability of oxygen and the surface will reach higher temperatures. *Figure 30.2* shows the ageing index of bitumen after mixing, storage, transport, paving and subsequent service.

30.1.4 BITUMEN AGEING TESTS

Tests related to the ageing of bitumen can be broadly divided into tests performed on neat bitumen and tests performed on asphalt mixtures. Most laboratory ageing of bitumen utilises thin film oven ageing to age the bitumen in an accelerated manner. Typically,

these tests are used to simulate the relative hardening that occurs during the mixing and laying process ('short-term ageing'). To include 'long-term hardening' in the field, thin film oven ageing is typically combined with pressure oxidative ageing.

The most commonly used short-term ageing test is the rolling thin film oven test (RTFOT), standardised in BS EN 12607-1. The RTFOT involves rotating eight glass bottles each containing 35 g of bitumen in a vertically rotating shelf, while blowing hot air into each sample bottle at its lowest travel position. During the test, the bitumen flows continuously around the inner surface of each container in relatively thin films of 1.25 mm at a temperature of 163°C for 75 minutes. The vertical circular carriage rotates at a rate of 15 revolutions/minute and the air flow is set at a rate of 4000 ml/minute. The method ensures that all the bitumen is exposed to heat and air and the continuous movement ensures that no skin develops to protect the bitumen. The conditions in the test are not identical to those found in practice, but experience has shown that the amount of hardening in the RTFOT correlates reasonably well with that observed in a conventional batch mixer.

Long-term ageing of bitumen can be achieved using the pressure ageing vessel (PAV), which was developed to simulate the in-service oxidative ageing of bitumen in the field. The method involves hardening of bitumen in the RTFOT followed by oxidation of the residue in a pressurised ageing vessel. The PAV procedure (AASHTO R28-06) entails ageing 50 g of bitumen in a 140 mm diameter pan (approximately 3.2 mm binder film thickness) within the heated vessel,

Table 30.1   Classification of voids in terms of permeability for asphalt mixtures (after Chen *et al.*, 2004)

Permeability, k *(cm/s)*	Permeable condition	Voids	Mixture
10^{-4} or lower	Impervious	Impermeable	Dense
10^{-4} to 10^{-2}	Poor drainage	Semi-effective	Stone mastic asphalt
10^{-2} or higher	Good drainage	Effective	Porous asphalt

pressurised with air to 2.07 MPa for 20 hours at temperatures between 90 and 110°C.

30.2 Permeability

Permeability is an important parameter of a bituminous mixture because it controls the extent to which both air and water can migrate into the material. The significance of exposure to air (ageing) was described in the previous section. Water may also bring about deterioration by causing the bitumen to strip from the aggregate particles (adhesive failure), or causing weakening (damage) of the bitumen and bitumen/filler/fines mastic (cohesive failure).

30.2.1 MEASUREMENT AND VOIDS ANALYSIS
The measurement of permeability is, in essence, a simple task, achieved by applying a fluid under pressure to one side of a specimen of a bituminous mixture and measuring the resulting flow of fluid at the opposite side. Both air and water have been used as the permeating fluid. *Table 30.1* presents typical ranges of permeability for three common types of asphalt mixture (Chen *et al.*, 2004).

30.2.2 FACTORS AFFECTING PERMEABILITY
The permeability of a bituminous mixture depends on a large number of factors. Of particular importance are the quantity of voids, the distribution of void size and the continuity of the voids. *Figure 30.3* shows how permeability varies with total voids in the mixture for a range of typical asphalt concrete, hot rolled asphalt, stone mastic asphalt and porous asphalt mixtures. It can be seen that there are significant differences in air void size, distribution and connectivity, and therefore permeability, in mixtures with the same proportion of air voids. *Figure 30.3* shows that the relationship between permeability and air voids is in all cases exponential.

The voids are also affected by the nature of the aggregate. The shape, texture and grading of the particles will govern the packing and hence void content at a particular bitumen content. The amount of compactive effort employed is also important.

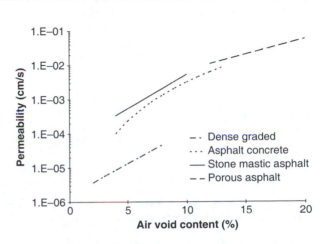

Fig. 30.3 *Relationship between the coefficient of permeability, k, and air voids for a range of asphalt mixtures (after Caro* et al., *2008).*

30.3 Adhesion

The quality of the adhesion of a bitumen to an aggregate is dependent on a complex assemblage of variables. *Table 30.2* identifies a number of factors that have an influence on the adhesion performance of bituminous mixtures. Although some of these relate to the ambient conditions and aspects of the mixture as a whole, the principal factors are the nature of the aggregate and, to a lesser extent, the bitumen.

30.3.1 THE NATURE OF THE AGGREGATE
The mineralogical and physical nature of the aggregate particles has an important bearing on adhesion, the adhesive capacity being a function of chemical composition, shape and structure, residual valence, surface energy and the surface area presented to the bitumen. Generalisations about the effect of mineralogy are difficult because the effects of grain size, shape and texture are also important. However, in general, the more siliceous aggregates such as granites, rhyolites, quartzites, siliceous gravel and cherts tend to be more prone to moisture-related adhesive failures. The facts that good performance with these materials has also been reported, and that failures in supposedly

Table 30.2 Material properties and external influences that can act singularly or together to affect the adhesion and stripping resistance of a bituminous mix

Aggregate properties	Bitumen properties	Interactive mix properties	External influences
Mineralogy	Composition and source	Compaction	Annual precipitation*
Surface texture	Durability and weathering	Grading	Relative humidity*
Porosity	Viscosity	Permeability	pH of water*
Surface coatings and dust	Curing time	Binder content	Presence of salts*
Mechanical durability	Oxidation	Cohesion	Temperature*
Surface area	Electrical polarity	Film thickness	Temperature cycling*
Absorption	Use of additives	Filler type	Light, heat and radiation*
Moisture content		Type of mix	Traffic
Abrasion		Method of production	Construction practice
pH		Use of additives	Design
Weathering grade			Workmanship
Exposure history			Drainage
Shape			
Additives			

* Factors considered uncontrollable.

good rock types such as limestones and basic igneous rocks have occurred, emphasise the complexity of the various material interactions. Therefore caution should be exercised when attempting to make generalisations on the adhesion performance of aggregates of different or even similar mineralogy.

The surface character of each individual aggregate type is important, particularly in relation to the presence of a residual valence or surface charges. Aggregates with unbalanced surface charges possess a surface energy, which can be attributed to a number of factors including broken co-ordination bonds in the crystal lattice, the polar nature of minerals, and the presence of adsorbed ions. Such surface energy will enhance the adhesive bond if the aggregate surface is coated with a liquid of opposite polarity.

Absorption of bitumen into the aggregate depends on several factors, including the total volume of permeable pore space and the size of the pore openings. The presence of a fine microstructure of pores, voids and microcracks can bring about an enormous increase in the absorptive surface available to the bitumen. This depends on the petrographic characteristics of the aggregate as well as its quality and state of weathering.

It is generally accepted that rougher surfaces exhibit a greater degree of adhesion. A balance is, however, required between the attainment of good wettability of the aggregate (smooth surfaces being more easily wetted), and a rougher surface that holds the binder more securely once wetting has been achieved. The presence of a rough surface texture can mask the effects of mineralogy.

30.3.2 THE NATURE OF THE BITUMEN

The important characteristics of bitumen affecting its adhesion to aggregate are its viscosity and surface tension, and its polarity. The viscosity and surface tension will govern the extent to which bitumen is absorbed into the pores at the surface of the aggregate particles. Both these properties change with temperature, and mixing of aggregate and bitumen is always done at high temperature (up to 180°C for 40/60 pen bitumen) in order that the bitumen coats the aggregate surface readily.

Bitumen will also chemically adsorb on to aggregate surfaces. Strongly adsorbed bitumen fractions have been identified at the bitumen–aggregate interface, forming a band on the order of 180 Å thick. Ketones, dicarboxylic anhydrides, carboxylic acids, sulphoxides and nitrogen-bearing components have been found in this layer (Ensley, 1975). The strongly adsorbed components have been found to have sites capable of hydrogen bonding to the aggregate, though in the presence of water the available bonds prefer the more active water. Migration of some bitumen components to the interface is inferred and therefore a dependence on binder composition, mixing temperature and viscosity. *Figure 30.4* illustrates the process, with molecules of bitumen at the surface aligned in the

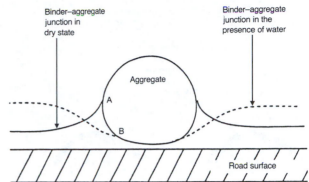

Fig. 30.5 *Retraction of the binder–water interface over the aggregate surface in the presence of water (after Majidzadeh and Brovold, 1968).*

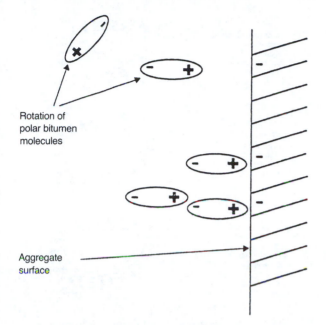

Fig. 30.4 *Adsorption of bitumen molecules to the aggregate surface (after Ensley, 1975).*

direction of polarity of the substrate (aggregate), usually a negative surface. The zone of orientation of bitumen molecules extends for a thickness of several thousand molecules.

30.3.3 MECHANISMS FOR LOSS OF ADHESION

Breakdown of the bond between bitumen and aggregate, known as stripping, may occur for a number of reasons. However, the principal agencies are the action of traffic and moisture, and these often act in combination.

The effect of moisture is significant since it causes loss of adhesion in a number of ways (Caro *et al.*, 2008). A number of mechanisms have been postulated for loss of adhesion, most of which involve the action of water. These are described below and each may occur depending on the circumstances.

Displacement

This occurs when the bitumen retracts from its initial equilibrium position as a result of contact with moisture. *Figure 30.5* illustrates the process in terms of an aggregate particle embedded in a bituminous film (Majidzadeh and Brovold, 1968). Point A represents the equilibrium contact position when the system is dry. The presence of moisture will cause the equilibrium point to shift to B, leaving the aggregate particle effectively displaced to the surface

of the bitumen. The positions of points A and B will depend on the type of bitumen and its viscosity.

Detachment

This occurs when the bitumen and aggregate are separated by a thin film of water or dust, though no obvious break in the bitumen film may be apparent. Although the bituminous film coats the aggregate particle, no adhesive bond exists and the bitumen can be cleanly peeled from the surface.

Film rupture

This occurs when the bitumen fully coats the aggregate but where the bitumen film thins, usually at the sharp edges of the aggregate particles (*Fig. 30.6*).

Blistering and pitting

If the temperature of the bitumen at the surface of a road rises, its viscosity falls. This reduced viscosity allows the bitumen to creep up the surface of any water droplets that fall on the surface, and it may eventually form a blister (*Fig. 30.7a*). With further heating the blister can expand and cause the bitumen film to rupture, leaving a pit (*Fig. 30.7b*).

Spontaneous emulsification

Water and bitumen have the capacity to form an emulsion with water as the continuous phase. The emulsion formed has the same (negative) charge as the aggregate surface and is thus repelled. The formation of the emulsion depends on the type of bitumen, and is assisted by the presence of finely divided particulate material such as clay materials, and the action of traffic.

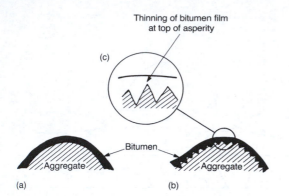

Fig. 30.6 *Thinning of the bitumen film on an aggregate with rough surface texture (b, c). Smooth aggregates (a) retain an unstressed and even film.*

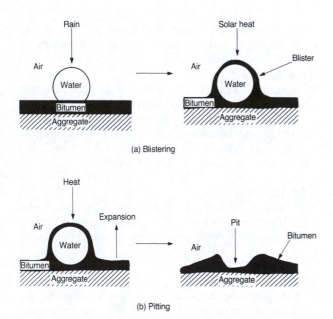

Fig. 30.7 *Formation of blisters and pits in a bituminous coating (after Thelan, 1983).*

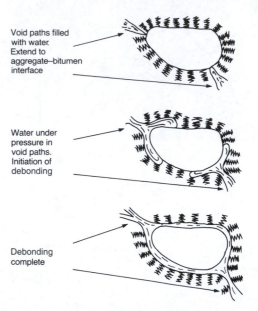

Fig. 30.8 *Pore pressure debonding mechanism (after McGennis et al., 1984)*

ficking. Subsequent trafficking acts on this trapped water and high pore pressures can result. This generates channels at the interface between bitumen and aggregate (*Fig. 30.8*) and eventually leads to debonding.

Hydraulic scouring

This is due principally to the action of vehicle tyres on a wet road surface. Water can be pressed into small cavities in the bitumen film in front of the tyre and, on passing, the action of the tyre sucks up this water. Thus a compression–tension cycle that can cause debonding is invoked.

Pore pressure

This mechanism is most important in open or poorly compacted mixtures. Water can become trapped in these mixtures as densification takes place by traf-

References

Caro S, Masad E, Bhasin A and Little DN (2008). Moisture susceptibility of asphalt mixtures, Part 1: mechanism. *International Journal of Pavement Engineering*, 9, 81–98.

Chen J, Lin K and Young S (2004). Effects of crack width and permeability on moisture-induced damage of pavements. *Journal of Materials in Civil Engineering*, ASCE, 16, 276–282.

Ensley EK (1975). Multilayer absorption with molecular orientation of asphalt on mineral aggregate and other substrates. *Journal of Applied Chemistry and Biotechnology*, 25, 671–682.

Majidzadeh K and Brovold FN (1968). *State of the art: Effect of water on bitumen-aggregate mixtures*. Highway Research Board, Publication 1456, Special Report 98, 77.

McGennis RB, Kennedy TW and Machemehl RB (1984). *Stripping and moisture damage in asphalt mixtures*. Research Report 253.1, Project 39-79-253, Center for Transportation Research, University of Texas, USA.

Thelan E (1958). Surface energy and adhesion properties in asphalt-aggregate systems. Highway Research Board Bulletin, 192, 63–74.

Chapter 31

Design and production of bituminous materials

31.1 Bituminous mixtures

There are a very large number of bituminous mixtures, which vary according to density, bitumen content, bitumen grade, aggregate size and aggregate grading. Previously in the UK these mixtures used to be classified into two groups, namely 'asphalts' and 'macadams'. The move to European standardisation has meant that bituminous mixtures are now classified into seven material specifications as described in the BS EN 13108 series. Of these seven only five have any relevance to the UK and of these five only two have major relevance.

Under BS EN 13108, bituminous mixtures are classified according to the grading of the aggregate (mixture type), their upper sieve size (maximum nominal aggregate size), the intended use of the material and the binder used in the mixture. *Table 31.1* summarises the designation of the bituminous mixtures.

31.1.1 ASPHALT CONCRETES

Asphalt concretes (previously termed dense macadams) are characterised by relatively low binder content and a continuously graded aggregate. They rely on the packing and interlock of the aggregate particles for their strength and stiffness. The binder coats the aggregate, and acts as a lubricant when hot and an adhesive and waterproofer when cold. The grade of binder used is softer than for hot rolled asphalts, being 20/30 to 160/220 pen inclusively. *Figure IV.1b* of the Introduction to Part 4 illustrates these features, and it can be seen that the material transmits load through the aggregate structure. Because of their lower binder content, asphalt concretes are cheaper than hot rolled asphalts (HRA) and stone mastic asphalts (SMA). In general asphalt concretes have a higher void content than HRA and SMA and are therefore more permeable and less durable.

Asphalt concrete is used for surface courses, binder courses and bases and is specified in BS EN 13108 Part 1. A summary of asphalt concrete material options for the UK is listed in *Table 31.2*.

31.1.2 HOT ROLLED ASPHALTS

Hot rolled asphalts (previously termed 'asphalts') are dense materials that are characterised by their high bitumen content and high filler/fines content. They derive their strength and stiffness from a dense stiff mortar of bitumen, filler and fines. The coarse

Table 31.1 Designation of bituminous mixtures under BS EN 13108

Mixture type	Size	Pavement layer	Bitumen grade, e.g. 40/60
Asphalt concrete	Aggregate size	Base/binder course/ surface course	Full bitumen grade designation (xx/yy)
Hot rolled asphalt	Grading designation	Base/binder course/ surface course	Full bitumen grade designation (xx/yy)
Stone mastic asphalt	Aggregate size	Base/binder course/ surface course	Full bitumen grade designation (xx/yy)
Porous asphalt	Aggregate size	Surface course	Full bitumen grade designation (xx/yy)

Table 31.2 Asphalt concrete mixtures complying with BS EN 13108-1

Material description	EN 13108 designation	Bitumen grade option
6 mm Medium graded surface course	AC 6 med surf	160/220
6 mm Dense surface course	AC 6 dense surf	100/150
10 mm Close graded surface course	AC 10 close surf	100/150
14 mm Close graded surface course	AC 14 close surf	100/150
20 mm Dense, heavy duty and high modulus binder course	AC 20 dense bin	100/150
20 mm Dense, heavy duty and high modulus binder course	AC 20 HDM bin	40/60
20 mm Dense, heavy duty and high modulus binder course	AC 20 HMB bin	30/45
32 mm Dense, heavy duty and high modulus base	AC 32 dense base	100/150
32 mm Dense, heavy duty and high modulus base	AC 32 HDM base	40/60
32 mm Dense, heavy duty and high modulus base	AC 32 HMB base	30/45

bin, binder course; surf, surface course.

Table 31.3 Hot rolled asphalt (HRA) mixtures complying with BS EN 13108-4

Material description	EN 13108 designation	Bitumen grade option
HRA 50% 14 mm for binder course	HRA 50/14 bin	40/60
HRA 60% 32 mm for base or binder course	HRA 60/32 base	40/60
	HRA 60/32 bin	40/60
HRA 30% 14 mm Type F surface course	HRA 30/14 F surf	40/60
HRA 35% 14 mm Type F surface course	HRA 35/14 F surf	40/60
HRA 30% 14 mm Type C surface course	HRA 30/14 C surf	40/60
HRA 35% 14 mm Type C surface course	HRA 35/14 C surf	40/60

bin, binder course; surf, surface course.

aggregate content is relatively low so the overall particle size distribution is gap-graded. *Figure IV.1a* of the Introduction illustrates these features, and it can be seen that the material transmits load through the mortar continuum. This mortar, being rich in bitumen, is expensive and the coarse aggregate serves to increase the volume of the mortar with a relatively cheap material, thereby reducing the overall cost. The binder used for hot rolled asphalt will normally be between 30/45 and 100/150 penetration grade bitumen.

Hot rolled asphalts are used in surface courses, binder courses and bases, and are specified in BS EN 13108 Part 4. A summary of hot rolled asphalt material options for the UK is listed in *Table 31.3*.

Surface course mixtures may be either type F, incorporating fine sand, or type C, incorporating crushed rock or slag fines that are more coarsely graded. *Table 31.4* shows the grading specification for the group of preferred hot rolled asphalt mixtures. It can be seen that each mixture is designated according to the coarse aggregate content and its nominal size. Thus a 50/14 mixture has 50% coarse aggregate with a nominal size of 14 mm.

Freshly laid surface course hot rolled asphalt presents a smooth surface with coarse aggregate particles submerged within the mortar. In order to provide a skid-resistant surface, coated chippings are rolled into the surface, which adds to the cost. Hot rolled asphalts have a very low permeability, and are capable of transmitting high stresses whilst providing some ductility. They are therefore very durable and normally used where traffic loads are high or durability is important.

31.1.3 POROUS ASPHALT

Porous asphalt is a bituminous material designed to provide a large volume (at least 20%) of interconnected

Table 31.4 Grading of target composition for hot rolled asphalt mixtures

Column number	1	2	3	4	5	6
Designation	50/14	60/32	30/14F	30/14C	35/14F	35/14C
			Passing sieve (% by mass)			
45	–	100	–	–	–	–
31.5	–	97	–	–	–	–
20	100	59–71	100	100	100	100
14	95	39–56	95	95	95	95
10	76–93	–	67–83	67–83	62–81	62–81
2	35–45	32	65	65	61	56
0.5	17–45	13–32	48–65	29–41	44–61	24–41
0.25	10–27	9–21	22–58	24–36	19–51	19–31
0.063	5.5	4.0	9.0	9.0	8.0	8.0

air voids so that water can drain through the material and run off within the thickness of the layer. This requires the underlying binder course to be impermeable. It is used exclusively for surface courses and can be laid in more than one layer. The very high content of interconnected voids not only allows the passage of water but also allows the movement of air, thereby providing noise reducing characteristics.

Porous asphalts are specified in BS EN 13108 Part 7. The aggregate grading consists predominantly of coarse aggregate: about 75% retained on the 2 mm sieve. Fine aggregate fractions are added to enhance the cohesion and stiffness of the mixture but in sufficiently small quantity so as not to interfere with the interlock of the coarse particles, and to leave enough voids to provide a pervious structure. Because of its porous nature, the material is vulnerable to ageing through oxidation of the bitumen. To counteract this, the bitumen content must be sufficient to provide a thick coating on the coarse aggregate particles. The advantages of porous asphalt are that it minimises spray in wet weather, reduces surface noise, improves skidding resistance, and offers lower rolling resistance than dense mixtures. However it is weaker than denser mixtures and in the UK has limited relevance and dwindling use owing to the increased use of SMA and thin surfacings on highways (see BRE Special Digest 1).

31.1.4 STONE MASTIC ASPHALT

Stone mastic asphalt (SMA) is specified in BS EN 13108 Part 5. SMA has a coarse aggregate skeleton but the voids are filled with a mortar of fines, filler and bitumen. It thus resembles hot rolled asphalt, particularly the high-stone-content mixtures, but it may best be considered as having a coarse aggregate structure similar to that of porous asphalt but with the voids filled. SMA differs from hot rolled asphalt in that the quantity of mortar is just sufficient to fill the voids in the coarse aggregate structure. It therefore provides high stiffness owing to the interlock of the coarse aggregate particles, and good durability because of a low void content.

Stone mastic asphalts are used in surface courses and binder courses, and a summary of SMA material options for the UK is listed in *Table 31.5*.

31.2 Recipe and designed mixtures

The majority of bituminous mixtures are recipe mixtures. In other words, the mixtures are put together according to prescribed proportions laid down in the appropriate European standard. These mixture proportions have been derived through experience in use and, provided the separate ingredients meet their specifications, the mixture will provide the required performance in most situations. This approach is consistent with the empirical method for the structural design of roads that predominated until relatively recently. Thus the design chart used to determine the thickness of the base layer requires that the base material has the same mixture proportions as the materials in the roads from which the design chart was originally established.

Recipe mixtures provide a satisfactory performance in many cases and there is some advantage in the simplified approach that recipe mixtures offer. However, there are limitations to the use of recipe

Table 31.5 Stone mastic asphalt (SMA) mixtures complying with BS EN 13108-5

Material description	EN 13108 designation	Bitumen grade option
6 mm SMA surface course	SMA 6 surf	40/60
10 mm SMA surface course	SMA 10 surf	40/60
14 mm SMA surface course	SMA 14 surf	40/60
14 mm SMA binder course	SMA 14 bin	40/60
20 mm SMA binder course	SMA 20 bin	40/60

bin, binder course; surf, surface course.

mixtures that match the limitations of empirical structural design of roads. These are as follows:

1. Non-specified materials cannot be used. For example, locally available sand may not meet the grading requirements of the specification but may produce a satisfactory mixture. Recipe mixtures preclude any assessment of the properties of a mixture containing that sand.
2. No procedure is available to assess causes of failure.
3. No procedure is available to optimise the mixture proportions. This is particularly important as far as the bitumen is concerned because this is the most expensive ingredient and has a strong bearing on the performance of mixtures, especially those which are denser.

These drawbacks have led to the development of a procedure for the design of bituminous mixtures, which has occurred in parallel with the development of analytical procedures for the structural design of roads. An analytical approach to road design enables the determination of the thickness of the road structure through an analysis of its behaviour under the applied load. This clearly requires knowledge of certain properties of the materials and it follows that materials will have to be produced with particular characteristics.

The most commonly used procedure for mixture design (BS 598: Part 107: 2004) is based upon the Marshall test, which was originally developed in the USA for designing mixtures for use on airfield runways. The objective of the procedure is to determine an optimum binder content from a consideration of mixture strength (stability), mixture density, and mixture deformability (flow).

Test samples of binder/aggregate mixtures are prepared using the materials to be used in the field. The aggregate grading is kept constant and samples with a range of binder content are produced. The samples are prepared and compacted in a standard way into moulds 101.6 mm in diameter and 70 mm high. The state of compaction achieved is determined by measuring the bulk density and calculating the compacted aggregate density. At low binder contents the mixture will lack workability and the densities will be correspondingly low. At high binder contents, aggregate will effectively be displaced by bitumen, and again the densities will be low. Each of these measures of density will thus produce an optimum binder content, as shown in *Fig. 31.1a* and *b*.

To test the strength and resistance to deformation of the material, the specimens are heated to 60°C

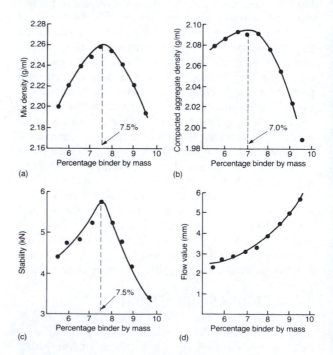

Fig. 31.1 Analysis of mix design data from the Marshall test.

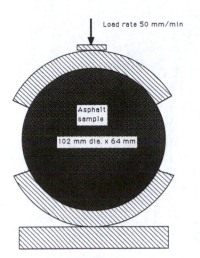

Load rate 50 mm/min

Asphalt
sample

102 mm dia. x 64 mm

Fig. 31.2 Testing arrangement for a Marshall asphalt design.

and subjected to a compression test using special curved jaws that match the curved sides of the specimens, as shown in *Fig. 31.2*. Thus the load is applied radially. The jaws of the machine are driven together at a constant rate of 50 mm per minute until the maximum load is obtained, which is termed the 'stability'. The deformation of the sample at this maximum load is also recorded and termed the 'flow'. Typical plots of stability and flow against binder content are shown in *Fig. 31.1c* and *d*. The stability plot gives a third optimum binder content, and the design binder content is obtained from the average of this and the optima from the two density plots. The flow at this design binder content can then be read off. Minimum values of stability and flow are specified according to the amount of traffic that the road will carry.

In evaluating mixtures it is helpful to consider the Marshall quotient, Q_m, which is derived from the stability and flow:

$$Q_m = \text{stability/flow} \qquad (31.1)$$

Thus Q_m bears some resemblance to a modulus (ratio of stress to strain) and may be taken as a measure of mixture stiffness.

More recently new approaches to bituminous mixture design have been proposed. The Superior Performing Asphalt Pavement (Superpave™) mixture design process was developed in the USA through the Strategic Highway Research Program (SHRP) in the early 1990s (Harrigan *et al.*, 1994). The Superpave™ method incorporates a pavement design methodology that takes into account the environ-mental conditions that the pavement could expect to experience during its design life. The design method involves the selection of the bitumen based on a performance grading (PG) system together with volumetric measurements (density, air voids content, etc.). Asphalt mixtures on heavily trafficked roads are also subjected to permanent deformation and fatigue tests as part of the mixture design process.

31.3 Methods of production

The process of manufacture of bituminous materials involves three stages. Firstly, the aggregate must be proportioned to give the required grading, secondly the aggregate must be dried and heated, and thirdly the correct amount of binder must be added to the aggregate and mixed to thoroughly coat the aggregate particles and produce a homogeneous material.

The most common type of plant in the UK is the indirectly heated batch mixing plant. A schematic diagram of this type of plant is shown in *Fig. 31.3*. The aggregate is blended from cold bins and passed through a rotary drier/heater. Here the moisture is driven off and the aggregate temperature raised to the prescribed mixing temperature for the type of material being produced. The aggregate is then transported by hot elevator to hot storage bins, where it is separated into fractions of specified size. Aggregates are released into the weigh box in the desired proportions and then released into the mixer. Bitumen heated to the prescribed temperature is also introduced to the mixer, the quantity being determined using a weigh bucket or volumetrically using a flow meter. The mixing time varies up to 60 seconds but should be as short as possible in order to limit oxidation of the binder. After mixing, the material is discharged directly into a wagon.

This type of plant is very versatile, being capable of producing a wide range of different asphalt mixture types, and being able to easily adjust to a wide range of mixture specifications.

A variation on this type of plant is to dry and heat the aggregate in batches before it is charged into the mixer. This eliminates the need for hot aggregate storage, but the proportioning of the cold aggregate needs to be very carefully controlled.

An alternative type of mixer is the drum mixer, which gives continuous rather than batch produc-tion. Here the cold aggregates are proportioned and conveyed directly into a drum mixer that has two zones. The first zone is where drying and heating occur, and in the second zone bitumen is introduced and mixing takes place. The advantages of this type

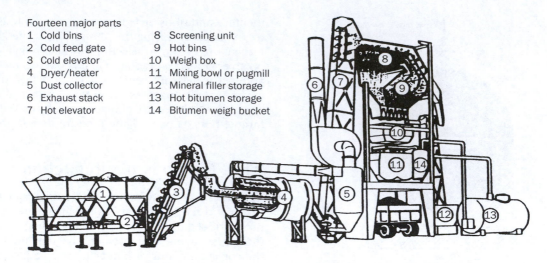

Fourteen major parts

1	Cold bins	8	Screening unit
2	Cold feed gate	9	Hot bins
3	Cold elevator	10	Weigh box
4	Dryer/heater	11	Mixing bowl or pugmill
5	Dust collector	12	Mineral filler storage
6	Exhaust stack	13	Hot bitumen storage
7	Hot elevator	14	Bitumen weigh bucket

Fig. 31.3 Schematic diagram of an indirectly heated batch mixing plant (from The Shell Bitumen Handbook, 2003).

of plant are that the amount of dust emission is reduced, the process is simpler and, above all, the rate of production can be very high – up to 500 tonnes per hour. This is advantageous where large quantities of the same type of material are required, but it is difficult to change production to a different mixture.

References

Harrigan ET, Leahy RB and Youtcheff JS (1994). *The SUPERPAVE mix design system manual of specifications*. SHRP-A-379, Strategic Highways Research Program, National Research Council, Washington DC.

Shell Bitumen UK (2003). *The Shell Bitumen Handbook*, Shell Bitumen UK.

Chapter 32

Recycling of bituminous materials

The recycling and reuse of bituminous materials is not a recent development, the first asphalt pavement recycling project being recorded in 1915 (Epps *et al.*, 1980). Recycled (or reclaimed) asphalt pavement (RAP) material consists of aggregate and bitumen that has been removed and/or reprocessed from an asphalt pavement. Although old asphalt mixtures were historically removed and disposed of in landfills, the use of RAP in new asphalt mixtures has increased substantially since the 1930s.

RAP material is produced when old, damaged pavement materials are milled and crushed for subsequent addition as a component in new asphalt mixtures. With the availability of high-quality virgin materials declining and landfilling becoming less practical and more expensive, the increase in the addition of RAP to pavement mixtures is not surprising. The benefits of recycling asphalt pavements can be attributed to a reduction in the costs of new construction and rehabilitation projects, environmental conservation of energy, mineral aggregates and bituminous binder, and the preservation of road geometry (Kandhal and Mallick, 1997).

Recycling of pavement material can be undertaken as an in-place or a central plant process. In addition, recycling can be grouped into hot, warm and cold processes depending on the virgin binder deployed in the recycling operation. RAP contents in new asphalt mixtures vary from approximately 10% to as high as 100%, although generally RAP contents of 30% can be considered to be an acceptable maximum (Planche, 2008). Owing to the benefits of pavement recycling techniques, up to 80% of asphalt pavements removed each year from widening and resurfacing projects are put back into roads, highway shoulders and embankments.

32.1 In-plant asphalt recycling

In-plant recycling involves stockpiling RAP at the asphalt plant before mixing with virgin material to produce a new asphalt mixture. Reclaimed pavements are transported to the asphalt production plant, ripped, milled, and crushed into required sizes before being blended with virgin bitumen and aggregate. Depending on the mixing temperatures required of the recycled mixture, in-plant recycling is further divided into cold, warm, and hot recycling (Karlson and Isacsson, 2006). In-plant hot recycling tends to be the most advantageous, hot recycled asphalt mixtures with 10–30% of RAP having a performance that is similar to, or even better than, that of virgin asphalt mixtures (Kandhal and Mallick, 1997).

One of the disadvantages of recycling material in an asphalt plant is the need to transport the RAP material to the plant and the subsequent transport of the recycled asphalt mixture to site. This need for transport results in added expense as well as pavement damage associated with delivery vehicles, which can only be alleviated by an *in-situ* process.

32.1.1 HOT IN-PLANT OPERATIONS

In the hot recycling process, RAP material is blended with superheated virgin aggregate before mixing with virgin bitumen. The role of the superheated aggregate is to transfer heat to the RAP to soften or liquefy the RAP binder and to break the RAP material into smaller lumps. To aid the mixing of the RAP binder with virgin bitumen, rejuvenating agents are also added during the production stage. The role of the rejuvenator is to diffuse into and regenerate the properties of the aged binder. A wide range of rejuvenators – for example, soft bitumens, fluxing oils, extender oils and aromatic oils – can be used (Karlson and Isacsson, 2006).

Batch and drum plants are conventionally used to produce hot asphalt mixtures, but in order to produce hot recycled asphalt mixtures, modifications to the conventional mixing plant are needed (Roberts *et al.*, 1991).

In a batch mixing plant, if the RAP material is dried in the same way as the virgin aggregate, the RAP bitumen will be burned directly by the flame

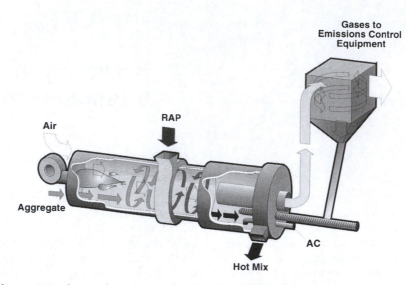

Fig. 32.1 *Parallel drum mixer designed to incorporate recycled asphalt pavement (RAP) (after Brock and Richmond, 2005).*

in the burner, causing blue smoke. RAP is therefore stored separately before being introduced to super-heated virgin aggregate in the hot bin, the excessive thermal energy being used to heat the RAP material from ambient temperature to mixing temperature.

The procedure in the drum facility is quite different from that of the batch plant. In a conventional drum facility, aggregate is heated and mixed with bitumen in the same drum mixer. The drum facility therefore needs to be modified to incorporate RAP material for the production of hot recycled asphalt mixture. *Figure 32.1* shows a modified drum mixer in which RAP is introduced in the middle section of the drum. The RAP is therefore mixed with superheated virgin aggregate before being mixed with virgin binder.

32.1.2 COLD IN-PLANT PROCESSES

Cold recycled asphalt mixtures are produced using bitumen emulsion rather than conventional bitumen. Cold mixtures have two very significant advantages over hot asphalt mixtures:

1. The energy consumption of the plant is typic-ally 10–20% that for hot asphalt mixture production.
2. Cold asphalt mixtures can be stored for long periods before use.

The first point leads to clear economic and environ-mental benefits, offsetting the loss of performance when compared to hot mixtures. The second point means that cold asphalt mixtures make ideal patching and repair materials. There is a long workability

period following placement during which compac-tion can be achieved, so trench backfills and other small areas of construction can be completed to a high standard. The arrival of cold-mix technology has also spawned a new term, '*ex-situ*'. This refers to the practice of setting up a mobile mixing plant at or near to a work site, which takes recycled materials from the site and incorporates them into a new mixture. The relative simplicity of cold mix production means that this is an economically attractive option, saving on transport and making re-use of materials practical.

In practice cold recycled asphalt mixtures tend to be used almost exclusively on relatively lightly trafficked roads, because of the need for the material to cure and the danger of damage during early life.

32.2 *In-situ* asphalt recycling

In-situ methods involve processing on site and are divided into hot *in-situ* recycling, which includes the remixing and repaving processes, and cold *in-situ* recycling, which includes full depth reclamation. *In-situ* recycling is also known as in-place recycling.

32.2.1 HOT *IN-SITU* ASPHALT RECYCLING

Hot *in-situ* asphalt recycling is used to transform an aged and partially failed asphalt surface layer into an as-good-as-new product. The process requires that the surface of the pavement be heated to a

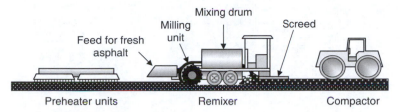

Labels: Feed for fresh asphalt · Milling unit · Mixing drum · Screed · Preheater units · Remixer · Compactor

Fig. 32.2 Schematic of hot in-situ recycling of asphalt (remix process) (after Thom, 2008).

temperature at which it can be reworked, healing all cracks and reshaping any deformed profile. This process is usually applicable to depths between 20 and 60 mm.

Repave

In the so-called 'repave' process, the uppermost materials are heated to 140–180°C and scarified, i.e. loosened, to a depth of approximately 20 mm before being re-profiled. While the scarified asphalt material is still hot (above approximately 90°C), a new asphalt mixture surfacing layer is laid on top of the old material and both layers are compacted together. The heated old material therefore blends with the new material to form a single contiguous surface course with fully restored properties.

Remix

The remix process is similar to, although slightly more sophisticated than, the repave process. In the remix process, the heated existing surface material is physically scraped up and mixed together with new material in the remixing machine before being re-laid as a new surface layer (*Fig. 32.2*). This ensures a fully integrated and blended layer.

The bitumen grade and mixture composition of the added material is selected to balance out any defects in the composition of the existing material. The re-mixed material, when compacted, forms a single new 40-mm surface course. This treatment is suitable when the existing surface course has started to deteriorate, by either rutting or cracking, but the underlying layers are still in a sound condition. Limitations are as follows:

- A fairly consistent existing surface course is necessary.
- A careful mixture design is required to establish the proportions of the added material needed.
- Material deeper than 20 mm below the original pavement surface is not heated sufficiently to be significantly remoulded or improved.
- The process can only be considered as partial recycling because of the need for new material.

- An increased surface level has to be accommodated.
- There is a danger of toxic fumes being emitted during the heating process if it is not carefully controlled.

Both the repave and remix hot recycling processes are excellent in that they transform a failing surface into an intact new layer. However, they have their limitations:

- The existing surface material must be suitably uniform in composition (i.e. not too many patches or other changes of surface material).
- The transformation is only effective to the depth to which the asphalt can be heated. This means that the treatment is ideal as a cure for rutting, top-down cracking or ravelling, but that it does nothing to overcome deeper problems.

32.2.2 COLD *IN-SITU* PROCESSES

In full-depth reclamation, the existing pavement (usually all the surface and part of the base) is ripped and milled to a certain depth, which can extend up to 350 mm. The broken fragments are then mixed with a new binder, consisting of bitumen emulsion, foamed bitumen or soft bitumen to produce a stabilised base, and compacted. A new surface is then laid and compacted on the new recycled base (Kandhal and Mallick, 1997). *Figure 32.3* provides an illustration of the process.

The result, if design and construction are carried out sensibly, is a strong new pavement, although it will always be necessary to apply a new surface course, since it is unlikely that the requisite surface properties will be present from the recycled material.

There is no need for expensive and disruptive transportation of materials (other than binder) and there is no requirement for quarrying virgin rock. All that is needed is the addition of an effective binder. Naturally there are difficulties, and the following are seen as the major issues confronting practitioners and designers:

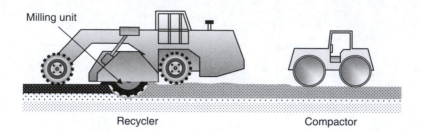

Milling unit

Recycler   Compactor

Fig. 32.3 Cold in-situ *recycling (after Thom, 2008).*

- Binder is unlikely to be uniformly mixed into the material.
- All pavements are to a certain extent non-homogeneous, notably in layer thickness. In many cases this means that the milled material will present varying proportions of asphalt, base (including hydraulically-bound materials), sub-base and even subgrade.
- Water content can only be controlled approximately.
- Aggregate gradation is likely to be low in fines content if the RAP percentage is high.
- Compaction beyond about 250 mm depth will be increasingly ineffective.

Practical advice on cold asphalt recycling can be found in Wirtgen's *Cold Recycling Manual* (Wirtgen, 2004).

32.3 Issues related to asphalt recycling

32.3.1 BLACK ROCK

If the binder content of the RAP is to be utilised in the final recycled asphalt mixture, then there is a need to activate the binder in the RAP. It will usually have aged over a period of many years and will therefore be much harder than new bitumen or bitumen emulsion. RAP in this state is often referred to as 'black rock'. Nevertheless, with appropriate addition of chemical rejuvenators it has been shown that it is possible to induce a degree of blending between old and new binder and therefore to enhance mixture properties (see, e.g., Walter, 2002).

32.3.2 MATERIAL VARIABILITY

The gradation of RAP is quite different from the gradation of the original asphalt mixture, since it consists of agglomerates rather than individual particles, and the key difference is a much reduced fines content. This material variability needs to be considered when designing recycled asphalt mixtures (Sherwood, 2001).

References

Brock JD and Richmond JL (2005). *Milling and Recycling*, Technical Paper T-127, Astec Inc., Chattanooga, TN, USA.

Epps JA, Little DN, Holmgreen RJ and Terrel RL (1980). *Guidelines for Recycling Pavement Materials. NHCRP 224*, National Highway Cooperative Research Program, USA.

Kandhal PS and Mallick RB (1997). *Pavement Recycling Guidelines for State and Local Government*. National Center for Asphalt Technology, Auburn University, AL 36849, USA.

Karlsson R and Isacsson U (2006). Material-related aspects of asphalt recycling – state-of-the-art. *Journal of Materials in Civil Engineering*, **18** (No. 1), 81–92.

Planche JP (2008). *European survey on the use of RAP*. Proceedings of the ISAP Symposium on Asphalt Pavements and Environment, Zurich, Switzerland, pp. 140–149.

Roberts FL, Kandhal PS, Brown ER, Lee DY and Kennedy TW (1991). *Hot Mix Asphalt Materials, Mixture Design and Construction*, NAPA Education Foundation, Lanham, Maryland, US.

Sherwood P (2001). *Alternative Materials in Road Construction*, 2nd edition, Thomas Telford, London.

Thom NH (2008). *Principles of Pavement Engineering*, Thomas Telford, London.

Walter J (2002). *Factors controlling RAP cold mix modulus*. Proceedings of the Third World Congress on Emulsions, Lyon, 4-EO57.

Wirtgen (2004). *Cold Recycling Manual*, 2nd edition, Wirtgen GmbH, Germany.

Further reading for Part 4 Bituminous materials

BOOKS

Hunter RN (ed.) (2000). *Asphalts in Road Construction*, Thomas Telford, London.

This book gives an excellent coverage of asphalts including recent developments in asphalt technology such as stone mastic asphalt, thin surfacings, and high-modulus bases. As well as covering the materials themselves, it deals with the design and maintenance of pavements, laying and compaction, surface dressing and other surface treatments, and failure of surfacings. The book is an updated version of Bituminous Materials in Road Construction. *This too is worth reading, since it deals with all types of bituminous material.*

Read JM and Whiteoak CD (2003). *The Shell Bitumen Handbook*, Shell UK Oil Products Limited.

Essential reading if you want to get to grips with the fundamentals of bitumen. It deals with the structure and constitution of bitumen and goes on to link this to its engineering properties. The book also covers the design and testing of bituminous mixes as well as pavement design.

Thom NH (2008). *Principles of Pavement Engineering*, Thomas Telford, London.

This excellent book sets out the material needed to equip practising engineers as well as students with the fundamental understanding needed to undertake the design and maintenance of all pavement types. The book contains theoretical concepts that are presented in a way that is easily understood together with quantitative material to illustrate the key issues.

O'Flaherty CA (2000). *Highways: The Location, Design, Construction and Maintenance of Road Pavements*, Elsevier/Butterworth Heinemann.

A comprehensive textbook on all aspects of road engineering from the planning stages through to the design, construction and maintenance of road pavements.

STANDARDS

American Association of State Highway and Transportation Officials. AASHTO R28-06. *Standard Practice for Accelerated Aging of Asphalt Binder Using a Pressurized Aging Vessel (PAV)*. Washington, DC.

American Society for Testing and Materials. ASTM D 4402-02. *Standard Test Method for Viscosity Determinations of Unfilled Asphalt at Elevated Temperatures Using a Rotational Viscometer*. ASTM, Philadelphia.

British Standard BS 434: Part 1: 1984: *Bitumen road emulsions (anionic and cationic). Part 1. Specification for bitumen road emulsions.*

British Standard, BS 594-1: 2005: *Hot rolled asphalt for roads and other paved areas – Part 1: Specification for constituent materials and asphalt mixtures.*

British Standard, BS 594-2: 2003: *Hot rolled asphalt for roads and other paved areas – Part 2: Specification for transport, laying and compaction of hot asphalt.*

British Standard, BS 594987: 2008: *Asphalt for roads and other paved areas – Specification for transport, laying and compaction and design protocols.*

British Standard 598: Part 107: 2004: *Method of test for the determination of the composition of design wearing course asphalt.*

British Standard 812: Part 110: 1990: *Methods for determination of aggregate crushing value.*

British Standard 812: Part 112: 1990: *Methods for determination of aggregate impact value.*

British Standard 812: Part 114: 1989: *Method for determination of polished stone value.*

British Standard 2000: Part 72: 1993: *Viscosity of cutback bitumen and road oil.*

British Standard 3690: Part 1: 1989: *Bitumens for building and civil engineering. Part 1: Specification for bitumens for road purposes.*

British Standard, BS 4987-1: 2005: *Coated macadam (asphalt concrete) for roads and other paved areas – Part 1: Specification for constituent materials and for mixtures.*

British Standard, BS 4987-2: 2003: *Coated macadam for roads and other paved areas – Part 2: Specification for transport, laying and compaction.*

BS EN 1097-2: 1998: *Tests for mechanical and physical properties of aggregates – Part 2: Methods for the determination of resistance to fragmentation.*

BS EN 1426: 2000: *Methods of test for petroleum and its products, bitumen and bituminous binders. Determination of needle penetration.*

BS EN 1427: 2000: *Methods of test for petroleum and its products, bitumen and bituminous binders. Determination of softening point. Ring and ball method.*

BS EN 12591: 2000 *Bitumen and bituminous binders – Specifications for paving grade bitumens.*

BS EN 12592: 2007 *Bitumen and bituminous binders – Determination of solubility.*

BS EN 12593: 2000 *Bitumen and bituminous binders. Determination of the Fraass breaking point.*

BS EN 12607-1: 2007 *Bitumen and bituminous binders – Part 1: Determination of the resistance to hardening under influence of heat and air – RTFOT method.*

BS EN 13036-1: 2002: *Road and airfield surface characteristics – Test methods – Part 1: Measurement of pavement surface macrotexture depth using a volumetric patch technique.*

BS EN 13108-1: year: *Bituminous mixtures – Material specifications – Part 1: Asphalt concrete.*

BS EN 13108-4: year: *Bituminous mixtures – Material specifications – Part 4: Hot rolled asphalt.*

BS EN 13108-5: year: *Bituminous mixtures – Material specifications – Part 5: Stone mastic asphalt.*

BS EN 13108-7: year: *Bituminous mixtures – Material specifications – Part 7: Porous asphalt.*

BS EN 13303: 2003: *Methods of test for petroleum and its products – BS 2000-506: Bitumen and bituminous binders – Determination of the loss in mass after heating of industrial bitumens.*

BS EN 13358: 2004: *Methods of test for petroleum and its products – BS 2000-525: Bitumen and bituminous binders – Determination of the distillation characteristics of petroleum cut-back bitumen products.*

BS EN 13808: 2005: *Bitumen and bituminous binders – Framework for specifying cationic bituminous emulsions.*

Highways Agency (2006). Pavement design, HD26, *Design manual for roads and bridges Vol. 7: Pavement design and maintenance – pavement design and construction.* Stationery Office, London.

PART 5

MASONRY: BRICKWORK, BLOCKWORK AND STONEWORK

Bob de Vekey

Introduction

Over the last three decades the term 'masonry' has been widened from its traditional meaning of structures built of natural stone to encompass all structures produced by stacking, piling or bonding together discrete chunks of rock, fired clay, concrete, etc. to form the whole. 'Masonry' in this wider sense is what these chapters are about. In contemporary construction most masonry in the UK is built from man-made materials such as bricks and blocks. Stone, because of its relatively high cost and the environmental disadvantages of quarrying, is mainly used as thin veneer cladding or in conservation work on listed buildings and monuments.

Second to wood, masonry is probably the oldest building material used by man; it certainly dates from the ancient civilisations of the Middle East and was used widely by the Greeks and Romans. Early cultures used mud building bricks, and very little of their work has survived, but stone structures such as the Egyptian pyramids, Greek temples and many structures made from fired clay bricks have survived for thousands of years. The Romans used both fired clay bricks and hydraulic (lime/pozzolana) mortar and spread this technology over most of Europe.

The basic principle of masonry is of building stable bonded (interlocked) stacks of handleable pieces. The pieces are usually chosen or manufactured to be of a size and weight that one person can place by hand but, where additional power is available, e.g. the pyramids in Egypt, larger pieces may be used, which give potentially more stable and durable structures. This greater stability and durability is conferred by the larger weight and inertia, which increase the energy required to remove one piece and make it more resistant to natural forces such as winds and water as well as human agency.

There are four main techniques for achieving stable masonry:

1. Irregularly shaped and sized but generally laminar pieces are selected and placed by hand in an interlocking mass (e.g. dry stone walls).
2. Medium to large blocks are made or cut very precisely to one or a small range of interlocking sizes and assembled to a basic grid pattern either without mortar or with very thin joints (e.g. ashlar or thin-joint).
3. Small to medium units are made to normal precision in a few sizes and assembled to a basic grid pattern, and the inaccuracies are taken up by use of a packing material such as mortar (e.g. normal brickwork).
4. Irregularly shaped and sized pieces are both packed apart and bonded together with adherent mortar (e.g. random rubble walls).

Type (4) structures and thin-joint systems depend significantly on the mortar for their stability; all the other types rely largely on the mechanical interlocking of the pieces. *Figure V.1* shows typical examples.

These descriptions are given to emphasise that most traditional masonry owes much of its strength and stability to interlocking action, weight and inertia while the mortar, when present, is not acting as a glue but as something to fill in the gaps resulting from the imperfect fitting together of the pieces. Most contemporary masonry is type (3) and although modern mortars do have an adhesive role much of the strength still derives from the mass and friction between interlocking shapes; it is important to remember this in design.

It is also important to remember that, although the wall is the most useful and effective masonry structure, many other structural forms such as columns, piers, arches, tunnels, floors and roads are used. Normal plain masonry must be designed such

Fig. V.1 *The main types of masonry: (a) dry stone wall, (b) ashlar stonework, (c) jointed brick and blockwork, (d) rubble masonry.*

that the predominant forces put it into compression since it cannot be relied on to resist tensile forces. If, however, tension structures such as cantilevers, earth-retaining walls and beams are required, masonry may be reinforced or post-tensioned in the same way as concrete.

This part of the book starts with a chapter on terminology, raw materials, composition and manufacturing processes for the components of masonry. In subsequent chapters we then consider: structural forms, architecture and detailing; structural behaviour and response to actions such as wind, imposed load and movement; other key properties such as thermal and noise insulation, and resistance to fire; and finally durability in relation to the environment and how to conserve and repair weathered masonry and maintain an attractive appearance. The important issues of energy efficiency and sustainability are covered in Chapter 62 along with the other materials described throughout this book.

Chapter 33

Materials and components for masonry

In this chapter we start with some of the terminology that will be important throughout this part of the book. The majority of the chapter then describes the raw materials, manufacture and properties of the components of masonry – the various brick and block units and the mortar that separates them in the final structure. It will become apparent as you read this that there is some overlap with the descriptions of cement and concrete in Part 3; although this part contains sufficient detail to be read independently you may find it helpful to refer to the relevant sections of Part 3 as you go.

33.1 Basic terminology

Units: brick or block-sized pieces of stone, fired clay, concrete or calcium silicate bonded aggregate that are assembled to make masonry. Usually, but not invariably, these are in the form of rectangular prisms, but many special shapes are manufactured and stones are often used either as found or shaped to be partly squared, faced or split faced. Manufactured units are produced in the following standard forms: solid, frogged, cellular, perforated, hollow, key-faced, fair-faced and split faced, as illustrated in *Fig. 33.1*. They are also available as 'standard specials' with a range of curves, non-right-angled corners, plinths, cappings, etc. These are described and specified in BS 4729 (2005); some typical examples are shown in *Fig. 34.7*.

Mortar: a material that is plastic and can flow when fresh but sets hard over a period of hours to days. Its purpose is to fill the gaps caused by variations in the size and shape of units such that the masonry is stable and resists the flow of air and water. Mortar is compounded from a binder (e.g. cement) and a filler/aggregate (usually sand).

Binder: a finely ground material which when mixed with water reacts chemically and then sets hard and binds aggregates into solid masses to form either

units or mortars. Non-hydraulic binders such as resins are occasionally used.

Work size: the size of a masonry unit specified for its manufacture, to which its actual size should conform within specified permissible deviations. As

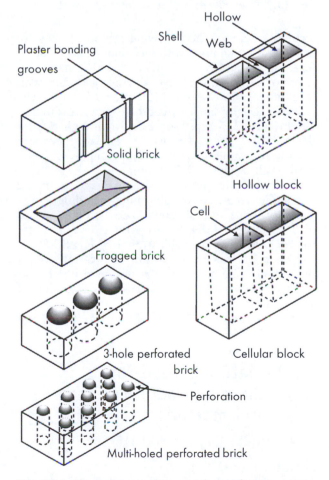

Fig. 33.1 Shape types and terminology of masonry units (based on Fig. 1 *in BRE Digest 441).*

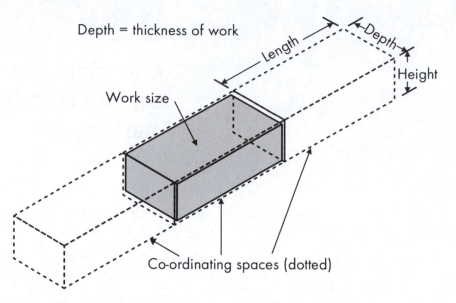

Fig. 33.2 *Size of units and co-ordinating spaces.*

a rough guide for the following sections, bricks are considered to be units with face dimensions of up to 337.5 mm long by 112.5 mm high with a maximum depth of 225 mm, while blocks are larger units, with face dimensions of up to 1500 mm by 500 mm. UK units are usually smaller than these limits. A standard UK brick is 215 mm long by 65 mm high, with a depth of 102.5 mm. There is no standard block size but the commonest size is 440 mm long by 215 mm high, with a depth of 100 mm.

Co-ordinating size: the size of a co-ordinating space allocated to a masonry unit, including allowances for joints and tolerances. The coordination grid into which they fit is usually around 10 mm larger for each dimension, as illustrated by *Fig. 33.2*.

Fair-faced: masonry, within the variability of individual units, precisely flat on the visible face.

This is normally only possible on one side of solid walls.

Other definitions can be found in BS 6100-6 (2008).

33.2 Materials used for the manufacture of units and mortars

33.2.1 ROCKS, SAND AND FILLERS

Rock (or stone)
The main types of rock used for masonry in the UK are:

- sedimentary rocks, formed from compressed sediments on the bottom of ancient seas e.g. limestones and sandstones
- metamorphic rocks, formed by the action of pressure and high temperature on other rock deposits, e.g. marbles and slates
- igneous rocks, formed by melting of rock during volcanic activity, e.g. granites and basalts.

Both the strength and durability of rocks are very variable and will depend on the porosity and the distribution of the pores. Generally strength increases and porosity decreases from sedimentary through metamorphic to igneous. Most sedimentary rocks have a layer structure and will be significantly stronger normal to the bedding plane than in the other two directions (i.e. they are anisotropic). Igneous rocks and some fine-grained sedimentary and metamorphic rocks are fairly isotropic, i.e. they have similar properties in all directions. Such rocks are termed 'freestones' because they can be cut in any plane and are usually suitable for carving to elaborate shapes and polishing.

Rock or stone is used in three main ways – as thin sheet cladding, as solid building units and in the form of crushed aggregate to make concretes and mortars.

Sand – nature and composition
Sand is used widely as a constituent of masonry in mortar, in concrete units and sandlime units, and in grouts and renders. It is a mixture of rock particles

of different sizes from about 10 mm diameter down to 75 μm diameter. Most sand is a naturally occurring rock powder derived from recent naturally occurring alluvial deposits such as the beds of rivers and sea beaches or from older deposits formed by alluvial or glacial action. In some areas it may be derived from dunes or by crushing quarried rocks. The chemical and geological composition will reflect the area from which it is derived. The commonest sands are those based on silica (SiO_2), partly because of its wide distribution in rocks such as sandstones and the flint in limestones, and partly because silica is hard and chemically resistant. Other likely constituents are clay, derived from the decomposition of feldspars, calcium carbonate ($CaCO_3$), in the form of chalk or limestone from shells in some marine sands, and micas in sands from weathered granites. Crushed rocks such as crushed basalts and granites will reflect their origins.

Sands should be mostly free of particles of clay (with a size of between 75 and 30 micrometres), which causes unsatisfactorily high shrinkage characteristics and chemical interactions with binders. Most of the constituents of sand are relatively chemically inert to environmental agents, but chalk or limestone particles will be dissolved slowly by mild acids and clays may react in time with acids or alkalis. Most sand constituents are also fairly hard and are resistant, in themselves, to mechanical abrasion and erosion by windblown dust or waterborne particles.

MORTAR AND RENDERING SANDS

Mortar sand must not contain particles with a diameter greater than about half the thinnest joint thickness, e.g. around 5 mm for masonry with 10 mm joints. It should also have a good range of particle sizes from the largest to the smallest (an even grading) since this leads to good packing of the particles to give a dense, strong mass resistant to erosion, permeation and chemical attack. Many naturally occurring alluvial deposits fall naturally into the required grading and may be used as dug or just with a few coarser particles screened off. These are usually termed pit sands. Sands that are outside the normal range must be sieved to remove coarse fractions and washed to remove excess clay particles.

The shape of the particles is also important for mortar sands. Very flaky materials such as slates and micas are not very suitable as it is difficult to make them workable. Very porous absorbent materials are also unsatisfactory for dry-mixed mortars since they cause rapid falls in workability during use by absorbing the mixing water. They may be suitable for mixes based on wet premixed lime-sand 'coarse stuff'. Sand may be sieved into fractions and regraded, but this is rarely done for a mortar sand. *Figure 33.3* shows the grading curves for the sands allowed under the previous standard BS 1200 (1976). This gives two allowed grades, S for structural use and G, with slightly wider limits, for general purposes. The current standard BS EN 13139 (2002) gives similar grading limits and tolerances plus limits on sulphates, chlorides and materials that modify setting-rate. As it is fairly complex there is an associated guidance document PD6682-3 (2003).

Rendering mixes require sands with broadly similar characteristics to those of mortars, but a good grading is even more important to avoid shrinkage cracking and spalling and to give good bond to the substrate.

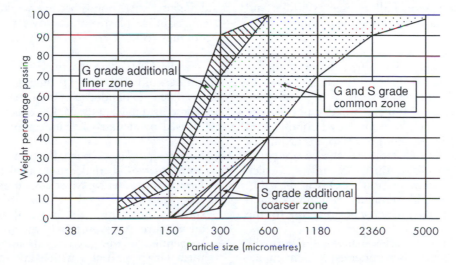

Fig. 33.3 Grading curves for mortar sands.

Table 33.1 Chemical compositions of some representative clays (weight%)

Oxide constituent	SiO_2	Al_2O_3	Fe_3O_4	CaO	MgO	X_2O	CO_2	H_2O	Organic
Broad type									
London brick	49.5	34.3	7.7	1.4	5.2	–	–	–	1.9
Blue clay	46.5	38	1	1.2	–	–	–	–	13.3
Loam	66.2	27	1.3	0.5	–	–	–	–	5
Fletton clay	50	16	7	10	1	3	–	6	6
Marl	33	10	3	26	3.5	–	20.5	4	–
Burnham clay	42.9	20.9	5	10.8	0.1	0.3	8.1	6.9	5
Red brick clay	49	24	8	7	–	1	11	–	–
Gault clay	44	15	6	17	–	–	–	18	–
Washed china clay	46	40	–	1	–	–	–	13	–
Stourbridge fireclay	65	22	2	1	–	–	–	10	–

X, Na/K.

CONCRETING SANDS

The requirement for sands for concrete units have been discussed in Chapter 17.

GROUND SAND

Finely ground silica sand is used particularly in the manufacture of aircrete (AAC) materials and as an inert filler.

Fly ash (pulverised fuel ash; pfa)

Fly ash, the main by-product of modern coal-fired electricity generating stations, is a chemically active filler often used in concrete units. It has been discussed in Chapter 15 when considering additions for concrete.

Chalk (calcium carbonate; $CaCO_3$)

In a finely ground state chalk is used as a filler and plasticity aid in masonry cement and some grouts.

33.2.2 CLAYS

Clay is a very widely distributed material that is produced by weathering and decomposition of acid alumino-silicate rocks such as the feldspars, granites and gneisses. Typical broad types are the kaolin group, of which kaolinite has a composition $Al_2O_3.2SiO_2.2H_2O$, the montmorillonite group, of which montmorillonite itself has the composition $Al_2O_3.4SiO_2.nH_2O$, and the clay micas, which typically have a composition $K_2O.MgO.4Al_2O_3$. $7SiO_2.2H_2O$. They will frequently contain iron and other transition metals, which can substitute for the aluminium.

The clays used for clay brick manufacture normally comprise only partly clay minerals, which impart plasticity when wetted, the balance being made up of other minerals. Brick earths, shales, marls, etc. mostly contain finely divided silica, lime and other materials associated with the particular deposit, e.g. carbon in coal-measure shales. Most brick clays contain iron compounds, which give the red, yellow, and blue colours to fired bricks. *Table 33.1* gives the compositions of some typical clays in terms of their content of oxides and organic matter (coal, oil, etc.).

The properties of clays result from their layer structure, which comprises SiO_4 tetrahedra bonded via oxygen to aluminium atoms, which are also bonded to hydroxyl groups to balance the charge. The layers form loosely bound flat sheet-like structures that are easily parted and can adsorb and bond lightly to varying amounts of water between the sheets. As more water is adsorbed the clay swells and the inter-sheet bonds become weaker, i.e. the clay becomes more plastic and allows various shaping techniques to be used.

33.2.3 LIGHTWEIGHT AGGREGATES

Manufactured lightweight aggregates, including sintered pfa, expanded clay and foamed slag have been described in Chapter 25. Other lightweight aggregates used particularly for unit manufacture are:

- *Furnace clinker*, a partially fused ash from the bottom of solid fuelled industrial furnaces.
- *Furnace bottom ash*. Most large modern furnaces, especially those used to raise steam in power stations, burn finely ground coal dust as a dust/air mixture. A proportion of the fine ash particles suspended in the gas stream sinter together to form larger particles which fall to the base of the furnace.

- *Perlite*, volcanic ash that is deposited as a fine glassy dust that can be converted to a lightweight aggregate by hot sintering.
- *Pumice*, a light foamed rock formed when volcanic lava cools. It is normally imported from volcanic regions such as Italy.

33.2.4 BINDERS

The binder is the component that binds together mixtures of sands, aggregates, fillers, plasticisers, pigments, etc. used to make mortars, concrete units, sandlime units and grouts. Widely used binders are based on one of:

- hydraulic cements, which react chemically with water at normal factory/site temperatures
- lime–silica mixtures, which react only in the presence of high-pressure steam
- lime–pozzolan mixtures, which set slowly at ambient temperatures, or pure lime which sets slowly in air by carbonation.

Portland cement

Currently, the most popular binder for general purposes is a CEM I Portland cement or a sulphate-resisting Portland cement. These have been described in some detail in Chapter 13.

Masonry cement

This is a factory prepared mixture of Portland cement with a fine inert filler/plasticiser (around 20%) and an air-entraining agent to give additional plasticity. It is intended solely for mixing with sand and water to make bedding mortars. When the fine powder is lime the appropriate standard is BS EN 197-1 Notation CEM1 (2000) while for any other filler, e.g. ground chalk or pfa etc., the relevant standard is BS EN 413-1, Class MC (2004).

Lime and hydraulic lime

Lime (CaO) is widely used as an ingredient in mortars, plasters and masonry units. The pure oxide form, called quicklime, was used widely in the past for mortars for stonework. It is produced by heating pure limestone to a high temperature and then 'slaking' with water to produce hydrated lime, $Ca(OH)_2$. Since it does not have any setting action in the short term it may be kept wet for days or weeks provided it is covered and prevented from drying out. The wet mix with sand is termed 'coarse stuff'. Contemporary lime mortar may be made from pre-hydrated lime but is otherwise similar. The initial setting action of this mortar depends only on dewatering by contact with the units so it is not

suitable for the construction of slender structures that require rapid development of flexural strength. Over periods of months or years the lime in this mortar carbonates and hardens to form calcium carbonate as in equation (33.1), but it is never as hard or durable as properly specified hydraulic cement mortars:

$$Ca(OH)_2 + CO_2 \rightarrow CaCO_3 + H_2O \quad (33.1)$$

Hydraulic lime was widely used in the past and is frequently specified for repairing historic buildings to match the original mortar. It is basically a quicklime – calcium oxide – produced by heating impure limestone to a high temperature. The impurities, usually siliceous or clay, lead to the formation of a proportion of hydraulically active compounds such as calcium silicates or aluminates. The binder is made by partial hydrolysis (slaking) of the lime with water. The high temperatures and steam caused by the reaction help to break down the mass to a powder. The mortar is made as normal by gauging (mixing in prescribed proportions) the finely ground binder with sand and water. The classic reference works are Vicat (1837) and Cowper (1927). More recent information is given by Ashurst (1983) and the BRE Good Building Guide 66 (2005).

Most of the hydraulically active cements and limes may be blended with pure hydrated lime in various proportions to make hybrid binders, which give mortars with a lower strength and rigidity but still maintain the plasticity of the 1:3 binder:sand ratio. This leads to mortars that are more tolerant of movement and more economical.

Sandlime

The binder used for sandlime bricks and aircrete (autoclaved aerated concrete – AAC) blocks is lime (calcium hydroxide, $Ca(OH)_2$), which reacts with silica during autoclaving to produce calcium silicate hydrates. The reaction, in a simplified form, is:

$$Ca(OH)_2 + SiO_2 \rightarrow CaSiO_3 + H_2O \quad (33.2)$$

The lime is usually added directly as hydrated calcined limestone or may be derived in part from Portland cement incorporated in small quantities to give early age strength to the unit.

33.3 Other constituents and additives

33.3.1 ORGANIC PLASTICISERS

Many organic compounds improve the plasticity, or workability, of mortars, rendering mortars, infilling

Table 33.2 Effect of latex polymer additives on properties of cement pastes with water:cement ratio of 0.3 and polymer:cement ratio of 0.1 (from de Vekey, 1975; de Vekey and Majumdar, 1977)

| | | Properties of polymer-cement with a W/C of 0.3 and a polymer solids: cement ratio of 0.1 as a proportion of neat cement paste after 2 yrs storage | | | |
| | | Elastic Modulus (GPa) | | Flexural strength (MPa) | |
Monomer	Reaction product with cement slurry	Storage in air, 65% RH	Storage in water at 20°C	Storage in air, 65% RH	Storage in water at 20°C
Vinyl acetate	Acetate ions	0.62*	0.59*	1.60*	0.66*
Vinyl propionate	None	0.62*	0.74	1.24*	1.22*
Butadiene and styrene	None	0.51	0.71	1.00*	1.63†
Vinylidene dichloride	Chloride ions	0.66†	0.84†	1.52†	1.36†
Acrylic acid and styrene	None	0.59	0.69	1.56	1.95
Acrylic acid	None	0.4	0.65	1.29	1.91
Acrylic acid and methacrylic acid	None	0.32	0.73	1.44	1.68

*Based on means of two products; †based on means of four products. RH, relative humidity.

grouts and concrete used for the manufacture of units. All the classic mortar plasticisers operate by causing air to be entrained as small bubbles. These bubbles fill the spaces between the sand grains and induce plasticity. Typical materials are based on Vinsol resin, a by-product of cellulose pulp manufacture, or other naturally available or synthetic detergents. They are surfactants and alter surface tension and other properties. Mortar plasticisers should conform to BS EN 934-3. Superplasticisers, used only for concrete and grout mixes, plasticise by a different mechanism that does not cause air entrainment. These have been described more fully in Chapter 14.

33.3.2 LATEX ADDITIVES

A number of synthetic copolymer plastics may be produced in the form of a 'latex', a finely divided dispersion of the plastic in water usually stabilised by a surfactant. Generally the solids content is around 50% of the dispersion. At a temperature known as the film-forming temperature, they dehydrate to form a continuous polymer solid. When combined with hydraulic cement mixes these materials have a number of beneficial effects: they increase adhesion of mortar to all substrates; increase the tensile strength and durability; and reduce the stiffness and permeability. Because of these effects they are widely used in flooring screeds and renders but are also used to formulate high-bond mortars and waterproof mortars. The better polymers are based on copoly-

merised mixtures of butadiene, styrene and acrylics. Polyvinylidene dichloride (PVDC) has also been marketed for this application but it can give off chlorine, which can attack buried metals. Polyvinyl acetate (PVA) is only suitable for use in dry conditions as it is unstable in moist conditions. Polyvinyl propionate has been found to give less satisfactory flow properties than the acrylic copolymers. These materials should never be used with sands containing more than 2% of clay or silt particles. Dosage is usually in the range of 5–20% of the cement weight. *Table 33.2* gives some properties of common types (de Vekey, 1975; de Vekey and Majumdar, 1977).

33.3.3 PIGMENTS

Through-coloured units and mortars of particular colours may be manufactured either by selecting suitably coloured ingredients or by adding pigments. Units may also be coloured by applying surface layers but this is more common for fired clay than for concrete or calcium silicate units. Pigments are in the form of inert coloured powders of a similar fineness to the binder, so they tend to dilute the mix and reduce strength. Most pigments should be limited to a maximum of 10% by weight of the binder in mortars and carbon black to 3%. Some typical pigments, from information in ASTM task group C09.03.08.05 (1980), are synthetic red iron oxide, Fe_2O_3; yellow iron oxide; black iron oxide, FeO (or Fe_3O_4) and brown iron oxide, $Fe_2O_3.xH_2O$; natural brown iron oxide, $Fe_2O_3.xH_2O$; chromium oxide green, Cr_2O_3; carbon black (concrete grade); cobalt blue; ultramarine blue; copper phthalocyanine; and dalamar (hansa) yellow. Only pigments resistant to alkali attack and wettable under test mix conditions are included. All but the last two are not faded by light. Pigments for mortar should conform to BS EN 12878 (2005).

33.3.4 RETARDERS

Retarders are used to delay the initial set of hydraulic cement mortars. They are generally polyhydroxy-carbon compounds. Typical examples are sugar, lignosulphonates and hydroxycarboxylic acids, as discussed in Chapter 14.

33.3.5 ACCELERATORS

Accelerators based on calcium chloride ($CaCl_2$) have, in the past, been used in small amounts in concrete block manufacture and for mortars (see Chapter 14). All current codes of practice and standards do not permit the addition of chlorides because they are corrosion accelerators for embedded steel fixings (see Chapter 24). Alternatives such as calcium formate ($Ca(CHO_2)_2$) may be satisfactory. Accelerators are not effective when building with mortar in frosty weather and are no substitute for proper protection of the work.

33.4 Mortar

Mortar has to cope with a wide range of sometimes conflicting requirements. To obtain optimum performance the composition must be tailored to the application. The broad principles are as follows:

1. Mortars with a high content of hydraulic cements are stronger, denser, more impervious and more durable, bond better to units under normal circumstances and harden rapidly at normal temperatures. They also lead to a high drying shrinkage and rigidity of the masonry. They are likely to cause shrinkage cracks if used with shrinkable low-strength units particularly for long lightly loaded walls such as parapets and spandrels.
2. Mortars with decreased or no content of hydraulic cements are weaker and more ductile and thus more tolerant of movement. They are matched better to low-strength units but at the cost of a reduction in strength, durability and bond. There is a corresponding reduction in shrinkage and hardening rate.
3. Mortars made with sharp, well-graded sands can have very high compressive strength, low permeability and generally good bond but poor workability. Fine loamy sands give high workability but generally with reduced compressive strength and sometimes reduced bond.
4. Lime confers plasticity and, particularly for the wet stored mixes, water retentivity (the ability of the mortar to retain its water in contact with highly absorbent bricks), which facilitates the laying process and makes sure that the cement can hydrate. Lime mortars perform poorly if subjected to freezing while in the green (unhardened) state but, when hardened, are very durable. Lime is white and thus tends to lighten the colour of the mortar. In some circumstances it can be leached out and may cause staining.
5. Air entrainment improves the frost resistance of green mortar and allows lower water:cement ratios to be used, but such plasticised mixes may be less durable and water-retentive than equivalent lime mixes. Air-entrained mixes also need careful manufacture and control of use since over-mixing gives very high air contents, and retempering (adding more water and reworking the mix) can lead to very poor performance owing to the high porosity of the set dry mortar.
6. Pigments weaken mortar and their content should never exceed the doses given in section 33.3.3.
7. Polymer latex additives can markedly improve some properties (see section 33.3.2), but they are expensive.
8. Retarders are widely used in the manufacture of ready-mixed mortars, delivered to site in the same way as ready-mixed concrete. The retarder is dosed to give a 'pot life' of between 1 and 3 days. Care is needed in the use of retarded mortars, especially in hot dry weather, because if they dry out too rapidly the curing process does not take place and the mortar only hardens when finally wetted.
9. Mortar has a relatively high thermal conductivity and thus causes heat loss in walls of insulating units. Insulating mortars, which use a low-density replacement for the coarse sand particles, can give improved performance. Another option is to use 1–5 mm thick joints of high-bond 'thin-bed mortars' with units with close tolerances on their size (see BRE Digest 432 (1998)).

33.4.1 PROPERTIES OF FRESHLY-MIXED (UNSET) MORTAR

Test methods for fresh mortar are given in the BS EN1015 series of standards (1998–1999) and a few in BS 4551 (1980), RILEM (1978) and ISO (1991). A key property of unset mortar is the workability, i.e. how easy it is to handle and place on to the masonry. This is covered either by the flow test, measured on a standard 254 mm diameter (ASTM) flow table covered by BS EN1015-3 (1999) or by consistence, measured by the plunger penetration test in EN1015-4 (1999). Other useful measurements are: the pot life and the setting time, i.e. how long it may be used for after mixing (see EN1015-9 (1999)); the water retentivity, i.e. how good it is at retaining water against the suction of the units measured in accordance with BS 4551 (1980); and the hardening rate. Associated parameters are the cement content, water content (often expressed as the water:cement ratio, w/c) and the air content, measured by weighing a known volume (0.5 litres) of mortar and then calculating, using the relative densities and proportions of the constituents, from equation 33.3 (see BS EN1015-7, 1999):

$$A = 100(1 - k\rho) \qquad (33.3)$$

where A is the air content, ρ is the relative density of the mortar and K is derived from:

$$K = \frac{\left[\dfrac{M_1}{d_1} + \dfrac{M_2}{d_2} + \cdots + M_w \right]}{M_1 + M_2 + \cdots + M_w} \qquad (33.4)$$

where $M_1, M_2, \ldots$, etc. are the relative masses of the constituents of the mortar of relative densities (specific gravities) d_1, d_2, etc., and M_w is the relative mass of water. The sum of all the values of M_1, $M_2, \ldots$ and M_w will be equal to 1. For precise measurements it is necessary to measure the relative densities of all the individual constituents by use of the density bottle method. If this cannot be done the following default values are suggested: Portland cement, 3.12; masonry cement, 3.05; silica sand, 2.65; white hydrated building lime, 2.26; and grey hydrated lime, 2.45. Alternatively the pressure method can be used.

The water content can be determined independently on fresh mortar by rapid oven drying of a weighed quantity. A simple site test to independently measure the cement content of fresh and unhardened mortars has been described by Southern (1989) and de Vekey (2001). A quantity of mortar is weighed and dissolved in a pre-diluted and measured volume of acid held in an insulated container. The cement content can be calculated from the temperature rise measured with an electronic thermometer.

33.4.2 PROPERTIES OF HARDENED MORTAR
The important properties of hardened mortar are its density, permeability, compressive plus flexural

strength and bond strength to units; methods of measuring these are described in BS EN1015-6 (1999), BS EN1015-19 (1999), BS EN1015-11 (1999) and BS EN1052-5 (2005), respectively. The compressive strength is measured by a cube-crushing test and the flexural strength is measured by the three-point bend (or modulus of rupture) test, both of which have been described in Chapter 21. Other useful indicators of performance are the drying shrinkage and Young's modulus, which may be measured by tests given in BS 4551 (1980), RILEM (1978) or ISO (1991). Durability is influenced by the combination of other properties and may be determined using the methods proposed by Harrison (1981, 1986), Harrison and Bowler (1990) and Bowler *et al.* (1995), now standardised by RILEM (1998).

The bond strength to typical units can be measured by the parallel wallette test. In this test a small wall is built and tested in the vertical attitude by a four-point bend test using articulated loading arms and supports, to prevent the application of any twisting moments, and resilient bolsters to prevent uneven loading.

The alternative bond wrench, which has been included in some codes and standards (such as ASTM C1072-86 (1986) and Standards Association of Australia (1988)), is now the basis of BS EN1052-5 (2005). It is a simpler way to measure bond strength than the wallette test although it may give slightly higher values. This device measures the moment required to detach a single unit from the top of a wall or stack-bonded prism using a lever clamped to the unit (*Fig. 33.4*). The load may be applied in a variety of ways, most simply by filling a container with lead shot. An electronically gauged version is described in BRE Digest 360 (1991) where the load is applied via a load cell transducer. Its main advantage is that it can be used both on site for quality control and diagnosis of problems and failures and as a laboratory tool.

Table 33.3 gives the common formulations of mortars in contemporary use and also lime mortars for restoration work. The table gives some ranges for the performance of mortars in terms of the range of their compressive, flexural and bond strengths. A 'designation' is a term used for a group of prescribed mortars giving approximately similar performance. The table also includes the equivalent performance measures used by the Eurocode and associated standards. This is based on compressive strength and expressed as MXX (where XX is the target value).

It is clear from the table that a very wide range of strengths is possible for any nominal mix ratio and that parameters other than just the binder type

Table 33.3 Strength ranges for mortar

Strength EN1996-1-1: 1996 Nat. Appn. Doc. Table 5	Designation: BS5628: part 1: 2005, Table 1	Proportions by volume of ingredients			Observed strength ranges at 28 days for cement mortars and 90 days for lime mortars		
		Cement: lime:sand	Masonry – cement:sand	Cement – sand:plasticiser	Compressive (MPa)	Flexural (MPa)	Bond (MPa)
M12	(i)	1:0–¼:3	–	–	8–30	2.8–6.6	0.6–1.6
M6	(ii)	1:½:4½	1:2.5–3.5	1:3–4	5–18	1.8–4.5	0.3–1.0
M4	(iii)	1:1:5–6	1:4–5	1:5–6	2–12	0.7–3.7	0.2–1.1
M2	(iv)	1:2:8–9	1:5.5–6.5	1:7–8	0.8–5.5	(0.7–1.7)	(0.36–0.5)
M1	(v)	1:3:10–12	1:6.5–7	1:8	0.5–1.0	(0.7–0.9)	no data
M1	(vi)*	0:1:2–3	–	–	0.5–1.0	no data	no data
M1	(vii)§	0:1:2–3	–	–	0.5–1.0	no data	no data
Varies	Thin bed	–	–	†	5–10	–	0.14–0.66
M2	Lightweight	1:1:5†	–	1:5†	1.7–3.6	0.7–1.4	0.05

*Hydraulic lime mortars; †The exact composition of thin bed and lightweight mortars is a commercial secret; §Pure (air) lime mortars.

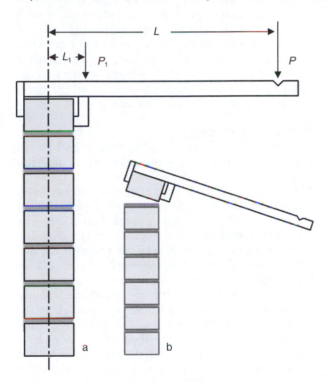

Fig. 33.4 Bond wrench (a) before and (b) after test.

and content influence the strength, including the water:cement ratio, sand grading and air content. A further factor, which only affects the properties of the mortar in the bed (and not, of course, mortar specimens cast in impervious moulds), is the amount of dewatering and compaction by the units. Where dewatering occurs it generally increases the intrinsic strength and density of the mortar but may reduce the bond by also causing shrinkage and local microcracking. Other key parameters are porosity (indicative of permeability), Young's modulus and drying shrinkage. Density (ρ) is obtained by dividing the weight of a prism or cube by its volume, which may be obtained by measuring all the dimensions; alternatively the density can be obtained by weighing the prism/cube submerged in water and then calculating:

$$\rho = m/bdL \quad \text{or} \quad \rho = m/(m - m_s) \quad (33.5)$$

where m is weight in air, b is the width, d is the depth and L is the length of the prism/cube, and m_s is its weight when submerged. The density will vary with moisture content and W is measured as saturated or oven dry. Typical densities for mortars are usually in the range 1500–2000 kg/m^3.

The water porosity is obtained by measuring the water absorption by evacuating the weighed dry specimen then immersing it in water at atmospheric pressure and weighing. The gain in weight of water can be converted to volume and then divided by the volume of the specimen to give the percentage porosity. The Young's modulus may be derived from the flexural strength test provided deflection is measured. Drying shrinkage is measured by attaching precision reference points to the ends of a prism, saturating the prism with water and measuring the length with a micrometer screw gauge or other precision length-measuring device. After a drying regime the length is again measured and the change is expressed as a percentage of the overall length. Various drying regimes have been used (BRE Digest 35 (1963), RILEM (1975) and CEN (2002)). Some data on these properties are given in *Table 35.4*.

33.4.3 THIN-BED AND LIGHTWEIGHT MORTARS

Thin-bed mortars (see BRE Digest 432, 1998; Fudge and Barnard, 1998; and Phillipson *et al.*, 1998) are proprietary formulas but normally contain some fine sand and cement plus bond-improving additives. They are normally supplied as a dry premix that just needs thorough mixing with water before use. The application technique is similar to that used for bedding tiles by use of serrated spreaders to produce thin ribbed layers of the order of 2–5 mm thick. Accurate levelling is crucial for good quality work.

Lightweight mortars (see Stupert *et al.*, 1998) are thermally more efficient replacements for normal mortars and contain cement, optionally lime and low-density fine aggregates such as pumice, pfa and perlite.

33.5 Fired clay bricks and blocks

Fired clay units are made by forming the unit from moist clay by pressing, extrusion or casting followed by drying and firing (burning) to a temperature usually in the range 850–1300°C. During the firing process complex chemical changes take place and the clay and other particles that go to make up the brick are bonded together by sintering (transfer of ions between particles at points where they touch) or by partial melting to a glass. During the drying and the firing processes the units generally shrink by several per cent from their first-made size and this has to be allowed for in the process. Some clays contain organic compounds, particularly the coal-measure shales and the Oxford clay used to make Fletton bricks. Some clays are deliberately compounded with waste or by-product organic compounds since their oxidation during firing contributes to the heating process and thus saves fuel. The burning out of the organic material leaves a more open, lower density structure. The ultimate example of this is 'Poroton', which is made by incorporating fine polystyrene beads in the clay. Wood and coal dust can be used to achieve a similar effect in some products.

33.5.1 FORMING AND FIRING

Soft mud process

The clay is dug, crushed and ground then blended with water using mixers to make a relatively sloppy mud. A water content of 25–30% is required for this method. In some plants other additives may be incorporated such as a proportion of already fired clay from crushed reject bricks (grog), lime, pfa, crushed furnace clinker and organic matter to act as fuel. In the well-known yellow or London Stock brick, ground chalk and ground refuse are added. The mud is formed into lumps of the size of one brick and the lump is dipped in sand to reduce the stickiness of the surface. In the traditional technique the lump is thrown by hand into a mould and the excess is cut off with a wire. This gives rise to the characteristic 'folded' appearance of the faces of the brick caused by the dragging of the clay against the mould sides as it flows. Nowadays most production is by machine, which mimics the hand-making process. These bricks usually have a small frog (depression) formed by a raised central area on the bottom face of the mould. Because of the high drying shrinkage of such wet mixes and the plasticity of the unfired (green) brick the size and shape of such units are fairly variable. This variability adds to their 'character' but means that precision brickwork with thin mortar beds is not feasible. The finished brick is also fairly porous, which improves its insulation properties, and, paradoxically, its effectiveness as a rain screen, but limits its strength.

Stiff plastic process

The clay is dug, crushed and ground then blended with water using mixers to make a very stiff but plastic compound with a water content of 10–15%. This is then extruded from the mixer and cut into roughly brick-shaped pieces and allowed to dry for a short period before being pressed in a die. The clay is very stiff so when ejected straight from the mould it retains very precisely the shape of the die. The low moisture content means that the shrinkage is low and therefore the size is easier to control and the drying time is relatively short. Another advantage is that the unfired brick is strong enough to be stacked in the kiln or on kiln cars without further drying. This type of unit will usually have at least one shallow frog and may have frogs in both bed faces. The process is used to produce engineering bricks, facing bricks, bricks with very accurate dimensions and pavers.

Wire cut process

Clay of intermediate consistency is used, with a moisture content of 20–25%, and the clay is extruded from a rectangular die with the dimensions of the length and breadth of the finished unit. The ribbon of clay, the 'column', is then cut into bricks by wires

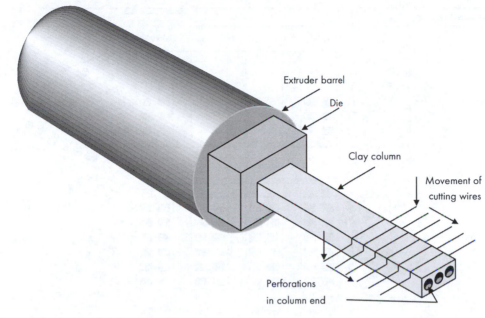

Extruder barrel

Die

Clay column

Movement of
cutting wires

Perforations
in column end

Fig. 33.5 Extruded wire-cut brick production.

set apart by the height of the unit plus the allowance for process shrinkage. The cutting machines are usually arranged such that the group of cutting wires can travel along at the same speed as the column while a multiple cut is made. This means that the process is fully continuous and the cut is perpendicular to the face and ends of the unit. A plain die produces a solid column with just the characteristic wire-cut finish and these bricks will have no depressions in their bed faces. In this process it is easy, however, to include holes or perforations along the length of the column by placing hole-shaped blockages in the die face. This has the following advantages:

1. Reduction in the weight of clay required per unit so transport costs at every stage of the production and use of the units and all clay preparation costs, i.e. for shredding, grinding, mixing, etc., are also reduced.
2. A reduction in the environmental impact by reducing the rate of use of clay deposits and therefore the frequency of opening up new deposits.
3. Reduction in the mass and opening up of the structure of the units thus speeding up drying and firing, cutting the fuel cost for these processes and reducing the capital cost of the plant per unit produced.
4. Oxidation of organic matter in the clay is facilitated by increasing the surface area to volume

ratio, and reduces the chance of blackhearting (see below).
5. The thermal insulation is improved. This has a modest effect for UK-Standard size bricks but the improvement can be substantial for large clay blocks.
6. The units are less tiring to lay because of their lower weight.

Because of these factors and the very large proportion of the production cost spent on fuel, most clay units are perforated at least to the extent of 10–25% by volume. It should be stressed that there is a penalty in that the clay must be very well ground and uniform in consistence for successful production of perforated units. Any lumps or extraneous air pockets can ruin the column. To improve consistency the mixers are commonly heated and the front of the extruder is de-aired (evacuated) to minimise air bubbles. *Figure 33.5* illustrates the production of a typical three-hole perforated brick by stiff plastic extrusion and wire-cutting.

Semi-dry pressing
This is one of the simplest processes for forming bricks. In the UK only Lower Oxford clay (or shale) is used; this comes from the Vale of Aylesbury and runs in a band towards the east coast. It contains about 7% natural shale oil, which reduces the cost of firing but does give rise to some pollution problems.

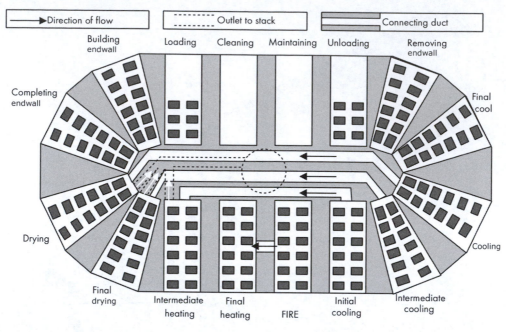

Fig. 33.6 *Principles of a Hoffman kiln.*

The clay is dug and then milled and ground to pass through a 2.4 mm or 1.2 mm sieve without altering the water content markedly from that as-dug (8–15%). The coarser size is used for common bricks and the finer for facings. The powdered clay is then fed into powerful automated presses, which form a deep-frogged, standard size brick known as a Fletton (named after one of the early manufacturers). Un-modified Flettons are limited to a single barred pink/cream colour. Flettons intended as facing bricks will either be mechanically deformed to give a patterned surface (rusticated) or have an applied surface layer, such as coloured sand, fired on.

Drying and firing in Hoffman kilns

The Hoffman kiln is a multi-chamber kiln in which the bricks remain stationary and the fire moves. It is mainly used for the manufacture of Flettons. In the classic form it consists of a row of chambers built of firebricks in the form of short tunnels or arches. In the most efficient form the tunnels form a circle or oval shape and are connected together and to a large central chimney by a complex arrange-ment of ducts. A single 'fire' runs round the circle and at any one time one chamber will be being loaded or 'set' ahead of the fire and one will be being unloaded behind the fire. The bricks are stacked in the kiln in groups of pillars termed 'blades', which leaves lots of spaces between units to enable the gases to circulate freely. Chambers immediately in front of the fire will be heating up using the exhaust gases from the hottest chamber, and those further ahead will be being dried or warmed by gases from chambers behind the fire that are cooling down.

Figure 33.6 illustrates the broad principles of the system, showing only the ducts in use for the fire in one position. In practice the ducts are positioned to ensure a flow through from the inlets to the outlets. The whole process is very efficient, particularly as a large proportion of the fuel is provided by the oil in the bricks themselves. Because of the organic content of the bricks the firing has to be carried out under oxidising conditions during the last phases in order to burn out the oil. If this is not done the bricks have a dark unoxidised central volume, known as a blackheart, which can give rise to deleterious soluble salts. During this phase some fuel is added to maintain the temperature. This is essentially a batch process and the average properties of the contents will vary a little from chamber to chamber. Addi-tionally, the temperature and oxidising condition will vary with position in the chamber, thus some selection is necessary to maintain the consistency of the product.

Drying and firing in tunnel kilns

Tunnel kilns are the complement of Hoffman kilns in that the fire is stationary and the bricks move

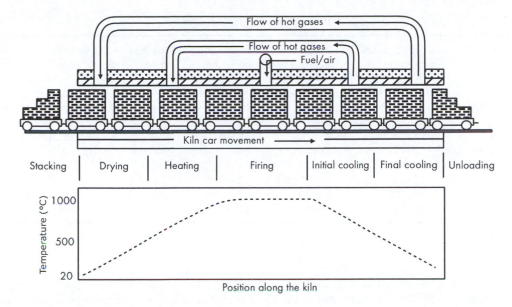

Fig. 33.7 *Principles of tunnel kiln.*

through the kiln as stacks on a continuous train of cars on rails (*Fig. 33.7*). In practice a long insulated tunnel is heated in such a way that the temperature rises along its length, reaches a maximum in the centre and falls off again on the other side. To maximise efficiency only the firing zone is fuelled and hot gases are recycled from the cooling bricks and used to heat the drying and heating-up zones of the kiln. Most extruded wire-cuts and stiff plastic bricks are now fired in such kilns, which are continuous in operation. Stocks and other mud bricks may also be fired this way after a pre-drying phase to make them strong enough to withstand the stacking forces.

Clamps
Clamps are the traditional batch kilns, comprising a simple insulated refractory beehive-shaped space with air inlets at the base and a chimney from the top and fired using solid fuel.

Intermittent kilns
These are the modern version of the clamp, where the units are fired in batch settings using oil or gas as a fuel. They are now only used for the production of small runs of specially shaped brick 'specials', some examples of which are shown in *Fig. 34.7*.

33.5.2 PROPERTIES
Clay bricks probably have the widest range of strengths of any of the manufactured masonry materials, with compressive strengths ranging from 10 MPa for an under-fired soft mud brick to as much as 200 MPa for a solid engineering brick. The common shapes and terminology are shown in *Fig. 33.1*. The compressive strength is measured by a crushing test on whole units with the stress applied in the same direction as the unit would be loaded in a wall. Solid and perforated units are tested as supplied, but frogs are normally filled with mortar as they would be in a wall. The European test method EN772-1 (2000) uses either mortar capping or face grinding to achieve even loading. The quoted strength is the average of six to ten determinations of stress based on the load divided by the area of the bed face. The flexural strength and modulus of elasticity are not normally designated test parameters in unit standards, but it is important to obtain data for finite element analysis models. A standard three-point bending method is included in RILEM LUM A.2 (1994), with linear elastic analysis used for the calculation of the maximum flexural stress.

Other important properties are the dimensions, water absorption and porosity, initial rate of water absorption, density and soluble salts content, which can be measured by methods specified in EN772-16 (2000), EN772-7 (1998), EN772-11 (2000), EN772-13 (2000) and EN772-5 (2001), respectively. In these standards the dimensions are measured for individual units with allowed individual tolerances for each replicate. Water absorption and porosity are measured in the same way as for mortar (see section 33.4.2), except that the preferred saturation technique is to

Table 33.4 Properties (typical ranges) for UK fired clay brick types

Brick type	Compressive strength (N/mm²)	Water absorption (weight%)	Water porosity (volume%)	Suction rate (IRWA) (kg/m²/min)	Bulk density (kg/m³)	Flexural strength (MPa)
Handmade facing	10–60	9–28	19–42	–	–	–
London Stock	5–20	22–37	36–50	–	1390	1.6
Gault wire-cut	15–20	22–28	38–44	–	1720	–
Keuper marl wire-cut	30–45	12–21	24–37	–	2030	–
Coalmeasure shale	35–100	1–16	2–30	–	2070	–
Fletton	15–30	17–25	30–40	1.0–2.0	1630	2.8
Perforated wire-cut	35–100	4.5–17	–	0.2–1.9	1470–2060	7
Solid wire-cut	20–110	4–21	10–35	0.25–2.00	1700–2400	6.5

IRWA, initial rate of water absorption.

boil the units in water for 5 hours. The initial rate of water absorption (IRWA or suction rate) is measured to give some idea of the effect of the unit on the mortar. Units with high suction rates need very plastic, high water:cement ratio mortars whereas units with low suction rates need stiff mortars. IRWA is determined by standing the unit in 3 mm depth of water and measuring the uptake of water in 60 seconds.

The IRWA is calculated from:

$$w_i = (m_2 - m_1)/Lb \qquad (33.6)$$

where: w_i = initial rate of absorption (IRWA);
m_1 = initial mass of the unit/specimen;
m_2 = mass after 60 seconds of water absorption;
L = length of the bed (mortar) face to ± 0.5%; and
b = width of the bed (mortar) face to ± 0.5%.

The result is normally given in units of kg/m²/min.

The content of water soluble salts is measured by standard wet chemical analysis techniques or by modern instrumental techniques such as flame photometry. The elements and compounds of concern are sulphates, sodium, potassium, calcium and magnesium. *Table 33.4* gives typical values and ranges for some of the key properties of clay bricks.

In most brickwork, bricks are loaded upon their normal bed face but often they are loaded on edge or on end. Typical examples are headers and 'soldiers' in normal walls (see *Fig. 34.3*), stretchers in arches and reinforced beams and headers in reinforced beams. While unfrogged solid bricks show a small variation in strength for loading in different directions, owing to the change in aspect ratio (height:thickness), perforated, hollow or frogged units may show marked differences, as illustrated in *Table 33.5*.

Taking the simplest geometry as an example it can be seen from *Fig. 33.8* that the minimum cross-sectional area of the 5-slot unit resisting the load will be 80% on bed but 71% on edge and 25% on end. The ratios of the strengths in *Table 33.5* follow approximately the ratios of the areas. Other factors such as the slenderness of the load-bearing sections and the effect of high local stresses at rectangular slot ends complicate the behaviour, and may explain the variations between different types. It can also be shown that porosity in the form of vertical perforations results in a smaller reduction in the strength of a material than does generally distributed porosity, and is the more efficient way of reducing the weight. This is illustrated in *Fig. 33.9*. More detailed information on clay brick properties is contained in BRE Digest 441 parts 1 and 2 (1999).

33.6 Calcium silicate units

Calcium silicate units are manufactured from firm mixtures of lime, silica-sand and water. Aggregates such as crushed rocks or flints may be incorporated to alter the performance and appearance, and pigments may be used to vary the colour. Common colours are whites, blacks, buffs and grey-blues. Reds are produced but they seldom have the richness of fired clay units. There is only one basic process, in which the mixture is pressed to high pressures in a die in a static press, ejected, set on cars and then placed

Table 33.5 Properties of some UK fired clay brick types in various orientations (from Lenczner, 1977; Davies and Hodgkinson, 1988; Sinha and de Vekey, 1990)

Type	Ref.*	Perforation (Volume%)	Compressive strength on bed (MPa)	Percentage of the on-bed strength			Water absorption (weight%)	Suction rate (IRWA) (kg/m²/min)
				On bed	On edge	On end		
23 hole	(a)	?	65.5	100	29	11	6.9	0.6
14-hole	(b)	21.3	74.3	100	35	14	3.9	–
14 hole	(c)	?	44.8	100	29	11	4.7	1.8
10 hole	(d)	30.1	70.2	100	42	31	5.4	–
3 hole	(e)	12.2	82	100	65	49	4.2	–
3 hole	(f)	?	57.8	100	36	25	8.5	1.5
5 cross slots	(g)	20	64.1	100	81	22	3.4	–
16 hole	(A)	20.1	64.7	100	31	13	5.5	0.35
Frogged	(B)	15.1	32.4	100	41	36	21.7	2.86
Frogged	(C)	6.2	33.7	100	49	51	14.4	1.06
Frogged	(D)	8.6	31.7	100	92	50	11.9	0.54
Solid	(E)	0	43.5	100	67	65	22.8	3.31

*The reference letter links this unit strength data to masonry strength data in *Table 35.1*. IRWA, initial rate of water absorption.

Fig. 33.8 *Area of 5-slot brick resisting load in the three orientations.*

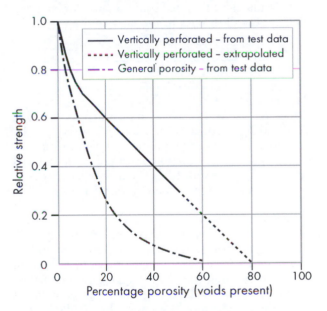

Fig. 33.9 *Compressive strength of ceramic bodies as a function of general porosity and aligned perforations of constant diameter.*

in autoclaves and cured in high-pressure steam for several hours. The mix is invariably fully compacted and makes a very precisely shaped, low tolerance unit with sharp, well-defined corners (arrises) and a fairly smooth finish. The ranges of properties often measured are given in *Table 33.6*.

33.7 Concrete and manufactured stone units

Concrete units have been made since the 1920s and were widely used, in the form of the 'breeze block',

Table 33.6 Property ranges of calcium silicate bricks (after West *et al.*, 1979)

Format	Compressive strength (MPa)	Water absorption (weight%)	Initial rate of water absorption (kg/m²/min)	Bulk density (kg/m³)	Frog volume (%)
Test ref. BS-EN	772-1 (2000)	*	772-18 (2000)†	772-13 (2000)	772-9 (2000)
Solid	20–50	8–22	0.25–2.0	1750–2000	0
Frogged	20–55	13–20	0.5–1.2	1650–1950	4–7

*Measured by vacuum absorption method; †clay brick test BS EN 772-11 actually used.

to build partitions in houses in the building boom of the 1930s. In the past 40 years, however, the range of products has expanded enormously to cover facing bricks and blocks, high-strength units, simulated stone units, thermally insulating blocks and pavers. The performance and behaviour of concrete are discussed in Chapters 20 and 21; in this section we will discuss units formed from concrete containing aggregates, which are produced by pressing specially designed mixes. Other processes are used to produce so-called concrete that does not contain aggregates, such as autoclaved aerated concrete (AAC or aircrete), which we will discuss in the next section.

33.7.1 PRODUCTION PROCESSES FOR CONCRETE UNITS

Casting concrete
Concrete blocks can be manufactured by pouring or vibrating a concrete mix into a mould and de-moulding after a curing process. While this method is used, particularly for some types of manufactured stone and stone-faced blocks, it is not favoured because of its slowness and labour demands.

Pressing of concrete
This is a widely used method for producing solid and frogged bricks and solid, cellular and hollow blocks either in dense concrete or as a porous open structure by using gap-graded aggregates of varying density and not compacting fully. The machine is basically a static mould (or die) that is filled automatically from a mixer and hopper system, and a dynamic press-head that compacts the concrete into the die. After each production cycle the green block is ejected on to a conveyor system and taken away to cure. The press-head may have multiple dies. A variation of the method is the 'egg layer'. This performs the same basic function as a static press but ejects the product straight on to the surface on which it is standing and then moves itself to a new position for the next production cycle.

Curing
All aggregate concrete products may be cured at ambient temperature or at elevated temperature. Elevated temperatures are usually achieved by the use of steam chambers, and allow the manufacturer to decrease the curing period or increase the strength or both. Products cured externally should not be made when the temperature is near or below 0°C since they will be damaged by freezing while in the green state.

33.7.2 CONCRETE PRODUCTS

Dense aggregate concrete blocks and concrete bricks
These are generally made from well-graded natural aggregates, sands, pigments, and CEM I or white Portland cement by static pressing to a well-compacted state. *Figure 33.10a* illustrates the principle of such materials, in which the voids between large particles are filled with smaller particles. They are strong, dense products and are often made with a good surface finish suitable for external facing masonry. They are also suitable for engineering applications. Bricks are produced mainly at the standard size (215 × 102 × 65 mm) in the UK but in a wide range of sizes in continental Europe. Blocks are produced as solid, cellular or hollow by varying the quantity of mix and the shape of the press platen. In order to facilitate demoulding the hollows will always have a slight taper. The hollows or cells in UK products are all designed to run vertically in the finished masonry as this gives the optimum strength to weight ratio. The face size of UK units is generally 440 mm long by 215 mm high but the thickness may vary from 50 mm to 300 mm. Some of the important properties are summarised in *Table 33.7*. Manufacture should comply with BS EN 771-3 (2003), and testing to the relevant BS EN 772-XX series of standards. BRE Digest 460 Parts 1 and 2 (2001) give some useful guidance and background.

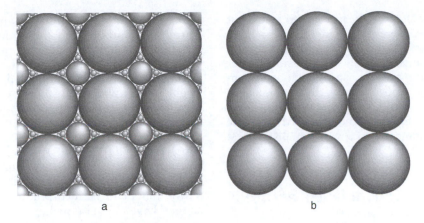

Fig. 33.10 *Types of aggregate: (a) well graded; (b) gap graded (schematic).*

Table 33.7 Properties of some typical dense aggregate (DA) and lightweight aggregate (LWA) concrete blocks

Unit type	Size L × H × T (mm)	Void (%)	Bulk density (kg/m³)	Concrete density (kg/m³)	Compressive strength (MPa)*	Flexural strength (MPa)†	Young's modulus (GPa)	Water absorption (weight%)‡
Brick	215 × 65 × 103	0	2160	2160	32.5	–	–	6.3
DA block	438 × 213 × 98	0	2140	2140	15.5	2.59	–	10.9
DA Block	390 × 190 × 140	41.6	1350	2320	31.6	–	42.3	–
LWA Block	390 × 190 × 140	22.1	1630	2090	21.5	–	32.8	–
LWA Block	390 × 190 × 140	0	2170	2170	29.9	–	30.6	–
LWA Block	390 × 190 × 90	0	2060	2060	21.6	–	17.5	–
LWA Block	390 × 190 × 90	19.9	1100	1380	8.1	–	9	–
LWA block	439 × 215 × 98	0	2190	2190	6.6	0.5	–	35

*Tested wet and mortar capped; †measured to BS6073 Appendix C; ‡measured by vacuum absorption.

Manufactured stone masonry units

These have an essentially similar specification to dense aggregate concrete blocks, except that the main aggregate will be a crushed natural rock such as limestone or basalt and the other materials will be chosen such that the finished unit mimics the colour and texture of the natural stone. Also the production will often be by casting. The relevant standard is BS EN 771-5 (2003).

Lightweight aggregate concrete blocks

These are generally produced as load-bearing building blocks for housing, small industrial buildings, in-fill for frames and partition walling. High strength and attractive appearance are rarely the prime consideration but handling, weight, thermal properties and economy are important. Inherently low-density aggregates are used and are often deliberately gap-graded, as illustrated by *Fig. 33.10b*, and only partly compacted to keep the density down. They will frequently be made hollow as well to reduce the weight still further. The aggregates used are sintered pfa nodules, expanded clay, furnace clinker, furnace bottom ash, pumice or foamed slag together with sand and binder. Breeze is a traditional term for a lightweight block made from furnace clinker. Often low-density fillers or aggregates such as sawdust, ground bark or polystyrene beads are incorporated to further reduce the density. They are produced either by static pressing or in egg-layer plants. Some of the important properties are summarised in *Table 33.7*. The properties commonly measured include compressive strength with the test face ground flat or capped. Flexural strength has also been used to evaluate partition blocks that bear lateral loads, but only self-weight compressive loads. It is a simple, three-point bend test of the type described in section 33.5.2 for clay units with an aspect ratio greater than 4. Other properties measured include dimensions, water absorption by the method of vacuum absorption, density and drying shrinkage. These are manufactured to the same standard as dense aggregate blocks and are also covered by BRE Digest 460 (2001).

33.8 Aircrete (autoclaved aerated concrete – AAC)

Aircrete is the current term for AAC and it is made by a process, developed originally in Scandinavia, that produces solid microcellular units that are light and have good insulating properties. Fine sand or pulverised fuel ash or mixtures thereof is used as the main ingredient. The binder is a mixture of Portland cement (A CEM I), to give the initial set to allow cutting, and lime, which reacts with the silica during the autoclaving to produce calcium silicate hydrates and gives the block sufficient strength for normal building purposes.

33.8.1 MANUFACTURING PROCESS
The method involves mixing a slurry containing a fine siliceous base material, a binder, some lime and the raising agent, aluminium powder, which reacts with the alkalis (mainly calcium hydroxide) to produce fine bubbles of hydrogen gas:

$$Ca(OH)_2 + 2Al + 2H_2O \rightarrow CaAl_2O_4 + 3H_2\uparrow$$
$$(33.7)$$

This mixture is poured into a mould maintained in warm surroundings and the hydrogen gas makes the slurry rise like baker's dough and set to a weak 'cake'. The cake is then cured for several hours at elevated temperature, de-moulded, trimmed to a set height and cut with two orthogonal sets of oscillating parallel wires to the unit size required using automatic machinery. The cut units are then usually set, as cut, on to cars which are run on rails into large autoclaves. The calcium silicate binder forms by reaction under the influence of high-pressure steam. Additional curing after autoclaving is not necessary all the units can be incorporated in work as soon as they have cooled down.

33.8.2 PROPERTIES
The binder reaction is conventional, as given for sandlime (equation 33.2). The structure, of small closed spherical cells with walls composed of a fine siliceous aggregate bound together by calcium silicate hydrates, gives the product a good resistance to permeation by water, good thermal properties and a high strength:density ratio. The nature of the principal siliceous material is identifiable from the colour: ground sand produces a white material and pulverised fuel ash a grey material.

Because of its light weight (low density) the product can be made into large blocks while remaining handleable. Some units are available with (double) face size, 447 mm by 447 mm, designed for building thin-joint masonry (see BRE Digest 432 (1998)). The most common size for normal work is 440 × 215 × 100 mm (or thicker). Some of the key properties are summarised in *Table 33.8*. Manufacture is to BS EN 771-4 (2003), testing to the BS EN 772 series or RILEM (1975). BRE Digest 468 Parts 1

Table 33.8 Properties of Aircrete

Type	Typical dry density (kg/m³)	Compressive strength tested to BS6073:1981		Typical flexural strength (MPa)	Typical tensile strength (MPa)	Young's modulus (GPa)*	Thermal conductance @ 3% moisture (W/m/°K)
		Nominal MPa	Typical MPa				
Low density	450	2.8	3.2	0.65	0.41	1.6	0.12
	525	3.5	4	0.75†	0.52	2.00†	0.14
Standard	600	4	4.5	0.85	0.64	2.4	0.16
	675	5.8	6.3	1.00†	0.76	2.55	0.18
High density	750	7	7.5	1.25	0.88	2.7	0.2

Nominal = as declared; typical = values achieved by modern production plants. *Extrapolation of a range of splitting tests (after Grimer and Brewer, 1965). †Interpolated value.

Table 33.9 Properties of some types of stone unit

Type of stone	Density (kg/m³)	Water absorption (%w/w)	Porosity (%v/v)	Compressive strength (MPa)	Young's modulus (GPa)
Typical ranges					
Limestone	1800–2700	0.1–17	0.3–30	20–240	1–8
Sandstone	2000–2600	0.4–15	1–30	20–250	0.3–8
Marble	2400–2800	0.4–2	0.4–5	40–190	–
Slate	2600–2900	0.04–2	0.1–5	50–310	–
Granite	2500–2700	0.04–2	0.1–4	80–330	2–6
Basalt	2700–3100	0.03–2	0.1–5	50–290	6–10
Individual examples					
York sandstone	2560	2.62	7.6	72.6	–
Portland limestone	2209	5.33	11.8	32	–

and 2 (2002) give some useful guidance and background.

33.9 Natural stone units

Stone units are either naturally occurring flints from chalk deposits (used to make rubble masonry), pieces formed by weathering from the original deposit or partly trimmed pieces used widely for domestic coursed or semi-coursed masonry or precisely cut blocks used to make ashlar stonework, usually for prestige or heritage buildings.

To ensure optimum performance the layered rocks are usually cut to maintain the bedding plane perpendicular to the compressive stress field in the building, e.g. horizontal in normal load-bearing masonry. This also gives the optimum durability. Specification and testing should comply with BS EN 771-6 (2005). Sourcing the stone and a useful set of references are covered in BRE Digest 420 (1997), and the Geological Society SP16 (1999) has comprehensive coverage. Specification and use of stone for conservation work is covered in BRE Digest 508 Part 1 (2008). Some performance characteristics of well-known types are given in Table 33.9 and movement data are given in Table 35.4.

33.10 Ancillary devices

In order to ensure their stability, masonry elements need to be connected either to other masonry elements to form stable box structures or to other structural elements such as frames, floors, roofs, beams and partitions. There is a huge range of ties and other connecting devices, summarised in Table 33.10 together with information on the current standards and guidance literature. Most of these devices are made from metal, predominantly galvanised mild steel, austenitic stainless steel and bronzes. A few light-duty tie products are made from plastic.

Masonry walls require openings for doors, windows and services. The masonry over the openings has to be supported by constructing a masonry arch, by use of a lintel (beam), or by reinforcing the masonry in-situ with bed-joint reinforcement to form a beam. Lintels are prefabricated beams made from steel, timber, concrete and clay-ware designed and sized to co-ordinate with and support masonry in walls. They are covered by BS EN845-2 (2003). Bed-joint reinforcement is typically prefabricated metal meshwork elements sized to be embedded in the mortar joints to increase the overall strength of the masonry. It is illustrated in the top left part of Fig. 34.2. Joint reinforcement is also used to resist out-of-plane loading and to tie masonry together to resist accidental damage.

References

(British standards and the BRE series of publications are listed in 'Further reading' at the end of Part 5.)

Ashurst J (1983). Mortars, plasters and renders in Conservation, Ecclesiastical Architects and Surveyors Association, London.

ASTM C1072-86 (1986). Standard method for measurement of masonry flexural bond.

ASTM Task group Subcommittee section C09.03.08.05 (1980). Pigments for integrally coloured concrete. Cement, Concrete and Aggregates, CCAGDP, 2 (No. 2), 74–77.

Table 33.10 Ancillary devices – manufactures can supply data on specification and installation

Ancillary device	Purpose	Specification document	Guidance documents
Cavity wall ties	To link the two leaves of cavity walls and share tension, compression and shear loads – see Fig. 34.4	BS EN845-1 (2003) BS1243, BSDD140,	BRE IP11/00 (2000), BS5628: Part 1 (2005), BS EN 8103-1: (2003),
Support ties	To link walls to frames	BS EN845-1 (2003)	BRE GBG 29 (1997),
Shear ties	To give shear connections between elements illustrated by Fig. 31.5	BS EN845-1 (2003)	BRE GBG 62 (2004), BRE GBG 41 (2000)
Slip ties	To give shear connections between elements while allowing in-plane movement	BS EN845-1 (2003)	
Straps	To tie walls to roofs and floors	BS EN845-1 (2003), BS8103	
Joist hangers	To support floors on unperforated walls via the joists	BS EN845-1 (2003)	
Brackets	To support the load of walls on frames	BS EN845-1 (2003)	
Angles	To support walls on frames	Specialist devices	
Anchors and fixings	Fixings devices such as screws, nails, screw-plug systems, resin-anchors, expanding anchors, for attaching fittings to walls	General-purpose fixings or specialist patent devices	BRE Digest 329 (2000), CIRIA TN 137 (1991)
Lintels	Prefabricated beams made from steel, timber, concrete and clay-ware designed and sized to co-ordinate with and support masonry in walls	BS EN845-2 (2003)	BS5628: Part 1 (2005), BS EN 1996 – 2 (2006)
Bed joint reinforcement	Prefabricated metal meshwork elements sized to be embedded in mortar joints to reinforce masonry illustrated at the top left of Fig. 31.2	BS EN845-3 (2003)	BS5628: Part 1 (2005), BS EN 1996 – 2 (2006) BS5628-2: (2005) BS EN 1996-1-1: (2005)

Bowler GK, Harrison WH, Gaze ME and Russell AD (1995). Mortar durability: an update. *Masonry International – Journal*, 8 (No. 30), 85–90.

CEN (2002). BS EN 772-14:2002, Methods of test for masonry units. Determination of moisture movement of aggregate concrete and manufactured stone masonry units.

Cowper AD (1927). *Lime and Lime Mortars* (reprinted 1998 BRE Ltd).

Davies S and Hodgkinson HR (1988). *The stress–strain relationships of brickwork when stressed in directions other than normal to the bed face: Part 2, RP755*, British Ceramic Research Establishment, Stoke-on-Trent.

de Vekey RC (1975). *The properties of polymer modified cement pastes*. Proceedings of the First International Polymer Congress, London.

de Vekey RC and Majumdar AJ (1977). Durability of cement pastes modified by polymer dispersions. *Materials and Structures*, 8 (No. 46), 315–321.

de Vekey RC (2001). *Bremortest for site control of mortar composition*. Proc. Intl. Workshop on Onsite control and evaluation of masonry structures, Proceedings PRO 26, RILEM Publications, Bagneux, France, pp. 57–66.

Fudge CA and Barnard M (1998). Development of AAC masonry Units with thin joint mortar for house-building in the UK, Proceedings of the 5th International Masonry Conference, London, 1998 (BMS Proc. 8), pp. 384–387.

Geological Society (1999). *Building stone, rock fill and armourstone in construction*, Special Publication No. 16, Geological Society, London.

Grimer FJ and Brewer RS (1965). *The within cake variation of autoclaved aerated concrete*. Proceedings of symposium on Autoclaved Calcium Silicate Building Products, Society of Chemical Industry, 163–170.

Harrison WH (1981). Conditions for sulphate attack on brickwork, *Chemistry and Industry*, 19 (September 1981), pp. 636–639.

Harrison WH (1986). Durability tests on building mortars – Effect of sand grading. *Magazine of Concrete Research*, 38 (No. 135), 95–107.

Harrison WH and Bowler GK (1990). Aspects of mortar durability. *Transactions of the British Ceramic Society*, **89**, 93–101.

Lenczner D (1977). *Strength of bricks and brickwork prisms in three directions*, Report No. 1, University of Wales Institute of Science and Technology, Cardiff.

Phillipson MC, Fudge CA, Garvin SL and Stupert AW (1998). Construction with thin joint mortar systems and AAC blockwork. Proc. 5th International Mas. Conf., London, 1998. Thermal and strength performance of two lightweight mortar products, Proc. 5th International Mas. Conf., London, 1998, (BMS Proc. 8), pp. 388–390.

RILEM (1975). Recommendations for testing methods of aerated concrete. *Materials and Structures*, **8** (No. 45), pp. 211–220.

RILEM (1978). Recommendations for testing of mortars and renderings, *Materials and Structures*, **11** (No. 63), 207.

RILEM (1994). *Technical recommendations for the testing and use of construction materials*, E&F Spon, London, LUM A.2 Flexural strength of masonry units, pp. 459–461.

RILEM (1998). de Vekey RC (ed.), RILEM recommendations of durability tests for masonry: MS.A.3. Unidirectional freeze–thaw test for masonry units and wallettes. *Materials and Structures*, **31** (No. 212), 513–519.

Sinha BP and de Vekey RC (1990). A study of the compressive strength in three orthogonal directions of brickwork prisms built with perforated bricks. *Masonry International*, **3** (No. 3), 105–110.

Southern JR (1989). *BREMORTEST: A rapid method of testing fresh mortars for cement content*, BRE Information Paper IP8/89, Building Research Establishment, Watford.

Standards Association of Australia (1988). Masonry code: Masonry in buildings. Appendix A7: Flexural strength by bond wrench, AS 3700-1988.

Stupert A, Skandamoorthy JS and Emerson F (1998). *Thermal and strength performance of two lightweight mortar products*. Proceedings of the 5th International Masonry Conference, London, 1998 (BMS Proc.8), pp. 103–106.

Vicat LJ (1837). *Treatise on Calcareous Mortars and Cements*, Translated by Smith JT, Reprinted by Donhead Publishing, Shaftesbury 1997.

West HWH, Hodgkinson HR, Goodwin JF and Haseltine BA (1979). The resistance to lateral loads of walls built of calcium silicate bricks, BCRL technical Note 288, BCRL, Penkhull, Stoke on Trent.

Chapter 34

Masonry construction and forms

This chapter is concerned with how masonry is built, the architectural forms used and the resulting appearance. The basic structural form of many types of masonry is expressed on the surface of buildings and other structures and can be a very attractive and reassuring aspect of these. Appearance is a synthesis of the size, shape and colour of the units, the bond pattern, the mortar colour and finish, the masonry elements – walls, piers, columns, corbels, arches, etc. – and the scale and proportion of the whole structure. Other key aspects are the workmanship, accuracy, the detailing in relation to other features and the use of specially shaped units.

The basic method of construction has hardly changed for several thousand years: units are laid one on top of another in such a way that they form an interlocking mass in at least the two horizontal dimensions. It is not practical to achieve interlocking in a third dimension with normal rectangular prismatic units, but a degree of such interlocking is sometimes used in ashlar stonework. Most practical masonry employs a mortar interlayer to allow for small to large inaccuracies of size between units and to make walls watertight, airtight and soundproof.

34.1 Walls and other masonry forms

Walls are built by laying out a plan at foundation level and bringing the masonry up layer by layer. To maximise the strength and attractiveness it is important to make sure that all the foundation levels are horizontal, are accurate to the plan and allow multiples of whole units to fit most runs between returns, openings, etc. It is also essential to maintain the verticality (plumb), the level of bed joints and the straightness of the masonry within reasonable limits. The thickness of the mortar joints must be kept constant within a small range, otherwise the

masonry will look untidy. The standard technique used is to generate reference points by building the corners (quoins) accurately using a plumb bob and line, a builder's level, a straight edge and a rule. Any openings are then filled with either a temporary or permanent frame placed accurately in the plan position. Lines are stretched between the reference points and the intervening runs of masonry are built up to the same levels. Columns, piers and chimneys are built in the same way but need plumbing in two directions and more care because of their small dimensions.

Some masonry built with precisely sized bricks, cut stone or terracotta ware is built to a higher accuracy, usually with mortar joints on the order of 3–5 mm thick. This is termed 'gauged masonry' and demands a higher standard of workmanship but can look very attractive. There is also a tradition for the use of pebbles, rubble and partly-squared stones combined with wider and more variable mortar joints, particularly in East Anglia, UK. This can also look very pleasing, especially when combined with more accurate units at corners and openings.

Arches and tunnels must follow a curved shape defined by the architect or engineer and are traditionally built on timber formwork. Adjustable reusable metal formwork systems are also available. Some arches use special tapered units called voussoirs but large radius or shallow arches may be built with standard units and tapered joints. *Figure 34.1* illustrates the main elements of arches and their associated vocabulary.

Reinforced and post-tensioned masonry are used to a limited extent in the UK, mainly for civil engineering structures, high single-storey halls, retaining walls and lintels within walls. Masonry lintels may sometimes be constructed by laying special bed-joint reinforcement in the mortar. This acts as tensile reinforcement for a masonry beam. Most other reinforced masonry is formed by building masonry boxes in the form of hollow piers, walls with cavities

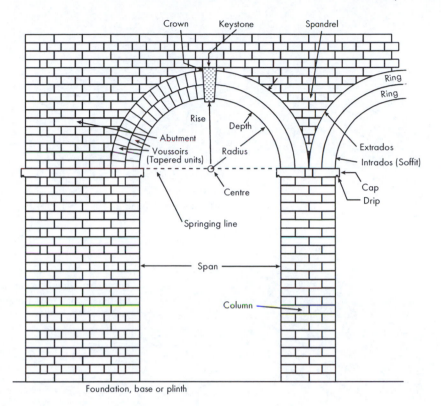

Fig. 34.1 *Structural elements and terminology of arches.*

or walls with slots in them, and then locking the reinforcing elements into the voids using a concrete grout. Post-tensioned masonry may be built in the same way but the reinforcement is then passed through the cavities and stressed against end plates, which removes any necessity to fill with grout. *Figure 34.2* shows some typical reinforced masonry forms.

34.2 Bond patterns

Most modern masonry is the thickness of a single unit breadth and is built by overlapping half the length with the next unit. This is known as *stretcher bond* or *half bond* and is shown in *Fig. 34.3*. Variations of stretcher bond may be achieved by using third or quarter bond (also shown). Soldier courses, where all the units stand on their ends, may be incorporated as a decorative feature but reduce the strength and robustness of the masonry. Much of this stretcher bonded work is used as cladding to frame structures where the strength is less important because of the presence of the supporting structure. In occupied structures it is widely used in the form

of the cavity wall, as illustrated in *Fig. 34.4*, which comprises two such walls joined with flexible metal ties across a space that serves principally to keep out rain and keep the inner wall dry. Blockwork is almost universally built with this bond, and broader units are used to achieve thicker walls. Stretcher-bonded walls may be built thicker by linking two or more layers with strong metal ties and filling the vertical 'collar' joint with mortar, as shown by *Fig. 34.5*.

In thicker walls built in multiples of a single unit breadth there are a large number of possible two-dimensional bonding patterns available, known by their traditional names. A few of the widely practised bonds are shown in *Fig. 34.6*. More are given in BS 5628 Part 3(2005), BS EN 1996-1-2 (2006) and Handisyde and Haseltine (1980).

34.3 Use of specials

It has always been possible to make structures more interesting by using specially shaped units to vary angles from 90° to generate tapers, plinths, curves,

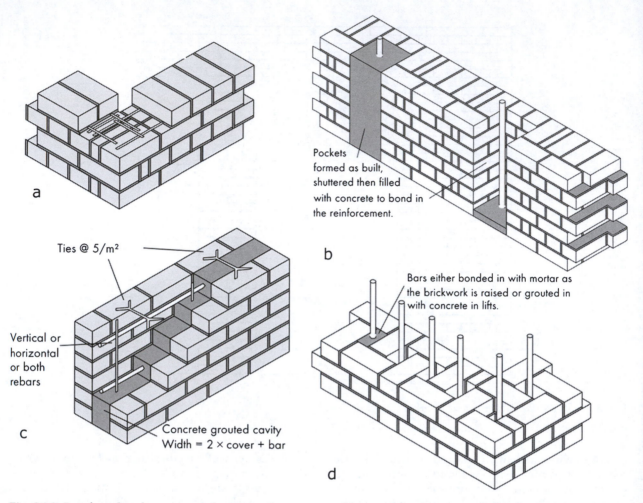

Fig. 34.2 Reinforced and post-stressed masonry forms. (a) Bed-joint reinforced wall; (b) Reinforced pocket wall; (c) Grouted cavity wall. (d) Quetta bond wall.

etc. In recent years such features have, if anything, become more popular. A very large range of shapes is available on a regular basis called 'standard specials'. Additionally, it is possible to get almost any shape manufactured to order, although it is inevitably quite expensive for small quantities. As an alternative, some specials can be made by gluing cut pieces of standard bricks with high performance adhesives. Some of the typical varieties are illustrated in *Fig. 34.7*.

34.4 Joint-style

It is often not realised how much the joint colour and shape influence the appearance and performance

of masonry. Obviously the colour contrast between the mortar and the units must have a profound effect on the appearance but so does the shape of the finished joint. The common joint styles are shown in *Fig. 34.8*. Recessed and weathered joints cast shadows and increase the contrast between mortars and light-coloured bricks in most lighting conditions.

34.5 Workmanship and accuracy

Standards of good workmanship in terms of how to lay out work and avoid weather problems by protection of new work against rain, wind and frost are covered in BS EN 1996-2 (2005) and BS 5628-3

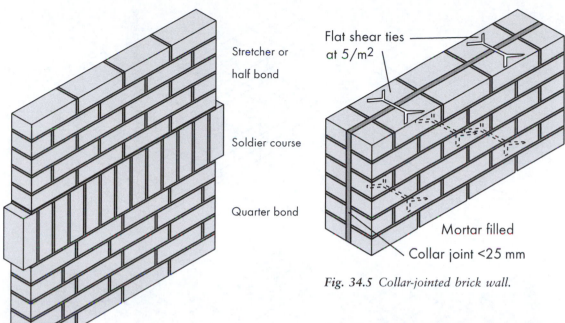

Stretcher or
half bond

Soldier course

Quarter bond

Fig. 34.3 *Half-brick bonds.*

Flat shear ties
at 5/m²

Mortar filled

Collar joint <25 mm

Fig. 34.5 *Collar-jointed brick wall.*

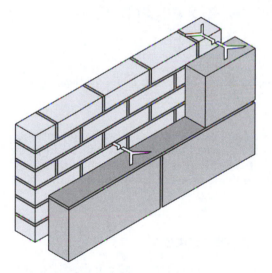

Fig. 34.4 *Typical block / brick cavity wall.*

(2005). Realistic tolerances for position on plan, straightness, level, height and plumb are given in BS EN1996-2 and in BS8000: Part 3 (2005), both based on the principles of BS5606 (1990).

34.6 Buildability, site efficiency and productivity

The process of constructing masonry has traditionally been regarded as difficult to mechanise and 'bring into the 21st century'. This is partly because it is a skilled craft and often has to be adapted to compensate for inaccuracies in other components. There is also a tendency to a high wastage rate because of the use of mortar that has a limited life and because of poor handling and storage conditions on site and loss and damage of materials between stores and the workpoint. Many of these problems have been reduced by innovations such as shrink wrapping of materials, crane delivery of packs direct to the work points and the use of retarded ready-to-use mortar with a long shelf life.

Despite many attempts it has not yet been possible to economically automate the site construction of masonry. Bricklaying machines have been developed but are only suitable for building the simplest of walls. For a brief period in the 1960/70s prefabricated brickwork cladding panels became popular in the USA and some degree of mechanised construction was feasible for relatively simple factory-built panels. Recent continental innovations are the use of very large but precisely sized masonry units, with cranes to lift them if necessary, and the use of thin-joint mortar to bond them. The larger size and simple technique necessary can improve productivity and speed to similar levels to that of timber frame for building the structural core of small to medium-sized

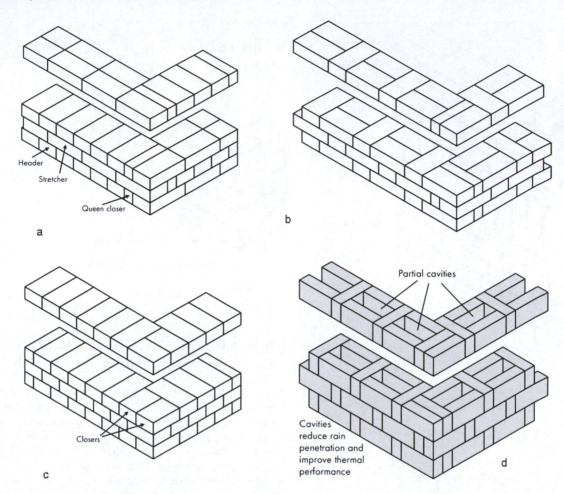

Fig. 34.6 *Common bonded wall types thicker than half brick. (a) English bond; (b) Flemish bond; (c) Heading bond; (d) 'Rat trap' bond.*

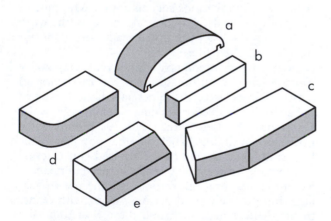

Fig. 34.7 *Examples of standard special bricks: (a) coping; (b) queen closer; (c) external angle; (d) single bull nose; (e) plinth.*

buildings. They can then be clad with conventional brickwork or other finishes. Since the cladding is off the critical path it is possible to simultaneously finish the outside and the inside in the same way as framed buildings.

34.7 Appearance

This is very much a matter of taste and expectation but there are some general rules to follow. Precisely shaped bricks with sharp arrises demand accurate layout with perpends lined up vertically and evenly sized mortar joints throughout, otherwise they tend to look untidy. Less accurate uneven bricks will tolerate some variation in joint size and position without looking ugly. Except with very accurate bricks, walls can only be made fair faced on one

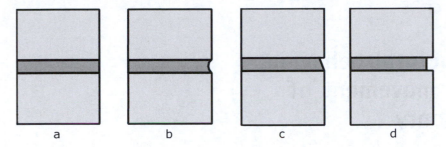

Fig. 34.8 *Joint styles; (a) Struck flush or wiped; (b) bucket handle or ironed; (c) weathered; (d) recessed.*

side while the other side has to suffer from any variability in thickness or length of the units. If a solid 220 mm thick wall is required to have two fair faces it should be built as a collar jointed wall (*Fig. 34.5*). If an internal half-brick wall is exposed on both faces it is probably best to use a recessed joint that is tolerant of some inaccuracy.

Exposed external walls should be protected as much as possible from run-off of rain. Any detail which causes large amounts of rain water to course down the wall in one spot or leach out through mortar joints at damp-proof membranes will eventually lead to discoloration due to staining by biological growths, e.g. lichen, lime or efflorescence.

References

(British standards and BRE series are listed in 'Further reading' at the end of Part 5.)

Handisyde CC and Haseltine BA (1980). *Bricks and Brickwork*, Brick Development Association.

Chapter 35

Structural behaviour and movement of masonry

Like any structural material, masonry must resist loads or forces due to a variety of external influences (or actions) and in various planes. *Figure 35.1* illustrates the various forces that can arise and the likely actions. In this chapter we will discuss the behaviour of masonry under all these types of action, an understanding of which is an essential prerequisite for successful structural design.

35.1 General considerations

Like plain concrete, unreinforced masonry is good at resisting compression forces, moderate to bad

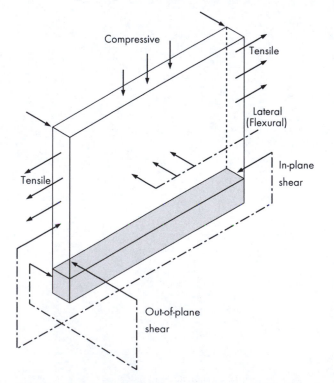

Fig. 35.1 *Forces on walls (from BRE Digest 246, 1981).*

at resisting shear and bending but very poor when subjected to direct tension. Masonry structures that are required to resist significant tensile forces should be reinforced by adding steel or other tension components. Unlike concrete, however, masonry is highly anisotropic because of its layer structure and this must always be borne in mind in design.

Masonry is quite effective at resisting bending forces when spanning horizontally between vertical supports but it is somewhat less effective at resisting bending forces when spanning vertically or cantilevering from a support (*Fig. 35.2*) because the resistance of a lightly loaded wall in that direction is dependent solely on the mass and the adhesion of the units to the mortar. Much of the resistance to bending and collapse, especially of simple cantilever masonry structures, is simply due to self-weight. Masonry is a heavy material, usually with a density in the range 500–2500 kg/m³, i.e. between 0.5 and 2.5 tonnes per cubic metre. In relatively squat structures such as some chimneys, parapets or low or thick boundary walls the force needed to overcome gravity to rotate the wall to a point of instability is sufficient to resist normal wind forces and minor impacts. Any masonry under compressive stress also resists bending since the compressive pre-stress in the wall must be overcome before any tensile strain can occur.

There is much empirical knowledge about how masonry works, and many small structures are still designed using experience-based rules. The Limit State Codes of Practice in the UK – BS EN 1996-1-1 (2005) supported by BMS (1997) or BS 5628: Part1 (2005) give a calculated design procedure, but much of this code is based on empirical data such as given by Davey (1952), Simms (1965), de Vekey *et al.* (1986, 1990) and West *et al.* (1968, 1977, 1986). Broadly, it predicts the characteristic strength of masonry elements, such as walls, from data on the characteristic strength or other characteristics of the materials using various engineering models for the different loading conditions. A check is then made

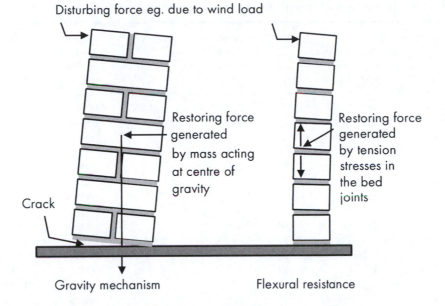

Disturbing force eg. due to wind load

Restoring force generated by mass acting at centre of gravity

Restoring force generated by tension stresses in the bed joints

Crack

Gravity mechanism

Flexural resistance

Fig. 35.2 Mechanisms for resisting bending forces in cantilevered masonry walls.

that the predicted strength is greater than the expected worse loading based on data about wind, dead, snow and occupancy loads. To allow for statistical uncertainty in loading data a safety factor, γ_f, is applied, which increases the design load level, and to allow for uncertainty about the strength of the masonry a further factor, γ_m, which reduces the design strength value for the masonry, is used. The combination of these partial factors of safety (FOS) gives an overall safety factor against failure of the structure that is usually in the range 3.5–5. This relatively high FOS is used because of the high variability in the properties of masonry and its brittle failure mode, which gives very little warning of failure. An alternative simplified design procedure is also available for smaller buildings such as houses (BS EN 1996-3, 2006).

35.2 Compressive loading

35.2.1 AXIAL LOADS

Masonry is most effective when resisting axial compressive loads normal to the bedding plane (*Fig. 35.1*). This is, clearly, the way in which most load-bearing walls function but also the way that arches and tunnels resist load since an inward force on the outside surface of a curved plate structure such as a tunnel will tend to put the material into radial compression, as in *Fig. 35.3*. If a load or force is put on a wall at a point, it would logically spread outward from the point of application in a stretcher-bonded wall since each unit is supported by the two units below it. This mechanism, shown diagrammatically in *Fig. 35.4* by representing the

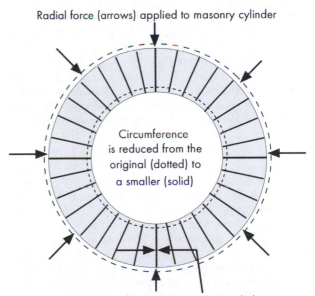

Radial force (arrows) applied to masonry cylinder

Circumference is reduced from the original (dotted) to a smaller (solid)

The smaller circumference generates a radial strain (contraction in thickness of units and mortar joints), which is converted elastically to a radial stress.

Fig. 35.3 Forces on curved masonry (e.g. arches, tunnels, sewers).

Table 35.1 Masonry prism strengths with bricks loaded in various directions

Type	Ref.*	Mortar Mix C:L:S	MPa	Prism compressive strength on-bed MPa	Proportion of the on-bed strength On bed	On edge	On end
23 hole	(a)	1:¼:3	30.1	22.4	100%	71%	48%
14-hole	(b)	1:¼:3	33.6	28.9	100%	29%	51%
14 hole	(c)	1:¼:3	30.1	19.9	100%	58%	57%
10 hole	(d)	1:½:4½	8.3	15.2	100%	104%	111%
10 hole	(d)	1:¼:3	33.6	22	100%	68%	91%
3 hole	(e)	1:¼:3	33.6	37.6	100%	81%	58%
3 hole	(e)	1:½:4½	7.8	20.3	100%	96%	83%
3 hole	(f)	1:¼:3	30.1	30.2	100%	61%	29%
5 cross slots	(g)	1:¼:3	33.6	34.1	100%	85%	41%
5 cross slots	(g)	1:½:4½	7.7	18	100%	132%	54%
16 hole	(A)	1:¼:3	–	26	100%	20%	29%
Frog	(B)	1:¼:3	–	9.7	100%	55%	55%
Frog	(C)	1:¼:3	–	10.8	100%	133%	122%
Frog	(D)	1:¼:3	–	19.2	100%	93%	86%
Solid	(E)	1:¼:3	–	16	100%	73%	64%

*The reference letter links this to the data in *Table 33.5*.

magnitude of the force by the width of the arrows, leads to some stress being spread at 45° in a half-bond wall, but the stress still remains higher near the axis of the load for a height of 2 m or more. Such a compressive force causes elastic shortening of the masonry. As a result of Poisson's ratio effects, a tension strain and hence a stress is generated normal to the applied stress. In bonded masonry the overlapping of the units inhibits the growth of cracks, which are generated in the vertical joints by tension, until the stress exceeds the tensile resistance of the units.

The compressive strength of masonry is measured by subjecting small walls or prisms or larger walls of storey height (2–3 m) to a force in a compression test machine. Loading is usually axial but may be made eccentric by offsetting the wall and loading only part of the thickness.

Masonry is not so good at resisting compression forces parallel to the bedding plane because there is no overlapping and the bed joints fail easily under the resultant tensile forces. Additionally, most bricks with frogs, perforations or slots are usually weaker when loaded on end than on bed because the area of material resisting the load is reduced and the stress distribution is distorted by the perforations. Data for bricks are given in *Table 33.5*. The equivalent data in *Table 35.1* for prisms show

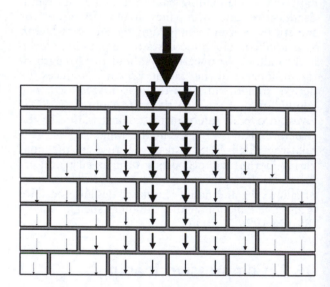

Fig. 35.4 Load spreading in stretcher-bonded walls.

that the strength does vary with loading direction, although not to the same extent as for units, because the aspect ratios of the masonry specimens are all similar.

The axial strength of masonry might be expected to depend on the strength of the units and of the mortar and, to a first approximation, the contribution

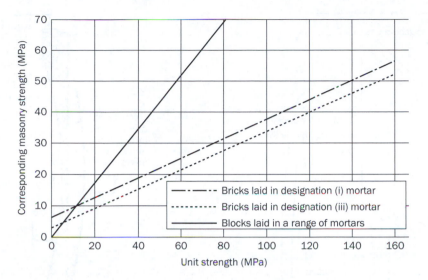

Fig. 35.5 *Influence of mortar and units on compressive strength of masonry (after de Vekey et al., 1990; West et al., 1990).*

of each to the overall strength should be related in some way to the volume proportion of each. This gives a reasonable model for the behaviour of squat structures. A complication is that most masonry comprises units that are much stronger than the mortar and the three-dimensional confining restraint increases the effective strength of the thin mortar beds. *Figure 35.5*, based on data for wallettes tested in compression, shows the minor effect of mortar strength (i) and (iii) on the compressive strength of masonry made with a range of strengths of bricks, but the much greater influence of the brick strength is clear. The third line shows the different behaviour of blocks tested to BS EN 772-1.

As we discussed in Chapter 2, section 2.6.2, platen friction in compression tests results in restraint of the material nearest to the platen and therefore in the standard test method inhibits tension failure in the zones shown shaded in *Fig. 35.6*, thus enhancing the measured strength of squat units (units wider than they are high) more than slender units. This anomaly is taken account of in BS EN 1996-1-1 by correcting all strengths measured using units of varying shape to a value representing the strength of a specimen with a square cross section (with a height:breadth ratio of 1) using a conversion table based on test data. The process, called normalisation, also corrects to a standard air-dry state whatever the test condition. The former British Code, BS5628: Part 1, uses a different table for each unit shape.

To calculate masonry strength, BS EN1996-1-1 uses an equation of the form:

$$f_k = K.f_b{}^a.f_m{}^b \qquad (35.1)$$

where f_k, f_b and f_m are the strength of the masonry, the normalised strength of the units and the strength of the mortar, respectively, K is a factor that may vary to take into account the shape or type of the units, a is a fractional power of the order of 0.7–0.8 and b is a fractional power of the order 0.2–0.3.

35.2.2 STABILITY: SLENDER STRUCTURES AND ECCENTRICITY

If a structure in the form of a wall or column is squat, so that the ratio of height to thickness (slenderness ratio) is small, then the strength will depend largely on the strength of the constituent materials. In real structures the material will be stiffer on one side than the other, the load will not be central and other out-of-plane forces may occur. This means that if the slenderness ratio is increased, at some point the failure mechanism will become one of instability and the structure will buckle and not crush.

Loads on walls, typically from floors and roofs, are commonly from one side and thus stress the wall eccentrically. *Figure 35.7* illustrates, in an exaggerated form, the effect of an eccentric load in reducing the effective cross-section bearing the load and putting part of the wall into tension. This is recognised in practice and usually a formula is used to calculate a reduction factor for the design load capacity, which is a function of the slenderness ratio and the net eccentricity of all the applied loads. BS EN 1996-1-1 gives a formula for the reduction

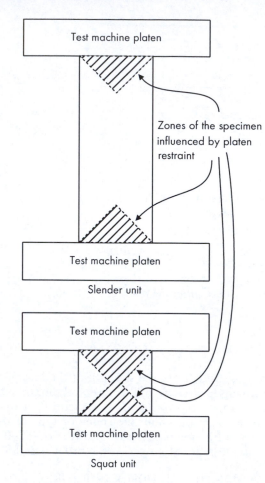

Fig. 35.6 Platen restraint (shown as shaded area), which enhances the measured compressive resistance of squat units more than slender units.

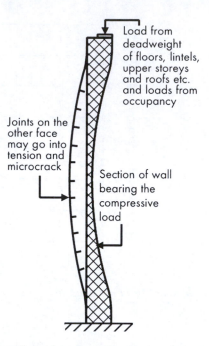

Fig. 35.7 Effect of eccentric load on masonry (exaggerated).

factor, ϕ, which varies between the bottom, middle and top of the wall.

35.2.3 CONCENTRATED LOAD

Many loads are fairly uniform in nature, being derived from the weight of the super-structure or more locally on floors. There will always be, however, some point loads, termed concentrated loads, in structures at the ends of beams, lintels, arches, etc. In general the masonry must be capable of withstanding the local stresses resulting from the concentrated loads, but the designer may assume that the load will spread out in the manner of *Fig. 35.4* so that it is only critical in the first metre or so below the load. Additionally, because the area loaded is restrained by adjacent unloaded areas, some local enhancement can be assumed.

Figure 35.8a shows the condition for a small isolated load applied via a pad where there is restraint from four sides and *Fig. 35.8b, c* and *d* show further conditions with reducing restraint. The local load capacity of the masonry in the patches compared to the average load capacity (the enhancement factor) can vary from 0.8 for case (*d*) to as much as 4 for case (*a*) (Ali and Page, 1986; Arora, 1988; Malek and Hendry, 1988). The enhancement factor decreases as the ratio of the area of the load to the area of the wall increases, as the load moves nearer the end of a wall and as the load becomes more eccentric. Formulae describing this behaviour are proposed in many of the references at the end of this chapter, and are codified in BS EN 1996-1-1 Clause 6.1.3.

35.2.4 CAVITY WALLS IN COMPRESSION

If the two leaves of a cavity wall share a compression load equally then the combined resistance is the sum of the two resistances provided that their elastic moduli are approximately the same. If one leaf is very much less stiff than the other, the stress is all likely to predominate in the stiffer wall and be applied eccentrically and then the combined wall may have less capacity than the single stiff wall loaded axially. It is common practice to put all the load on the inner leaf and use the outer leaf as a rain screen. In this load condition more stress is allowed on the loaded

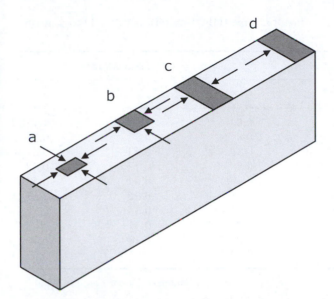

Fig. 35.8 *Concentrated loads on masonry (restraint indicated by arrows).*

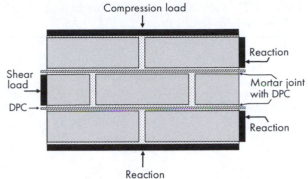

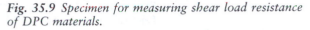

Fig. 35.9 *Specimen for measuring shear load resistance of DPC materials.*

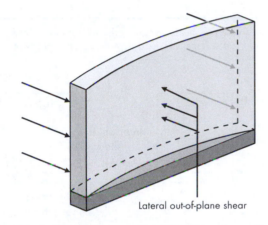

Fig. 35.10 *Shear failure on the DPC line of a wall laterally loaded out of plane (e.g. by wind).*

35.3 Shear load in the bed plane

If a wall is loaded by out-of-plane forces, e.g. wind, impact, or seismic (earthquake) action, the force will act to try to slide the wall sideways (like a piston in a tube). In practice the action can be at any angle in the 360° plane, although it is most commonly parallel or normal to the wall face. This is a very complex loading condition and the result is rarely a pure shearing failure. For small test pieces measured in an idealised test, the shear strength f_v can be shown to follow a friction law with a static threshold 'adhesion' f_{v0} and a dynamic friction term K dependent on the force normal to the shearing plane, s_a. The formula is simply:

$$f_v = f_{v0} + s_a.K \qquad (35.2)$$

Measurement of pure shear is very difficult because of the tendency to induce rotation in virtually any physical test arrangement. The simple double-shear test of the type shown in *Fig. 35.9* is suitable for measuring shear resistance of damp-proof course (DPC) materials as described in EN 1052-4 (1999) but is unsatisfactory for mortar joints, where a much shorter specimen is preferred of the type described in BS EN 1052-3. *Table 35.2* gives some typical shear data.

In the example sketched in *Fig. 35.10* the ends are supported so the wall tends to adopt a curved shape. In this case it is shown as failing by shear at the line of the DPC. If a wall is loaded by lateral forces acting on the end as in *Fig. 35.11* (which can arise from wind load on a flank wall at right angles) the force will initially tend to distort the wall to a parallelogram shape as in the top right hand of the diagram and induce compression forces where shaded. Failure if it occurs can be by a number of alternative mechanisms, as shown and listed in *Fig. 35.11*.

35.4 Flexure (bending)

Traditionally, masonry was made massive or made into shapes that resisted compression forces. Such structures do not depend to any great extent on the bond of mortar to units. Much of the masonry built

wall because it is propped (its buckling resistance is increased) by the outer leaf.

Table 35.2 Shear data for two mortars and some damp proof course (DPC) materials (after Hodgkinson and West, 1981)

Brick	DPC	1:¼:3 Mortar		1:1:6 Mortar	
		s_a	K	s_a	K
16 hole w/c	Blue brick*	0.72	0.82	0.44	1.14
	Bituminous	0.4	0.8	0.31	0.91
	Permagrip†	0.8	0.58	0.43	0.61
	Hyload‡	0.21	0.8	0.17	0.58
	Vulcathene§	0.06	0.47	0.04	0.54
Frogged semi-dry pressed	Blue brick*	0.73	0.95	0.36	1.57
	Bituminous	0.45	0.84	0.36	0.72
	Permagrip†	0.55	0.93	0.41	0.72
	Hyload‡	0.17	0.82	0.18	0.79
	Vulcathene§	0.09	0.52	0.06	0.49

*For this case the shear strength is that of the mortar beds. †Permagrip is a trade name for a reinforced bitumen product with a coarse sand surface. ‡Hyload is probably pitch/polymer. §Vulcathene is polyethythene based.

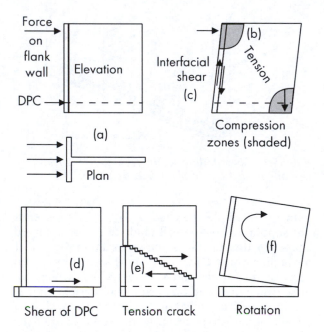

Fig. 35.11 *Stresses and strains resulting from in-plane shear forces. The wall may fail by: (top right) crushing in zones of high compressive stress or by shear at the vertical junction between the two walls at right angles; (bottom left) horizontal shear at the DPC line; (bottom middle) by tension splitting; or (bottom right) by rotation.*

in the last few decades has, however, been in the form of thin walls, for which the critical load condition can result from lateral forces, e.g. wind loads. This phenomenon was largely made possible by the use of ordinary Portland cement (OPC) mortars, which give a positive bond to most units and allow the masonry to behave as an elastic plate. There are two distinct principal modes of flexure about the two main orthogonal planes:

1. The vertically spanning direction shown in *Fig. 35.2*, which is commonly termed the parallel (or p) direction because the stress is applied about the plane parallel to the bed joints.
2. The horizontally spanning direction, shown in *Fig. 35.12*, which is commonly termed the normal (or n) direction because the stress is applied about the plane normal to the bed joints.

Clearly the strength is likely to be highly anisotropic since the stress in the parallel direction is only resisted by the adhesion of the units to the mortar while the stress in the normal direction is resisted by:

- the shear resistance of the mortar beds (a)
- the elastic resistance of the head joints to the rotation of the units (b)
- the adhesion of the head joints (c)
- the flexural resistance of the units themselves (d).

Generally the limiting flexural resistance will be the lesser of (a)+(b) or (c)+(d), giving two main modes

Table 35.3 Flexural strength ranges (MPa) for common masonry units bedded in designation (iii) mortar

Material	Normal (strong) direction	Parallel (weak) direction
Clay brick (0–7% water absorption)	1.8–4.7	0.35–1.1
Clay brick (7–12% water absorption)	1.9–3.2	0.3–1.3
Clay brick (> 12% water absorption)	1.0–2.1	0.3–0.8
Concrete brick (25–40 N/mm² strength)	1.9–2.4	0.5–0.9
Calcil brick (25–40 N/mm² strength)	0.7–1.5	0.05–0.4
Aircrete (AAC) block (100 mm thick)	0.3–0.7	0.3–0.6
Lightweight aggregate concrete block (100 mm thick)	0.7–1.3	0.3–0.5
Dense aggregate concrete block (100 mm thick)	0.7–1.7	0.2–0.7

There is some variation of flexural strength with mortar strength and thickness of blocks.

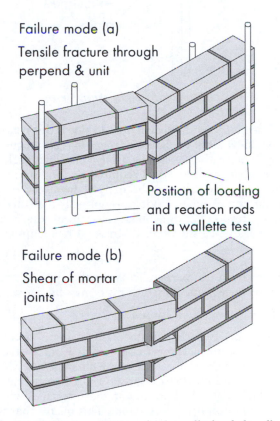

Fig. 35.12 *Modes of failure for laterally loaded walls in the strong (n) direction.*

of horizontal spanning failure – shearing, shown in the lower part of *Fig. 35.12* and snapping, shown in the upper part of *Fig. 35.12*.

Using small walls (wallettes), either as shown in *Fig. 35.12* for measuring horizontal bending or 10

courses high by 2 bricks wide for vertical bending, and tested in four-point bending mode, the flexural strength of a large range of combinations of UK units and mortars has been measured for the two orthogonal directions; typical ranges are given in *Table 35.3*, with further data in Hodgkinson *et al.* (1979), West *et al.* (1986), de Vekey *et al.* (1986) and de Vekey *et al.* (1990).

The ratio of the strength in the two directions, expressed as p-direction divided by n-direction, is termed the orthogonal ratio and given the symbol μ. In cases where only the bond strength (p-direction) is required a simpler and cheaper test is the bond wrench (discussed in section 30.3.2).

The flexural strengths given in *Table 35.3* are for simply supported pieces of masonry spanning in one direction. If the fixity of the masonry at the supports (the resistance to rotation) is increased the load resistance will increase in accordance with normal structural principles (*Fig. 35.13*). Again, if one area of masonry spans in two directions the resistances in the two directions are additive. Seward (1982), Haseltine *et al.* (1977), Lovegrove (1988), Sinha (1978) and Sinha *et al.* (1997) cover some aspects of the resistance of panels. Further, in cases where the edge supports for a masonry panel allow no outward movement arching occurs and may be the dominant flexural resistance mechanism for thicker walls.

35.5 Tension

Masonry made with conventional mortars has a very limited resistance to pure tension forces and, for the purposes of design, the tensile strength is usually

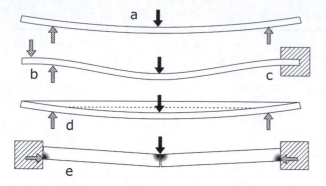

Fig. 35.13 *Effect of edge support conditions on flexural behaviour of masonry: (a) wall with simple edge support; (b) wall with fixed edge support; (c) wall with fixed edge by encastre condition; (d) wall spanning both horizontally and vertically sectioned at half height; and (e) wall arching horizontally between two rigid end buttresses after initial crack.*

taken to be zero. In practice it does have some resistance in the horizontal direction and somewhat less in the vertical direction. In an attempt to make a viable prefabricated masonry panel product for use as a cladding material, polymer latex additives can be used to improve the tensile strength. Panels of storey height and a metre or more wide have been manufactured and could be lifted and transported without failure.

The horizontal tensile strength of masonry has been measured but there is no standard test and virtually no data has been published in the public domain. Tensile bond strength is usually measured using a simple two-brick prism test, as illustrated in *Fig. 35.14*. Data from such tests indicate that the direct tensile strength across the bond is between one-third and two-thirds of the parallel flexural strength (see *Table 35.3*). Other tests have been developed along similar lines including one in which one unit is held in a centrifuge and the bond to another unit is stressed by centrifuging. A useful review is give by Jukes and Riddington (1998).

35.6 Elastic modulus

The stiffness or elastic modulus of masonry is an important parameter required for calculations of stresses resulting from strains arising from loads, concentrated loads, constrained movement and also for calculating the required area of reinforcing and post-stressing bars.

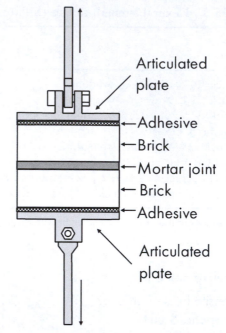

Fig. 35.14 *Stack-bonded couplet tensile test.*

The most commonly measured value is the Young's modulus (E), but the Poisson's ratio (v) is also required for theoretical calculations using techniques such as finite element analysis. If required the bulk (K) and shear (G) moduli may be derived from the other parameters. Young's modulus is normally measured in a compression test by simultaneously measuring strain (ε_p) parallel to the applied stress (s) whereupon:

$$E = s/\varepsilon_p \qquad (35.3)$$

If the strain (ε_n) perpendicular to the applied stress is also measured, Poisson's ratio may be derived:

$$v = \varepsilon_n/\varepsilon_p \qquad (35.4)$$

Masonry is not an ideally elastic material because it is full of minor imperfections such as microcracks in the bond layers and because the differences between the unit and mortar stiffnesses and Poisson's ratios produce high local strains at the interface, which results in non-linear behaviour. This means that the stress–strain curve is typically of a broadly parabolic form with an early elastic region, i.e. it is similar to that of concrete (Chapter 20) for similar reasons. An instantaneous value of E can be obtained from the tangent to the curve at any point but for some calculations, such as creep loss of post-stressed masonry, the effective E derived from the secant value is required. *Figure 35.15* illustrates this behaviour. Data on elastic properties in compression are given in

Table 35.4   Movement data and elastic modulus of masonry materials

Masonry component material	Thermal expansion coefficient (per C deg × 10⁻⁶)	Reversible moisture movement (± mm/m)	Irreversible moisture movement (mm/m)	Modulus of elasticity, E (GPa)
Granite	8–10	–	–	20–60
Limestone	3–4	0.1	–	10–80
Marble	4–6	–	–	35
Sandstone	7–12	0.7	–	3–80
Slate	9–11	–	–	10–35
Mortar	10–13	0.2–0.6	−0.4 to −1	20–35
Dense concrete	6–12	0.2–0.4	−0.2 to −0.6	10–25
LWAC blockwork	8–12	0.3–0.6	−0.2 to −0.6	4–16
AAC blockwork	8	0.2–0.3	−0.5 to −0.9	3–8
Calcil brickwork	8–14	0.1–0.5	−0.1 to −0.4	14–18
Clay brickwork	5–8	0.2	−0.02 to +0.10	4–26

LWAC, lightweight aggregate concrete.

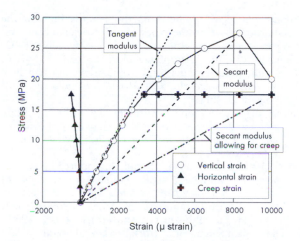

Fig. 35.15 Stress–strain behaviour of masonry.

Davies and Hodgkinson (1988). The elastic modulus is also important in estimating the deflections of walls out of plane due to lateral loads such as wind. In this case the modulus can be measured by using load deflection data for small walls tested in four-point bending, with E given by:

$$E = 8Wa(3L^2 - 4a^2)/384Id \qquad (35.5)$$

where W is the applied force, L is the support span, a is the distance from the supports to the loading points, I is the moment of inertia and d is the deflection.

In compression tests the value of E has generally been found to be in the range 500–1000 times the compressive strength. For typical materials this is likely to be around 2–30 GPa. In flexure the early tangent modulus has been found to be in the range 2–4 GPa for tests in the strong (normal) direction and 1–2 GPa for equivalent tests in the weak (parallel) direction. Some more data are given in *Table 35.4*.

35.7 Movement and creep

Unrestrained masonry is subject to cyclic movement due to moisture and temperature changes, irreversible creep due to dead loads or post stress loads, irreversible shrinkage due to drying/carbonation of mortar, concrete and calcium silicate materials and expansion due to adsorption of moisture by fired clay materials. *Table 35.4* contains some typical ranges for common masonry components derived mainly from Digest 228 (1979).

In simple load-bearing masonry elements vertical movements are accommodated by the structure going up and down as required and are no problem. Problems can arise, however, where materials with different movement characteristics are joined or where thick elements have differential temperature/moisture gradients through them. Elements with a vertical pre-stress much larger than the differential stresses will tolerate the movement, but lightly stressed and unrestrained elements will need a slip plane or soft joint between the elements to avoid problems. A classic case is (expanding) clay masonry on (shrinking) concrete frames where soft joints are used to stop stress transfer to the masonry.

Where restraint is present horizontal shrinkage will be converted into tensile forces and expansion into compressive forces. Since walls are probably two orders of magnitude weaker in tension than in compression the result is that restrained walls in tension usually crack while those in compression usually just build up stress. Where walls are unrestrained the reverse is usually the case: the shrinking wall simply contracts, but the expanding wall interacts with flank walls (those at right angles) and causes cracking by generating rotation at the corner. Design strategies to cope with movement and avoid cracking are contained in BS EN 1996-2 and BS 5628-3. Because there is nearly always differential restraint the pure tensile or compressive forces will also usually give rise to some associated shear forces.

References

(British standards and BRE series are listed in 'Further reading' at the end of Part 5.)

Ali S and Page AW (1986). An elastic analysis of concentrated loads on brickwork. *International Journal of Masonry Construction*, No. 6, Edinburgh, pp. 9–21.

Arora SK (1988). *Review of BRE research into performance of masonry walls under concentrated load*. Proceedings of the 8th International Brick/Block Masonry Conference, Dublin, p. 446.

British Masonry Society (1997). *Eurocode for masonry, ENV 1996-1-1: Guidance and worked examples*. Special Publication No. 1, 1997. British Masonry Society, Stoke-on-Trent.

Davey N (1952). Modern research on loadbearing brickwork. *The Brick Bulletin*, 1–16.

Davies S and Hodgkinson HR (1988). *The stress–strain relationships of brickwork, Part 2*. BCRL research paper 7S5.

de Vekey RC, Bright NJ, Luckin KR and Arora SK (1986). Research results on autoclaved aerated concrete blockwork. *The Structural Engineer*, **64A** (No. 11), 332–341.

de Vekey RC, Edgell GJ and Dukes R (1990). The effect of sand grading on the performance and properties of masonry. *Proceedings of the British Masonry Society*, No. 4, 152–159.

Haseltine BA, West HWH and Tutt JN (1977). Design of walls to resist lateral loads. *The Structural Engineer*, 55 (No. 10), 422–430.

Hendry AW, Sinha BB and Davies SR (1996). *An Introduction to Load Bearing Brickwork Design*, 3rd edition, Taylor and Francis.

Hodgkinson HR and West HWH (1981). *The shear resistance of some damp-proof course materials*. Technical Note 326, British Ceramic Research Ltd, Stoke-on-Trent.

Jukes P and Riddington JR (1998). A review of masonry tensile bond strength test methods. *Masonry International*, **12** (No. 2), 51–57.

Lovegrove R (1988). The effect of thickness and bond pattern upon the lateral strength of brickwork. *Proceedings of the British Masonry Society*, No. 2, 95–97.

Malek MH and Hendry AW (1988). Compressive strength of brickwork masonry under concentrated loading. *Proceedings of the British Masonry Society*, No. 2, 56–60.

Seward DW (1982). A developed elastic analysis of lightly loaded brickwork walls with lateral loading. *International Journal of Masonry Construction*, **2** (No. 2), 129–134.

Simms LG (1965). The strength of walls built in the laboratory with some types of clay bricks and blocks. *Proceedings of the British Ceramic Society*, July, 81–92.

Sinha BP (1978). A simplified ultimate load analysis of laterally loaded model orthotropic brickwork panels of low tensile strength. *The Structural Engineer*, **56B** (No. 4), 81–84.

Sinha BP, Ng CL and Pedreschi RF (1997). Failure criteria and behaviour of brickwork in biaxial bending. *Journal of Materials in Civil Engineering*, **9** (No. 2), 61–66 May.

West HWH, Hodgkinson HR and Davenport STE (1968). *The performance of walls built of wirecut bricks with and without perforations*. Special Publication No. 60, British Ceramic Research Ltd, Stoke-on-Trent.

West HWH, Hodgkinson HR and Haseltine BA (1977). The resistance of brickwork to lateral loading. *The Structural Engineer*, 55 (No. 10), 411–421.

West HWH, Hodgkinson HR, Goodwin JF and Haseltine BA (1979). *The resistance to lateral loads of walls built of calcium silicate bricks*. BCRL, Technical Note 288, BCRL, Penkhull, Stoke-on-Trent, UK.

West HWH, Hodgkinson HR, Haseltine BA and de Vekey RC (1986). Research results on brickwork and aggregate blockwork since 1977. *The Structural Engineer*, **64A** (No. 11), 320–331.

Chapter 36

Non–structural physical properties of masonry

The main non-structural function of masonry elements is as a cladding to buildings and the key function of such elements is to maintain habitable conditions within the building. It is therefore important to know how effective masonry wall systems are at preventing heat loss in winter and maintaining comfortable conditions in summer, preventing ingress of wind and rain, reducing noise transmission and limiting the spread of fire should it break out. Another key concern that is becoming increasingly important is the effect of a product on the environment – the sustainability or green credentials of a product. Increasingly, materials will be chosen not just for their inherent performance, appearance or economy but also taking into account their energy cost and other effects on the global environment such as emission of greenhouse gases and land destruction due to quarrying. This topic will be covered for all the materials discussed in this book in Chapter 62.

36.1 Thermal performance

The rate of heat flow through a given material is controlled by the thermal conductivity. As discussed in Chapter 7, metals generally have higher conductivities and very-low-density porous materials (containing a lot of air or other gas) have lower values. Masonry materials fall in a band between the two extremes. The property is important in that it affects the winter heat loss from a building through the walls and thus the energy efficiency of the structure. Surprisingly, although it is often of lower density and quite porous, normal bedding mortar is frequently a poorer insulator than many bricks and blocks, as shown by *Fig. 36.1*. This has led to the development of insulating mortars containing low-density aggregate particles and thin-joint mortar. The latter type reduces heat loss because of its much smaller area proportion of the wall face. Because porosity is a key parameter the thermal conductivity (k) is bound to be partially related to material density, and general equations for dry solid porous building materials tend to be a function of density (ρ) with regression equations of the form:

$$k = 0.0536 + 0.000213\rho - 0.0000002\rho^2 \quad (36.1)$$

The presence of moisture increases the conductivity of porous materials because evaporation and condensation heat-transfer mechanisms become possible.

Hollow and perforated products give some improvement over plain solid products, although the potential gain from the trapped air pockets is compromised by the convection of the air within them. Units with many small perforations perform better than hollow units because the smaller size of the holes reduces convection and the smaller solid cross-section reduces conduction. There is also an improvement if the holes are staggered such that the direct conduction

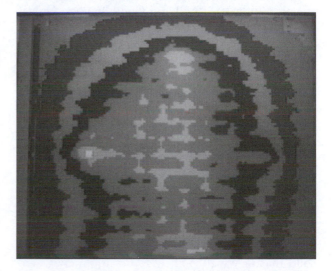

Fig. 36.1 Infra-red photograph showing greater heat loss (lighter colour) through mortar joints than through the bricks.

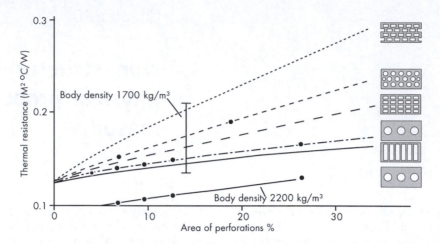

Fig. 36.2 *Effect of brick perforation pattern on thermal resistance of walls.*

path through the solid material is as long as possible. *Figure 36.2* illustrates the effects of different perforation patterns on thermal resistance. If convection is prevented by filling the hollows with foamed plastic materials such as urea–formaldehyde there is a further substantial improvement.

The properties of walls used as thermal barriers are normally expressed in the form of the 'U value', the overall heat transfer coefficient, which is a synthesis of the *k* value of the actual materials and the heat transfer coefficients at the hot and cold sides. More detail on thermal insulation is given in Diamant (1986), BRE Digest 273 (1983), BRE Digest 355 (1990), BR 262 (2002 edition), Good Repair Guide 26 (1999) and Good Building Guides 44-1 and 44-2.

36.2 Resistance to damp and rain penetration

From the earliest use of built housing one of the primary requirements has been that the walls will keep the occupants dry and thus most masonry forming the perimeter walls of houses and other buildings is called upon to resist the ingress of rain. All masonry component materials are porous, however, and there are always some imperfections in the bond and seal of the mortar joints (de Vekey *et al.*, 1997) and the workmanship (Newman and Whiteside, 1981; Newman, 1988) that will admit some water so no solid masonry wall is likely to be absolutely watertight (de Vekey *et al.*, 1997). Dampness rising through porous masonry with no damp proof course (DPC) has never been shown to occur under laboratory

conditions, and the current view is that cases diagnosed as such are really falling or horizontal damp from faults or retained soil (Howell, 2008).

Paradoxically, it is normally easier under UK conditions to make a rain-resistant wall from porous bricks. This is because some leakage always occurs at the joints, which is mopped up by high-absorption units but allowed free passage by low-absorption units. Provided the rain does not persist until the units are saturated, they can dry out in the intervening dry spell and do not actually leak significant amounts of moisture to the inner face. Under similar conditions, some modern, low-absorption facings may leak quite seriously during a moderate storm. As would be expected, resistance is greater for thicker walls or if a water-resistant coating is applied over the exterior. Typical coatings are renders and paints or overcladding systems such as vertical tiling.

The commonest technique for avoiding rain penetration in the UK is the cavity wall. This is a twin layer wall with an enclosed air space between the two leaves. Some leakage through the outer layer (leaf) is anticipated and any such water is directed back out through weepholes before it reaches the inner leaf by the use of damp-proof membranes and cavity trays. This is sometimes thought of as a very recent wall form but it was probably used in ancient Greece and has been in use in the damper parts of the UK for nearly 200 years. It has given remarkably good service and is quite tolerant of workmanship variations. The main problems have been leakage of rainwater, which affects a small percentage of cases, and the corrosion of the steel ties used to ensure shared structural action of the two leaves. Useful references are Newman *et al.* (1982a, 1982b),

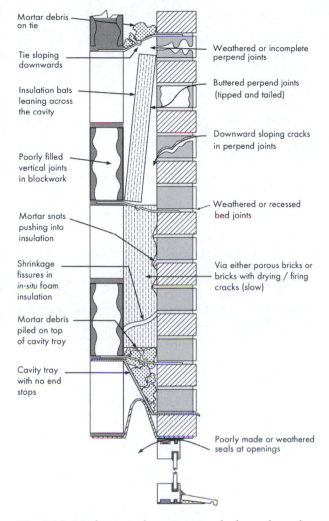

Labels on the figure:
- Mortar debris on tie
- Tie sloping downwards
- Insulation bats leaning across the cavity
- Poorly filled vertical joints in blockwork
- Mortar snots pushing into insulation
- Shrinkage fissures in *in-situ* foam insulation
- Mortar debris piled on top of cavity tray
- Cavity tray with no end stops
- Weathered or incomplete perpend joints
- Buttered perpend joints (tipped and tailed)
- Downward sloping cracks in perpend joints
- Weathered or recessed bed joints
- Via either porous bricks or bricks with drying / firing cracks (slow)
- Poorly made or weathered seals at openings

Fig. 36.3 Mechanisms for rain water leakage through the external leaf and for tracking across cavities.

BRE Digest 380 (1993), Good Repair Guide 5 (1997) and Good Building Guide 33. *Figure 36.3* illustrates some routes for moisture and dampness to penetrate firstly the outer leaf (or a solid masonry wall) and then to reach the inner leaf, usually because of bad design or workmanship in the construction of the cavity.

36.3 Moisture vapour permeability

Vapour permeability is important since vapour trapped within cold walls will form condensation on surfaces leading to mould and rot and causing health risks. Vapour absorbed within the wall will condense and cause damp walls. Permeable materials will allow the damp to escape provided there is ventilation to carry it away. It is measured in accordance with BS EN 772-15 (2000) for AAC or BS EN ISO 12572 (2001) for other materials.

36.4 Sound transmission

Sound transmission is another parameter that is very dependent on density, because it is the mass of the wall that is critical. Generally, the greater the mass, m, of a wall the more effective it is at attenuating (absorbing) the sound passing through it. A typical equation for the resistance, R, in decibels, dB, given in BRE Digest 273 (1983) is:

$$R = 14.3 \log m + 1.4 \, dB \qquad (36.2)$$

Any holes will allow the sound to bypass the wall so that wet plastered walls where any minor perforations or imperfections are repaired by the plaster layer are more effective than dry-lined walls (walls covered with plasterboard). There are additional techniques to try to cut out sound such as air gaps and cavities with fireproof blankets hung in them, which will damp out some frequencies. More details on principles and basic values are given in Diamant (1986), BRE Digest 337 (1988), and BRE Digest 338 (1989).

36.5 Fire resistance

Fire resistance of masonry is an important characteristic since it has long been recognised that it is a very effective material for resisting and preventing the spread of fire. In the UK this is now enshrined in various building regulations dating from the Great Fire of London. Masonry's effectiveness in this role is due to the following characteristics:

- a relatively low thermal conductivity, which prevents the fire spreading by inhibiting the rise in temperature of the side of a wall away from the fire
- a relatively high heat capacity, which also inhibits the rise in temperature of the side of a wall away from the fire (this is especially true of concrete-based products that contain a lot of loosely bound water, which absorbs heat while being boiled away)
- zero flammability and surface spread of flame
- refractory properties that mean that it retains its strength and integrity up to very high temperatures, approaching 1000°C in some cases.

289

These properties mean that it does not catch fire itself, it inhibits the spread of fire by conduction and radiation and it is not easily breached by the fire. Fire-resistant insulating finishes such as vermiculite plaster improve performance still further. It has been shown to resist fire for between half an hour and six hours depending on material, thickness and finishes. The classic data on masonry walls are contained in Davey and Ashton (1953). Relevant Codes of Practice are BS EN1996-1-2 (2005) and BS5628: Part 3: 2005 and BRE Digest 487-3 (2004).

References

(British standards and BRE series are listed in 'Further reading' at the end of Part 5.)

Davey N and Ashton LA (1953). *Investigation of building fires. Part V. Fire tests on structural elements*, National Building Studies No. 12, HMSO, London.

de Vekey RC (1993). Cavity walls – Still a good solution, BMS Proc. M(5), 35–38.

de Vekey RC, Russell AD, Skandamoorthy J and Ferguson A (1997). *Bond and water resistance of masonry walls*. Proceedings of 11th International Brick/Block Masonry Conference, Shanghai, vol 2, pp. 836–844.

de Vekey RC (1999). *BRE Digest 441: Clay bricks and clay brick masonry*, Part 1 and Part 2.

Diamant RME (1986). *Thermal and Acoustic Insulation*, Butterworth, London.

Howell J (2008). *The rising damp myth*, Nosecone Publications, Woodbridge.

Newman AJ and Whiteside D (1981). Water and air penetration through brick walls – A theoretical and experimental study. *Transactions of the British Ceramic Society*, **80**, 17–26.

Newman AJ, Whiteside D, Kloss PB and Willis W (1982a). Full-scale water penetration tests on twelve cavity fills – Part I. Nine retrofit fills. *Building and Environment*, **17** (No. 3), 175–191.

Newman AJ, Whiteside D and Kloss PB (1982b). Full-scale water penetration tests on twelve cavity fills – Part II. Three built-in fills. *Building and Environment*, **17** (No. 3), 193–207.

Newman AJ (1988). *Rain penetration through masonry walls: diagnosis and remedial measures*. BR 117. Garston, Construction Research Communications Ltd.

Chapter 37

Deterioration and conservation of masonry

There are a large number of mechanisms by which masonry structures can deteriorate; these can be categorised into:

1. Chemical/biological attack on either the mortar or the units or both, due to water and waterborne acids, sulphates, pollution and chemicals released by growing plants.
2. Corrosion of embedded metal (usually steel) components, particularly ties, straps, reinforcing rods, hangers, etc. – a special case of chemical attack.
3. Erosion of units or mortar by particles in flowing water and wind, by frost attack and by salt crystallisation.
4. Stress-related effects due to movement of foundations, movement/consolidation/washout of in-fill, vibration, overloading, moisture movement of bricks and blocks, thermal movement, growth of woody plants.
5. Staining due to efflorescence, lime, iron, silica, vanadium and biological growth.

After discussing each of these, we will consider some of the factors relating to conservation, which is the art and science of maintaining, repairing and cleaning buildings, ranging from ancient monuments to humble dwellings, such that they maintain their appearance and continue to perform a useful function or can be adapted for a new role.

37.1 Chemical attack

Because they must be finely divided to be able to bind together the sand grains in aggregates and to react fairly rapidly to set to a hard adhesive mass, binders are usually more chemically reactive than the other components of masonry. Their chemical reactivity is their weakness, in that they often react with chemicals in the environment with resultant deterioration.

Mortar is generally the least durable of the concrete-like materials because it contains binders, usually has a relatively high connected-porosity that allows water to percolate through it, and usually has only a modest hardness and abrasion resistance. Dense concrete units are durable because they are hard and resist percolation and lightweight concrete units because they have a mixture of large and fine pores. Well-fired clay units are generally very resistant to chemical attack. The durability of natural stone is very variable, ranging from the highest performance – given by dense impermeable granites and marbles – to the quite poor performance of porous limestones and lime and clay-bound sandstones.

37.1.1 WATER AND ACID RAIN

Water percolating into masonry is always a potential source of damage and, where possible, the structure should be designed to throw falling rain away from façades and to channel absorbed water away, or at least to allow it to escape via weepholes then drip away from the face. Absolutely pure water will have no direct chemical effect but some of the constituents of mortar are very slightly soluble and will dissolve very slowly. Rainwater containing dissolved carbon dioxide is a very mild acid that dissolves calcium carbonate by formation of the soluble bicarbonate via the reactions:

$$CO_2 + H_2O \rightarrow H_2CO_3 \qquad (37.1)$$
$$CaCO_3 + H_2CO_3 \rightarrow Ca(HCO_3)_2 \qquad (37.2)$$

This means that lime mortars, weak ordinary Portland cement (OPC), lime mortars, porous limestones, porous lime-bonded sandstones and porous concrete blocks made with limestone aggregate will eventually be destroyed by percolating rain water because calcium carbonate is a key constituent. Strong OPC mortars with well-graded sand and most concrete and sandlime units are less susceptible partly because the calcium silicate binder is less soluble but mainly because they are less permeable, so free

Fig. 37.1 Contour scaling of a stone baluster.

percolation is prevented. Typical visible effects of water leaching on mortar are loose sandy or friable joints, loss of mortar in the outside of the joints giving a raked joint appearance, and in serious cases the loss of units from the outer layer of masonry, particularly from tunnel/arch heads. The process will sometimes be accompanied by staining due to re-precipitation of the dissolved materials. Stones lose their surface finish and may develop pits or rounded arrises.

Sulphur dioxide reacts with water to form initially sulphurous acid but can oxidise further in air to sulphuric acid:

$$SO_2 + H_2O \rightarrow H_2SO_3 \qquad (37.3)$$
$$2H_2SO_3 + O_2 \rightarrow 2H_2SO_4 \qquad (37.4)$$

There is no systematic evidence that rain acidified by sulphur dioxide from flue gases at the normal levels has a particular effect on mortar, but very clear evidence that sulphur dioxide and its reaction products do attack limestones, usually with the formation of black crusts in partly-sheltered smoky environments followed by surface spalling. Lime-bound sandstones are also attacked and suffer contour scaling, as shown in *Fig. 37.1*. One mechanism of failure is the expansive conversion of calcium carbonate to gypsum:

$$CaCO_3 + H_2SO_4 + 2H_2O \rightarrow CaSO_4.2H_2O + H_2CO_3 \qquad (37.5)$$

37.1.2 CARBONATION

Gaseous carbon dioxide (CO_2) at humidities between about 30 and 70% neutralises any alkalis present. This process occurs for all lime and Portland cement binders with the conversion of compounds such as sodium hydroxide (NaOH) and, most commonly, calcium hydroxide ($Ca(OH)_2$) to their respective carbonates. In lime mortars this process probably increases the strength and durability. In Portland cement-based materials the key effect of the process is to reduce the pH from around 12–13 down to below 7, i.e. converting the material from highly alkaline to slightly acid. This can have a profound effect on the durability of embedded steel components, as discussed in Chapter 24. There is also some evidence that there is a slight associated shrinkage that may reduce the strength of very lightweight concrete units. Very dense concrete and calcium silicate units will carbonate slowly and may take 50–100 years or more to carbonate through.

37.1.3 SULPHATE ATTACK

Sulphate attack is the next most common problem, and is due to the reaction between sulphate ions in aqueous solution and the components of hardened Portland cement to form ettringite. We have given a detailed coverage of the physical/chemical mechanisms of this in Chapter 24. The resulting expansion, which can be on the order of several per cent, causes both local disruption of mortar beds and stresses in the brickwork, but only in wet or saturated conditions and where there is a source of a water-soluble sulphate compound. It will never occur in dry or slightly damp masonry. The common sulphates found in masonry are the freely soluble sodium, potassium and magnesium salts and calcium sulphate, which is less soluble but will diffuse in persistently wet conditions. Sulphates may be present in groundwater and can affect masonry in contact with the ground such as in foundations, retaining walls, bridges and tunnels. In this situation porous concrete units are also at risk – see BRE Special Digest 1 (2005). Soluble sulphates are also present in some types of clay brick and will be transported to the mortar in wet conditions. Examples are any clay bricks with unoxidised centres ('blackhearts'), some Scottish composition bricks and semi-dry pressed bricks made from Oxford clay (Flettons), which have high levels of calcium sulphate. Sulphates may also attack lime mortars by conversion of the lime to gypsum in a similar reaction to that shown in equation 37.5. Sulphate-resisting Portland cement is deliberately formulated to have a low C_3A content but may be attacked in very extreme conditions. Another compound, thaumasite Thaumasite Experts Group (1999), may form by reaction between dicalcium and tricalcium silicate, sulphate, carbonate and water, as also discussed in Chapter 24. This process can disrupt mortar beds.

Fig. 37.2 Sulphate attack on mortar.

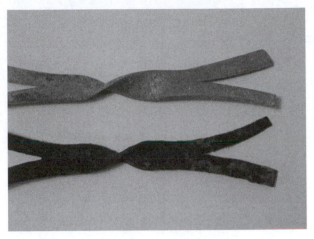

Fig. 37.3 Uncorroded (upper) and corroded cavity wall tie (lower) showing white rust (ZnO) and red rust ($Fe_2O_3.H_2O$).

Visible effects of sulphate attack on mortar are expansion of the masonry where it is unrestrained and increase in stress where it is restrained. Surface spalling is common. Typically the mortar is affected more within the body of the wall than on the surface, so small horizontal cracks are sometimes visible in the centre of each bed joint – as in *Fig. 37.2* – and vertical cracks may appear on the external elevations of thick masonry. Rendered masonry often exhibits a network of cracks termed 'map-cracking' or cracking that follows the mortar joints. The susceptibility of mortar to sulphate attack (and frost attack or a combination of the two) can be tested using the Harrison technique, which has been standardised by RILEM (1998a).

Sulphates rarely affect the units but precautions are advisable when building in ground containing sulphates (BRE Special Digest 1 (1996)) or constructing flumes and tunnels to carry contaminated effluents (WRC, 1986). A special type of 'engineering quality' concrete brick is available which is designed to be stable in effluents (Concrete, 1986). These units are manufactured to a high strength and low permeability with sulphate resisting Portland cement as the binder. Mundic concrete blocks made using tin-mine tailings as aggregate in southwest England have suffered attack from indigenous sulphates (Bomley and Pettifer, 1997; RICS, 1997).

37.1.4 ACIDS
The effects of acids, e.g. rain run-off from peat moors, industrial or agricultural pollution, on cement-based products has been discussed in Chapter 24. Fired clay products are normally resistant to acids.

37.1.5 CHLORIDES
Chlorides can have a weakening effect on calcium silicate units but have little effect on mortars, clay units or concrete masonry units. They also catalyse the rusting of embedded steel even in alkaline conditions (see below).

37.1.6 CORROSION OF EMBEDDED METALS
Chapter 24 includes a detailed consideration of the corrosion of steel in concrete, and the principles involved apply equally to steel in mortar in masonry construction. *Figure 37.3* shows a corroded masonry wall tie; de Vekey (1990a, 1990b) has given specific coverage of wall-tie corrosion.

More details on all the chemical processes described above are given in Yu and Bull (2006).

37.2 Erosion

Erosion processes such as wind and water scour attack both units and mortar but erode the softer of the two at a faster rate. Freeze–thaw frost attack and salt crystallisation are complex cyclic erosion processes where the susceptibility is dependent on pore size distribution and the number of cycles, and not simply on hardness, strength and overall porosity.

37.2.1 FREEZE–THAW ATTACK
Freeze–thaw damage is one of the principal eroding agents of porous materials, including masonry units and mortar, exposed to normal exterior conditions.

Fig. 37.4 Erosion due to frost attack.

Quite clearly it will not affect masonry buried more than a few feet and so will not affect foundations, the insides of tunnels away from portals or buried culverts. It may affect any masonry exposed on the exterior of structures but is more likely to affect exposed components that become saturated. Typical problem areas are civil engineering structures such as bridges, canal locks, earth-retaining walls and exposed culverts and parts of buildings such as parapets, copings, chimneys, freestanding walls and masonry between the ground and the damp-proof course.

Freeze–thaw attack is due to the stresses created by the 8% expansion of water on freezing to ice in the pore systems of units and mortars and thus only occurs in water-saturated or near-saturated masonry. A pictorial explanation of one of the classic mechanisms is given in Yu and Bull (2006).

Typical effects are the spalling (scaling) of small pieces of either the unit or the mortar or both forming a layer of detritus at the foot of the wall, as in *Fig. 37.4*. Clay bricks, particularly lightly fired examples of types made from some shales and marls, are especially susceptible and tend to delaminate. Semi-dry pressed bricks made from Oxford clay tend to break down into grains the same size as they were made from (e.g. around 2–4 mm diameter). Old types of solid clay brick with ellipsoidal drying micro-cracks tend to spall from the boundary between the heart and the outside of the brick. Some natural stones with a large proportion of fine pores are also susceptible. Modern perforated clay bricks are generally more frost-resistant because the more open structure allows more even drying and firing with less chance of drying/firing cracks forming.

It can be puzzling as to why some building materials are susceptible to freeze–thaw attack while other, apparently similar, materials are not. Clearly materials such as glass, plastic and metals, which are totally non-porous to water, are not affected. Materials such as well-fired 'glassy' engineering clay bricks, well-compacted concretes, low-porosity granites, marbles and slates are also affected very little by frost since the tiny volume of water that can penetrate will cause only a trivial level of stress on freezing. Materials with water absorptions ranging from about 4 to 50% tend to suffer damage but do not do so invariably. Closed-pore materials such as aircrete (AAC) and air-entrained mortars, which are not easy to saturate, are generally resistant as are materials such as stock bricks and lightweight concrete blocks, with a wide range of pore sizes from very large to fine. This is probably because it is difficult to fully saturate the mixed pore system, the water filling the finer but not the larger pores. Providing around 10% of the pore system remains air filled there is sufficient space for the ice crystals to expand into without damaging the structure. Materials – particularly some mortars, clay bricks and natural stones – having a limited range of pore sizes, usually of the finer sizes, tend to fail.

Most of the older data on susceptibility to frost attack are based on experience in use but this is a very slow and inefficient way of evaluating new products. To try to speed up the process, accelerated freezing tests have been developed. A typical example is the panel test (West *et al.*, 1984), which was published as an international standard by RILEM (1998a) and as a draft CEN standard prEN 772-20 (1999). Work has also been done by Beardmore and Ford (Beardmore and Ford, 1986; Beardmore, 1990) and Stupart (1996) to develop maps that indicate the average number of days per year of combined driving rain and freeze–thaw cycling affecting different areas of the UK.

37.2.2 CRYPTO-EFFLORESCENCE (SUB-FLORESCENCE) DAMAGE

This is basically the same process as efflorescence (see below) but at certain temperature/humidity conditions it occurs just below the surface of the masonry unit. The hydrated crystals of compounds such as magnesium and sodium sulphates growing in the pore structure result in a compression force in the surface layers and consequently shear and tensile forces at the boundary with the unaffected core. There needs to be a source of water or water containing salts and a surface that is sufficiently warm or well ventilated (or both) to encourage drying of the salt solution. In UK conditions it is more likely to affect clay brick or natural stonework than other units or mortars.

Fig. 37.5 *Salt crystallisation damage to masonry together with associated soluble salt crystals (efflorescence).*

Fig. 37.6 *Lime staining of brickwork.*

The typical appearance is similar to that of freeze–thaw damage but it will usually be associated with soluble salt crystals, as shown by *Fig. 37.5*. Accelerated test methods have been standardised by RILEM (1998b) and by BS EN 12370 (1999).

37.2.3 ABRASION

Abrasion by particles in wind and water probably acts more in concert with other processes than in isolation. Likely areas for such erosion are bridge columns founded in riverbeds and buildings near road surfaces where splash up can occur from vehicles. All types of marine/hydraulic structures such as dams, culverts, lock walls, flumes, etc., where high-velocity flows can occur, may suffer from localised abrasion/cavitation damage known as 'scour'.

37.3 Stress effects

Stress effects normally cause cracking of varying types but the effect is on the masonry composite and not on the individual components. There are some problems related to faults in manufacture of the units, particularly under- or over-firing of clay bricks and the inclusion of foreign particles in bricks or other types of unit. A good range of coloured illustrations of problems and corrective strategies is given in BRE Digest 361 (1991) and Cracking in Buildings (1995).

37.4 Staining

37.4.1 EFFLORESCENCE

This is a staining process caused by dissolution of soluble salts such as sodium, potassium, magnesium and calcium sulphates within the masonry pores by rain, groundwater or construction water from within brickwork which then crystallise on the surface as an off-white powder or encrustation. The surface may be any surface from which drying occurs. It is commonly the external facade but may be the interior of solid walls particularly those in contact with earth, e.g. cellar walls and ground-floor walls in older structures without damp courses. The salts commonly derive from clay bricks, but may also come from groundwater or stored materials. *Figure 37.6* shows typical effects.

37.4.2 LIME STAINING

This is caused by calcium hydroxide leaching from mortar or Portland cement-based concretes and being carbonated at the surface to form a white deposit of calcium carbonate crystals. It may also result from the dissolution of calcium carbonate in carbonated rainwater to form calcium bicarbonate, which reverts to the carbonate at the surface (stalactites grow by this mechanism). It is most commonly seen as white 'runs' from mortar joints on earth-retaining walls, see *Fig. 37.6* but can occur on the walls of buildings. It is often seen on hollow/perforated unit masonry and accumulates in the holes as the rainwater seeps slowly out via porous mortar or concrete units.

37.4.3 IRON STAINING

Iron compounds can be present in bricks, concrete units or mortars. They give little problem if well distributed, as in many red-brown sands, but will give ugly brown run-off stains if present as larger discrete particles. Iron may be present in clay deposits in the form of pyrites and is sometimes a constituent of aggregates and additives to clay.

37.4.4 BIOLOGICAL STAINING

Coloured deposits, usually white/green to brown/black or orange, can build up owing to the growth of algae, fungi, lichens, mosses etc. Such deposits only form if the masonry is wet for significant periods and is frequently the result of blocked downpipes or leaking gutters. Differential deposition can result from water streaming unequally off features such as windows and mullions, and this can become noticeable owing to the colour contrast. More information and illustrations of staining are given in BRE Digests 441, 460 and 508.

37.5 Conservation of masonry

37.5.1 INTRODUCTION

Conservation of the built environment is a difficult subject to cover comprehensively because it involves a complex interplay between politics, history, legislation, sustainability, the construction arts and crafts, materials science and structural engineering. In this short section only a flavour can be conveyed and an indication of where to look for more information. *Figure 37.7* shows an example of how masonry can be conserved given sufficient attention to all the principles involved.

37.5.2 PRINCIPLES

Over time and in different countries, districts etc. the principles behind the conservation of what has been built in the past have varied enormously between:

- complete indifference, where unwanted buildings were simply demolished for their materials and new ones built over the rubble (a fate of many Roman examples in the UK)
- complete negativity (such as the destruction of the monasteries by King Henry VIII)

Fig. 37.7 Well-conserved flint-work masonry in Norfolk.

- often misguided prettification/upgrading
- the current more-considered and responsible policy.

The philosophy applied currently in the UK in respect of important monuments is:

- minimal intervention
- protection against disaster and neglect
- use of identical replacement materials or the best compromise
- additions designed to be in keeping with the style of the original building or alternatively clearly distinguishable from it
- changes are recorded in a log
- the philosophy and management must be agreed with the regulating body such as a local authority, English Heritage, Heritage Scotland etc. before the start of projects.

Put more colloquially this is:

- if it ain't broke don't fix (or alter) it
- protect against fire, flood, rain, frost, vandalism, theft and neglect
- don't replace Portland limestone with concrete blocks
- don't change from art nouveau to art deco
- record what was done
- try to get all affected parties on-board including the public.

BRE Digest 502 (2007) gives a brief overview and a useful reference list.

37.5.3 REPLACEMENT MATERIALS: STONE

As explained in Chapter 33, stone is a very variable natural product formed from a wide range of minerals by different consolidation processes. Because its properties, particularly durability, can have such a wide range, even within each sub-class, it is important to be able to accurately identify the type and source of the original rock when choosing replacements. This is particularly important for the sedimentary rocks limestone and sandstone because they are the most widely used and have the greatest variability.

It is best to start by trying to obtain information from the building owner's own knowledge and records. As stone is heavy and expensive to transport it is often quarried nearby and often whole villages and towns are built from the local material. Typical examples are the villages surrounding the Ham stone in Dorset. If available, original stone from redundant parts of the same building/s or rescued from local architectural salvage yards is ideal provided that it has not deteriorated.

Visual identification can be facilitated by comparison with examples held in stone libraries as listed in BRE Digest 508: Part 1 using information such as grain (fine to coarse), overall colour or patterning, degree of stratification (layering), orientation of any layers in the work, size and shape of uncut rubble stones and the degree to which the stone can be worked or polished. If visual data are inconclusive then data from petrology – examination of thin sections under a microscope or X-ray diffraction – can be used to identify the minerals present and chemical analysis can determine the constituent elements. Other properties such as density, strength, hardness and porosity may give further clues.

Having identified the source quarry or an alternative, newly quarried stone may be suitable if an equivalent material is still in production. Stone can vary quite markedly with quarry, position in the quarry and depth from which it is cut, so it will have to be tested in accordance with the methods given in standards such as BS EN 771-6 (2005) and BS EN 1467 (2003) to check the match with the original.

37.5.4 REPLACEMENT MATERIALS: CLAY BRICKS, TERRACOTTA WARE, CONCRETE AND CALCIUM SILICATE UNITS

Even though such bricks and blocks are man-made, the process of identification and/or determination of a suitable replacement product is broadly similar to that for stone, as are the sources of original material. Important visual clues are makers' marks (usually impressed in the frog of mud or pressed bricks or sometimes stamped on a face of an extruded unit), overall colour, colour variation, appearance of any coatings or impressed patterning, perforation pattern, frog size and shape. The size and proportion of clay bricks has varied since their original introduction by the Romans. Terracotta, which is hollow moulded clay ware, often of elaborate and made-to-measure shapes, was imported during the 15–17th centuries and made in England in the 19–20th centuries for use in prestige buildings. Concrete and calcium silicate units will only be found in unmodified buildings dating from around 1900 onwards.

If new replacement units are necessary they should be matched in accordance with the test methods given in standards such as BS EN 771-1 to 5 (2003). BRE Digest 508 Part 2 (2009) gives some helpful guidance on man-made units.

37.5.5 REPLACEMENT MATERIALS: MORTARS

Mortars are normally identified on the basis of six main characteristics:

1. The type of binder, e.g. natural hydraulic lime.
2. The main type of aggregate, e.g. silica sand.
3. Any other discernible ingredients such as plasticisers, added fine material, pozzolans and pigments.
4. The hydraulicity of the mixture.
5. The particle shape and size range (grading) of the aggregates (from sieve analysis).
6. The porosity of the cured mortar.

The state of carbonation and the cube strength may also be helpful.

Where required measurements should conform to tests specified in BS EN 998-2 (2003) or BS4551 (2005). The composition of most of the mortar types likely to be encountered in conservation work and guideline data on their identification are covered in more detail in BRE GBG66 (2005).

37.5.6 SELECTION OF REPLACEMENT MATERIALS

Ightham Moat, a National Trust property, is typical of the problems that might arise when choosing replacements for the complex cocktail of materials present. Having identified the original units the broad rules for replacing them are:

1. Try to use un-weathered identical examples if available.
2. Similar but stronger and/or harder replacements may be used for additions or wholesale replacement of elements.

3. If anything, select slightly softer, lower-strength examples for patching walls because stronger harder materials often accelerate the weathering of the surviving original materials.
4. Do not mix materials that have widely differing chemical compositions or movement characteristics without consulting specialist materials scientists. The wrong choice could cause a 10-fold increase in weathering damage or disfigurement.

Mortar should be easier to reinstate as it is always placed as a plastic material that sets to a solid mass where placed. Mortar should have as near the same composition and be of compatible texture, colour, porosity, hardness, vapour permeability, strength and durability as the 'original', providing that the original was of acceptable durability. To avoid stress-related damage the hardened, particularly repointing, mortar should never be stronger/harder than the units, and to minimise shrinkage cracking, washed well-graded sand should be used. BRE GBG 66 (2005) lists most of the applicable mortar types and there is useful guidance on selection in Tables 2 and 3 of BS8221-2.

37.5.7 REPAIR METHODS
The main methods used (in order of severity) are:

- demolition and reconstruction of whole elements, e.g. walls, arches, or parts thereof
- removal and replacement of whole units, e.g. blocks, bricks
- cutting out the whole face of a unit and replacement with a slip
- cutting out of areas of deteriorated units and replacement with a shaped piece
- plastic repairs with mortars formulated to match the characteristics of the units.

In all cases, except at ground level, safe access must be provided using scaffolding or lift systems, a building professional/engineer should evaluate any collapse hazards, propping should be used if necessary and, ideally, a trial of the proposed technique should be carried out before the main contract starts. Techniques for partial demolition, removal of units and cutting out of parts of units are given in BRE Digest 508 together with further references.

37.5.8 CLEANING OF MASONRY
Exposed masonry has always been subject to staining (discussed in section 37.4) and defacement processes by interaction with the environment, water (moisture) being the main agent and probably industrial smoke the second. Since staining or surface deterioration changes the appearance from the original architectural concept, various cleaning methods have been developed based mainly on: mechanical action using brushes or abrasives; various water washing techniques; and treatment with weak to strong acids and alkalis. This is a very complex topic and only the basic rules can be covered here, which are:

1. Before industrialisation in the 17th century and more recently following the clean air acts of the 1950s, biological growths are the most likely staining agent.
2. During the most industrialised period (1750–1950) smoke and sulphur dioxide were probably the worst offenders especially affecting limestones, lime-bound sandstones and porous mortars. Such staining/erosion may still be encountered.
3. From 1950 to date there have been increased low-level effects of road splash and diesel exhausts.
4. It is essential to be sure of both the masonry materials and the staining agent before choosing a cleaning method.
5. Water is a relatively harmless cleaner but may cause further staining/damage if used in excess on, and not dried out from, porous materials.
6. Strong chemicals, such as acids, should not be allowed to penetrate and remain in pore systems, as they almost certainly will cause accelerated soiling and deterioration.
7. Most strong acids pose serious health and safety risks to operatives and should be avoided if at all possible.
8. Research by Historic Scotland (2003) on sandstones has shown an up to 10-fold increase in deterioration rate following cleaning.
9. Light brushing and grit blasting with softer abrasives such as crushed nut-shells is usually safe.
10. Laser-based ablation cleaning is probably safe but it is expensive and slow.
11. Do not use mechanical or chemical methods that remove surface layers from materials that depend on protective skins such as terracotta ware and weathered limestones.
12. If there is any doubt or known risk, do not use any cleaning process, particularly on stones or mortars with poor durability. Trials of cleaning systems may give some guidance but are rarely of sufficient duration.

Some useful and more detailed information is given in the extended reference list below and in the British

Standards and BRE guides listed in 'Further Reading', which follows this chapter.

References

DURABILITY AND DETERIORATION

Beardmore C and Ford RW (1986). *Winter weather records relating to potential frost failure of brickwork*. Technical note TN372, Ceram, Stoke-on-Trent.

Beardmore C (1990). *Winter weather records relating to potential frost failure of brickwork (2)*. Research Paper 781, British Ceramic Research Ltd.

Bomley AV and Pettifer K (1997). Sulfide-related degradation of concrete in Southwest England. Royal Institute of Chartered Surveyors (1997) The Mundic problem.

Bonshor RS and Bonshor LE (1995). *Cracking in buildings*, IHS BRE Press, Bracknell.

BRE Report BR 262 (2002). Thermal insulation: avoiding risks. IHS BRE Press, Bracknell.

Concrete (1986). The concrete engineering quality brick, *Concrete*, April.

de Vekey RC (1990a). *Corrosion of steel wall ties: background, history of occurrence and treatment*. BRE Information Paper IP12/90, Building Research Establishment, Watford.

de Vekey RC (1990b). *Corrosion of steel wall ties: recognition and assessment*. BRE Information Paper IP13/90, Building Research Establishment, Watford.

Harrison WH (1987). Durability of concrete in acid soils and groundwaters. *Concrete* (magazine), **21** (No. 2), 18–24.

RILEM recommendations of durability tests for masonry (1998a). MS.A.3. Unidirectional freeze–thaw test for masonry units and wallettes. *Materials and Structures*, **31** (No. 212), 513–519.

RILEM recommendations of durability tests for masonry (1998b). MS.B.5 Determination of the damage to wallettes caused by acid rain. *Materials and Structures*, **31** (No. 212), 519–521.

Royal Institute of Chartered Surveyors (RICS) (1997). The 'Mundic' Problem – A Guidance Note, 2nd edition.

Stupart AW (1996). Possible extensions to developing a frost index. *Masonry International*, **7** (No. 1), 4–9.

Thaumasite Experts Group, Department of the Environment, Transport and the Regions (1999). *The thaumasite form of sulfate attack: risks, diagnosis, remedial works and guidance on new construction*, DETR, London.

West HWH, Ford RW and Peake FA (1984). A panel freezing test for brickwork. *Proceedings of the British Ceramic Society*, **83**, 112–121.

WRC Information and Guidance Note IGN 4-10-01 (1986). *Bricks and Mortar*, Water Research Council, Oct.

Yu CW and Bull JW (2006). Durability of materials and structures in building and civil engineering, Whittles Publishing, Caithness, Chapters 8 to 10.

CONSERVATION AND CLEANING

Ashurst J and Ashurst N (1988). *Practical building conservation. Volume 1: Stone masonry*. English Heritage Technnical Handbook, Gower Publishing, London.

Ashurst J and Ashurst N (1988). *Practical building conservation, Volume 2: Brick, terracotta and earth*. English Heritage Technnical Handbook, Gower Publishing, London.

Ashurst J and Ashurst N (1988). *Practical building conservation. Volume 3: Mortars, plasters and renders*. English Heritage Technnical Handbook, Gower Publishing, London.

Ashurst J (2002). *Mortars, plasters and renders in conservation*, 2nd edition, Ecclesiastical Architects and Surveyors Association, Sheffield.

Ashurst N (1994). *Cleaning historic buildings, Vol.1: Substrates, soiling and investigation*, Donhead, London.

Ashurst N (1994). *Cleaning historic buildings, Vol.2: Cleaning materials and processes*, Donhead, London.

Martin Cooper (ed.) (1998). *Laser Cleaning in Conservation: an Introduction*, Butterworth Heinemann, Oxford.

Historic Scotland (1997). *Stone cleaning of Sandstone Buildings*, Crown Copyright, Edinburgh.

Historic Scotland (2003). *The consequences of past stone cleaning intervention on future policy and resources*, Masonry Conservation Research Group, The Robert Gordon University, Crown Copyright, Historic Scotland, Edinburgh.

Optoelectronics Research Centre, The Robert Gordon University and The Building Research Establishment (2005). *Laser stone cleaning in Scotland*, Crown Copyright, Historic Scotland, Edinburgh.

Urquhart D, Young M and Cameron S (1997). *Stone cleaning of Granite Buildings. Advice on the soiling, decay and cleaning of granite buildings and related testing, specification and execution of the work*, Crown Copyright, Historic Scotland, Edinburgh.

Webster GM (ed.) (1992). *Stone cleaning and the nature of soiling and decay mechanisms of stone*, Donhead, London.

Young ME, Ball J, Laing RA and Urquhart DCM (2003). *Maintenance and repair of cleaned stone buildings*, Crown Copyright Historic Scotland, Edinburgh.

Further reading for Part 5 Masonry: brickwork, blockwork and stonework

Publications on masonry construction are extensive, with much useful information contained in documents from trade and government organisations, particularly the Building Research Establishment and the Brick Development Association. Those most relevant to the contents of this part of the book are listed below, starting with BRE Digests.

BUILDING RESEARCH ESTABLISHMENT (BRE) SERIES PUBLICATIONS

Digests (DG)

DG35 (1963). *Shrinkage of natural aggregates in concrete*, BRE, Watford (revised 1968).

DG108 (1975). *Standard U-values*, BRE, Watford

DG157, *Calcium silicate (sandlime, flintlime) brickwork*, BRE, Watford

DG228 (1979). *Estimation of thermal and moisture movements and stresses: Part 2*, BRE, Watford (also Digests 227 and 229)

DG240 (1993). *Low rise buildings on shrinkable clay soils: Part 1*. BRE, Watford

DG241 (1990). *Low rise buildings on shrinkable clay soils: Part 2*. BRE, Watford

DG246, *Strength of brickwork and blockwork walls: design for vertical load*. BRE, Watford

DG273 (1983). *Perforated clay bricks*. BRE, Watford

DG298 (1999). *The influence of trees on house foundations in clay soils*. BRE, Watford

DG329 (2000). *Installing wall ties in existing construction*, BRE, Watford

DG333 (1988). *Sound insulation of separating walls and floors. Part 1: Walls*, BRE, Watford

DG337 (1994). *Sound insulation: basic principles*, BRE, Watford

DG338 (1989). *Insulation against external noise*, BRE, Watford

DG355 (1990). *Energy efficiency in dwellings*, BRE, Watford

DG359 (1991). *Repairing brickwork*, BRE, Watford

DG360 (1991). *Testing bond strength of masonry*, BRE, Watford

DG361 (1991). *Why do buildings crack?* BRE, Watford.

DG362 (1991). *Building mortar*. BRE, Watford

DG370 (1992). *Control of lichens and similar growths*, BRE, Watford

DG380 (1993). *Damp proof courses*

DG418 (1996). *Bird, bee and plant damage to buildings*, BRE, Watford

DG420 (1997). *Selecting natural building stones*, BRE, Watford

DG421 (1997). *Measuring the compressive strength of masonry materials: the screw pull-out test*, IHS BRE Press, Bracknell.

DG432 (1998). *Aircrete: thin joint mortar systems*, BRE, Watford

DG433 (1998). *Recycled aggregates*

DG441, Part 1 (1999). *Clay bricks and clay brick masonry*, BRE Watford

DG441, Part 2 (1999). *Clay bricks and clay brick masonry*, BRE Watford

DG448 (2000). *Cleaning buildings Legislation and good practice*

DG449 (2000). *Cleaning exterior masonry, Part 1 Developing and implementing a strategy*

DG449 (2000). *Cleaning exterior masonry, Part 2 Methods and materials*

DG460, Part 1 (2001). *Bricks blocks and masonry made from aggregate concrete: Part 1 – performance requirements*

DG460, Part 2 (2001). *Bricks blocks and masonry made from aggregate concrete: Part 2 – Appearance and environmental aspects*, BRE, Watford

DG468, Part 1 (2002). *Autoclaved aerated concrete 'aircrete' blocks and masonry: Part 1 – performance requirements*, BRE

DG468, Part 2 (2002). *Autoclaved aerated concrete 'aircrete' blocks and masonry: Part 2 – Appearance and environmental aspects*, BRE Watford
DG487-3 (2004). *Structural fire engineering design: Part 3: Materials behaviour – Masonry*
DG502 (2007). *Principles of masonry conservation management*, IHS BRE Press, Bracknell
DG508 (2007). *Conservation and cleaning of masonry, Part 1: Stonework*, IHS BRE Press, Bracknell
Special digest SD1 (2005). *Concrete in aggressive ground 2005 edition*
Special digest SD4 (2007). *Masonry walls and beam and block floors U-values and Building Regulations*, 2nd edition

Good Building Guides (GG)
GG 13 (1992). *Surveying brick or blockwork freestanding walls*
GG 14 (1994). *Building simple plan brick or blockwork freestanding walls*
GG 17 (1993). *Freestanding brick walls: repairs to copings and cappings*
GG 19 (1994). *Building reinforced, diaphragm and wide plan freestanding walls*
GRG 27 (2000). *Cleaning external walls of buildings, Part 1 Cleaning methods*
GRG 27 (2000). *Cleaning external walls of buildings, Part 2 Removing dirt and stains*
GG 29 *Connecting walls and floors. Parts 1 and 2*
GG 33 (1999). *Assessing moisture in building materials (three parts)*
GG 44-1 (2000). *Insulating masonry cavity walls: Part 1 Techniques and materials*
GG 44-2 (2000). *Insulating masonry cavity walls: Part 2 Principal risks and guidance*
GG 66 (2005). *Building masonry with lime based mortars*

Good Repair Guides (GR)
GR5 (1997). *Diagnosing the causes of dampness*
GR6 (1997). *Treating rising damp in houses*, IHS BRE Press
GR7 (1997). *Treating condensation in houses*, IHS BRE Press
GR8 (1997). *Treating rain penetration in houses*, IHS BRE Press
GR23 (1999). *Treating dampness in basements*, IHS BRE Press
GR26-1 (1999). *Improving energy efficiency: Part 1 Thermal insulation*
GR27-1 (2000). *Cleaning external walls of buildings, Part 1 Cleaning methods*
GR27-2 (2000). *Cleaning external walls of buildings, Part 2 Removing dirt and stains*

Other BRE series

Information Papers
BRE report BR466. *Understanding dampness; effects, causes, diagnosis and remedies*, IHS BRE Press
Performance specifications for wall ties – BRE report. BRE, Watford
BRE CP24/70 (1970). *Some results of exposure tests on durability of calcium silicate bricks*
BRE CP23/77 (1977). Chemical resistance of concrete, *Concrete*, **11** (No. 5), 35–37

Book
Harrison HW and de Vekey RC (1998). *Walls Windows and Doors*, in BRE Building Elements series, CRC, Watford

BRICK DEVELOPMENT ASSOCIATION PUBLICATIONS
Brick diaphragm walls in tall single storey buildings (and earth retaining walls)
BDA Design note 3, *Brickwork Dimensions Tables*
BDA Design note 7, *Brickwork Durability*

Ceram Research publications
Technical Note 368, *The performance of calcium silicate brickwork in high sulphate environments.*
SP56: 1980: *Model specification for clay and calcium silicate structural brickwork* (in process of updating)
Supplement No.1 to SP56, *Glossary of terms relating to the interaction of bricks and brickwork with water*
SP108, *Design guide for reinforced clay brickwork pocket-type retaining walls*
brickwork.SP109, *Achieving the functional requirements of mortar*

BRITISH CEMENT ASSOCIATION
Technical report TRA/145, *The effects of sulphates on Portland cement concretes and other products*

CONCRETE BRICK MANUFACTURERS ASSOCIATION PUBLICATIONS
CBMA Information Sheet 2, *Concrete bricks – product information*

BRITISH/EUROPEAN STANDARDS AND CODES OF PRACTICE REFERRED TO IN THE TEXT
The list below is of those standards, specifications and design codes published by the British Standards Institution that are mentioned in the text. It is not intended to be an exhaustive list of all those concerned with masonry materials, components and construction. These can be found be looking at the BSI website: www.bsi-global.com/

Codes of Practice and guides
BS 5628: Part 1 (2005). *Code of practice for the use of masonry Part 1: Structural use of unreinforced masonry* (covers design of walls, arches, tunnels, columns, etc. subject to compressive, lateral and shear loads)

BS 5628: Part 2 (2005). *Structural use of reinforced and prestressed masonry* (covers the design of earth-retaining walls, chamber covers, beams, floors, cantilevers etc.)

BS 5628: Part 3 (2006). *Materials and components, design and workmanship* (covers non-structural aspects of brickwork design, particularly the specification of units and mortars for durability over a wide range of applications and also workmanship, detailing, bonding patterns, fire resistance and resistance to weather conditions)

BS 5606 (1990). *Code of practice for accuracy in building*

BS 6100-6 (2008). *Glossary of building and civil engineering terms, Part 6: masonry*

BS8000: Part 3 (2001). *Workmanship in building, Part 3: masonry*

BS 8221-1 (2000). *Code of Practice for the cleaning and surface repair of buildings – Part 1: Cleaning of natural stones, brick, terracotta and concrete*

BS 8221-2 (2000). *Code of Practice for the cleaning and surface repair of buildings – Part 2: Surface repair of natural stones, brick and terracotta*

BS EN 1745 *Masonry and masonry products – Methods for determining design thermal values*

BS EN 1996-1-1 (2005). *Eurocode 6: Design of masonry structures – Part 1-1 General rules for reinforced and unreinforced structures*

BS EN 1996-1-2 (2005). *Eurocode 6: Design of masonry structures – Part 1-2 General rules Structural fire design*

BS EN 1996-2 (2005). *Eurocode 6: Design of masonry structures – Part 2: Design considerations, selection of materials and execution of masonry*

BS EN 1996-3 (2006). *Eurocode 6: Design of masonry structures – Part 3: Simplified calculation methods for unreinforced masonry structures*

Materials standards

BS 1200:1976	Sands from natural sources: sands for mortar for plain and reinforced brickwork, blockwork and stone masonry
BS 1243:1978	Specification for metal ties for cavity wall construction (Replaced by BS845-1)
BS 4729:2005	Clay and calcium silicate bricks of special shapes and sizes. Recommendations
BS 6677:Part 1:1985	Specification for clay and calcium silicate pavers. (withdrawn)
BS EN 197-1:2000	Cement. Composition, specifications and conformity criteria for common cements
BS EN 413-1:2004	Masonry cement. Composition, specifications and conformity criteria
BS EN 771-1:2003	Specification for masonry units. Clay masonry units
BS EN 771-2:2003	Specification for masonry units. Calcium silicate masonry units
BS EN 771-3:2003	Specification for masonry units. Aggregate concrete masonry units (dense and lightweight aggregate)
BS EN 771-4:2003	Specification for masonry units. Autoclaved aerated concrete masonry units
BS EN 771-5:2003	c for masonry units. Manufactured stone masonry units
BS EN 771-6:2005	Specification for masonry units. Natural stone masonry units
BS EN 845-1:2003	Specification for ancillary components for masonry, ties, straps, hangers and brackets
BS EN 845-2:2003	Specification for ancillary components for masonry, lintels
BS EN 845-3:2003	Specification for ancillary components for masonry, bed joint reinforcement of steel meshwork
BS EN 934-3:2003	Admixtures for concrete, mortar and grout. Admixtures for masonry mortar. Definitions, requirements, conformity, marking and labelling
BS EN 998-1:2003	Specification for mortar for masonry: Part 1 Rendering and plastering mortar
BS EN 998-2:2003	Specification for mortar for masonry: Part 2 Masonry mortar
BS EN 1467:2003	Natural stone rough blocks: requirements

BS EN 12878:2005	Pigments for the colouring of building materials based on cement and/or lime. Specifications and methods of test
BS EN 13139:2002	Aggregates for mortar
BS EN 1338:2003	Precast unreinforced concrete paving blocks. Requirements and test methods (was BS 6717-1)

Methods of test for masonry units

BS EN 772-1:2000	Determination of compressive strength
BS EN 772-2:1998	Determination of percentage area of voids in aggregate concrete masonry units (by paper indentation)
BS EN 772-3:1998	Determination of net volume and percentage of voids of clay masonry units by hydrostatic weighing
BS EN 772-4:1998	Determination of real and bulk density and of total and open porosity for natural stone masonry units
BS EN 772-5:2001	Determination of the active soluble salts content of clay masonry units
BS EN 772-6:2001	Determination of bending tensile strength of aggregate concrete masonry units
BS EN 772-7:1998	Determination of water absorption of clay masonry damp course units by boiling in water
BS EN 772-9:1998	Determination of volume and percentage of voids and net volume of calcium silicate masonry units by sand filling
BS EN 772-10:1999	Determination of moisture content of calcium silicate and autoclaved aerated concrete units
BS EN 772-11:2000	Determination of water absorption of manufactured stone and natural stone masonry units due to capillary action and the initial rate of absorption of clay masonry units
BS EN 772-13:2000	Determination of net and gross dry density of masonry units (except for natural stone)
BS EN 772-14:2002	Determination of moisture movement of aggregate concrete and manufactured stone masonry units
BS EN 772-15:2000	Determination of water vapour permeability of autoclaved aerated concrete masonry units
BS EN 772-16:2000	Determination of dimensions
BS EN 772-18:2000	Determination of freeze–thaw resistance of calcium silicate masonry units
BS EN 772-19:2000	Determination of moisture expansion of large horizontally-perforated clay masonry units
BS EN 772-20:2000	Determination of flatness of faces of masonry units
BSI. prEN 772-22 (draft):1999	*Methods of test for masonry units – Determination of the frost resistance of clay masonry units*, London
BS EN ISO 12572 (2001)	Hygrothermal performance of building materials and products: determination of water vapour transmission properties

Methods of test for mortar for masonry

BS 4551:2005	Methods of testing mortars screeds and plasters: Chemical analysis and aggregate grading
BS EN 1015-1:1999	Determination of particle size distribution (by sieve analysis)
BS EN 1015-2:1999	Bulk sampling of mortars and preparation of test mortars
BS EN 1015-3:1999	Determination of consistence of fresh mortar (by flow table)
BS EN 1015-4:1999	Determination of consistence of fresh mortar (by plunger penetration)
BS EN 1015-6:1999	Determination of bulk density of fresh mortar
BS EN 1015-7:1999	Determination of air content of fresh mortar
BS EN 1015-9:1999	Determination of workable life and correction time of fresh mortar
BS EN 1015-10:1999	Determination of dry bulk density of hardened mortar
BS EN 1015-11:1999	Determination of flexural and compressive strength of hardened mortar
BS EN 1015-12:2000	Determination of adhesive strength of hardened rendering and plastering mortars on substrates
BS EN 1015-14:1999	Determination of durability of hardened mortar (review in progress)
BS EN 1015-17:2000	Determination of water-soluble chloride content of fresh mortars
BS EN 1015-18:2002	Determination of water absorption coefficient due to capillary action of hardened mortar
BS EN 1015-19:1999	Determination of water vapour permeability of hardened rendering and plastering mortars
BS EN 1015-21:2002	Determination of the compatibility of one-coat rendering mortars with substrates

Methods of test for ancillary components for masonry

BS EN 846-XX   (Tests cover aspects of the dimensional accuracy, strength, stiffness and durability of wall ties, straps, hangers, brackets, lintels and bed-joint reinforcement of steel meshwork)

Methods of test for masonry composites

BS EN 1052-1:1999   Methods of test for masonry, determination of compressive strength
BS EN 1052-2:1999   Methods of test for masonry, determination of flexural strength
BS EN 1052-3 (1999)   Methods of test for masonry, shear strength of unit-mortar joints
BS EN 1052-4 (1999)   Methods of test for masonry, shear strength of dpcs
BS EN 1052-5:1999   Methods of test for masonry, determination of bond strength

PART 6

POLYMERS

Len Hollaway

Introduction

Pioneers in the development of plastics include Alexander Parkes in the 1860s in the UK and Leo Baekeland in the USA, who developed 'Parkesine' and 'Bakelite', respectively. Much of the development and exploitation of polymers during the last century stemmed from the growth of the oil industry. Since the 1930s oil has been our main source of organic chemicals, from which synthetic plastics, fibres, rubbers and adhesives are manufactured. The by-products of the distillation of petroleum are called basic chemicals and they provide the building blocks from which many chemicals and products, including plastics, can be made.

A large variety of polymers, with a wide range of properties, have been developed commercially since 1955. For example, phenol formaldehyde (PF) is a hard thermosetting material, polystyrene is a hard, brittle thermoplastic; polythene and plasticised polyvinyl chloride (PVC) are soft, tough thermoplastic materials. Plastics can also exist in various physical forms: bulk solids, rigid or flexible foams, sheet or film. Many of these materials have found use in the construction industry, and this section describes the processing, properties and applications of those that have been used as sealants, adhesives, elastomers and geosynthetics.

Although many of the materials are relatively strong, their stiffness is too low for most structural applications. As we will see in the next section they can be combined with fibres of high stiffness and strength to form composites with improved structural properties.

Chapter 38

Polymers: types, properties and applications

38.1 Polymeric materials

Polymers are produced by combining a large number of small molecular units (monomers) by the chemical process known as polymerisation to form long-chain molecules. There are two main types. Thermoplastics consist of a series of long-chain polymerised molecules, in which all the chains are separate and can slide over one another. In thermosetting polymers the chains become cross-linked so that a solid material is produced that cannot be softened and that will not flow. Polymers are usually made in one of two polymerisation processes. In condensation-polymerisation the linking of molecules creates by-products, usually water, nitrogen or hydrogen gas. In addition-polymerisation no by-products are created. Both thermosetting and thermoplastic polymers can be manufactured by these processes.

38.1.1 THERMOPLASTIC POLYMERS

The long-chain molecules of a thermoplastic polymer are held together by relatively weak van der Waals forces (Chapter 1), but the chemical bonds along the chain are extremely strong (*Fig. 38.1a*) When the material is heated, the intermolecular forces are weakened and the polymer becomes soft and flexible; at high temperatures it becomes a viscous melt. When it is allowed to cool again it solidifies. The cycle of softening by heating and hardening by cooling can be repeated almost indefinitely, but with each cycle the material tends to become more brittle.

Thermoplastic materials can have either a semi-crystalline ordered structure or an amorphous random structure. Civil engineering materials such as polypropylene, nylon 66 and polycarbonate are examples of amorphous thermoplastic polymers. Developments in the field of engineering polymers include the introduction of high-performance polymers, such as polyethersulphone (PES), which is amorphous, and polyetheretherketone (PEEK), which is semicrystalline; these polymers offer properties far superior to those of the normal thermoplastic polymers. They are not normally employed in civil engineering owing to their high costs but they are used in the aerospace engineering industry (Hollaway and Thorne, 1990).

38.1.2 THERMOSETTING POLYMERS

The principal thermosetting polymers that are used in construction are polyesters, vinylesters and epoxies. They may be used for two different functions: firstly as a composite when combined with a fibrous material and secondly as an adhesive; in the latter case epoxies are generally used (see section 38.4.2). Thermosetting polymers are formed in a two-stage chemical reaction when a polymer (e.g. polyester, vinylester or epoxy) is reacted with a curing agent (e.g. triethylenetetramine (TETA)). Firstly, a substance consisting of a series of long-chain polymerised molecules, similar to those in thermoplastics, is produced, then the chains become cross-linked. This reaction can take place either at room temperature or under the application of heat and pressure. As the cross-linking is by strong chemical bonds, thermosetting polymers are rigid materials and their mechanical properties are affected by heat.

Two basic procedures are employed to polymerise the thermosetting polymers used in the civil engineering industry; these are the cold-cure systems

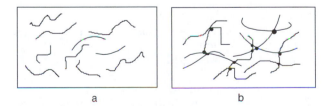

a   b

Fig. 38.1 *(a), The long-chain molecules of a thermoplastic polymer; (b), the attached and/or cross-linked molecules of a thermosetting polymer.*

that are cured at ambient temperatures and the hot-cure systems, in which the polymerisation is performed at elevated temperatures. Thus different resin and curing systems are required for site and factory manufacture of FRP composites (e.g. the cold-cure systems for site work and the hot-cure systems for factory fabrication). Consideration must be given to the glass transition temperature, Tg, of cold-cure epoxy polymers (see Chapter 42, section 4.2.1).

38.1.3 FOAMED POLYMERS

A rigid foam is a two-phase system of gas dispersed in solid polymer, and is produced by adding a blowing agent to molten resin. In the exothermic polymerisation reaction the gas is released and causes the polymer to expand, increasing its original volume many times by the formation of small gas cells. Like solid polymers, rigid foam polymers can be either thermoplastic or thermosetting and generally any polymer can be foamed.

38.2 Processing of thermoplastic polymers

Thermoplastic polymers may readily be processed into sheets or rods or complex shapes in one operation, which is often automated. Stages such as heating, shaping and cooling will ideally be a single event or a repeated cycle. The principal processing methods are extrusion, injection moulding, thermoforming and calendering.

The first is the most important method from the civil engineering viewpoint, and this is therefore outlined below. Powder or granules of thermoplastic polymer are fed from a hopper to a rotating screw inside a heated barrel; the screw depth is reduced along the barrel so that the material is compacted. At the end of the barrel the melt passes through a die to produce the desired finished article. Changing the die allows a wide range of products to be made, such as:

- profile products
- film-blown plastic sheet
- blow-moulded hollow plastic articles
- co-extruded items
- highly orientated grid sheets.

38.2.1 PROFILE PRODUCTS

With different extrusion dies, many profiles can be manufactured, such as edging strips, pipes, window-frames, etc. However, success depends upon the correct design of the die.

38.2.2 FILM-BLOWN PLASTIC SHEET

Molten plastic from the extruder passes through an annular die to form a thin tube; a supply of air inside the tube prevents collapse and when the film is cooled it passes through collapsing guides and nip rolls and is stored on drums. Biaxial orientation of the molecules of the polymer can be achieved by varying the air pressure in the polymer tube, which in turn controls the circumferential orientation. Longitudinal orientation can be achieved by varying the relative speeds of the nip roll and the linear velocity of the bubble; this is known as draw-down.

38.2.3 BLOW-MOULDED HOLLOW PLASTIC ARTICLES

A molten polymer tube, the Parison, is extruded through an annular die. A mould closes round the Parison and internal pressure forces the polymer against the sides of the mould. This method is used to form such articles as bottles and cold-water storage tanks. The materials commonly used are polypropylene, polyethylene and polyethylene terephthalate (PET).

38.2.4 CO-EXTRUDED ITEMS

A multilayered plastic composite is sometimes needed to withstand the end use requirements. Two or more polymers are combined in a single process by film blowing with an adhesive film between them. Reactive bonding processes to chemically cross-link the polymers are under development.

38.2.5 HIGHLY ORIENTATED GRID SHEETS

Polymer grids are used in civil engineering as the reinforcement for soil in reinforced earth. Continuous sheets of thermoplastic polymers, generally polypropylene or polyethylene, are extruded to very fine tolerances and with a controlled structure. A pattern of holes is stamped out in the sheet and the stampings are saved for re-use. The perforated sheet is stretched in the longitudinal and then in the transverse direction to give a highly orientated polymer in the two directions, with a tensile strength similar to that of mild steel. The low original stiffness of the material can be increased ten-fold. The stiffness of unorientated high-density polyethylene (HDPE), for instance, is initially only 1 GPa, but after forming into an orientated molecular structure it increases to 10 GPa. The use of these sheets is discussed later in this chapter. In injection moulding, softened thermoplastic polymer is forced through a nozzle into a clamped cold mould. When the plastic becomes cold, the mould is opened and the article is ejected; the operation is then repeated.

Table 38.1 Typical mechanical properties for some thermosetting and thermoplastic polymers

Material	Relative density	Ultimate tensile strength (MPa)	Modulus of elasticity in tension (GPa)	Coefficient of linear expansion $\times 10^{-6}/^{\circ}C$
Thermosetting				
Polyester	1.28	45–90	2.5–4.0	100–110
Vinylester (BASF Palatel A430-01)	1.07	90	4.0	80
Epoxy	1.03	90–110	3.5–7.0	48–85
Thermoplastic				
Polyvinyl chloride (PVC)	1.37	58.0	2.4–2.8	50
Acrylonitrile butadiene styrene (ABS)	1.05	17–62	0.69–2.82	60–130
Nylon	1.13–1.15	48–83	1.03–2.76	80–150
Polyethylene	0.96	30–35	1.10–1.30	120

38.3 Polymer properties

38.3.1 MECHANICAL PROPERTIES

Thermoplastic polymers, which are not cross-linked, derive their strength and stiffness from the properties of the monomer units and their high molecular weight. Consequently, in crystalline thermoplastic polymers there is a high degree of molecular order and alignment, and during any heating the crystalline phase will tend to melt and to form an amorphous viscous liquid. In amorphous thermoplastic polymers there is a high degree of molecular entanglement so they act like a cross-linked material. On heating, the molecules become disentangled and the polymer changes from a rigid solid to a viscous liquid.

The thermosetting polymers used in construction are reinforced with glass, aramid or carbon fibres to form the fibre/matrix composite for civil/structural utilisation. These polymers are cross-linked and form a tightly bound three-dimensional network of polymer chains; the mechanical properties are highly dependent upon the network of molecular units and upon the lengths of cross-linking chains. The characteristics of the network units are a function of the polymers used, their curing agents and the heat applied at the polymerisation stage. In addition, the length of the cross-linked chains is determined by the curing process. The most satisfactory way to cure polymers (and hence composites) is by the application of heat, thus achieving optimum cross-linking and hence enabling the substance to realise its potential. Shrinkage of the polymer during curing does occur, particularly with polyesters; thus contraction on cooling at ambient temperature can lead to stress build-up between the matrix and fibre of a composite. This effect is caused by the differences between the thermal expansion coefficients of the matrix and fibre, and it can have a major effect on the internal micro-stresses, which are sometimes sufficient to produce micro-packing, even in the absence of external loads; this will be discussed later. *Table 38.1* gives the most important mechanical properties of the common thermosetting and thermoplastic polymers used in civil engineering.

38.3.2 TIME-DEPENDENT CHARACTERISTICS

The deformation of a polymer material over time under the application of a load is referred to as the creep of that material; this deformation will continue as long as the load is applied. Upon removing the load the polymer will regain some of its original length but will not return completely to its original condition. *Figure 38.2* illustrates the total creep curve for a polymer under a given uniaxial tensile stress at constant temperature; it is divided into five regions, as shown on *Fig. 38.2*. The reason for this characteristic is that a polymer has both the properties of an elastic solid and a viscous fluid. Consequently, polymer materials have mechanical characteristics that lie somewhere between the ideal Hookean materials, where stress is proportional to strain, and the Newtonian materials, where stress is proportional to rate of strain. Thus they are classified as a viscoelastic material and their stress (σ) is a function of strain (ε) and time (t), as described by the equation:

$$\sigma = f(\varepsilon, t) \tag{38.1}$$

This non-linear viscoelastic behaviour can be simplified for design purposes:

$$\sigma = \varepsilon . f(t) \tag{38.2}$$

309

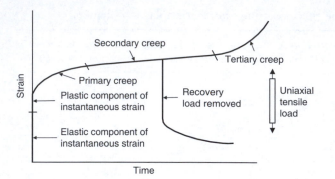

Fig. 38.2 *Total creep curve for a polymer under a uniaxial tensile stress.*

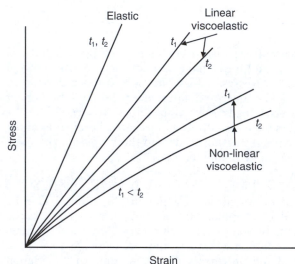

Fig. 38.3 *Stress–strain behaviour of elastic and viscoelastic materials at two values of time.*

This linear viscoelastic response indicates that, under sustained tensile stress, after a particular time interval the stress is directly proportional to strain. *Figure 38.3* illustrates schematically the various types of response discussed above.

The creep, flow and plastic deformation in polymeric materials result from the irreversible slippage, decoupling and disentanglements of the polymer chains in semi-crystalline polymers. An important consequence of the time-dependent behaviour of polymers is that when they are subjected to a particular strain, the stress necessary to maintain this strain decreases with time.

The creep of a polymeric material depends upon:

- the time-dependent nature of the micro-damage in the composite material subjected to stress
- the molecular characteristics and microstructure

- the loading history and the nature of the applied load
- the temperature and moisture environments into which it is placed.

In assessing the creep of a polymer material it is particularly important to ensure that the service temperature does not approach the glass transition temperature of the polymer (see Chapter 42, section 42.1). As the polymer approaches this temperature its mechanical characteristics change.

Figure 38.4 illustrates a family of creep curves consisting of an isostrain creep curve (*Fig. 38.4a*), an isostress creep curve (*Fig. 38.4b*) and a 100-second isochronous stress–strain creep curve (*Fig. 38.4c*). The first and third curves have been produced by cross-plotting, from the isostress creep curves, at constant times. By varying the stress as shown in *Fig. 38.4c* the isochronous stress–strain curves will correspond to a specific loading direction. Constant load tests are carried out under controlled conditions, as required by BS 4618-5.3: (1972), at durations of 60 s, 100 s, 1 h, 2 h, 100 h, 1 year, 10 years and 80 years, and the 100-s isochronous stress–strain curve is a plot of the total strain (at the end of 100 s) against the corresponding stress level (*Fig. 38.4c*). The three-dimensional creep graph shows the relationships of curves (a), (b) and (c) on one graph. The creep modulus of the material may then be found by measuring the slope at any specific stress level on the isochronous stress–strain curve; the creep modulus will vary for every stress level.

Leadermann (1943), was the first to suggest that temperature was an 'accelerating factor' in the time dependence of polymer properties and Aklonis and MacKnight (1983) developed the time–temperature superposition principle (TTSP). It is based upon the observation that the short-term behaviour of viscoelastic materials at higher temperatures is similar to the long-term behaviour at some lower reference temperature (Cardon *et al.*, 2000). The TTSP assumes that the effect of increasing temperature is equal to expanding the time of the creep response by a shift factor, such that creep curves made at different temperatures are superimposed by horizontal shifts along a logarithmic time scale to give a single curve that describes the creep response over a larger range of time, or the master curve (ASTM D2990, 2001; Goertzen and Kessler, 2006). Thus the TTSP allows for short-term creep curves at a range of temperatures to be used to generate a creep compliance master curve that is much longer than the short-term creep curve. The method can be used for polymers and polymer composites.

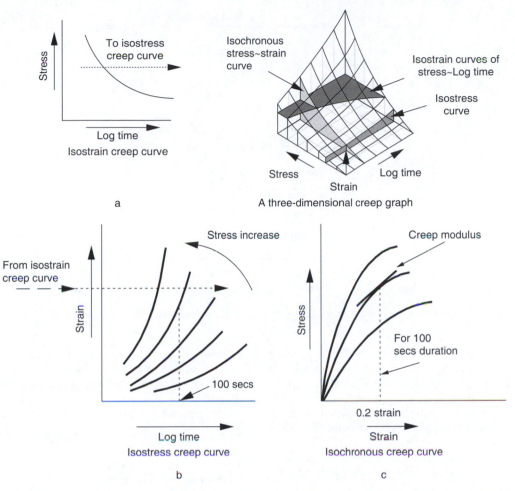

Fig. 38.4 *Typical creep curves for a non-linear viscoelastic material for varying stress values. Stress–log time curve and the build up to the stress–strain isochronous creep curve (Adapted from Hollaway, 2009).*

When applied to the generation of a creep master curve the following steps are required:

- A master specimen is subjected to a constant load at a certain temperature as in conventional creep tests.
- Similar experiments are performed for identical specimens at different temperature levels and the relevant creep curves drawn.
- An arbitrary reference temperature is selected.
- All the individual creep curves corresponding to different temperature levels are shifted along the log(time) co-ordinate to obtain the superimposed master curve.

The shift process is shown diagrammatically in *Fig. 38.5*.

Cheng and Yang (2005) developed the TTSP further by introducing a matched theoretical calculated curve from a supposed model of transition kinetics in which only time is involved as the independent variable. Cessna (1971) developed the time, applied stress superposition principle (TSSP), which can also estimate creep values for polymers and polymer composite materials.

38.4 Applications and uses of polymers

38.4.1 SEALANTS

Sealants are elastomeric materials that can be used for sealing joints against wind and water in construction. Thin curtain wall construction employs highly effective materials to provide the heat installation, but generally there is no cavity for the dispersion of water that may leak through the joints on the

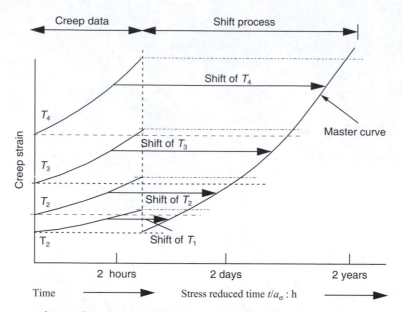

Fig. 38.5 *The shift process shown diagrammatically (Adapted from Hollaway, 2009).*

outside. In addition, in the event of air blowing directly to the inside of a joint there must be an effective material to provide heat insulation, consequently a baffling system to provide this must be installed. Therefore, adhesive and elastic sealants are required to enable this type of construction to be used efficiently.

The largest variety of sealants fall into the classification of solvent release and are composed of three component parts:

- the basic non-volatile vehicle (the liquid portion of the compound)
- the pigment component
- a solvent or thinner to make application easier.

The non-volatile vehicle can vary from a vegetable oil (e.g. linseed) to a synthetic elastomer. Opacity or colour will be introduced into the material by the pigment component. To enable the sealant to be applied easily and to ensure the correct thickness is achieved a solvent is introduced. The sealant is cured and its required viscosity is reached by the evaporation of the solvent. The butyl rubber solution and the acrylic copolymer solution fall into this category.

Another group of sealants comprises those that are chemically cured. Polysulphide compounds and silicone-based compounds are the main sealants under this heading. The latter, which is a two part sealant, is highly dependent upon the environmental conditions for its rate of cure, thus, if the temperature and humidity are low, the curing period could

be very long. The chemically cured compounds require adhesion additives in order to develop bond to a surface, as they do not generally contain much solvent.

The desired properties of a sealant are:

- a good adhesion with the joint
- low rate of hardening
- low rate of shrinkage
- permanent elasticity.

The choice of sealants is a compromise as no one product has all the above mentioned attributes.

38.4.2 ADHESIVES

Within the context of the construction industry the term adhesive embraces not only those materials that are used to bond together two components of a structure, but also those materials that provide a specific function in themselves (e.g. protection, decoration) and are at the same time self-adhesive to the substrate whose surface they modify. Thus, a mortar that may be used to bond bricks together may also be applied as a self-adhesive protection and often decorative rendering over the finished block work.

In this section adhesive bonded connections are the primary concern. The physical nature of fibre/matrix composites introduces problems that are not encountered in metals. The fibre type and arrangement, as well as the resin type and fibre volume fraction will influence the behaviour of the joint. In addition, composites are not generally homogeneous throughout their thickness as many thermosetting polymers have gel coats applied to the laminating resin to protect it against aggressive environmental influences met with in construction; the resin-rich surface layers will thus be brittle and, when overloaded, are liable to display a brittle fracture. An appropriate resin should therefore be chosen, such as a compliant one that will distribute the applied load over a large area, thus reducing the stress taken by the friable surface of the composite. There are two particular problems associated with adhesive bonding of fibre-reinforced polymer (FRP) materials:

- the attachment to the surface of a layered material
- the surface may be contaminated with mould-release agents remaining on it from the manufacturing procedure.

As the matrix material in a polymer composite is also an organic adhesive, the polymers that are used to join composite materials together are likely to be similar in terms of chemical composition and mechanical properties. Currently epoxy- and acrylic-based toughened adhesives are used for general application and have proved over the years to be very versatile and easy to use; they are durable, robust and relatively free from toxic hazards, and the toughened adhesives exhibit high pull strengths.

The basic requirements for the production of a satisfactory joint are:

- The adhesive should exhibit adequate adhesion to the materials involved.
- A two-part epoxy resin with a polyamine-based hardener should be used, which exhibits good moisture resistance and resistance to creep.
- The Tg of the adhesive should generally be greater than 60°C.
- The flexural modulus of the material should fall within the range of 2 to 10 GPa at 20°C.
- The equilibrium water content should not exceed 3% by weight after immersion in distilled water at 20°C. The coefficient of permeability should not exceed 5×10^{-14} m^2/s.
- It should possess gap-filling properties, be thixotropic, and be suitable for application to vertical and overhead surfaces.
- It should not be sensitive to the alkaline nature of concrete if this material is the adherent and it should not adversely affect the durability of the joint.

Adhesive formulations are, in general, complex. To the base resin is added one of a range of different types of curing agents (hardener) and additives, such as fillers, toughening agents, plasticizers, surfactants, anti-oxidants and any other required materials.

Curing agents are chosen depending upon whether the cure of the resin is to be at ambient or at elevated temperatures; the rate of a chemical reaction is approximately doubled for every 8°C degree rise in temperature. It will be clear that the properties of the adhesive will be altered with the large variety of additives that can be incorporated into the base resin.

If the adhesive is required to join two dissimilar materials, such as polymer composite and concrete or steel, the mechanical and thermal properties should be considered in relation to these two materials. The effects of environmental and other service conditions on the adhesive material and on the behaviour of bonded joints must be considered carefully.

With some bonding surfaces, such as steel, it will generally be necessary to apply an adhesive-compatible primer coat to generate a reliable and reproducible surface. With concrete surfaces it might be advisable to use a primer to give suitable conditions on which to apply a relatively viscous adhesive. It will not generally be necessary to prime a composite surface.

As with all resins it is necessary to keep a close check on the following items to ensure that the adhesive, when used in the field, is in pristine condition:

- the shelf-life is within the manufacturer's recommended time limit
- the viscosity and wetting ability are satisfactory
- the curing rate is correct
- the ambient temperature does not fall below the specification value
- post cure is complete before load is applied to the joint.

38.4.3 ELASTOMERS

The elastomer is another member of the polymer family. The material consists of long-chain molecules that are coiled and twisted in a random manner and the molecular flexibility is such that the material is able to undergo very large deformations. The material is cross-linked by a process known as vulcanisation, which prevents the molecules of the elastomer moving irreversibly relative to each other when under load. After a curing process, the molecules are crossed-linked like a thermosetting polymer. As the vulcanisation process does not change the form of the coiled molecules but merely prevents them from sliding, the elastomeric material will completely recover its original shape after the removal of an external force.

38.4.4 GEOSYNTHETICS

One area in which major advances in polymer science have been made in the last 35 years has been the burgeoning use of these materials in the geotechnical engineering industry. The most common material is the geotextile, a simple definition of which is that it is a textile material used in a soil environment.

In the early 1970s these materials were referred to as civil engineering fabrics or filter fabrics, their primary use being as filters. In the latter part of the decade they became known as geotechnical fabrics, as they were primarily used in geotechnical soil engineering applications. It was in the early 1980s that the term geotextile was suggested as a suitable name for this type of material. At the same time impermeable polymeric membranes were also being used increasingly. These materials became known as geomembranes. Thus in the mid-1980s many types of polymeric-based materials were being used in the geotechnical engineering industry, and many of these could not be classed as either a geotextile or a geomembrane. To encompass all these polymeric materials the new name 'geosynthetic' was derived, which is defined as a synthetic (polymeric) material used in a soil environment. Ingold and Miller (1988) discuss the different materials available for use in civil engineering, their properties and their measurement.

Geosynthetics, which are all thermoplastic polymers, can be divided into five broadly based categories:

1. *Geotextiles*: polymeric textile materials used in geotechnical engineering applications. These materials are essentially permeable to the passage of water.
2. *Geogrids*: open, mesh like polymeric structures.
3. *Geomembranes*: polymeric materials in sheet form that are essentially impermeable to the passage of water.
4. *Geolinear elements*: long, slender, polymeric materials normally used as reinforcing tendons in soils and rocks.
5. *Geocomposites*: covers all polymeric materials used in a soil environment not covered by the above four categories.

Each of these is now discussed.

Geotextiles

Geotextiles are usually classified by their method of manufacture and are made in two stages: the manufacture of the linear elements, such as fibres, tapes, etc. and the fabrication of those linear elements into geotextiles. The fibres are the basic load-bearing elements in the material and the forming technique determines the structure and hence the physical and mechanical characteristics of the system. The main fibres used in geotextiles are synthetic ones such as polyethylene, polypropylene, polyester and polyamide.

Geogrids

Geogrids are often grid-like structures of thermoplastic polymeric material, and in conjunction with the soil form a quasi-composite system, where the grid structure is the fibre and the soil is assumed to be the 'matrix' and forms an efficient bond with the fibre. Geogrids are of two forms: cross-laid strips and punched thermoplastic polymer sheets. The manufacturing techniques for the strips and polymer sheet are discussed in Hollaway (1993).

Geomembranes

Geomembranes are synthetic materials manufactured in impermeable sheet form from thermoplastic polymers or bituminous materials. Both materials can be reinforced or unreinforced; the former is manufactured on a production line and the latter can be produced on a production line or *in situ*. The matrix can be reinforced by textiles.

Geolinear elements

Geolinear elements are long, slender strips or bars consisting of a unidirectional filament fibre core that is made from a polyester, aramid or glass fibre in a polymer sheath of a low-density polyethylene or a resin. The components of the system form a composite, in that the fibre provides the strength and extension characteristics and the matrix protects the fibre from internal influences and provides the bonding characteristics with the soil.

Geocomposites

Geocomposites consist of two or more different types of thermoplastic polymer system combined into a hybrid material. Their main function is to form a drainage passage along the side of the water course, with a polymer core as the drainage channel and the geotextile skin as the filter.

As is apparent, many of the materials in each of the above groups are fibre composites of a parent polymer reinforced with polymer fibres and so will be discussed in more detail in the next section.

References

Aklonis JJ and MacKnight WJ (1983). *Introduction to Polymer Viscoelasticy*, 2nd edition, Wiley, New York, NY, pp. 36–56.

ASTM D2990 (2001). *Standard Test Methods for Tensile, Compressive and Flexural Creep and Creep Rupture of Plastics*, American Society for Testing and Materials, Pennsylvania.

Cardon AH, Qin Y, Van Vossole C and Bouquet P (2000). Prediction of residual structural integrity of a polymer matrix composite construction element. *Mechanics of Time-Dependent Materials*, **4**, 155–167.

Cessna LC (1971). Stress-time superposition for creep data for polypropylene and coupled glass reinforced polypropylene. *Polymer Engineering Science*, **13**, 211–219.

Cheng R and Yang H (2005). Application of time-temperature superposition principle to polymer transient kinetics. *Journal of Applied Polymer Science*, **99** (No. 4), 1767–1772.

Goertzen WK and Kessler MR (2006). Creep behavior of carbon fiber/epoxy matrix composites. *Material Science and Engineering A*, **421**, 217–225.

Hollaway LC and Thorne AM (1990). *High technology carbon-fibre/polyethersulphone composites for space applications*. Proceedings of the ESA Symposium: Space Applications of Advanced Structural Materials, ESTEC, Noordwik, The Netherlands, March 21–23, SP-303, June.

Hollaway LC (1993). *Polymer Composites for Civil and Structural Engineering*, Blackie Academic and Professional, Glasgow, New York, Tokyo, Victoria and Madras.

Hollaway LC (2009). 'Advanced Polymer Composites', Chapter 3 of Section 7, ICE Manual of Construction Materials, editor Forde MC, Vol. 2, published by Thomas Telford, London.

Ingold TS and Miller KS (1988). *Geotextiles Handbook*, Thomas Telford Limited, London.

Leadermann H (1943). 'Elastic and Creep Properties of Filamentous Materials and Other High Polymers', Washington, D.C.: The Textile Foundation.

Bibliography

Cook DI (2003). *Geosynthetics – RAPRA* Report 158, **14**, No. 2, pp. 1–120.

Horrocks AR and Anand SC (eds) (2000). *Handbook of Technical Textiles*, Woodhead Publishing, Cambridge, UK.

Michalowski RL and Viratjandr C (2005). *Combined Fiber and Geogrid Reinforcement for Foundation Soil Slabs*. Symposium on Geosynthetics and Geosynthetic-Engineered Structures, Louisiana, 1–3 June (eds Ling HI, Kaliakin VN and Leshchinsky D), Session 3, Paper 2.

PART 7

FIBRE COMPOSITES

Introduction

The history of fibre-reinforced composites as construction materials is more than 3000 years old. Well-known examples are the use of straw in clay bricks, mentioned in Exodus, and horsehair in plaster. Other natural fibres have been used over the ages to reinforce mud walls and give added toughness to rather brittle building materials.

This part is divided into two sections to reflect the difference in the mechanics of the reinforcing process between fibre-reinforced polymers and cementitious materials. The first, on fibre-reinforced polymers, is mostly concerned with pre-cracking behaviour, and the second, on fibre-reinforced cement and concrete, is mainly concerned with post-cracking performance.

In the last section we commented that polymers, although relatively strong, have low stiffness. They can, however, be combined with fibres of high stiffness and strength to form a range of composites with improved structural performance that have an increasing number of applications. These include 'hi-tech' uses of carbon fibres in resin systems for aircraft components and sports equipment and glass fibre reinforced systems (GFRP) for car bodies and ships' hulls. In civil engineering there has been a steadily increasing use of FRP composites for rehabilitation and retrofitting of reinforced concrete, metallic and timber structural components, large span all-composite footbridges and modest span all-composite road bridges, bridge decks, etc. The composition, manufacture, analysis, properties and uses of these systems are described in Section 1.

Fibre-reinforced composites can also be based on inorganic cements and binders. Most fibre-reinforced cement-based composites differ from fibre-reinforced polymers in that most of the reinforcing effects of the fibres occur after the brittle matrix has cracked either at the microscopic level or with visible cracks through the composite. This is the result of the relatively low strain to failure of the cement matrix (~0.01–0.05 per cent) compared with the high elongation of the fibres (~1–5 per cent). The fibres in cement-based systems often have a lower modulus of elasticity than the cement matrix and hence little or no increase in cracking stress is expected from the fibre reinforcement.

The most notable exception to this division is asbestos cement, which was the most commercially successful fibre-reinforced composite in the 20th century, in terms of both tonnage and turnover for many years. It was invented in about 1900 by Hatschek and its high tensile cracking stress and failure strain (in excess of 0.1 per cent) results in part from the suppression of cracks propagating from flaws. The use of asbestos fibres reduced after about 1980 owing to their carcinogenic effects and this stimulated the search for alternative fibre systems, chiefly glass and polypropylene, which have enabled the traditional products to be safely marketed even if with considerably changed properties.

From the 1960s there has been growing interest in modifying the properties of fresh and hardened concrete by the addition of fibres. These include both steel and polypropylene and, to a limited extent, glass fibres. Their use is now widespread in the construction industry, and the range of materials, their properties and most common applications are described in Section 2.

Section 1: Polymer composites

Len Hollaway

Introduction

Polymer composites are used to form engineering materials and consist of strong stiff fibres in a polymer resin; they require scientific understanding from which design procedures may be developed. The mechanical and physical properties of the composite are clearly controlled by their constituent properties and by their micro-structural configurations. It is, therefore, necessary to be able to predict properties when parameter variations take place.

An important aspect of composite material design is the property of anisotropy; it is necessary to give special attention to the methods of controlling this property and its effect on analytical and design procedures.

Research work has demonstrated that with the correct process control and a soundly based material design approach, it is possible to produce composites that can satisfy stringent structural requirements. However, in the construction industry, because it is difficult to establish the durability of the composite over a period of say 50 years, great care has to be exercised in predicting stress and deformations over this time span.

In this section we first consider the manufacture and properties of the range of fibres that are most commonly used, and then consider the analysis of the behaviour of the composites, which provides an essential basis for their design. We then describe manufacturing processes, which we follow with a discussion of durability and design issues. In the last chapter we describe a number of applications that illustrate the wide and increasing range of uses in both the building and civil engineering sectors.

Chapter 39

Fibres for polymer composites

When a load is applied to a fibre-reinforced composite consisting of a low-modulus matrix reinforced with high-strength, high-modulus fibres, the viscoelastic flow of the matrix under stress transfers the load to the fibre; this results in a high-strength, high-modulus material that determines the stiffness and strength of the composite and is in the form of particles of high aspect ratio (i.e. fibres), is well dispersed and bonded by a weak secondary phase (i.e. matrix).

Many amorphous and crystalline fibres can be used, including glass, and fibres produced from synthetic polymers, such as carbon fibre (made by elongating polyacrylonitrile and then placing its elongated form in various inert gases). Making a fibre involves aligning the molecules of the material, and the high tensile strength is associated with improved intermolecular attraction resulting from this alignment. Polymeric fibres are made from those polymers whose chemical composition and geometry are basically crystalline and whose intermolecular forces are strong. As the extensibility of the material has already been utilised in the process of manufacture, such fibres have a low elongation.

The following sections discuss the manufacture, make-up and properties of fibres that can be used to upgrade polymers, cements, mortars and concretes. These include glass, carbon and Kevlar fibres that are used in conjunction with thermosetting polymers such as polyesters, vinylesters and epoxies to form civil engineering composites. Some typical properties of a number of these composites are then given.

39.1 Fibre manufacture

39.1.1 GLASS FIBRES

Glass fibres are manufactured by drawing molten glass from an electric furnace through platinum bushings at high speed; *Fig. 39.1* illustrates the

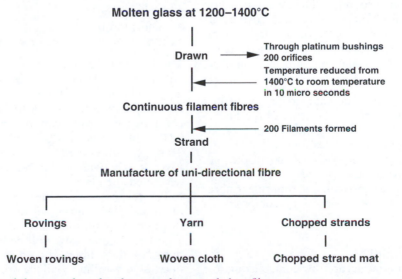

Fig. 39.1 Illustration of the procedure for the manufacture of glass fibre.

procedure. The filaments cool from the liquid state at about 1200°C to room temperature in 10^{-5} seconds. On emerging from the bushings the filaments are treated with a lubricant or size and 200 filaments are bundled together to form a strand. The main functions of the lubricant are to:

- facilitate the manufacturing of the strands and moulding of the composite
- reduce damage to the fibres during mechanical handling
- reduce the abrasive effect of the filaments against one another.

The following four types of glass fibre are the major ones used in construction:

- E-glass fibre of low alkali content is the commonest glass fibre on the market and is the major one used in composites in the construction industry. It was first used in 1942 with polyester resin to manufacture radomes (a structural weatherproof enclosure that protects radar equipment), and is now widely used with polyester, vinylester and epoxy resin.
- A-glass fibre of high alkali content was formerly used in the aircraft industry, and is now used for special manufactured articles in civil engineering.
- Z-glass (zirconia glass) was developed for reinforcing cements, mortars and concretes because of its high resistance to alkali attack.
- S2-glass fibre is used in extra-high-strength and high-modulus applications in aerospace and on occasions in civil engineering.

Strands of glass fibre are combined to form thicker parallel bundles called rovings which, when twisted, can form several different types of yarn; rovings or yarns can be used individually or in the form of woven fabric. Glass strands and rovings for reinforcing thermosetting polymers may be used in a number of different forms:

- woven rovings, to provide high strength and stiffness characteristics in the direction of the fibres
- chopped fibres, to provide a randomly orientated dispersion of fibres in the composite
- chopped strand mat, to provide a quasi-isotropic reinforcement to the composite
- surface tissues, to provide a thin glass-fibre mat when a resin-rich surface of composite is required.

39.1.2 CARBON FIBRES

Carbon fibres are manufactured by controlled pyrolysis and cyclisation of certain organic fibre precursors

(the raw material used to make carbon fibre). About 90% of carbon fibres are produced from polyacrylonitrile (PAN), the remaining 10% are made from rayon or petroleum pitch. The PAN precursor is manufactured by spinning to produce a round cross-section fibre; the yield is only 50% of the original precursor fibre. It can also be manufactured by a melt assisted extrusion as part of the spinning operation. I-type and X-type rectangular cross-section carbon fibre composites are produced with a closer fibre packing in the composite compared to that of the circular fibre. The PITCH precursor fibres are derived from petroleum, asphalt, coal tar and PVC, the carbon yield is high but the uniformity of the fibre cross-sections is not constant from batch to batch. This non-uniformity is acceptable to the construction industry although it must be said that it is not acceptable to the aerospace industry. The pitch fibre precursor is invariably used when ultra-high (European definition – see below) stiffness carbon fibres are used in construction.

The conversion process for carbon fibre includes stabilisation at temperatures up to 400°C, carbonisation at temperatures from 800 to 1200°C and graphitisation in excess of 2000°C. Surface treatment of the fibres, which includes sizing and spooling, is then undertaken. *Figure 39.2* shows a diagrammatic representation of the process.

Carbon fibres are organic polymers, characterised by long strings of molecules bound together by carbon atoms. The exact composition of each precursor varies from one manufacturing technique to another. They are classified according to their mechanical properties as:

- high strength
- high modulus
- ultra-high modulus fibres (European definition).

The last two are alternatively defined in the USA and in some Asian countries as normal modulus and high-modulus fibres. Carbon filaments are typically between 5 and 8 μm in diameter and are combined into tows containing between 3000 and 12 000 filaments. A tow count is typically between 200 tex and 900 tex. The tows are twisted into yarns and woven into fabrics analogous to those described for glass.

The techniques for the manufacture of the high modulus (HM) and the ultra-high modulus (UHM) (European definitions) are the same but the heat treatment temperature will be higher the higher the modulus of the fibres, thus, a more highly orientated fibre of crystallites will be formed for the UHM fibre.

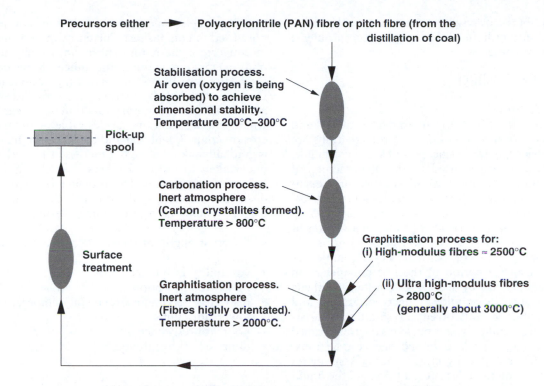

Precursors either → **Polyacrylonitrile (PAN) fibre or pitch fibre (from the distillation of coal)**

Stabilisation process.
Air oven (oxygen is being absorbed) to achieve dimensional stability.
Temperature 200°C–300°C

Pick-up spool

Carbonation process.
Inert atmosphere
(Carbon crystallites formed).
Temperature > 800°C

Graphitisation process for:
(i) High-modulus fibres ≈ 2500°C

Surface treatment

Graphitisation process.
Inert atmosphere
(Fibres highly orientated).
Temperasture > 2000°C.

(ii) Ultra high-modulus fibres > 2800°C
(generally about 3000°C)

Polyacrylonitrile fibres for production of high modulus fibres (constructionindustry) or production of high modulus or ultra-high modulus (aerospace industry).
Pitch fibres for production of ultra-high modulus carbon fibres (constructionindustry)
{Definitions of fibres used here are the European ones}

Fig. 39.2 Schematic representation of the production of carbon fibre (Adapted from Hollaway (2009)).

39.1.3 ARAMID FIBRES

Aramid (aromatic polyamide) fibres are produced by an extrusion and spinning process typically used to produce a thermoplastic acrylic fibre. A solution of the polymer in a suitable solvent at a temperature of between −50 and −80°C is extruded into a hot cylinder which is at a temperature of 200°C; this causes the solvent to evaporate and the resulting fibre is wound onto a bobbin. To increase its strength and stiffness properties, the fibre undergoes a stretching and drawing process, thus aligning the molecular chains, which are then made rigid by means of aromatic rings linked by hydrogen bridges. There are two grades of stiffness available; one has a modulus of elasticity of the order of 130 GPa – which is the one used in polymer composites for upgrading structural systems – and the other has a modulus of elasticity of 60 GPa and is used in bullet proof systems.

Aramid fibres are resistant to fatigue, both static and dynamic. They are elastic in tension but exhibit non-linear characteristics under compression and care must be taken when high strain compressive or flexural loads are involved. Aramid fibres exhibit good toughness and general damage tolerance characteristics.

39.1.4 LINEAR ORGANIC FIBRES

By orientating the molecular structure of simple thermoplastic polymers into one direction during their manufacture, a high-strength and high-modulus organic fibre can be produced. This fibre, in future, could be one of the major reinforcements for civil engineering structures. With a relative density of 0.97, high-modulus polyethylene fibres, produced in the USA and The Netherlands, have mechanical properties of the same order as those of aramid fibres, with modulus of elasticity and tensile strength values of 117 GPa and 2.9 GPa, respectively. These values were determined at ambient temperature but will decrease rapidly with increasing temperature. Furthermore, with non-cross linked thermoplastic polymer fibres, creep will be significant, but by cross-linking using radiation technology, creep problems can be overcome.

39.1.5 OTHER FIBRES

Synthetic fibres

The important fibres for upgrading cements and mortars or for use in reinforced earth situations are polypropylene, polyethylene, polyester and polyamide. The first two are utilised in the manufacture of cement/mortar composites; all are used in geosynthetics, especially to form geotextiles and geogrids. Synthetic fibres are the only ones that can be engineered chemically, physically and mechanically to suit particular geotechnical engineering applications.

The manufacture of synthetic fibres commences with the transformation of the raw polymer from solid to liquid either by dissolving or melting. Synthetic polymers such as acrylic, modacrylic, aramid and vinyl polymers are dissolved into solution, whereas the polyolefin and polyester polymers are transformed into molten liquid; chlorofibre polymers can be transformed into a liquid by either means. A spinneret consisting of many holes is used to extend the liquid polymer, which is then solidified into continuous filaments. The filaments undergo further extension in their longitudinal axes, thus further increasing the orientation of the molecular chain within the filament structure, with a consequent improvement in the stress–strain characteristics. Different types of synthetic fibre or yarn may be produced, including monofilament fibres, heterofilament fibres, multi-filament yarns, staple fibres, staple yarns, split-film tapes and fibrillated yarns.

Natural fibres

In recent years there has been interest in the natural fibre as a substitute for glass fibre because of the potential advantages of weight saving, lower raw material price, and potential for recycling and renewing. Natural fibres are used to reinforce conventional thermoplastics, for example, injection moulding and press moulding interior parts for the automobile industry. The fibres are generally short and randomly orientated. They are obtained from different parts of plants: jute, ramie, flax, kenaf and hemp are obtained from the stem whereas sisal, banana and pineapple from the leaf and cotton and kapok from seed. All plant species are built up of cells and the components of natural fibres are cellulose, hemicellulose, lignin, pectin, waxes and water-soluble substances; the first three components govern the physical properties of the fibre; Pickering

(2008) provides an overview of the types of natural fibre used in composites. Currently, there is interest in converting the natural fibres into long, aligned reinforcement to exploit the inherent mechanical properties of plants in structural applications. Diversification of the market in geotextiles, which are required for temporary functions – for example, where biodegradation is desirable – temporary erosion control, building and construction materials is gradually taking place but there is little work being undertaken to use these fibres in civil engineering structural components. This is owing to the following disadvantages compared with those fibre composites currently being used in civil engineering:

- lower strength properties, particularly impact strength
- variability in quality
- moisture absorption
- restricted maximum processing temperature
- lower durability
- poor fire resistance
- dimensional instability.

39.2 Fibre properties

The advantage of fibre/polymer composite materials over the more conventional civil engineering materials is that they have high specific strength and high specific stiffness, achieved by the use of low-density fibres with high strength and modulus values *Table 39.1*. Some strength and stiffness values of carbon, glass and Kevlar fibres have been mentioned in the previous sections. *Table 39.1* gives a more comprehensive set of values. The degree of alignment of the small crystalline units in the carbon fibres varies considerably with the manufacturing technique, which thus affects the stiffness of the three types of fibre. The arrangement of the layer planes in the cross-section of the fibre is also important, because it affects the transverse and shear properties.

The strength and modulus of elasticity of glass fibres are determined by the three-dimensional structure of the constituent oxides, which can be of calcium, boron, sodium, iron or aluminium. The structure of the network and the strength of the individual bonds can be varied by the addition of other metal oxides and so it is possible to produce glass fibres with different chemical and physical properties. The properties of carbon and glass fibres are anistropic and therefore the modulus of elasticity of both fibres along and transverse to the fibres will not be the same. The

Table 39.1 Typical tensile mechanical properties of glass, carbon and aramid fibres used in civil engineering

Material	Fibre	Elastic modulus (GPa)	Tensile strength (MPa)	Ultimate strain (%)
Glass fibre	E	69	2400	3.5
	A	69	3700	5.4
	S-2	86	3450	4.0
Carbon fibre				
Pan based fibre:	HM	300	5200	1.73
Hysol Grafil Apollo	UHM	450	3500	0.78
	HS	260	5020	1.93
Pan based fibre:	G-40-700	300	4960	1.66
BASF Celion	Gy 80	572	1860	0.33
Pan based fibre:	T300	234	3530	1.51
Torayca				
Pitch based fibres:	T-300	227.5	2758.0	1.76
Hysol Union carbide	T-500	241.3	3447.5	1.79
	T-600	241.3	4137.0	1.80
	T-700	248.2	4550.7	1.81
Aramid fibre	49	125	2760	2.2
	29	83	2750	3.3

HM, High modulus (European definition); UHM, ultra high modulus (European definition); HS, high strain.

main factors that determine the ultimate strength of glass fibres are the processing conditions and the damage sustained during handling and processing.

The manufacturing processes for Kevlar fibres align the stiff polymer molecules parallel to the fibre axes, and the high modulus achieved indicates that a high degree of alignment is possible. When the fibres have been incorporated into a matrix material, composite action takes place and as discussed in the next chapter, a knowledge of the fibre alignment, fibre volume fraction and method of manufacture is necessary to obtain the optimum mechanical characteristic of the material.

39.3 Polymer composite properties

There have been several definitions of the meaning of advanced polymer composites. A clear definition is essential to their understanding, and in 1989 a study group of the Institution of Structural Engineers, the Advanced Polymer Composites Group, defined an advanced polymer composite for the construction industry as follows:

'Composite materials consist normally of two discrete phases, a continuous matrix which is often a resin, surrounding a fibrous reinforcing structure. The reinforcement has high strength and stiffness whilst the matrix binds the fibres together, allowing stress to be transferred from one fibre to another producing a consolidated structure. In advanced or high performance composites, high strength and stiffness fibres are used in relatively high volume fractions whilst the orientation of the fibres is controlled to enable high mechanical stresses to be carried safely. In the anisotropic nature of these materials lies their major advantage. The reinforcement can be tailored and orientated to follow the stress patterns in the component leading to much greater design economy than can be achieved with traditional isotropic materials. The reinforcements are typically glass, carbon or aramid fibres in the form of continuous filament, tow or woven fabrics. The resins which confer distinctive properties such as heat, fibre or chemical resistance may be chosen from a wide spectrum of thermosetting or thermoplastic synthetic materials, and those commonly used are polyester, epoxy and phenolic resins. More advanced heat resisting types such as vinylester and bismaleimides are gaining useages in high performance applications and advanced carbon fibre/thermoplastic composites are well into a market development phase.'

Table 39.2 Typical mechanical properties of long directionally aligned epoxy fibre composites (fibre weight fraction 65%) used in civil engineering and manufactured by an automated process

Fibre	Relative density	Tensile strength (MPa)	Tensile modulus (GPa)	Flexural strength (MPa)	Flexural modulus (GPa)
E-glass	1.9	760–1030	41.0	1448	41.0
S–2 glass	1.8	1690	52	–	–
Aramid	1.45	1150–1380	70–107	–	–
Carbon (PAN)	1.6	2689–1930	130–172	1593	110.0
Carbon (Pitch)	1.8	1380–1480	331–440	–	–

The method of manufacture of polymer composites for construction are given in Chapter 41.

Table 39.3 Typical mechanical properties of glass fibre/vinylester polymer composites used in civil engineering and manufactured by different fabrication methods

Method of manufacture	Tensile strength (MPa)	Tensile modulus (GPa)	Flexural strength (MPa)	Flexural modulus (GPa)
Wet Lay-up	62–344	4–31	110–550	6–28
Spray-up	35–124	6–12	83–190	5–9
RTM	138–193	3–10	207–310	8–15
Filament winding	550–1380	30–50	690–1725	34–48
Pultrusion	275–1240	21–41	517–1448	21–41

Table 39.4 Typical mechanical properties of glass fibre/vinylester polymer manufactured by an automated process – randomly orientated fibres

Fibre:matrix ratio (%)	Relative density	Flexural strength (MPa)	Flexural modulus (GPa)	Tensile strength (MPa)	Tensile modulus (GPa)
67	1.84–1.90	483	17.9	269	19.3
65	1.75	406	15.1	214	15.8
50	1.8	332	15.3	166	15.8

Structural polymer composites have a wide spectrum of mechanical properties. These properties will be dependent upon:

- the relative proportions of fibre and matrix materials (the fibre/matrix volume or weight ratio)
- the method of manufacture (Chapter 41)
- the mechanical properties of the component parts (a carbon fibre array will give greater stiffness to the composite than an identical glass fibre array)
- the fibre orientation within the polymer matrix (the fibre orientations can take the form of unidirectional, bi-directional, various off-axis directions and randomly orientated arrays).

The fibre arrangement within the matrix will influence the type and the mechanical properties of the composite material. *Table 39.2* gives typical mechanical properties of composites manufactured using long directionally aligned fibre reinforcement of glass, aramid and carbon with a fibre:matrix ratio by weight of 65:35. *Table 39.3* shows typical mechanical properties of glass fibre composites manufactured by different techniques; it clearly illustrates the effect that the methods of fabrication have on the properties. *Table 39.4* shows the variation of the composite properties when the fibre:matrix ratio is changed, the method of manufacture and component parts of the composite remaining constant.

The methods of manufacture of polymer composites for construction are described in Chapter 44.

Reference

Pickering K (2008). *Properties and performance of natural-fibre composites*, Woodhead Publishing Ltd, Cambridge, England.

Bibliography

Hollaway LC and Head PR (2001). *Advanced Polymer Composites and Polymers in the Civil Infrastructure*, Elsevier, Oxford.

Hollaway LC (2009). 'Advanced Polymer Composites', Chapter 3 of Section 7, ICE Manual of Construction Materials, editor Forde MC, Vol. 2, published by Thomas Telford, London.

Kim DH (1995). *Composite structures for civil and architectural engineering*, E & F Spon, London.

Matthews FL and Rawlings RD (2002). *Composite materials: Engineering and Science*, Woodhead Publishing Ltd, Cambridge.

Chapter 40

Analysis of the behaviour of polymer composites

40.1 Characterisation and definition of composite materials

The mechanical properties of polymers can be greatly enhanced by incorporating fillers and/or fibres into the resin formulations. Therefore, for structural applications, such composite materials should:

- consist of two or more phases, each with their own physical and mechanical characteristics
- be manufactured by combining the separate phases such that the dispersion of one material in the other achieves optimum properties of the resulting material
- have enhanced properties compared with those of the individual components.

In fibre-reinforced polymer materials, the primary phase (the fibre) uses the plastic flow of the secondary phase (the polymer) to transfer the load to the fibre; this results in a high-strength, high-modulus composite. Fibres generally have both high strength and high modulus but these properties are only associated with very fine fibres with diameters on the order of 7–15 mm; they tend to be brittle. Conversely, polymers may be either ductile or brittle and will generally have low strength and stiffness. By combining the two components a bulk material is produced with a strength and stiffness that depend on the fibre volume fraction and the fibre orientation.

The properties of fibre/matrix composite materials are highly dependent upon the micro-structural parameters such as:

- fibre diameter
- fibre length

- fibre volume fraction of the composite
- fibre orientation and packing arrangement.

It is important to characterise these parameters when considering the processing of the composite material and the efficient design and manufacture of the composite made from these materials. The interface between the fibre and the matrix plays a major role in the physical and mechanical properties of the composite material. The transfer of stresses between fibre and fibre takes place through the interface and the matrix and in the analysis of composite materials a certain number of assumptions are made to enable solutions to mathematical models to be obtained:

- the matrix and the fibre behave as elastic materials
- the bond between the fibre and the matrix is perfect, consequently there will be no strain discontinuity across the interface
- the material adjacent to the fibre has the same properties as the material in bulk form
- the fibres are arranged in a regular or repeating array.

The properties of the interface region are very important in understanding the stressed composite. The region is a dominant factor in the fracture toughness of the material and in its resistance to aqueous and corrosive environments. Composite materials that have weak interfaces have low strength and stiffness but high resistance to fracture, and those with strong interfaces have high strength and stiffness but are very brittle. These effects are functions of the ease of de-bonding and pull-out of the fibres from the matrix material during crack propagation. Using the above assumptions, it is possible to calculate the distribution of stress and strain in a composite material in terms of the geometry of the component materials.

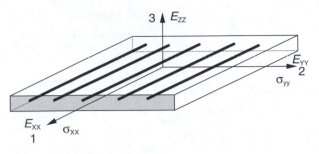

Fig. 40.1 Basic laminate.

40.2 Elastic properties of continuous unidirectional laminate

40.2.1 LONGITUDINAL STIFFNESS

A basic laminate is shown in *Fig. 40.1* and it is assumed that the orthotropic layer has three mutually perpendicular planes of property symmetry; it is characterised elastically by four independent elastic constants (see section 40.6 and section 40.8, *Fig. 40.5*). They are:

E_{11} = modulus of elasticity along fibre direction
E_{22} = modulus of elasticity in the transverse direction
v_{12} = Poisson's ratio, i.e. strains produced in direction 2 when specimen is loaded in direction 1
G_{12} = longitudinal shear modulus
and

$$E_{11}v_{21} = E_{22}v_{12}$$

where v_{21} = Poisson's ratio

If the line of action of a tensile or compressive force is applied parallel to the fibres of a unidirectional laminate, the strain ε_m in the matrix will be equal to the strain ε_f in the fibre, provided the bond between the two components is perfect. As both fibre and matrix behave elastically then:

$$\sigma_f = E_f\varepsilon_f \quad \text{and} \quad \sigma_m = E_m\varepsilon_m, \quad \text{where} \quad \varepsilon_f = \varepsilon_m$$

As $E_f > E_m$ the stress in the fibre must be greater than the stress in the matrix and will therefore bear the major part of the applied load. The composite load $P_c = P_m + P_f$ or:

$$\sigma_c A_c = \sigma_m A_m + \sigma_f A_f$$
$$\sigma_c = \sigma_m V_m + \sigma_f V_f \qquad (40.1)$$

where A = the area of the phase, V = the volume fraction of the phase, with V_c (the volume of composite) = 1. As the bond is perfect:

$$\varepsilon_c = \varepsilon_m = \varepsilon_f$$

and from eqn (40.1):

$$E_c\varepsilon_c = E_m\varepsilon_c V_m + E_f\varepsilon_c V_f \qquad (40.2)$$

Thus:

$$E_c = E_m V_m + E_f V_f$$

and

$$E_c = E_{11} = E_m(1 - V_f) + E_f V_f. \qquad (40.3)$$

This equation is often referred to as the *law of mixtures* equation.

40.2.2 TRANSVERSE STIFFNESS

The same approach can be used to obtain the transverse modulus of a unidirectional laminate E_{22}. The applied load transverse to the fibres acts equally on the fibre and matrix and therefore:

$$\sigma_f = \sigma_m$$
$$\varepsilon_f = \sigma_{22}/E_f \quad \text{and} \quad \varepsilon_m = \sigma_{22}/E_m \qquad (40.4)$$
$$\varepsilon_{22} = V_f\varepsilon_f + V_m\varepsilon_m \qquad (40.5)$$

Substituting equation (40.4) into equation (40.5):

$$\varepsilon_{22} = V_f\sigma_{22}/E_f + V_m\sigma_{22}/E_m \qquad (40.6)$$

Substituting $\sigma_{22} = E_{22}\varepsilon_{22}$ into equation (40.6) gives;

$$E_{22} = E_f E_m/[E_f(1 - V_f) + E_m V_f] \qquad (40.7)$$

Equation (40.7) predicts E_{22} with reasonable agreement when compared with experimental results. Equation (40.8) has been proposed and takes account of Poisson contraction effects:

$$E_{22} = E_f E_m'/[E_f(1 - V_f) + E_m' V_f] \qquad (40.8)$$

where

$$E' = E_m/(1 - v_m^2)$$

40.3 Elastic properties of in-plane random long-fibre laminate

A laminate manufactured from long randomly orientated fibres in a polymer matrix are, on a microscopic scale, isotropic in the plane of the laminate. The general expression (the proof is given in Hollaway, 1989) for the elastic modulus of laminate consisting of long fibres is:

$$1/E_\theta = (1/E_{11})(\cos^4\theta) + (1/E_{22})(\sin^4\theta)$$
$$+ [(1/G_{12}) - (2v_{12}/E_{11})]\cos^2\theta \sin^2\theta \qquad (40.9)$$

where θ = angle defining the direction of required stiffness. *Figure 40.2* shows the relationship of E_θ

when θ varies between 0° and 90°. It can be seen then that laminate can be made with a predetermined distribution of the orientation of the fibres, so that elastic and other mechanical properties can be designed to meet specific needs.

40.4 Macro-analysis of stress distribution in a fibre/matrix composite

The behaviour of composites reinforced with fibres of finite length l cannot be described by the above equations. As the aspect ratio, which is defined by the fibre length divided by the fibre diameter (l/d), decreases, the effect of fibre length becomes more significant. When a composite containing uniaxially aligned discontinuous fibres is stressed in tension parallel to the fibre direction there is a portion at the end of each finite fibre length, and in the surrounding matrix, where the stress and strain fields are modified by the discontinuity. The efficiency of the fibre to stiffen and to reinforce the matrix decreases as the fibre length decreases. The critical transfer length

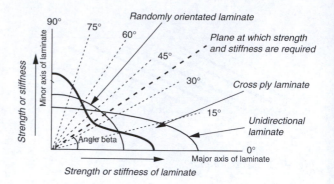

Fig. 40.2 *The relationship of E_θ and angle θ between 0° and 90°.*

over which the fibre stress is decreased from the maximum value, under a given laminate load, to zero at the end of the fibre is referred to as half the critical length of the fibre. To achieve the maximum fibre stress, the fibre length must be equal to or greater than the critical value l_c. *Figure 40.3* shows a schematic representation of a discontinuous fibre/matrix laminate subjected to an axial stress; the stress distributions of the tensile and shear components are shown.

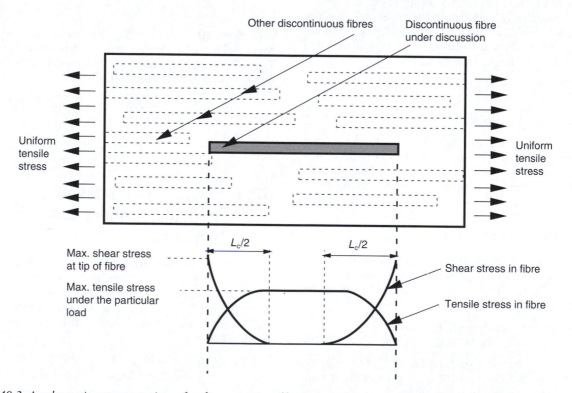

Fig. 40.3 *A schematic representation of a discontinuous fibre/matrix laminate subjected to an axial stress.*

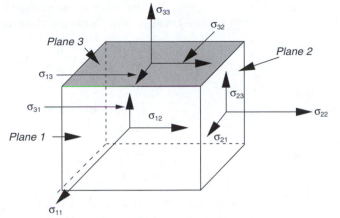

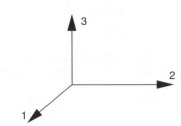

Through thickness stresses (on plane 3)
are zero for laminate construction

Fig. 40.4 *Components of stress acting on an elemental unit cube.*

40.5 Elastic properties of short-fibre composite materials

As discussed above the reinforcing efficiency of short fibres is less than that of long fibres. In addition the orientation of short fibres in a laminate is random and therefore the laminate can be assumed to be isotropic on a macro scale. The law of mixtures as given in equation 40.3 can be modified by the inclusion of a fibre orientation distribution factor η, thus the composite modulus of elasticity is given by:

$$E_c = E_{11} = E_m V_m + \eta E_f V_f \qquad (40.10)$$

Values of η have been calculated by Krenchel (1964) for different fibre orientations:

- $\eta = 0.375$ for a randomly oriented fibre array
- $\eta = 1.0$ for unidirectional laminate when tested parallel to the fibre
- $\eta = 0$ for unidirectional laminate when tested perpendicularly to the fibre
- $\eta = 0.5$ for a bidirectional fibre array.

The following sections are important for a complete understanding of composite theory, and you may find that you have to go back to study them in more detail after the first reading.

40.6 Laminate theory

Sections 40.2, 40.3, 40.4 and 40.5 discussed individual laminate properties; this section concentrates

upon laminates that are formed when two or more laminate are combined to produce a composite. It describes the methods used to calculate the elastic properties of the laminates, and briefly introduces the elasticity theory.

The stresses at a point in a body are generally represented by stress components acting on the surface of a cube; *Fig. 40.4* shows the three normal and the three shear stresses. The notation employed here is such that the first subscript refers to the plane upon which the stress acts and the second subscript is the coordinate direction in which the stress acts; the equivalent strains have the same notation. As laminate are assumed to be sufficiently thin the through thickness stresses are zero. Thus $\sigma_{33} = \sigma_{31} = \sigma_{13} = 0$ and plane stress conditions hold.

40.6.1 ISOTROPIC LAMINATE

For a homogeneous isotropic laminate the stress–strain relationship is:

$$\sigma_{11} = (E/(1 - \upsilon^2))(\varepsilon_{11} + \upsilon\varepsilon_{22})$$
$$\sigma_{22} = (E/(1 - \upsilon^2))(\varepsilon_{22} + \upsilon\varepsilon_{11}) \qquad (40.11a)$$
$$\sigma_{12} = (E/2(1 + \upsilon)(\varepsilon_{12})$$

or in matrix form:

$$\begin{pmatrix} \sigma_{11} \\ \sigma_{22} \\ \sigma_{12} \end{pmatrix} = \begin{pmatrix} Q_{11} & Q_{12} & 0 \\ Q_{21} & Q_{22} & 0 \\ 0 & 0 & Q_{33} \end{pmatrix} \begin{pmatrix} \varepsilon_{11} \\ \varepsilon_{22} \\ \varepsilon_{12} \end{pmatrix}$$

$$[\sigma] = [Q][\varepsilon] \qquad (40.11b)$$

where:

$$Q_{11} = E/(1 - v^2) = Q_{22}$$
$$Q_{12} = vE/(1 - v^2) = Q_{21}$$
$$Q_{33} = E/2(1 + v) = G$$

There are two independent constants in these equations; these are E and v, and this indicates isotropic material properties.

The corresponding set of equations to those in equation (40.11b), which relate strains to stresses, are:

$$\begin{pmatrix} \varepsilon_{11} \\ \varepsilon_{22} \\ \varepsilon_{12} \end{pmatrix} = \begin{pmatrix} S_{11} & S_{12} & 0 \\ S_{21} & S_{22} & 0 \\ 0 & 0 & S_{33} \end{pmatrix} \begin{pmatrix} \sigma_{11} \\ \sigma_{22} \\ \sigma_{12} \end{pmatrix} \qquad (40.12a)$$

$$[\varepsilon] = [S][\sigma] \qquad (40.12b)$$

where:

$$S_{11} = 1/E = S_{22}$$
$$S_{33} = 1/G$$
$$S_{12} = -v/E = S_{21}$$

40.6.2 ORTHOTROPIC LAMINATE

The orthotropic laminate can be assumed to be isotropic in plane 1, as shown in *Fig. 40.4* (i.e. the plane normal to the axis direction (1), as the properties are independent of direction in that plane. The stress–strain relationship for an orthotropic laminate is:

In matrix form

$$\begin{pmatrix} \sigma_{11} \\ \sigma_{22} \\ \sigma_{33} \end{pmatrix} = \begin{pmatrix} Q_{11} & Q_{12} & 0 \\ Q_{21} & Q_{22} & 0 \\ 0 & 0 & Q_{33} \end{pmatrix} \begin{pmatrix} \varepsilon_{11} \\ \varepsilon_{22} \\ \varepsilon_{33} \end{pmatrix}$$

$$[\sigma] = [Q][\varepsilon] \qquad (40.13)$$

where

$$Q_{11} = E_{11}/(1 - v_{12}v_{21}); \quad Q_{22} = E_{22}/(1 - v_{12}v_{21})$$
$$Q_{12} = v_{21}E_{11}/(1 - v_{12}v_{21}); \quad Q_{21} = v_{12}E_{22}/(1 - v_{12}v_{21})$$
$$Q_{33} = G_{12}.$$

As the Q matrix is symmetric we have $v_{21}E_{11} = v_{12}E_{22}$. The Poisson's ratio v_{12} refers to the strains produced in direction 2 when the laminate is loaded in direction 1. There are four independent constants in these equations, E_{11}, E_{22}, v_{12} and v_{21}, and this indicates orthotropic material properties. From the above equation it can be seen that the shear stress σ_{12} is independent of the elastic properties E_{11}, E_{22}, v_{12} and v_{21}, and therefore no coupling between tensile and shear strains takes place.

The corresponding set of equations to those in equation (40.13b), which relate strains to stresses are:

$$\begin{pmatrix} \varepsilon_{11} \\ \varepsilon_{22} \\ \varepsilon_{12} \end{pmatrix} = \begin{pmatrix} S_{11} & S_{12} & 0 \\ S_{21} & S_{22} & 0 \\ 0 & 0 & S_{33} \end{pmatrix} \begin{pmatrix} \sigma_{11} \\ \sigma_{22} \\ \sigma_{12} \end{pmatrix}$$

$$[\varepsilon] = [S][\sigma] \qquad (40.14)$$

where:

$$S_{11} = 1/E_{11}; \quad S_{22} = 1/E_{22}; \quad S_{33} = 1/G_{12}$$
$$S_{12} = -v_{21}/E_{22}; \quad S_{21} = -v_{12}/E_{11}$$

If the line of application of the load is along some axis other than the principal one, then the laminate principal axes do not coincide with the reference axes x, y of the load and the former axes must be transformed to the reference axes. *Figure 40.5* illustrates the orientation of the orthotropic laminate about the reference axis. Hollaway (1989) showed that the stress–strain relationship in the (x,y) coordinate system at angle θ to the principle material direction becomes:

$$\begin{pmatrix} \sigma_{xx} \\ \sigma_{yy} \\ \sigma_{xy} \end{pmatrix} = \begin{pmatrix} \bar{Q}_{11} & \bar{Q}_{12} & \bar{Q}_{13} \\ \bar{Q}_{21} & \bar{Q}_{22} & \bar{Q}_{23} \\ \bar{Q}_{31} & \bar{Q}_{32} & \bar{Q}_{33} \end{pmatrix} \begin{pmatrix} \varepsilon_{xx} \\ \varepsilon_{yy} \\ \varepsilon_{xy} \end{pmatrix}$$

or

$$[\sigma] = [\bar{Q}][\varepsilon] \qquad (40.15)$$

where

$$\bar{Q}_{11} = Q_{11}m^4 + Q_{22}n^4 + 2(Q_{12} + 2Q_{33})n^2m^2$$

$$\bar{Q}_{12} = \bar{Q}_{21} = (Q_{11} + Q_{22} - 4Q_{33})n^2m^2 + Q_{12}(n^4 + m^4)$$

$$\bar{Q}_{13} = \bar{Q}_{31} = (Q_{11} - Q_{12} - 2Q_{33})nm^3 + (Q_{12} - Q_{22} + 2Q_{33})n^3m$$

$$\bar{Q}_{22} = Q_{11}n^4 + Q_{22}m^4 + 2(Q_{12} + 2Q_{33})n^2m^2$$

$$\bar{Q}_{23} = \bar{Q}_{32} = (Q_{11} - Q_{12} - 2Q_{33})n^3m + (Q_{12} - Q_{22} + 2Q_{33})nm^3$$

$$\bar{Q}_{33} = (Q_{11} + Q_{22} - 2Q_{12} - 2Q_{33})n^2m^2 + Q_{33}(n^4 + m^4)$$

where Q_{11}, Q_{22}, Q_{12}, Q_{21} and Q_{33} have been defined in equation (40.13) and $m = \cos θ$, $n = \sin θ$. The equivalent expression for strain components in the reference axis x, y in terms of the stress components in that axis becomes:

$$\begin{pmatrix} \varepsilon_{xx} \\ \varepsilon_{yy} \\ \varepsilon_{xy} \end{pmatrix} = \bar{S} \begin{pmatrix} \sigma_{xx} \\ \sigma_{yy} \\ \sigma_{xy} \end{pmatrix}$$

where $[\bar{S}]$ is a 3×3 compliance matrix whose components are:

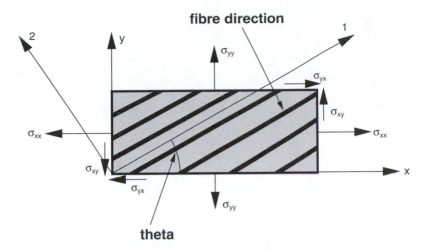

Fig. 40.5 *The orientation of the orthotropic laminate about the reference axis.*

$\bar{S}_{11} = S_{11}m^4 + S_{22}n^4 + (2S_{12} + S_{33})n^2m^2$

$\bar{S}_{12} = \bar{S}_{21} = (S_{11} + S_{22} - S_{33})n^2m^2 + S_{12}(n^4 + m^4)$

$\bar{S}_{13} = \bar{S}_{31} = (2S_{11} - 2S_{12} - S_{33})nm^3 - (2S_{22} - S_{12}$
$\qquad - S_{33})n^3m$

$\bar{S}_{23} = \bar{S}_{32} = (2S_{11} - 2S_{12} - S_{33})n^3m - (2S_{22} - 2S_{12}$
$\qquad - S_{33})nm^3$

$\bar{S}_{22} = S_{11}n^4 + S_{22}m^4 + (2S_{12} + S_{33})n^2m^2$

$\bar{S}_{33} = 2(2S_{11} + 2S_{22} - 4S_{12} - S_{33})n^2m^2 + S_{33}(n^4 + m^4)$

where S_{11}, S_{22}, S_{12}, S_{21} and S_{33} have been defined in equation (40.14).

40.7 The strength characteristics and failure criteria of composite laminate

In the two preceding sections, the stiffness relationships in terms of stress and strain were presented for isotropic and orthotropic materials. It is now necessary to have an understanding of the ultimate strengths of the laminate to enable a complete characterisation of the composite material to be made. The stress–strain relationship stated in the previous sections described the actual stresses occurring at any point in a laminate, and the strength characteristics may be considered as describing the allowable stress at any point.

When the formulation of the stiffness characteristics of the laminate was developed, properties in both tension and compression were assumed. However, the ultimate strength behaviour of composite systems may be different in tension and compression and the characteristics of the failure mode will be highly dependent upon the component materials. Therefore,

a systematic development of the strengths of these materials is not possible; consequently a series of failure criteria for composite materials will be given.

40.7.1 STRENGTH THEORIES FOR ISOTROPIC LAMINATES

In isotropic materials both normal and shear failure can occur, but it is usual to equate the combined stress situation to the experimentally determined uniaxial tension or compression value. When a tensile load is applied to a specimen in a uniaxial test it is possible for failure in the specimen to be initiated by either an ultimate tensile stress or a shear stress, because a tensile stress of s (the maximum principal stress in this type of test) on the specimen produces a maximum shear value of s/2. Consequently the failure theories are related to the applied tensile or compressive stress that causes failure, irrespective of whether it was a normal or a shear stress failure.

Many theories and hypotheses have been developed to predict the failure surface for composite materials under tensile loads, and probably the best known theories that have been used to predict failure – and that are discussed in Holmes and Just (1983) – are as follows.

The maximum principal stress theory

$$\sigma_{xx} = \sigma_t^* \qquad (40.17)$$

where

$\qquad \sigma_{xx}$ = maximum principal stress
$\qquad \sigma_t^*$ = failure stress in a uniaxial tensile test
or $\quad \sigma_{zz} = \sigma_c^*$
$\qquad \sigma_{zz}$ = minimum principal stress
$\qquad \sigma_c^*$ = failure stress in a uniaxial compressive test.

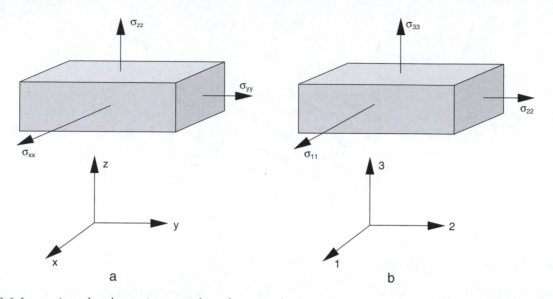

Fig. 40.6 *Isotropic and orthotropic materials under normal stresses: (a) isotropic element under three principal stresses $\sigma_{xx} > \sigma_{yy} > \sigma_{zz}$; (b) orthotropic material under normal stress.*

Figure 40.6 shows the principal stresses acting on an element of material.

The maximum principal strain theory

$$\varepsilon_{xx} = \varepsilon_t^* \qquad (40.18)$$

where:

ε_{xx} = maximum principal tensile strain
ε_t^* = tensile strain at failure

in terms of stress:

$$(\sigma_{xx} - \upsilon\sigma_{yy} - \upsilon\sigma_{zz})/E = \sigma_t^*/E$$
or $\quad (\sigma_{xx} - \upsilon\sigma_{yy} - \upsilon\sigma_{zz}) = \sigma_t^*$
$(\upsilon = \text{Poission's ratio})$

Similarly:

$$\varepsilon_{zz} = \varepsilon_c^*$$
ε_{zz} = minimum principal strain
ε_c^* = compression strain at failure.
or $\quad \sigma_{zz} - \upsilon(\sigma_{xx} + \sigma_{yy}) = \sigma_c^*$

Both of the above theories assume failure to be due to normal stresses and ignore any shear stress present. Consequently the theories are relevant to the failure of brittle materials under tension.

The total strain energy theory

The above theories express the failure criterion as either limiting stress or limiting strain; the total strain energy theory attempts to combine these two theories. The development of the theory, which is based upon strain energy principles, has been discussed in Hollaway (1989) and only the final solution will be given here. The laminate theory gives the solution as:

$$\sigma^{*2} = \sigma_{xx}^2 + \sigma_{yy}^2 - 2\sigma_{xx}\sigma_{yy}\upsilon \qquad (40.19)$$

The theory applies particularly to brittle materials in which the ultimate tensile stress is less than the ultimate shear stress.

Deviation strain energy theory

This theory is known as the von Mises criterion and in it the principal stresses σ_{xx}, σ_{yy} and σ_{zz} can be expressed as the sum of two components, namely the hydrostatic stress, which causes only a change in volume, and the deviation stress, which causes distortion of the body. The system is shown in *Fig. 40.7*.

The hydrostatic stress components produce equal strains in magnitude and are consistent in the three directions and therefore produce equal strain in these directions. The system, therefore, undergoes change in volume but not change in shape. The stress deviation system will cause the body to undergo changes in shape but not in volume. Again the theory has been developed in Hollaway (1989) and will not be repeated here; the laminate theory gives the solution as:

$$\sigma_{xx}^2 + \sigma_{yy}^2 - \sigma_{xx}\sigma_{yy} = \sigma^{*2} \qquad (40.20)$$

The above failure criterion is most relevant to ductile materials. It is not obvious which of the

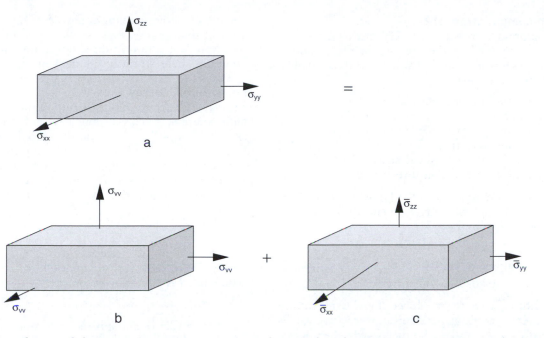

Fig. 40.7 *Volume and deviation stress system: (a) principal stress; (b) volume stress system; (c) stress deviation (or distortion) system.*

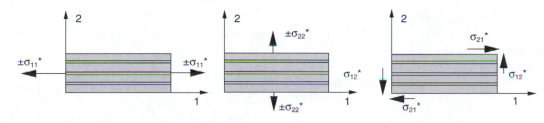

Fig. 40.8 *Critical stress values in the principal material axes.*

above failure criteria is most relevant to composites as the fibre volume fraction and orientation of the fibres in the polymer will influence their strength and ductility properties. However, the last theory has been applied to quasi-isotropic composites with some success.

40.7.2 STRENGTH THEORIES FOR ORTHOTROPIC LAMINATE

The theories based upon the strength characteristics of orthotropic materials are considerably more complicated than those for the isotropic ones. As with these latter materials the strength hypothesis is based upon simple fundamental tests, but because orthotropic materials have different strengths in different directions a more intensive set of data is required than for isotropic materials. Three uniaxial tests are

required, one in each of the principal axis directions to determine the three moduli of elasticity, Poisson's ratio and the strength characteristics; tests in these directions will eliminate any coupling effects of shearing and normal strains that would occur if the laminate were tested in any other direction. The shearing strengths with respect to the principal directions must be determined from independent experiments.

Figure 40.8 shows schematically the critical stress values in the principal material axes.

For orthotropic materials the stress condition at a point is resolved into its normal and shearing components relative to the principal material axis at the point. Consequently the failure criteria in these materials become functions of the basic normal and shearing strengths described for isotropic materials.

The maximum stress theory

The maximum stress theory of failure assumes that failure occurs when the stresses in the principal material axes reach a critical value. The three possible modes of failure are:

$\sigma_{11} = \sigma_{11}^*$ the ultimate tensile or
    compressive stress in direction 1
$\sigma_{22} = \sigma_{22}^*$ the ultimate tensile or
    compressive stress in direction 2   (40.21)
$\sigma_{12} = \sigma_{12}^*$ the ultimate shear stress
    action in plane 1 in direction 2

If the load were applied to the laminate at an angle θ to the principal axis direction shown in *Fig. 40.5* then by transformation:

$$\sigma_{11} = \sigma_{xx} \cos^2\theta = \sigma_\theta \cos^2\theta$$
$$\sigma_{22} = \sigma_{xx} \sin^2\theta = \sigma_\theta \sin^2\theta \qquad (40.22)$$
$$\sigma_{12} = -\sigma_{xx} \sin\theta \cos\theta = -\sigma_\theta \sin\theta \cos\theta$$

The failure strength produced by the maximum stress theory would depend upon the relative values of σ_{11}, σ_{22} and σ_{12} and would therefore be the smallest value of the following:

$$\sigma_{11} = \sigma_{11}^*/\cos^2\theta$$
$$\sigma_{22} = \sigma_{22}^*/\sin^2\theta \qquad (40.23)$$
$$\sigma_{12} = \sigma_{12}^*/\sin\theta \cos\theta$$

The maximum strain theory

The maximum strain theory of failure assumes that failure occurs when the strains in the principal material axes reach a critical value. Here again there are three possible modes of failure:

$$\varepsilon_{11} = \varepsilon_{11}^*$$
$$\varepsilon_{22} = \varepsilon_{22}^* \qquad (40.24)$$
$$\varepsilon_{12} = \varepsilon_{12}^*$$

Where ε_{11}^* is the maximum tensile or compressive strain in direction 1, ε_{22}^* is the maximum tensile or compressive strain in direction 2 and ε_{12}^* is the maximum shear strain on plane 1 in direction 2.

The Tsai–Hill energy theory

The Tsai–Hill criterion is based upon the von Mises failure criterion, which was originally applied to homogeneous isotropic bodies. It was then modified by Hill to suit anisotropic bodies, and finally applied to composite materials by Tsai. Hollaway (1989) has discussed the derivation of the equation that describes the failure envelope and this may be expressed as:

$$\sigma_{11}^2/\sigma_{11}^{*2} - \sigma_{11}\sigma_{22}/\sigma_{11}^{*2} + \sigma_{22}^2/\sigma_{22}^{*2} + \sigma_{12}^2/\sigma_{12}^{*2} = 1 \qquad (40.25)$$

For most composite materials $\sigma_{11}^* \gg \sigma_{22}$; consequently, the second term of equation (40.25) is negligible and this equation becomes:

$$\sigma_{11}^2/\sigma_{11}^{*2} + \sigma_{22}^2/\sigma_{22}^{*2} + \sigma_{12}^2/\sigma_{12}^{*2} = 1 \qquad (40.26)$$

Equations (40.25) and (40.26) are only apply to orthotropic laminate under in-plane stress conditions. To enable a prediction to be made of the failure strength in direction θ to the principal axes (*Fig. 40.5*) on unidirectional laminate, equations (40.25) and (40.23) can be combined to give:

$$\sigma_{xx} = \sigma_\theta = [\cos^4\theta/\sigma_{11}^{*2} + (1/\sigma_{12}^{*2} - 1/\sigma_{11}^{*2}) \sin^2\theta \cos^2\theta + \sin^4\theta/\sigma_{22}^{*2}]^{-1/2} \qquad (40.27)$$

Hull and Clyne (1996) have stated that when equation (40.27) has been fitted to the results of experimental tests on carbon fibre–epoxy resin laminates, the predicted values are much better than those for the maximum stress theory (equation 40.21).

Finally, *Fig. 40.9* shows a laminate made from three laminate. Providing laminates 1 and 3 have the

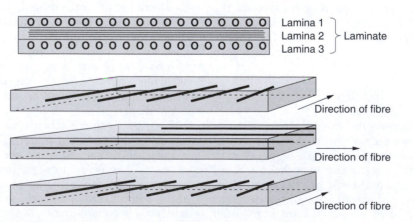

Fig. 40.9 Laminate arrangements.

same thickness, the laminate would be described as symmetric; laminate 2 could have any value of thickness. If, however, the thicknesses of laminates 1 and 3 were different, the laminate would be described as non-symmetric. Under a thermal and mechanical load, coupling forces are introduced into a non-symmetric laminate because of the different mechanical properties of the individual laminate. For this reason it is common practice in many applications to use symmetric laminates that are not subjected to this type of coupling.

References

Holmes M and Just DJ (1983). *GRP in Structural Engineering*, Applied Science Publishers, London and New York.

Hollaway L (1989). Design of composites. In *Design with Advanced Composite Materials* (ed. Phillips L), The Design Council, London.

Hull D and Clyne TW (1996). *An introduction to Composite Materials*, 2nd edition, Cambridge University Press, Cambridge.

Krenchel H (1964). *Fibre reinforcement*, Akademisk Forlag, Copenhagen.

Bibliography

Hollaway LC (1993). *Polymer Composites for Civil and Structural Engineering*, Blackie Academic & Professional, London, Glasgow, New York Tokyo, Melbourne, Madras.

Hollaway LC and Head PR (2001). *Advanced Polymer Composites and Polymers in the Civil Infrastructure*, Elsevier, Oxford.

Matthews FL and Rawlings RD (2004). *Composite Materials: Engineering and Science*, Woodhead Publishing, Cambridge, England.

Chapter 41

Manufacturing techniques for polymer composites used in construction

The two parts of this chapter concentrate upon the manufacturing techniques for civil engineering fibre-reinforced thermosetting and thermoplastic polymer composites, respectively.

41.1 Manufacture of fibre-reinforced thermosetting composites

There are three basic techniques used to manufacture advanced polymer composites for the civil engineering industry; each technique will have an influence on the mechanical properties of the final component. The techniques and their sub-divisions are:

- Manual
  - wet lay-up (either factory or site fabricated) and cold cured
  - pressure bag methods, fabricated and cold cured in a factory or on site
  - the vacuum assisted resin-transfer procedures (RTM, site fabricated and cold cured).
- Semi-automated
  - resin injection (cold cured)
  - low-temperature mould factory made impregnated fibre (prepreg) site fabricated and cured under pressure and elevated temperature.
- Automated
  - pultrusion (hot cured)
  - filament winding (cold cured)
  - the cold melt prepreg factory manufactured, cured under a vacuum assisted pressure of 1 bar and an elevated temperature
  - injection moulding (cold cured).

41.1.1 THE MANUAL PROCESSES
The manual processes currently used in construction are variations of the general *wet lay-up method*. The commercial methods of manufacture of fibre–polymer composites by this process are:

- the REPLARK™ method
- the Dupont method
- the Tonen Forca method.

These techniques are basically the same with minor variations. The wet lay-up process consists of *in-situ* wetting of dry fibres in the form of sheets or fabrics impregnated *in-situ* with a polymer. These are wrapped around the structural member during rehabilitation (see Chapter 43, section 43.5.1) or placed on a mould of the desired geometric shape. Their size will depend upon the size of the member or mould; the fibre reinforcements are generally of widths varying between 150 and 1500 mm.

The REPLARK process (REPLARK is a trade name used by Sumotomo Corporation, Europe PLC) is a commercially available method of rehabilitating (or retrofitting) a structural member to strengthen structures in flexure and shear by bonding the material to their tensile and/or shear faces; in these cases the surface of the structure forms the mould for the composite (see section 43.5). Furthermore, planar and non-planar composites can be manufactured independently and used as structural units. Pressure- and vacuum-bag moulding are similar lay-up systems, but pressure or vacuum is applied to the mould through a rubber membrane for compaction before curing commences (the voids in the composite material are considerably reduced, thus providing a glass fibre:polymer weight ratio of up to 55% for vacuum-bag mouldings and 65% for pressure-bag mouldings). To protect fibres from exposure to the atmosphere and especially to moisture penetration of the interface of the fibre and matrix, a resin-rich coat, known as a gel coat, is sometimes applied to the surface of the composite. The thickness of the gel coat is generally about 0.35 mm. Sometimes a surface tissue mat is used to reinforce the gel coat.

A second method for rehabilitating/retrofitting procedures is known as the Dupont method, which uses Kevlar fibres; it is marketed as a repair system for concrete structures.

In addition to manufacturing structural components made from polymer composites by the wet lay-up or spray-up techniques, commercially available procedures to rehabilitate composite materials to existing structural members, to improve their tensile and shear strengths, do exist.

The autoclave is a modification of the pressure-bag method; pressures of up to 6 bar are developed within the autoclave and the system produces a high-quality composite with a fibre:matrix weight ratio of up to 70%. The cost of production also increases.

These methods have been described in Hollaway (1993) and Hollaway and Head (2001).

41.1.2 THE SEMI-AUTOMATED PROCESS FOR THE REHABILITATION OF A STRUCTURAL MATERIAL

The semi-automated processes used currently are:

* the resin infusion under flexible tooling (RIFT) process
* the resin transfer moulding process (RTM)
* the low-temperature factory-made pre-impregnated fibre (prepreg), cured under pressure and elevated temperature.

Resin-infusion moulding. The semi-automated resin infusion under flexible tooling (RIFT) process has been developed by DML, Devonport, Plymouth, to allow quality composites to be formed. In this process dry fibres are preformed in a mould in the fabrication shop and the required materials are attached to the preform before packaging. The preform is taken to site and is attached to the structure; a resin supply is then channelled to the prepreg. The prepreg and resin supply is then enveloped in a vacuum-bagging system. As the resin flows into the dry fibre preform it develops both the composite material and the adhesive bond between the carbon fiber-reinforced polymer (CFRP) and the structure. The process provides composites with high fibre volume fractions on the order of 55%; these have high strength and stiffness values.

XXsys Technologies, Inc., San Diego, California, has developed a wrapping system for seismic retro-fitting to columns. The technology associated with the technique is based upon filament winding (described in section 41.1.3) of prepreg carbon fibre tows around the structural unit, thus forming a carbon fibre jacket; currently, structural units to be upgraded would be columns. The polymer is then cured by a controlled temperature oven and can, if desired, be coated to match the existing structure.

Resin-transfer moulding is a low-pressure, closed mould semi-mechanised process. In the RTM process, several layers of dry continuous strand mat, woven roving or cloth are placed in the bottom half of a two-part closed mould and a low-viscosity catalysed resin is injected under pressure into the mould cavity, and cured. Flat reinforcing layers, such as a continuous strand mat, or a 'preform' that has already been shaped to the desired product, can be used as the starting material in this process. The potential advantages of RTM are the rapid manufacture of large, complex, high-performance structures with good surface finish on both sides, design flexibility and the capability of integrating a large number of components into one part. This method can be employed to form large components for all composite bridge units but it is not often used.

The low-temperature mould factory-made pre-impregnated fibre (prepreg) is cured either in the factory, for the production of pre-cast plates, or on site if the prepreg composite is to be fabricated onto a structural member. In the latter case a compatible film adhesive is used and the film adhesive and the prepreg components are cured in one operation under an elevated temperature of 65°C applied for 16 hours or 80°C applied for 4 hours; a vacuum-assisted pressure of 1 bar is applied for simultaneous compaction of the composite and the film adhesive. This method is new but it is estimated that it will be used increasingly for rehabilitating degraded structural members (see Chapter 43, section 43.5). In the UK the manufacturing specialist in the production of hot-melt factory made pre-impregnated fibre for the construction industry is ACG, of Derbyshire.

41.1.3 AUTOMATED PROCESSES

The automated processes that are available to the construction industry are:

* filament winding
* the pultrusion technique.

The filament winding technique is a highly mechanised and sophisticated technique for the manufacture of pressure vessels, pipes and rocket casings when exceptionally high strengths are required. In the construction industry filament winding has been used to form high-pressure pipes and pressure vessels. Sewerage pipes have also been manufactured by filament winding.

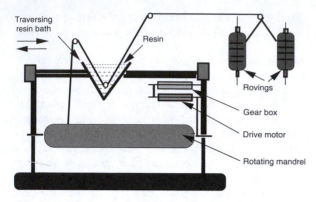

Fig. 41.1 *Schematic representation of the filament winding technique.*

Continuous reinforcement, usually rovings, is fed through a traversing bath of activated resin and is then wound on to a rotating mandrel. If resin pre-impregnated reinforcement is used, it is passed over a hot roller until tacky and is then wound on to the rotating mandrel. *Figure 41.1* illustrates the process and it is evident that the angle of the helix is determined by the relative speeds of the traversing bath and the mandrel. After completion of the initial polymerisation, the composite is removed from the mandrel and cured, for which process the composite unit is placed in an enclosure at 60°C for 8 hours.

The pultrusion technique and the pull-winding technique are used within a closed-mould system,

utilising heat to produce high-quality units. Owing to the high capital equipment outlay, particularly for the manufacture of the metal moulds and the initial set-up of runs, it is essential that large production runs are performed. Only a small skilled workforce is required owing to the mechanisation of the system. The technique consists of impregnating continuous strands of a reinforcing material with a polymer and drawing them through a die, as shown in *Fig. 41.2*. Thermosetting polymers are used in this process, although research is currently being undertaken to pultrude thermoplastic materials. Curing of the thermosetting composite component is undertaken when the die is heated to about 135°C. A glass content of between 60 and 80% by weight can be achieved. Composites manufactured by this method tend to be reinforced mainly in the longitudinal direction with only a relatively small percentage of fibres in the transverse direction. A technique was developed (Shaw-Stewart, 1988) to 'wind' fibres in the transverse direction simultaneously with the pultrusion operation. The process is known as *pull-winding* and gives the designer greater flexibility in the production of composites, particularly those of circular cross-section.

The pultrusion technique is the process used extensively in the civil engineering industry and is an important technique to form flexural/shear structural units and also to manufacture high-pressure water and sewerage system pipes (using the pull-winding procedure). The finished pultrusion sections are generally straight and dies can be manufactured to give

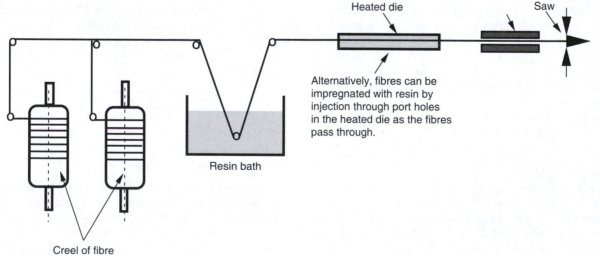

Fig. 41.2 *Schematic diagram of the pultrusion technique.*

most geometrical shapes; the most common of these are I, L, Tee, and circular sections. Curved-in-plan pultrusion sections can also be manufactured.

41.2 Manufacture of fibre-reinforced thermoplastic composites

Reinforced thermoplastic composites can be manufactured by most of the thermoplastic processing techniques such as extrusion, blow-moulding and thermoforming of short-fibre reinforced thermoplastics. However, the most important technique for civil engineering industry use is injection moulding. It is a similar technique to the manufacture of un-reinforced thermoplastics but the melt viscosity is higher in the reinforced polymer process, consequently injection pressures are higher. With all the techniques, production difficulties can occur because the reinforced composite is stiffer than the un-reinforced one. The cycle time is less but the increased stiffness can affect ejection from the mould, so the mould design has to be modified from that of the un-reinforced polymer mould.

One of the problems of thermoplastic composites is that they use short fibres (typically 0.2–0.4 mm long) and consequently their full strength is not developed. Continuous fibre tapes and mats in the form of prepregs can help to overcome this problem. The best known examples of these systems are the aromatic polymer composites (APC) and the glass-mat-reinforced thermoplastic composites (GMT). The systems use unidirectional carbon fibre in a matrix of polyethersulphone (PES) or polyetheretherketone (PEEK). The material for APCs comes in a prepreg form of unidirectional or 0°/90° fibre, and for GMT in a tape prepreg form. The composite is manufactured by the film-stacking process, with the prepregs arranged in the desired directions.

Film-stacking products can be made from prepreg reinforcement, and one system uses a polyethersulphone polymer content of about 15% by weight. The final volume fractions of fibre and resin are obtained by adding matrix in the form of a polymer film. The film-stacking process, therefore, consists of alternating layers of fibre impregnated with insufficient matrix, with polymer films of complementary mass to bring the overall laminate to the correct fibre volume ratio. The required stacked sequence is rolled around a central mandrel (PTFE material) and is placed into one part of a split steel mould; the two half moulds are joined and placed in an oven for a specific time. The difference in thermal expansion of the PTFE and steel causes pressure to be applied to the curing polymer.

This technique is used mainly for high-technology composites in the aerospace and space industries.

References

Hollaway LC (1993). *Polymer Composites for Civil and Structural Engineering*, Blackie Academic & Professional, London, Glasgow, New York, Tokyo, Melbourne, Madras.

Hollaway LC and Head PR (2001). *Advanced Polymer Composites and Polymers in the Civil Infrastructure*, Elsevier, Oxford.

Shaw-Stewart D (1988). *Pullwinding*. Proceedings of the Second International Conference on Automobile Composites 88, Noordwijkerhout, The Netherlands.

Bibliography

Bank LC (2006). *Composites for Construction Structural Design with FRP Materials*, John Wiley and Sons Inc., Hoboken, New Jersey.

EPSRC Innovative Manufacturing Initiative (1994). 'Construction as a Manufacturing Process' Workshop, *New Builder*, 18 March.

EPSRC (1994). '*Materials for Better Construction Programme*' Brochure, September, 1994.

Hollaway LC (1986). 'Pultrusion' Chapter 1 of *Developments in Plastic Technology* – 3 (eds Whelan A and Craft JL), Elsevier Applied Science, Oxford.

Philips LN (1989), (editor), 'Design with Advanced Composite Materials, The Design Council, published Springer-Verlag, Berlin, Heidelberg, New York, London, Paris and Tokyo.

Chapter 42

Durability and design of polymer composites

It will be clear from the other parts of this book that all engineering materials are prone to degradation over time and FRP materials are no exception to this rule. However we will see in this chapter that they do offer some significant durability advantages over the more conventional construction materials, although they are not without their own problems.

The ageing of a polymer can be defined as a slow and irreversible variation of the material structure, morphology and/or composition, leading to a deleterious change. This slow degradation is caused by the interaction of the material with the environment into which it is placed. Clearly knowledge of the processes and the controlling factors is a prerequisite for successful composite design.

The most significant factors that cause polymer composites to degrade with time under various environments are:

* moisture and aqueous solutions, particularly alkaline environments
* corrosion
* fire
* thermal effects
* ultra-violet radiation.

All of these are considered in this chapter, but before doing so we will make some general points.

Some field surveys undertaken on FRP composite materials throughout the world have been described by Hollaway (2007) and indicate the relative importance of degradation mechanisms. For example, in-service factors that contribute to the degradation of GFRP composites include temperature, ultraviolet rays from the sun, moisture absorption and freeze–thaw cycles, the latter two factors being considered the most critical. Exposure to alkaline environments and UV radiation also affect long-term durability, but to a lesser extent. Furthermore, UV degradation resistance of most composites is being improved by

applying protective coats and additives during the manufacturing process.

It is difficult to analyse the problems owing to their very slow progress (typically over a number of years), consequently accelerated testing is sometimes undertaken. This requires specimens being exposed to an accelerated test regime, which generally involves an environment many times more severe than that which would be experienced in practice. Furthermore, these test samples are sometimes also exposed to elevated temperatures to further increase the degradation. This accelerated test regime to obtain durability data in one environmental situation will generally not be relevant to the more gradual degradation effect that would have taken place had the conditions been less rigorous. An accelerated test programme can, however, be undertaken to build kinetic models that describe the changes over time of the behaviour of the material; these models are then used to predict the durability from a conventional lifetime criterion. It is then necessary to prove the pertinence of the choice of accelerated ageing conditions by a mathematical form of the kinetic model. Some investigators use empirical models but these are highly questionable because they have to be used in extrapolations for which they are not appropriate. It is therefore important to treat the results of accelerated experimental testing with caution and care, particularly if the polymer is heated to increase the rate of degradation. Heating would change the characteristics of the specimen, and in addition the mechanical properties of polymers degenerate with increase in heat.

It is particularly important to appreciate the effects that heat has on thermosetting polymers and hence on thermosetting polymer composites. Therefore, before considering the durability factors it is necessary to discuss what happens as the temperature of a polymer rises above a certain temperature termed the glass transition temperature, T_g.

42.1 The glass transition temperature of a polymer (T_g)

The T_g is the temperature below which a wholly or a partially amorphous polymer behaves in a similar way to that of a solid phase (a glassy state) and above which it behaves in a manner similar to that of a liquid (a rubbery state). The T_g is the mid-point of a temperature range of a few degrees in which the wholly or partially amorphous polymer gradually becomes less viscous and changes from being in a solid to being in a liquid state. The epoxy and vinylester polymers used in construction are generally in the amorphous state, with a small amount of crystalline structure. Thermoplastic polymers are crystalline solids and have a melting temperature, T_m, which spans a range of a few degrees. Above this temperature they lose their crystalline structure. Furthermore, they have a T_g below the T_m value and at temperatures below the T_g they are rigid and brittle and can crack and shatter under stress. Most crystalline polymers possess some degree of amorphous structure and most amorphous polymers have some degree of crystallinity, thus they can have both a glass transition temperature (the amorphous portion) and a heat distortion temperature (the crystalline portion).

In summary, all polymers below the T_g are rigid; they therefore have both stiffness and strength. Above the T_g, amorphous polymers are soft elastomers or viscous liquids and have no stiffness or strength. Crystalline polymers will range in properties from a soft to a rigid solid depending upon the degree of crystallinity.

42.2 Moisture and aqueous solutions

Polymers are not impervious to moisture and aqueous solutions, and if such solutions do penetrate the polymer they could do damage to the fibres. For example, all glass fibres are very susceptible to alkaline environments, and when incorporated as rebar reinforcement for concrete, the susceptibility is primarily due to the presence of silica in the glass fibres. This problem does not affect carbon fibres, but aramid fibres do suffer some reduction in tensile strength when exposed to an alkaline environment (Balazs and Borosnyoi, 2001). However, there are now glass fibres that are more resistant to an alkaline environment and therefore will increase the durability of a GFRP composite. Ownes Corning manufactures

a glass fibre known as Advantex; when this is embedded in a polymer matrix and immersed in a simulated concrete pore-water solution and loaded to 30% of its ultimate load capacity, Benmokrane *et al.* (2002) have reported that it retains 100% of its strength after 140 days of immersion; conventional E-glass fibres exposed to the same conditions resulted in a 16% loss of tensile strength.

As the matrix protects the fibre from external influences, the long-term properties of the matrix of CFRP and AFRP composites are of importance to the overall properties of the composite. In construction there are many different polymers on the market, some of which have been modified by chemists over the years to improve their in-service performance. Furthermore, additives are on occasion incorporated into cold-setting polymers to enhance curing. Each time these polymers are changed/modified the durability will be affected.

42.3 Corrosion resistance

In comparison with other construction materials FRP composites do not 'rust'; this makes them attractive in applications where corrosion is a concern. For example FRP composites are used in:

- rebars and grids for reinforcing concrete
- cables for pre- and post-tensioning concrete
- cable stays to bridges
- the upgrading of components of reinforced concrete and steel structures where the material is exposed to salt solutions, e.g. marine waterfront structures, cladding panels, pipe lines and walkways in harsh environments, and de-icing solution during winter snow storms.

A fully cured polymer exhibits good resistance to acidic and alkaline attack if selected and designed properly; resin manufacturers should be consulted when choosing a resin to be utilised for a specific corrosive environment.

42.4 Behaviour of polymer composites in fire

The polymer component of a composite is an organic material and is composed of carbon, hydrogen and nitrogen atoms; these materials are flammable to varying degrees. Consequently, a major concern for the building construction engineer using polymers is the problem associated with fire; the major health hazard derived from polymers and composites in a fire is from the toxic combustion products produced during burning. Smoke toxicity plays an important role during fire accidents in buildings, where the majority of people who die do so from inhalation of smoke. Improved methods of assessment need to be developed if toxicity is to be included as part of the fire hazard risk identification. The degree of toxicity generated depends on:

- the phase of burning of the fire
- oxidative pre-ignition
- flaming combustion or fully developed combustion
- ventilation controlled fires.

When a composite material is specified, it must meet the appropriate standards of fire performance. It is usually possible to select a resin system that will meet the requirements of BS Specification BS 476. The UK Building Regulations require that, depending upon their use, building components or structures should conform to given standards of fire safety. The fire tests as defined in BS 476: Parts 4–7 and in BS 476: Parts 3–8, respectively, fall into two categories:

- reaction to fire – tests on materials
- fire resistance – tests on structures.

Virtually all composites used in structural engineering will have high fibre volume fractions and thus the rate of progress of the fire through the composite is slow; carbon, glass or aramid fibres do not burn. To enable the flame-retardant properties of the composites to be improved additives are incorporated into the resin formulations, but in so doing an impurity is added to the polymer and some mechanical and/or in-service property of the polymer may be compromised. The chemical structure of the polymer could be altered, thereby modifying the burning behaviour and producing a composite with an enhanced fire property. Aluminium trihydrate and antimony trioxide may be used as fillers for both lamination and gel-coated resins, but the use of flame retardants can affect the colour retention of the polymer; a pigment is then added to produce a particular colour in a structural component.

Het-acid-based resins can be used where flame-retardant characteristics are required. Nano-clay particles will give some protection against fire and may be added to the pristine polymer, but the process is complicated and at present is expensive for the civil engineering industry (Hackman and Hollaway, 2006).

Nevertheless, modification of the polymer can only aid the fire resistance of the composite to a certain degree; eventually fire will damage composites and indeed all civil engineering materials.

42.5 Thermal effects

Thermal effects can be divided into thermal expansion and thermal conductivity.

42.5.1 THERMAL EXPANSION

The coefficients of thermal expansion of polymers, which range from 50 to 100×10^{-6}/C degree, are much higher than those of the fibre component of the fibre/matrix composites, e.g. 8.6×10^{-6}/C degree for glass fibre and from 1.6×10^{-6}/ to 2.1×10^{-6}/C degree for carbon fibres, depending on the fibre's structural properties. The thermal expansion of an FRP composite system is thus reduced from the high value of the polymer to a value near to that of conventional materials; this reduction is due to the stabilising effect that the fibres have on the polymer. The final value will depend upon:

- the type of fibre
- the fibre array
- the fibre volume fraction of the composite
- the temperature and the temperature range into which the composite is placed
- the degree of cross-linking of the polymer will also influence the rate of thermal expansion.

42.5.2 THERMAL CONDUCTIVITY

The thermal conductivity of polymers is low, consequently they are good heat insulators. This property is particularly important when FRP composites are exposed directly to the sun's rays. An example where this effect is particularly relevant is in FRP bridge decks that are incorporated into the superstructure of a reinforced concrete bridge (see Chapter 43, section 43.7).

42.6 Temperature effects

The effects of temperature on polymers can be separated into short-term and long-term effects. Short-term effects are generally physical and reversible when the temperature returns to ambient, while long-term effects are generally dominated by chemical change and are not reversible.

As the temperature varies both physical and mechanical properties of polymers change, therefore it will be necessary to fully characterise a material over a range of temperatures. These remarks on the selection of properties apply equally to measuring the ageing effects of long-term exposure. Certain short-term effects such as glass transition temperature, thermal expansion and melting point, are thought of as separate properties, although they are particular cases of the effects of temperature.

Constantly fluctuating temperatures have a greater deleterious effect on all composites but, particularly GFRP. At a micro scale, the difference in the coefficients of thermal expansion of the glass and of the resin may contribute to progressive de-bonding and weakening of the materials, although the extensibility of the resin system will usually accommodate differential movement. When GFRP composites are exposed to high temperatures a discoloration of the resin may occur; this is noticed by the composite's becoming yellow. Both polyester and epoxy show this effect and the problem will be aggravated if flame retardants are added to the resin during manufacture of the composite. Furthermore, as a result of the exposure to high temperatures, the composite will become brittle. These effects are not noticed when carbon fibre composites are used.

42.7 Ultraviolet radiation

The ultraviolet (UV) component of sunlight degrades polymers and therefore composites to varying degrees by either causing discolouration of the material causing it to become brittle; the short wavelength band at 330 nm has the most effect upon polymers. It is manifested by a discoloration of the polymer and a breakdown of the surface of the composite. Ultraviolet stabilisers are incorporated into polymer resin formulations to obviate this problem. The inclusion of stabilisers in epoxy resin formulations seems to have little effect regarding the discoloration but there is no evidence that continuous exposure to sunlight affects the mechanical properties of epoxy polymers. A gel coat surface coating can also be applied to the composite for increased UV and weather protection.

42.8 Design with composites

Designing with composites is an interactive process between the designer and the production engineer responsible for the manufacturing technique. It is essential that a design methodology is selected and rigorously used, because many different composite materials are on the market and they can be affected by the quality of their manufacture and the environment into which they are placed. It is also important that the designer recognises the product cost, because the constituents of composite materials (the fibre and the matrix) can vary significantly in price and the manufacturing process can range from simple compact moulded units cured at room temperature to sophisticated high-temperature- and pressure-cured composites.

The design process can be divided into five main phases:

- the design brief and an estimation of cost
- the structural, mechanical and in-service details
- the manufacturing processes and cost details
- the material testing and specification information
- the quality control and structural testing information.

The choice of design factors of safety is an important aspect of the work; these are likely to be given in the relevant code of practice. However, if the design is unique, it may be necessary for the designer/analyst to select specific factors of safety, bearing in mind the exactness of the calculations, the manufacturing processes, the in-service environment, the life of the product and the loading. The selection of these design factors follows the pattern for other materials but, with the variation in properties, owing to the anisotropic nature and the different manufacturing techniques of composites, a more involved calculation and a greater reliance upon the design factors will result.

In recent years a significant number of design guides, design codes and specifications have been published by technical organisations in several countries throughout the world; these provide guidance for design with FRP materials for civil engineering. As a considerable volume of FRP composites has been concerned with bridge work these design guides are mainly directed to bridge engineering. A list of some of these design guides has been given in the bibliography to this chapter.

References

Balazs GL and Borosnyoi A (2001). *Long term behaviour of FRP*. International Workshop Composites in Construction: a reality (eds Cosenza E, Manfredi G, Nanni A), American Society of Civil Engineers, Reston, 2001, pp. 84–91.

Benmokrane B, Wang P, Ton-That TM, Rahman H and Robert J-F (2002). Durability of glass fiber-reinforced polymer reinforcing bars in concrete environment. *Journal of Composites for Construction*, **6** (No. 3), 143–153.

Hackman I and Hollaway LC (2006). Epoxy-layered silicate nanocomposites in civil engineering. *Composites Part A*, **37** (No. 8), 1161–1170.

Hollaway LC (2007). Fibre-reinforced polymer composite structures and structural components: current applications and durability issues. Chapter 10 of *Durability of composites for civil structural applications* (ed. Karbhari V), Woodhead Publishing, Cambridge, UK.

Manfredi, A Nanni, *American Society of Civil Engineers*, Reston, 84–91.

Bibliography

ASTM standards that are concerned with the measurement of gases present or generated during fires and their smoke density are *ASTM E 800-01, ASTM E1678-02 and ASTM E176-04. The British Standard Codes* that are concerned with smoke toxicity in fire hazards and risk assessment are *BSS 7239-88 the Boeing Toxicity Test and BSS 179-03, the use of bench-scale toxicity data in fire hazards.*

Building Research Establishment (1963) Internal records, Building Research Establishment, Garston, UK.

Demers M, Labossière P and Neale K (2005). *Ten years of Structural Rehabilitation with FRPs – A Review of Quebec Applications.* Proceedings of Composites in Construction 2005 – 3rd International Conference, Lyon, France, July 11–July 13, 2005.

Engineering Science Data Unit (1987a). *Stiffnesses and properties of laminated plates, ESDU 20–22,* ESDU International, London.

Engineering Science Data Unit (1987b). *Failure of composite laminates, ESDU 20–33,* ESDU International, London.

Farhey DN (2005). Long-term performance monitoring of the Tech 21 all-composite bridge. *Journal of Composites for Construction,* **9** (No. 3), 255–262.

Hollaway LC and Head P (2001). *Advanced Polymer Composites and Polymers in the Civil Infrastructure,* Elsevier, Oxford.

ISO/CD standard 6721–11 (2001). *Plastics – Determination of dynamic mechanical properties – Part 11: Glass transition temperature.*

ISO standard 11357-1 (1997). *Plastics – Differential scanning calorimetry (DSC) – Part 1: General principles.*

Kajorncheappunngam S, Gupta RK and GangaRao HVS (2002). Effect of aging environment on degradation of glass-reinforced epoxy. *Journal of Composites for Construction,* **6** (No. 1), 61–69.

Kootsookos A and Mouritz AP (2004). Seawater durability of glass- and carbon-polymer composites. *Composites Science and Technology,* **64** (No. 10–11), 1503–1511.

DESIGN GUIDES

In recent years a significant number of design codes and specifications have been published by technical organisations that provide guidance for design with FRP materials for civil engineering. The key publications are listed below.

Europe

1. Structural Design of Polymer Composites, Eurocomp Design Code and handbook (ed. Clarke JL), 1996.
2. Fib Task Group 9.3, *FRP Reinforcement for Concrete Structures,* Federation Internationale du Beton, 1999.
3. Concrete Society Technical Report (2000). *Design Guidance for Strengthening Concrete Structures Using Fibre Composite Materials,* TR55, 2nd edition, Camberley, UK.
4. Fib Bulletin 14, *Design and use of Externally Bonded FRP Reinforcement for RC Structures,* Federation Internationale du Beton 2001.
5. Concrete Society Technical Report (2003). *Strengthening Concrete Structures using Fibre Composite Materials: Acceptance, Inspection and Monitoring,* TR57, Camberley, UK.
6. Cadei JM, Stratford TK, Hollaway LC and Duckett WG (2004). *Strengthening Metallic Structures Using Externally Bonded Fibre-Reinforced Polymers,* CIRIA Report C595.
7. Eurocrete Modifications to NS3473 – *When Using FRP Reinforcement,* Report No. STF 22 A 98741, Norway, 1998.

USA

1. ACI (2004). *Prestressing Concrete structures with FRP Tendons,* ACI 440.4R-04, American Concrete Institute, Farmington Hills, MI.
2. ACI (2006). *Guide for the Design and Construction of Structural Concrete Reinforced with FRP Bars,* 440.1R-06, American Concrete Institute, Farmington Hills, MI.
3. ACI (2002). *Report on Fibre Plastic Reinforcement for Concrete Structures,* 440.R-96 (Re-approved 2002).
4. ACI (2004). *Guide Test Methods for Fibre-Reinforced Polymers (FRP) for reinforcing or Strengthening Concrete Structures,* 440.3R-04, American Concrete Institute, Farmington Hills, MI.
5. ACI (2002). *Guide for the Design and Construction of Externally Bonded FRP Systems for Strengthening Concrete Structures,* 440.2R-02 American Concrete Institute, Farmington Hills, MI.

Canada

1. AC 125 (1997). *Acceptance Criteria for Concrete and Reinforced and Unreinforced Masonry Strengthening Using Fibre Reinforced Polymer Composite Systems.* ICC Evaluation Service, Whittier, CA.
2. AC 187 (2001). *Acceptance Criteria for Inspection and Verification of Concrete and Reinforced and Unreinforced Masonry Strengthening Using Fibre Reinforced Polymer Composite Systems.* ICC Evaluation Service, Whittier, CA, Canada.
3. CSA (2000). *Canadian Highway Bridge Design Code,* CSA-06-00, Canadian Standards Association, Toronto, Ontario, Canada.

4. CSA (2002). *Design and Construction of Building Components with Fiber-Reinforced Polymers*, Canadian Standards Association, Toronto, Ontario, Canada, CSA S806-02.

5. ISIS Canada, Design Manual No. 3, *Reinforcing Concrete Structures with Fiber Reinforced Polymers*, Canadian Network of Centers of Excellence on Intelligent Sensing for Innovative Structures, ISIS Canada Corporation, Winnipeg, Manitoba, Canada, Spring 2001.

Japan

1. Japan Society of Civil Engineers (JSCE). *Recommendation for Design and Construction of Concrete Structures Using Continuous Fiber Reinforced Materials*, Concrete Engineering Series 23 (ed. Machida A), Research Committee on Continuous Fiber Reinforcing Materials, Tokyo, Japan, 1997.

2. BRI (1995). *Guidelines for Structural Design of FRP Reinforced Concrete Building Structures*, Building Research Institute, Tsukuba, Japan.

3. JSCE (1997). *Recommendation for Design and Construction of Concrete Structures using Continuous Fiber Reinforcing Materials*, Concrete Engineering Series 23, Japan Society of Civil Engineers, Tokyo.

4. JSCE (2001). *Recommendations for Upgrading of Concrete Structures with Use of Continuous Fibre Sheets*, Concrete Engineering Series 41, Japan Society of Civil Engineers, Tokyo.

Australia

Oehlers DJ, Seracino R and Smith S (2007). 'Design Guideline for RC structures retrofitted with FRP and metal plates: beams and slabs', Publishers SAI Global Limited, published under auspices of Standards Australia, 108 pages.

British standards

BS 476-3: 1958 External fire exposure roof tests.

BS 476-4: 1970 Non-combustibility tests for materials.

BS 476-6: 1981 Method of test for fire propagation for products.

BS 476-7: 1987 Method of classification of the surface spread of flame for materials.

BS 476-8: 1972 Test methods and criteria for the fire resistance of elements of building construction.

Chapter 43

Applications of FRP composites in civil engineering

During the introduction of FRP composites into the building and construction industry in the 1970s glass fibres were used in a polyester matrix as a construction material. Skeletal frames constructed from reinforced concrete (RC) or steel columns and beams were in-filled with non-load-bearing or semi-load-bearing GFRP panels manufactured by the wet lay-up process or by the spray-up technique to form structural buildings. Several problems developed owing to a lack of understanding of the FRP material, mainly arising from insufficient knowledge of its in-service properties relating to durability and the enthusiasm of architects and fabricators for developing geometric shapes and finding new outlets for their products without undertaking a thorough analysis of them. Consequently, to improve certain physical properties of the FRP some additives were incorporated into the polymers by the fabricators without a full understanding of their effect on the durability of the FRP material, or indeed were omitted in cases where additives should have been added.

Advanced polymer composites did not enter the civil engineering construction industry until the middle to late 1980s; polyester and epoxy polymers were used initially and vinylester was introduced in the 1990s. From the 1970s, universities, research institutes and industrial firms have been involved in researching the in-service, mechanical properties of FRPs and in the design and testing of structural units manufactured from fibre/polymer composites. This was followed by the involvement of interested civil engineering consultants undertaking industrial research and the utilisation of the structural material in practice. The application of advanced polymer composites, over the past 35 years for the building industry and the past 25 years for the civil engineering industry, can be conveniently divided into some specific areas, which will be discussed briefly in this chapter:

- Building industry:
  - infill panels and new building structures.

- Civil engineering industry:
  - civil engineering structures, fabricated entirely from advanced polymer composite material, known as *all-polymer/fibre composite structures*
  - bridge enclosures and fairings
  - bridge decks
  - external reinforcement rehabilitation and retrofitting to RC structures (including FRP confining of concrete columns)
  - external reinforcement rehabilitation and retrofitting to steel structures
  - internal reinforcement to concrete members
  - FRP/concrete duplex beam construction
  - polymer bridge bearings and vibration absorbers

All these, other than the first, involve a combination of advanced polymer composites and conventional construction materials and are therefore often termed *composite construction*.

FRP composites are durable and lightweight and consequently they can fulfil many of the requirements of structural materials for many forms of construction. Ideally when new civil engineering structures are manufactured from polymer composite systems the component parts should be modular to provide rapid and simple assembly. An example of the importance of this is in the installation of highway infrastructure, where any construction or long maintenance period of the infrastructure will cause disruption to traffic flow and will be expensive.

The examples of the applications of polymer fibre composites in those areas that we will discuss in this chapter have been chosen to illustrate all the areas of use listed above.

43.1 The building industry

During the 1970s two sophisticated and prestigious GFRP buildings were developed and erected in the

UK, Mondial House, the GPO Headquarters in London (Berry, 1974) and the classroom of the primary school in Thornton Clevelys, Lancashire; these are discussed below. Other FRP buildings that were erected during this period were Covent Garden Flower Market (Roach, 1974; Berry, 1974), the American Express Building in Brighton (Southam, 1978), and Morpeth School, London (Leggatt, 1974, 1978). These structures played a major role in the development of polymer composite materials for construction. Because of the relatively low modulus of elasticity of the material, all except one of these buildings were designed as folded plate systems and erected as a composite modular system, with either steel or reinforced concrete units as the main structural elements and the GFRP composite as the load-bearing infill panels. The exception to this is the classroom of the primary school, Thornton Clevelys, Lancashire, UK (Stephenson, 1974), which is entirely manufactured from GFRP material.

43.1.1 MONDIAL HOUSE, ERECTED ON THE NORTH BANK OF THE THAMES IN LONDON 1974

This building was clad above the upper ground floor level and the panels were manufactured from glass fibre polyester resin. The outer skin of the panel included a gel coat that used isophthalic resin, pigmented white, with an ultraviolet stabiliser backed up with a glass fibre reinforced polymer laminate; the latter used a 3 oz per square foot chopped strand mat and a self-extinguishing laminating resin reinforced with 9 oz per square foot glass fibre chopped strand mat reinforcement. Some degree of rigidity was obtained from a core material of rigid polyurethane foam bonded to the outer skin and covered on the back with a further glass-reinforced laminate; this construction also provided thermal insulation. Further strength and rigidity were obtained by the use of lightweight top-hat section beams, manufactured as thin formers and incorporated and over-laminated into the moulding as manufacture proceeded. The effect of the beams was transferred to the front of the panel by means of glass-fibre reinforced ties or bridges formed between the polyurethane foam at the base of each beam. The face of the beam was reeded on the vertical surfaces in order to mask any minor undulations and to provide channels off which the water ran and thereby cleaned the surface. The reeding also gave the effect of a matt panel without reducing the high surface white finish.

The structure was visually inspected in 1994 by Scott Bader and the University of Surrey and the degradation was found to be minimal. It was

Fig. 43.1 The 'all-polymer composite' classroom of the primary school, Thornton Clevelys, Lancashire, UK.

demolished in 2007 to allow for redevelopment of that area. A part of the composite material from the demolished structure was analysed at the University of Surrey for any variations in the mechanical properties due to the degradation of the composite material during its life (Sriramula and Chryssanthopoulos, 2009).

43.1.2 AN 'ALL-POLYMER COMPOSITE' CLASSROOM OF PRIMARY SCHOOL, THORNTON CLEVELYS, LANCASHIRE, UK, 1974

The classroom, shown in *Fig. 43.1*, is an 'all-composite' FRP building in the form of a geometrically modified icosahedron, and is manufactured from 35 independent self-supported tetrahedral panels of chopped strand glass-fibre reinforced polyester composite. Twenty eight panels have a solid single skin GFRP composite and in five of these panels circular apertures were constructed to contain ventilation fans. In the remaining seven panels non-opening triangular windows were inserted. The wet lay-up method was utilised to manufacture the E-glass fibre/polyester composite skins. The inside of the panels has a 50 mm thick integral skin phenolic foam core acting as a non-load bearing fire protection lining to the GFRP composite skins. The icosahedron structure is separated from the concrete base by a timber hardwood ring. The FRP panels were fabricated onto a mould lining of Perspex with an appropriate profile to give a fluted finish to the flat surfaces of the panels. The edges of the panels were specially shaped to provide a flanged joint, which formed the connection with adjacent panels. Sandwiched between two adjacent flanges is a shaped hardwood batten, which provides the correct geometric angle between the panels; the whole is bolted together using galvanised steel bolts placed at 450 mm intervals. The external joint surfaces between the adjacent panels were sealed with polysulphide mastic. The glass windows were fixed in position on site by means of neoprene gaskets. The classroom was designed by Stephenson (1974).

When the classroom structure was under construction in 1974 a fire test at the BRE Fire Research Station was undertaken on four connected GFRP panels, with the integral skin phenolic foam in place. At the same time, tests were also undertaken on an identical geometrically shaped school system used at that time. The results demonstrated that the GFRP classroom had over 30 minutes fire rating whereas the existing school system had only 20 minutes.

These two descriptions of the Mondial House and the school classroom at Thornton Clevelys have been based on Hollaway (2009).

43.2 The civil engineering industry

The 'all-polymer composite' structure systems – like those of the building industry produced to date – have tended to be single prestigious structures, manufactured from 'building blocks', Hollaway and Head (2001). The advantages of this are:

- the controlled mechanised or manual factory manufacture and fabrication of identical structural units
- the transportation to site of the lightweight units, which can be readily stacked; it is more economical to transport lightweight stacked FRP units than the heavier steel and concrete units.

McNaughton (2006) said: 'The majority of the Network Rail's bridges in the UK are 100 years old and are constructed in a variety of materials, for example cast iron, wrought iron, steel, reinforced concrete, brick, masonry and timber. Future construction is likely to use more complex forms of composite construction, in particular fibre reinforced polymers, which are already being used to strengthen bridges'.

Examples of some of these 'more complex structures' are the Aberfeldy Footbridge, Scotland (1993), the Bonds Mill Single Bascule Lift Road Bridge, Oxfordshire (1994) (Head, 1994), Halgavor Bridge (2001) (Cooper, 2001), the road bridge over the River Cole at West Mill, Oxfordshire (2002) (Canning *et al.*, 2004), the Willcott Bridge (2003) (Faber Maunsell, 2003), the New Chamberlain Bridge, Bridgetown, Barbados (2006) and the Network Rail

footbridge which crosses the Paddington–Penzance railway at St Austell, UK (2007). An innovative £2 million Highways Agency super-strength FRP composite bridge (The Mount Pleasant Bridge) was installed in 2006 over the M6, between Junctions 32 and 33; the structure won the National Institution of Highways and Transportation Award for Innovation in June 2007.

All these structures were of modular construction, manufactured utilising advanced composite materials; for the construction to be successful the material had to be durable, and assembly of the units had to be rapid and simple with reliable connections. As we have already seen advanced polymer composite materials are durable and lightweight and consequently they fulfil these requirements, provided that the initial design of the basic building modular system is properly undertaken and the material properly installed.

A number of bridges have used the concept of the Maunsell structural plank, shown in *Fig. 43.2*.

43.3 Bridge enclosures and fairings

It is a requirement that all bridge structures have regular inspection and maintenance, which will often cause disruption to travellers, particularly if closure of roads and interruption to railway services are required. Furthermore, increasingly stringent standards are causing the cost of closures to be high, particularly if maintenance work is over or beside busy roads and railways. Most bridges that have been designed and built over the last 30 years do not have good access for inspection, and in Northern Europe and North America deterioration caused by de-icing salts is creating an increasing maintenance workload.

The function of 'bridge enclosures' is to erect a 'floor' underneath the girder of a steel composite bridge to provide access for inspection and maintenance. The concept was developed jointly by the Transport Research Laboratory (TRL, formerly TRRL) and Maunsell (now AECOM) in 1982 to provide a solution to the problems. Most bridge enclosures that have been erected in the UK have utilised polymer composites. These materials are ideal because they add little weight to the bridge, are highly durable, and as they are positioned on the soffit of the bridge they are protected from direct sunlight.

The floor is sealed on to the underside of the edge girders to enclose the steelwork and to protect it

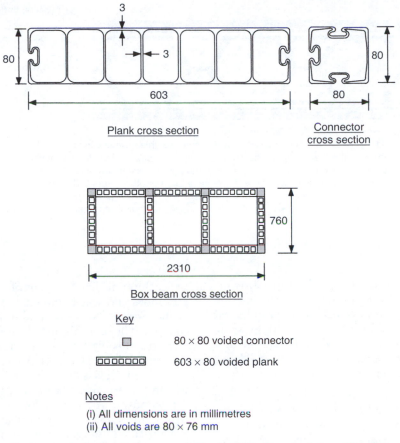

Plank cross section

Connector cross section

Box beam cross section

Key

⬜ 80 × 80 voided connector

▭▭▭▭▭▭▭ 603 × 80 voided plank

Notes
(i) All dimensions are in millimetres
(ii) All voids are 80 × 76 mm

Fig. 43.2 *The Maunsell structural plank (Hollaway and Head, 2001, by permission, Elsevier).*

from further corrosion. Once the enclosures have been erected the rate of corrosion of uncoated steel in the protected environment within the enclosure is 2–10% of that of painted steel in the open (McKenzie, 1991; 1993). The enclosure space has a high humidity; chloride and sulphur pollutants are excluded by seals and when condensation does occur (as in steel girders) the water drops onto the enclosure floor, which is set below the level of the steel girders from where it escapes through small drainage holes.

Figure 43.3 shows an example of the enclosure on the approach span of the Dartford River Bridge (QE2) where it passes over the Channel Tunnel rail link (CTRL) (before the train rails were laid).

43.4 Bridge decks

The development of FRP deck structures has been based generally on the pultruded systems, but occasionally on moulded structures. Recently FRP deck

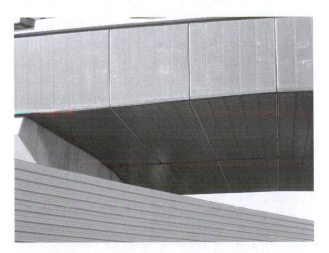

Fig. 43.3 *Photograph of the enclosure on the approach span of the Dartford River Bridge (QE2) where it passes over the CTRL (before the train rails were laid) (Courtesy of AECON).*

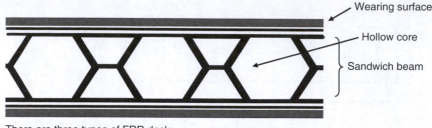

There are three types of FRP deck:

1. *Honeycomb*: core construction provides considerable flexibility in tailored depth, however the wet lay-up method now employed requires painstaking attention to quality control in the bonding of the top and bottom face material to the core.
2. *Solid core sandwich*: solid core decks have foam or other fillers in the core.
3. *Hollow core sandwich*: consists of pultruded shapes fabricated together to form deck sections. FRP decks typically have continuous hollow core patterns, as shown above.

Fig. 43.4 A typical cross-section of an FRP bridge deck.

replacements in conjunction with FRP superstructure replacement for road bridges have been carried out. This type of construction is becoming popular for replacement decks of bridges up to 20 m span. *Figure 43.4* illustrates a typical cross-section of a bridge deck. The reasons for FRP material being used in particular circumstances are:

- the bridge deck is the most vulnerable element in the bridge system because it is exposed to the direct actions of wheel loads, chemical attack, and temperature/moisture effects including freeze–thaw shrinkage and humidity; FRP material characteristics satisfy these requirements
- reduced future maintenance (FRP composites are durable materials)
- quick installation owing to pre-fabrication and easy handling.

In the USA over 100 concrete bridge decks have been replaced by FRP deck installations, most of which have been built using proprietary experimental systems and details. The lack of standardisation is a challenge to bridge engineers, who traditionally have been accustomed to standard shapes, sizes and material properties. The first FRP European bridge deck and superstructure replacement was conceived and developed under the innovative European ASSET Project led by Mouchel Consulting. It culminated in 2002 in the construction of the West Mill Bridge over the River Cole in Oxfordshire; the beam and deck structures were manufactured by the pultrusion technique.

The first vehicle-carrying FRP bridge deck in the UK to span over a railway replaced the existing over-line bridge at Standen Hey, near Clitheroe,

Lancashire; it has a span of 10 metres, weighs 20 tonnes and was completed in March 2008. This is the first of Network Rail's six trial sites in the country. The consultants Tony Gee and Partners were responsible for the design of the deck, which comprises three layers of ASSET panel deck units made from E-glass fibres in the form of biaxial mats within a UV-resistant resin matrix.

Composite Advantage (CA) has recently built (April 2008) a new 'drop-in-place' GFRP composite prefabricated integral beams and deck bridge super-structure, 6.77 m long by 19.0 m wide (22 feet by 62 feet) in Hamilton County, Ohio, USA. No heavy lifting equipment was required and it took one day to install (Composite Advantage, 2008).

A new single carriageway road bridge over the M6 motorway (UK) has recently been completed by the UK Highways Agency. The superstructure comprises a novel pre-fabricated FRP deck spanning transversely over, and adhesively bonded to, two longitudinal steel plate girders. The Mouchel Group designed the FRP bridge deck, which provides general vehicular access to an equestrian centre (*Fig. 43.5*); this was designed for unrestricted traffic loading (Canning, 2008).

43.5 External reinforcement to reinforced concrete (RC) structural members

The repair, upgrading and strengthening/stiffening of deteriorated, damaged and substandard infra-structure has become one of the fastest growing and

Fig. 43.5 Craning in the 100-tonne FRP deck onto the supports of the bridge over the M6 (Courtesy of Mouchel).

- *New loading requirements.* For example, a structure may not have originally been designed to carry blast or seismic loads.
- *Material deterioration.* For example, concrete degradation by the alkali–silica reaction or corrosion of steel reinforcement in marine or industrial environments or from the de-icing salts used on highways, all of which were discussed in Chapter 24.
- *Structural deterioration.* The condition of a structure will deteriorate with time owing to the service conditions to which it is subjected. In some cases this deterioration might be slowed or rectified by maintenance, but if unchecked the structure will become unable to perform the purpose for which it was originally designed.
- *Fatigue.* This is a secondary cause of structural degradation, and it can govern the remaining life of a structure, as discussed in Chapter 2.

Structural degradation can also result from *hazard events*, such as impact (for example, 'bridge bashing' by over-height vehicles), vandalism, fire, blast loading or inappropriate structural alterations during maintenance. A single event may not be structurally significant, but multiple events could cause significant cumulative degradation to a structure.

The following discussions and examples illustrate the strengthening of members by external bonding of FRP plates or members. These will be considered as un-stressed at the time of bonding onto the structural beam. It is however possible to pre-stress the plate before bonding it onto the beam; this is known as active flexural strengthening. This topic will not be discussed here but further reading on it may be found in Teng *et al.* (2002) and De Lorenzis *et al.* (2008), and a practical example is cited in Hollaway (2008).

Many experimental and analytical research investigations have been undertaken on reinforced concrete beams strengthened by FRP composites; some of these are discussed in Triantafillou and Plevris (1991), Hollaway and Leeming (1999), Teng *et al.* (2002), Concrete Society Technical Reports (2000, 2003), Oehlers and Seracino (2004) and Hollaway and Teng (2008). Both flexural and shear upgrading can be undertaken using FRP composites.

43.5.1 REHABILITATION OF DEGRADED FLEXURAL RC STRUCTURAL BEAMS USING FRP PLATES

Within the scope of 'strengthening' concrete, it is essential to differentiate between the terms repair, rehabilitation, strengthening and retrofitting; these

most important challenges confronting the bridge engineer worldwide. It is generally much less expensive and less time consuming to repair a bridge or building structure than to replace it.

Civil infrastructure routinely has a serviceable life in excess of 100 years. It is inevitable that some structures will eventually be required to fulfil a role not envisaged in the original specification. It is often unable to meet these new requirements, and consequently needs strengthening. Changes in use of a structure include:

- *Increased live load.* For example, increased traffic load on a bridge; change in use of a building resulting in greater imposed loads.
- *Increased dead load.* For example, additional load on underground structures owing to new construction above ground.
- *Increased dead and live load.* For example, widening a bridge to add an extra lane of traffic.
- *Change in load path.* For example, by making an opening in a floor slab to accept a lift shaft, staircase or service duct.
- *Modern design practice.* An existing structure may not satisfy modern design requirements; for example, owing to the development of modern design methods or to changes in design codes.
- *Design or construction errors.* Poor construction workmanship and management, the use of inferior materials, or inadequate design, can result in deficient structures that are unable to carry the intended loads.

terms are often erroneously interchanged but they do refer to four different structural upgrading procedures.

- *Repair* to an RC structural member implies the filling of cracks by the injection of a polymer into the crack.
- *Rehabilitation* of a structural member (of any type) refers to the improvement of a functional deficiency of that member, such as caused by severe degradation, by providing it with additional strength and stiffness to return it to its original structural form.
- *Strengthening* of a structural member is specific to the enhancement of the existing designed performance level.
- *Retrofit* is used to relate to the upgrading of a structural member damaged during a seismic event.

Bonding of FRP plates to the adherend

As with all bonding operations the adherends must be free of all dust, dirt and surface grease. Consequently, the concrete or steel surface onto which the composite is to be bonded must be grit blasted to roughen and clean the surface. It will then be air blasted to remove any loose particles and wiped with acetone or equivalent to remove any grease before the bonding operation. The surface preparation of component materials of FRP composite plate bonding to concrete surfaces is described in Hutchinson (2008).

The thickness of the adhesive and FRP composite plate would generally be about 1.0–1.5 mm and about 1.2 mm, respectively; the total length of the FRP plate as delivered to site would be of the order of 18 metres. It is possible to roll the material into a cylinder of about 1.5 metre diameter for transportation and for bonding the plate onto the beam in one operation.

Power actuated (PA) fastening 'pins' for fastening FRP composites

This method, which has been recently developed, is known as the *Mechanically-fastened unbonded FRP* (MF-UFRP) method and is a viable alternative to the adhesive bonding of a preformed pultruded or a prepreg rigid plate. It mechanically fastens the FRP plate to the RC beam by using many closely spaced steel power-actuated (PA) fastening 'pins' and a limited number of steel expansion anchors. The process is rapid and uses conventional hand tools, lightweight materials and unskilled labour. In addition, the MF-UFRP method requires minimal

surface preparation of the concrete and permits immediate use of the strengthened structure. The advantage of using multiple small fasteners as opposed to large diameter bolts, which are generally used for anchorages, is that the load is distributed uniformly over the FRP strip and this reduces the stress concentrations that can lead to premature failure. The method was developed by researchers at the University of Wisconsin, Madison, USA (Bank, 2004). Bank *et al.* (2003a, 2003b) have discussed the strengthening of a 1930 RC flat-slab bridge of span 7.3 m by mechanically fastening the rigid FRP plates using the MF-UFRP method.

Unstressed FRP plates

Figure 43.6 shows an FRP composite flexural plate bonded in position. The plate material used for the bonding or the MF-UFRP operations is generally the high-modulus (European Definition) CFRP, AFRP (Kevlar 49) or GFRP composite. These will be fabricated by one of three methods:

- the pultrusion technique, in which the factory made rigid pre-cast FRP plate is bonded onto the degraded member with cold-cure adhesive polymer
- the factory made rigid fully cured FRP prepreg plate, which is bonded to the degraded member with cold-cure adhesive polymer
- the low-temperature mould prepreg FRP prepreg/adhesive film placed onto the structural member and both components are cured simultaneously on site under pressure and elevated temperature (see Chapter 41, section 41.1.2).

The third method for the bonding operation is superior to the precast plate and cold-cure adhesive systems (first and second methods) as the site compaction and cure procedure of the prepreg and film adhesive ensure a low void ratio in the composite and an excellent join to the concrete. The current drawback to this method is the cost; it is about twice as expensive as the other two, and the currently preferred manufacturing system for upgrading is either the first or the second method. With these systems the plate material cannot be reformed to cope with any irregular geometry of the structural member. In addition, a two-part cold-cure epoxy adhesive is used to bond the plate onto the substrate. This is the Achilles' heel of the system, particularly if it is cured at a low ambient temperature since without post cure the polymerisation of the polymer will continue over a long period of time; this incomplete polymerisation might affect the durability of the material.

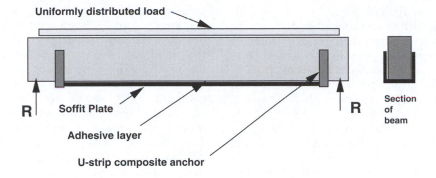

Plated RC beam with FRP U-strip end anchorage

Fig. 43.6 An FRP flexural plate bonded in position with cold-cure adhesive.

Near-surface mounted (NSM) FRP composite reinforcement technique

This is another method for the rehabilitation of RC structural members. CFRP, AFRP and GFRP composites can be utilised and generally the cross-section of the member is either circular or rectangular. Grooves are cut into the surface of the member, generally into the soffit of the concrete beam, but if the cover to the steel rebars is insufficient for this the grooves may be cut into the vertical side of the beam as near to the bottom of the section as is practical. The NSM FRP reinforcement is embedded and bonded into this groove with an appropriate binder (usually high-viscosity epoxy or cement paste). *Figure 43.7* shows the position of NSM bars in an RC structural member.

The NSM reinforcement can significantly increase the flexural capacity of RC elements. Bond may be the limiting factor to the efficiency of this technique as it is with externally bonded laminates. A review of the technique has been given by De Lorenzis and Teng (2007).

NSM FRP reinforcement has also been used to enhance the shear capacity of RC beams. In this case, the bars are embedded in grooves cut into the sides of the member at the desired angle to the axis. Utilising NSM round bars, De Lorenzis and Nanni (2001) have shown experimentally that an increase in capacity as high as 106% can be achieved, thus when stirrups are used a significant increase can be obtained.

Flexural strengthening of pre-stressed concrete members

Limited research has been undertaken on strengthening pre-stressed concrete (PC) members; the fib

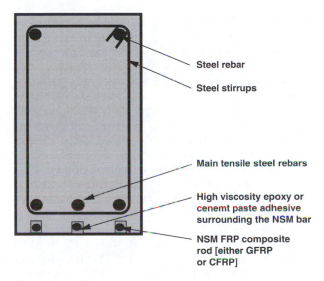

Fig. 43.7 Near-surface mounted (NSM) FRP composite reinforcement technique.

have reported that less than 10% of FRP-strengthened bridges as of 2001 are pre-stressed (fib Task Group 9.3, 2001). Strengthening usually takes place when all long-term phenomena (creep, shrinkage, relaxation) have fully developed, which may complicate the preliminary assessment of the existing condition. As in RC strengthening, the required amount of FRP will generally be governed by the ultimate limit state design in PC members. Additional failure modes controlled by rupture of the pre-stressing tendons must also be considered, and consideration should be given to limitations on cracking.

Seismic retrofit of RC columns

The properties of FRP composites (their light weight and tailorability characteristics) provide immense advantages for the development of structural components for bridges and buildings in seismic regions. The retrofit of RC structures improves the strength of those members that are vulnerable to seismic attack.

The seismic retrofit of RC columns tends to change the column failure mode from shear to flexural failure, or to transfer the failure criteria from column to joint and/or from joint to beam failure, depending upon the strengthening parameters. This technique is used in existing reinforced concrete columns where insufficient transverse reinforcement and/or seismic detailing are provided; three different types of failure mode can be observed under seismic input. These are:

- *Column shear failure mode*: This mode of failure is the most critical one. The modern seismic column designs contain detailed transverse or shear reinforcement, but the shear strength of existing substandard columns can be enhanced by providing external shear reinforcement or by strengthening the column through composite fibres in the hoop direction.
- *Confinement failure at the flexural plastic hinge*: Subsequent to flexural cracking, the cover-concrete will crush and spall; this is followed by buckling of the longitudinal steel reinforcement, or a compression failure of the concrete, which in turn initiates plastic hinge deterioration.
- *Confinement of lower ends of columns*: Some bridge columns have lap splices in the column reinforcement; these are starter bars used for ease of construction and are located at the lower column end to form the connection between the footings and the columns. This is a potential plastic hinge region and it is advantageous to provide confinement by external jacketing or continuous fibre winding in this area.

None of these failure modes and associated column retrofits can be viewed separately since retrofitting for one deficiency may shift the seismic problem to another location and a different failure mode without necessarily improving the overall deformation capacity.

The confinement of RC columns can be undertaken by fabricating FRP composites using techniques such as the wet lay-up, the semi-automated cold-melt factory-made pre-impregnated fibre or the automated filament-winding processes. The fib have discussed the use of prefabricated (pre-cured) elements in the form of shells or jackets that are bonded to the concrete and to each other to provide confinement (fib Task Group 9.3, 2001). The wet lay-up and the prefabricated systems are generally placed with the principal fibre direction perpendicular to the axis of the member. The concrete column takes essentially axial load therefore the ratio of the areas of the circumferential to axial fibres of the composite is large thus providing confinement to the concrete. This allows the tensile strength in the circumferential direction to be virtually independent of the axial stress value. A review of the effectiveness of FRP composites for confining RC columns has been given in De Lorenzis and Tepfers (2003).

43.5.2 SHEAR STRENGTHENING OF DEGRADED RC BEAMS

Shear strengthening of RC beams and columns may be undertaken by bonding FRP laminates to the sides of the member. The principal fibre direction is parallel to that of the maximum principal tensile stresses, which in most cases is at approximately 45° to the member axis. However, for practical reasons it is usually preferable to attach the external FRP reinforcement with the principal fibre direction perpendicular to the member axis. Various researchers – El-Hacha and Rizkalla (2004), Triantafillou (1998) – and current design recommendations – El-Refaie *et al.* (2003) and Ibell and Silva (2004) – have shown that an FRP-shear-strengthened member can be modelled in accordance with Mörsch's truss analogy. Further information on this topic can be found in Lu *et al.* (2009).

43.6 Upgrading of metallic structural members

Advanced polymer composite materials have not been utilised to upgrade metallic structures to the same extent as they have been for reinforced concrete structures. However, as a result of research into this subject, which commenced at the latter part of the 20th century (Mertz and Gillespie, 1996; Mosallam and Chakrabarti, 1997; Luke, 2001; Moy, 2001; Tavakkolizadeh and Saadatmanesh, 2003; Cadei *et al.*, 2004; Moy, 2004; Luke and Canning, 2004, 2005; Photiou *et al.*, 2006; Hollaway *et al.*, 2006; Zhang *et al.*, 2006), there have been a number of applications of CFRP to metallic structures that have shown that the technique can have significant benefits over alternative methods of strengthening.

The number of applications to date in the UK has led to the publication of two comprehensive guidance documents:

1. ICE Design & Practice Guide. *FRP Composites – Life Extension and Strengthening of Metallic Structures* (Moy, 2001).
2. CIRIA Report C595. *Strengthening Metallic Structures using Externally-Bonded FRP* (Cadei *et al.*, 2004).

Design guidance has also been published recently by the Italian National Research Council (CNR, 2006), Schnerch *et al.* (2006) and ISIS (Canada), 2007.

FRP strengthening can be used to address any of the structural deficiencies described in the concrete section. The reasons for using FRP to rehabilitate a metallic or concrete structure may be similar; however, the way in which the FRP works with an existing metallic structure can often be very different to that in a concrete structure.

The FRP composite plate material used for the bonding operation is either the ultra-high-modulus (European definition) or the high-modulus (European definition) CFRP, AFRP (Kevlar 49) or possibly GFRP composites and these will be fabricated by one of four methods:

1. The pultrusion technique, in which the factory made rigid pre-cast FRP plate is bonded onto the degraded member with cold-cure adhesive.
2. The factory made rigid fully cured FRP prepreg plate bonded to the degraded member with cold-cure adhesive.
3. The low-temperature mould prepreg FRP prepreg/adhesive film placed onto the structural member and both components compacted and cured under vacuum at an elevated temperature.
4. Vacuum infusion (The Resin Infusion under Flexible Tooling (RIFT) process).

Figure 43.8 shows the upgrading of a curved steel structural beam by a carbon fibre/epoxy composite prepreg.

It should be mentioned that the ultra high-modulus carbon fibre composite has a low strain to failure, of the order of 0.4% strain, and a modulus of elasticity of the composite of about 40 GPa, so the system will fail with a small inelastic characteristic. The high-modulus CFRP composites have a value of ultimate strain of the order of 1.6% strain for modulus of elasticity of 28 GPa. This implies that the material is ductile and is unlikely to fail in a rehabilitation situation by ultimate strain but by some other method (Photiou, 2006).

Fig. 43.8 The upgrading of a curved steel structural beam by the carbon fibre/epoxy composite low-temperature mould prepreg (Courtesy of Taylor Woodrow, UK, and ACG Derbyshire, UK).

43.7 Internal reinforcement to concrete members

FRP rebars for reinforcing concrete members are generally fabricated by the pultrusion method (Nanni, 1993; ACI, 1996; Pilakoutas, 2000; Bank, 2006). The rebars can be manufactured from carbon, aramid and glass fibres using epoxy or vinylester polymers. The surfaces of pultruded composites are smooth and therefore it is necessary to post-treat them to develop a satisfactory bond characteristic between the concrete and the rebar. Several techniques are used for this, including:

- applying a peel-ply to the surface of the pultruded bar during the manufacturing process; the peel ply is removed before encasing the bar with concrete, thus leaving a rough surface on the pultruded rebar
- over-winding the pultruded rebar with additional fibres
- bonding a layer of sand with epoxy adhesive to the surface of the pultruded rod; this is a secondary operation at the end of the pultrusion line.

The features and benefits of using FRP rebars are:

- they are non-corrosive – they will not corrode when exposed to a wide variety of corrosive elements, including chloride ions, and are not susceptible to carbonation-initiated corrosion in a concrete environment

- they are non-conductive – they provide good electrical and thermal insulation
- they are fatigue resistant – they perform well in cyclic loading situations
- they are impact resistant – they resist sudden and severe point loading
- they have magnetic transparency – they are not affected by electro-magnetic fields.

FRP rebars manufactured from a thermosetting resin (viz. vinlyesters or epoxies) are unable to be reshaped once they are polymerised and therefore cannot be bent on site. If bends are required, for instance anchorages or stirrups, they must be produced by the FRP rebar manufacturer as a special item, but their strength at the bend will be considerably reduced. One option would be to use thermoplastic polymers as spliced bends; this material can be bent on site but the system is still in its development stage.

Although carbon and aramid fibre composites are higher in cost than are glass composites, they are inert to alkaline environment degradation and can be used in the most extreme cases. We discuss the behaviour of these materials in an alkaline environment in section 42.2.

43.8 FRP confining of concrete columns

The confinement of concrete enhances its durability and strength. In the past it was usual to enhance reinforced concrete columns by the addition of longitudinal steel bars and concrete around existing columns. A further method consisted of placing a steel jacket around a column. However, these two methods are difficult to apply.

Numerous experiments since the 1980s have demonstrated the effectiveness of FRP composites for confining RC columns by external wrapping with composite sheets (De Lorenzis and Tepfers, 2003). Confinement with polymer composite strands or sheets of composite prepreg have shown many advantages in compression over the above confinement methods. These include:

- high specific strength and stiffness
- relative ease of applying the composite materials in construction site situations
- with the large ratio of the areas of the circumferential to axial fibres, the modulus of elasticity of the FRP axial composite is small, thus allowing the concrete to take essentially the entire axial load
- the tensile strength in the circumferential direction is very large and essentially independent of the value of the axial stress

- ease and speed of application result from the FRP's low weight
- their minimal thickness does not alter the shape and size of the strengthened elements
- the good corrosion behaviour of FRP materials makes them suitable for use in coastal and marine structures.

Composite wrapping systems have been used throughout the world on a number of bridges, mainly for seismic loading, predominately in Japan and the USA. The available composite systems include epoxy with glass fibre, aramid fibre or carbon fibre fabric materials. Both wet lay-up and prefabricated systems are normally used with the principal fibre direction perpendicular to the axis of the member. The wrapping can be applied either continuously over the surface (which poses the problem of moisture migration) or as strips with a particular width between them (the spaced confining devices provide reduced effectiveness compared to the equal continuous device, as portions of the column between adjacent strips remain unconfined). The FRP confinement action is passive, that is, it arises as a result of the lateral expansion of the concrete core under axial load, and the confining reinforcement develops a tensile stress balanced by pressures reacting against the concrete's lateral expansion. An FRP confined column can deform longitudinally much more under an extreme stress state than a conventional material system before failure. The lateral confinement of the concrete provides an order of magnitude improvement in the ultimate compressive strain.

Confinement is most effective for circular columns, as the confinement pressure is in this case uniform. Both strength and ductility can be significantly enhanced. In the case of rectangular columns, the confining action is less efficient. The achievable increase in strength is usually modest or negligible, but a ductility enhancement can still be obtained. The effectiveness decreases as the cross-sectional aspect ratio increases.

The REPLARK and the XXsys (sections 41.1.1 and 41.1.2, respectively) are the two main systems available for site work.

43.9 FRP/concrete duplex beam construction

The combination of a fibre matrix composite and a conventional civil engineering material, i.e. concrete, to form a 'duplex' beam was researched by Triantafillou and Meier (1992). *Figure 43.9a* shows the basic 'duplex' beam that they conceived. The

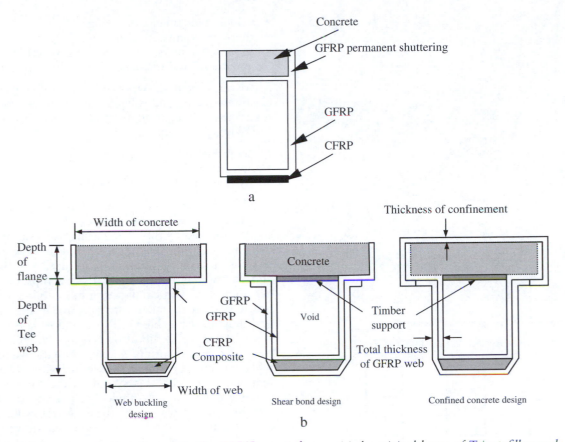

Fig. 43.9 Diagrammatic elevations of duplex FRP/Concrete beams: (a) the original beam of Triantafillou and Meier (1992); (b) the Tee beam of Hulatt et al. *(2003) (adapted from ICE Manaul of Bridge Engineering, Hollaway, Fig. 4).*

emphasis of their work was to use the concrete in the compressive and APCs in the tensile regions of a beam. Thus the two materials are used to their best structural advantage.

Hulatt *et al.* (2003a, 2003b, 2004) further developed the idea by testing and numerically analysing a Tee system under various geometries and loading configurations. *Figure 43.9b* shows diagrammatic elevations of FRP/concrete beams; the items shown are a design for web buckling; a design for shear bond; and a design for confining the concrete (which aids the compression strength of the concrete). The FRP composite materials used in this work were prepregs of glass and carbon supplied by Advanced Composites Group (ACG), Heanor, UK; this material is described in Chapter 41, section 41.1.1. As a result of the above work NECSO Entrecanales Cubiertas, Madrid, Spain and ACG have developed an equivalent beam; an element consisting of this beam and the completed bridge are shown in *Fig. 43.10*. This utilises the high compressive strength of concrete

and the high tensile strength of the carbon fibre. Load testing at 80% of ultimate load demonstrated that the beam behaved as a typical steel girder, indicating that the traditional principles of flexural design can be utilised. Analysis indicated that the manufacturing cost of a duplex beam is comparable to that of long-span concrete beams. The real benefit is in the significant cost savings provided owing to the lower weight and reduction of life cost of the beam. Obvious opportunities for this technology are more site installations and refurbishment of infrastructure in developing countries or war-damaged regions.

43.10 Polymer bridge bearings and vibration absorbers

43.10.1 BRIDGE BEARINGS

Bridge bearings are used to transfer loads from the deck of a bridge to its sub-structure and thence to its foundation and to avoid damage from:

Foam core (also acting as permanent shuttering)

Concrete flange

Permanent shuttering

u.d. CFRP fibre prepreg

±45° CFRP fibre prepreg

Diagramatic section through bridge beam

Fig. 43.10 The equivalent (Duplex) beam developed by NECSO Entrecanales Cubiertas, Madrid, Spain and ACG Ltd. Heanor, UK (Courtesy of ACG Derbyshire, UK).

- vehicle movements on bridges
- thermal expansion of bridges
- loading to piers, thus reducing reaction forces and rotational movement to within safe limits.

There are broadly two types of bridge bearing, elastic materials and roller bearings. In elastomeric materials, the bearing is made from one of the following polymers:

- neoprene polymer
- natural rubber
- styrene butadiene rubber (SBR).

The bearings are either plain pad-and-strip bearings or laminates with steel plates. Movements are accommodated by the basic mechanisms of internal deformation. The bearings allow the deck to be flexible in shear to accommodate deck translation and rotation but they are stiff in compression to accommodate vertical loads. The stability of the bearings must be taken into account in the design and they must be able to absorb and isolate energy from impacts and vibrations. *Figure 43.11a* illustrates a typical bearing. A diagrammatic sketch of a plane sliding bearing is shown in *Fig. 43.11b*; the material used with this system is the low-friction polymer polytetrafluoroethylene (PTFE), which slides against a metal plate. This bearing resists loads in the vertical direction but not rotational movements in the longitudinal or transverse directions; the rotational and transverse loads are resisted by providing mechanical keys.

A typical multi-roller bearing is shown in *Fig. 43.11c*. Vertical loads only can generally be resisted by these bearings, but large longitudinal movements can be accommodated. The roller material tends to be steel and therefore this type of bearing is outside the scope of this chapter.

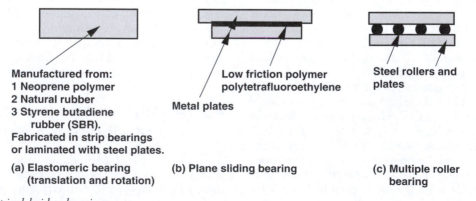

Manufactured from:
1 Neoprene polymer
2 Natural rubber
3 Styrene butadiene rubber (SBR).
Fabricated in strip bearings or laminated with steel plates.

Metal plates

Low friction polymer polytetrafluoroethylene

Steel rollers and plates

(a) Elastomeric bearing (translation and rotation)

(b) Plane sliding bearing

(c) Multiple roller bearing

Fig. 43.11 Typical bridge bearings.

43.10.2 SEISMIC ISOLATION SYSTEMS

Seismic isolation systems have two functions:

- to introduce flexibility at the base of a building structure in the horizontal plane
- to provide damping elements to restrict the amplitude of the motion caused by the earthquake.

There are three basic elements in a system, which have to provide:

- a damper or energy dissipator to control relative deflections between a building and the ground. Elastomers with high damping characteristics could be used for this element
- a flexible mounting so that the period of vibration of the total system is lengthened sufficiently to reduce the force response
- rigidity under low service load levels (e.g. wind or minor earthquakes).

The rubber-based isolation system is manufactured from:

- a high damping steel–rubber member
- a lead–rubber laminations member of thickness between 160 mm and 200 mm.

The typical residential buildings of reinforced concrete frame or wall construction of more than five stories high use the lead–rubber type. Other systems use elastomeric pads constructed of neoprene layers in series and are available with alternating raised diagonal ribs or square-cell pattern.

43.10.3 ANTI-VIBRATION AND STRUCTURAL ISOLATION SYSTEMS

In areas of ground-borne vibrations due to low-frequency rumble from underground and surface trains there is a risk of sound transmission from the rolling stock. Building structures might require isolation systems to be installed in their foundations. In these cases the isolation of the structure can be effected by placing elastomeric bearings (such as polyurethane-bound rubber granulate, polyurethane mixed-cell structure foam, a medium-density closed-cell structural foam such as isolation sheets, etc.) under the foundations of the building. The isolators are installed on top of the basement walls or columns. Applications of this technology include:

- foundation isolation
- column heads
- pile cap
- perimeter isolation
- floating floor systems.

43.11 Use of geosynthetics

The use of geosynthetics in civil engineering is a wide subject and cannot be competently covered here. However, some of the uses of this family of materials are in the form of:

- *Geotextiles* to prevent intermixing of the soft subgrade with granular material during the passage of lorries on civil engineering construction sites.
- *Geotextile overlay* to prevent existing cracks in pavements migrating into new overlay surfaces during the maintenance of asphalt roads.
- *Geolinear* elements used as anchors to stabilise an RC retaining wall.
- *Geogrids* acting as reinforced earth to reinforce slopes and retaining walls.
- *Geomembranes* to prevent loss of liquid from containment structures, such as water courses.
- *Geocomposites*, which have a wide range of applications, e.g. prefabricated drains, flexible skins, etc.

Further information on this topic may be found in Hollaway (1993), Akagi (1996), Cook (2003) and Giroud (2005).

References

ACI (1996). State-of-the-Art Report on Fiber Reinforced Plastic (FRP) Reinforcement for concrete structures. ACI 440R-96, American Concrete Institute, Farmington Hills, Michigan.

Akagi T (1996). *Application of Geosynthetics in Roads, Railways and Ground Improvements*. Proceedings of the International Conference on Environmental Geotechnology with Geosynthetics, New Delhi, pp. 171–180.

Bank LC, Arora D, Borowicz DT and Oliva M (2003a). *Rapid strengthening of reinforced concrete bridges*. Wisconsin Highway Research Program, Report No. 03-06. 166 pages.

Bank LC, Borowicz DT, Arora D and Lamanna AJ (2003b). *Strengthening of concrete beams with fasteners and composite material strips – Scaling and anchorage Issues*. US Army Corps of Engineers. Draft Final Report Contract Number DACA42-02-P-0064.

Bank LC (2004). *Mechanically-fastened FRP (MF-FRP) – a viable alternative for strengthening RC members*. Proceedings of the FRP 2nd International Conference on Composites in Civil Engineering – CICE 2004 – (ed. Seracino R), AA Balkema Publishers, Leiden, London, New York, Philadelphia, Singapore.

Bank LC (2006). *Composites for Construction Structural Design with FRP Materials*, John Wiley and Sons Inc., Hoboken, New Jersey.

Berry DBS (1974). *Tests on full size plastics panel components*. Proceedings of The Use of Plastics for Load bearing and Infill Panels (ed. Hollaway LC), University of Surrey, 12th and 13th September 1974, pp. 148–155.

Cadei JM, Stratford TK, Hollaway LC and Duckett WG (2004). CIRIA Report C595 'Strengthening Metallic Structures Using Externally Bonded Fibre-Reinforced. Published by CIRIA, London.

Canning L, Luke S, Taljsten B and Brown P (2004). *Field testing and long term monitoring of West Mill Bridge*. Proceedings of the 2nd International Conference, Advanced Polymer Composites for Structural Applications in Construction (eds Hollaway LC, Chryssanthopoulos MK and Moy SSJ), University of Surrey, Guildford, Surrey, UK, 20–22 April 2004.

Canning L (2008). *Mount Pleasant FRP bridge deck over M6 motorway*. Proceedings of the 4th International Conference on FRP Composites in Civil Engineering (CICE 2008), Zurich, Switzerland, 22–24 July 2008, pp. 243–249.

Composite Advantage (2008). Composite Advantage Builds New Drop-in-Place FRP Superstructure. Press Release, *Composite Advantage Newsletter* 1st May.

Concrete Society Technical Report (2000). *Design Guidance for Strengthening Concrete Structures Using Fibre Composite Materials*. TR55, 2nd edition, Concrete Society, Camberley, UK.

Concrete Society Technical Report (2003). *Strengthening Concrete Structures using Fibre Composite Materials: Acceptance, Inspection and Monitoring*. TR57, Concrete Society, Camberley, UK.

Cook DI (2003). Geosynthetics. *RAPRA Review Report*, 14 (No. 2), Report 158.

Cooper D (2001). GRP bridge cuts traffic disruption. *Reinforced Plastics*, 45 (No. 6), 4.

De Lorenzis L and Nanni A (2001). Shear strengthening of RC beams with near surface mounted FRP rods. *ACI Structural Journal*, 98 (No. 1), 60–68.

De Lorenzis L and Tepfers R (2003). A comparative study of models on confinement of concrete cylinders with FRP composites. *Journal of Comparative Construction*, ASCE, 7 (No. 3), 219–237.

De Lorenzis L and Teng JG (2007). Near-surface mounted reinforcement: an emerging technique for structural strengthening. *Composites Part B: Engineering*, 38 (No. 2), 119–143.

De Lorenzis L, Stafford T and Hollaway LC (2008). Structurally deficient civil engineering infrastructure: concrete, metallic masonry and timber structures. Chapter 1 of *Strengthening and Rehabilitation of Civil Infrastructures using Advanced Fibre/Polymer Composites* (eds Hollaway L and Teng J-G), Woodhead Publishing Ltd, Cambridge, UK.

El-Hacha R and Rizkalla SH (2004). Near-surface-mounted fibre-reinforced polymer reinforcements for flexural strengthening of concrete structures. *ACI Structural Journal*, 101 (No. 5), 717–726.

El-Refaie SA, Ashour AF and Garrity SW (2003). Sagging and hogging strengthening of continuous reinforced concrete beams using CFRP sheets. *ACI Structural Journal*, 100 (No. 4), 446–453.

Faber Maunsell (2003). FRP footbridge in place. *Reinforced Plastics*, 47 (No. 6), 9.

Federation Internationale du Beton, FIB Task Group 9.3 (2001). *Externally bonded FRP reinforcement for RC structures*, FIB, Lausanne.

Giroud JP (2005). Quantification of geosynthetic behavior. *Geosynthetics International*, Special Issue on the Giroud Lectures, 12 (No. 1), 2–27.

Head P (1994). *The world's first advanced composite road bridge*. Symposium on short- and medium-span bridges, Calgary, Canada.

Hollaway LC (1993). *Polymer Composites for Civil and Structural Engineering*, Blackie Academic & Professional, London, Glasgow, New York Tokyo, Melbourne, Madras.

Hollaway LC and Leeming MB (eds) (1999). *Strengthening of Reinforced Concrete Structures*, Woodhead Publishing, Abingdon, UK.

Hollaway LC and Head PR (2001). *Advanced Polymer Composites and Polymers*, Elsevier, London, Amsterdam, 223.

Hollaway LC (2008). Case Studies. Chapter 13 of *Strengthening and Rehabilitation of Civil Infrastructures using Advanced Fibre/Polymer Composites* (eds Hollaway L and Teng J-G), Woodhead Publishing Ltd, Cambridge, UK.

Hollaway LC, Zhang L, Photiou NK, Teng JG and Zhang SS (2006). Advances in adhesive joining of carbon fibre/polymer composites to steel members for repair and rehabilitation of bridge structures. *Advances in Structural Engineering*, 9 (No. 6), 791–803.

Hollaway LC and Teng JG (2008). *Strengthening and rehabilitation of civil infrastructures using fiber-reinforced polymer (FRP) composites*, Woodhead Publishing Ltd, Cambridge, UK.

Hollaway LC (2009). Applications. Chapter 58, Section 7 of *ICE Manual of Construction Materials* (ed. Forde MJ), Institution of Civil Engineers, London.

Hulatt J, Hollaway L and Thorne A (2003a). The use of advanced polymer composites to form an economic structural unit. *Construction and Building Materials*, 17 (No. 1), 55–68.

Hulatt J, Hollaway L and Thorne A (2003b). Short term testing of a hybrid T-beam made from a new prepreg material. *ASCE Journal of Composites for Construction*, 7 (No. 2), 135–145.

Hulatt J, Hollaway LC and Thorne AM (2004). A novel advanced polymer composite/concrete structural element. *Proceedings of the Institution of civil engineers*, Special Issue: *Advanced Polymer Composites for Structural Applications in Construction*. February, pp. 9–17.

Hutchinson AR (2008). Surface Preparation of Component Materials, Chapter 3 of *Strengthening and Rehabilitation of Civil Infrastructures using Advanced Fibre/Polymer Composites* (eds Hollaway L and Teng J-G), Woodhead Publishing Ltd, Cambridge, UK.

Ibell TJ and Silva PF (2004). A theoretical strategy for moment redistribution in continuous FRP-strengthened concrete structures. *Proceedings of the 2nd International Conference on Advanced Polymer Composites for Structural Applications in Construction*, edited by Hollaway LC, Chryssanthopoulos MK and Moy SSJ. Published by Woodhead Publishing Ltd., Cambridge.

ISIS (Canada). *Design guidance for strengthening steel structures using FRP*, to be published in 2010.

Italian National Research Council (CNR, DT200/2004). *Guidelines for the Design and Construction of Externally Bonded FRP Systems for Strengthening Existing Structures 2004 – Metallic Structures*. (English translation, 2006).

Leggatt AJ (1974). *Contribution given on 'Convent Garden Flower Market'*. Proceedings of The Use of Plastics for Load bearing and Infill Panels (ed. Hollaway LC), University of Surrey, 12th and 13th September 1974, p. 207.

Leggatt AJ (1978). *The role of the engineer*. Proceedings of the Conference on Design and Specification of GRP Cladding (ed. Hollaway LC), Royal Institute of British Architects, 19th October 1978, pp. 21–30.

Lu XZ, Chen JF, Ye LP, Teng JG and Rotter JM (2009). RC beams shear-strengthened with FRP: Stress distributions in the FRP reinforcement. *Construction and Building Materials*, **23** (No. 4), 1544–1554.

Luke S (2001). *Strengthening structures with carbon fibre plates. Case histories for Hythe Bridge, Oxford and Qafco Prill Tower*. NGCC first annual conference and AGM – Composites in Construction through life performance, Watford, UK, 30–31 October 2001.

Luke S and Canning L (2004). *Strengthening Highway and Railway Bridge Structures with FRP Composites – Case Studies*. Proceedings of the 2nd International Conference, Advanced Polymer Composites for Structural Applications in Construction (eds Hollaway LC, Chryssanthopoulos MK and Moy SJ), University of Surrey, Guildford, Surrey, UK, 20–22 April 2004, pp. 747–754.

Luke S and Canning L (2005). Strengthening and Repair of Railway Bridges Using FRP Composites. In *Bridge Management 5* (eds Parke GAR and Disney P), Thomas Telford Ltd, London, 684 pages.

McNaughton A (2006). Foreword to 'The Maintenance and renewal of Bridges', Network Rail sponsored Supplier Conference, Marriott Hotel Centre, Bristol, UK, 22–23 November 2006.

McKenzie M (1991). *Corrosion protection: The environment created by bridge enclosure*, Research Report 293, TRRL, Crowthorne, 1991.

McKenzie M (1995). *The corrosivity of the environment inside the Tees Bridge Enclosure: Final year results*, Project Report PR/BR/10/93, TRRL, Crowthorne, 1993.

Mertz D and Gillespie J (1996). *Rehabilitation of steel bridge girders through the application of advanced composite material*. NCHRP 93-ID11, Transportation Research Board, Washington, DC, 1–20.

Mosallam AS and Chakrabarti PR (1977). 'Making connection', Civil Engineering, ASCE, pp. 56–59.

Moy SSJ (ed.) (2001). *FRP composites – Life Extension and strengthening of Metallic Structures*, Institution of Civil Engineers, London, 33–35.

Moy SSJ (2004). *The Strengthening of Wrought Iron Using Carbon Fibre Reinforced Polymer Composites*, Proceedings of the 2nd International Conference on Advanced Polymer Composites for Structural Applications in Construction, edited by Hollaway LC, Chryssanthopoulos MK and Hoy SS. Published by Woodhead Publishing Ltd., Cambridge.

Nanni A (ed.) (1993). *Fibre Reinforced Plastics (FRP) for Concrete Structures: Properties and Applications*, Elsevier Science, New York.

Oehlers DJ and Seracino R (2004). *Design of FRP and Steel Plated RC Structures – Retrofitting Beams and Slabs for Strength, Stiffness and Durability*, Elsevier, Amsterdam, London, New York, Sydney.

Pilakoutas K (2000). Composites in Construction. Chapter 10 of *Failure Analysis of Industrial Composite Materials* (eds Gdoutos EE, Pilakoutas K and Rodopoulos CA), McGraw-Hill, New York.

Photiou NK, Hollaway LC and Chryssanthopoulos MK (2006). Strengthening of an artificially degraded steel beam utilising a carbon/glass composite system. *Construction and Building Materials*, **20** (Nos. 11–21).

Photiou N (2006). 'Rehabilitation of Steel Members Utilising Hybrid FRP Composite Materials Systems', PhD thesis, University of Surrey, Guildford, UK.

Roach EC (1974). *The manufacturer's view of the general use of plastics panels as structural and non-load bearing units*. Proceedings of The Use of Plastics for Load bearing and Infill Panels (ed. Hollaway LC), University of Surrey, 12th and 13th September 1974, pp. 24–35.

Schnerch D, Dawood M and Rizkalla S (2005). 'Strengthening steel-concrete composite bridge with high modulus carbon fiber reinforced polymer (CFRP) laminates', Proceedings of the Third International Conference on Composites in Construction (CCC 2005), Lyon, France, July 11–13, 2005, pp. 283–290.

Reinforced Polymer (CFRP) Strips, Technical Report No. IS-06-02. Constructed Facilities Laboratory, North Carolina State University.

Southam NLF (1978). *The role of the architect*. Proceedings of the Conference on Design and Specification of GRP Cladding (ed. Hollaway LC), Royal Institute of British Architects, 19th October 1978, pp. 8–19.

Sriramula S and Chryssanthopoulos MK (2009). Probabilistic models for spatially varying mechanical properties of in-service GFRP cladding panels. *Journal of Composites for Construction*, **13**, 159–167.

Stephenson B (1974). *The architect's view of the general use of plastics panels as structural and non-load bearing units*. Proceedings of The Use of Plastics for Load bearing and Infill Panels (ed. Hollaway LC), University of Surrey, 12th and 13th September 1974, pp. 17–23.

Tavakkolizadeh M and Saadatmanesh H (2003). Strengthening of steel-concrete composite girders using carbon fibre reinforced polymer sheets. *Journal of Structural Engineering, ASCE*, **129** (No. 1), 30–40.

Teng JG, Chen JF, Smith ST and Lam L (2002). *FRP Strengthened RC Structures*, John Wiley, England, USA, Germany, Australia, Canada, Singapore.

Triantafillou TC and Plevris N (1991). *Post-strengthening of RC beams with epoxy bonded fibre composite materials*. Proceedings of the Specialty Conference on Advanced Composites Materials in Civil Engineering, Nevada, pp. 245–256.

Triantafillou TC and Meier U (1992). *Innovative design of FRP combined with concrete*. Proceedings of the First International Conference on Advanced Composite Materials for Bridges and structures (ACMBS), Sherbrooke, Quebec, pp. 491–499.

Triantafillou TC (1998). Shear strengthening of reinforced concrete beams using epoxy-bonded FRP composites. *Structural Journal*, **95** (No. 2), 07–115.

Zhang L, Hollaway LC, Teng J-G and Zhang SS (2006). *Strengthening of Steel Bridges under Low Frequency Vibrations*. Proceedings of the 3rd International Conference on FRP Composites in Civil Engineering (CICE 2006), Miami, Florida, USA, 13–15 December 2006.

Bibliography

Hollaway LC and Leeming MB (eds) (1999). *Strengthening of Reinforced Concrete Structures – using externally bonded FRP composites in structural and civil engineering*, Woodhead Publishing Ltd, Cambridge, UK.

Kelly A, Buresch FE and Biddulph RH (1987). 'Composites for the 1990s' Philosophical Transactions of the Royal Society, London. July 27, 1987, Vol. 322, pp. 409–423.

McKenzie M (1995). *The corrosivity of the environment inside the Tees Bridge enclosure*. Final Year results, Project Report PR/BR/10/93, TRRL, Crowthorne.

Meier U (1987). Proposal for a carbon fibre reinforced composite bridge crossing the Strait of Gibraltar at its narrowist point, *Proceedings Institution Mechanical Engineers*, **201**, Issue B2, 73–78.

Meier U and Kaiser HP (1991). *Strengthening of structures with CFRP laminates*. Proceedings of the Speciality Conference Advanced Composites Materials in Civil Engineering Structures, Las Vegas, Nevada, editors Iyer SL and Sen R, American Society of Civil Engineers, pp. 224–232.

Priestley MJN, Seible F and Calvi M (1996). *Seismic Design and Retrofit of Bridges*, John Wiley & Sons, Inc., New York.

Richmond BR and Head PR (1988). *Alternative materials in long-span bridge structures*. Proceedings of the 1st Oleg Kerensky Memorial Conference, London, June 1988.

Seible F, Priestley MJN, Hegemier GA and Innamorato D (1997). Seismic retrofit of RC columns with continuous carbon fiber jackets. *ASCE Journal of Composites for Construction*, Vol. 1, pp. 52–62.

Triantafillou TC and Plevris N (1995). Reliability analysis of reinforced concrete beams strengthened with CFRP laminates. In *Non-metallic (FRP) Reinforcement for Concrete Structures* (ed. Taerwe L), E & FN Spon, London, pp. 576–583.

Section 2: Fibre–reinforced cements and concrete

Phil Purnell

Introduction

Almost every publication on fibre-reinforced cements and concretes (FRC) opens by reminding us that man has formed useful composites by combining brittle materials with more ductile fibres for millennia. I see no reason why I should digress:

> Ye shall no more give the people straw to make brick, as heretofore: let them go and gather straw for themselves. And the tale of the bricks, which they did make heretofore, ye shall lay upon them; ye shall not diminish ought thereof: for they be idle . . . (Exodus 5:7–8 KJV)

Although some progress has been made since in the handling of stakeholder productivity issues, the basic technical problem remains the same. Cementitious materials are relatively cheap and, as we have discussed is some detail in Part 3, are easy to form and strong in compression but have poor tensile strength, impact strength and toughness. This makes them susceptible to cracking and intolerant of local transient overloads, especially those developed at points of fixing or caused by installation. Surrounding a more ductile fibre with a concrete or mortar matrix can produce a material with a degree of 'pseudo-ductility' or toughness orders of magnitude greater than that of the plain concrete. Thus the mode of reinforcement is very different to that generally found in fibre-reinforced polymers (FRP) where, as we discussed in the last section, stiff strong fibres reinforce a weak but ductile matrix. Paradigms for analysing FRP are not generally valid for FRC; hence this chapter is required.

The most widely used FRC in recent times has been asbestos cement, where a natural fibrous silicate was used as the reinforcement in automated production of thin sheets (the Hatschek process). These once-ubiquitous, often corrugated sheets were used for roofing, cladding and fireproofing. Since the widely publicised health concerns surrounding the use and disposal of asbestos cement have come to light (e.g. Health & Safety Commission, 1979), its use has declined rapidly, often under legislative decree. However, as the failure strain of cementitious materials is low (generally around 0.03%), a wide range of other fibres – glass, carbon, polymer, steel and so on – are potentially suitable alternative reinforcements. Many of them have technical advantages over asbestos; for example, glass and carbon can be used to provide primary reinforcement, increasing the ultimate strength of the plain matrix, as well as providing increased crack resistance and toughness.

The most common matrix used for FRC is Portland cement (PC) concrete or mortar. The microstructural environment within a PC concrete is typically highly alkaline, changes with time as the matrix continues to hydrate and is influenced by external factors such as humidity and temperature (see Part 3). The interaction of the fibres with this environment means that FRC properties are time-dependent, generally on timescales involving years, with important implications for durability. Modifying this fibre–matrix interaction to improve durability has formed the main thrust of FRC research over the past 20 years, mainly focused on the development of alternative matrices. Advances in the use of fibres have also been made. Previously exotic fibres, such as carbon and aramids, are beginning to drift downwards in price sufficiently for researchers to consider using them in cement composites. Specialist textile reinforcements are also being developed to allow the production of structural, load-bearing FRC components with high fibre contents. Eventually, such composites may compete with traditional reinforced concrete (RC) components. However, at the moment the distinction between applications for FRC and RC remains fairly clear. FRC is preferred where thin sections (i.e. of thickness insufficient to provide cover for rebars) are required, such as roofing and cladding products. It is also often preferable where localised deformations are considerable and/or unpredictable, such as tunnel linings, industrial floors, marine structures and blast-resistant structures. In particular, FRC excels in restraining cracking caused by secondary effects, such as shrinkage, humidity changes, creep, and temperature fluctuations, owing to the distributed nature of the reinforcement. Fibres are sometimes also added to help control plastic shrinkage cracking during setting and curing.

We will look at applications later in the section, but a flagship example of what can uniquely be achieved using FRC (in this case, glass-fibre reinforced concrete) is the 37 m, 11-storey Merlion on Sentosa Island, Singapore (1996) (*Fig. VII.2.1*). Housing a visitors' centre, studded with fibre optic illumination and with a viewing platform on top, FRC was considered the only choice of material that could combine flexibility, durability, surface finish and form to create this unique structure.

In the first chapter in this section we introduce some terminology specific to FRC and which is useful for the subsequent discussions. The properties of FRC are dictated by the properties of its constituent materials (i.e. fibres and matrices) and these are dealt with in Chapter 45. The nature of the fibre–matrix interface is discussed in Chapter 46, followed by a description of reinforcement layouts in Chapter 47. Chapter 48 then outlines the mechanisms by which fibres modify the properties of the composite and the various models that describe this. Manufacturing processes, typical applications and examples are then covered in Chapters 49 and 50, respectively, followed by a discussion of the durability and time-dependent behaviour and, briefly, recycling of FRC in Chapter 51.

Fig. VII.2.1 *The Merlion, Sentosa Island, Singapore. Photo taken by Sengkang, Singapore (2006). Taken from the Wikipedia Commons resource at http://commons.wikimedia.org/wiki/Sentosa.*

Chapter 44

Terminology for FRC

44.1 FRC – cement or concrete?

In the wider literature, one finds references to both fibre-reinforced cement and fibre-reinforced concrete. The first normally refers to thin-sheet material with high fibre content, no coarse aggregate and a matrix with markedly higher cement content than normal concrete. The fibres are intended to provide primary reinforcement, i.e. substantially enhancing a key property such as bending strength or toughness and the composite containing them is more correctly referred to as *primary FRC* (other authors (Bentur and Mindess, 2007, p. 3) use the term *high-performance FRC*). The second normally refers to more traditional concrete to which fibres are added either to provide post-failure integrity in the event of accidental overload or spalling (*secondary FRC*), or to provide control of shrinkage-related cracking (*tertiary FRC*). Bentur and Mindess (2007) refer to both these latter types as *conventional FRC*. Many authors use the terms indiscriminately and there is significant overlap between the classes; in this chapter, where the difference is important it has been clarified.

44.2 Key parameters in FRC

Several FRC parameters pertain to the matrix, fibre and composite. Where this is the case, subscripts m, f and c are used, respectively. Where 'ultimate' parameters are discussed (i.e. failure strength or strain), the subscript u is added. For example, the failure strength of the fibres would be σ_{fu} and the failure strain of the matrix would be ε_{mu}.

44.2.1 VOLUME FRACTIONS

Many properties of FRC are functions of the volume fraction of fibres V_f, defined as the volume of fibres divided by the volume of the composite, usually expressed as a percentage. A typical value of V_f for primary FRC is ~5%; for secondary FRC it will be lower than this down to a minimum of ~0.2%. In very high performance textile FRC, or asbestos FRC, it might be as high as 15%. Of particular interest is the critical volume fraction, V_{fcrit}, which is the minimum V_f at which primary reinforcement action can be observed in the composite. We occasionally refer also to V_m (the volume fraction of matrix), which is of course $1 - V_f$.

44.3 Reinforcement elements

44.3.1 MULTIFILAMENT/MICROFIBRE FRC

The unit reinforcement element in FRC is often not a single filament but some grouping or bundle of fibres. The bundle is not necessarily intended to disperse such that each individual filament is completely surrounded by matrix (in contrast to FRP); fibres in the centre of the bundle may remain uncoated by matrix, which allows the bundle to remain flexible. Glass-, carbon- natural- and most polymer-FRC fall into this category. We will call this *multifilament FRC* and we require terminology to describe the various configurations. In this chapter, a single monolithic fibre will be called a *filament*. These tend to be relatively thin, < 0.1 mm in diameter, which leads to *multifilament FRC* often being referred to as *microfibre FRC*. A group or bundle of filaments will be called a *strand* (e.g. for glass-FRC, a strand of 204 filaments is typically used). The filaments in a strand are normally loosely held together by a soluble coating known as a 'size'. A group of strands will be called a *roving* (in glass-FRC, rovings of between 1 and 64 strands are common). Strands may be twisted into a roving (in which case it is normally called a 'tow') or they may remain parallel.

44.3.2 MONOFILAMENT/MACROFIBRE FRC

In *monofilament FRC*, the matrix fully surrounds each individual filament and/or the filaments are dispersed rather than grouped. This is the case for all steel-FRC and some polymer-FRC. Fibres in monofilament FRC tend to be larger, > 0.1 mm effective diameter i.e. *macrofibres* and are used at relatively low volume fractions as secondary or tertiary reinforcement.

Reference

Bentur A and Mindess S (2007). *Fibre Reinforced Cementitious Composites*, 2nd edition (Modern Concrete Technology 15), Taylor and Francis, Oxford, UK, 601 pages.

Chapter 45

Component materials

45.1. Fibres

Almost every imaginable type of fibre has been used at some time or another with cement or concrete. The most commercially significant types are described below and their properties are summarised in *Table 45.1*. More details are given in Chapter 39; here we focus on the properties particularly relevant to FRC.

45.1.1 POLYMER FIBRES

Polymer fibres have been used very successfully in FRC for many years. Since the modulus of elasticity of polymers tends to be rather less than that of the matrix (~5 GPa vs. ~20 GPa) they normally only provide a secondary or tertiary reinforcement action (providing 'post-peak' strength and toughness, or controlling shrinkage cracking) unless quite high values of $V_f > 5\%$ are used. Polypropylene (PP) fibres are the most commonly used polymer fibre for FRC and are made by extruding high-molecular-weight PP into either monofilaments or films. These are then stretched in order to orient the polymer molecules, which improves the fibre modulus from around 2 GPa to > 5 GPa. Short monofilaments are used at low V_f (< 0.5%) and dispersed throughout concrete

Table 45.1 Typical properties of cement-based matrices and fibres

Material or fibre	Relative density	Diameter or thickness (microns)	Length (mm)	Elastic modulus (GPa)	Tensile strength (MPa)	Failure strain (%)	Volume in composite (%)
Mortar matrix	1.8–2.0	300–5000	–	10–30	1–10	0.01–0.05	85–97
Concrete matrix	1.8–2.4	10 000–20 000	–	20–40	1–4	0.01–0.02	97–99.9
Aromatic polyamide (aramid)	1.45	10–15	5–continuous	70–130	2900	2–4	1–5
Asbestos	2.55	0.02–30	5–40	164	200–1800	2–3	5–15
Carbon	1.16–1.95	7–18	3–continuous	30–390	600–2700	0.5–2.4	3–5
Cellulose	1.5	20–120	0.5–5.0	10–50	300–1000	20	5–15
Glass	2.7	12.5	10–50	70	600–2500	3.6	3–7
Polyacrylonitrile (PAN)	1.16	13–104	6	17–20	900–1000	8–11	2–10
Polyethylene pulp	0.91	1–20	1	–	–	–	3–7
HDPE filament	0.96	900	3–5	5	200	–	2–4
High modulus	0.96	20–50	continuous	10–30	> 400	> 4	5–10
Polypropylene							
Monofilament	0.91	20–100	5–20	4	–	–	0.1–0.2
Chopped film	0.91	20–100	5–50	5	300–500	10	0.1–1.0
Continuous nets	0.91–0.93	20–100	continuous	5–15	300–500	10	5–10
Polyvinyl alcohol (PVA, PVOH)	1–3	3–8	2–6	12–40	700–1500	–	2–3
Steel	7.86	100–600	10–60	200	700–2000	3–5	0.3–2.0

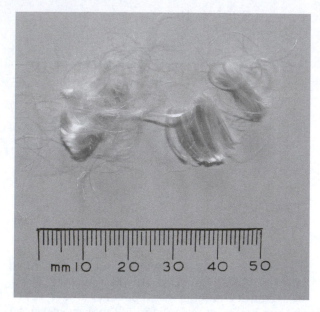

Fig. 45.1 Network of polypropylene fibres for use in FRC.

as tertiary reinforcement. Films are 'fibrillated' (*Fig. 45.1*), with tape-like cross-sections of effective diameter ~0.05 to 0.5 mm and used as secondary (0.5% $< V_f < 5\%$) or more rarely primary ($V_f > 5\%$) microfilament reinforcement in thin sheet FRC. Since the PP fibre surface is chemically hydrophobic, it is normally treated in some way in order that a reasonable bond with cement paste can be formed.

The use of other polymer fibres in FRC is much less common. Acrylic and polyester fibres are not stable in the high pH environment of most cement-based matrices and are thus not suitable for FRC. There is some interest in using nylon fibres as a replacement for PP but commercial use is not widespread. Polyethylene fibres have been used as short dispersed fibres in concrete (V_f ~4%) and in fibrillated 'pulp' form as a direct replacement for asbestos fibre (V_f ~10%). Their slightly higher modulus compared with that of PP (5–30 GPa depending on processing) and improved mechanical bond with the cement matrix mean that they have greater potential to produce primary FRC than PP. Certain polyvinyl alcohol (PVA) fibres are also suitable to act as asbestos replacement in FRC. They have relatively high modulus and strength (20–30 GPa and 1200–1500 MPa, respectively) and bond chemically to the cement matrix, so at $V_f = 3\%$ they can provide a significant primary reinforcement action.

Aramid fibres – such as Kevlar® – are special polymer fibres with aligned chain microstructures,

imparting much higher mechanical properties than other polymers (moduli of ~70–120 GPa and tensile strengths of ~3000 MPa). The unit 'fibre' is actually a bundle (called a tow or roving) of several hundred filaments, 0.010–0.015 mm in diameter. The price of these high-performance fibres is currently too high for serious commercial FRC application but they could potentially provide very effective primary reinforcement.

45.1.2 STEEL FIBRES

Historically, the use of steel–FRC has probably outweighed that of any other FRC; it is still very widely used, although polymer–FRC is catching up fast. In all cases, steel fibres can only provide secondary or tertiary monofilament reinforcement, since workability issues prevent V_f from exceeding 1.5–2% and the fibres are normally dispersed in a random 3D manner throughout the concrete matrix, which minimises their reinforcement efficiency. Fibres for general use are made from ordinary carbon steel, although where enhanced corrosion resistance is required (e.g. marine applications) stainless alloy steel or galvanised fibres may be used. There are many manufacturing routes, including simple wire drawing, melt spinning or manufacture from cut sheet steel. This leads to a variation in tensile strength from around 350 MPa to > 1000 MPa, although the modulus remains constant at about 200 GPa.

All modern steel fibres intended for use with FRC have complex cross sections (e.g. crimped, twisted or flattened) and/or deformed axial shape (e.g. hooked ends or a 'wavy' shape) to maximise the anchorage between the fibre and the matrix (*Fig. 45.2*). Physical and chemical surface treatments may also be applied to enhance the bond. Equivalent diameters and lengths vary from 0.1 to 1 mm and 10 to 60 mm, respectively.

45.1.3 NATURAL FIBRES

Natural fibres – by which we generally mean those of vegetable origin – are the oldest form of reinforcement for cementitious materials. The motivation for use of a particular natural fibre normally stems from the desire to use a cheap, locally available and sustainable resource. Technical performance issues are of secondary interest, since the use of most viable natural fibres will lead to similar FRC properties, i.e. inferior to those of glass-, steel- or polymer-FRC. Since the fibres also have a poor tolerance for the highly alkaline FRC matrix, they also degrade quickly. However, since natural fibres are normally used to provide tertiary reinforcement, permitting easier shaping of green forms, or short-term secondary

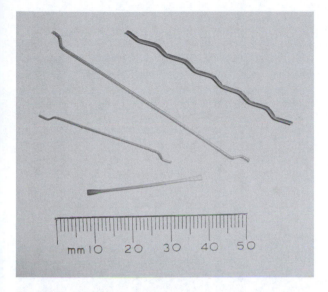

Fig. 45.2 Steel fibres showing a variety of types of mechanical deformation.

toughness to permit installation handling, their use is widespread, particular in less economically developed countries. The volume fraction rarely rises above 5% except in specialised cases where a natural fibre is used as a direct replacement for asbestos in the Hatschek process.

Fibres derived from plant stems (e.g. jute and flax), leaves (e.g. sisal), or woody parts (waste structural wood or bamboo) are processed to extract the cellulose-rich fibres from the organic matrices. The degree of processing applied will determine the quality of the fibres; jute, flax and sisal are relatively easily processed using natural methods called 'retting' to produce medium-quality fibres. To extract fibres from timber, it must be chipped and heated in industrial processing plants, but high-quality cellulose fibres can be obtained. Other fibres such as cotton and coir (coconut husk) can be extracted with minimal processing, but tend to be of low quality.

However obtained, all natural fibres have the same basic multifilament structure (*Fig. 45.3a*). A single fibre is effectively a roving, with a variable number of strands of ~100 cellular filaments. Each of these cells is a tube, around 1–2 mm long and < 50 μm across, with a hollow central core called a lumen. The walls of the tube are a composite of crystalline cellulose fibres in a hemi-cellulose–lignin matrix (*Fig. 45.3b*); the roving itself is held together by more lignin. The lumen has a tendency to absorb water, and the fibre properties are strongly dependent on the water content. As the degree of processing increases, the roving is progressively separated into strands, removing the low-strength lignin and concentrating the high-strength cellulose fibres. The properties are highly variable; strength and stiffness

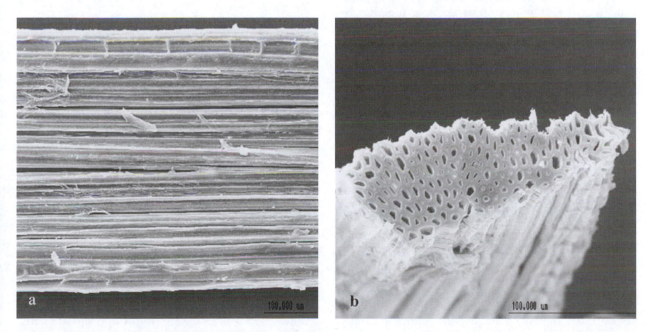

Fig. 45.3 Scanning electron micrograph of sisal fibre used for FRC. (a) Side view, (b) end-on view. (Reprinted from Toledo Filho et al. © 2005 with permission from Elsevier.)

may range from 200 to 800 MPa and 10 to 80 GPa, respectively, depending on fibre type, quality and processing parameters.

45.1.4 GLASS

As with polymer matrices, glass fibres are frequently used to reinforce cement composites. Standard 'E' glass (from the SiO_2–CaO–Al_2O_3–B_2O_3 system) is not stable in most cement-based matrices owing to their highly alkaline nature, so special alkali-resistant (AR) glass fibres (SiO_2–Al_2O_3–ZrO_2–alkali oxide) are used in commercial glass–FRC. The addition of zirconia reduces (but does not completely halt) the overall rate of reaction between the glass and the alkaline matrix and also causes a protective zirconia-rich layer to form on the fibre surface. Since the modulus and strength of the glass are around 70 GPa and 1500 MPa, respectively, it can be used to manufacture primary FRC; in fact, glass–FRC is probably the most common primary FRC, generally formed in thin sheets by spraying, with $V_f \sim 5\%$. It is also sometimes used as secondary and/or tertiary reinforcement in larger concrete components. It is always used in multifilament mode.

Glass fibres are normally supplied as rovings of up to 64 strands, each with ~200 filaments of ~14 μm diameter. They may be supplied as a continuous roving on a roll and chopped to length of 10–40 mm during primary FRC manufacture, or come pre-chopped to 3–15 mm for secondary/tertiary applications (*Fig. 45.4*). The strands are held together by a size (typically based on polyhydroxyphenol), which modifies the hydration of the matrix at the interface to help promote durability. Although the strain to failure is relatively low (~2%), this is significantly more than that of the cement matrix, so very useful toughness can also be provided by glass fibres.

45.1.5 CARBON

Carbon fibres were commercially developed in the USA in the 1960s. Their very high strength- and stiffness-to-weight ratios quickly found them applications in high-performance polymer composites and, as with most fibres, it was only a matter of time before they were investigated for use with FRC. As their price falls, they are becoming increasingly attractive and several commercial carbon–FRC applications have now emerged. They are particularly attractive for FRC since the carbon is virtually inert with respect to the alkaline environment in the cementitious matrix. A single carbon fibre naturally forms a tow of around 10 000 filaments, each about 7–15 μm in diameter. The filaments consist of stacked

Fig. 45.4 Glass fibres for FRC. Foreground – chopped strand mat. Background – continuous roving for chopping.

layers of graphite, oriented more-or-less parallel with the fibre axis. Within the layers strong covalent bonds prevail, while the layers are held together by weak van der Waals bonding (see Chapter 1). The strength of the fibre varies depending on the manufacturing method; as the degree of distortion of the layers decreases and thus a greater proportion of the layers are oriented parallel to the fibre axis, the strength increases. Pitch-based fibres (modulus ~30 GPa and strength ~600 MPa) are normally used for FRC, as they are rather cheaper than the polyacrylonitrile-based fibres, which are about 10 times stiffer and 4 times stronger. The fibres can be formed into a convenient fabric for use in FRC (*Fig. 45.5*).

45.1.6 ASBESTOS

Although the use of asbestos–FRC is now banned in most developed countries, it is still very commonly encountered during refurbishment of older buildings so a brief description is appropriate here.

Asbestos fibres are derived from naturally occurring crystalline fibrous silicate minerals. The two most

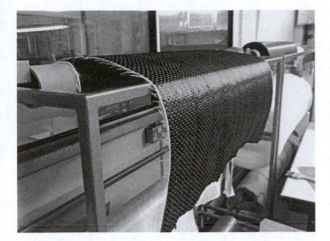

Fig. 45.5 Roll of unidirectional carbon fibre fabric for FRC.

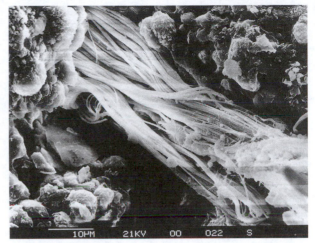

Fig. 45.6 Asbestos fibre bundle in cement paste after natural weathering for more than 10 years.

commonly encountered are white asbestos (chrysotile, a serpentine mineral) and blue asbestos (crocidolite, an amphibole mineral). Both have a microstructure based on silicate sheets; a planar banded structure interspersed with iron and sodium compounds in blue asbestos, and a 'roll of carpet' structure interspersed with brucite $(Mg(OH)_2)$ in white asbestos. These chemical structures are highly stable in the alkaline cement environment, and their surface chemistry and particle size/shape allows a stiff, felt-like slurry of asbestos fibres and cement to be produced in an automated, continuous process, resulting in a final hardened thin sheet product with V_f ~10% (*Fig. 45.6*). Fibre diameters are typically 20–200 nm (i.e. 2–3 orders of magnitude smaller than glass or carbon, hence the damaging bioactivity). The length varies with the processing method but rarely exceeds 20 μm. The strength and stiffness of such small fibres is very difficult to measure so reported values vary widely, but values of 1000–2000 MPa and 160 GPa, respectively, are typical.

45.2 Matrices

There are a wide variety of matrices for FRC. At one end of the scale (tertiary FRC), fibres are simply added to a standard concrete in order, for example, to control plastic shrinkage. At the other extreme (primary FRC), a cementitious matrix may be specially developed for use with a particular type of FRC, with mix design, admixtures and additives chosen to optimise e.g. textile penetration, composite durability, manufacturability and/or surface finish. However,

the majority of these matrices are based on Portland cement (though frequently heavily modified), which are described in Part 3, is only aspects particular to FRC are discussed here.

45.2.1 MIX DESIGN PARAMETERS
The basic physical requirements of a matrix for a typical FRC (i.e. < 40 mm thickness, primary or secondary reinforcement, typically glass, carbon or polypropylene fibres at $V_f > 3.5$%) are:

- small aggregate particle size and high binder and/ or fines content to ensure that unit reinforcement elements (strands or rovings) are fully surrounded
- sufficient filler/aggregate content to prevent shrinkage
- sufficient fluidity to ensure compaction, fibre encapsulation and quality surface finish
- low water:cement ratio to ensure good matrix strength (which is often used as the serviceability limit design parameter).

Thus FRC matrices tend to be more like mortars than traditional concretes. Traditionally, simple 0:1, 1:1 or 2:1 sand:binder mixes have been used depending on the application. Recent work on optimising matrices for textile reinforced concrete (TRC) suggests a typical binder:aggregate ratio of about 3:5 w/w, a maximum aggregate size of 0.6 mm and a water:cement ratio (w/c) of 0.3–0.4 (Brameshuber, 2006). The use of admixtures such as superplasticisers, accelerators or retarders (see Chapter 14) to modify the fresh mix is common.

For secondary/tertiary FRC (larger components, for example slabs or tunnel linings, normally steel–FRC but occasionally glass or polymer fibres at lower V_f), matrix recipes may differ little from normal concrete mix designs, although the adverse effect of the addition of fibres on the workability of the mix often needs to be taken into account (normally by increasing the cement content and/or fine:coarse aggregate ratio, or using plasticisers).

45.2.2 MECHANICAL PROPERTIES

As with normal concrete, the mechanical properties of FRC matrices are mainly a function of the water:cement ratio used and the age of the component. In contrast to normal concrete, however, tensile and flexural strength are of more interest than compressive strength. The high cement content and low water:cement ratio of FRC matrices lead to relatively high 28-day values of flexural strength, generally around 8–15 MPa with a modulus of 18–30 GPa. Strain to failure is generally less than 0.03%.

45.2.3 BINDERS

Portland cement (PC, i.e. CEM I in BS EN 197-1 – see Chapter 13) is the most common binder for FRC, but the use of additions (see Chapter 15) and alternative binders is more widespread than in normal concrete practice. This is driven mainly by durability concerns, which are discussed in more detail later in this section. All FRC matrices must be chemically compatible with the fibres, but un-modified PC is highly alkaline (with pH > 13), which may degrade some fibres, particularly natural fibres, some polymers and glass. It also continues to hydrate for many years, causing calcium hydroxide crystals to be precipitated between the filaments in fibre strands; this can also have an effect on properties, even in FRC where the fibre itself is less susceptible to alkalis (e.g. carbon or polypropylene). Using additions such as fly ash, ggbs, metakaolin and microsilica reduces both the alkalinity and calcium hydroxide content of the matrix. Alternative binders for FRC include calcium–aluminate and calcium–sulphoaluminate cements, blast-furnace slag cements and phosphate cements (Vubonite®), most of which have low alkali content and little or no calcium hydroxide in the matrix.

Much FRC, particularly glass–FRC, uses polymer-modified concrete or mortar as the matrix. Acrylic polymer dispersions, equivalent to ~5% polymer by matrix volume, are typically added to help prevent surface water evaporation and associated shrinkage cracking, and to improve workability and mould finish. They may also confer some durability benefits.

For steel–FRC, the alkalinity of PC protects the steel in the same way as it does for normal reinforced concrete (see section 24.4), so modifications to the PC–concrete matrix are normally made for other reasons, e.g. adding fly ash for increased consistence or ggbs to reduce the heat of hydration.

References

Brameshuber W (ed.) (2006). *Textile Reinforced Concrete*, State of the art report of RILEM technical committee 201-TRC edition, RILEM Publications SARL, Bagneux, France, pp. 187–210.

Toledo Filho RD, Ghavami K, Sanjuán MA and England GL (2005). Free, restrained and drying shrinkage of cement mortar composites reinforced with vegetable fibres. *Cement and Concrete Composites*, **27**, 537–546.

Chapter 46

Interface and bonding

As with all fibre composites, the properties of the fibre–concrete interface have a crucial bearing on the properties of FRC. The interface in FRC is uniquely complex for two key reasons:

- Bond strength, the type of bonding and interface morphology/chemistry can change significantly as the matrix continues to hydrate over many years.
- In multi-filament FRC, the interface is between a bundle of fibres and concrete; not all the filaments are necessarily completely surrounded by the matrix.

Both these factors can pose unique challenges to characterising such seemingly simple parameters as fibre diameter and bond strength. In particular, they also have serious implications for assessing durability, so some detailed discussion of the interface is also included in Chapter 51.

46.1 Interfacial morphology and properties

In monofilament/macrofibre FRC (especially steel–FRC), the fibre–matrix interface is generally considered to be very similar to the interface between clean rebars and concrete in normal reinforced concrete. The zone close to the fibre, within about < 10 μm, is analogous to the transition or interface zone in normal concrete (see Chapter 21, section 21.2.1), which has locally enhanced calcium hydroxide content (~30% compared to 12% in the bulk HCP). It also has higher porosity, at least double that of the bulk HCP, and can thus be expected to have lower strength. Recent work has also shown that the nature of the interface is strongly dependent on the orientation of the rebar with respect to the casting direction (Horne *et al.*, 2007), thus we can expect the interface in a normal 3D random monofilament/macrofibre FRC to be highly variable.

The interface in multifilament/microfibre FRC is more complex. *Figure 46.1* shows the cross-section

of a single reinforcing strand in typical multifilament glass–FRC at 28 days. The cement matrix is in close contact with most (but not all) of the perimeter filaments but does not penetrate into the centre of the strand. As the composite ages and the cement matrix continues to hydrate, contact between the perimeter fibres and the matrix will become more intimate, and hydration products will gradually begin to be deposited between the filaments, changing the bond with time.

46.2 Bonding

Three types of bond are available in FRC:

- elastic bonding – the fibres adhere to the matrix
- frictional bonding – friction between the fibres and the matrix provides resistance to pullout
- mechanical bonding – the fibres are purposely deformed along their length such that they are mechanically interlocked with the matrix. This can be thought of as 'enhanced' frictional bond.

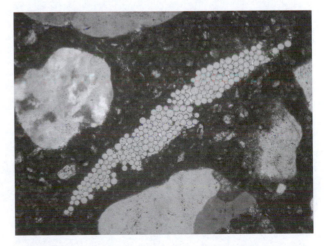

Fig. 46.1 Strand cross-section in glass–FRC. (Thin section petro-micrograph, horizontal field of view 730 microns.)

Frictional bonding dominates the important part of FRC stress–strain behaviour – the post cracking region (see Chapter 48) – and so is of most interest. Pull-out tests are used to estimate frictional bond. These are difficult to interpret, and reported results vary widely, but typical values for the average bond stress (τ) in monofilament FRC with smooth cylindrical fibres are around 0.1, 0.5 and 1 MPa for polyethylene, polypropylene and steel fibres, respectively. Crimping the fibres to provide mechanical bonding can effectively increase this value by a factor of around 4 (Bentur and Mindess, 2007, p. 62). Evaluating pull-out tests for multifilament fibres is even more difficult, not least because the contact perimeter between the fibre bundle and the matrix is not known. Typical reported values of τ are around 0.6 MPa for carbon, and 0.5–1 MPa for fresh and aged glass–FRC, respectively (Bentur and Mindess, 2007; Purnell

et al., 2000). Adding microsilica, ggbs or other additions will affect the interfacial properties and increase the bond strength, as they densify the matrix.

References

Bentur A and Mindess S (2007). *Fibre Reinforced Cementitious Composites*, 2nd edition (Modern Concrete Technology 15), Taylor and Francis, Oxford, UK, 601 pages.

Horne AT, Richardson IG and Bryson RMD (2007). Quantitative analysis of the microstructure of interfaces in steel-reinforced concrete. *Cement and Concrete Research*, 37, 1613–1623.

Purnell P, Buchanan AJ, Short NR, Page CL and Majumdar AJ (2000). Determination of bond strength in glass fibre reinforced cement using petrography and image analysis. *Journal of Material Science*, 35, 4653–4659.

Chapter 47

Reinforcement layouts

The manner in which fibres are distributed throughout an FRC component is determined largely by the manufacturing method. These are discussed in more detail in Chapter 49, but some generic issues are discussed here. As with FRP, choosing the correct fibre layout or *fibre architecture* for a given loading case is crucial in order that the fibre can be used most efficiently. There are two key parameters: fibre length and fibre orientation.

As an FRC component is loaded, an embedded fibre can 'fail' in one of two ways; either it can break in two, or it can be pulled out of the cement or concrete matrix. Either mode of failure may be desirable, depending on the particular FRC application. There is a *critical length* of fibre, l_c, which is the minimum length of fibre required such that on failure of the matrix, the full strength of the fibre is mobilised (i.e. the fibre will break rather than pull out of the matrix, as illustrated in *Fig. 47.1*).

At the critical length:

$$\sigma_{fu}A_f = l_c P_f \tau \quad \therefore \quad l_c = \frac{A_f}{P_f} \cdot \frac{\sigma_{fu}}{\tau} \qquad (47.1)$$

where σ_{fu} = ultimate strength of the fibres, A_f = cross-sectional area of the fibre, P_f = bond area and τ = bond shear strength.

For a cylindrical fibre, $A_f/P_f = r/2$ where r is the radius of the fibre. We therefore see that the critical length decreases with increased bond strength and increases with increased fibre strength. Fibre length l is expressed in terms of critical length l_c when analysing FRC behaviour. The threshold for long-fibre behaviour is ~5 l_c, but in practical terms the difference is usually between effectively continuous fibres and short chopped fibres. This change in mode of failure is critically important to FRC designers, who need to select the correct mode of failure (or combination of modes) for a given application by careful specification of fibre type, fibre length and bond parameters. Critical length is much lower in multifilament/microfibre FRC compared with macrofibre FRC. Typical calculated values for plain, uncrimped steel fibres and glass strands would be 100 mm and 7 mm, respectively (which is why almost all steel fibres used for modern FRC have crimped profiles and/or hooked ends to decrease the effective value of l_c).

Fibre orientation can be 1-dimensional (i.e. aligned fibres), random 2-dimensional (normally in thin sheets, random in plan view but aligned when viewed edge-on) or random 3-dimensional (dispersed throughout a bulk concrete), as illustrated in *Fig. 47.2*.

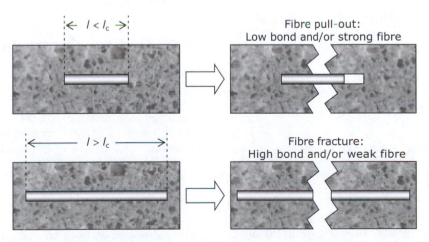

$\leftarrow l < l_c \rightarrow$

Fibre pull-out:
Low bond and/or strong fibre

$\leftarrow\quad l > l_c \quad\rightarrow$

Fibre fracture:
High bond and/or weak fibre

Fig. 47.1 Critical fibre length (after Purnell, 2007; Fig. 9.3).

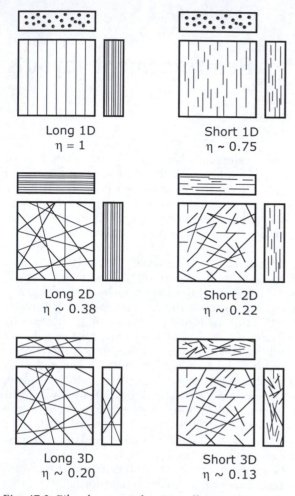

Fig. 47.2 *Fibre layouts (after Purnell, 2007; Fig. 9.1).* Note: η, Efficiency factor.

Aligned layouts allow fibres to be placed parallel to the applied stress, optimising reinforcement efficiency at the expense of creating anisotropic properties (a 'grain' effect, as in timber). Random layouts give uniform properties in all directions at the expense of peak strength or toughness. Combinations of layouts can be used to give intermediate behaviour.

47.1 Efficiency factors

Efficiency factors ($0 \leq \eta \leq 1$) are used in analysis to account for this variation in composite properties with fibre architecture. For a given architecture, η will be a function of l, l_c and the fibre orientation. For long (i.e. continuous) fibres aligned in the direction of applied stress $\eta = 1$, i.e. the fibres are fully

mobilised. However, this is a very unusual layout and for most applications $\eta < 1$ and must be calculated. There are various analytical approaches to determining η, derived from statistical considerations of fibres crossing cracks within an FRC, but most return similar values. Three scenarios are of interest.

For short, aligned fibres (Bentur and Mindess, 2007, p. 109):

$$l > l_c, \quad \eta = l - \frac{l_c}{2l}; \quad l < l_c, \quad \eta = \frac{l}{2l_c} \quad (47.2)$$

For 2D random fibres (the most common scenario; Laws, 1971):

$$l > \tfrac{5}{3}l_c, \quad \eta = \frac{3}{8}\left(1 - \frac{5}{6} \cdot \frac{l_c}{l}\right); \quad l < \tfrac{5}{3}l_c, \quad \eta = \frac{9}{80} \cdot \frac{l}{l_c}. \quad (47.3)$$

For 3D random fibres many authors just use a value of 0.2, but more detailed relationships have been derived e.g. (Laws 1971):

$$l > \tfrac{10}{7}l_c, \quad \eta = \frac{1}{5}\left(1 - \frac{5}{7} \cdot \frac{l_c}{l}\right); \quad l < \tfrac{10}{7}l_c, \quad \eta = \frac{7}{100} \cdot \frac{l}{l_c} \quad (47.4)$$

We can use these factors to define an effective volume fraction $V_f' = \eta V_f$. It is useful to think of this as the volume fraction of fibres in the composite that are aligned with the direction of applied stress. Typical efficiency factors (for $l = 2l_c$ where relevant) are included in *Fig. 47.2* for each layout. For example, using a 3D short-fibre layout would require three times more fibre to achieve the same effective volume fraction (and hence FRC properties) as using a 2D long-fibre layout.

47.2 Textile reinforcement

In order to provide better control over the fibre architecture, engineered textiles are increasingly being used. In primary FRC applications where significant continuous structural loads are to be carried, textiles allow the reinforcement to be placed to optimise load resistance for a given V_f, rather than be equally distributed throughout the cross-section, as with simple FRCs. By varying V_f throughout the thickness, resistance to bending can also be increased (e.g. by concentrating textile on the tension side of the neutral axis). Most FRC textile is made from glass fibres, although carbon, aramid and hybrid textiles containing mixed fibre, often including polymers,

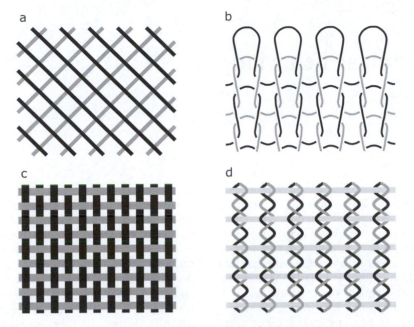

Fig. 47.3 *Textile configurations: a, Scrim; b, Knit; c, Plain weave; d, Leno weave.*

are also available. There is a huge variety of textile forms available, but for FRC purposes the textiles must have a relatively open structure (to permit ingress of the cementitious matrix) and displacement stability (to prevent distortion during composite manufacture). Such textiles fall into three categories, illustrated in *Fig. 47.3*:

- scrim, a 'loose' bi- or multi-axial textile, in which superimposed threads are not fixed at the crossing points (*Fig. 47.3a*)
- warp knits, in which crossing threads are knotted and/or looped to provide stability (*Fig. 47.3b*)
- weave, a 'tight' bi-axial textile with two orthogonal thread systems – the warp and weft – that cross alternately (*Fig. 47.3c,d*).

Combinations of basic forms can also be made. Modern textiles can be produced in multiple layers interlinked with binding or spacer warp thread to produce 3D reinforcement architectures such as tubes or sandwich structures (*Fig. 47.4*).

The efficiency factors applicable to simple textiles can easily be estimated (e.g. for the simple scrim in *Fig. 47.3a* it would be 0.5) but for more complex textiles experimental values would need to be obtained.

Fig. 47.4 *3D textile (Courtesy of Institut für Textiltechnik, ITA, RWTH Aachen University, Germany).*

References

Bentur A and Mindess S (2007). *Fibre Reinforced Cementitious Composites*, 2nd edition (Modern Concrete Technology 15), Taylor and Francis, Oxford, UK, 601 pages.

Laws V (1971). The efficiency of fibrous reinforcement in brittle matrices. *Journal of Physics D: Applied Physics*, **4**, 1737–1746.

Purnell P (1998). The durability of glass-fibre reinforced cements made with new cementitious matrices. PhD Thesis, Aston University, UK.

Chapter 48

Mechanical behaviour of FRC

At a conceptual level, FRC is simple. By adding small amounts of expensive, relatively ductile fibres to a cheap, easily formed but brittle matrix we form a tough composite. Putting numbers to this idea is rather trickier; the response of FRC to mechanical loading – especially bending or impact loading, for which most FRC is designed – is complex and still not fully understood. *Figure 48.1* shows six schematic, idealised tensile stress–strain curves that cover most FRC materials. Although tensile behaviour is not strictly representative of much FRC loading, the key features of the tensile stress curve illustrate important aspects of FRC response. The particular curve that will apply to a given FRC is a function of volume fraction, fibre length, fibre, matrix and interfacial properties, fibre architecture and manufacturing quality control. In theory, either a composite materials approach or the application of fracture mechanics can be used to quantify the mechanical behaviour of FRC. In practice, neither can fully describe all aspects of FRC behaviour and elements of both approaches are required.

48.1 Composite materials approach

In curves a–e in *Fig. 48.1*, the segment OA represents the initial, linear, uncracked behaviour of the FRC under load. This is often termed the *pre-cracking* behaviour (region I). The fibres and matrix act together with full composite action and the modulus of the FRC (E_c) in this region (i.e. the slope of line OA) is given by the rule of mixtures, derived in the previous section on FRP, modified using the efficiency factor:

$$E_c = E_m V_m + E_f V_f' = E_m(1 - V_f) + E_f \eta V_f \quad (48.1)$$

(The symbols are as defined in Chapter 44)

Substituting typical values into equation 48.1 (see 44.2.1 above) shows that E_c is not significantly

greater than E_m except in specialised cases discussed below with reference to curve f in *Fig. 48.1*. (Strictly speaking, we should use different efficiency factors in the pre-cracking and post-cracking regions but the difference is negligible in this context.) At point A – defined by the failure strain of the matrix ε_{mu} and '*first crack stress*' $\sigma_{c,A}$ – the matrix will crack (this point is also referred to as a 'bend-over point' (BOP) or 'loss of proportionality' (LOP) in other literature). After this point, the curves diverge. The key parameter determining which type of post-cracking behaviour will occur is the critical fibre volume fraction, V_{fcrit}.

48.1.1 CRITICAL FIBRE VOLUME FRACTION

The stress carried by the composite immediately before the matrix cracks ($\sigma_{c,A}$) is shared by the matrix and the fibre. When the matrix cracks, this stress must initially be transferred to the fibres. The fibre volume fraction that is just sufficient to carry this load is the critical fibre volume fraction. With reference to equation 48.1 at point A just before the matrix cracks:

$$\sigma_{c,A} = E_c \varepsilon_{mu} = \varepsilon_{mu} E_m(1 - V_f) + \varepsilon_{mu} E_f \eta V_f$$
$$= \sigma_{m,A}(1 - V_f) + \sigma_{f,A} \eta V_f \quad (48.2)$$

Immediately after the matrix cracks (and assuming that $\varepsilon_{fu} \gg \varepsilon_{mu}$, as should be the case for all FRC), $\sigma_{m,A} = 0$. If we have just sufficient fibres to carry the load then $\sigma_{f,A} = \sigma_{fu}$ and $V_f = V_{fcrit}$ and we can write:

$$V_{fcrit} = \frac{\sigma_{c,A}}{\eta \sigma_{fu}} \quad (48.3)$$

For most FRC, E_c up to point A is not significantly different from E_m, thus $\sigma_{c,A} \approx \sigma_{mu}$ and so:

$$V_{fcrit} \approx \frac{\sigma_{mu}}{\eta \sigma_{fu}} \quad (48.4)$$

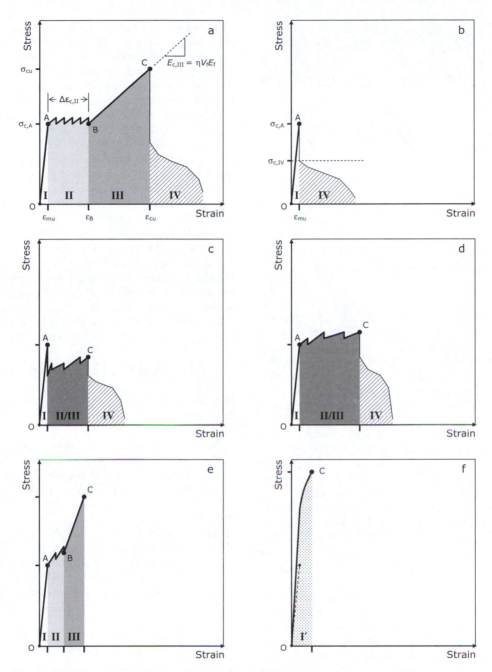

Fig. 48.1 *Idealised tensile stress–strain curves for FRC with different composite parameters.*

48.1.2 PRIMARY FRC: ACK THEORY AND MULTIPLE CRACKING

If the volume fraction of fibres is comfortably above the critical value then the composite failure stress will be significantly higher than the first crack stress. The FRC will behave as a primary FRC, with a stress–strain curve similar to *Fig. 48.1a*. Two important sectors of the curve can be identified (labelled II and III). Region II is the *multiple cracking* region. After the first crack has formed, if the load on the composite is increased, stress is transferred back into the matrix owing to the fibre–matrix

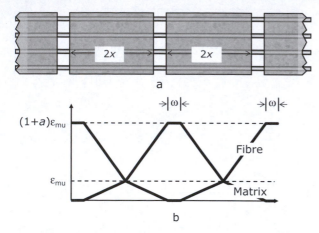

Fig. 48.2 *Multiple cracking (a) and strain distribution (b) in FRC (after Aveston* et al., *1971, Fig. 4).*

bond. Further increases in stress will cause further matrix cracking until the matrix forms a network of closely-spaced cracks. The first formal analysis of this was published by Aveston and co-workers (1971, 1973, 1974) and the following, derived from their work, is known as *ACK theory*.

Figure 48.2 shows an 'edge-on' view of a section through a thin sheet 1-dimensional long-fibre FRC. The matrix has been broken into blocks by parallel multiple cracks. The stress in the matrix must be zero at the crack faces at each end of a block, and it is assumed to vary linearly with distance away from the crack up to the maximum possible value – the matrix strength. Thus the maximum length available for stress transfer (x) is half the block length, i.e. the maximum block length is $2x$. The maximum force that can be transferred (per unit plan area) by the fibres into the block depends on the bond strength, τ, the number of fibres per unit plan area, N, and the contact area between the fibres and matrix, $P_f x$, where P_f is the perimeter of the fibre cross section and x is the distance over which the load is transferred (i.e. half the block length). The maximum force that can be transferred from the matrix to the fibre (again, per unit plan area) is limited by the failure stress of the matrix. Balancing the two forces:

$$NP_f x\tau = V_m\sigma_{mu} \qquad (48.5)$$

Since $N = V_f/a_f$, where a_f is the cross-sectional area of a fibre, then x is given by:

$$x = \frac{V_m}{V_f} \cdot \frac{\sigma_{mu}}{\tau} \cdot \frac{a_f}{P_f} \qquad (48.6)$$

The strain distribution in the matrix and fibre is also shown in *Fig. 48.2*. Note that any further increase in composite stress will cause the block of length $2x$ to break in half as the failure strain of the matrix is reached in the centre. This means that x is a lower bound for the block length. The block length – i.e., the crack spacing – thus varies between x and $2x$ and will in fact average about $1.364x$ owing to statistical concerns (Aveston *et al.*, 1974). For 2D and 3D random fibre layouts, the crack spacings can be derived using $x_{2D} = (\pi x)/2$ and $x_{3D} = 2x$, respectively (Aveston and Kelly, 1973).

If the average block length is $1.364x$ then the additional stress transferred to the fibre during region II on the stress–strain curve varies from $\sigma_{mu}V_m/V_f$ at the crack face to $(1 - 1.364/2)(\sigma_{mu}V_m/V_f) = 0.318\sigma_{mu}(V_m/V_f)$ in the centre of the block, i.e. the average additional fibre stress throughout the block is $0.659\,\sigma_{mu}(V_m/V_f)$. Thus the average additional strain imparted to the composite $\Delta\varepsilon_{c,II}$ – i.e., the 'length' of the multiple cracking region, $\varepsilon_B - \varepsilon_{mu}$ (*Fig. 48.1a*) can be determined as:

$$\Delta\varepsilon_{C,II} = 0.659\sigma_{mu}\frac{V_m}{V_f} \cdot \frac{1}{E_f} = 0.659\alpha\varepsilon_{mu} \qquad (48.7)$$

Where $\alpha = V_m E_m/V_f E_f$. As before, we would expect $\Delta\varepsilon_{c,II}$ to be $\frac{1}{2}\pi$ and 2 times greater for 2D and 3D fibre layouts, respectively.

Since the entire multiple cracking region develops at an approximately constant stress of $\sigma_{c,A} \approx \sigma_{mu}$ we can estimate the area of region II and compare it to the area of region I. The area under a stress–strain curve is directly related to the toughness of the material under test (more precisely, it is the strain energy absorbed per unit volume, W). This will allow us to graphically estimate the increase in toughness imparted by multiple cracking behaviour and thus evaluate the benefit of fibre addition. This is the approach taken by published standards for the testing of FRC (e.g. ASTM C1018).

$$W_I = \tfrac{1}{2}\varepsilon_{mu}\sigma_{c,A} \approx \tfrac{1}{2}\varepsilon_{mu}\sigma_{mu} \approx \tfrac{1}{2}E_m\varepsilon_{mu}^2$$
$$W_{II} \approx \Delta\varepsilon_{C,II}\sigma_{c,A} \approx 0.659\alpha \cdot E_m\varepsilon_{mu}^2 \qquad (48.8)$$

Thus the ratio of the energy absorbed by the composite at the end of the multiple cracking region to the toughness of the unreinforced matrix (sometimes called the *multiple cracking toughness index*, I_{MC}) is given by:

$$I_{MC} = \frac{W_I + W_{II}}{W_I} = 1 + 1.318\alpha \qquad (48.9)$$

Inserting typical values (see *Table 45.1*) for $V_f = 2\%$ gives I_{MC} of about 7, 20 and 100 for 1D carbon, glass and polypropylene–FRC, respectively. The factors to apply to this to account for 2D and 3D reinforcement remain ½π and 2, respectively, as above. For a more realistic FRC ($V_f = 5\%$, 2D long fibres), I_{MC} would then be about 5, 10 and 80 for carbon, glass and polypropylene–FRC, respectively. However, it is critically important to note that equations 48.5 to 48.9 are *only* valid if $\eta V_f > V_{fcrit}$ by a significant margin, allowing multiple cracking to be fully mobilised (this would be marginal for the polypropylene–FRC at this volume fraction, for example).

The crack width, ω, can be estimated by multiplying the average block length (crack spacing) by the *total* strain at the end of multiple cracking:

$$\omega \approx 1.364(0.659\alpha + 1) \cdot \varepsilon_{mu}x \qquad (48.10)$$

For example, crack spacings of ~5 mm are typical for glass–FRC (Purnell, 1998), which suggests crack widths of around 0.01 mm, barely visible to the naked eye. This is rather lower than the 0.3 mm maximum allowed crack width in traditional reinforced concrete (RC) (e.g. Kong and Evans, 1987) and for this reason fibres are sometimes added to RC matrices to reduce crack widths and promote durability in severe environments.

48.1.3 POST CRACKING BEHAVIOUR

Region III (*Fig. 48.1a*) – the post-cracking region – begins when no further multiple cracking can take place. In this region, assuming that the effective fibre volume fraction is sufficiently above the critical value, the fibres alone carry any further load until failure at point C. The post-cracking modulus of the composite is given by:

$$E_{c,III} = \eta V_f E_f \qquad (48.11)$$

and the ultimate strength by

$$\sigma_{cu} = \eta V_f \sigma_{fu} \qquad (48.12)$$

although this tends to be an over-estimate owing to progressive fibre damage towards failure. The ultimate strain capacity is slightly less than the fibre failure strain:

$$\varepsilon_{cu} = \varepsilon_{fu} - 0.341\alpha\varepsilon_{mu} \qquad (48.13)$$

Once again, we can graphically estimate the extra contribution to toughness of region III. We could define a peak toughness index I_{PC} in various ways, but here we will determine the ratio of failure toughness (the area under the stress–strain curve up to peak stress) to the area under the curve up to the end of multiple cracking. This helps us assess the extra contribution to toughness gained in the post-cracking region.

$$W_{III} = \tfrac{1}{2}(\sigma_{c,A} + \sigma_{cu}) \cdot (\varepsilon_{cu} - \varepsilon_B)$$
$$\approx \tfrac{1}{2}(\sigma_{mu} + \eta V_f\sigma_f) \cdot (\varepsilon_{fu} - [1 + \alpha]\varepsilon_{mu}) \qquad (48.14)$$

$$I_{PC} = \frac{W_I + W_{II} + W_{III}}{W_I + W_{II}} \qquad (48.15)$$

Using typical values (lower bounds for fibre strength) from *Table 45.1*, I_{PC} for a 2%, 1D fibre architecture would be 20, 20 and 5 for carbon, glass and polypropylene–FRC, respectively. Adjusting the strain terms in equation 48.14 accordingly, I_{PC} for a 5%, 2D long-fibre architecture would be 30, 30 and 9 for carbon, glass and polypropylene–FRC, respectively. In practice, these values would be reduced considerably (perhaps by factors of 2–5) owing to unavoidable fibre damage during FRC manufacture, non-linear stress–strain behaviour close to the failure point and excessive deflection. For this reason, in design work we tend to use experimentally defined values for efficiency factors in the post-cracking region. Nonetheless, equations 48.14 and 48.15 serve to show the considerable potential for increased toughness that can be accessed by ensuring that the effective fibre volume fraction is high enough to mobilise region III, post-cracking behaviour.

48.1.4 FAILURE, POST-PEAK BEHAVIOUR AND SECONDARY FRC

Many FRC materials will retain some residual strength and toughness after the peak stress has been reached at point C; this is labelled region IV in *Fig. 48.1*. For secondary FRC, $\eta V_f < V_{fcrit}$ so multiple cracking and post-cracking behaviour cannot be mobilised. Post-peak behaviour is thus the only way by which the toughness of secondary FRC is improved over that of the unreinforced matrix. A typical stress–strain curve for secondary FRC is given in *Fig 48.1b* and determination of the area under the curve in region IV (W_{IV}) is clearly important.

The nature of post-peak behaviour in secondary FRC is difficult to model using composite theory, but some key parameters can be identified. On failure of the matrix at point A (*Fig. 48.1b*) the load carried by the matrix will be transferred to the fibres. If the composite strain is subsequently increased past ε_{mu} then what happens next depends on the length of the fibres compared to the critical length. If $l \gg l_c$ then the fibres will simply break and no post-peak toughness will be evident. If $l \approx l_c$ or $l < l_c$ then it is not possible for all the fibres to break as some or all will pull out of the matrix before their breaking stress can be mobilised (see *Fig. 47.1*). The load required to overcome the frictional fibre–matrix bond and progressively pull the fibres out of the matrix will provide the composite with some residual strength $(\sigma_{c,IV})$, which tends to reduce as the strain increases since a smaller total length of fibre is embedded in the matrix. The upper bound for $\sigma_{c,IV}$ at ε_{mu} will be equal to $\eta V_f \sigma_{fu}$. The lower bound can be estimated by considering the fibre–matrix bond and the aspect ratio of the fibres. The mean embedded length of a fibre crossing a crack will be $l/2$, thus the mean pull-out load per fibre will be $\frac{1}{2}\eta_{IV} l P_f \tau$ (where η_{IV} is a post-cracking efficiency factor, equivalent to the length-independent terms in equations 2–4 i.e. 1, 0.375 and 0.2 for 1D, 2D and 3D layouts, respectively). The number of fibres crossing a unit area of composite is given by V_f/a_f and thus:

$$\eta V_f \sigma_{fu} \geq \sigma_{c,IV} \geq \tfrac{1}{2}\varphi \eta_{IV} \tau V_f;$$
$$\tfrac{1}{2}\varphi \eta_{IV} \tau \leq \eta \sigma_{fu}, \quad \eta V_f < V_{fcrit} \qquad (48.16)$$

Where φ = the aspect ratio of the fibres (length/diameter). The higher the aspect ratio, the higher the pull-out force will be up to the limit imposed by the fibre strength. Careful consideration of fibre aspect ratio, strength and bond is required to design FRC with pull-out toughness. Unfortunately, bond (and for many microfibres, aspect ratio) can be difficult to measure or define, which significantly complicates matters.

Predicting the shape of the post-peak stress–strain curve and the toughness represented by region IV is very difficult. The exact initial value of $\sigma_{c,IV}$ is difficult to establish and its rate of change with increasing strain is complex and poorly understood. In particular, region IV is very sensitive to the strain rate (or more commonly, displacement rate) at which the stress–strain testing is carried out. Typically, fast tests ($\sim 10^{-4}$–10^{-3} s^{-1}) give high values of $\sigma_{c,IV}$ but rapid decay and low apparent toughness; slower tests ($\sim 10^{-5}$–10^{-6} s^{-1}) give lower values of $\sigma_{c,IV}$ but much slower decay and higher apparent toughness. In a load-controlled test (where load – rather than displacement – is continuously increased) region IV behaviour cannot be produced at all since the derivative of the stress–strain curve is nominally negative; failure will occur when the matrix fails and the sample is broken in two. Post-peak toughness normally must be evaluated by experimentation and it is critically important that standard procedures (particularly with regard to displacement rates) are used to compare different secondary FRC materials.

A design based approach to describing post-peak behaviour involves defining an 'equivalent strength'.

An experimental stress–strain curve is obtained and the area under the curve up to a predefined strain is recorded. The equivalent strength is then defined as the area divided by the strain. For flexural testing of steel–FRC, strains (defined in terms of mid-span deflections divided by the total span rather than true bending strain) of 1/150 and 1/300 are often used. This equivalent flexural strength is used to calculate ultimate limit state behaviour, assuming the formation of a plastic hinge and subsequent redistribution of stresses (Nemegeer, 2002).

For primary FRC, region IV is frequently ignored or of minor importance. In textile FRC or other primary FRC types where the fibres are effectively continuous, region IV cannot be mobilised at all since the aspect ratio φ is effectively infinite. This leads to curves similar to those in *Fig. 48.1e*.

48.1.5 INTERMEDIATE BEHAVIOUR

Figures 48.1c and *48.1d* illustrate the behaviour of FRC where $\eta V_f \approx V_{fcrit}$. This is an extremely common situation. For many FRC types, the maximum attainable volume fraction that can be attained using simple manufacturing methods is close to the effective critical volume fraction. This holds for steel–FRC, natural fibre–FRC and many polymer–FRCs. In such cases, regions II and III tend to become blurred together (as the low fibre content prevents full multiple cracking from being realised, causing premature transition to post-cracking behaviour). If V_f is marginally below V_{fcrit} then the behaviour in *Fig. 48.1c* will be observed; if it is marginally above, then *Fig. 48.1d* will apply. Different samples of the same FRC may show either *Fig. 48.1c* or *48.1d* type behaviour, since the transition between the two can be caused by inherent minor variations in V_f and/or σ_{mu}. This can cause such FRC to exhibit a wide variation in failure toughness, which may cause quality-control problems.

48.1.6 HIGH MODULUS/HIGH V_f BEHAVIOUR

Figure 48.1f shows the behaviour of FRC where $\eta V_f E_f > E_m$ and $V_f \gg V_{fcrit}$. Although rarely encountered in the past, recent developments in carbon textile–FRC (where V_f may comfortably exceed 10%) could lead to such composites being manufactured.

Stress–strain curves similar to those in *Fig. 48.1f* are also generated for asbestos–FRC and certain other FRC where high volume fractions (> 10%) of very small microfibres are used. In these cases, the composite cracking strength $\sigma_{c,A}$ becomes significantly greater than σ_{mu} and/or the strain at first crack $\varepsilon_{c,A}$ becomes significantly greater than ε_{mu}. Composite theory as outlined above cannot explain this and it is generally attributed to the suppression and/or modification of crack growth in the composite by the fibres. A fracture mechanics approach is thus required, and is in fact considered by many investigators a superior approach for all FRC.

48.2 Fracture mechanics approach

Fracture mechanics concerns modelling the behaviour of materials using an energy-balance approach (see Chapter 4). For FRC, we need to consider this in the context of the concrete matrix. This contains 'stable' flaws in many forms – capillary porosity, aggregate–paste interfaces, air bubbles, microcracking etc. – that act as crack initiators. When the strain energy input exceeds the energy required for new surfaces to be formed within the concrete – either by formation of a crack from a flaw or microcrack, or by extension of an existing crack – then crack propagation will occur. Failure will occur when the speed at which the crack propagates becomes such that a crack sufficient to critically weaken a component forms within its service life. In practice, cracks tend to be either stable or to propagate extremely fast, and defining the threshold parameters between these types of behaviour is the essence of fracture mechanics.

In FRC, there are three ways in which the presence of fibres can prevent, retard or modify the propagation of cracks:

- *Crack suppression.* The presence of the dispersed fibres suppresses the formation of cracks in the matrix (i.e. the extension of small stable flaws/microcracks into macroscopic cracking) by increasing the energy required for crack initiation.
- *Crack stabilisation.* Once cracks are formed, fibres suppress further crack growth both by continuing to provide crack suppression at the crack tip and, by bridging the crack, the fibres provide a 'closing force' that resists crack opening and increases the energy required for further propagation.
- *Fibre–matrix debonding.* This can be modelled as the growth of cracks at the interface and/or the diversion of propagating matrix cracks along the interface, effectively arresting them.

The mathematics of modelling such concepts is generally elegant, but diverse and fearsome. Here we will focus on general concepts and their implications for FRC design rather than the detail of the various models.

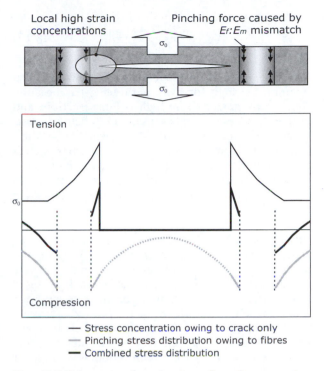

Fig. 48.3 Schematic of mechanism of crack suppression by closely spaced fibres.

48.2.1 CRACK SUPPRESSION

As illustrated in *Fig. 48.3*, at the tip of an existing flaw in the matrix, stress concentration causes the local strain to be greater than the bulk composite strain. If the tip is adjacent to a fibre, the fibre will resist this enhanced strain (since it is stiffer than the matrix), applying an opposing 'pinching' force and reducing the stress concentration. In theory, the cracking strength of an FRC is found to be inversely proportional to the spacing of the fibres S, since more closely spaced fibres increase the chance that a flaw will have a fibre in its vicinity. Since $S = (\pi r^2/V_f)^{1/2}$ for cylindrical fibres, then for a given fibre volume fraction, using narrower fibres will increase the cracking strength.

It has been suggested that for steel–FRC under ideal conditions, the minimum spacing for this effect to be significant is around 10 mm and at a spacing of around 3 mm the cracking strength is approximately double that of FRC with widely spaced wires at the same volume fraction (Romualdi and Batson, 1963).

The energy extension to the ACK theory suggests that the cracking strain of the matrix is actually given by:

$$\varepsilon_{mu} = \left[\frac{12\tau\gamma_m E_f V_f^2}{E_c E_m^2 r V_m} \right]^{\frac{1}{3}} \qquad (48.17)$$

Where γ_m is the surface energy of the matrix, which is not a straightforward parameter to measure. Again, this suggests that high volume fractions and narrower fibres (as well as good bond) help suppress cracking. Many more models are available but their complexity increases and further exotic parameters are required as input. In any case, for most FRC (except asbestos–FRC), the crack suppression mechanism is not considered a primary toughening mechanism.

48.2.2 CRACK STABILISATION

Linear elastic fracture mechanics (LEFM) defines a surface as a 'traction-free' area of matrix where the normal stress is zero. Since a crack involves two surfaces immediately opposite one another, it assumes that no load is transferred across a crack. Clearly, for a macroscopic crack in FRC, fibres bridging the crack will resist crack opening (*Fig. 48.4*) and so this assumption is no longer valid.

Most modelling approaches start from the idealisation of a crack in FRC into three distinct zones: the traction-free zone, the fibre-bridging zone and the 'process zone' at the crack tip, where both fibre and aggregate bridging have an effect (*Fig. 48.5a*). The cement paste fraction of the matrix is assumed to behave according to LEFM, i.e. its critical stress intensity factor (or fracture toughness, K_{IC}) is assumed to be a size-independent material property. For crack propagation to occur, the critical stress intensity at the crack tip K_{tip} is assumed to be equal to the fracture toughness of the cement paste, but consists of the sum of two contributions (*Fig. 48.5 b, c* and *d*):

Fig. 48.4 *Fibres bridging a crack in steel–FRC.*

- K_a from the external load σ_a
- K_b, the bridging force supplied by the fibres (and to a much lesser extent, aggregate bridging σ_b, which is a function of the crack opening displacement δ) (Zhang and Li, 2004).

Both these parameters are expressed in terms of their variation along the axis of a crack (x) and the initial unbridged flaw size (a):

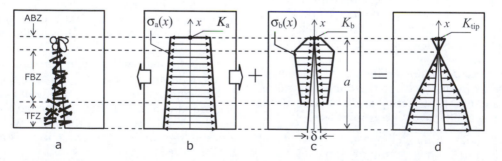

Fig. 48.5 *Modelling of crack stabilisation: (a) Ideal crack model (after Wecharatana and Shah, 1983, Fig. 1); (b), (c), (d) superposition procedure for fracture mechanics modelling of crack propagation (after Zhang and Li, 2004).* Note: *ABZ, aggregate-bridging zone (process zone); FBZ, fibre-bridging zone; TFZ, traction-free zone.*

$$K_{tip} = K_a + K_b; \quad K_a = 2 \int_0^a G\sigma_a \cdot dx; \quad K_b = -2 \int_0^a G\sigma_b \cdot dx$$

$$G = f(x, a, h); \quad \sigma_a = f(x); \quad \sigma_b = f(\delta); \quad \delta = f(x)$$

$$(48.18)$$

G is a weighting function that relates the crack tip stress intensity factor to a unit force on the crack surface and will depend on the geometry (e.g. depth of beam h), loading type (e.g. bending/flexure or direct tension) and crack configuration. The integrations are then performed numerically. Most of the input parameters are straightforward to obtain except the non-linear function $\sigma_b(\delta(x))$, referred to as the *crack bridging law*, which must be experimentally derived by considering stress vs. crack-width curves and is considered to be a material property. The 'initial flaw unbridged flaw length', a_0, must also be either estimated or derived, but the solutions for post-cracking behaviour are not overly sensitive to its value within sensible limits. The model can be used to predict the critical external load capacity as a function of applied stress and $\delta(x)$. *Figure 48.6* compares the fracture model predictions with experimental results for two types of steel–FRC. The model predicts the general form of the curves extremely well, can cope directly with bending or tensile loads and represents a promising advance over previous fracture mechanics models, which often required large numbers of exotic input parameters.

48.2.3 FIBRE–MATRIX DEBONDING

Pull-out and debonding – the major mechanisms adding post-peak, region IV toughness – can be modelled using fracture mechanics as crack growth along the fibre–matrix interface (*Fig. 48.7a*). Detailed analysis of the problem is complex and usually undertaken using finite element analysis, but most investigators have come to the same conclusions, namely that the strain energy release rate associated with debonding (3–7 N m^{-1}) is less than that associated with forming new cracks in the matrix (5–12 N m^{-1}). This means that cracks preferentially propagate along the fibre–matrix interface (Bentur and Mindess, 2007, p. 137). Cracks propagating through the matrix that encounter a fibre are also likely to bifurcate, rapidly reducing their growth rate as their effective tip radius is increased (*Fig. 48.7b* and *c*). Thus the fibres also act indirectly as crack stoppers.

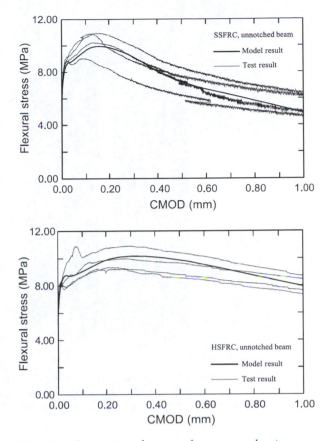

Fig. 48.6 *Comparison between fracture mechanics model predictions and experimental results for two types of steel–FRC tested in flexure. SSFRC, straight steel fibres ø 0.4 × 25 mm, HSFRC = hooked steel fibres ø 0.5 × 30 mm; V_f = 1%, CMOD = crack mouth opening displacement. (Reprinted from Zhang and Li, 2004, with permission from Elsevier.)*

References

ASTM C1018-97 Standard Test Method for Flexural Toughness and First-Crack Strength of Fiber-Reinforced Concrete (Using Beam with Third-Point Loading). ASTM International, PA, USA.

Aveston J, Cooper GA and Kelly A (1971). Single and multiple fracture. In *The Properties of Fibre Composites: Proceedings of the NPL Conference*, November 1971, IPC Science and Technology Press, UK, pp. 15–26.

Aveston J and Kelly A (1973). Theory of multiple fracture of fibrous composites. *Journal of Material Science*, 8, 352–362.

Aveston J, Mercer RA and Sillwood JM (1974). *Fibre reinforced cements – scientific foundations for specifications*. Proceedings of the NPL Conference on

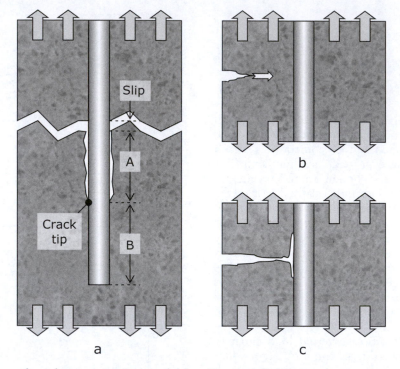

Fig. 48.7 *Debonding and crack arrest. (a) Onset of debonding modelled as interface cracking: (after Morrison et al., 1988); (b), (c) crack growth arrested by bifurcation at fibre–matrix interface.*
Note: *A, debonded length (crack length); B, bonded length.*

Composites – Standards, Testing and Design, IPC Science and Technology Press, Guildford, Surrey, UK, pp. 93–103.

Bentur A and Mindess S (2007). *Fibre Reinforced Cementitious Composites*, 2nd edition (Modern Concrete Technology 15), Taylor and Francis, Oxford, UK, 601 pages.

Kong FK and Evans RH (1987). *Reinforced and Prestressed Concrete*, 3rd edition, Chapman and Hall, London.

Morrison JK, Shah SP and Jenq Y-S (1988). Analysis of fiber debonding and pullout in composites. *Journal of Engineering Mechanics*, **114** (No. 2), 277–294.

Nemegeer D (ed.) (2002). *Design guidlelines for Dramix® steel wire fibre reinforced concrete*, NV Bekaert SA, Zwevegem, Belgium, 23 pages.

Purnell P (1998). The durability of glass-fibre reinforced cements made with new cementitious matrices'. PhD Thesis, Aston University, UK.

Purnell P (2007). Degradation of fibre-reinforced cement composites. Chapter 9 of *Durability of Concrete and Cement Composites* (eds Page CL and Page MM), Woodhead Publishing, Cambridge, UK, pp. 316–363.

Romualdi JP and Batson GB (1963). Mechanics of crack arrest in concrete. *Journal of Engineering Mechanics – Proceedings of the American Society of Civil Engineers*, **89** (EM3), 147–162.

Toledo Filho RD, Ghavami K, Sanjuán MA and England GL (2005). Free, restrained and drying shrinkage of cement mortar composites reinforced with vegetable fibres. *Cement and Concrete Composites*, **27**, 537–546.

Toledo Filho RD, Scrivener K, England GL and Ghavami K (2000). Durability of alkali-sensitive sisal and coconut fibres in cement mortar composites. *Cement and Concrete Composites*, **22**, 127–143.

Wecharatana M and Shah SP (1983). A model for predicting fracture resistance of fiber reinforced concrete. *Cement and Concrete Research*, **13** (No. 6), 819–829.

Zhang J and Li VC (2004). Simulation of crack propagation in fiber-reinforced concrete by fracture mechanics. *Cement and Concrete Research*, **34**, 333–339.

Chapter 49

Manufacturing of FRC

49.1 Cast premix

By mass of FRC produced, casting of premixed FRC is probably the most common method, especially for secondary or tertiary FRC. *Figure 49.1* shows a relatively small-scale casting operation of a glass fibre-reinforced cladding panel.

Fibres are normally supplied in 'pre-batched' bags, with each bag suitable for direct addition to one cubic meter of concrete to provide the correct volume fraction V_f. Steel fibres for traditional secondary FRC and polypropylene fibres intended for plastic cracking control (tertiary FRC) are almost exclusively supplied this way. Glass fibres may also be supplied in pre-batched bags for many applications. For applications where large quantities of ready-mix or pre-cast FRC are to be manufactured on an ongoing basis, or more precise control of V_f is required, continuous batching equipment is available. Once the fibres have been added to the concrete, it can be installed using any of the normal concrete placing methods (pouring, pumping, vibration etc.).

The performance of cast premix FRC is limited by the effect of increasing V_f on workability and compaction. Steel–FRC suppliers place a typical limit on V_f of 2% (equivalent to 160 kg fibres m^{-3} concrete (Nemegeer, 1998)), above which balling of fibres in the mixer, or poor compaction leading to air voids and decreased strength and durability, will occur. Guidelines for premix glass–FRC manufacture suggest an upper limit of 3.5% (GRCA, 2006) and the use of two-stage mixing in high-shear mixers to ensure fibre dispersion. Polypropylene fibres for tertiary FRC are normally added at 0.91 kg m^{-3} concrete (i.e. V_f ~0.1%), although some commercial products that offer a degree of secondary reinforcement may be added at up to 5 kg m^{-3} concrete.

49.2 Sprayed premix

In many systems, after premixing the fresh FRC slurry is placed by spraying onto a mould or substrate (*Fig. 49.2*). For steel–FRC, standard concrete spraying/shotcreting equipment and methods are used, with few restrictions except that the nozzle diameter should be at least 1.5 times the fibre length. Both dry-mix and wet-mix systems can be used (see Chapter 25, section 25.1). Mix design guidelines suggest that rather lower fibre contents are typically used than in premix, up to 70 kg m^{-3} concrete (Vandewalle, 2005). For glass–FRC, specialised equipment (including peristaltic pumps, high-shear mixing and purpose-designed spray guns) are used to spray premix. Fibre volume fractions of up 5.5% are claimed to be attainable (Peter, 2008).

49.3 Dual–spray systems

Some spray systems deliver the matrix and the fibre to the spray gun separately, using either a twin or

Fig. 49.1 Manufacture of a cast premix glass–FRC cladding panel. (Supplied by Iain D. Peter, Powersprays Ltd, UK (www.power-sprays.co.uk).)

Fig. 49.2 *Spraying of premix glass–FRC. (Supplied by Iain D. Peter, Powersprays Ltd., UK (www.power-sprays.co.uk).)*

concentric double nozzle (*Fig. 49.3*). This allows closer control and monitoring of V_f during manufacture. Fibre is delivered as a continuous roving to the gun and cut to the required length by an internal chopper. Mixing and consolidation occur as the sprayed constituents impinge on the mould or substrate. The sprayed layer is then generally further compacted by hand rollers (for *in-situ* work) or automated production machinery (for factory pre-fabricated components). This arrangement is the main system in use for producing primary glass–FRC, and volume fractions well in excess of 5% are attainable by skilled operatives. The strength of dual-sprayed GRC is generally superior to that of premix by 50–100%, owing to the higher V_f attainable and better compaction. Production rates of up to several tons of glass–FRC per day can be achieved.

49.4 Hand lay-up

Textile reinforcement lends itself to hand lay-up techniques similar to those used for FRP production. It allows more control of fibre placing than the spray methods and can potentially produce very high volume fractions, perhaps > 10%, but it is labour intensive and comparatively low output. The steps are very similar to FRP lay-up. First a thin 'gel coat' of modified matrix slurry is used to coat the inside of the mould and provide good adhesion and surface finish. Next, alternate layers of matrix and textile are added, each layer being compacted using hand rollers, until the required component thickness is

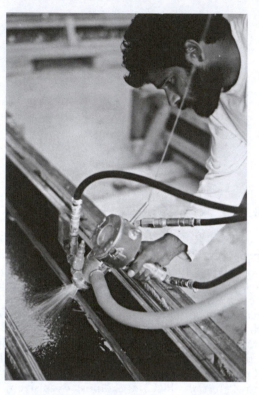

Fig. 49.3 *Dual-spray concentric nozzle system for glass–FRC. The large pipe at the bottom right of the spray-gun supplies matrix; the small straight metal pipe at the top left is the roving intake. The black pipes supply compressed air both to power the roving chopper and to propel the chopped fibre and matrix. (Supplied by Iain D. Peter, Powersprays Ltd., UK (www.power-sprays.co.uk).)*

obtained. After preliminary curing, the component is then released from the mould and finished.

49.5 Automated systems

Several automated systems for FRC manufacture are available. The most well established is the continuous 'Hatschek process', developed for asbestos–FRC sheets from paper-making processes (*Fig. 49.4*). Fibre, fillers and cement are formed into a dilute slurry with water (about 6% solids by weight). The slurry is drained through a continuously rotating porous roller, depositing a layer of wet solids on the surface, which are then picked up by a continuous loop of permeable mat called a 'felt'. This felt, with the layer of green FRC attached, passes over a vacuum dewatering machine, where most of the remaining water is sucked out, consolidating the wet

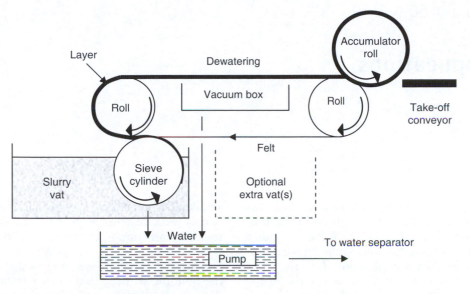

Fig. 49.4 The Hatschek process (schematic).

solids into a dense but flexible green-sheet product. Several layers are built up before it is taken off the felt mat. It is then cut to size and further shaped (e.g. into corrugated forms) if required, before curing at either normal or elevated temperatures. Fibre contents of V_f = 9–30% can be achieved depending on the application.

Once the equipment has been installed, the production process is very cheap. The widespread outlawing of asbestos–FRC has spurred interest in using other fibres within the Hatschek process. Polyethylene fibres and cellulose fibres (i.e. processed natural fibres derived from woody materials), both in pulp form, are the most widely investigated for this application. In principle, any fibre that can be dispersed within cementitious slurry to produce a dense fibre mat can be used. The controlling factor tends to be whether the fibre size and surface morphology are suitable to 'trap' cement particles in the slurry and prevent washout (Coutts, 2005). Since the properties of such fibres tend to be inferior to those of asbestos fibre, the FRC properties are correspondingly reduced.

Several systems are available that use robotic or quasi-robotic systems to control sprayed FRC nozzles. These range from factory-based prefabrication systems for glass–FRC components to computer controlled steel–FRC guns for *in-situ* tunnel lining fabrication. Several other methods, mainly arising from FRP production technology, are also in development, including (Brameshuber, 2006):

- *Pultrusion*: textiles are passed through baths of slurry and then pulled through shaped rollers to consolidate the matrix and fibre and form FRC laminates.
- *Filament winding*: rovings or tapes are passed through slurry baths and wound onto a mandrel to form, e.g., pipes and other cylindrical vessels.
- *Extrusion*: FRC premix is extruded under pressure into a die, either to directly form components or to produce green shapes for compression moulding.

References

Brameshuber W (ed.) (2006). *Textile Reinforced Concrete*, State of the art report of RILEM technical committee 201-TRC edition, RILEM Publications SARL, Bagneux, France, pp. 187–210.

Coutts RSP (2005). A review of Australian research into natural fibre cement composites. *Cement and Concrete Composites*, **27**, 518–526.

Glassfibre Reinforced Concrete Association (GRCA) (2006). *GRC in Action*. GRCA/Concrete Society, Camberley, UK, 23 pages. (Retrieved from http://www.grca. co.uk, October 2008.)

Nemegeer D (ed.) (1998). *The properties of Dramix® steel fibre concrete*, NV Bekaert SA, Zwevegem, Belgium, 11 pages.

Peter ID (2008). Sprayed premix – the new GRC. *Concrete*, February 2008, pp. 13–14.

Vandewall M (2005). *Tunnelling is an art*, NV Bekaert SA, Zwevegem, Belgium, 400 pages.

Chapter 50

Applications

In this chapter we describe some typical applications for the various types of FRC discussed in the preceding chapters. The list is not exhaustive – other FRC could be used for the applications given, and there are countless other applications for FRC – but they are intended to give you an idea of the major uses of FRC in building and construction.

50.1 Architectural cladding: glass–FRC

One of the highest volume semi-structural applications for glass–FRC is architectural cladding, particularly where complex surface mouldings or faithful restoration of heritage features such as capitals or cornices are required. Its major competition is traditional pre-cast concrete. Since glass–FRC has no steel reinforcement and thus no cover concrete is required, elements can be made very thin (> 6 mm), making glass–FRC cladding components extremely light in comparison with pre-cast concrete elements (which generally must be at least 50 mm thick). As well as reducing structural loads – often important in renovation works – this can significantly reduce installation costs, handling complexity and erection time. The thin sections can also form a wider range of shapes than traditional pre-cast elements, are less susceptible to visible cracking and do not contain any steel to corrode. The low weight reduces both transport costs and cement usage, reducing environmental impact. Glass–FRC for architectural cladding is manufactured using dual-spray systems, giving a high V_f, quality surface finish and good dimensional tolerance. Panels are normally fixed to the supporting substructure using L-shaped flexible steel anchors bonded to the rear of the panels (GRCA, 2006).

The Newcastle Council Chambers building, Australia (*Fig. 50.1*) was refurbished using glass–FRC panels in the 1990s. The original pre-cast reinforced concrete panels had deteriorated, with dangerous spalling occurring on the panel surfaces. Glass–FRC panels were designed to fit over the existing façade to cover and contain the spalling. Their light weight allowed them be installed with simple scaffolding and manual handling equipment – thus not requiring the building to be closed during installation – and did not add sufficient additional structural load to require strengthening of the building. In addition, the new panels were designed to seal the building to allow more efficient operation of heating and air-conditioning systems (Glenn Industries). More advanced applications of the same basic system, but using *in-situ* spraying rather than factory prefabricated panels, can produce extremely complex 'megasculptural' structures such as the Merlion (*Fig. VII.2.1*) and the UK's Millennium Dome 'Body Zone' (*Fig. 50.2*). The Body Zone structure had to transport up to 3500 people per hour through its interior. Since the resultant live load varied throughout the body of the structure, the FRC skin thickness

Fig. 50.1 Newcastle Council Chambers, Newcastle, Australia. (Photo by Kate Farquharson, Australia (2008), reproduced from flickr.com/photos/zigwamp under Attribution-No Derivative Works 2.0 Generic License (see creativecommons.org for license details).)

Fig. 50.2 Body Zone Figure, Millennium Dome, UK.

over the steel sub-frame had to be continuously varied (Glenn Industries).

50.2 Tunnel linings: steel–FRC and polymer–FRC

Tunnel boring machines are now increasingly used in preference to other tunnelling methods. Robust and rapidly deployable linings are required to prevent ground settlement, especially in urban areas. Pre-cast RC segments are often used, which are jacked into place after tunnel excavation has finished. Manufacture of the steel reinforcing cage for these segments is expensive, and during installation the cover concrete at edges and vertices often spalls under the jacking forces, leading to durability and finishing problems. Using pre-cast steel–FRC reduces

cost and weight, and eases installation of the lining (Vandewalle, 2002).

In larger tunnels, the lining is normally placed *in-situ*. An initial lining of rapid setting/hardening concrete or steel–FRC, around 75 mm thick, is sprayed on to support and stabilise the fresh excavation (*Fig. 50.3* and see Chapter 25, section 25.4). The inner, structural liner (about 300–350 mm thick) is then either cast or sprayed in place. Waterproofing membranes may be placed between the two, depending on the system in use. Cast RC liners require temporary lattice girders to be placed at a set distance from the tunnel roof/walls to orient and support the reinforcing steel (since it cannot be attached to the inner liner without disrupting the membrane or other waterproofing system). This is cumbersome, and installation can present a health and safety hazard as it occurs in an unsupported excavation (Eddie and Neumann, 2004). Using sprayed FRC for both layers can reduce costs and complexity. Advances in admixtures and placing technologies that allow a reduction in water:cement ratio to < 0.4 can also render the FRC sufficiently impermeable to remove the need for waterproof membranes. Modern semi-automated remote spraying equipment (*Fig. 50.3*) allows linings to be applied without personnel entering an unexcavated area, and laser guidance systems do away with the orientation role of the temporary lattice girders.

Polymer fibres, particularly polypropylene, are often added to tunnel lining concrete (at ~1–3 kg m^{-3})

to help protect against explosive spalling during tunnel fires. Since most tunnel-lining concrete has a low water:cement ratio the capillary porosity is 'segmented', i.e. it does not consist of a continuous network of pores, and the permeability of the concrete is very low (see Chapter 13). During a fire, water vaporised by the intense heat cannot easily escape and large pressures build up, which can lead to catastrophic explosive spalling of the concrete. By adding polypropylene fibres, which melt at relatively low temperatures (130–160°C), pathways for vapour to escape are provided, reducing spalling without sacrificing strength.

50.3 Industrial flooring: steel–FRC and polymer–FRC

A typical *in-situ* placing operation of a concrete industrial floor is shown in *Fig. 50.4*. Installation of such floors is a demanding application, since the large exposed surface area (per m^3 of concrete) can lead to increased probability of cracking caused by plastic shrinkage and/or drying shrinkage (see Chapters 19 and 20). Also floors are required to deal with concentrated dynamic and static loads from, e.g. fork-lift trucks or warehouse shelving and so need enhanced resistance to localised cracking. Thus a form of distributed crack-control reinforcement (in addition to any structural reinforcement required, e.g. for floors laid over pile caps) is required. Traditionally, this has been provided by a double layer of steel mesh. Large sheets of mesh – typically of ~10 mm diameter bars crossing in a 100–200 mm grid – are laid on spacers to ensure that the layers are positioned at the correct depth. The concrete is then poured over the mesh. Using mesh in this way is cumbersome, especially on multi-storey projects. Large sheets of mesh must be craned and manoeuvred into the building and fixed into place; the pour must be closely supervised to ensure that the spacers are not accidentally knocked out by personnel necessarily walking on or 'between' the mesh during concrete pouring; the mesh itself is a trip hazard.

Replacing the mesh with steel fibres provides several advantages:

- All handling and placing cost/time associated with the steel mesh is eliminated.
- The reinforcement is distributed throughout the full thickness of the slab, rather than just concentrated in one or two layers.
- Since the shear resistance of steel–FRC is greater than that of mesh–RC, and no cover to the steel

Fig. 50.4 Steel–FRC floor installation. (Supplied by J. Greenhalgh, Bekaert Ltd, UK (www.bekaert.com).)

is required, the floor thickness can be reduced, reducing dead weight.
- FRC can be used more easily in conjunction with composite floor construction.

The technical performance (with regard to strength and deflection) of FRC is generally claimed to be as good as or better than that of mesh–RC. Fibre manufacturers also claim that 10–40% time savings and 10–30% cost savings can be achieved (Stadlober, 2006).

While steel fibres are the usual choice for mesh replacement, polymer fibres are also used. Some hybrid systems combine steel macro-fibres to control service cracking with a relatively small amount of polypropylene micro-fibres (12 mm long, 18 µm diameter at ~1 kg/m^3 concrete) to control plastic shrinkage cracking. Other systems use a combination of crimped polypropylene or PVA macrofibres (2–7 kg m^{-3} concrete) with polypropylene micro-fibres (as above) to obtain a similar effect. In lightly loaded floors (or other applications where shrinkage control is of enhanced importance), polypropylene microfibres might be used on their own; independent tests (by the British Board of Agrément, see BBA certificate no. 06/4373 details sheet 3) have shown that such fibres can reduce plastic-shrinkage cracking in slabs by a factor of almost 10.

50.4 Sheet materials for building: natural–FRC

The most widespread use for natural–FRC is probably 'siding' – external cladding for domestic and

Fig. 50.5 Natural–FRC sidings being installed. (Photo by Mark Zimmerman, US (2007), reproduced from flickr.com/photos/pdz_house under Attribution 2.0 Generic License (see creativecommons.org for license details).)

light commercial buildings (*Fig. 50.5*). In this application it competes with plastics, plywood and metallic sheeting, and is a direct replacement for asbestos sheeting. The sheets are made using the Hatschek process (*Fig. 49.4*). Wood-pulp fibres obtained using the kraft process (where wood chips are treated with sodium hydroxide and sodium sulphide at < 180°C to extract the cellulose fibres from the lignin matrix) are mechanically treated to internally and externally fibrillate the surfaces in order that the fibres can flocculate and retain the cement particles in the green felt (Coutts, 2005). Products may be air-cured or autoclaved. The strength of typical boards can vary considerably according to the moisture state (typically 10.0 MPa dry strength, 7.0 MPa wet strength; see Anon, 2008).

The siding market is currently dominated by wood-based and polymer products, but natural–FRC now has a ~9% share of a US market estimated at around $500 million (Coutts, 2005). As with glass–FRC, panel products can be made to a wide range of finishes to re-create the look of traditional timber siding, but with much lower maintenance requirements.

References

Anon (2008). Technical Data Sheet, *Artisan® Matrix™ panel*. James Hardie International Finance BV. (Retrieved from http://www.jameshardiecommercial.com/pdf/artisan-matrix-panel-td.pdf, October 2008.)

British Board of Agrément certificate no 06/4373 Detail sheet 3: Fibrin 23. BBA, Watford, UK, 2006.

Coutts RSP (2005). A review of Australian research into natural fibre cement composites. *Cement and Concrete Composites*, **27**, 518–526.

Eddie CM and Neumann C (2004). *Development of the LaserShell method of tunnelling*. In Proceedings of the North American Tunnelling Conference, Atlanta 2004 (eds Levent O and Ozdemir O). Taylor and Francis, UK.

Glassfibre Reinforced Concrete Association (GRCA) (2006). *GRC in Action*. GRCA/Concrete Society, Camberley, UK, 23 pages. (Retrieved from http://www.grca.co.uk, October 2008.)

Glenn Industries (unknown date) project information sheets (*The Project: Newcastle Council Chambers, Newcastle, Australia*; *The Project: The Body Zone, London, UK*). Glenn Industries Pty Ltd, ABN 46 007 654 024. (Retrieved from http://www.glenn.com.au/, October 2008.)

Stadlober H (ed.) (2006). *Dramix® Steel fibres for industrial floors*, NV Bekaert SA, Zwevegem, Belgium, 12 pages.

Vandewall M (ed.) (2002). *Steel wire reinforced segments for tunnel linings*, NV Bekaert SA, Zwevegem, Belgium 2002, 6 pages.

Chapter 51

Durability and recycling

Combining a continuously hydrating, highly alkaline matrix with a wide variety of types, shapes and sizes of reinforcing fibres will lead to complex, time-dependent fibre–matrix interactions. These interactions lead to the engineering properties of most FRC varying markedly with time. This was recognised by even the earliest modern FRC researchers, so the durability of FRC has been extensively studied for several decades. A good deal more is probably known about the mechanisms and magnitudes of FRC durability issues than those of any other composite material. It is important that the FRC designer is aware of these issues from the outset to avoid unwarranted failures in FRC components.

51.1 Time–dependent behaviour

51.1.1 MULTIFILAMENT/MICROFIBRE FRC

In *Fig. 51.1*, the residual bending strength for various common types of multifilament, PC matrix–FRC is plotted as a function of time of exposure to UK outdoor weathering conditions. The strength is expressed as a fraction of the original, as-supplied value. The fibres in each case are intended to provide primary reinforcement. The main observations to note are:

- There are appreciable losses in strength with time for most FRC types, and these need to be accounted for in design.

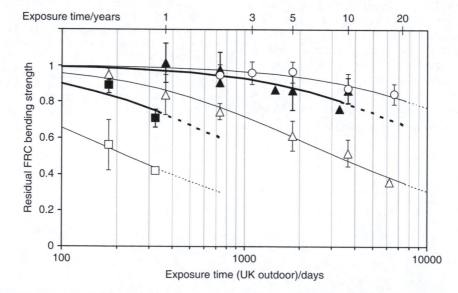

Fig. 51.1 Residual FRC bending strength (fraction of original value) vs. time in UK outdoor environment for various types of fibre and PC-based matrix (after Purnell, 2007, Fig. 9.4). FRC fibre type: □ Sisal (plant leaf fibre), ■ Coir (coconut husk fibre), ▲ 2nd generation AR glass, △ 1st generation AR glass, ○ Polypropylene.
Note: *(Data sources: glass – Majumdar and Laws, 1991; polypropylene – Hannant, 1998; sisal and coir – Tolêdo et al., 2000. Error bars, ±1 standard deviation where available.)*

- The magnitude of the strength loss is very different not only between different fibre types, e.g. polymer compared to natural fibres, but also within different fibre types, e.g. coir compared to sisal and 1st compared to 2nd generation AR glass. Thus we cannot talk about overall FRC durability; it varies according to the particular composite being used.
- In terms of durability, we can rank the various fibre types as polypropylene > 2nd generation (i.e. modern) AR glass >> 1st generation AR glass >> unprocessed natural fibres (Purnell, 2007).

Primary FRC will degrade until the composite strength is reduced to the matrix strength, at which point all region II and III behaviour will be lost (*Fig. 48.1*). In secondary FRC, we have seen that the benefit provided by the fibres – region IV toughness – is not related to the composite strength, which is generally less than the matrix strength (*Fig. 48.1*). In fact, as the composite ages, the matrix tends to increase in strength. This can give misleading results. For example, the strength of some cellulose-fibre secondary FRC can increase by half over five years of exposure as the matrix hydrates, increasing matrix strength and bonding. Yet the strain to failure (an indicator of region IV performance) over this period decreases from 3% to < 0.1% (Akers and Studinka, 1989). Thus in many cases, it is more appropriate to monitor toughness or strain to failure instead of strength in durability studies. However, since there is no accepted standard toughness test for FRC, and strain-to-failure results often have large scatter, strength remains the most popular metric.

There is not much information available on FRC made with fibres other than those shown in *Fig. 51.1*. Studies on PVA–FRC report no strength loss after about 7 years of weathering, but since V_f is almost always less than V_{fcrit} this does not necessarily prove its durability. No long-term weathering studies of carbon–FRC have been reported. Carbon fibres are generally assumed to be inert and immune to any corrosion-related strength loss, but it is possible that other mechanisms (see below) could affect their long-term properties.

51.1.2 MONOFILAMENT/MACROFIBRE FRC
Time-dependent behaviour in steel–FRC has been less extensively studied. Its performance is more frequently compared with that of traditional RC, so attention has tended to focus on RC-related issues such as surface spalling, cracking and structural integrity rather than directly measuring strength loss. Even after several months of the most severe lab-based environments (cyclic wetting/drying in hot saline solutions), steel–FRC retains 70% of its strength and 40% of its toughness (Kosa and Naaman, 1990). In normal service only superficial corrosion of fibres at the surface is to be expected unless significant structural cracking is present (in which case, the engineer generally has more pressing concerns).

Other monofilament/macrofibre FRCs are not common enough for durability data to have been reported.

51.2 Property loss mechanisms

There are two types of process that can cause loss of mechanical properties in FRC: weakening of the fibre and processes associated with the progressive hydration and development of the cementitious matrix. Both types can reduce the strength of primary FRC, change the failure mode of FRC from primary to secondary by eliminating region II/III behaviour, or reduce the toughness of secondary FRC (i.e. reduce W_{IV}). The severity and relative importance of each type will be different for each FRC, and affected by fibre type, matrix chemistry, V_f and intended composite action.

51.2.1 FIBRE WEAKENING
Most cement-based matrices are highly alkaline, the pore water having a pH of 12–13. Many fibres can degrade in these environments. Cellulose, common glass (E-glass) and some polymer fibres may suffer gross corrosion, i.e. the effective cross-sectional area of the fibres is significantly reduced (*Fig. 51.2*). Alkali-resistant (AR) glass fibres do not suffer gross corrosion in normal service, but sub-microscopic flaws form, which act as Griffiths stress concentrations; these grow, slowly weakening the fibre. Steel fibres are of course passivated (covered with a tightly adhering molecular layer of protective oxide; see Chapter 24, section 24.3.1) in these alkaline conditions in the same way as reinforcing bars in normal reinforced concrete. They will only corrode if that passivation is disrupted by carbonation or chloride ions. Natural fibres also tend to weaken as they absorb water from the surroundings.

Reduction of fibre strength contributes to four mechanisms for loss of durability:

1. In primary FRC, it reduces the composite strength since σ_{cu} is proportional to σ_{fu} (see section 48.1.3).
2. Since V_{fcrit} is inversely proportional to σ_{fu} (equation 48.4) then a loss of fibre strength increases

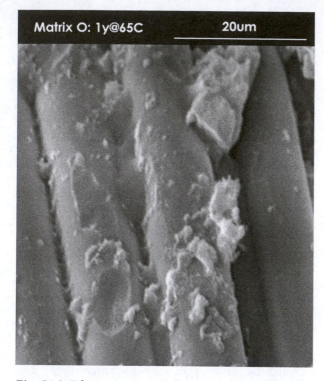

Fig. 51.2 Fibre corrosion in glass–FRC after exposure to an extremely aggressive environment (1 year at 65°C, equivalent to ~500 years of UK weathering). (Taken from Purnell, 1998.)

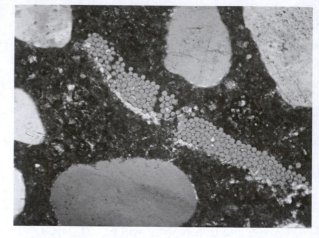

Fig. 51.3 Microstructure in aged glass–FRC. The bright areas at the interface and between the fibres are calcium hydroxide crystals. (Thin section petromicrograph, horizontal field of view 730 microns.)

the critical volume fraction. If V_f is not sufficiently high to accommodate this, the failure mode can change from primary to secondary by eliminating the multiple cracking and post-cracking behaviour (regions II and III, *Fig. 48.1a*). This is often of greater concern than simple strength reduction since it is accompanied by a drastic loss of toughness, especially if $l \gg l_c$.

3. In secondary FRC, a reduction in σ_{fu} reduces the upper bound for region IV (*Fig. 48.1b* and equation 48.17) and thus potentially W_{IV}.

4. It also leads to a reduction in the critical length l_c (equation 47.1). This increases the likelihood that the dominant failure mode in secondary FRC will be fibre fracture rather than fibre pull-out, significantly reducing the toughness of the composite (i.e. W_{IV} (*Fig. 48.1b* and section 48.1.4), will be significantly reduced or eliminated).

51.2.2 CONTINUED MATRIX HYDRATION

In common with most pre-cast cementitious products, FRC is generally supplied or installed after having been cured for periods ranging from about 7 to 28 days. A significant proportion of unhydrated cement

will remain available for hydration at this time (see Chapter 13) and its continued hydration can cause a number of durability issues.

The critical fibre volume fraction is proportional to the matrix strength (equation 48.4). If the matrix strength increases sufficiently after installation (increases of 10–30% are typical) and if V_{fcrit} was borderline to start with, then the failure mode may change from primary to secondary with a corresponding loss of toughness.

The initial interface between the fibres and the matrix tends to be quite 'loose' and porous (see *Fig. 46.1*). The matrix at the interface is relatively weak and the bond strength is low. As the matrix hydrates, this interface becomes denser, and in multi-filament FRC the spaces within the fibre bundles may also become partially filled with various hydration products, particularly calcium hydroxide crystals (*Fig. 51.3*). The direct effect of this densification is to significantly increase the bond strength, perhaps by a factor of up to 3 (Purnell *et al.*, 2000). Increased bond strength τ is generally welcomed in most composites as it increases the efficiency of the fibres. In secondary FRC, though, since increased bond strength also decreases the critical length it may decrease the toughness, W_{IV}, in the same manner as described in section 48.2, by changing the fibre failure mode from pull-out to fracture.

Indirect effects of interfacial densification may include:

• Localised aggravation of fibre degradation – 'notching' – by calcium hydroxide crystals

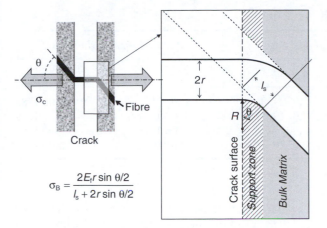

$$\sigma_B = \frac{2E_f r \sin \theta/2}{l_s + 2r \sin \theta/2}$$

Fig. 51.4 *Crack bridging (after Purnell, 2007, Fig. 9.12).*

growing at the fibre–matrix interface, which can weaken fibres.

- Loss of flexibility of multifilament strands as the spaces between the filaments are filled with hydration products. Strands bridging cracks rarely do so at right angles and thus have to be able to bend freely (*Fig. 51.4*). This also affects natural fibres in a process known as mineralisation, where the lumens within the fibres become filled with calcium hydroxide precipitates, causing similar effects.
- Decreased radii of curvature – and thus increased bending stress σ_b – in monofilament fibres crossing cracks, as the 'support zone' becomes stronger and less able to yield locally under a fibre (see *Fig. 51.4*).
- Decreased contrast in fracture toughness, K_{IC}, between the interface and the bulk matrix, which prevents the 'crack blunting' mechanism (*Fig. 48.7b, c*).

51.3 Designing durable FRC

There are three approaches to designing durable FRC and the approach, or combination of approaches, to be taken in any given case will depend on the fibre type, installed application and commercial considerations.

51.3.1 GOOD DESIGN AND MANUFACTURING PRACTICE

The most important approach is to encourage high quality with regard to FRC manufacture. It is good practice to specify a mean fibre volume fraction to be used that is somewhat in excess of that indicated

by design calculations. This will give some protection against many of the mechanisms outlined above. It is important to then apply a sensible factor of safety to this value, and to carefully monitor quality control during manufacture both to ensure that the volume fraction remains high and that the fibres are placed and consolidated properly to maximise their efficiency. 'Skimping' on fibre content to save a few pennies at the manufacture stage can have expensive (and possibly litigious) consequences in the longer term. Good practice with regard to the matrix – good compaction, high cement contents, low water:cement ratio etc. – will also protect against ingress of outside agents (chlorides, water) that can damage e.g. natural or steel fibres.

51.3.2 INCREASING FIBRE RESISTANCE

Making fibres less susceptible to attack by the matrix will increase durability. For example, virtually all glass fibres now used in FRC are 'second-generation' fibres, which combine alkali-resistant glass with a soluble coating that reduces the precipitation of calcium hydroxide crystals at the interface (see section 45.1.4). In very demanding environments (e.g. marine structures), galvanised or alloy steel fibres may be used to help prevent corrosion of the steel. Multifilaments such as natural fibres may be treated in various ways to increase their alkali resistance, often by pre-impregnation with fine materials such as microsilica, or polymers, which block calcium hydroxide precipitation. Carbon and polypropylene fibres are generally assumed to be more-or-less inert and so other approaches are required.

51.3.3 MATRIX MODIFICATION

Another approach is to modify the matrix so that it is less aggressive towards the fibres. This can be achieved by reducing the alkalinity of the matrix, and/or its propensity to precipitate calcium hydroxide at the interface. Additions (see Chapter 15) are invaluable in this respect, since they react with both the free alkali and the calcium hydroxide in the matrix. Waste materials such as microsilica, blast-furnace slag or fly ash, plus manufactured materials such as metakaolin (a calcined china clay) are all routinely used in most multifilament FRC where durability is a prime concern, and can significantly increase the predicted service life of the materials, especially in warm service conditions (e.g. Purnell and Beddows, 2005). Additions are also added to steel–FRC matrices to provide the same durability benefits as in reinforced and pre-stressed concrete, i.e. increased resistance to carbonation, ingress of chloride, and penetration of water (see Chapter 24).

Table 51.1 k values (days^{-1}) for glass–FRC strength prediction

Service temperature (°C)	CEM1 (PC) matrix glass–FRC		Modified matrix glass–FRC	
	$n = 1$	$n = 0.5$	$n = 1$	$n = 0.5$
10	0.000170	0.000438	0.000150	0.000383
20	0.000662	0.00190	0.000351	0.000959
30	0.00235	0.00747	0.000778	0.00226
40	0.00772	0.0270	0.00164	0.00506

In glass–FRC, the matrix is frequently modified by adding acrylic polymer dispersions. As well as acting as a curing and workability aid, these polymers enhance durability, reducing the degree to which strain-to-failure is degraded over about 20 years by 75% (Ball, 2003). The mechanism by which it works is not clear, but it probably involves disrupting the precipitation of calcium hydroxide at the interface rather than providing a protective coating on the fibres.

An alternative approach is to use non-Portland cement systems such as calcium aluminate cement or sulpho-aluminate cements (see Chapter 16), which have lower intrinsic alkalinity and develop little or no calcium hydroxide during hydration. This is in its infancy but some such matrix formulations are commercially available, especially in China. Such matrices often have other advantages such as lower embodied energy/CO_2 and rapid strength development, which will see them being used more widely in the future.

51.4 Modelling and service life prediction

Predicting strength loss in FRC is important, as it is the most widely used property in specifying the performance of the material. Several models of the strength vs. time relationship have been proposed, mainly for glass–FRC, that take into account one or more of the various parameters that affect the service life (e.g. service temperature and humidity, matrix chemistry and hydration, fibre type and so on). The most recent relates normalised strength S (i.e. the ratio of the strength at a given time to the original strength) to time (t) using a relationship of the form:

$$S = \frac{1}{\sqrt{(1 + kt)^n}}$$ (51.1)

The parameter n is normally taken as either 1 or 0.5 depending on the assumptions made, but 0.5 is probably the more correct value. The rate constant k depends on the service temperature, the particular fibre/matrix combination concerned, and the value of n used. *Table 51.1* gives some typical values of k for glass–FRC. Using this relationship, a critical normalised strength can be defined (usually the original value of $\sigma_{c,A}/\sigma_{cu}$, i.e. the threshold at which region II/III toughness is lost) and service lifetimes predicted. Values of 60 and 80 years for PC-matrix and modified matrix glass–FRC have been suggested (Purnell, 2007).

51.5 Recycling

Recycling of any composite component in an assembly involves one of three options:

- disassembly and consequent reuse of entire components in a new structure or other application
- reduction of the composite into its component phases (i.e. fibre and matrix) and separate recycling of each phase
- crushing and recycling of the composite component as a lower-grade material.

When considering FRC, all these options are problematic. Recycling of entire components is the most promising option, but is rarely carried out. FRC panels tend to be used in external applications, and in common with all other types of external panel, are subjected to more weathering than other structural components. Thus it is unlikely in general that they will outlast the rest of the structure.

Reduction of FRC into fibre and matrix is neither economically nor technically feasible. Chemical separation processes involving dissolution of the matrix would most likely damage the fibre and produce large quantities of waste. Physical separation is not possible, except conceivably for some steel–FRC.

Regulations surrounding the use of crushed and recycled construction materials (i.e. BS8500-2) as aggregate limit the content of 'other foreign material' such as glass or plastics to <1.0% by mass, and the mass fraction of 'fines' (i.e. particles passing the 0.063-mm sieve) must be less than 5%. The high fibre and cement content of most primary FRC, plus the possibility of microfibres passing the sieve, will exclude it from being recycled in this way. However, some secondary FRC with suitably low V_f may be able to be recycled as either aggregate for structural concrete, i.e. as recycled concrete aggregate (RCA), or for low-grade concrete, structural sub-base for roads or to make-up ground level, i.e. as recycled aggregate, RA (see Chapter 26).

References

Akers SAS and Studinka JB (1989). Ageing behaviour of cellulose fibre cement composites in natural weathering and accelerated test. *International Journal of Cement Composites and Lightweight Concrete*, **11**, 93–97.

Ball H (2003). *Durability of naturally aged gfrc mixes containing forton polymer and SEM analysis of the fracture interface*. In Proceedings of the 13th Congress of the Glass Fibre Reinforced Concrete Association, October 2003, Barcelona, Spain (eds Clarke JN and Ferry R). Concrete Society, UK, paper 17, 30 pages.

Hannant DJ (1998). Durability of polypropylene cement composites: 18 years of data. *Cement and Concrete Research*, **28**, 1809–1817.

Kosa K and Naaman AE (1990). Corrosion of steel fiber-reinforced concrete. *ACI Materials Journal*, **87** (No. 1), 27–37.

Majumdar A and Laws V (1991). *Glass Fibre Reinforced Cement*. BSP Professional Books, Oxford, 197 pages.

Purnell P (1998). The durability of glass-fibre reinforced cements made with new cementitious matrices. PhD Thesis, Aston University, UK.

Purnell P, Buchanan AJ, Short NR, Page CL and Majumdar AJ (2000). Determination of bond strength in glass fibre reinforced cement using petrography and image analysis. *Journal of Material Science*, **35**, 4653–4659.

Purnell P (2007). Degradation of fibre-reinforced cement composites. Chapter 9 of *Durability of Concrete and Cement Composites* (eds Page CL and Page MM), Woodhead Publishing, Cambridge, UK, pp. 316–363.

Purnell P and Beddows J (2005). Durability and simulated ageing of new matrix glass fibre reinforced concrete. *Cement and Concrete Composites*, **27**, 875–884.

Tolêdo Filho RD, Scrivener K, England GL and Ghavami K (2000). Durability of alkali-sensitive sisal and coconut fibres in cement mortar composites. *Cement and Concrete Composites*, **22**, 127–143.

PART 8

TIMBER

John Dinwoodie

Introduction

In the industrial era of the 19th century timber was used widely for the construction not only of roofs but also of furniture, waterwheels, gearwheels, rails of early pit railways, sleepers, signal poles, bobbins and boats. The 20th century saw an extension of its use in certain areas and a decline in others, owing to its replacement by newer materials. Despite competition from lightweight metals and plastics, whether foamed or reinforced, timber continues to be used on a massive scale.

World production of timber for industrial and constructional purposes in 2005 (the last year for which complete data are available) was $635.2 \times 10^6 \, m^3$, made up of softwood timber, hardwood timber and wood-based panels (see Chapter 56) as set out in *Table VIII.1*.

UK consumption of timber and wood-based panels for industrial and constructional purpose in 2005 was $17.1 \times 10^6 \, m^3$ that was valued at £2.5 billion (excluding secondary processing), comprising £1.5 billion for timber and £1 billion for wood-based panels. UK production of timber and panels in 2005 was equivalent to 37% of total timber consumption

by volume. In 2006 the value of timber and panels consumed in the UK had risen to £2.7 billion, comprising £1.5 billion for timber and £1.2 billion for panels (TTF, 2008).

The Office of National Statistics reveals that the value of the total output of the wood-based industry from the harvesting to end use in the UK in 2004 (the latest year for which official figures are available) was £11.1 billion; this results in placing the wood-based industry at 42nd in the valuation of the top 100 industries in the country (TTF 2008).

In the UK timber and timber products are consumed by a large range of industries, but the bulk of the material continues to be used in construction, either structurally, such as roof trusses or floor joists (about 43% of total consumption), or non-structurally, e.g. doors, window frames, skirting boards and external cladding (about 9% of total consumption). On a volume basis, annual consumption continues to increase slightly and there is no reason to doubt that this trend will be maintained in the future, especially with the demand for more houses, the increasing price of plastics, the favourable strength:weight and strength:cost ratios of timber and panel products, and the increased emphasis on environmental

Table VIII.1 World production, UK consumption and UK production of timber and wood-based panels in 2005. (Sources: Wadsworth, 2007 and TTF, 2007.)

Product	World production ($\times 10^6 \, m^3$)	World production (%)	UK consumption ($\times 10^6 \, m^3$)	UK production ($\times 10^6 \, m^3$)	UK production as percentage of UK consumption
Softwood	323.1	51	10.0	2.8	28
Hardwood	102.7	16	0.7	0.05	7
Panels	209.4	33	6.4	3.4	53
Total	635.2	100	17.1	6.25	37

performance and sustainability, in which timber is the only renewable construction material.

Timber is cut and machined from trees, themselves the product of nature and time. The structure of the timber of trees has evolved through millions of years to provide a most efficient system that supports the crown, conducts mineral solutions and stores food material. Since there are approximately 30000 different species of tree, it is not surprising to find that timber is an extremely variable material. A quick mental comparison of the colour, texture and density of a piece of balsa and a piece of lignum vitae, formerly used to make playing bowls, will illustrate the wide range that occurs. Nevertheless, man has found timber to be a cheap and effective material and, as we have seen, continues to use it in vast quantities. However, he must never forget that the methods by which he utilises this product are quite different from the purpose that nature intended and many of the criticisms levelled at timber as a material are a consequence of man's use or misuse of nature's product. Unlike so many other materials, especially those used in the construction industry, timber cannot be manufactured to a particular specification. Instead the best use has to be made of the material already produced, though it is possible from the wide range available to select timbers with the most desirable range of properties. Timber as a material can be defined as a low-density, cellular, polymeric composite, and as such does not conveniently fall into any one class of material, rather tending to overlap a number of classes. In terms of its high strength performance and low cost, timber remains the world's most successful fibre composite.

Four orders of structural variation can be recognised – macroscopic, microscopic, ultrastructural and molecular – and in subsequent chapters the various physical and mechanical properties of timber will be related to these four levels of structure. In seeking correlations between performance and structure, it is tempting to describe the latter in terms of smaller and smaller structural units. While this desire for refinement is to be encouraged, a cautionary note must be recorded, for it is all too easy to overlook the significance of the gross features. This is particularly so where large sections of timber are being used under practical conditions; in these situations gross features such as knots and grain angle are highly significant factors in reducing performance.

References

TTF (2007). *Timber in Context. A statistical Review of the Timber Industry 2005*, The Timber Trade Federation, London.

TTF (2008). *The Timber Industry Statistical Review 2006*, The Timber Trade Federation, London.

Wadsworth J (2007). *Panel market developments – a personal view*. Proceedings of The International Panel Products Symposium, Cardiff.

Chapter 52

Structure of timber and the presence of moisture

52.1 Structure at the macroscopic level

The trunk of a tree has three physical functions to perform: firstly, it must support the crown, a region responsible for the production not only of food, but also of seed; secondly, it must conduct the mineral solutions absorbed by the roots upwards to the crown; and thirdly it must store manufactured food (carbohydrates) until required. As will be described in detail later, these tasks are performed by different types of cell.

Whereas the entire cross-section of the trunk fulfils the function of support, and increasing crown diameter is matched with increasing diameter of the trunk, conduction and storage are restricted to the outer region of the trunk. This zone is known as *sapwood*, while the region in which the cells no longer fulfil these tasks is termed the *heartwood*. The width of sapwood varies widely with species, rate of growth and age of the tree. Thus, with the exception of very young trees, the sapwood can represent from 10 to 60% of the total radius, though values from 20 to 50% are more common (*Figs 52.1 and 52.2*); in very young trees, the sapwood will extend across the whole radius. The advancement of the heartwood to include former sapwood cells results in a number of changes, primarily chemical in nature. The acidity of the wood increases slightly, though certain timbers have heartwood of very high acidity. Substances collectively called *extractives* are formed in small quantities and these impart not only colouration to the heartwood, but also resistance to both fungal and insect attack. Different substances are found in different species of wood and some timbers are devoid of them altogether: this explains the very wide range in the natural durability of wood, about which more will be said in section 55.2.1. Many timbers develop gums and resins in the heartwood, while the moisture content of the heartwood

of most timbers is appreciably lower than that of the sapwood in the freshly felled state. However, in exceptional cases high moisture contents can occur in certain parts of the heartwood. Known as *wetwood* these zones are frequently of a darker colour than the remainder of the heartwood and are thought to be due to the presence of micro-organisms, which produce aliphatic acids and gases (Ward and Zeikus, 1980; Hillis, 1987).

With increasing radial growth of the trunk by division of the cambial cells (see later), commensurate increases in crown size occur, resulting in the enlargement of existing branches and the production of new ones; crown development is not only outwards but upwards. Radial growth of the trunk must accommodate the existing branches and this is achieved by the structure that we know as the *knot*. If the cambium of the branch is still alive at the point where it fuses with the cambium of the trunk, continuity in growth will arise even though there will be a change in orientation of the cells. The structure so formed is termed a *green* or *live* knot (*Fig. 52.3*). If, however, the cambium of the branch is dead – and this frequently happens to the lower branches – there will be an absence of continuity, and the trunk will grow round the dead branch, often complete with its bark. Such a knot is termed a *black* or *dead* knot (*Fig. 52.4*), and will frequently drop out of planks on sawing. The direction of the grain in the vicinity of knots is frequently distorted, and in a later section the loss of strength due to different types of knot will be discussed.

52.2 Structure at the microscopic level

The cellular structure of wood is illustrated in *Figs 52.5* and *52.6*. These three-dimensional blocks are produced from micrographs of samples of wood

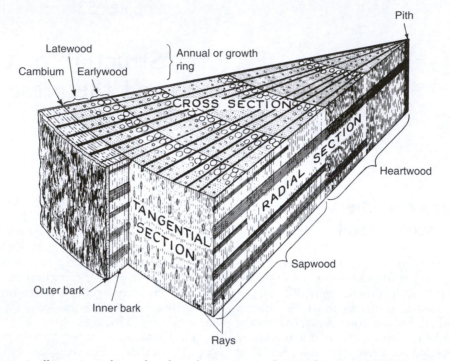

Fig. 52.1 *Diagramatic illustration of a wedge-shaped segment cut from a five year-old hardwood tree, showing the principal structural features (© Building Research Establishment).*

Fig. 52.2 *Cross-section through the trunk of a Douglas fir tree showing the annual growth rings, the darker heartwood, the lighter sapwood and the bark (© Building Research Establishment).*

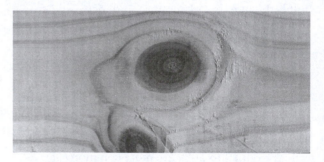

Fig. 52.3 *Green or live knots showing continuity in structure between the branch and tree trunk (© Building Research Establishment).*

8 × 5 × 5 mm in size removed from a coniferous tree, known technically as a *softwood* (*Fig. 52.5*), and a broadleaved tree, known as a *hardwood* (*Fig. 52.6*). In the softwoods about 90% of the cells are aligned in the vertical axis, while in the hardwoods there is a much wider range in the percentage of cells that are vertical (80–95%); the remaining percentage is present in bands, known as *rays*, aligned in one of the two horizontal planes known as the radial plane or quartersawn plane or more loosely, as the radial section (*Fig. 52.1*). This means that there is

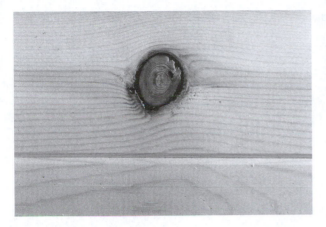

Fig. 52.4 Black or dead knot surrounded by the bark of the branch and hence showing discontinuity between branch and tree trunk (© Building Research Establishment).

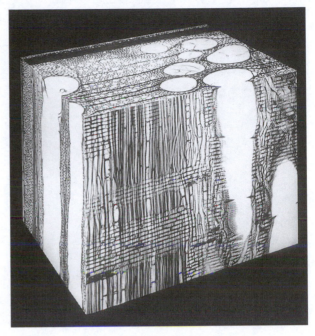

Fig. 52.6 Cellular arrangement in a ring-porous hardwood (Quercus robur – European oak) (© Building Research Establishment).

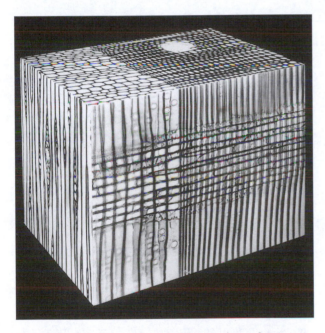

Fig. 52.5 Cellular arrangement in a softwood (Pinus sylvestris – Scots pine, redwood) (© Building Research Establishment).

a different distribution of cells on the three principal axes, which is one of the two main reasons for the high degree of anisotropy present in timber.

It is popularly believed that the cells of wood are living cells, but this is certainly not the case. Wood cells are produced by division of the *cambium*, a zone of living cells that lies between the bark and the woody part of the trunk and branches (*Fig. 52.1*). In winter the cambial cells are dormant and generally consist of a single circumferential layer. With the onset of growth in the spring, the cells in this single layer subdivide radially to form a cambial zone some ten cells in width. This is achieved by the formation within each dividing cell of a new vertical wall called the primary wall. During the growing season these cells undergo further radial subdivision to produce what are known as daughter cells. Some of these will remain as cambial cells while others will either develop into bark if on the outside of the zone or change into wood if on the inside. There is thus a constant state of flux within the cambial zone with the production of new cells and the relegation of existing cambial cells to bark or wood. Towards the end of the growing season the emphasis is on relegation, and a single layer of cambial cells is left for the period when growth does not occur.

To accommodate the increasing diameter of the tree the cambial zone must increase circumferentially, which is achieved by the periodic tangential division of the cambial cells. In this case, the new wall is sloping and subsequent elongation of each half of the cell results in cell overlap, often frequently at shallow angles to the vertical axis, giving rise to the formation of spiral grain in the timber. The rate

Table 52.1 The functions and wall thicknesses of the various types of cell found in softwoods and hardwoods

Cell	Softwood	Hardwood	Function	Wall thickness, schematic
Parenchyma	✓	✓	Storage	
Tracheids	✓	✓	Support Conduction	
Fibres		✓	Support	
Vessels (pores)		✓	Conduction	

at which the cambium divides tangentially has a significant effect on the average cell length of the timber produced.

The daughter cells produced radially from the cambium undergo a series of changes extending over a period of about three weeks; this process is known as *differentiation*. Changes in cell shape are paralleled with the formation of the secondary wall, the final stages of which are associated with the death of the cell; the degenerated cell contents are frequently to be found lining the cell cavity. It is during the process of differentiation that the standard daughter cell is transformed into one of four basic cell types (*Table 52.1*).

Chemical dissolution of the lignin–pectin complex cementing the cells together will result in their separation, and this is a useful technique for separating and examining individual cells. In softwoods two types of cell can be observed (*Fig. 52.7*). Those present in greater number are known as *tracheids*, which are some 2–4 mm in length with an aspect ratio (length:diameter) of about 100:1. These cells, which lie vertically in the tree trunk, are responsible for both supporting and conducting roles. The small block-like cells some 200 × 30 µm in size, known as *parenchyma*, are mostly located in the *rays* and are responsible for the storage of food material.

In contrast, in hardwoods four types of cell are present albeit that one, the tracheid, is present in small amounts (*Fig. 52.8*). The role of storage is again

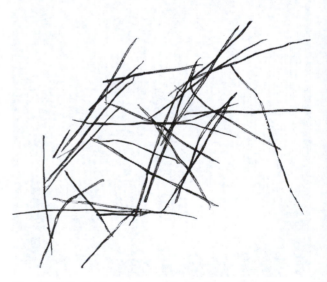

Fig. 52.7 *Individual softwood cells (magnification × 12) (© Building Research Establishment).*

primarily taken on by the parenchyma, which can be present horizontally in the form of a ray, or vertically, either scattered or in distinct zones. Support is effected by long thin cells with very tapered ends, known as *fibres*; these are usually about 1–2 mm in length with an aspect ratio of about 100:1. Conduction is carried out in cells whose end walls have been dissolved away either completely or in

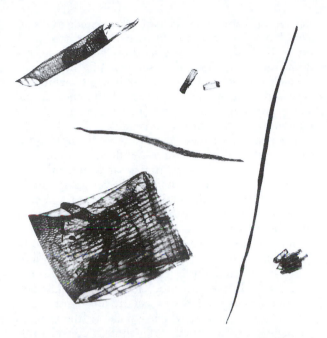

Fig. 52.8 Individual cells from a ring-porous hardwood (original magnification × 40) (© Building Research Establishment).

part. These cells, known as *vessels* or *pores*, are usually short (0.2–1.2 mm) and relatively wide (up to 0.5 mm) and when situated above one another form an efficient conducting tube. Tracheids can also be present in some hardwoods but represent a very small percentage of the total cell count. It can be seen, therefore, that while in softwoods the three functions are performed by two types of cell, in hardwoods each function is performed primarily by a single type of cell (*Table 52.1*).

Although all cell types develop a secondary wall this varies in thickness, being related to the function that the cell performs. Thus, the wall thickness of fibres is several times that of the vessel (*Table 52.1*). Consequently, the density of the wood, and hence its quality (Butterfield, 2003) and many of its strength properties will be related to the relative proportions of the various types of cell (as will be discussed later). Density, of course, will also be related to the absolute wall thickness of any one type of cell, for it is possible to obtain fibres of one species of wood with a cell wall several times thicker than those of another. The range in density of timber is from about 120 to 1200 kg/m^3, corresponding to pore volumes of from 92 to 18% (see section 52.6.1).

Growth may be continuous throughout the year in certain parts of the world and the wood formed tends to be uniform in structure. In the temperate and

subarctic regions and in parts of the tropics growth is seasonal, resulting in the formation of *growth rings*; where there is a single growth period each year these rings are referred to as *annual rings* (*Fig. 52.1*).

When seasonal growth commences, the dominant function appears to be conduction, while in the latter part of the year the dominant factor is support. This change in emphasis manifests itself in the softwoods with the presence of thin-walled tracheids (about 2 μm) in the early part of the season (the wood being known as *earlywood*) and thick-walled (up to 10 μm) and slightly longer (10%) in the latter part of the season (the *latewood*) (*Fig. 52.5*).

In some of the hardwoods, but certainly not all of them, the earlywood is characterised by the presence of large-diameter vessels surrounded primarily by parenchyma and tracheids; only a few fibres are present. In the latewood, the vessel diameter is considerably smaller (about 20%) and the bulk of the tissue comprises fibres. It is not surprising to find, therefore, that the technical properties of the earlywood and latewood are quite different from one another. Timbers with this characteristic two-phase system are referred to as having a *ring-porous* structure (*Fig. 52.6*).

The majority of hardwoods, whether of temperate or tropical origin, show little differentiation into earlywood and latewood. Uniformity across the growth ring occurs not only in cell size, but also in the distribution of the different types of cell (*Fig. 52.9*): these timbers are said to be *diffuse-porous*.

In addition to determining many of the technical properties of wood, the distribution of cell types and their sizes is used as a means of timber identification.

Interconnection by means of pits occurs between cells to permit the passage of mineral solutions and food in both longitudinal and horizontal planes. Three basic types of pit occur. *Simple pits*, generally small in diameter and taking the form of straight-sided holes with a transverse membrane, occur between parenchyma and parenchyma, and also between fibre and fibre. Between tracheids a complex structure known as the *bordered pit* occurs (*Fig. 52.10*; see also *Fig. 52.27a* for a sectional view). The entrance to the pit is domed and the internal chamber is characterised by the presence of a diaphragm (the *torus*), which is suspended by thin strands (the *margo strands*). Differential pressure between adjacent tracheids will cause the torus to move against the pit aperture, effectively stopping flow. As will be discussed later, these pits have a profound influence on the degree of artificial preservation of the timber. Similar structures are to be found interconnecting vessels in a horizontal plane.

Fig. 52.9 Cellular arrangement in a diffuse-porous hardwood (Fagus sylvatica – beech) (© Building Research Establishment).

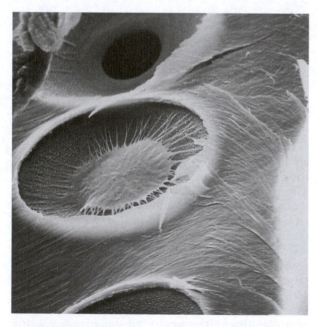

Fig. 52.10 Scanning electron micrograph of softwood bordered pits on the radial wall of a spruce tracheid. The arched upper dome of the pits has been removed in specimen preparation, and in the central pit the torus and supporting margo strands are revealed; these have been torn out of the lower and upper pits during the preparation process (magnification × 3000) (© Building Research Establishment).

Between parenchyma cells and tracheids or vessels, *semi-bordered pits* occur and are often referred to as ray pits. These are characterised by the presence of a dome on the tracheid or vessel wall and the absence of such on the parenchyma wall: a pit membrane is present, but the torus is absent. Differences in the shape and size of these pits provide an important diagnostic feature in the softwoods.

The general arrangement of the vertically aligned cells is referred to as *grain*. While it is often convenient when describing timber at a general level to regard these cells as lying truly vertically, this is not really true in the majority of cases; these cells generally deviate from the vertical axis in a number of different patterns.

In many timbers, and certainly in most softwoods, the direction of the deviation from the vertical axis is consistent and the cells assume a distinct *spiral* mode, which may be either left- or right-handed. In young trees the helix is usually left-handed and the maximum angle, which is near to the core, is frequently of the order of 4°, though considerable variability occurs both within a species and also between different species. As the tree grows, so the helix angle in the outer rings decreases to zero and quite frequently in very large trees the angle in the outer rings subsequently increases, but the spiral has changed direction. Spiral grain has very significant technical implications; strength is lowered, while the degree of twisting on drying and the amount of pick-up on machining increase as the degree of spirality of the grain increases (Brazier, 1965).

In other timbers the grain can deviate from the vertical axis in a number of more complex ways, of which *interlocked* and *wavy* are perhaps the two most common and best known. Since each of these types of grain deviation gives rise to a characteristic decorative figure, further discussion on grain is reserved until section 52.5.

52.3 Molecular structure and ultrastructure

52.3.1 CHEMICAL CONSTITUENTS

Chemical analysis reveals the existence of four constituents and provides data on their relative proportions. This information may be summarised as in *Table 52.2*: proportions are for timber in general and slight variations in these can occur between timber of different species, or in different parts of a single tree. Thus, Bertaud and Holmbom (2004), for example, record that the heartwood of Norway

Table 52.2   Chemical composition of timber

Component	Per cent mass		Polymeric state	Molecular derivatives	Function
	Softwood	Hardwood			
Cellulose	42 ± 2	45 ± 2	Crystalline Highly oriented Large linear molecule	Glucose	Fibre
Hemicelluloses	27 ± 2	30 ± 5	Semi-crystalline Smaller molecule	Galactose, mannose, xylose	Matrix
Lignin	28 ± 3	20 ± 4	Amorphous Large 3D molecule	Phenylpropane	
Extractives	3 ± 2	5 ± 4	Generally compounds soluble in organic solvents	Terpenes, polyphenols, stilbenoids	Extraneous

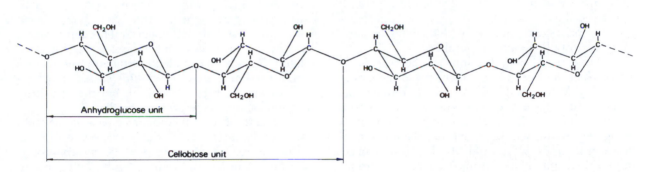

Fig. 52.11 Structural formula for the cellulose molecule in its 'chair' configuration (© Building Research Establishment).

spruce contained significantly more lignin and less cellulose than the sapwood; differences in the amounts of specific hemicelluloses were also found to vary not only between heartwood and sapwood, but also between earlywood and latewood.

Cellulose

Cellulose $(C_6H_{10}O_5)_n$ occurs in the form of long slender filaments or chains, these having been built up within the cell wall from the glucose monomer $(C_6H_{12}O_6)$. While the number of units per cellulose molecule (the degree of polymerisation) can vary considerably even within one cell wall, it is thought that a value of 8000–10 000 is a realistic average for the secondary cell wall, while the primary cell wall has a degree of polymerisation of only 2000–4000 (Simson and Timell, 1978). The anhydroglucose unit $C_6H_{10}O_5$, which is not quite flat, is in the form of a six-sided ring consisting of five carbon atoms and one oxygen atom (*Fig. 52.11*); the side groups play an important part in intra- and intermolecular bonding as will be noted later. Successive glucose units are covalently linked in the 1,4 positions giving rise

to a potentially straight and extended chain; i.e. moving in a clockwise direction around the ring it is the first and fourth carbon atoms after the oxygen atom that combine with adjacent glucose units to form the long-chain molecule. The anhydroglucose units comprising the molecule are not flat, as noted above; rather they assume a 'chair' configuration, with the hydroxyl groups (one primary and two secondary) in the equatorial positions and the hydrogen atoms in the axial positions (*Fig. 52.11*).

Glucose, however, can be present in one of two forms depending on the position of the –OH group attached to carbon 1. When this group lies above the ring, i.e. on the same side as that on carbon 4, the unit is called α-glucose and when this combines with an adjacent unit with the removal of H–O–H (known as a condensation reaction) the resulting molecule is called starch, a product which is manufactured in the crown and stored in the parenchyma cells.

When the –OH group lies below the ring, the unit is known as β-glucose, and on combining with adjacent units, again by a condensation reaction, a molecule of cellulose is produced in which alternate anhydroglucose units are rotated through 180°: it is this product that is the principal wall-building constituent of timber.

Cellulose chains may crystallise in many ways, but one form, namely cellulose I, is characteristic of natural cellulosic materials. Over the years there have been various attempts to model the structure of cellulose I. The model that has gained widest acceptance, is that proposed by Gardner and Blackwell (1974). Using X-ray diffraction methods on the cellulose of *Valonia*, these authors proposed an eight-chain unit cell with all the chains running in the same direction. Forty-one reflections were observed in their X-ray diffractions and these were indexed using a monoclinic unit cell having dimensions $a = 1.634$ nm, $b = 1.572$ nm and $c = 1.038$ nm (the cell axis) with $\beta = 97°$; the unit cell therefore comprises a number of whole chains or parts of chains totalling eight in number.

All but three of the reflections can be indexed by a two-chain unit cell almost identical to the earlier model by Meyer and Misch (1937), though this model had adjacent chains aligned in opposite directions. These three reflections are reported as being very weak, which means that the differences between the four Meyer and Misch unit cells making up the eight-chain cell must be small. Gardner and Blackwell therefore take a two-chain unit cell ($a = 0.817$ nm, $b = 0.786$ nm and $c = 1.038$ nm) as an adequate approximation to the real structure. Their proposed model for cellulose I is shown in *Fig. 52.12*, which shows the chains lying in a parallel configuration, the centre chain staggered by $0.266 \times c$ (= 0.276 nm).

Cellulose that has regenerated from a solution displays a different crystalline structure, and is known as cellulose II; in this case there is complete agreement that the unit cell possesses an antiparallel arrangement of the cellulose molecule.

Within the structure of cellulose I, both primary and secondary bonding are represented and many of the technical properties of wood can be related to the variety of bonding present. Covalent bonding both within the glucose rings and linking together the rings to form the molecular chain contributes to the high axial tensile strength of timber. There is no evidence of primary bonding laterally between the chains; rather this seems to be a complex mixture of (fairly strong) hydrogen bonds and (weak) van der Waals forces. The same –OH groups that take part in this hydrogen bonding are highly attractive to water molecules, which explain the affinity of cellulose for water. Whereas some earlier workers,

though placing the intermolecular hydrogen bonds in the *ac* plane, recorded that the intramolecular hydrogen bonds were on a diagonal plane, thereby linking different layers, Gardner and Blackwell (1974) identified the existence of both intermolecular and intramolecular hydrogen bonds, all of which, however, are interpreted as lying only on the *ac* plane (*Fig. 52.12*); they consider the structure of cellulose as an array of hydrogen-bonded sheets held together by van der Waals forces across the *cb* plane.

The degree of crystallinity of cellulose is usually assessed by X-ray and electron diffraction techniques, though other methods have been employed. Generally, a value of about 60% is obtained, though values as high as 90% are recorded in the literature. This wide range in values is due in part to the different techniques employed in the determination of crystallinity and in part to the fact that wood is comprised not just of crystalline and noncrystalline constituents, but rather of a series of substances of varying crystallinity. Regions of complete crystallinity and regions with a total absence of crystalline structure (amorphous zones) can be recognised, but the transition from one state to the other is gradual.

The length of the cellulose molecule is about 5000 nm (0.005 mm) whereas the average size of each crystalline region determined by X-ray analysis is only 60 nm in length, 5 nm in width and 3 nm in thickness. This means that any cellulose molecule will pass through several regions of high crystallinity – known as *crystallites* or *micelles* – with intermediate non-crystalline or low-crystalline zones in which the cellulose chains are in only loose association with each other (*Fig. 52.12c*). Thus, the majority of chains emerging from one crystallite will pass to the next, creating a high degree of longitudinal coordination; this collective unit is termed a *microfibril* and has 'infinite' length. It is clothed with chains of cellulose mixed with chains of sugar units other than glucose (see below), which lie parallel but are not regularly spaced. This brings the microfibril in timber to about 10 nm in breadth, and in some algae, such as *Valonia*, to 30 nm. The degree of crystallinity will therefore vary along its length and it has been proposed that this could be periodic.

A more comprehensive account of the structure of cellulose can be fond in Chapter 1 of Dinwoodie (2000).

Hemicelluloses and lignin

In *Table 52.2* reference is made to the other constituents of wood besides cellulose. Two of these, the hemicelluloses and lignin, are regarded as cementing materials contributing to the structural integrity of

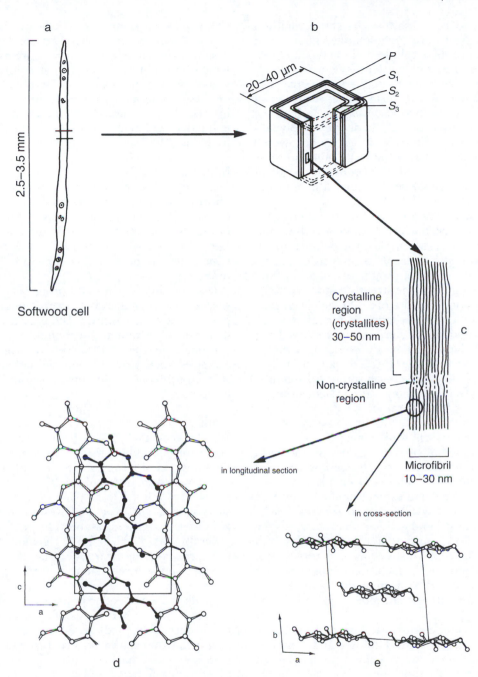

Fig. 52.12 *Relationship between the structure of timber at different levels of magnitude. (a), low power microscopic level; (b), high power microscopic level; (c), ultrastructural (electron microscopic) level; (d), (e), molecular level.*
(d), (e), Projections of the Gardner and Blackwell two-chain cell used as an approximation to the eight-chain unit cell of the real structure. (d), The projection viewed perpendicular to the ac plane; (e), the projection viewed perpendicular to the ab plane (i.e. along the cell axis). Planes are characterised according to North American rather than European terminology. Note that the central chain (in black) has the same orientation as the other chains and is staggered vertically with respect to them by an amount equal to ca. c/4. ((a), (b), (c), Adapted from Siau, 1971, reproduced by permission of Syracuse University Press; (d), (e), from Gardner and Blackwell, 1974, by permission of John Wiley and Sons, Inc.)

wood and also to its high stiffness. The hemicelluloses, like cellulose itself, are carbohydrates built up of sugar units, but unlike cellulose in the type of units they comprise; these units differ between softwoods and hardwoods and generally, the total percentage of the hemicelluloses present in timber is greater in hardwoods than in softwoods (*Table 52.2*). Both the degree of crystallisation and the degree of polymerisation of the hemicelluloses are generally low, the molecule containing fewer than 200 units; in these respects, and also in their lack of resistance to alkali solutions, the hemicelluloses are quite different from true cellulose (Siau, 1984).

Lignin, present in about equal proportions to the hemicelluloses, is chemically dissimilar to these and to cellulose. Lignin is a complex, three-dimensional, polymeric, aromatic molecule composed of phenyl groups with a molecular weight of about 11 000. It is non-crystalline and the structure varies between wood from a conifer and from a broadleaved tree. About 25% of the total lignin in timber is to be found in the middle lamella, an intercellular layer composed of lignin and pectin together with the primary cell wall. Since this compound middle lamella is very thin, the concentration of lignin is correspondingly high (about 70%). Deposition of the lignin in this layer is rapid.

The bulk of the lignin (about 75%) is present within the secondary cell wall, having been deposited following completion of the cellulosic framework. Initiation of lignification of the secondary wall commences when the compound middle lamella is about half completed and extends gradually inwards across the secondary wall (Saka and Thomas, 1982). Termination of the lignification process towards the end of the period of differentiation coincides with the death of the cell. Most cellulosic plants do not contain lignin and it is the inclusion of this substance within the framework of timber that is largely responsible for the stiffness of timber, especially in the dried condition.

A recent and comprehensive account of the structure and influence of the hemicelluloses and lignin in determining certain aspects of wood quality is given by Pereira *et al.* (2003).

Extractives

Before leaving the chemical composition of wood, mention must be made of the presence of *extractives* (*Table 52.2*). This is a collective name for a series of highly complex organic compounds that are present in certain timbers in relatively small amounts. Some, like waxes, fats and sugars, have little economic significance, but others, for example rubber and resin (from which turpentine is distilled), are of considerable economic importance. The heartwood of timber, as described previously, generally contains extractives which, in addition to imparting coloration to the wood, bestow on it its natural durability, since most of these compounds are toxic to both fungi and insects. Readers desirous of more information on extractives are referred to the comprehensive text by Hillis (1987).

Minerals

Elements such as calcium, sodium, potassium, phosphorus and magnesium are all components of new growth tissue, but the actual mass of these inorganic materials is small and constitutes on the basis of the oven-dry mass of the timber less than 1% for temperate woods and less than 5% for tropical timbers.

Certain timbers show a propensity to conduct suspensions of minerals that are subsequently deposited within the timber. The presence of silica in the rays of certain tropical timbers, and calcium carbonate in the cell cavities of iroko, are two examples where large concentrations of minerals cause severe problems in log conversion and subsequent machining.

Acidity

Wood is generally acidic in nature, the level of acidity being considerably higher in the heartwood than in the sapwood of the same tree. The pH of the heartwood varies in different species of timber, but is generally about 4.5 to 5.5; however, in some timbers such as eucalypt, oak, and western red cedar, the pH of the heartwood can be as low as 3.0. Sapwood generally has a pH at least 1.0 units higher than the corresponding heartwood, i.e. the acidity is at least ten times lower than that of the corresponding heartwood.

Acidity in wood is due primarily to the generation of acetic acid by hydrolosis of the acetyl groups of the hemicelloses in the presence of moisture; this acidity in wood can cause severe corrosion of certain metals and care has to be exercised in the selection of metallic fixings, especially at higher relative humidities.

52.3.2 THE CELL WALL AS A FIBRE COMPOSITE

In the introductory remarks wood was defined as a natural composite, and the most successful model used to interpret the ultrastructure of wood from the various chemical and X-ray analyses ascribes the role of 'fibre' to the cellulosic microfibrils while the lignin and hemicelluloses are considered as separate components of the 'matrix'. The cellulosic microfibril, therefore, is interpreted as conferring high tensile strength to the composite owing to the presence of covalent bonding both within and between the anhydroglucose units. Experimentally it has been shown that reduction in chain length following gamma irradiation markedly reduces the tensile strength of timber; the significance of chain length in determining strength has been confirmed in studies of wood with inherently low degrees of polymerisation. While slippage between the cellulose chains was previously considered to be an important contributor to the development of ultimate tensile strength, this is now thought to be unlikely owing to the forces involved in fracturing large numbers of hydrogen bonds.

Preston (1964) has shown that the hemicelluloses are usually intimately associated with the cellulose, effectively binding the microfibrils together. Bundles of cellulose chains are therefore seen as having a polycrystalline sheath of hemicellulose material, consequently the resulting high degree of hydrogen bonding would make chain slippage unlikely: rather it would appear that stressing results in fracture of the C—O—C linkage.

The deposition of lignin is variable in different parts of the cell wall, but it is obvious that its prime function is to protect the hydrophilic (water-seeking) non-crystalline cellulose and the hemicelluloses, which are mechanically weak when wet. Experimentally, it has been demonstrated that removal of lignin markedly reduces the strength of wood in the wet state, though its removal results in an increase in strength in the dry state, calculated on a net cell wall area basis. Consequently, the lignin is regarded as lying to the outside of the microfibril, forming a protective sheath.

Since the lignin is located only on the exterior it must be responsible for cementing together the fibrils and in imparting shear resistance in the transference of stress throughout the composite. The role of lignin in contributing towards the stiffness of timber has already been mentioned.

There has been great debate over the years as to the juxtaposition of the cellulose, hemicellulose and lignin in the composition of a microfibril, and to the size of the basic unit. One of the many models proposed is illustrated in *Fig. 52.13*. In this widely accepted model, the crystalline core is considered to be about 5×3 nm containing about 48 chains in either 4- or 8-chain unit cells. Passing outwards from the core of the microfibril, the highly crystalline cellulose gives way first to the partly crystalline layer containing mainly hemicellulose and non-crystalline cellulose chains, and then to the amorphous lignin:

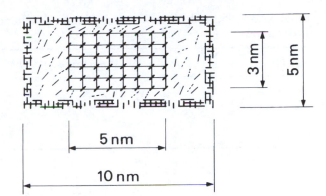

Fig. 52.13 A model of the cross-section of a microfibril in which the core is regarded as being homogeneous. (Adapted from Preston, 1974 and reproduced by permission of Chapman and Hall.) ⟋ Cellulose chains, ⋰ Hemicellulose chains, ⊞ Lignin.

this gradual transition of crystallinity from fibre to matrix results in high inter-laminar shear strength, which contributes considerably to the high tensile strength and toughness of wood.

52.3.3 CELL WALL LAYERS

When a cambial cell divides to form two daughter cells a new wall is formed comprising the middle lamella and two primary cell walls, one to each daughter cell. These new cells undergo changes within about three days of their formation and one of these developments will be the formation of a secondary wall. The thickness of this wall will depend on the function that the cell will perform, as described earlier, but its basic construction will be similar in all cells.

Early studies on the anatomy of the cell wall used polarisation microscopy, which revealed the direction of orientation of the crystalline regions. These studies indicated that the secondary wall could be subdivided into three layers, and measurement of the extinction position was indicative of the angle at which the microfibrils were orientated. Subsequent studies with transmission electron microscopy confirmed these findings and provided some additional information, with particular reference to wall texture and variability of angle. However, much of our knowledge on microfibrillar orientation has been derived using X-ray diffraction analysis. Most of these techniques yield only mean angles for any one layer of the cell wall, but recent analysis has indicated that it may be possible to determine the complete microfibrillar angle distribution of the cell wall (Cave, 1997).

The relative thickness and mean microfibrillar angle of the layers in a sample of spruce timber are illustrated in *Table 52.3*.

Table 52.3 Microfibrillar orientation and percentage thickness of the cell wall layers in spruce timber (*Picea abies*)

Wall layer	Approximate % thickness	Angle to longitudinal axis
P	3	Random
S_1	10	50–70°
S_2	85	10–30°
S_3	2	60–90°

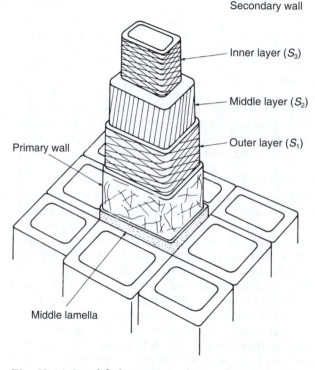

Fig. 52.14 Simplified structure of the cell wall showing mean orientation of microfibrils in each of the major wall layers (© Building Research Establishment).

The middle lamella, a lignin–pectin complex, is devoid of cellulosic microfibrils, while in the primary wall (P) the microfibrils are loosely packed and interweave at random (*Fig. 52.14*); no lamellation is present. In the secondary wall layers the microfibrils are closely packed and parallel to each other. The outer layer of the secondary wall, the S_1, is again thin and is characterised by having from four to six concentric lamellae, the microfibrils of each alternating between a left- and right-hand spiral (S and Z helix), both with a pitch to the longitudinal axis of from 50° to 70° depending on the species of timber.

The middle layer of the secondary wall (S_2) is thick and is composed of 30–150 lamellae, the closely packed microfibrils of which all exhibit a similar orientation in a right-hand spiral (Z helix) with a pitch of 10–30° to the longitudinal axis, as illustrated in *Figs 52.14* and *52.15*. Since more than three-quarters of the cell wall is composed of the S_2 layer, it follows that the ultrastructure of this layer will have a very marked influence on the behaviour of the timber. In later sections, anisotropic behaviour, shrinkage, tensile strength and failure morphology will all be related to the microfibrillar angle in the S_2 layer.

Kerr and Goring (1975) were among the first workers to question the extent of these concentric lamellae in the S_2 layer; these workers found that though there was a preferred orientation of lignin and carbohydrates in the S_2 layer, the lamellae were certainly not continuous. Thus, the interrupted lamellae model proposed by them embraced lignin and carbohydrate entities that were bigger in the tangential than in the radial direction. Cellulose microfibrils were envisaged as being embedded in a matrix of hemicelluloses and lignin.

The existence within the S_2 layer of concentric lamellae has been questioned again later in the century. Evidence has been presented (Sell and Zimmermann, 1993) from both electron and light microscopy that indicates radial, or near radial orientations of the transverse structure of the S_2 layer. The transverse thickness of these agglomerations of microfibrils is 0.1–1.0 nm and they frequently extend the entire width of the S_2 layer. A modified model of the cell wall of softwoods has been proposed (Sell & Zimmerman, 1993).

The S_3 layer, which may be absent in certain timbers, is very thin, with only a few concentric lamellae; it is characterised, as is the S_1 layer, by alternate lamellae possessing microfibrils orientated in opposite spirals with a pitch of 60–90°, though the presence of the right-handed spiral (Z helix) is disputed by some workers. Generally, the S_3 has

*Fig. 52.15 Electron micrograph of the cell wall of Norway spruce timber (*Picea abies*), showing the parallel and almost vertical microfibrils of an exposed portion of the S$_2$ layer (© Building Research Establishment).*

a looser texture than the S$_1$ and S$_2$ layers and is frequently encrusted with extraneous material. The S$_3$, like the S$_1$, has a higher concentration of lignin than does the S$_2$ (Saka & Thomas, 1982). Electron microscopy has also revealed the presence of a thin warty layer overlaying the S$_3$ layer in certain timbers.

Investigations have indicated that the values of microfibrillar angle quoted in *Table 52.3* are only average for the layers and that systematic variation in angle occurs within each layer. Thus, Abe *et al.* (1991) have shown that in *Abies sachalinensis* the microfibrillar angle of the secondary wall, as seen from the lumen, changed in a LH direction from

the outermost S$_1$ to the middle of the S$_2$ and then in a RH direction to the innermost S$_3$. This resulted in the boundaries between the three principal layers being very indistinct, confirming reports by previous workers on other species and suggesting that the wall structure can be viewed as a systematically varying continuum.

Microfibrillar angle appears to vary systematically along the length of the cell as well as across the wall thickness. Thus, the angle of the S$_2$ layer has been shown to decrease towards the ends of the cells, while the average S$_2$ angle appears to be related to the length of the cell, itself a function of the rate of growth of the tree. Systematical differences in microfibrillar angle have been found between radial and tangential walls and this has been related to differences in degree of lignification between these walls. Openings occur in the walls of cells and many of these pit openings are characterised by localised deformation of the microfibrillar structure.

Further information on the variability of microfibrillar angle and its importance in determining wood quality is to be found in Dinwoodie (2000) and Butterfield (2003).

52.4 Variability in structure

Variability in performance of wood is one of its inherent deficiencies as a material. It will be discussed later how differences in mechanical properties occur between timbers of different species and how these are manifestations of differences in wall thickness and distribution of cell types. However, superimposed on this genetic source of variation are both a systematic and an environmental one.

There are distinct patterns of variation in many features within a single tree. Length of the cells, thickness of the cell wall and hence density, angle at which the cells are lying with respect to the vertical axis (spiral grain) and angle at which the microfibrils of the S$_2$ layer of the cell wall are located with respect to the vertical axis, all show systematic trends outwards from the centre of the tree to the bark and upwards from the base to the top of the tree. This pattern results in the formation of a core of wood in the tree with many undesirable properties, including low strength and high shrinkage. This zone, usually regarded as some ten to twenty growth rings in width, is known as *core wood* or *juvenile wood* as opposed to the *mature wood* occurring outside this area. The boundary between juvenile and mature wood is usually defined in terms of the change in slope of the variation in magnitude of one anatomical feature (e.g. cell length, density) when plotted against ring number from the pith.

Environmental factors have considerable influence on the structure of wood and any environmental influence, including forest management, which changes the rate of growth of the tree will affect the technical properties of the wood. However, the relationship is a complex one; in softwoods, increasing growth rate generally results in an increase in the width of early-wood with a resulting decrease in density and mechanical properties. In diffuse-porous hardwoods increasing growth rate, provided it is not excessive, has little effect on density, while in ring-porous hardwoods, increasing rate of growth, again provided it is not excessive, results in an increase in the width of latewood and consequently in density and strength.

There is a whole series of factors that may cause defects in the structure of wood and consequent lowering of its strength. Perhaps the most important defect with regard to its utilisation is the formation of *reaction wood*. When trees are inclined to the vertical axis, usually as a result of wind action or of growing on sloping ground, the distribution of growth-promoting hormones is disturbed, resulting in the formation of an abnormal type of tissue. In softwoods, this reaction tissue grows on the compression side of the trunk and is characterised by having a higher than normal lignin content, a higher microfibrillar angle in the S_2 layer resulting in increased longitudinal shrinkage, and a generally darker appearance (*Fig. 52.16*); this abnormal timber, known as *compression wood*, is also considerably more brittle than normal wood. In hardwoods, reaction wood forms on the tension side of trunks and large branches and is therefore called *tension wood*. It is characterised by the presence of a gelatinous celulosic layer (the *G layer*) to the inside of the cell wall; this results in a higher than normal cellulose content to the total cell wall, which imparts a rubbery characteristic to the fibres, resulting in difficulties in sawing and machining.

A more comprehensive description of reaction wood is given by Barnett and Jeronimidis (2003).

One other defect of considerable technical significance is *brittleheart*, which is found in many low-density tropical hardwoods and is one manifestation of the presence of longitudinal growth stresses in large diameter trees. Yield of the cell wall occurs under longitudinal compression with the formation of shear lines through the cell wall and throughout the core wood; compression failure will be discussed in greater detail in section 54.6.1.

More information on the variability in structure and its influence on the technical performance of

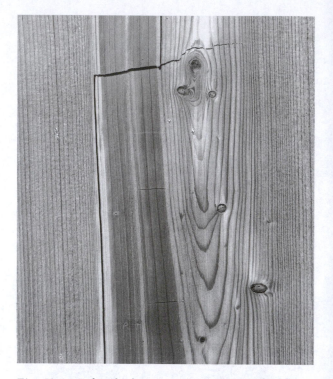

Fig. 52.16 *A band of compression wood (centre left) in a Norway spruce plank, illustrating the darker appearance and higher longitudinal shrinkage of the reaction wood compared with the adjacent normal wood (© Building Research Establishment).*

timber can be found in Chapters 5 and 12 of Desch and Dinwoodie (1996).

52.5 Appearance of timber in relation to its structure

Most of you will agree that many timbers are aesthetically pleasing and the various and continuing attempts to simulate the appearance of timber in the surface of synthetic materials bear testament to the very attractive appearance of many timbers. Although a very large proportion of the timber consumed in the UK is used within the construction industry, where the natural appearance of timber is of little consequence, excepting the use of hardwoods for flush doors, internal panelling and wood-block floors, a considerable quantity of timber is still utilised purely on account of its attractive appearance, particularly for furniture and various sports goods. The decorative appearance of many timbers is due to the *texture*, or to the *figure*, or to the *colour* of the material and, in many instances, to combinations of these.

52.5.1 TEXTURE

The texture of timber depends on the size of the cells and on their arrangement. A timber such as boxwood in which the cells have a very small diameter is said to be *fine-textured*, while a *coarse-textured* timber such as keruing has a considerable percentage of large-diameter cells. Where the distribution of the cell-types or sizes across the growth ring is uniform, as in beech, or where the thickness of the cell wall remains fairly constant across the ring, as in some of the softwoods, e.g. yellow pine, the timber is described as being *even-textured*; conversely, where variation occurs across the growth ring, either in distribution of cells as in teak or in thickness of the cell walls as in larch or Douglas fir, the timber is said to have an *uneven texture*.

52.5.2 FIGURE

Figure is defined as the 'ornamental markings seen on the cut surface of timber, formed by the structural features of the wood', but the term is also frequently applied to the effect of marked variations in colour. The four most important structural features inducing figure are *grain*, *growth rings*, *rays* and *knots*.

Grain

Mention was made in section 52.2 that the cells of wood, though often described as vertically orientated, frequently deviate from this convenient arrangement. In the majority of cases this deviation takes the form of a spiral, the magnitude of the angle varying with distance from the pith. Although of considerable technical importance because of loss in strength and induced machining problems, the common form of spiral grain has no effect on the figure presented on the finished timber. However, two other forms of grain deviation do have a very marked influence on the resulting figure of the wood. Thus, in certain hardwood timbers, and the mahoganies are perhaps the best example, the direction of the spiral in the longitudinal–tangential plane alternates from left to right hand at very frequent intervals along the radial direction; grain of this type is said to be *interlocked*. Tangential faces of machined timber will be normal, but the radial face will be characterised by the presence of alternating light and dark longitudinal bands produced by the reflection of light from the tapered cuts of fibres inclined in different directions (*Fig. 52.17*). This type of figure is referred to as *ribbon* or *stripe* and is desirous in timber for furniture manufacture.

If instead of the grain direction alternating from left to right within successive layers along the radial direction as above, the grain direction alternates at right angles to this pattern, i.e. in the longitudinal–

Fig. 52.17 *Illustration of 'ribbon' or 'stripe' figure on the cut longitudinal–radial plane in mahogany timber. The fibres in successive radial zones of the timber are inclined in opposite directions in the longitudinal–tangential plane (© Building Research Establishment).*

radial plane, a *wavy* type of grain is produced. This is very conspicuous in machined tangential faces where it shows up clearly as alternating light and dark horizontal bands (*Fig. 52.18*); this type of figure is described as *fiddleback*, since timber with this distinctive type of figure has been used traditionally for the manufacture of the backs of violins. It is also found on the panels and sides of expensive wardrobes and bookcases.

Growth rings

Where variability occurs across the growth ring, either in the distribution of the various cell types or in the thickness of the cell walls, distinct patterns will appear on the machined faces of the timber. Such patterns, however, will not be regular like many of the man-made imitations, but will vary according to changes in width of the growth ring and in the relative proportions of early and latewood.

On the radial face the growth rings will be vertical and parallel to one another, but on the tangential face a most pleasing series of concentric arcs is

produced as successive growth layers are intersected. In the centre part of the plank of timber illustrated in *Fig. 52.19*, the growth rings are cut tangentially forming these attractive arcs, while the edge of the board with parallel and vertical growth rings reflects timber cut radially. In the case of ring-porous timbers, it is the presence of the large earlywood vessels that makes the growth ring so conspicuous, while in timbers like Douglas fir or pitch pine, the striking effect of the growth ring can be ascribed to the very thick walls of the latewood cells.

Rays

Another structural feature that may add to the attractive appearance of timber is the ray, especially where, as in the case of oak, the rays are both deep and wide. When the surface of the plank coincides with the longitudinal–radial plane, these rays can be seen as sinuous light-coloured ribbons running across the grain.

Knots

Knots, though troublesome from the mechanical aspects of timber utilisation, can be regarded as a decorative feature; the fashion for knotty-pine furniture and wall panelling in the early seventies is a very good example of this. However, as a decorative feature, knots do not possess the subtlety of variation in grain and colour that arises from the other structural features described above.

Fig. 52.18 '*Fiddleback' figure on the longitudinal–tangential plane in* Terminalia amazonia *due to wavy grain development in the longitudinal–radial plane (© Building Research Establishment).*

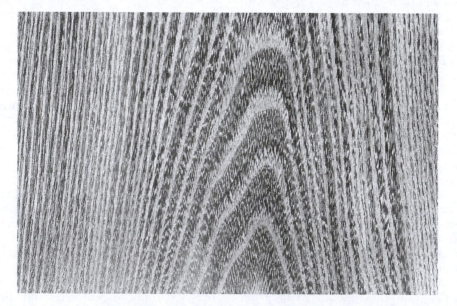

Fig. 52.19 *The effect of growth rings on figure in robinia timber; in the centre the growth rings have been cut tangentially to produce a series of overlapping cones, while at the edges of this plank the rings are cut radially to produce parallel vertical bands (© Building Research Establishment).*

Exceptionally, trees produce a cluster of small shoots at some point on the trunk and the timber subsequently formed in this region contains a multitude of small knots. Timber from these *burrs* is highly prized for decorative work, especially if walnut or yew.

52.5.3 COLOUR

In the absence of extractives, timber tends to be a rather pale straw colour, which is characteristic of the sapwood of almost all timbers. The onset of heartwood formation in many timbers is associated with the deposition of extractives, most of which are coloured, thereby imparting coloration to the heartwood zone. In passing, it should be recalled that although a physiological heartwood is always formed in older trees, extractives are not always produced; thus, the heartwood of timbers such as ash and spruce is colourless.

Where coloration of the heartwood occurs, a whole spectrum of colours exists among the different species. The heartwood may be yellow, e.g. boxwood; orange, e.g. opepe; red, e.g. mahogany; purple, e.g. purpleheart; brown, e.g. African walnut; green, e.g. greenheart; or black, e.g. ebony. In some timbers the colour is fairly evenly distributed throughout the heartwood, while in other species considerable variation in the intensity of the colour occurs. In zebrano, distinct dark brown and white stripes occur, while in olive wood patches of yellow merge into zones of brown. Dark gum-veins, as present in African walnut, contribute to the pleasing alternations in colour. Variations in colour such as these are regarded as contributing to the 'figure' of the timber.

It is interesting to note in passing that the non-coloured sapwood is frequently coloured artificially to match the heartwood, thereby adding to the amount of timber converted from the log. In a few rare cases, the presence of certain fungi in timber in the growing tree can result in the formation of very dark coloured heartwood; the activity of the fungus is terminated when the timber is dried. Both *brown oak* and *green oak*, produced by different fungi, have always been prized for decorative work.

52.6 Mass–volume relationships

52.6.1 DENSITY

The *density* of a piece of timber is a function not only of the amount of wood substance present, but also of the presence of both extractives and moisture. In a few timbers extractives are completely absent, while in many they are present, but only in small amounts and usually less than 3% of the dry mass of the timber. In some exceptional cases, the extractive content may be as high as 10% and in these cases it is necessary to remove the extractives prior to the determination of density.

The presence of moisture in timber not only increases the mass of the timber, but also results in swelling of the timber, and hence both mass and volume are affected. Thus, in the determination of density where:

$$\rho = \frac{m}{v} \tag{52.1}$$

both the mass (m) and volume (v) must be determined at the same moisture content. Generally, these two parameters are determined at zero moisture content. However, as density is frequently quoted at a moisture content of 12% – since this level is frequently experienced in timber in use – the value of density at zero moisture content is corrected to 12% if volumetric expansion figures are known, or else the density determination is carried out on timber at 12% moisture content.

Thus, if:

$$m_x = m_0(1 + 0.01\mu) \tag{52.2}$$

where m_x is the mass of timber at moisture content x, m_0 is the mass of timber at zero moisture content, and μ is the moisture content %, and:

$$v_x = v_0(1 + 0.01s_v) \tag{52.3}$$

where v_x is the volume of timber at moisture content x, v_0 is the volume of timber at zero moisture content, and s_v is the volumetric shrinkage/expansion %, it is possible to obtain the density of timber at any moisture content in terms of the density at zero moisture content, thus:

$$\rho_x = \frac{m_x}{v_x} = \frac{m_0(1 + 0.01\mu)}{v_0(1 + 0.01s_v)} = \rho_0 \left(\frac{1 + 0.01\mu}{1 + 0.01s_v} \right) \tag{52.4}$$

As a very approximate rule of thumb, the density of timber increases by approximately 0.5% for each 1.0% increase in moisture content up to 30%. Density therefore will increase, slightly up to moisture contents of about 30% as both total mass and volume increase; however, at moisture contents above 30%, density will increase rapidly and curvilinearly, with increasing moisture content, since, as will be explained later in this chapter, the volume remains constant above this value, while the mass increases.

The determination of density by measurement of mass and volume takes a considerable period of time and over the years a number of quicker techniques have been developed for use where large numbers of density determinations are required. These methods range from the assessment of the opacity of a photographic image that has been produced by either light or β-irradiation passing through a thin section of wood, to the use of a mechanical device (the Pilodyn) that fires a spring-loaded bolt into the timber after which the depth of penetration is measured. In all these techniques, however, the method or instrument has to be calibrated against density values obtained by the standard mass/volume technique.

In section 52.2, timber was shown to possess different types of cell that could be characterised by different values of the ratio of cell-wall thickness to total cell diameter. Since this ratio can be regarded as an index of density, it follows that the density of timber will be related to the relative proportions of the various types of cell. Density, however, will also reflect the absolute wall thickness of any one type of cell, since it is possible to obtain fibres of one species of timber the cell-wall thickness of which can be several times greater than that of fibres of another species. The influence of various growth factors on determining density is provided by Saranpää (2003).

Density, like many other properties of timber, is extremely variable; it can vary by a factor of ten, ranging from an average value at 12% moisture content of 176 kg/m³ for balsa, to about 1230 kg/m³ for lignum vitae (*Fig. 52.20*). Balsa, therefore, has a density similar to that of cork, while lignum vitae has a density slightly less than half that of concrete or aluminium. The values of density quoted for different timbers, however, are merely average values, as each timber will have a range of densities reflecting differences between early and latewood, between the pith and outer rings, and between trees on the same site. Thus, for example, the density of balsa can vary from 40 to 320 kg/m³. In certain publications, reference is made to the *weight* of timber, a term widely used in commerce; it should be appreciated that the quoted values are really densities.

52.6.2 SPECIFIC GRAVITY

The traditional definition of *specific gravity* (*G*, also known as *relative density*) can be expressed as:

$$G = \frac{\rho_t}{\rho_w} \qquad (52.5)$$

where ρ_t is the density of timber, and ρ_w is the density of water at 4°C (1.0000 g/ml). *G* will therefore vary with moisture content, consequently the specific gravity of timber is usually based on the oven-dry mass, and volume at some specified moisture

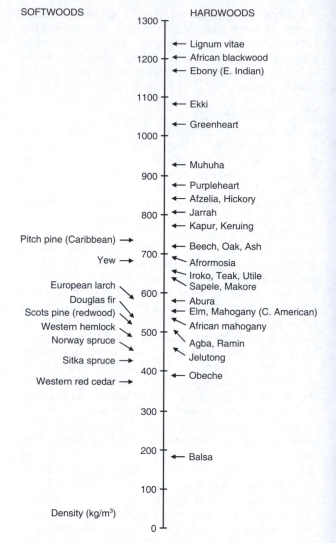

Fig. 52.20 Mean density values at 12% moisture content for some common hardwoods and softwoods (© Building Research Establishment).

content. This is frequently taken as zero though, for convenience, green or other moisture conditions are sometimes used, when the terms *basic specific gravity* and *nominal specific gravity* are applied, respectively. Hence:

$$G_\mu = \frac{m_0}{V_\mu \rho_w} \qquad (52.6)$$

where m_0 is the oven-dry mass of timber, V_μ is the volume of timber at moisture content μ, ρ_w is the density of water, and G_μ is the specific gravity at moisture content μ.

At low moisture contents, specific gravity decreases slightly with increasing moisture content up to 30%, thereafter remaining constant. In research activities specific gravity is defined usually in terms of oven-dry mass and volume. However, for engineering applications specific gravity is frequently presented as the ratio of oven-dry mass to volume of timber at 12% moisture content; this can be derived from the oven-dry specific gravity, thus:

$$G_{12} = \frac{G_0}{1 + 0.01\mu G_0/G_{s12}} \qquad (52.7)$$

where G_{12} is the specific gravity of timber at 12% moisture content, G_0 is the specific gravity of timber at zero moisture content, μ is the moisture content %, and G_{s12} is the specific gravity of bound water at 12% moisture content.

The relationship between density and specific gravity can be expressed as:

$$\rho = G(1 + 0.01\mu)\rho_w \qquad (52.8)$$

where ρ is the density at moisture content μ, G is the specific gravity at moisture content μ, and ρ_w is the density of water. Equation 52.8 is valid for all moisture contents. When $\mu = 0$ the equation reduces to:

$$\rho = G_0 \qquad (52.9)$$

i.e. density and specific gravity are numerically equal.

52.6.3 DENSITY OF THE DRY CELL WALL

Although the density of different timbers may vary considerably, the density of the actual cell wall material remains constant for all timbers, with a value of approximately 1500 kg/m^3 when measured by volume-displacement methods. The exact value for cell-wall density depends on the liquid used for measuring the volume; densities of 1525 and 1451 kg/m^3 have been recorded for the same material using water and toluene, respectively.

52.6.4 POROSITY

In section 52.2 the cellular nature of timber was described in terms of a parallel arrangement of hollow tubes. The *porosity (p)* of timber is defined as the fractional void volume and is expressed mathematically as:

$$p = 1 - V_f \qquad (52.10)$$

where V_f is the volume fraction of cell-wall substance. The calculation of porosity is set out in Chapter 3 of Dinwoodie (2000).

52.7 Moisture in timber

52.7.1 EQUILIBRIUM MOISTURE CONTENT

Timber is hygroscopic, that is it will absorb moisture from the atmosphere if it is dry and correspondingly yield moisture to the atmosphere when wet, thereby attaining a moisture content that is in equilibrium with the water vapour pressure of the surrounding atmosphere. Thus, for any combination of vapour pressure and temperature of the atmosphere there is a corresponding moisture content of the timber such that there will be no inward or outward diffusion of water vapour; this moisture content is referred to as the *equilibrium moisture content* (EMC). Generally, it is more convenient to use relative humidity rather than vapour pressure. Relative humidity is defined as the ratio of the partial vapour pressure in the air to the saturated vapour pressure, expressed as a percentage.

The fundamental relationships between moisture content of timber and atmospheric conditions have been determined experimentally, and the average equilibrium moisture content values are shown graphically in *Fig. 52.21*. A timber in an atmosphere of 20°C and 22% relative humidity will have a moisture content of 6% (see below), while the same timber if moved to an atmosphere of 40°C and 64% relative humidity will double its moisture content. It should be emphasised that the curves in *Fig. 52.21* are average values for moisture in relation to relative humidity and temperature, and that slight variations in the equilibrium moisture content will occur owing

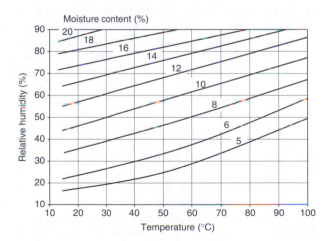

Fig. 52.21 *Chart showing the relationship between the moisture content of timber and the temperature and relative humidity of the surrounding air; approximate curves based on values obtained during drying from green condition (© Building Research Establishment).*

to differences between timbers and to the previous history of the timber with respect to moisture.

52.7.2 DETERMINATION OF MOISTURE CONTENT

It is customary to express the moisture content of timber in terms of its oven-dry mass using the equation:

$$\mu = \frac{m_{init} - m_{od}}{m_{od}} \times 100 \qquad (52.11)$$

where m_{init} = initial mass of timber sample (g), m_{od} = mass of timber sample after oven-drying at 105°C (g) and μ = moisture content of timber sample (%).

Expression of the moisture content of timber on a dry-mass basis is in contrast to the procedure adopted for other materials, where moisture content is expressed in terms of the wet mass of the material.

Determination of the moisture content of timber is usually carried out using the basic gravimetric technique above, though it should be noted that at least a dozen different methods have been recorded in the literature. Suffice it here to mention only two of these alternatives. First, where the timber contains volatile extractives, which would normally be lost during oven drying, thereby resulting in erroneous moisture content values, it is customary to use a distillation process, heating the timber in the presence of a water-immiscible liquid such as toluene, and collecting the condensed water vapour in a calibrated trap. Second, where ease and speed of operation are preferred to extreme accuracy, moisture content is assessed using electric moisture meters. The type most commonly used is known as the *resistance meter*, though this battery-powered hand-held instrument actually measures the conductance or flow (the reciprocal of resistance) of an electric current between two probes. Below the fibre saturation point (about 27% moisture content; see later) an approximately linear relationship exists between the logarithm of conductance and the logarithm of moisture content. However, this relationship, which forms the basis for this type of meter, changes with species of timber, temperature, and grain angle. Thus, a resistance-type meter is equipped with a number of alternative scales, each of which relates to a different group of timber species; it should be used at temperatures close to 20°C. with the pair of probes inserted parallel to the direction of the grain.

Although the measurement of moisture content is quick with such a meter, there are, however, two drawbacks to its use. First, moisture content is measured only to the depth of penetration of the two probes, a measurement that may not be representative of the moisture content of the entire depth of the timber member; the use of longer probes can be beneficial though these are difficult to insert and withdraw. Second, the working range of the instrument is only from 7 to 27% moisture content.

52.7.3 THE MOISTURE CONTENT OF GREEN TIMBER

In the living tree, water is to be found not only in the cell cavity, but also within the cell wall. Consequently the moisture content of green (newly felled) wood is high, usually varying from about 60% to nearly 200%, depending on the location of the timber in the tree and the season of the year.[1] However, seasonal variation is slight compared to the differences that occur within a tree between the sapwood and heartwood regions. The degree of variation is illustrated for a number of softwoods and hardwoods in *Table 52.4*; within the former group the sapwood may contain twice the percentage of moisture to be found in the corresponding heartwood, while in the hardwoods this difference is appreciably smaller or even absent. However, pockets of 'wet' wood can be found in the heartwood, as described in section 52.1.

Green timber will yield moisture to the atmosphere with consequent changes in its dimensions. At moisture contents above 20% many timbers, especially their sapwood, are susceptible to attack by fungi; the strength and stiffness of green wood are considerably lower than for the same timber when dry. For all these reasons it is necessary to dry or *season* timber following felling of the tree and prior to its use in service.

52.7.4 REMOVAL OF MOISTURE FROM TIMBER

Drying or seasoning of timber can be carried out in the open, preferably with a top cover. However, it will be appreciated from the previous discussion on equilibrium moisture content that the minimum moisture content that can be achieved is determined by the lowest relative humidity of the summer period. In the UK it is seldom possible to achieve moisture contents of less than 16% by air seasoning. The planks of timber are separated in rows by stickers (usually 25–30 mm across) that permit air currents to pass through the pile; nevertheless it may take from two to ten years to air-season timber, depending on the species of timber and the thickness of the timber members.

---

[1] Values greater than 100% arise because the percentage is expressed as percentage of the *dry* weight of the wood.

Table 52.4   Average green moisture content of the sapwood and heartwood

| Botanical name | Commercial name | Moisture content (%) | |
		Heartwood	Sapwood
Hardwoods			
Betula lutea	Yellow birch	64	68
Fagus grandifolia	American beech	58	79
Ulmus americana	American elm	92	84
Softwoods			
Pseudotsuga menziesii	Douglas fir	40	116
Tsuga heterophylla	Western hemlock	93	167
Picea sitchensis	Sitka spruce	50	131

The process of seasoning may be accelerated artificially by placing the stacked timber in a drying kiln, basically a large chamber in which the temperature and humidity can be controlled and altered throughout the drying process; control may be carried out manually or programmed automatically. Humidification is sometimes required in order to keep the humidity of the surrounding air at a desired level when insufficient moisture is coming out of the timber; it is frequently required towards the end of the drying run, and is achieved either by the admission of small quantities of steam or by the use of water atomisers or low-pressure steam evaporators. Various designs of kiln are used, which are reviewed in detail by Pratt (1974).

Drying of softwood timber in a kiln can be accomplished in from four to seven days, the optimum rate of drying varying widely from one timber to the next; hardwood timber usually takes about three times longer than softwood of the same dimensions. Following many years of experimentation, kiln schedules have been published for different groups of timber, which provide wet- and dry-bulb temperatures (maximum of 70°C) for different stages in the drying process, and their use should result in the minimum amount of degrade in terms of twist, bow, spring, collapse and checks (Pratt, 1974). Most timber is now seasoned by kilning, and little air drying is carried out. Dry stress-graded timber in the UK must be kiln-dried to a mean value of 20% moisture content with no single piece greater than 24%. However, UK and some Swedish mills are now targeting 12% ('superdried'), as this level is much closer to the moisture content in service.

Recently, *solar kilns* have become commercially available and are particularly suitable for use in developing countries to season many of the difficult slow-drying tropical timbers. These small kilns are very much cheaper to construct than conventional kilns and are also much cheaper to run. They are capable of drying green timber to about 7% moisture content in the dry season and about 11% in the rainy season.

52.7.5 INFLUENCE OF STRUCTURE

As previously mentioned in section 52.7.3, water in green or freshly felled timber is present both in the cell cavity and within the cell wall. During the seasoning process, irrespective of whether this is by air or within a kiln, water is first removed from within the cell cavity; this holds true down to moisture contents of about 27–30%. Since the water in the cell cavities is *free*, not being chemically bonded to any part of the timber, it can readily be appreciated that its removal will have no effect on the strength or dimensions of the timber. The lack of variation of the former parameter when moisture content is reduced from 110 to 27% is illustrated in *Fig. 52.22*.

However, at moisture contents below 27% water is no longer present in the cell cavity, but is restricted to the cell wall, where it is chemically bonded (hydrogen bonding) to the matrix constituents, to the hydroxyl groups of the cellulose molecules in the noncrystalline regions and to the surface of the crystallites; as such, this water is referred to as *bound water*. The uptake of water by the lignin component is considerably lower than that by either the hemicellulose or the amorphous cellulose; water may be present as a monomolecular layer, though frequently up to six layers can be present. Water cannot penetrate the crystalline cellulose since the hygroscopic hydroxyl groups are mutually satisfied by the formation of both intra- and intermolecular bonds within

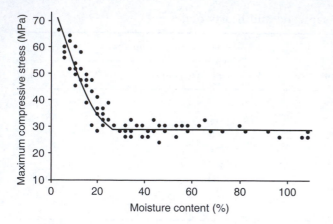

Fig. 52.22 *Relationship between longitudinal compressive strength and moisture content (© Building Research Establishment).*

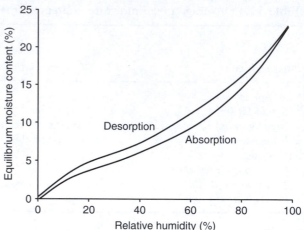

Fig. 52.23 *Hysteresis loop resulting from the average adsorption and desorption isotherms for six species of timber at 40°C (© Building Research Establishment).*

the crystalline region, as described in section 52.3.1. This view is confirmed by X-ray analysis, which indicates no change of state of the crystalline core as timber gains or loses moisture.

However, the percentage of noncrystalline material in the cell wall varies between 8 and 33% and the influence of this fraction of cell-wall material as it changes moisture content on the behaviour of the total cell wall is very significant. The removal of water from these areas within the cell wall results first in increased strength and second in marked shrinkage. Both changes can be accounted for in terms of drying out of the water-reactive matrix, thereby causing the microfibrils to come into closer proximity, with a commensurate increase in inter-fibrillar bonding and decrease in overall dimensions. Such changes are reversible, or almost completely reversible.

52.7.6 FIBRE SATURATION POINT

The increase in strength on drying is clearly indicated in *Fig. 52.22*, from which it will be noted that there is an approximately three-fold increase in strength as the moisture content of the timber is reduced from about 27% to zero. The moisture content corresponding to the inflexion in the graph is termed the *fibre saturation point*, where in theory there is no free water in the cell cavities while the walls are holding the maximum amount of bound water. In practice this situation rarely exists; a little free water may still exist while some bound water is removed from the cell wall. Consequently, the fibre saturation 'point', while a convenient concept, should really be regarded as a 'range' in moisture contents over which the transition occurs.

The fibre saturation point therefore corresponds in theory to the moisture content of the timber when placed in a relative humidity of 100%; in practice, however, this is not so since such a situation would result in total saturation of the timber (Stamm, 1964). Values of EMC above 98% are unreliable. It is generally found that the moisture content of hardwoods at this level is from 1 to 2% higher than for softwoods. At least nine different methods of determining the fibre saturation point are recorded in the literature; the value of this parameter depends on the method used.

52.7.7 SORPTION

Timber, as already noted in section 52.7.1, assumes with the passage of time a moisture content that is in equilibrium with the relative vapour pressure of the atmosphere. This process of water sorption is typical of solids with a complex capillary structure, and this phenomenon has also been observed in concrete. The similarity in behaviour between timber and concrete with regard to moisture relationships is further illustrated by the presence of S-shaped isotherms when moisture content is plotted against relative vapour pressure. Both materials have isotherms that differ according to whether the moisture content is reducing (desorption) or increasing (adsorption), thereby producing a *hysteresis loop* (*Fig. 52.23*).

For more information on sorption and diffusion, especially on the different theories of sorption, you should consult the comprehensive text by Skaar (1988).

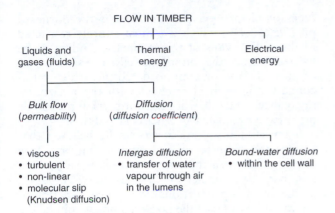

FLOW IN TIMBER

Liquids and gases (fluids) — Thermal energy — Electrical energy

Bulk flow (permeability) — Diffusion (diffusion coefficient)

- viscous
- turbulent
- non-linear
- molecular slip (Knudsen diffusion)

Intergas diffusion
- transfer of water vapour through air in the lumens

Bound-water diffusion
- within the cell wall

Fig. 52.24 The different aspects of flow in timber that are covered in this chapter.

52.8 Flow in timber

The term *flow* is synonymous with the passage of liquids through a porous medium such as timber, but the term is also applicable to the passage of gases, thermal energy and electrical energy; it is this wider interpretation of the term that is applied in this chapter, albeit that the bulk of the chapter is devoted to the passage of both liquids and gases (i.e. fluids).

The passage of *fluids* through timber can occur in one of two ways, either as *bulk flow* through the interconnected cell lumens or other voids, or by *diffusion*. The latter term embraces both the transfer of water vapour through air in the lumens and the movement of bound water within the cell wall (*Fig. 52.24*). The magnitude of the bulk flow of a fluid through timber is determined by its *permeability*.

Looking at the phenomenon of flow of moisture in wood from the point of view of the type of moisture, rather than the physical processes involved as described above, it is possible to identify the involvement of three types of moisture:

1. *Free water* in the cell cavities giving rise to bulk flow above the fibre saturation point (see section 52.7.5).
2. *Bound water* within the cell walls, which moves by diffusion below the fibre saturation point (see section 52.7.5).
3. *Water vapour*, which moves by diffusion in the lumens both above and below the fibre saturation point.

It is convenient when discussing flow of any type to think of it in terms of being either constant or variable with respect to either time or location within the specimen; flow under the former conditions is referred to as *steady-state flow*, whereas when flow

is time and space dependent it is referred to as *unsteady-state flow*. Because of the complexity of the latter, only the former is covered in this text; for more information you should consult Siau (1984).

One of the most interesting features of steady-state flow in timber in common with many other materials is that the same basic relationship holds irrespective of whether one is concerned with liquid or gas flow, diffusion of moisture, or thermal and electrical conductivity. The basic relationship is that the flux or rate of flow is proportional to the pressure gradient:

$$\frac{\text{Flux}}{\text{Gradient}} = k \qquad (52.12)$$

where flux is the rate of flow per unit cross-sectional area, gradient is the pressure difference per unit length causing flow, and k is a constant, dependent on form of flow, e.g. permeability, diffusion or conductivity.

52.8.1 BULK FLOW AND PERMEABILITY

Permeability is simply the quantitative expression of the bulk flow of fluids through a porous material. Flow in the steady-state condition is best described in terms of Darcy's law. Thus:

$$\text{Permeability} = \frac{\text{Flux}}{\text{Gradient}} \qquad (52.13)$$

and for the flow of liquids, this becomes:

$$k = \frac{QL}{A\Delta P} \qquad (52.14)$$

where k is the permeability (cm²/atm s), Q is the volume rate of flow (cm³/s), ΔP is the pressure differential (atm), A is the cross-sectional area of the specimen (cm²), and L is the length of the specimen in the direction of flow (cm).

Because of the change of pressure of a gas and hence its volumetric flow rate as it moves through a porous medium, Darcy's law for the flow of gases has to be modified as follows:

$$k_g = \frac{QLP}{A\Delta P\bar{P}} \qquad (52.15)$$

where k_g is the superficial gas permeability and Q, L, A and ΔP are as in equation (52.14), P is the pressure at which Q is measured, and $\bar{P}$ is the mean gas pressure in the sample (Siau, 1984).

Of all the numerous physical and mechanical properties of timber, permeability is by far the most variable; when differences between timbers and differences between the principal directions within a timber are taken into consideration, the range is of the order of 10^7. Not only is permeability important in the impregnation of timber with artificial preservatives, fire retardants and stabilising chemicals, but it is also significant in the chemical removal of lignin in the manufacture of wood pulp and in the removal of *free* water during drying.

Flow of fluids

The bulk of flow occurs as *viscous (or laminar) flow* in capillaries where the rate of flow is relatively low and when the viscous forces of the fluid are overcome in shear, thereby producing an even and smooth flow pattern. In viscous flow Darcy's law is directly applicable, but a more specific relationship for flow in capillaries is given by the Poiseuille equation, which for liquids is:

$$Q = \frac{N\pi r^4 \Delta P}{8\eta L} \qquad (52.16)$$

where N is the number of uniform circular capillaries in parallel, Q is the volume rate of flow, r is the capillary radius, ΔP is the pressure drop across the capillary, L is the capillary length and η is the viscosity. For gas flow, the above equation has to be modified slightly to take into account the expansion of the gas along the pressure gradient. The amended equation is:

$$Q = \frac{N\pi r^4 \Delta P \bar{P}}{8\eta L P} \qquad (52.17)$$

where $\bar{P}$ is the mean gas pressure within the capillary and P is the pressure of gas where Q was measured. In both cases:

$$Q \propto \frac{\Delta P}{L} \qquad (52.18)$$

or flow is proportional to the pressure gradient, which conforms with the basic relationship for flow.

Other types of flow can occur, e.g. turbulent, non-linear and molecular diffusion, the last mentioned being of relevance only to gases. Information on all these types can be found in Siau, 1984.

Flow paths in timber

SOFTWOODS

Because of their simpler structure and their greater economic significance, much more attention has been paid to flow in softwood timbers than in hardwood timbers. It will be recalled from section 52.2 that both tracheids and parenchyma cells have closed ends and that movement of liquids and gases must be by way of the pits in the cell wall. Three

types of pit are present. The first is the bordered pit (*Fig. 52.10*), which is almost entirely restricted to the radial walls of the tracheids, tending to be located towards the ends of the cells. The second type of pit is the ray or semi-bordered pit, which interconnects the vertical tracheid with the horizontal ray parenchyma cell, while the third type is the simple pit between adjacent parenchyma cells.

For very many years it was firmly believed that – since the diameter of the pit opening or of the openings between the margo strands is very much less than the diameter of the cell cavity, and since permeability is proportional to a power function of the capillary radius – the bordered pits would be the limiting factor controlling longitudinal flow. However, it has been demonstrated that this concept is fallacious and that at least 40% of the total resistance to longitudinal flow in *Abies grandis* sapwood that had been specially dried to ensure that the torus remained in its natural position could be accounted for by the resistance of the cell cavity (Petty and Puritch, 1970).

Both longitudinal and tangential flowpaths in softwoods are predominantly by way of the bordered pits, as illustrated in *Fig. 52.25*, while the horizontally aligned ray cells constitute the principal pathway for radial flow, though it has been suggested that very fine capillaries within the cell wall may contribute slightly to radial flow. The rates of radial flow are found to vary very widely between species.

It is not surprising to find that the different pathways to flow in the three principal axes result in anisotropy in permeability. Permeability values quoted in the literature illustrate that for most timbers longitudinal permeability is about 10^4 times the transverse permeability; mathematical modelling of longitudinal and tangential flow supports a degree of anisotropy of this order. Since both longitudinal and tangential flow in softwoods are associated with bordered pits, a good correlation is to be expected between them; radial permeability is only poorly correlated with permeability in either of the other two directions, and is frequently found to be greater than tangential permeability.

Not only is permeability directionally dependent, but it is also found to vary with moisture content, between earlywood and latewood, between sapwood and heartwood (*Fig. 52.26*) and between species. In the sapwood of *green* timber the torus of the bordered pit is usually located in a central position and flow can be at a maximum (*Fig. 52.27a*). Since the earlywood cells possess larger and more frequent bordered pits, the flow through the earlywood is considerably greater than that through the latewood.

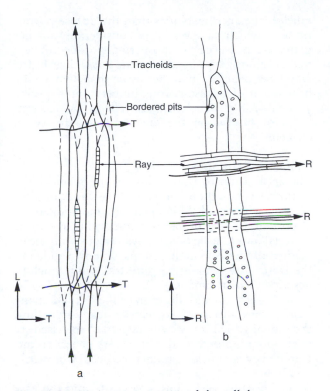

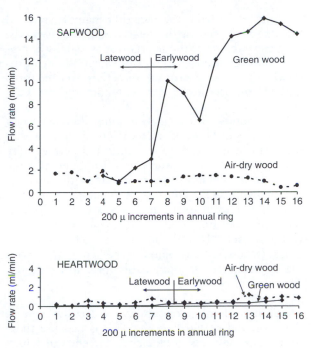

Fig. 52.25 *(a) A representation of the cellular structure of a softwood in a longitudinal–tangential plane illustrating the significance of the bordered pits in both longitudinal and tangential flow; (b) softwood timber in the longitudinal–radial plane, indicating the role of the ray cells in defining the principal pathway for radial flow (© Building Research Establishment).*

Fig. 52.26 *The variation in rate of longitudinal flow through samples of green and dry earlywood and latewood of Scots pine sapwood and heartwood (From Banks, 1968, © Building Research Establishment).*

However, on drying, the torus of the earlywood cells becomes aspirated (*Fig. 52.27b*), owing, it is thought, to tension stresses set up by the retreating water meniscus (Hart and Thomas, 1967). In this process the margo strands obviously undergo very considerable extension, and the torus is rigidly held in a displaced position by strong hydrogen bonding.

This displacement of the torus effectively seals the pit and markedly reduces the level of permeability of dry earlywood. In the latewood, the degree of pit aspiration on drying is very much lower than in the earlywood, a phenomenon that is related to the smaller diameter and thicker cell wall of the latewood pit. Thus, in dry timber, in marked contrast to green timber, the permeability of the latewood is at least as high as that of the earlywood and may even exceed it (*Fig. 52.26*). Rewetting of the timber causes only a partial reduction in the number of aspirated pits, and it appears that aspiration is mainly irreversible.

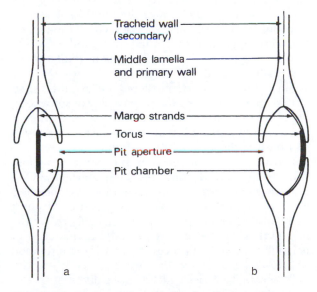

Fig. 52.27 *Cross-section of a bordered pit in the sapwood of a softwood timber: (a) in timber in the green condition with the torus in the 'normal' position; (b) in timber in the dried state with the torus in an aspirated position (© Building Research Establishment).*

Quite apart from the fact that many earlywood pits are aspirated in the heartwood of softwoods, the permeability of the heartwood is usually appreciably lower than that of the sapwood owing to the deposition of encrusting materials over the torus and margo strands and also within the ray cells (*Fig. 52.26*).

Permeability varies widely among different species of softwood. Thus, Comstock (1967) found that the ratio of longitudinal-to-tangential permeability varied between 500:1 and 80 000:1. Generally, the pines are much more permeable than the spruces, firs, or Douglas fir. This can be attributed primarily, though not exclusively, to the markedly different type of semi-bordered pit present between the vertical tracheids and the ray parenchyma in the pines (fenestrate or pinoid type) compared with the spruces, firs, or Douglas fir (piceoid type).

HARDWOODS

The longitudinal permeability is usually high in the sapwood of hardwoods. This is because these timbers possess vessel elements, the ends of which have been either completely or partially dissolved away. Radial flow is again by way of the rays, while tangential flow is more complicated, relying on the presence of pits interconnecting adjacent vessels, fibres and vertical parenchyma; however, intervascular pits in sycamore have been shown to provide considerable resistance to flow (Petty, 1981). Transverse flow rates are usually much lower than in the softwoods, but somewhat surprisingly a good correlation exists between tangential and radial permeability; this is owing, in part, to the very low permeability of the rays in hardwoods.

Since the effects of bordered pit aspiration, so dominant in controlling the permeability of softwoods, are absent in hardwoods, the influence of drying on the level of permeability in hardwoods is very much less than is the case with softwoods.

Permeability is highest in the outer sapwood, decreasing inwards and reducing markedly with the onset of heartwood formation as the cells become blocked either by the deposition of gums or resins or, as happens in certain timbers, by the ingrowth into the vessels of cell-wall material of neighbouring cells, a process known as the formation of *tyloses*.

Permeability varies widely among different species of hardwood. This variability is due in large measure to the wide variation in vessel diameter that occurs among hardwood species. Thus, the ring-porous hardwoods, which are characterised as having earlywood vessels of large diameter, generally have much higher permeabilities than the diffuse-porous timbers, which have vessels of considerably smaller diameter; however, in those ring–porous timbers that develop tyloses (e.g. the white oaks) their heartwood permeability may be lower than that of the heartwood of diffuse-porous timbers. Inter-specific variability in permeability also reflects the different types of pitting on the end walls of the vessel elements.

Timber and the laws of flow

The application of Darcy's law to the permeability of timber is based on a number of assumptions not all of which are upheld in practice. Among the more important are that timber is a homogeneous porous material and that flow is always viscous and linear; neither of these assumptions is strictly valid, but the Darcy law remains a useful tool with which to describe flow in timber.

By de-aeration and filtration of their liquid, many workers have been able to achieve steady-state flow, the rate of which is inversely related to the viscosity of the liquid, and to find that in very general terms Darcy's law is upheld in timber (see, e.g., Comstock, 1967).

Gas, because of its lower viscosity and the ease with which steady flow rates can be obtained, is a most attractive fluid for permeability studies. However, at low mean gas pressures, owing to the presence of slip flow, deviations from Darcy's law have been observed by a number of investigators. At higher mean gas pressures, however, an approximately linear relationship between conductivity and mean pressure is expected and this, too, has been observed experimentally. However, at even higher mean gas pressures, flow rate is sometimes less than proportional to the applied pressure differential owing, it is thought, to the onset of non-linear flow. Darcy's law may thus appear to be valid only in the middle range of mean gas pressures.

52.8.2 MOISTURE DIFFUSION

Flow of water below the fibre saturation point embraces both the diffusion of water vapour through the void structure comprising the cell cavities and pit membrane pores and the diffusion of bound water through the cell walls (*Fig. 52.24*). In passing, it should be noted that because of the capillary structure of timber, vapour pressures are set up and vapour can pass through the timber both above and below the fibre saturation point; however, the flow of vapour is usually regarded as being of secondary importance to that of both bound and free water.

Moisture diffusion is another manifestation of flow, conforming with the general relationship between flux and pressure. Thus, it is possible to express diffusion of moisture in timber at a fixed temperature in terms of Fick's first law, which states that the flux of moisture diffusion is directly proportional to the gradient of moisture concentration. As such, it is analogous to Darcy's law on the flow of fluids through porous media.

The total flux F of moisture diffusion through a plane surface under isothermal conditions is given by:

$$F = \frac{dm}{dt} = -D \cdot \frac{dc}{dx} \qquad (52.19)$$

where dm/dt is the flux (rate of mass transfer per unit area), dc/dx is the gradient of moisture concentration (mass per unit volume) in the x direction, and D is the moisture diffusion coefficient, which is expressed in m^2/s (Siau, 1984; Skaar, 1988).

Under steady-state conditions the diffusion co-efficent is given by:

$$D = \frac{100mL}{tA\rho\Delta M} \qquad (52.20)$$

where m is the mass of water transported in time t, A is the cross-sectional area, L is the length of the wood sample, and ΔM is the moisture content difference driving the diffusion.

The vapour component of the total flux is usually much less than that for the bound water. The rate of diffusion of water vapour through timber at moisture contents below the fibre saturation point has been shown to yield coefficients similar to those for the diffusion of carbon dioxide, provided that corrections are made for differences in molecular weight of the gases. This means that water vapour must follow the same pathway through timber as does carbon dioxide, and implies that diffusion of water vapour through the cell walls is negligible in comparison to that through the cell cavities and pits (Tarkow and Stamm, 1960).

Diffusion of bound water occurs when water molecules bound to their sorption sites by hydrogen bonding receive energy in excess of the bonding energy, thereby allowing them to move to new sites. At any one time the number of molecules with excess energy is proportional to the vapour pressure of the water in the timber at that moisture content and temperature. The rate of diffusion is proportional to the concentration gradient of the migrating molecules, which in turn is proportional to the vapour pressure gradient.

The most important factors affecting the diffusion coefficient of water in timber are temperature, moisture content and density of the timber. Thus, Stamm (1959) showed that the bound-water diffusion coefficient of the cell-wall substance increases with temperature approximately in proportion to the increase in the saturated vapour pressure of water, and increases exponentially with increasing moisture content at constant temperature. The diffusion coefficient has also been shown to decrease with increasing density and to differ according to the method of determination at high moisture contents. It is also dependent on grain direction; the ratio of longitudinal to transverse coefficients is approximately 2.5:1.

Various alternative ways of expressing the potential that drives moisture through wood have been proposed. These include percentage moisture content, relative vapour pressure, osmotic pressure, chemical potential, capillary pressure and spreading pressure, the last mentioned being a surface phenomenon derivable from the surface sorption theory of Dent, which in turn is a modification of the Brunauer–Emmet–Teller (BET) sorption theory (Skaar & Babiak, 1982). Although all this work has led to much debate on the correct flow potential, it has no effect on the calculation of flow; moisture flow is the same irrespective of the potential used, provided the mathematical conversions between transport coefficients, potentials and capacity factors are carried out correctly (Skaar, 1988).

As with the use of the Darcy equation for permeability, so with the application of Fick's law for diffusion there appears to be a number of cases in which the law is not upheld and the model fails to describe the experimental data. Claesson (1997) in describing some of the failures of Fickian models claims that this is due to a complicated, but transient sorption in the cell wall; it certainly cannot be explained by high resistance to flow of surface moisture.

The diffusion of moisture through wood has considerable practical significance since it relates to the drying of wood below the fibre saturation point, the day-to-day movement of wood through diurnal and seasonal changes in climate, and in the quantification of the rate of transfer of vapour through a thin sheet such as the sheathing used in timber-frame construction.

What is popularly called 'vapour permeability' of a thin sheet, but is really vapour diffusion, is determined using the *wet-cup* test, and its reciprocal is now quantified in terms of the *water vapour resistance factor* (μ). Values of μ for timber range from 30

431

to 50, while for wood-based panels, μ ranges from 15 for particleboard to 130 for oriented strand board (see also Dinwoodie, 2000).

52.8.3 THERMAL CONDUCTIVITY

The basic law for flow of thermal energy is ascribed to Fourier, and when described mathematically it is:

$$K_h = \frac{HL}{tA\Delta T} \qquad (52.21)$$

where K_h is the thermal conductivity for steady state flow of heat through a slab of material, H is the quantity of heat, t is time, A is the cross-sectional area, L is the length, and ΔT is the temperature differential. This equation is analogous to that of Darcy for fluid flow.

Compared with permeability, where the Darcy equation was shown to be only partially valid for timber, thermal flow is explained adequately by the Fourier equation, provided the boundary conditions are defined clearly.

Thermal conductivity will increase slightly with increased moisture content, especially when calculated on a volume-fraction-of-cell-wall basis. However, it appears that conductivity of the cell-wall substance is independent of moisture content (Siau, 1984); at 12% moisture content, the average thermal conductivity of softwood timber parallel to the grain is of the order of 0.38 W/mK. Conductivity is influenced considerably by the density of the timber, i.e. by the volume-fraction-of-cell-wall substance, and various empirical and linear relations between conductivity and density have been established. Conductivity will also vary with timber orientation owing to its anisotropic structure; the longitudinal thermal conductivity is about 2.5 times the transverse conductivity. Values of thermal conductivity are given in Siau (1984) and Dinwoodie (2000).

Compared with metals, the thermal conductivity of timber is extremely low, although it is generally up to eight times higher than that of insulating materials. The average transverse value for softwood timber (0.15 W/mK) is about one seventh that for brick, thereby explaining the lower heating requirements of timber houses compared with the traditional brick house.

Thermal insulation materials in the UK are usually rated by their U-value, where U is the conductance or the reciprocal of the thermal resistance. Thus:

$$\text{U-value} = K_h/L \qquad (52.22)$$

where K_h is the thermal conductivity and L is the thickness of the material.

References

Abe H, Ohtani J and Fukazawa K (1991). FE-SEM observations on the microfibrillar orientation in the secondary wall of tracheids. *IAWA Bulletin new series*, **12** (No. 4), 431–438.

Banks WB (1968). A technique for measuring the lateral permeability of wood. *Journal of the Institute of Wood Science*, **4** (No. 2), 35–41.

Barnett JR and Jeronimidis G (2003). Reaction Wood. In *Wood quality and its biological basis* (eds Barnett JR and Jeronimidis G), Blackwell Publishing, Oxford, pp. 30–52.

Bertaud F and Holmbom B (2004). Chemical composition of earlywood and latewood in Norway spruce heartwood, sapwood and transition zone wood. *Wood Science and Technology*, **38**, 245–256.

Brazier J (1965). An assessment of the incidence and significance of spiral grain in young conifer trees. *Forest Products Journal*, **15** (No. 8), 308–312.

Butterfield BG (2003). Wood anatomy in relation to wood quality. In *Wood quality and its biological basis* (eds Barnett JR and Jeronimidis G), Blackwell Publishing, Oxford, pp. 118–136.

Cave ID (1997). Theory of x-ray measurement of microfibril angle in wood. Part 2: The diffraction diagram, X-ray diffraction by materials with fibre type symmetry. *Wood Science and Technology*, **31**, 225–234.

Claesson J (1997). *Mathematical modelling of moisture transport*. Proceedings of the International conference on wood–water relations, Copenhagen (ed. P Hoffmeyer). Published by the management committee of EC COST Action E8, pp. 61–68.

Comstock GL (1967). Longitudinal permeability of wood to gases and nonswelling liquids. *Forest Products Journal*, **17** (No. 10), 41–46.

Desch HE and Dinwoodie JM (1996). *Timber – structure, properties, conversion and use*, 7th edition, Macmillan, Basingstoke, 306 pages.

Dinwoodie JM (2000). *Timber – Its nature and behaviour*, 2nd edition, E & F N Spon, London, 257 pages.

Gardner KH and Blackwell J (1974). The structure of native cellulose. *Biopolymers*, **13**, 1975–2001.

Hart CA and Thomas RJ (1967). Mechanism of bordered pit aspiration as caused by capillarity. *Forest Products Journal*, **17** (No. 11), 61–68.

Hillis WE (1987). *Heartwood and tree exudates*, Springer-Verlag, Berlin, 268 pages.

Kerr AJ and Goring DAI (1975). Ultrastructural arrangement of the wood cell wall. *Cellulose Chemistry and Technology*, **9** (No. 6), 563–573.

Meyer KH and Misch L (1937). Position des atomes dans le nouveau modele spatial de la cellulose. *Helvetica Chimica Acta*, **20**, 232–244.

Pereira H, Graca J and Rodrigues C (2003). Wood chemistry in relation to quality. In *Wood quality and its biological basis* (eds Barnett JR and Jeronimidis G), Blackwell Publishing, Oxford, pp. 53–86.

Petty JA (1981). Fluid flow through the vessels and intervascular pits of sycamore woods. *Holzforschung*, **35**, 213–216.

Petty JA and Puritch GS (1970). The effects of drying on the structure and permeability of the wood of *Abies grandis*. *Wood Science and Technology*, **4** (No. 2), 140–154.

Pratt GH (1974). *Timber drying manual*, HMSO, London, 152 pages.

Preston RD (1964). Structural and mechanical aspects of plant cell walls. In *The Formation of Wood in Forest Trees* (ed. Zimmermann HN), Academic Press, New York, 169–188.

Preston RD (1974). *The physical biology of plant cell walls*, Chapman & Hall, London, 491 pages.

Saka S and Thomas RJ (1982). A study of lignification in Loblolly pine tracheids by the SEM-EDXA technique. *Wood Science and Technology*, **12**, 51–62.

Saranpää P (2003). Wood density and growth. In *Wood quality and its biological basis* (eds Barnett JR and Jeronimidis G), Blackwell Publishing, Oxford, pp. 87–117.

Sell J and Zimmermann T (1993). Radial fibril agglomeration of the S_2 on transverse fracture surfaces of tracheids of tension-loaded spruce and white fir. *Holz als Roh- und Werkstoff*, **51**, 384.

Siau JF (1971). *Flow in Wood*, Syracuse University Press, Syracuse, USA, 99 pages.

Siau J (1984). *Transport processes in wood*, Springer-Verlag, Berlin, 245 pages.

Simson BW and Timell TE (1978). Polysaccharides in cambial tissues of *Populus tremuloides* and *Tilia americana*. V. Cellulose. *Cellulose Chemistry and Technology*, **12**, 51–62.

Skaar C (1988). *Wood-water relations*, Springer-Verlag, Berlin, 283 pages.

Skaar C and Babiak M (1982). A model for bound-water transport in wood. *Wood Science and Technology*, **16**, 123–138.

Stamm AJ (1959). Bound water diffusion into wood in the fiber direction. *Forest Products Journal*, **9** (No. 1), 27–32.

Stamm AJ (1964). *Wood and cellulose science*, Ronald, New York, 549 pages.

Tarkow H and Stamm AJ (1960). Diffusion through the air filled capillaries of softwoods: I, Carbon dioxide; II, Water vapour. *Forest Products Journal*, **10**, 247–250 and 323–324.

Ward JC and Zeikus JG (1980). Bacteriological, chemical and physical properties of wetwood in living trees. In *Natural variations of wood properties* (ed. Bauch J). *Mitt Bundesforschungsanst Forst-Holzwirtsch*, **131**, 133–166.

Chapter 53

Deformation in timber

Timber may undergo dimensional changes solely on account of variations in climatic factors; on the other hand, deformation may be due solely to the effects of applied stress. Frequently stress and climate interact to produce enhanced levels of deformation. This chapter commences by examining the dimensional changes that occur in timber following variations in its moisture content and/or temperature. The magnitude and consequently the significance of such changes in the dimensions of timber are much greater in the case of alterations in moisture content than in changes in temperature. Consequently, the greater emphasis in this first section is placed on the influence of changing moisture content. Later in the chapter the effect of stress on deformation is examined in detail.

53.1 Dimensional change due to moisture

In timber it is customary to distinguish between those changes that occur when green timber is dried to very low moisture contents (e.g. 12%), and those that arise in timber of low moisture content due to seasonal or daily changes in the relative humidity of the surrounding atmosphere. The former changes are called *shrinkage* while the latter are known as *movement*.

53.1.1 SHRINKAGE

As explained in Chapter 52, removal of water from below the fibre saturation point occurs within the amorphous region of the cell wall and manifests itself by reductions in strength and elastic modulus, as well as inducing dimensional shrinkage of the material.

Anisotropy in shrinkage

The reduction in dimensions of the timber, technically known as *shrinkage*, can be considerable but, owing to the complex structure of the material, the degree of shrinkage is different on the three principal axes; in other words, timber is anisotropic in its water relationships. The variation in degree of shrinkage that occurs between different timbers and, more important, the variation among the three different axes within any one timber are illustrated in *Table 53.1*. It should be noted that the values quoted in the table represent shrinkage on drying from the green state (i.e. > 27%) to 12% moisture content, a level which is of considerable practical significance. At 12% moisture content, timber is in equilibrium with an atmosphere having a relative humidity of 60% and a temperature of 20°C; these conditions would be found in buildings having regular but intermittent heating.

From *Table 53.1* it will be observed that shrinkage ranges from 0.1% to 10%, i.e. a 100-fold range. Longitudinal shrinkage, it will be noted, is always an order of magnitude less than transverse, while in the transverse plane radial shrinkage is usually some 60–70% of the corresponding tangential figure.

The anisotropy between longitudinal and transverse shrinkage, amounting to approximately 40:1, is due in part to the vertical arrangement of cells in timber and in part to the particular orientation of the microfibrils in the middle layer of the secondary cell wall (S_2). Thus, since the microfibrils of the S_2 layer of the cell wall are inclined at an angle of about 15° to the vertical, the removal of water from the matrix and the consequent movement closer together of the microfibrils will result in a horizontal component of the movement considerably greater than the corresponding longitudinal component (see *Table 53.1*).

Various theories have been developed over the years to account for shrinkage in terms of microfibrillar angle. The early theories were based on models that generally consider the cell wall to consist of an amorphous hygroscopic matrix in which

Table 53.1   Shrinkage (%) on drying from green to 12% moisture content

| Botanical name | Commercial name | Transverse | | Longitudinal |
		Tangential	Radial	
Chlorophora excelsa	Iroko	2.0	1.5	< 0.1
Tectona grandis	Teak	2.5	1.5	< 0.1
Pinus strobus	Yellow pine	3.5	1.5	< 0.1
Picea abies	Whitewood	4.0	2.0	< 0.1
Pinus sylvestris	Redwood	4.5	3.0	< 0.1
Tsuga heterophylla	Western hemlock	5.0	3.0	< 0.1
Quercus robur	European oak	7.5	4.0	< 0.1
Fagus sylvatica	European beech	9.5	4.5	< 0.1

are embedded parallel crystalline microfibrils that restrain swelling or shrinking of the matrix. One of the first models considered part of the wall as a flat sheet consisting only of an S_2 layer, in which microfibrillar angle had a constant value (Barber and Meylan, 1964). This model treated the cells as square in cross-section and there was no tendency for the cells to twist as they began to swell. An improved model (Barber, 1968) treated the cells as circular in cross-section and embraced a thin constraining sheath outside the main cylinder, which acted to reduce transverse swelling; experimental confirmation of this model was carried out by Meylan (1968). Later models have treated the cell wall as comprising two layers of equal thickness, having microfibrillar angles of equal and opposite sense, and these two-ply models have been developed extensively over the years to take into account the layered structured of the cell wall, differences in structure between radial and tangential walls, and variations in wall thickness. The principal researcher using this later type of model was Cave, whose models are based on an array of parallel cellulose microfibrils embedded in a hemicellulose matrix, with different arrays for each wall layer; these arrays of basic wall elements were bonded together by lignin microlayers. Earlier versions of the model included consideration of the variation in the elastic modulus of the matrix with changing moisture content (Cave, 1972, 1975). The model was later modified to take account of the amount of high-energy water absorbed rather than the total amount of water (Cave, 1978a, b). Comparison with previously obtained experimental data was excellent at low moisture contents, but poorer at moisture contents between 15 and 25%. All these theories are extensively presented and discussed by Skaar (1988)

and more recently by Pang (2002) who, in addition, has modified the Barber and Meylan model to accommodate changes in lumen shape during shrinkage and the presence of layers of variable shrinkage in the wood.

The influence of microfibrillar angle on the degree of longitudinal and transverse shrinkage described for normal wood is supported by evidence derived from experimental work on compression wood, one of the forms of reaction wood described in section 52.4. Compression wood is characterised by possessing a middle layer to the cell wall, the microfibrillar angle of which can be as high as 45°, though 20–30° is more usual. The longitudinal shrinkage is much higher and the transverse shrinkage correspondingly lower than in normal wood, and it has been demonstrated that the values for compression wood can be accommodated on the shrinkage–microfibrillar angle curve for normal wood.

Differences in the degree of transverse shrinkage between tangential and radial planes (*Table 53.1*), are usually explained in terms of: first, the restricting effect of the rays on the radial plane; second, the increased thickness of the middle lamella on the tangential plane compared with the radial; third, the difference in degree of lignification between the radial and tangential cell walls; fourth, the small difference in microfibrillar angle between the two walls; and fifth, the alternation of earlywood and latewood in the radial plane which, because of the greater shrinkage of latewood, induces the weaker earlywood to shrink more tangentially than it would if isolated. Considerable controversy reigns as to whether all five factors are actually involved and their relative significance. Comprehensive reviews of the evidence supporting these five possible explanations of differential shrinkage in the radial and tangential planes is to be found in Boyd (1974) and Skaar (1988).

Recently Kifetew (1997) demonstrated that the gross transverse shrinkage anisotropy of Scots pine timber, with a value approximately equal to 2, can be explained primarily in terms of the earlywood–latewood interaction theory by using a set of mathematical equations proposed by him; the gross radial and transverse shrinkage values were determined from the isolated early and latewood shrinkage values taken from the literature.

Volumetric shrinkage, s_v, is slightly less than the sums of the three directional components.

Practical significance

In order to avoid shrinkage of timber after fabrication, it is essential that it be dried down to a moisture content that is in equilibrium with the relative humidity of the atmosphere in which the article is to be located. A certain latitude can be tolerated in the case of timber frames and roof trusses, but in the production of furniture, window frames, flooring and sports goods it is essential that the timber is seasoned to the expected equilibrium conditions, namely 12% for regular intermittent heating and 10% in buildings with central heating, otherwise shrinkage in service will occur with loosening of joints, crazing of paint films, and buckling and delamination of laminates. An indication of the moisture content of timber used in different environments is presented in *Fig. 53.1*.

53.1.2 MOVEMENT

So far only those dimensional changes associated with the initial reduction in moisture content have been considered. However, dimensional changes, albeit smaller in extent, can also occur in seasoned or dried wood owing to changes in the relative humidity of the atmosphere. Such changes certainly occur on a seasonal basis and frequently also on a daily basis. Since these changes in humidity are usually fairly small, inducing only slight changes in the moisture content of the timber, and since a considerable delay occurs in the diffusion of water vapour into or out of the centre of a piece of timber, it follows that these dimensional changes in seasoned timber are small, considerably smaller than those caused by shrinkage.

To quantify such movements for different timbers, dimensional changes are recorded over an arbitrary range of relative humidities. In the UK, the standard procedure is to condition the timber in a chamber at 90% relative humidity and 25°C, then to measure its dimensions and to transfer it to a chamber at

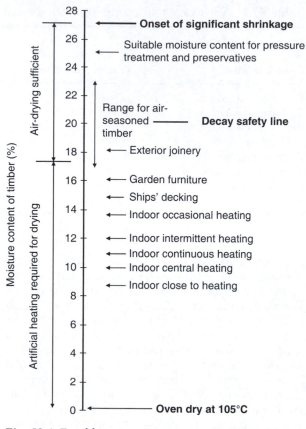

Fig. 53.1 *Equilibrium moisture content of timber in various environments. The data for different species vary, and the chart shows only average values (© Building Research Establishment).*

60% relative humidity and 25°C, allowing it to come to equilibrium before re-measuring it; the corresponding average change in moisture content is from 21 to 12%. Movement values in the tangential and radial planes for those timbers listed in *Table 53.1* are presented in *Table 53.2*. The timbers are recorded in the same order, thus illustrating that although a broad relationship holds between values of shrinkage and movement, individual timbers can behave differently over the reduced range of moisture content associated with movement. Since movement in the longitudinal plane is so very small, it is generally ignored. Anisotropy within the transverse plane can be accounted for by the same set of variables that influence shrinkage.

Where timber is subjected to wide fluctuations in relative humidity, care must be exercised to select a species that has low movement values.

Moisture in timber has a very pronounced effect not only on its strength (*Fig. 52.22*), but also

Table 53.2 Movement (%) on transferring timber from 90 to 60% relative humidity at 25°C

| Botanical name | Commercial name | Transverse | |
		Tangential	Radial
Chlorophora excelsa	Iroko	1.0	0.5
Tectona grandis	Teak	1.2	0.7
Pinus strobus	Yellow pine	1.8	0.9
Picea abies	Whitewood	1.5	0.7
Pinus sylvestris	Redwood	2.2	1.0
Tsuga heterophylla	Western hemlock	1.9	0.9
Quercus robur	European oak	2.5	1.5
Fagus sylvatica	European beech	3.2	1.7

on its elastic modulus, toughness and fracture morphology; elastic modulus is discussed later in this chapter, while the other two parameters are discussed in Chapter 54.

Aware of the technological significance of the instability of wood under changing moisture content, many attempts have been made over the years to find a solution to the problem; although it has not been possible to achieve complete dimensional stabilisation, it has been possible through the effects of either heat treatment or chemical modification to reduce the dimensional movement in wood by about 50%. Thermal modification is discussed in section 56.3 and chemical treatment in section 56.2.1.

53.2 Thermal movement

Timber, like other materials, undergoes dimensional changes commensurate with increasing temperature. This is attributed to the increasing distances between the molecules as they increase the magnitude of their oscillations with increasing temperature. Such movement is usually quantified for practical purposes as the *coefficient of linear thermal expansion*, and values for certain timbers (and other substances) are listed in *Table 53.3*. Although differences occur between species these appear to be smaller than those occurring for shrinkage and movement. The coefficient for transverse expansion is an order of magnitude greater than that in the longitudinal

Table 53.3 Coefficient of linear thermal expansion of various woods and other materials

| Material | Coefficient of thermal expansion ($\times 10^{-6}$/C degree) | | |
	Longitudinal		Transverse
Timber			
Picea abies (Whitewood)	5.41		34.1
Pinus strobus (Yellow pine)	4.00		72.7
Quercus robur (European oak)	4.92		54.4
Other			
GRP, 60/40, unidirectional	10.0		10.0
CFRP, 60/40 unidirectional	10.0		−1.00
Mild steel		12.6	
Duralumin (aluminium alloy)		22.5	
Nylon 66		125.0	
Polypropylene		110.0	

GRP, glass-reinforced plastic; CFRP, carbon fibre-reinforced plastic.

direction. This degree of anisotropy (10:1 on average) can be related to the ratio of length to breadth dimensions of the crystalline regions within the cell wall. Transverse thermal expansion appears to be correlated with specific gravity, but somewhat surprisingly this relationship is not sustained in the case of longitudinal thermal expansion, where the values for different timbers are roughly constant (Weatherwax and Stamm, 1946).

The expansion of timber with increasing temperature appears to be linear over a wide temperature range; the slight differences in expansion that occur between the radial and tangential planes are usually ignored and the coefficients are averaged to give a transverse value, as recorded in *Table 53.3*. For comparative purposes the coefficients of linear thermal expansion for glass- and carbon fibre-reinforced plastic, two metals and two plastics are also listed. Even the transverse expansion of timber is considerably less than those of the plastics.

The dimensional changes in timber caused by differences in temperature are small when compared with changes in dimensions resulting from the uptake or loss of moisture. Thus for timber with a moisture content greater than about 3%, the shrinkage due to moisture loss on heating will be greater than the thermal expansion, with the result that the net dimensional change on heating will be negative. For most practical purposes thermal expansion or contraction can be safely ignored over the range of temperatures in which timber is generally employed.

53.3 Deformation under load

This section is concerned with the type and magnitude of the deformation that results from the application of external load. As in the case of both concrete and polymers, the load–deformation relationship in timber is exceedingly complex, resulting from the facts that:

- timber does not behave in a truly elastic mode, rather its behaviour is time-dependent
- the magnitude of the strain is influenced by a wide range of factors; some of these are property dependent, such as density of the timber, angle of the grain relative to direction of load application, and angle of the microfibrils within the cell wall; others are environmentally dependent, such as temperature and relative humidity.

Under service conditions timber often has to withstand an imposed load for many years, perhaps even centuries; this is particularly relevant in construction applications. When loaded, timber will deform and a generalised interpretation of the variation in deformation with time together with the various components of this deformation is illustrated in *Fig. 53.2*. This is essentially similar behavior to that discussed in Chapter 2, section 2.7. The instantaneous (and reversible) deformation on load application at time t_0 is true elastic behaviour. On maintaining the load to time t_1 the deformation increases by creep, although the rate of increase is continually decreasing. On load removal at time t_1

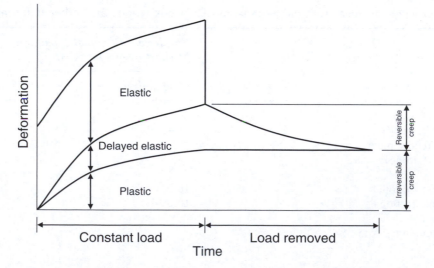

Fig. 53.2 *The various elastic and plastic components of the deformation of timber under constant load (© Building Research Establishment).*

an instantaneous recovery is approximately equal in magnitude to the initial elastic deformation. With time, the remaining deformation will decrease at an ever-decreasing rate until at time t_2 no further reduction occurs. The creep that has occurred during loading can be conveniently subdivided into a *reversible* component, which disappears with time and which can be regarded as *delayed elastic* behaviour, and an *irreversible* component, which results from *plastic or viscous* flow. Therefore, timber on loading possesses three forms of deformation behaviour – elastic, delayed elastic and viscous. Like so many other materials, timber can be treated neither as a truly elastic material where, by Hooke's law, stress is proportional to strain but independent of the rate of strain, nor as a truly viscous liquid where, according to Newton's law, stress is proportional to rate of strain, but independent of strain itself. Therefore timber is, like many other materials, viscoelastic (see Chapter 5, section 5.2).

Having defined timber as such (i.e. as a visco-elastic material), you will no doubt be surprised to find that half of what follows in this chapter is devoted to the elastic behaviour of timber. The following section will indicate how, at low levels of stressing and short periods of time, there is considerable justification for treating the material as such. Perhaps the greatest incentive for this viewpoint is the fact that classical elasticity theory is well established and, when applied to timber, has been shown to work very well. The question of time in any stress analysis can be accommodated by the use of safety factors in design calculations.

Consequently, we will deal first with elastic deformation as representing a very good approximation of what happens in practice, and then follow this with viscoelastic deformation, which embraces both delayed elastic and irreversible deformation. Although technically more applicable to timber, viscoelasticity is certainly less well understood and developed in its application than is the case with elasticity theory.

53.3.1 ELASTIC DEFORMATION

When a sample of timber is loaded in tension, compression or bending, the instantaneous deformations are approximately proportional to the values of the applied load. *Figure 53.3* illustrates that this approximation is certainly truer of the experimental evidence in longitudinal tensile loading than in the case of longitudinal compression. In both modes of loading, the approximation appears to become a reality at the lower levels of loading. The point of inflection of the stress–strain curve, i.e. the *limit of propor-*

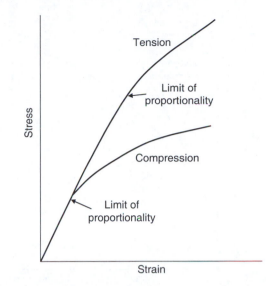

Fig. 53.3 *Stress–strain graphs for timber stressed in tension and compression parallel to the grain. The assumed limit of proportionality for each graph is indicated.*

tionality, is analogous to the yield or proof stress of metals (Chapter 2, section 2.3). Generally, the limit of proportionality in longitudinal tension is found to occur at about 60% of the ultimate load to failure, while in longitudinal compression the limit is considerably lower, varying from 30 to 50% of the failure value.

As with other materials, at stresses below the limit of proportionality we can describe the behaviour with the *modulus of elasticity* (E), which is the slope of the straight line (Chapter 2, section 2.2). It is worth recalling that E is also referred to as the elastic modulus, Young's modulus, or simply and frequently, though incorrectly, as stiffness.

The apparent linearity at the lower levels of loading is really an artefact introduced by the rate of testing. At fast rates of loading, a very good approximation to a straight line occurs, but as the rate of loading decreases, the stress–strain line assumes a curvilinear shape (*Fig. 53.4*). Such curves can be treated as linear by introducing a straight line approximation, which can take the form of either a tangent or a secant, giving the tangent or secant modulus (Chapter 2, section 2.2). This is common practice for timber and wood-fibre composites, with the tangent modulus determined at zero stress and the lower stress level for the secant modulus being zero (*Fig. 53.4*).

Thus, while in theory it should be possible to obtain a true elastic response, in practice this is

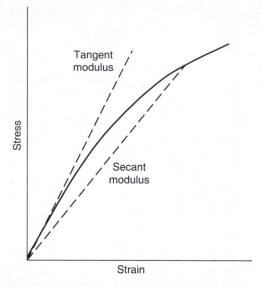

Fig. 53.4 The approximation of a curvilinear stress–strain curve for timber stressed at low loading rates by linear tangents or secants (© Building Research Establishment).

rarely the case, though the degree of divergence is frequently very low. It should be appreciated in passing that a curvilinear stress–strain curve must not be interpreted as an absence of true elastic behaviour. The material may still behave elastically, though not linearly elastically; the prime criterion for elastic behaviour is that the stress–strain curve is truly reversible, i.e. no permanent deformation occurs on release of the load.

The modulus of elasticity or elastic modulus in the longitudinal direction is one of the principal elastic constants of our material. You can find information on the different methods used to determine the value of the modulus under both static and dynamic loading in Dinwoodie (2000).

As we discussed in Chapter 2, section 2.2, the elastic behaviour of a material is characterised by two further constants:

- the *modulus of rigidity* or the *shear modulus*, G – the slope of the shear stress vs. shear strain graph
- the *Poisson's ratio*, ν – the ratio of the strain in the direction perpendicular to the applied stress to that in the direction of the stress.

Because of timber's anisotropic behaviour and its treatment as a rhombic system, six Poisson's ratios occur.

Orthotropic elasticity and timber

In applying the elements of orthotropic elasticity to timber, the assumption is made that the three principal elasticity directions coincide with the longitudinal, radial and tangential directions in the tree. The assumption implies that the tangential faces are straight and not curved, and that the radial faces are parallel and not diverging. However, by dealing with small pieces of timber removed at some distance from the centre of the tree, the approximation of rhombic symmetry for a system possessing circular symmetry becomes more and more acceptable.

The nine independent constants required to specify the elastic behaviour of timber are the three moduli of elasticity, one in each of the longitudinal (L), radial (R) and tangential (T) directions; the three moduli of rigidity, one in each of the principal planes longitudinal–tangential (LT), longitudinal–radial (LR) and tangential–radial (TR); and three Poisson's ratios, namely ν_{RT}, ν_{LR} and ν_{TL}. These constants, together with the three dependent Poisson's ratios ν_{RL}, ν_{TR} and ν_{LT}, are presented in *Table 53.4* for a selection of hardwoods and softwoods. The table illustrates the high degree of anisotropy present in timber. Comparison of E_L with either E_R or E_T, and G_{TR} with G_{LT} or G_{LR} will indicate a degree of anisotropy that can be as high as 60:1; usually the ratio of E_L to E_{HORIZ} is of the order of 40:1. Note should be taken that the values of ν_{TR} are frequently greater than 0.5.

Factors influencing the elastic modulus

The elastic modulus of timber is influenced by many factors, some of which are properties of the material while others are components of the environment.

GRAIN ANGLE

Figure 53.5, in addition to illustrating the marked influence of grain angle on elastic modulus, shows the degree of fit between experimentally derived values and the line obtained using transformation equations to calculate theoretical values (see equation 54.5 in section 54.5.1).

DENSITY

Elastic modulus is related to the density of timber, a relationship that is apparent in *Table 53.4* and that is confirmed by the plot of over two hundred species of timber contained in Bulletin 50 of the former Forest Products Research Laboratory (*Fig. 53.6*); the correlation coefficient was 0.88 for timber at 12% moisture content and 0.81 for green timber, and the relation is curvilinear. A high

Table 53.4 Values of the elastic constants of six hardwoods and four softwoods determined on small clear specimens (from Hearmon (1948), but with different notation for the Poisson's ratios)

Species	Density (kg/m³)	Moisture content (%)	E_L	E_R	E_T	ν_{TR}	ν_{LR}	ν_{RT}	ν_{LT}	ν_{RL}	ν_{TL}	G_{LT}	G_{LR}	G_{TR}
Hardwoods														
Balsa	200	9	6300	300	106	0.66	0.018	0.24	0.009	0.23	0.49	203	312	33
Khaya	440	11	10 200	1130	510	0.60	0.033	0.26	0.032	0.30	0.64	600	900	210
Walnut	590	11	11 200	1190	630	0.72	0.052	0.37	0.036	0.49	0.63	700	960	230
Birch	620	9	16 300	1110	620	0.78	0.034	0.38	0.018	0.49	0.43	910	1180	190
Ash	670	9	15 800	1510	800	0.71	0.051	0.36	0.030	0.46	0.51	890	1340	270
Beech	750	11	13 700	2240	1140	0.75	0.073	0.36	0.044	0.45	0.51	1060	1610	460
Softwoods														
Norway spruce	390	12	10 700	710	430	0.51	0.030	0.31	0.025	0.38	0.51	620	500	23
Sitka spruce	390	12	11 600	900	500	0.43	0.029	0.25	0.020	0.37	0.47	720	750	39
Scots pine	550	10	16 300	1100	570	0.68	0.038	0.31	0.015	0.42	0.51	680	1160	66
Douglas fir*	590	9	16 400	1300	900	0.63	0.028	0.40	0.024	0.43	0.37	910	1180	79

*Listed in original as Oregon pine. E, modulus of elasticity in a direction indicated by the subscript (MPa); G, modulus of rigidity in a plane indicated by the subscript (MPa); ν_{ij}, Poisson's ratio for an extensional stress in j direction = compressive strain in j direction/extensional strain in j direction.

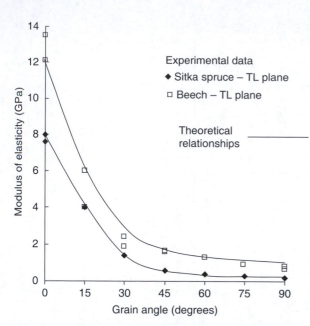

Fig. 53.5 *Effect of grain angle on the modulus of elasticity (© Building Research Establishment).*

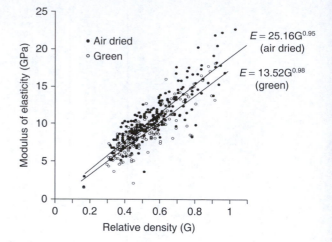

Fig. 53.6 *Effect of specific gravity on the longitudinal modulus of elasticity for over 200 species of timber tested in the green and dry states (© Building Research Establishment).*

correlation is to be expected, since density is a function of the ratio of cell wall thickness to cell diameter; consequently, increasing density will result in increasing elastic modulus of the cell.

Owing to the variability in structure that exists between different timbers, the relationship between density and elastic modulus will be higher where only a single species is under investigation. Because of the reduced range in density, a linear regression is usually fitted.

Similar relationships with density have been recorded for the modulus of rigidity in certain species; in others, however, for example spruce, both the longitudinal–tangential and longitudinal–radial shear moduli have been found to be independent of density. Most investigators agree, however, that the Poisson's ratios are independent of density.

KNOTS

Since timber is anisotropic in behaviour, and since knots are characterised by the occurrence of distorted grain, it is not surprising to find that the presence of knots in timber results in a reduction in the elastic modulus. The relationship is difficult to quantify since the effect of the knots will depend not only on their number and size, but also on their distribution both along the length of the sample and across the faces. Dead knots, especially where the knot has fallen out, will result in larger reduc-

tions in elastic modulus than will green knots (see section 52.1).

ULTRASTRUCTURE

Two components of the fine or chemical structure have a profound influence on both the elastic and rigidity moduli. The first relates to the existence of a matrix material with particular emphasis on the presence of lignin. In those plants devoid of lignin, e.g. the grasses, or in wood fibres that have been delignified, the stiffness of the cells is low and it would appear that lignin, apart from its hydrophilic protective role for the cellulosic crystallites, is responsible to a considerable extent for the high elastic modulus found in timber.

The fact that lignin is significant determining the elastic modulus does not imply that the cellulose fraction plays no part; on the contrary, it has been shown that the angle at which the microfibrils are lying in the middle layer of the secondary cell wall, S_2, also plays a significant role in controlling the elastic modulus (*Fig. 53.7*).

A considerable number of mathematical models have been devised over the years to relate elastic modulus to microfibrillar angle. The early models were two-dimensional in approach, treating the cell wall as a planar slab of material, but later the models became much more sophisticated, taking into account the existence of cell-wall layers other than the S_2, the variation in microfibrillar angle between the radial and tangential walls and consequently the probability that they undergo different strains, and,

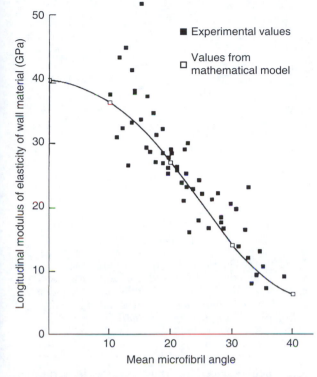

Fig. 53.7 *Effect of the mean microfibrillar angle of the cell wall on the longitudinal modulus of elasticity of the wall material in Pinus radiata. Calculated values from a mathematical model are also included (from Cave, 1968) by permission of Springer-Verlag.*

cellulose sheath and the surrounding matrix. The model predicts orthotropic elastic properties that are in good agreement with recorded values (Astley *et al.*, 1998). The predicted variation of axial elastic modulus with the S_2 microfibrillar angle is consistent with observed behaviour and aligned with results from other cell wall models.

The elastic modulus of a material is very dependent on the type and degree of chemical bonding within its structure; the abundance of covalent bonding in the longitudinal plane and hydrogen bonding in one of the transverse planes contributes considerably to the moderately high levels of elastic modulus characteristic of timber.

MOISTURE CONTENT

The influence of moisture content on elastic modulus is similar to, though not quite as sensitive as, that for strength (see *Fig. 52.22*). Early experiments in which elastic modulus was measured on a specimen of Sitka spruce as it took up moisture from the dry state clearly indicated a linear loss in modulus as the moisture content increased to about 27%, corresponding to the fibre saturation point as discussed in Chapter 52; further increase in moisture content has no influence on the elastic modulus. These results for the variation in longitudinal moduli with moisture content under static loading have been confirmed using simple dynamic methods. Measurement of the frequency of vibration – from which the moduli can be deduced – was carried out at regular intervals as samples of Sitka spruce were dried from 70% to zero moisture content (*Fig. 53.8*).

Confirmation of the reduction in modulus of elasticity with increasing moisture content is forthcoming first, from *Fig. 53.6*, in which the regression lines of elasticity against density for over 200 species of timber at 12% moisture content and in the green state are presented, and second, from the review of more recent results by Gerhards (1982). Moisture increase has a far greater effect on the modulus perpendicular to the grain than on the modulus along the grain (Gerhards, 1982).

TEMPERATURE

In timber, like most other materials, increasing temperature results in greater oscillatory movement of the molecules and an enlargement of the crystal lattice. These in turn affect the mechanical properties, and the elastic modulus and strength of the material decrease. Although the relationship between modulus and temperature has been shown experimentally to be curvilinear, the degree of

lastly, the possibility of complete shear restraint within the cell wall. These three-dimensional models are frequently analysed using finite element techniques. The early models were reviewed by Dinwoodie (1975).

Later modelling of the behaviour of timber in terms of its structure has been reviewed by Astley *et al.* (1998); it makes reference to the work of Cave (1975), who used the concept of an elastic fibre composite consisting of an inert fibre phase embedded in a water-reactive matrix. The constitutive equation is related to the overall elastic modulus of the composite, the volume fraction, and the elastic modulus and sorption characteristics of the matrix. Unlike previous models, the equation can be applied not only to elasticity, but also to shrinkage and even moisture-induced creep.

Recently, Harrington *et al.* (1998) have developed a model of the wood cell wall based on the homogenisation first, of an amorphous lignin–polyose matrix, and then of a representative volume element comprising a cellulose microfibril, its polyose–

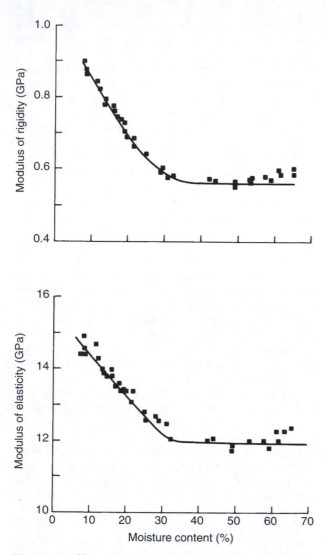

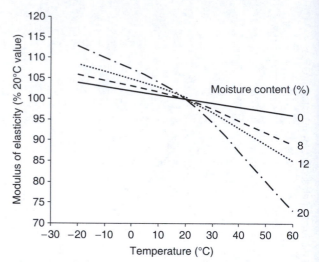

Fig. 53.9 *The interaction of temperature and moisture content on the modulus of elasticity. Results are averaged for six species of timber and the modulus at 20°C and 0% moisture content is taken as unity (© Building Research Establishment).*

Fig. 53.8 *Effect of moisture content on the longitudinal modulus of elasticity and the modulus of rigidity in the L–R plane in Sitka spruce. Both moduli were determined dynamically (© Building Research Establishment).*

curvature is usually slight at low moisture contents and the relation is frequently treated as linear thus:

$$E_T = E_t[1 - a(T - t)] \qquad (53.1)$$

where E is the elastic modulus, T is a higher temperature, t is a lower temperature, and a is the temperature coefficient. The value of a for the longitudinal modulus has been shown to lie between 0.001 and 0.007 for low moisture contents. The effect of a temperature increase is greater on the

perpendicular modulus than on the longitudinal modulus (Gerhards, 1982).

At higher moisture contents the relationship between elastic modulus and temperature is markedly curvilinear and the interaction of moisture content and temperature in influencing modulus is clearly shown in *Fig. 53.9*. At zero moisture content the reduction in modulus between −20 and +60°C is only 6%; at 20% moisture content the corresponding reduction is 40%. This increase in the effect of temperature with increasing moisture content has been confirmed by Gerhards (1982).

Long-term exposure to elevated temperature results in a marked reduction in elastic modulus as well as strength and toughness, the effect usually being greater in hardwoods than in softwoods. Even exposure to cyclic changes in temperature over long periods will result in a loss of elastic modulus (Moore, 1984).

53.3.2 VISCOELASTIC DEFORMATION

In the introduction to section 53.3, timber was described as being neither truly elastic in its behaviour nor truly viscous, but rather a combination of both states; such behaviour is usually described as viscoelastic and, in addition to timber, materials such as concrete, bitumen and the thermoplastics are also viscoelastic in their response to stress.

Viscoelasticity implies that the behaviour of the material is time dependent; at any instant in time under load its performance will be a function of its past history. Now if the time factor under load is reduced to zero, a state that we can picture in concept, but never attain in practice, the material will behave truly elastically, and we have seen in section 53.3.1 how timber can be treated as an elastic material and how the principles of orthotropic elasticity can be applied. However, where stresses are applied for a period of time, viscoelastic behaviour will be experienced and, while it is possible to apply elasticity theory with a factor covering the increase in deformation with time, this procedure is at best only a first approximation.

In a material such as timber, time-dependent behaviour manifests itself in a number of ways, of which the more common are *creep, relaxation, damping capacity* and the dependence of strength on *duration of load*. When the load on a sample of timber is held constant for a period of time, the increase in deformation over the initial instantaneous elastic deformation is called creep, and *Fig. 53.2* illustrates not only the increase in creep with time, but also the subdivision of creep into a reversible and an irreversible component – of which more will be said in a later section.

Most timber structures carry a considerable dead load and the component members of these will undergo creep; the dip towards the centre of the ridge of the roof of very old buildings bears testament to the fact that timber does creep. However, compared with thermoplastics and bitumen, the amount of creep in timber is appreciably lower.

Viscoelastic behaviour is also apparent in the form of relaxation, where the load necessary to maintain a constant deformation decreases with time; in timber utilisation this has limited practical significance and the area has attracted very little research. Damping capacity is a measure of the fractional energy converted to heat compared with that stored per cycle under the influence of mechanical vibrations; this ratio is time dependent. A further manifestation of viscoelastic behaviour is the apparent loss in strength of timber with increasing duration of load; this feature is discussed in detail in section 54.6.12 and illustrated in *Fig. 54.7*.

Creep

CREEP PARAMETERS

It is possible to quantify creep by a number of time-dependent parameters, of which the two most common are *creep compliance* (known also as

specific creep) and relative creep (known also as the *creep coefficient*); both parameters are a function of temperature. Creep compliance (c_c) is the ratio of increasing strain with time to the applied constant stress, i.e.:

$$c_c(t, T) = \frac{\text{strain variation}}{\text{applied constant stress}} \quad (53.2)$$

while relative creep (c_r) is defined as either the deflection or, more usually, the increase in deflection, expressed in terms of the initial elastic deflection, i.e.:

$$c_r(t, T) = \frac{\varepsilon_t}{\varepsilon_0} \quad \text{or} \quad \frac{\varepsilon_t - \varepsilon_0}{\varepsilon_0} \quad (53.3)$$

where ε_t is the deflection at time t, and ε_0 is the initial deflection. Relative creep has also been defined as the change in compliance during the test expressed in terms of the original compliance.

CREEP RELATIONSHIPS

In both timber and timber products such as plywood or particleboard (chipboard), the rate of deflection or creep slows down progressively with time (*Fig. 53.10*); creep is frequently plotted against log(time) and the graph assumes an exponential shape. The results of creep tests can also be plotted as relative creep against log(time) or as creep

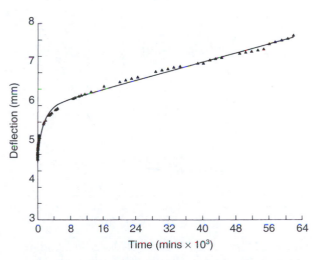

Fig. 53.10 The increase in deformation with time of urea–formaldehyde-bonded particleboard (chipboard): the regression line has been fitted to the experimental values using equation 53.7 (© Building Research Establishment).

445

compliance against stress as a percentage of the ultimate short-time stress.

In section 53.4.1 it was shown that the degree of elasticity varies considerably between the horizontal and longitudinal planes. Creep, as one particular manifestation of viscoelastic behaviour, is also directionally dependent. In tensile stressing of longitudinal sections produced with the grain running at different angles, it was found that relative creep was greater in the direction perpendicular to the grain than it was parallel to the grain.

Timber and wood-based panels, therefore, are viscoelastic materials, the time dependent properties of which are directionally dependent. The next important question is whether they are linearly viscoelastic in behaviour. For viscoelastic behaviour to be defined as linear, the instantaneous, recoverable and non-recoverable components of the deformation must vary directly with the applied stress. An alternative definition is that the creep compliance or relative creep must be independent of stress and not a function of it.

Timber and wood-based panels exhibit linear viscoelastic behaviour at lower levels of stressing, but at higher stress levels this behaviour reverts to being non-linear. Examples of this transition in behaviour are illustrated in *Figs 53.11* and *53.12*, where for both redwood timber and UF bonded particleboard, respectively, the change from linear to non-linear behaviour occurs between the 45 and 60% stress levels.

The linear limit for the relationship between creep and applied stress varies with mode of testing, with species of timber or type of panel, and with both temperature and moisture content. In tension parallel to the grain at constant temperature and moisture content, timber has been found to behave as a linear viscoelastic material up to about 75% of the ultimate tensile strength, though some workers have found considerable variability and have indicated a range of from 36 to 84%. In compression parallel to the grain, the onset of non-linearity appears to occur at about 70%, though the level of actual stress will be much lower than in the case of tensile strength, since the ultimate compression strength is only one third that of the tensile strength. In bending, non-linearity seems to develop very much earlier – at about 50–60% (*Figs 53.11* and *53.12*); the actual stress levels will be very similar to those for compression.

In both compression and bending, the divergence from linearity is usually greater than in the case of tensile stressing; much of the increased deformation occurs in the non-recoverable component of creep and is associated with progressive structural changes, including the development of incipient failure (see later).

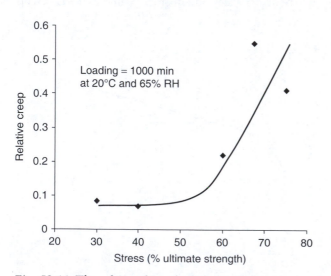

Fig. 53.11 *The relationship of relative creep to stress level at fixed time periods, illustrating the transition from linear to non-linear viscoelastic behaviour in redwood timber (Pinus sylvestris) (© Building Research Establishment).*

RH, relative humidity.

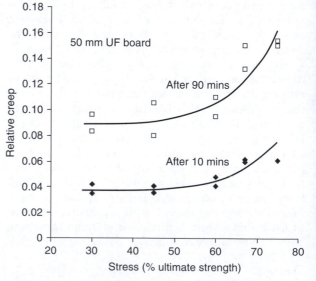

Fig. 53.12 *The relationship of relative creep to stress level at fixed time periods illustrating the transition from linear to non-linear viscoelastic behaviour in 50 mm urea–formaldehyde-bonded particleboard (© Building Research Establishment).*

Increases not only in stress level, but also in temperature to a limited extent, and in moisture content to a considerable degree, result in an earlier onset of non-linearity and a more marked departure from linearity. For most practical purposes, however, working stresses are only a small percentage of the ultimate, rarely approaching even 50%, and it can be safely assumed that timber, like concrete, will behave as a linear viscoelastic material under normal service conditions.

PRINCIPLE OF SUPERPOSITION

Since timber behaves as a linear viscoelastic material under conditions of normal temperature and humidity and at low to moderate levels of stressing, it is possible to apply the *Boltzmann's principle of superposition* to predict the response of timber to complex or extended loading sequences. This principle states that the creep occurring under a sequence of stress increments is taken as the superposed sum of the responses to the individual increments. This can be expressed mathematically in a number of forms, one of which for linear materials is:

$$\varepsilon_c(t) = \sum_1^n \Delta\sigma_i c_{ci} \qquad (53.4)$$

where n is the number of load increments, $\Delta\sigma_i$ is the stress increment, c_{ci} is the creep compliance for the individual stress increments applied for differing times, $t - \tau_1, t - \tau_2, \ldots, t - \tau_n$ and $\varepsilon_c(t)$ is the total creep at time t; or in integrated form:

$$\varepsilon_c(t) = \int_{\tau_1}^t c_c(t - \tau)\frac{d\sigma}{d\tau}(\tau)d\tau \qquad (53.5)$$

In experiments on timber it has been found that in the comparison of deflections in beams loaded either continuously or repeatedly for two or seven days in every fourteen, for six months at four levels of stress, the applicability of the Boltzmann's principle of superposition was confirmed for stress levels up to 50% (Nakai and Grossman, 1983). The superposition principle has been found to be applicable even at high stresses in both shear and tension in dry samples. However, at high moisture contents, the limits of linear behaviour in shear and tension appear to be considerably lower, thereby confirming views expressed earlier on the non-linear behaviour of timber subjected to high levels of stressing and/or high moisture content.

MATHEMATICAL MODELLING OF STEADY-STATE CREEP

The relationship between creep and time has been expressed mathematically using a wide range of equations. It should be appreciated that such expressions are purely empirical, none of them possessing any sound theoretical basis. Their relative merits depend on how easily their constants can be determined and how well they fit the experimental results.

One of the most successful mathematical descriptions for creep in timber under constant relative humidity and temperature appears to be the power law, of general form:

$$\varepsilon(t) = e_0 + at^m \qquad (53.6)$$

where $\varepsilon(t)$ is the time-dependent strain, e_0 is the initial deformation, a and m are material-specific parameters to be determined experimentally ($m = 0.33$ for timber), and t is the elapsed time. The prime advantage of using a power function to describe creep is its representation as a straight line on a log/log plot, thereby making onward prediction on a time basis that much easier than using other models. The shape of the viscoelastic creep curve was predicted by Van der Put (1989) based on deformation kinetic theory.

Alternatively, creep behaviour in timber, like that of many other high polymers, can be interpreted with the aid of mechanical (rheological) models comprising different combinations of springs and dashpots (piston in a cylinder containing a viscous fluid). The springs act as a mechanical analogue of the elastic component of deformation, while the dashpot simulates the viscous or flow component. When more than a single member of each type is used, these components can be combined in a wide variety of ways, though only one or two will be able to describe adequately the creep and relaxation behaviour of the material.

The simplest linear model that successfully describes the time-dependent behaviour of timber under constant humidity and temperature for short periods of time is the four-element model illustrated in *Fig. 53.13*; the central part of the model will be recognised as a Kelvin element. To this unit has been added in series a second spring and dashpot. The strain at any time t under a constant load is given by the equation:

$$Y = \frac{\sigma}{E_1} + \frac{\sigma}{E_2}\left[1 - \exp\left(\frac{-tE_2}{\eta_2}\right)\right] + \frac{\sigma t}{\eta_3} \qquad (53.7)$$

where Y is the strain at time t, E_1 is the elasticity of spring 1, E_2 is the elasticity of spring 2, σ is the stress applied, η_2 is the viscosity of dashpot 2 and η_3 is the viscosity of dashpot 3.

The first term on the right-hand side of equation (53.7) represents the instantaneous deformation,

447

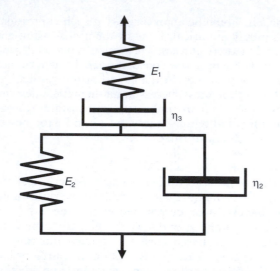

Fig. 53.13 Mechanical analogue of the components of creep: the springs simulate elastic deformation and the dashpots viscous flow. The model corresponds to equation 53.7 (© Building Research Establishment).

while the second term describes the delayed elasticity and the third term the plastic flow component. Thus, the first term describes the elastic behaviour while the combination of the second and third terms accounts for the viscoelastic or creep behaviour. The response of this particular model will be linear and it will obey the Boltzmann superposition principle.

The degree of fit between the behaviour described by the model and experimentally derived values can be exceedingly good; an example is illustrated in *Fig. 53.10*, where the degree of correlation between the fitted line and experimental results for creep in the bending of urea–formaldehyde particleboard (chipboard) beams was as high as 0.941.

A much more demanding test of any model is the prediction of long-term performance from short-term data. For timber and the various board materials, it has been found necessary to make the viscous term non-linear in these models where accurate predictions of creep (±10%) are required for long periods of time (> 10 years) from short-term data (6–9 months) (Dinwoodie *et al.*, 1990a). The deformation of this non-linear mathematical model is given by the equation:

$$Y = \beta_1 + \beta_2[1 - \exp(-\beta_3 t)] + \beta_4 t^{\beta_5} \qquad (53.8)$$

where $\beta_1 = \sigma/E_1$, $\beta_2 = \sigma/E_2$, $\beta_3 = E_2/\eta_2$, $\beta_4 = \sigma/\eta_3$, β_5 is the viscous modification factor, with a value $0 < b < 1$. An example of the successful application

of this model to predict the deflection of a sample of cement-bonded particleboard after ten years from the first nine months of data is given in Dinwoodie *et al.* (1990a).

Various non-linear viscoelastic models have been developed and tested over the years, ranging from the fairly simple early approach by Ylinen (1965) – in which a spring and a dashpot in this rheological model are replaced by non-linear elements – to the much more sophisticated model by Tong and Ödeen (1989a), in which the linear viscoelastic equation is modified by the introduction of a non-linear function either in the form of a simple power function, or by using the sum of an exponential series corresponding to ten Kelvin elements in series with a single spring.

All this modelling provides further confirmation of the non-linear viscoelastic behaviour of timber and wood-based panels when subjected to high levels of stressing, or to lower levels at high moisture contents. The development of models for unsteady-state moisture content are described later.

REVERSIBLE AND IRREVERSIBLE COMPONENTS OF CREEP
In timber and many of the high polymers, creep under load can be subdivided into reversible and irreversible components; passing reference to this was made in section 53.4 and the generalised relationship with time was depicted in *Fig. 53.2*. The relative proportions of these two components of total creep appear to be related to stress level and to prevailing conditions of temperature and moisture content.

The influence of level of stress is clearly illustrated in *Fig. 53.14*, where the total compliance at 70% and 80% of the ultimate stress for hoop pine in compression is sub-divided into the separate components. At 70%, the irreversible creep compliance accounts for about 45% of the total creep compliance, while at 80% of the ultimate, the irreversible creep compliance has increased appreciably, to 70% of the total creep compliance at the longer periods of time, though not at the shorter durations. Increased moisture content and increased temperature will also result in an enlargement of the irreversible component of total creep.

Reversible creep is frequently referred to in the literature as delayed elastic or primary creep and in the early days was ascribed to either polymeric uncoiling or the existence of a creeping matrix. Owing to the close longitudinal association of the molecules of the various components in the amorphous regions, it appears unlikely that uncoiling of the polymers under stress can account for much of the reversible component of creep.

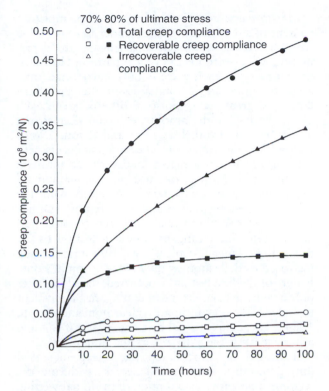

70% 80% of ultimate stress
○ ● Total creep compliance
□ ■ Recoverable creep compliance
△ ▲ Irrecoverable creep compliance

Fig. 53.14 The relative proportions of the recoverable and irrecoverable creep compliance in samples of hoop pine (Araucaria cunninghamii) *stressed in bending (from Kingston and Budgen (1972) by permission of Springer-Verlag).*

The second explanation of reversible creep utilises the concept of time-dependent two-stage molecular motions of the cellulose, hemicellulose and lignin constituents. The pattern of molecular motion for each component is dependent on that of the other constituents, and it has been shown that the difference in directional movement of the lignin and non-lignin molecules results in considerable molecular interference, such that stresses set up in loading can be transferred from one component (a creeping matrix) to another component (an attached, but non-creeping structure). It is postulated that the lignin network could act as an energy sink, maintaining and controlling the energy set up by stressing (Chow, 1973).

Irreversible creep, also referred to as viscous, plastic or secondary creep, has been related to either time-dependent changes in the active number of hydrogen bonds, or to the loosening and subsequent remaking of hydrogen bonds as moisture diffuses through timber with the passage of time (Gibson, 1965). Such diffusion can result directly from

stressing; thus early work indicated that when timber was stressed in tension it gained in moisture content, and conversely when stressed in compression its moisture content was lowered. It is argued, though certainly not proven, that the movement of moisture by diffusion occurs in a series of steps from one adsorption site to the next, necessitating the rupture and subsequent re-formation of hydrogen bonds. The process is viewed as resulting in loss of elastic modulus and/or strength, possibly through slippage at the molecular level. Recently, however, it has been demonstrated that moisture movement, while affecting creep, can account for only part of the total creep obtained, and this explanation of creep at the molecular level warrants more investigation; certainly not all the observed phenomena support the hypothesis that creep is due to the breaking and remaking of hydrogen bonds under a stress bias. At moderate to high levels of stressing, particularly in bending and compression parallel to the grain, the amount of irreversible creep is closely associated with the development of kinks in the cell wall (Hoffmeyer and Davidson, 1989).

Boyd (1982) in a lengthy paper demonstrates how creep under both constant and variable relative humidity can be explained quite simply in terms of stress-induced physical interactions between the crystalline and non-crystalline components of the cell wall. Justification of his viewpoint relies heavily on the concept that the basic structural units develop a lenticular trellis format containing a water sensitive gel, which changes shape during moisture changes and load applications, thereby explaining creep strains. Attempts have been made to describe creep in terms of the fine structure of timber, and it has been demonstrated that creep in the short term is highly correlated with the angle of the microfibrils in the S_2 layer of the cell wall, and inversely with the degree of crystallinity. However, such correlations do not necessarily prove any causal relationship, and it is possible to explain these correlations in terms of the presence or absence of moisture that would be closely associated with these particular variables.

ENVIRONMENTAL EFFECTS ON RATE OF CREEP

Temperature: Steady-state. In common with many other materials, especially the high polymers, the effect of increasing temperature on timber under stress is to increase both the rate and the total amount of creep. *Figure 53.15* illustrates a two-and-a-half-fold increase in the amount of creep as the temperature is raised from 20 to 54°C; there is a marked increase in the irreversible component of

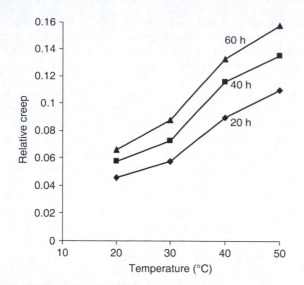

Fig. 53.15 *The effect of temperature on relative creep of samples of hoop pine (Araucaria cunninghamii) loaded in compression for 20, 40 and 60 hours (from Kingston and Budgen (1972) by permission of Springer-Verlag).*

creep at the higher temperatures. Various workers have examined the applicability to wood of the time/temperature superposition principle; results have been inconclusive and variable and it would appear that caution must be exercised in the use of this principle.

Temperature: Unsteady-state. Cycling between low and high temperatures will induce in stressed timber and panel products a higher creep response than would occur if the temperature was held constant at the higher level; however, this effect is most likely due to changing moisture content as temperature is changed, rather than to the effect of temperature itself.

Moisture content: Steady-state. The rate and amount of creep in timber of high moisture content are appreciably higher than those of dry timber; an increase in moisture content from 6 to 12% increases the deflection in timber at any given stress level by about 20%. It is interesting to note the occurrence of a similar increase in creep in nylon when wet. Hunt (1999) recorded that the effect of humidity can be treated by the use of an empirical humidity-shift factor curve to be used with an empirical master-creep curve. At high moisture contents this logarithmic shift factor increases rapidly; creep at 22% moisture content compared with that at 10% was found to be $10^{1.5}$ or 32 times as fast.

Moisture content: Unsteady-state. If the moisture content of small timber beams under load is cycled from dry to wet and back to dry again, the deformation will also follow a cyclical pattern. However, the recovery in each cycle is only partial and over a number of cycles the total amount of creep is very large; the greater the moisture differential in each cycle, the higher the amount of creep (Armstrong and Kingston, 1960; Hearmon and Paton, 1964). *Figure 53.16* illustrates the deflection that occurs with time in matched test pieces loaded to 3/8 ultimate short-term load where one test piece is maintained in an atmosphere of 93% relative humidity, while the other is cycled between 0 and 93% relative humidity. After 14 complete cycles the latter test piece had broken after increasing its initial deflection by 25 times; the former test piece was still intact, having increased its deflection by only twice its initial value. Failure of the first test piece occurred, therefore, after only a short period of time and at a stress equivalent to only 3/8 of its ultimate (Hearmon and Paton, 1964). The figure also shows the effect of loading a matched test piece to 1/8 its maximum load.

It should be appreciated that creep increased during the drying cycle and decreased during the wetting cycle with the exception of the initial wetting when creep increased. It was not possible to explain the negative deflection observed during absorption, though the energy for the change is probably provided by the heat of absorption. The net change at the end of a complete cycle of wetting and drying was considered to be a redistribution of hydrogen bonds, which manifests itself as an increase in deformation of the stressed sample (Gibson, 1965).

More early work showed that the rate of moisture change affects the rate of creep, but not the amount of creep; this appears to be proportional to the total change in moisture content (Armstrong and Kingston, 1962).

This complex behaviour of creep in timber when loaded under either cyclic or variable changes in relative humidity has been confirmed by a large number of research workers (e.g. Schniewind, 1968; Ranta-Maunus, 1973; Hunt, 1982; Mohager, 1987). However, in board materials the cyclic effect appears to be somewhat reduced (Dinwoodie *et al.*, 1990b). Later test work, covering longitudinal compression and tension stressing as well as bending, indicates that the relationship between creep behaviour and moisture change was more complex than first thought. The results of this work (e.g. Ranta-Maunus 1973, 1975; Hunt, 1982) indicate that there are three separate components to this form of creep; these are illustrated in *Fig. 53.17* and are:

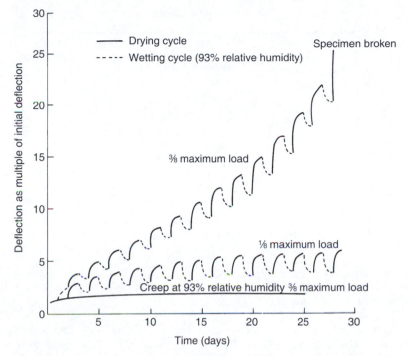

Fig. 53.16 *The effect of cyclic variations in moisture content on relative creep of samples of beech loaded to 1/8 and 3/8 of ultimate load (© Building Research Establishment).*

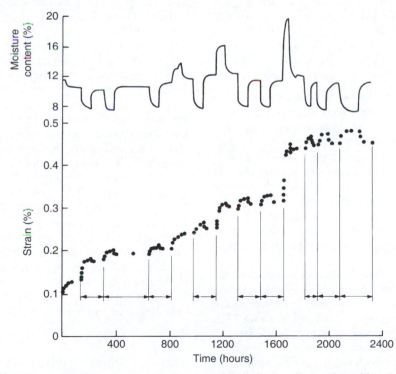

Fig. 53.17 *Creep deflection under changing moisture content levels for small samples of beech stressed in tension. Note the different responses to increasing moisture content (from Hunt (1982) by permission of the Institute of Wood Science).*

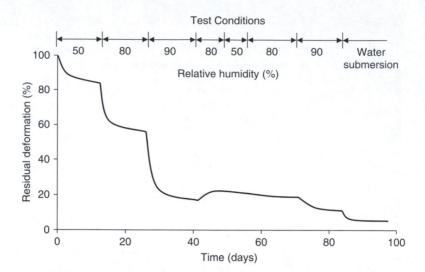

Fig. 53.18 *The amount of recovery of both visco-elastic and mechano-sorptive deflection that occurred when dried bent beams were subjected to a sequence of humidity changes (adapted from Arima and Grossman (1978) by permission of the Institute of Wood Science).*

1. An increase in creep, which follows a decrease in moisture content of the sample, as described previously.
2. An increase in creep, which follows any increase in moisture content above the previous highest level reached after loading; three examples can be seen in the middle of the graph in *Fig. 53.16*.
3. A decrease in creep, which follows an increase in moisture content below the previous highest level reached after loading, as described previously.

It follows from 1 and 2 above that there will always be an initial increase in creep during the initial change in moisture content, irrespective of whether adsorption or desorption is taking place.

Further experimentation has established that the amount of creep that occurs depends on the size and rate of the moisture change, and is little affected by its duration or by whether such change is brought about in one or more steps (Armstrong and Kingston, 1962). These findings were to cast doubt on the previous interpretation that such behaviour constituted true creep.

Reinforcement of that doubt occurred with the publication of work by Arima and Grossman (1978). Small beams of *Pinus radiata*, 680 × 15 × 15 mm, were cut from green timber and stressed to about 25% of their short-term bending strength. While held in their deformed condition, the beams were allowed to dry for 15 days, after which the retaining clamps were removed and the initial recovery measured. The unstressed beams were then subjected to changes in relative humidity followed by water immersion, and

Fig. 53.18 shows the changes in recovery with changing humidity. Most important is the fact that total recovery was almost achieved; what was thought to have been viscous deformation in the post-drying and clamping stage turned out to be reversible.

These two phenomena – that creep is related to the magnitude of the moisture change and not to time, and that deformation is reversible under moisture change – cast very serious doubts on whether the deformation under changes in moisture content constituted true creep.

It was considered necessary to separate these two very different types of deformation, namely the true viscoelastic creep that occurs under constant moisture content and is directly a function of time, from that deformation that is directly related to the interaction of change in moisture content and mechanical stressing, which is a function of the history of moisture change and is relatively uninfluenced by time. A term of convenience was derived to describe this latter type of deformation, namely *mechano-sorptive behaviour* (Grossman, 1976).

Changing levels of moisture content, however, will result in changes in the dimensions of timber and an allowance for this must be taken into account in the calculation of mechano-sorptive deformation. Thus:

$$\varepsilon_m = \varepsilon_{vc} + \varepsilon_{ms} + \varepsilon_s \qquad (53.9)$$

where ε_m = total measured strain, ε_{vc} is the normal time (constant moisture content) viscoelastic creep, ε_{ms} is the mechano-sorptive strain under changing moisture content and ε_s is the swelling or shrinkage strain of a matched, zero-loaded control test piece. The swelling/shrinkage strain (ε_s) (referred to as *pseudo-creep* by some workers), which is manifest during moisture cycling by an increase in deflection during desorption and a decrease during adsorption, has been ascribed to differences in the normal longitudinal swelling and shrinkage of wood – a tensile strain resulting in a smaller shrinkage coefficient and a compression strain resulting in a larger one (Hunt and Shelton, 1988).

Mechano-sorptive behaviour is linear only at low levels of stress; in the case of compression and bending the upper limit of linear behaviour is of the order of 10%, while in tension it is slightly higher (Hunt, 1980). The most accepted theoretical explanation of mechano-sorptive behaviour is that changes in moisture content result in the rupture of bonds with consequent slippage between the two new surfaces under stress, and the re-formation of bonds at different locations (Grossman, 1976).

Susceptibility to mechano-sorptive behaviour is positively correlated with elastic compliance, microfibrillar angle of the cell wall and dimensional change rates (Hunt, 1994). Thus, both juvenile wood and compression wood have been shown to creep much more (up to five times more) than adult wood.

MODELLING OF DEFORMATION UNDER VARIABLE
MOISTURE CONTENT

Mathematical modelling of viscoelastic creep has been described in section 53.4.2, to which reference should be made. The requirement for a model of mechano-sorptive creep was originally set out by Schniewind (1966) and later developed by Grossman (1976) and Mårtensson (1992). More recently, the list of requirements that must be satisfied for such a model has been reviewed and considerably extended by Hunt (1994); aspects of sorption were included on the list.

Many attempts have been made over the last three decades to develop a model for mechano-sorptive behaviour. A few of these models have been explanatory or descriptive in nature, seeking relationships at the molecular, ultrastructural or microscopic levels. Most of them, however, have been either purely mathematical, with the aim of producing a generalised constitutive equation, or partly mathematical, where the derived equation is linked to some physical phenomenon, or change in structure of the timber under stress. More information on the parameters that have been included in these mathematical models as well as the types of models that have been developed over the years are to be found in Ranta-Maunus (1973, 1975), Tong and Ödeen (1989b), Hunt (1994), Morlier and Palka (1994) and Hanhijärvi (1995).

It may be recalled that it is fifty years since the concept of mechano-sorptive deformation was first established (Armstrong and Kingston, 1960). Since then, the concept of mechano-sorptive deformation – which is primarily a function of the amount of moisture change and associated dimensional changes – has been clearly separated from that of viscoelastic creep, which is present under both constant and changing levels of moisture and which is primarily a function of time. However, in complex cases involving environmental changes, there appears in certain circumstances to be a small interaction between viscoelastic creep and mechano-sorptive creep, albeit that this is probably not a common mechanism (Hanhijärvi and Hunt, 1998; Matar and Hunt, 2003).

Hunt (1999) employed a new way to characterise wood creep by plotting data in the form of strain rate against strain; solution of this differential equation led to the more normal relation of strain against time. It was then found that normalisation of both the ordinate and abscissa resulted in a single master creep curve for both juvenile and mature wood from a single sample and, more important, approximately also for all test humidities.

The traditional approach to the quantitative modelling of creep under variable humidity conditions incorporated four separate components, three of which relate to mechano-sorptive behaviour (as described earlier and illustrated in *Fig. 53.16*), while the fourth component deals with viscoelastic creep. However, owing to the difficulty of analysing the three separate components of mechano-sorptive creep, some workers have resorted to a different approach by taking only two components, one reversible and one irreversible, which are then added to the viscoelastic component to quantify total creep (Matar and Hunt, 2003). It should be appreciated that such a model is based almost entirely on the convenience of handling of the experimental data.

Using their new model, Matar and Hunt regard the gradually decreasing increments of additional mechano-sorptive creep during humidity changes as the irreversible component (while the sample is still under load) and this component can then be quantified by an exponential-decay type of equation. Such analysis implies that there is an eventual creep limit, a conclusion that is not shared by many workers (e.g. Navi *et al.*, 2002).

The remaining creep/recovery component in Matar's and Hunt's new model is treated as reversible, and the term *pseudocreep* is applied to it since this component is not a creep component, but rather a differential expansion or shrinkage component, the magnitude of which depends on the level of strain (Hunt, 2004). The real merit of such a model is that it allows the estimation of a practical upper-bound value for creep in Service Classes 1 and 2 of Eurocode 5. Equally important, the amount of mechano-sorptive limiting creep was nearly the same at 63 and 90% relative humidity. This implies that the mechano-sorptive creep occurs mainly at low and intermediate rather than high moisture contents; its importance, therefore, is that it correlates with swelling and shrinkage, which are also greater at lower moisture contents. This is a very significant conclusion.

Following on from the theoretical explanation of mechano-sorptive behaviour mentioned earlier, the idea has been developed of using the concept of physical ageing developed for polymers to explain mechano-sorptive effects in timber (Hunt and Gril, 1996). Thus it is argued that the effects of humidity changes require the additional measurement of the increased activity associated with the molecular destabilisation and its relaxation-time constraint associated with the physical ageing phenomenon (thermodynamic equilibrium); the application of this concept suggests that the speed of moisture change might be important in mechano-sorptive deformation. This concept has been taken up and developed by Ishimaru and others in Japan (Ishimaru, 2003).

In a further quest to refine the modelling of creep, especially in its relationship to the anatomy of the wood, Hunt and Gril (2006) have recently focused attention on the anomalies that take place during swelling and shrinkage and of the dangers in assuming that these processes can be characterised by a single value. From modelling, it was demonstrated that the restraint produced by the S_1 layer of the cell wall is most important and that the level of this restraint relaxes with time and with changes in moisture content. Such findings complement the work of Esteban *et al.* (2004), who found that repeated moisture cycling resulted in a considerable reduction in moisture sorption and rate of shrinkage in various species, a phenomenon that they termed *swelling fatigue*, and is a further example of the physical ageing process.

Although much work has been undertaken to try to understand and model mechano-sorptive creep, especially in trying to increase the accuracy of the model, much still remains to be done. While the primary variables have been identified, great difficulty arises in quantifying how these vary with the number and magnitude of the changes that take place in moisture content and their associated effects on the physical and chemical structure of wood.

References

Arima T and Grossman PUA (1978). Recovery of wood after mechano-sorptive deformation. *Journal of the Institute of Wood Science*, 8 (No. 2), 47–52.

Armstrong LD and Kingston RST (1960). Effect of moisture changes on creep in wood. *Nature*, 185 (No. 4716), 862–863.

Armstrong LD and Kingston RST (1962). The effect of moisture content changes on the deformation of wood under stress. *Australian Journal of Applied Science*, 13 (No. 4), 257–276.

Astley RJ, Stol KA and Harrington JJ (1998). Modelling the elastic constraints of softwood. Part II: The cellular microstructure. *Holz als Roh-und Werkstoff*, 56, 43–50.

Barber NF (1968). A theoretical model of shrinking wood. *Holzforschung*, 22, 97–103.

Barber NF and Meylan BA (1964). The anisotropic shrinkage of wood. A theoretical model. *Holzforschung*, 18, 146–156.

Boyd JD (1974). Anisotropic shrinkage of wood. Identification of the dominant determinants. *Mokuzai Gakkaishi*, 20, 473–482.

Boyd JD (1982). An anatomical explanation for viscoelastic and mechano-sorptive creep in wood, and effects of loading rate on strength. In *New Perspectives in Wood Anatomy* (ed. Bass P), Martinus Nijhoff/Dr W Junk, The Hague, 171–222.

Cave ID (1968). The anisotropic elasticity of the plant cell-wall. *Wood Science and Technology*, 2, 268–278.

Cave ID (1972). A theory of shrinkage of wood. *Wood Science and Technology*, 6, 284–292.

Cave ID (1975). Wood substance as a water-reactive fibre-reinforced composite. *Journal of Microscopy*, 104 (No. 1), 47–52.

Cave ID (1978a). Modelling moisture-related mechanical properties of wood. I. Properties of the wood constitutents. *Wood Science and Technology*, 12, 75–86.

Cave ID (1978b). Modelling moisture-related mechanical properties of wood. II. Computation of properties of a model of wood and comparison with experimental data. *Wood Science and Technology*, 12, 127–139.

Chow S (1973). Molecular rheology of coniferous wood tissues. *Transactions of the Society of Rheology*, 17, 109–128.

Dinwoodie JM (1975). Timber – a review of the structure–mechanical property relationship. *Journal of Microscopy*, 104 (No. 1), 3–32.

Dinwoodie JM (2000). *Timber: its nature and behaviour*, 2nd edition, E & F N Spon, London, 257 pages.

Dinwoodie JM, Higgins JA, Paxton BH and Robson DJ (1990a). Creep in chipboard. Part 7: Testing the efficacy of models on 7–10 years' data and evaluating optimum period of prediction. *Wood Science and Technology*, **24**, 181–189.

Dinwoodie JM, Higgins JA, Paxton BH and Robson DJ (1990b). Creep research on particle board – 15 years' work at UK BRE. *Holz als Roh und Werkstoff*, **48**, 5–10.

Esteban LG, Gril J, Casasus AG and de Palacios PP (2004). Reduction of wood hygroscopicity and associated dimensional response by repeated humidity cycles. *Annals of Forest Science*, **62** (No. 3), 275–284.

Gerhards CC (1982). Effect of moisture content and temperature on the mechanical properties of wood. An analysis of immediate effects. *Wood and Fiber*, **14** (No. 1), 4–36.

Gibson E (1965). Creep of wood: role of water and effect of a changing moisture content. *Nature (London)*, **206**, 213–215.

Grossman PUA (1976). Requirements for a model that exhibits mechano-sorptive behaviour. *Wood Science and Technology*, **10**, 163–168.

Hanhijärvi A (1995). *Modelling of creep deformation mechanism in wood*. Technical Research Centre of Finland Publication 231, 144 pages.

Hanhijärvi A and Hunt D (1998). Experimental indication of interaction between viscoelastic and mechanosorptive creep. *Wood Science and Technology*, **32**, 57–70.

Harrington JJ, Booker R and Astley RJ (1998). Modelling the elastic properties of softwood. Part 1: The cell-wall lamellae. *Holz als Roh-und Werkstoff*, **56**, 37–41.

Hearmon RFS (1948). Elasticity of wood and plywood. *Special Report 7*, Forest Products Research, HMSO, London.

Hearmon RFS and Paton JM (1964). Moisture content changes and creep in wood. *Forest Products Journal*, **14**, 357–359.

Hoffmeyer P and Davidson R (1989). Mechano-sorptive creep mechanism of wood in compression and bending. *Wood Science and Technology*, **23**, 215–227.

Hunt DG (1980). *A preliminary study of tensile creep of beech with concurrent moisture changes*. Proceedings of the Third International Conference on Mechanical Behaviour of Materials, Cambridge, 1979 (eds Miller KJ and Smith RF), Vol. 3, pp. 299–308.

Hunt DG (1982). Limited mechano-sorptive creep of beech wood. *Journal of the Institute of Wood Science*, **9** (No. 3), 136–138.

Hunt DG (1994). Present knowledge of mechano-sorptive creep of wood. In *Creep in Timber Structures* (ed. Morlier P), RILEM Report 8, E & F N Spon, London, 73–97.

Hunt DG (1999). A unified approach to creep in wood. *Proceedings of the Royal Society of London, A*, **455**, 4077–4095.

Hunt DG (2004). *Some questions regarding time, moisture and temperature as applied to dimensional changes and creep of wood*. Proceedings of the 3rd International Conference of ESWM, Portugal, pp. 239–245.

Hunt DG and Gril J (1996). Evidence of a physical ageing phenomenon in wood. *Journal of Material Science Letters*, **15**, 80–92.

Hunt DG and Gril J (2006). *Anomalies of shrinkage and swelling as related to cell-wall anatomy*. To be published in the Proceedings of the of 5th International Conference of ESWM, Florence.

Hunt DG and Shelton CF (1988). Longitudinal moisture-shrinkage coefficents of softwood at the mechano-sorptive creep limit. *Wood Science and Technology*, **22**, 199–210.

Ishimaru Y (2003). *Mechanical properties of wood in unstable states caused by changes in temperature and/or swelling*. Proceedings of the 2nd International Conference of ESWM, Stockholm, Sweden, pp. 37–46.

Kifetew G (1997). *Application of the early–latewood interaction theory of the shrinkage anisotropy of Scots pine*. Proceedings of the International Conference on Wood–Water Relations, Copenhagen (ed. Hoffmeyer P), The Management Committee of EC COST Action E8, pp. 165–171.

Kingston RST and Budgen B (1972). Some aspects of the rheological behaviour of wood, Part IV: Non-linear behaviour at high stresses in bending and compression. *Wood Science and Technology*, **6**, 230–238.

Mårtensson A (1992). *Mechanical behaviour of wood exposed to humidity variations*. Report TVBK-1066, Lund Institute of Technology, Sweden, 189 pages.

Matar A and Hunt DG (2003). *Progress in the modelling of reversible and irreversible mechano-sorptive creep of softwood*. Proceedings of the 2nd International Conference of EWSM, Stockholm, pp. 25–28.

Meylan BA (1968). Cause of high longitudinal shrinkage in wood. *Forest Products Journal*, **18** (No. 4), 75–78.

Mohager S (1987). *Studier av krypning hos trä (Studies of creep in wood)*. Report 1987-1 of the Department of Building Materials, The Royal Institute of Technology, Stockholm, 140 pages.

Moore GL (1984). The effect of long term temperature cycling on the strength of wood. *Journal of the Institute of Wood Science*, **9** (No. 6), 264–267.

Morlier P and Palka LC (1994). Basic Knowledge. In *Creep in Timber Structures* (ed. P. Morlier), RILEM Report 8, E and FN Spon, London, 9–42.

Nakai T and Grossman PUA (1983). Deflection of wood under intermittent loading. Part 1: Fortnightly cycles. *Wood Science and Technology*, **17**, 55–67.

Navi P, Pittet V and Plummer CJG (2002). Transient moisture effects on wood creep. *Wood Science and Technology*, **36** (No. 6), 447–462.

Pang S (2002). Predicting anisotropic shrinkage of softwood. Part 1: Theories. *Wood Science and Technology*, **36**, 75–91.

Ranta-Maunus A (1973). *A theory for the creep of wood with application to birch and spruce plywood*, Technical Research Centre of Finland, Building

Technology and Community Development, Publication 4, 35 pages.

Ranta-Maunus A (1975). The viscoelasticity of wood at varying moisture content. *Wood Science and Technology*, 9, 189–205.

Schniewind AP (1966). Uber den Einfluss von Feuchtigkeitsanderungen auf das Kriechen von Buchenholz quer zur Faser unter Berucksichtigung von Temperatur und temperaturanderungen. *Holz als Roh-und Werkstoff*, 24, 87–98.

Schniewind AP (1968). Recent progress in the study of rheology of wood. *Wood Science and Technology*, 2, 189–205.

Skaar C (1988). *Wood-water relations*, Springer-Verlag, Berlin, 283 pages.

Tong L and Ödeen K (1989a). *A non-linear viscoelastic equation for the deformation of wood and wood structure*. Report No. 4 of the Department of Building Materials, Royal Institute of Technology, Stockholm, 6 pages.

Tong L and Ödeen K (1989b). *Rheological behaviour of wood structures*. Report No. 3 of the Department of Building Materials, Royal Institute of Technology, Stockholm, 145 pages.

Van der Put T (1989). *Deformation and damage processes in wood*, PhD thesis, Delft University Press, Delft, The Netherlands, 70 pages.

Weatherwax RG and Stamm AJ (1946). *The coefficents of thermal expansion of wood and wood products*. US Forest Products Laboratory, Madison, Report No. 1487.

Ylinen A (1965). Prediction of the time-dependent elastic and strength properties of wood by the aid of a general non-linear viscoelastic rheological model. *Holz als Roh und Werkstoff*, 5, 193–196.

Chapter 54

Strength and failure in timber

While it is easy to appreciate the concept of deformation primarily because it is something that can be observed, it is much more difficult to define in simple terms what is meant by the *strength* of a material. Perhaps one of the simpler definitions of strength is that it is a measure of the resistance to *failure*, providing of course that we are clear in our minds what is meant by failure.

Let us start with a recap of some of the ideas discussed in Chapters 2 and 4 and consider how we can define failure. In those modes of stressing where a distinct break occurs with the formation of two fracture surfaces, failure is synonymous with rupture of the specimen. However, in certain modes of stressing, fracture does not occur and failure must be defined in some arbitrary way such as the maximum stress that the sample will endure or, in exceptional circumstances such as compression strength perpendicular to the grain, the stress at the limit of proportionality.

Having defined our end point, it is now easier to appreciate our definition of strength as the natural resistance of a material to failure. But how do we quantify this resistance? This may be done by calculating either the stress necessary to produce failure or the amount of energy consumed in producing failure. Under certain modes of testing it is more convenient to use the former method of quantification, as the latter tends to be more limited in application.

54.1 Determination of strength

54.1.1 TEST PIECE SIZE AND SELECTION

Although in theory it should be possible to determine the strength properties of timber independently of size, in practice this is found not to be the case. A definite though small size effect has been established, and in order to compare the strength of a timber sample with recorded data it is necessary to adopt the sizes set out in the standards.

The size of the test piece to be used will be determined by the type of information required. Where tests are required to characterise new timbers or for the strict academic comparison of wood from different trees or different species, small, knot-free, straight-grained, perfect test pieces representing the maximum quality of wood that can be obtained and known as 'small clears', should be used. However, where tests are required to determine the strength performance of structural-grade timber with all its imperfections, such as knots and distorted grain, generally large test pieces of structural timber are required.

Current testing standards still permit the derivation of structural stresses from small clear test pieces, but it is better to use structural-size test pieces as the effects of strength limiting characteristics are easier to quantify with such pieces.

Use of small clear test pieces

This size of test piece was originally used for the derivation of working stresses for timber, but in the 1970s this was superseded, though not exclusively, by the use of actual structural-size timber. However, the small clear test piece still remains valid for the derivation of structural stresses as well as characterising new timbers and the strict academic comparison of the strength of wood from different trees or different species.

Two standard procedures for testing small clear test pieces have been used internationally; the original was introduced in the USA as early as 1891 using a test sample 2 × 2 inches in cross section; the second, European in origin, employs a test specimen 20 × 20 mm in cross section. Before 1949 the former size was adopted in the UK, but after this date this larger sample was superseded by the smaller, thereby making it possible to obtain an adequate number of test specimens from smaller trees. Because of the difference in size, the results obtained from the two standard procedures are not strictly comparable and a series of conversion values has been determined (Lavers, 1969).

The early work in the UK on species characterisation employed a sampling procedure in which the test samples were removed from the log in accordance with a cruciform pattern. However, this was subsequently abandoned and a method devised applicable to the centre plank removed from a log; 20×20 mm sticks, from which the individual test pieces are obtained, are selected at random in such a manner that the probability of obtaining a stick at any distance from the centre of a cross section of a log is proportional to the area of timber at that distance. Test samples are cut from each stick eliminating knots, defects and sloping grain; this technique is described fully by Lavers (1969).

Methods of test for 'small clears' are given in BS 373[1] (retained as a national standard) for the UK and ASTM D143-52 for the USA. The use of small clears for the derivation of structural stresses is given in BS EN 384. Dinwoodie (2000) describes most of the tests in BS 373.

Use of structural-size test pieces

Use of these larger test pieces reproduces actual service loading conditions, and they are of particular value because they allow directly for defects such as knots, splits and distorted grain rather than indirectly (by applying a series of reduction factors), as is necessary with small clear test pieces.

54.1.2 STANDARDISED TEST PROCEDURES

Europe at the present time is in the final stages of a transition period in which national test procedures have been largely replaced by European procedures. It is interesting to note that many of the European Standards (ENs) on testing have now been adopted as International Standards (ISOs). Methods for the testing of structural timber in Europe are described in BS EN 408, and for small clears in BS EN 384. Some of these tests are described by Dinwoodie (2000).

54.2 Strength values

54.2.1 STRENGTH DERIVED USING SMALL CLEAR TEST PIECES

For a range of strength properties determined on small clear test pieces, with the exception of tensile strength parallel to the grain, the mean values and standard deviations (see below) are presented in *Table 54.1* for a selection of timbers covering the range in densities to be found in hardwoods and softwoods. Many of the timbers whose elastic constants were presented in *Table 53.4* are included. All values relate to a moisture content that is in equilibrium with a relative humidity of 65% at 20°C; this is of the order of 12% and the timber is referred to as 'dry'. Modulus of elasticity values have also been included in *Table 54.1*, which has been compiled from data presented in Bulletin 50 of the former Forest Products Research Laboratory (Lavers, 1969), which lists data for both the dry and green states for 200 species of timber.

In *Table 54.2* tensile strength parallel to the grain is listed for certain timbers and it is in this mode that clear timber is at its strongest. Comparison of these values with those for compression strength parallel to the grain in *Table 54.1* indicates that, unlike many other materials, the compression strength is only about one-third that of tensile strength along the grain.

54.2.2 STRENGTH DERIVED USING STRUCTURAL-SIZE TEST PIECES

After the end of October 2008 all structural test work and design within Europe should be carried out according to the new European standards by testing to EN 408 and deriving the *characteristic values* (see Chapter 2, section 2.10.2) according to EN 384. Within the European system, the characteristic value for the strength properties is taken as the 5-percentile value; for modulus of elasticity there are two characteristic values, one the 5-percentile, the other the mean or 50-percentile value. The design of structures must be carried out according to Eurocode 5 (BS EN 1995-1-1).

The sample 5-percentile value is determined for each sample by the equation:

$$f_{05} = f_r \qquad (54.1)$$

where f_{05} is the sample 5-percentile value, and f_r is obtained by ranking all the test values for a sample in ascending order. The 5-percentile value is the test value for which 5% of the values are lower. If this is not an actual test value (i.e. the number of test values is not divisible by 20) then interpolation between the two adjacent values is permitted.

The characteristic value of strength (f_k) is calculated from the equation:

$$f_k = \bar{f}_{05} k_s k_v \qquad (54.2)$$

where $\bar{f}_{05}$ is the mean (in MPa) of the adjusted 5-percentile values (f_{05}) for each sample (see above) weighted according to the number of pieces in each sample, k_s is a factor to adjust for the number of

---

[1] As in the other parts of this book, the details of the standards referred to in the text are listed in 'Further reading' at the end of the part.

Table 54.1 Estimated average and (standard deviation) of various mechanical properties of selected timbers at 12% moisture content from small clear test pieces (from Lavers, 1969)

Material	Dry density (kg/m³)	Static bending* Modulus of rupture (bending strength) (MPa)	Modulus of elasticity (MPa)	Energy to max. load (mm N/mm³)	Energy to facture (mm N/mm³)	Impact: drop of hammer (m)	Compression parallel to grain (MPa)	Hardness on side grain (N)	Shear parallel to grain (MPa)	Cleavage Radial plane (N/mm width)	Tangential plane (N/mm width)
Hardwoods											
Balsa	176	23 (7.3)	3200 (1060)	0.018 (0.007)	0.035 (0.017)		15.5 (4.43)		2.4 (0.62)		
Obeche	368	54 (6.5)	5500 (620)	0.058 (0.010)	0.095 (0.015)	0.48 (0.072)	28.2 (3.00)	1910 (268)	7.7 (0.67)	9.3 (1.82)	8.4 (1.58)
Mahogany (*Khaya ivorensis*)	497	78 (15.0)	9000 (1520)	0.070 (0.026)	0.128 (0.044)	0.58 (0.149)	46.4 (8.45)	3690 (816)	11.8 (2.56)	10.0 (2.08)	14.0 (2.90)
Sycamore	561	99 (11.0)	9400 (1160)	0.121 (0.028)	0.163 (0.049)	0.84 (0.136)	48.2 (4.83)	4850 (639)	17.1 (2.32)	16.8 (2.95)	27.3 (3.91)
Ash	689	116 (16.6)	11 900 (2170)	0.182 (0.045)	0.281 (0.097)	1.07 (0.216)	53.3 (7.73)	6140 (1158)	16.6 (2.52)		
Oak	689	97 (16.8)	10 100 (1960)	0.093 (0.026)	0.167 (0.051)	0.84 (0.209)	51.6 (7.98)	5470 (911)	13.7 (2.38)	14.5 (2.86)	20.1 (2.08)
Afzelia	817	125 (26.6)	13 100 (1760)	0.100 (0.043)	0.203 (0.087)	0.79 (0.215)	79.2 (12.02)	7870 (914)	16.6 (2.28)	10.5 (2.00)	13.3 (2.49)
Greenheart	977	181 (20.9)	21 000 (1990)	0.213 (0.047)	0.395 (0.088)	1.35 (0.207)	89.9 (8.49)	10 450 (1531)	20.5 (3.06)	17.5 (4.79)	22.2 (4.97)
Softwoods											
Norway spruce (European)	417	72 (10.2)	10 200 (2010)	0.086 (0.022)	0.116 (0.040)	0.58 (0.116)	36.5 (5.26)	2140 (353)	9.8 (1.44)	8.4 (1.07)	9.1 (1.20)
Yellow pine (Canada)	433	80 (10.9)	8300 (1440)	0.089 (0.015)	0.097 (0.019)	0.56 (0.100)	42.1 (6.14)	2050 (473)	9.3 (1.61)	8.2 (1.57)	11.6 (1.77)
Douglas fir (UK)	497	91 (16.9)	10 500 (2160)	0.097 (0.038)	0.172 (0.081)	0.69 (0.200)	48.3 (8.03)	3420 (865)	11.6 (2.29)	9.5 (1.90)	11.4 (2.17)
Scotts pine (UK)	513	89 (16.9)	10 000 (2130)	0.103 (0.032)	0.134 (0.053)	0.71 (0.167)	47.4 (9.25)	2980 (697)	12.7 (2.45)	10.3 (1.82)	13.0 (2.47)
Caribbean pitch pine	769	107 (14.5)	12 600 (1800)	0.126 (0.042)	0.253 (0.060)	0.91 (0.196)	56.1 (7.76)	4980 (1324)	14.3 (2.81)	12.1 (1.23)	13.3 (1.58)

*In three point loading.

Table 54.2 Tensile strength parallel to the grain of certain timbers derived from small clear test pieces

Timber	Moisture content (%)	Tensile strength (MPa)
Hardwoods		
Ash (UK grown)	13	136
Beech (UK grown)	13	180
Yellow poplar (imported)	15	114
Softwoods		
Scots pine (UK grown)	16	92
Scots pine (imported)	15	110
Sitka spruce (imported)	15	139
Western hemlock (imported)	15	137

samples and their size, and k_v is a factor to allow for the lower variability of f_{05} values from machine grades in comparison with visual grades; for visual grades $k_v = 1.0$, and for machine grades $k_v = 1.12$.

In order to permit a comparison between the strength values obtained from structural-size test pieces and small clear specimens, previous data obtained from the testing of structural test pieces to the now withdrawn BS 5820 and the results recorded as mean values are presented in *Table 54.3* for each of the major strength modes. Not only are these mean values considerably lower than the mean values derived from small clear test pieces (*Tables 54.1* and *54.2*), but the tensile strengths are now lower than the compression strengths. This is directly related to the presence of knots and associated distorted grain in the structural-size test pieces.

54.3 Variability in strength values

In Chapter 52 attention was drawn to the fact that timber is a very variable material and that for many

of its parameters, e.g. density, cell length and micro-fibrillar angle of the S_2 layer, distinct patterns of variation could be established within a growth ring, outwards from the pith towards the bark, upwards in the tree, and from tree to tree. The effects of this variation in structure are all too apparent when mechanical tests are performed.

Test data for small clear test pieces are usually found to follow a normal distribution and, as described earlier and as explained in Chapter 2, section 2.10.1, we can describe the variability with the *sample standard deviation*, SD, and, sometimes more usefully, the *coefficient of variation*, which relates SD to the mean strength value. The coefficient of variation for timber varies considerably, but is frequently under 15%. However, reference to *Table 54.1* will indicate that this value is frequently exceeded. For design purposes the two most important properties are the modulus of elasticity and bending strength, which have coefficients of variation typically in the range of 10–30%.

As a general rule, a normal distribution curve fits the data from small clear test pieces better than do the data from structural-size test pieces, for which – as noted earlier – the 5 percentile characteristic value is determined simply by ranking the results.

54.4 Interrelationships between the strength properties

54.4.1 MODULUS OF RUPTURE (BENDING STRENGTH) AND MODULUS OF ELASTICITY

A high correlation exists between the moduli of rupture and elasticity for a particular species, but it is doubtful whether this represents any causal relationship; rather it is more probable that the correlation arises as a result of the strong correlation

Table 54.3 Mean values for dry strength derived on structural-size test pieces (approximately 97 × 47 mm with moisture content of 15–18%) using the now withdrawn BS 5820

Timber	Mean values (MPa)		
	Bending	Tension	Compression
Sitka spruce (UK)	32.8	19.7	29.5
Douglas fir (UK)	35.7	21.4	32.1
Spruce/pine/fir (Canada)	43.9	26.3	39.5
Norway spruce (Baltic)	50.9	30.5	45.8

that exists between density and each modulus. Whether it is a causal relationship or not, it is nevertheless put to good use, for it forms the basis of the stress grading of timber by machine (see section 54.7).

54.4.2 IMPACT BENDING AND TOTAL WORK

Good correlations have been established between the height of drop in impact bending tests and both work to maximum load and total work; generally the correlation is higher with the latter property.

54.4.3 HARDNESS AND COMPRESSION PERPENDICULAR TO THE GRAIN

Correlation coefficients of 0.902 and 0.907 have been established between hardness and compression strength perpendicular to the grain of timber at 12% moisture content and timber in the green state, respectively. It is general practice to predict the compression strength from the hardness result using the following equations:

$$Y_{12} = 0.00147x_{12} + 1.103 \qquad (54.3)$$
$$Y_g = 0.00137x_g - 0.207 \qquad (54.4)$$

where Y_g and Y_{12} are the compression strengths perpendicular to the grain (MPa) for green timber and timber at 12% moisture content, respectively, and x_g and x_{12} are hardness in N.

54.5 Factors affecting strength

Many of the variables noted in Chapter 53 as influencing modulus of elasticity also influence the various strength properties of timber. Once again, these can be regarded as being either material dependent or manifestations of the environment.

54.5.1 ANISOTROPY AND GRAIN ANGLE

The marked difference between the longitudinal and transverse planes in both shrinkage and modulus of elasticity has been discussed in previous chapters.

Strength likewise is directionally dependent, and the degree of anisotropy present in both tension and compression is presented in *Table 54.4* for small clear test pieces of Douglas fir. Irrespective of moisture content, the highest degree of anisotropy is in tension (48:1); this reflects the fact that the highest strength of clear, straight-grained timber is in tension along the grain while the lowest is in tension perpendicular to it. A similar degree of anisotropy is present in the tensile stressing of both glass-reinforced plastics and carbon-fibre-reinforced plastics when the fibre is laid up in parallel strands.

Table 54.4 also demonstrates that the degree of anisotropy in compression is an order of magnitude less than in tension. While the compression strengths are markedly affected by moisture content, tensile strength appears to be relatively insensitive, reflecting the exclusion of moisture from the crystalline core of the microfibril; it is this crystalline core that imparts to timber its very high longitudinal tensile strength. Comparison of the data in *Table 54.4* for tension and compression strengths along the grain reveals that clear, straight-grained timber, unlike most other materials, has a tensile strength considerably greater than its compression strength. In structural timber containing knots and distorted grain, the opposite is the norm (see *Table 54.3*).

Anisotropy in strength is due in part to the cellular nature of timber and in part to the structure and orientation of the microfibrils in the wall layers. Bonding along the direction of the microfibrils is covalent, while bonding between microfibrils is by hydrogen bonds. Consequently, since the majority of the microfibrils are aligned at only a small angle to the longitudinal axis and the timber is cellular, it will be easier to rupture the cell wall if the load is applied perpendicular to than if applied parallel to the fibre axis.

Since timber is an anisotropic material, it follows that the angle at which stress is applied relative to the longitudinal axis of the cells will determine the

Table 54.4 Anisotropy in strength parallel to and perpendicular to the grain in small clear test pieces

Timber	Moisture content (%)	Tension			Compression		
		Parallel (MPa)	Perpendicular (MPa)	Ratio, parallel to perpendicular	Parallel (MPa)	Perpendicular (MPa)	Ratio, parallel to perpendicular
Douglas fir	> 27	131	2.69	48.7	24.1	4.14	5.8
Douglas fir	12	138	2.90	47.6	49.6	6.90	7.2

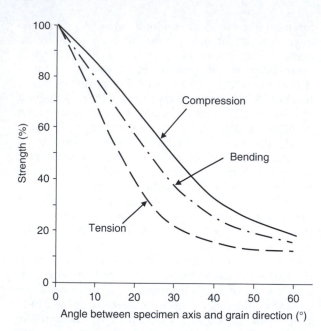

Fig. 54.1 Effect of grain angle on the tensile, bending and compression strength of timber.

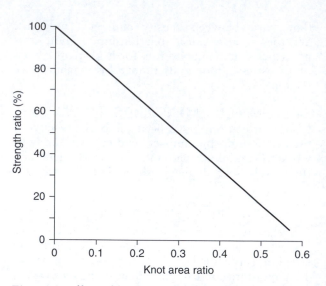

Fig. 54.2 Effect of knot-area ratio on the strength of timber; average of values for 200 Douglas fir boards (© Building Research Establishment).

ultimate strength of the timber. *Figure 54.1* illustrates that over the range 0–45° tensile strength is much more sensitive to grain angle than is compression strength. However, at angles as high as 60° to the longitudinal axis both tension and compression strengths have fallen to only about 10% of their value in straight-grained timber. The sensitivity of strength to grain angle in clear straight-grained timber is identical to that for fibre orientation in both glass-fibre- and carbon-fibre-reinforced plastics.

It is possible to obtain an approximate value of strength at any angle to the grain from a knowledge of the corresponding values of strength both parallel and perpendicular to the grain using the following formula which, in its original form, was credited to Hankinson:

$$f_\theta = \frac{f_L \cdot f_T}{f_L \sin^n\theta + f_T \cos^n\theta} \qquad (54.5)$$

where f_θ is the strength property at angle θ from the fibre direction, f_L is the strength parallel to the grain, f_T is the strength perpendicular to the grain, and n is an empirically determined constant; in tension $n = 1.5$–2 while in compression $n = 2$–2.5. The equation has also been used for the modulus of elasticity, for which a value of $n = 2$ has been adopted.

54.5.2 KNOTS

Knots are associated with distortion of the grain, and since even slight deviations in grain angle reduce the strength of the timber appreciably, it follows that knots will have a marked influence on strength. The significance of knots, however, will depend on their size and distribution both along the length of a piece of timber and across its section. Knots in clusters are more important than evenly distributed knots of a similar size, while knots on the top or bottom edge of a beam are more significant than those in the centre; large knots are much more critical than small knots.

It is very difficult to quantify the influence of knots; one of the parameters that have been successfully used is the *knot-area ratio*, which relates the sum of the cross-sectional area of the knots in a cross section to the cross-sectional area of the piece. The loss in bending strength that occurred with increasing knot area ratio in 200 UK grown Douglas fir boards is illustrated in *Fig. 54.2*. The very marked reduction in tensile strength of structural-size timber compared with small clear test pieces (*Tables 54.2* and *54.3*) is due primarily to the presence and influence of knots in the former.

54.5.3 DENSITY

In section 52.2 density was shown to be a function of cell-wall thickness and therefore dependent on the relative proportions of the various cell components and also on the level of cell-wall development

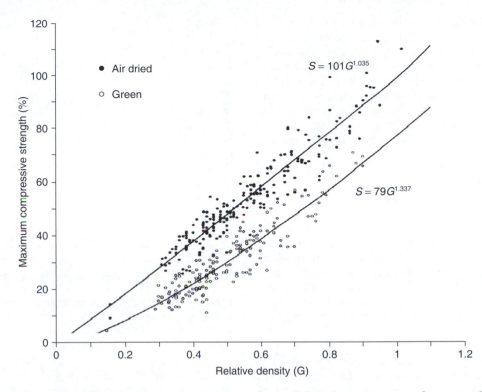

Fig. 54.3 *The relationship of maximum compression strength parallel to the grain to specific gravity for 200 species tested in the green and dry states (© Building Research Establishment).*

of any one component. However, variation in density is not restricted to different species, but can occur to a considerable extent within any one species and even within a single tree. Some measure of the inter-specific variation that occurs can be obtained from both *Fig. 52.20* and the limited amount of data in *Tables 53.1* and *54.1*. It will be observed from the latter that as density increases, so modulus of elasticity and the various strength properties increase. Density continues to be the best predictor of timber strength, since high correlations between strength and density are a common feature in timber studies.

Most of the relations that have been established throughout the world between the various strength properties and timber density take the form of:

$$f = kg^n \qquad (54.6)$$

where *f* is any strength property, *g* is the specific gravity, *k* is a proportionality constant differing for each strength property, and *n* is an exponent that defines the shape of the curve. An example of the use of this expression on the results of over 200 species tested in compression parallel to the grain is presented in *Fig. 54.3*; the correlation coefficient

between compression strength and density of the timber at 12% moisture content was 0.902.

Similar relationships have been found to hold for other strength properties, though in some the degree of correlation is considerably lower. This is the case in tension parallel to the grain, where the ultrastructure probably plays a more significant role. Over the range of density of most of the timbers used commercially, the relationship between density and strength can safely be assumed to be linear, with the possible exception of shear and cleavage; similarly, within a single species, the range is low and the relationship can again be treated as linear.

54.5.4 RING WIDTH

Since density is influenced by the rate of growth of the tree it follows that variations in ring width will change the density of the timber and hence the strength. However, the relationship is considerably more complex than it first appears. In the ring-porous timbers such as oak and ash (see Chapter 52), increasing rate of growth (ring width) results in an increase in the percentage of the latewood, which contains most of the thick-walled fibres;

consequently, density will increase and so will strength. However, there is an upper limit to ring width beyond which density begins to fall owing to the inability of the tree to produce the requisite thickness of wall in every cell. In the diffuse-porous timbers such as beech, birch and khaya, where there is uniformity in structure across the growth ring, increasing rate of growth (ring width) has no effect on density unless, as before, the rate of growth is excessive.

In the softwoods, however, increasing rate of growth results in an increased percentage of the low-density earlywood, consequently both density and strength decrease as ring width increases. Exceptionally, it is found that very narrow rings can also have very low density; this is characteristic of softwoods from the very northern latitudes where latewood development is restricted by the short summer period. Hence ring width of itself does not affect the strength of the timber, nevertheless, it has a most important indirect effect working through density.

54.5.5 RATIO OF LATEWOOD TO EARLYWOOD
Since the latewood comprises cells with thicker walls, it follows that increasing the percentage of latewood will increase the density and therefore the strength of the timber. Differences in strength of 150–300% between the late- and earlywood are generally explained in terms of the thicker cell walls of the former. However, some workers maintain that when strength is expressed in terms of the cross-sectional area of the cell wall the latewood cell is still stronger than the earlywood. Various theories have been advanced to account for the higher strength of the latewood wall material; the more likely are couched in terms of the differences in microfibrillar angle in the middle layer of the secondary wall, differences in degree of crystallinity and differences in the proportion of the chemical constituents.

54.5.6 CELL LENGTH
Since the cells overlap one another, it follows that there must be a minimum cell length below which there is insufficient overlap to permit the transfer of stress without failure in shear occurring. Some investigators have gone further and have argued that there must be a high degree of correlation between the length of the cell and the strength of cell wall material, since a fibre with high strength per unit of cross-sectional area would require a larger area of overlap in order to keep constant the overall efficiency of the system.

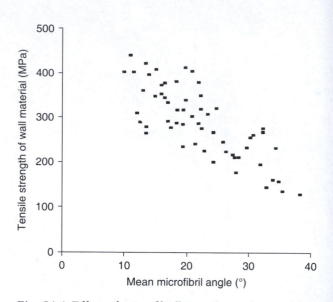

Fig. 54.4 *Effect of microfibrillar angle on the tensile strength of* Pinus radiata *blocks (from Cave (1969) by permission of Springer-Verlag).*

54.5.7 MICROFIBRILLAR ANGLE
The angle of the microfibrils in the S_2 layer has a most significant effect in determining the strength of wood. *Figure 54.4* illustrates the marked reduction in tensile strength that occurs with increasing angle of the microfibrils; the effect on strength closely parallels that which occurs with changing grain angle.

54.5.8 CHEMICAL COMPOSITION
In Chapter 52 the structure of the cellulose molecule was described and emphasis was placed on the existence in the longitudinal plane of covalent bonds both within the glucose units and also linking them together to form filaments containing from 5000 to 10000 units. There is little doubt that the high tensile strength of timber owes much to the existence of this covalent bonding. Certainly, experiments in which many of the β-1–4 linkages have been ruptured by gamma irradiation, resulting in a decrease in the number of glucose units in the molecule from over 5000 to about 200, resulted in a most marked reduction in tensile strength; it has also been shown that timber with inherently short molecules, e.g. compression wood, has a lower than normal tensile strength.

Until the 1970s it had been assumed that the hemicelluloses – which constitute about half of the matrix material – played little or no part in determining the strength of timber. However, it has now been demonstrated that some of the hemicelluloses are oriented within the cell wall, and it is now thought that these will be load-bearing.

It is known that lignin is less hydrophilic than either cellulose or hemicellulose and, as indicated earlier, at least part of its function is to protect the more hydrophilic substances from the ingress of water and consequent reduction in strength. Apart from this indirect effect on strength, lignin is thought to make a not too insignificant direct contribution. Much of the lignin in the cell wall is located in the primary wall and in the middle lamella. Since the tensile strength of a composite with fibres of a definite length will depend on the efficiency of the transfer of stress by shear from one fibre to the next, it will be appreciated that in timber lignin plays a most important role. Compression strength along the grain has been shown to be affected by the degree of lignification not between the cells, but rather within the cell wall, when all other variables have been held constant.

It would appear, therefore, that both the fibre and the matrix components of the timber composite contribute to its strength, as in fact they do in most composites, but the relative significance of the fibre and matrix roles will vary with the mode of stressing.

54.5.9 REACTION WOOD

Compression wood

The chemical and anatomical properties of this abnormal wood, which is found only in the softwoods, were described in section 52.4. When stressed, it is found that the tensile strength and toughness are lower and the compressive strength higher than those of normal timber. Such differences can be explained in terms of the changes in fine structure and chemical composition.

Tension wood

This second form of abnormal wood, which is found only in the hardwoods, has tensile strengths higher and compression strengths lower than normal wood. Again this can be related to changes in fine structure and chemical composition (see section 52.4).

54.5.10 MOISTURE CONTENT

The marked increase in strength on drying from the fibre saturation point to oven-dry conditions was described in detail in section 52.7.5 and illustrated in *Fig. 52.22*; experimentation has indicated the probability that at moisture contents of less than 2% the strength of timber may show a slight decrease rather than the previously accepted continuation of the upward trend.

Confirmatory evidence of the significance of moisture content on strength is forthcoming from *Fig. 54.3*, in which the regression line for the compression strength of green timber against density is lower than that for timber at 12% moisture content for over 200 species; strength data for timber are generally presented for these two levels of moisture content (Lavers, 1969). However, reference to *Table 54.4* indicates that the level of moisture has almost no effect on the tensile strength parallel to the grain. This strength property is determined by the strength of the covalent bonding along the molecule, and since the crystalline core is unaffected by moisture (Chapter 52), retention of tensile strength parallel to the grain with increasing moisture content is to be expected.

Within certain limits and excluding tensile strength parallel to the grain, the regression of strength, expressed on a logarithmic basis, and moisture content can be plotted as a straight line. The relationship can be expressed mathematically as:

$$\log_{10} f = \log_{10} f_s + k(\mu_s - \mu) \qquad (54.7)$$

where f is the strength at moisture content μ, f_s is the strength at the fibre saturation point, μ_s is the moisture content at the fibre saturation point, and k is a constant. It is possible, therefore, to calculate the strength at any moisture content below the fibre saturation point, assuming f_s to be the strength of the green timber and μ_s to be 27%. This formula can also be used to determine the strength changes that occur for a 1% increase in moisture content over certain ranges (*Table 54.5*); the table illustrates for small clear test pieces how the change in strength per unit change in moisture content is non-linear.

This relationship between moisture content and strength may not always apply when the timber contains defects, as is the case with structural-size timber. Thus, it has been shown that the effect of moisture content on strength diminishes as the size of knots increases.

The relationship between moisture content and strength presented above, even for knot-free timber, does not always hold for the impact resistance of timber. In some timbers, though certainly not all, impact resistance or toughness of green timber is considerably higher than it is in the dry state; the impact resistances of green ash, cricket bat willow and teak are approximately 10, 30 and 50% higher, respectively, than the values at 12% moisture content. In the case of structural timber, several types of model have been proposed to represent moisture–property relationships. These models reflect the

Table 54.5 Percentage change in strength and stiffness of Scots pine timber per 1% change in moisture content (after Lavers, 1969, 1983)

Property	Moisture range (%)		
	6–10	12–16	20–24
Modulus of elasticity (MOE – stiffness)	0.21	0.18	0.15
Modulus of rupture (MOR – bending strength)	4.2	3.3	2.4
Compression, perpendicular to the grain	2.7	2.0	1.40
Hardness	0.058	0.053	0.045
Shear, parallel to the grain	0.70	0.53	0.36

finding that increases in strength with drying are greater for high-strength structural timber than for low-strength material.

54.5.11 TEMPERATURE

At temperatures within the range –20 to +200°C and at constant moisture content, strength properties are linearly (or almost linearly) related to temperature, decreasing with increasing temperature. However, a distinction must be made between short- and long-term effects.

When timber is exposed for short periods of time to temperatures below 95°C the changes in strength with temperature are reversible. These reversible effects can be explained in terms of the increased molecular motion and greater lattice spacing at higher temperatures. For all the strength properties, with the possible exception of tensile parallel to the grain, a good rule of thumb is that an increase in temperature of 1°C produces a 1% reduction in their ultimate values (Gerhards, 1982).

At temperatures above 95°C, or at temperatures above 65°C for very long periods of time, there is an irreversible effect of temperature due to thermal degradation of the wood substance, generally taking the form of a marked shortening of the cellulose molecules and chemical changes within the hemicelluloses (see section 55.1.3). All the strength properties show a marked reduction with temperature, but toughness is particularly sensitive to thermal degradation. Repeated exposure to elevated temperature has a cumulative effect and usually the reduction is greater in the hardwoods than in the softwoods. Even exposure to cyclic changes in temperature over long periods of time has been shown to result in thermal degradation and loss in strength and especially toughness.

The effect of temperature is very dependent on moisture content, sensitivity of strength to temperature increasing appreciably as moisture content

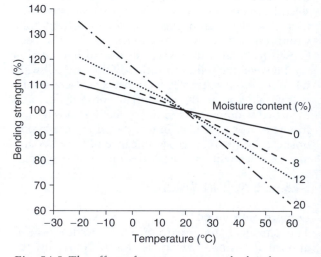

Fig. 54.5 The effect of temperature on the bending stength of Pinus radiata timber at different moisture contents.

increases (Fig. 54.5), as occurs also with modulus of elasticity (Fig. 53.9); these early results have been confirmed by Gerhards (1982). The relationships between strength, moisture content and temperature appear to be slightly curvilinear over the range 8–20% moisture content and –20 to 60°C. However, in the case of toughness, while at low moisture content it is found that toughness decreases with increasing temperature, at high moisture contents toughness actually increases with increasing temperature.

54.5.12 TIME

In Chapter 53 timber was described as a viscoelastic material and as such its mechanical behaviour will be time dependent. Such dependence will be apparent in terms of its sensitivity to both rate of loading and duration of loading.

Rate of loading

An increase in the rate of load application results in increased strength values, the increase in green timber being some 50% greater than that of timber at 12% moisture content; strain to failure, however, actually decreases. A variety of explanations have been presented to account for this phenomenon, most of which are based on the theory that timber fails when a critical strain has been reached and consequently at lower rates of loading viscous flow or creep is able to occur, resulting in failure at lower loads. The various standard testing procedures adopted throughout the world set tight limits on the speed of loading in the various tests. Unfortunately, recommended speeds vary throughout the world, thereby introducing errors in the comparison of results from different laboratories; the introduction of European standards (ENs) and the wider use of International standards (ISOs), should give rise to greater uniformity in the future.

Duration of load (DOL)

In terms of the practical use of timber, the duration of time over which the load is applied is perhaps the single most important variable. Many investigators have worked in this field and each has recorded a direct relationship between the length of time over which a load can be supported at constant temperature and moisture content and the magnitude of the load. This relationship appears to hold true for all loading modes, but is especially important for bending strength. The modulus of rupture (maximum bending strength) will decrease in proportion, or nearly in proportion, to the logarithm of the time over which the load is applied; failure in this particular time-dependent mode is termed *creep rupture* or *static fatigue* (see Chapter 2, section 2.7). Wood (1951) indicated that the relationship was slightly different for ramp and constant loading, was slightly curvilinear and that there was a distinct levelling off at loads approaching 20% of the ultimate short-term strength such that a critical load or stress level occurs below which failure is unlikely to occur. The hyperbolic curve that fitted Wood's data best for both ramp and sustained loading, and which became known as the *Madison curve*, is illustrated in *Fig. 54.6*.

Other workers have reported a linear relation, though a tendency to non-linear behaviour at very high stress levels has been recorded by some of them. Pearson (1972), in reviewing previous work in the field of duration of load and bending strength, plotted on a single graph the results obtained over a 30-year period and found that despite differences

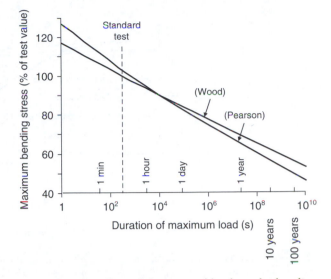

Fig. 54.6 *The effect of duration of load on the bending strength of timber (after Wood (1951) by permission of the Forest Products Laboratory, Madison, and Pearson (1972) by permission of Walter de Gruyter & Co., Berlin, New York).*

in method of loading (ramp or constant), species, specimen size, moisture content, or whether the timber was solid or laminated, the results showed remarkably little scatter from a straight line described by the regression:

$$f = 91.5 - 7 \log_{10} t \qquad (54.8)$$

where f is the stress level (%), and t is the effective duration of maximum load in hours. This regression is also plotted in *Fig. 54.6* to allow comparison with Wood's curvilinear line. Pearson's findings certainly threw doubt on the existence of a critical stress level below which creep rupture does not occur. These regressions indicate that timber beams that have to withstand a dead load for 50 years can be stressed to only 50% of their ultimate short-term strength. Although this type of log-linear relationship is still employed for the derivation of duration of load factors for wood-based panel products, this is certainly not the case for solid timber.

By the early 1970s there was abundant evidence available to indicate that the creep rupture response of structural timber beams differed considerably from the classic case for small clear test pieces described above. Between 1970 and 1985 an extensive amount of research was carried out in America, Canada and Europe; you can follow the historical development of the new concepts in DOL in comprehensive reviews by Tang (1994) and Barrett

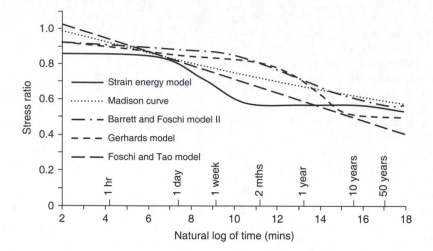

Fig. 54.7 *Comparison of duration of load predictions among four damage-accumulation models and one model based on strain energy (after Fridley, Tang and Soltis (1992a) with permission of the American Society of Engineers).*

(1996). This research confirmed that the DOL effect in structural timber was different to that in small clear test pieces and was also less severe than the Madison curve predicted for loading periods of up to one year. It also confirmed that high-strength timber possessed a larger DOL effect than low-strength timber.

The above test work clearly indicated that the DOL factors then in current use were conservative and in order to obtain a more realistic prediction of time to failure, attention moved to the possible application of reliability-based design principles for the assessment of the reliability of timber members under in-service loading conditions. In particular, this approach led to the adoption of the concept of *damage accumulation*.

It should be appreciated that in the application of this concept there does not exist any method for quantifying the actual damage; the development of damage is simply deduced from the time-to-failure data from long-term loading experiments under a given loading history. These models generally use the stress-level history as the main variable and are thus independent of material strength. In order to calculate time to failure for a given stress history under constant temperature and humidity, the damage rate is integrated from an assumed initial value of 0 to the failure value of 1 (Morlier *et al.*, 1994). Several types of damage accumulation models have been recorded, of which the most important are those listed by Morlier *et al.* (1994) and Tang (1994).

The dependence of these damage models on the stress ratio results is a logistical problem, since both the short-term and the long-term strength has to be known for the same structural test piece, but the test piece can be tested only once. This problem is usually resolved by using two side-matched test pieces and assuming equal strengths! A comparison among four of the damage-accumulation models, together with one model based on strain energy (Fridley *et al.*, 1992a), is presented in *Fig. 54.7*, from which it will be noted how large is the variability among them.

The levels of both moisture content and temperature have a marked effect on time to failure. Thus, increasing relative humidity results in reduced times to failure when stressed at the same stress ratios (Fridley *et al.*, 1991), while varying levels of humidity have an even greater effect in reducing time to failure (Hoffmeyer, 1990; Fridley *et al.*, 1992b).

54.6 Strength, toughness, failure and fracture morphology

There are two fundamentally different approaches to the concept of strength and failure. The first is the classical strength of materials approach, attempting to understand strength and failure of timber in terms of the strength and arrangement of the molecules, the fibrils, and the cells by thinking in terms of a theoretical strength and attempting to identify the reasons why the theory is never satisfied.

The second and more recent approach is much more practical in concept since it considers timber in its current state, ignoring its theoretical strength and its microstructure and stating that its performance will be determined solely by the presence of some defect, however small, that will initiate on stressing a small crack; the ultimate strength of the material will depend on the propagation of this crack. Many of the theories have required considerable modification for their application to the different fracture modes in an anisotropic material such as timber. Both approaches are discussed below for the more important modes of stressing.

54.6.1 CLASSICAL APPROACH

Tensile strength parallel to the grain

Over the years a number of models have been employed in an attempt to quantify the theoretical tensile strength of timber. In these models it is assumed that the lignin and hemicelluloses make no contribution to the strength of the timber; in the light of recent investigations, however, this may not be valid for some of the hemicelluloses. One of the earliest attempts modelled timber as comprising a series of endless chain molecules, and strengths of the order of 8000 MPa were obtained. More recent modelling has taken into account the finite length of the cellulose molecules and the presence of amorphous regions. Calculations have shown that the stress needed to cause chain slippage is generally considerably greater than that needed to cause chain scission, irrespective of whether the latter is calculated on the basis of potential energy function or bond energies between the links in the chain; preferential breakage of the cellulose chain is thought to occur at the C–O–C linkage. These important findings have led to the derivation of minimum tensile stresses of the order of 1000–7000 MPa (Mark, 1967).

The ultimate tensile strength of timber is of the order of 100 MPa, though this varies considerably between species. This figure corresponds to a value between of 0.1 and 0.015 of the theoretical strength of the cellulose fraction. Since this accounts for only half the mass of the timber (*Table 52.2*) and since it is assumed, perhaps incorrectly, that the matrix does not contribute to the strength, it can be said that the actual strength of timber lies between 0.2 and 0.03 of its theoretical strength.

In attempting to integrate these views of molecular strength with the overall concept of failure, it is necessary to examine strength at the next order of magnitude, namely the individual cells. It is possible to separate these by dissolution of the lignin–pectin complex cementing them together (section 53.2 and *Figs 52.7* and *52.8*). Using specially developed techniques of mounting and stressing, it is possible to determine their tensile strengths. Much of this work has been done on softwood tracheids, and mean strengths of the order of 500 MPa have been recorded by a number of investigators. The strengths of the latewood cells can be up to three times those of the corresponding earlywood cells.

Individual tracheid strength is therefore approximately five times greater than that of solid timber. Softwood timber also contains parenchyma cells, which are found principally in the rays and lining the resin canals, and which are inherently weak. Many of the tracheids tend to be imperfectly aligned and there are numerous discontinuities along the cell, consequently it is to be expected that the strength of timber is lower than that of the individual tracheids. Nevertheless, the difference is certainly substantial and it seems doubtful whether the features listed above can account for the total loss in strength, especially when it is realised that the cells rupture on stressing and do not slip past one another.

When timber is stressed in tension along the grain, failure occurs catastrophically with little or no plastic deformation (*Fig. 53.3*) at strains of about 1%. Visual examination of the sample usually reveals an interlocking type of fracture that can be confirmed by optical microscopy. However, as illustrated in *Fig. 54.8*, the degree of interlocking is considerably greater in the latewood than in the earlywood. Whereas in the former, the fracture plane is essentially vertical, in the latter the fracture plane follows a series of shallow zig-zags in a general transverse plane; it is now thought that these thin-walled cells contribute very little to the tensile strength of timber. Thus, failure in the stronger latewood region is by shear, while in the earlywood, though there is some evidence of shear failure, most of the rupture appears to be transwall or brittle.

Examination of the fracture surfaces of the latewood cells by electron microscopy reveals that the plane of fracture occurs either within the S_1 layer or, more commonly, between the S_1 and S_2 layers. Since shear strengths are lower than tensile strengths these observations are in accord with comments made previously on the relative superiority of the tensile strengths of individual fibres compared with the tensile strength of timber. Failing in shear implies that the shear strength of the wall layers is lower than the shear strength of the lignin–pectin material cementing together the individual cells.

Confirmation of these views is forthcoming from the work of Mark (1967), who calculated the theoretical strengths of the various cell-wall layers and

Fig. 54.8 Tensile failure in spruce (Picea abies) showing mainly transverse cross-wall failure of the earlywood (left) and longitudinal intra-wall shear failure of the latewood cells (right) (magnification × 200, polarized light) (© Building Research Establishment).

Thus, both the microscopic observations and the developed theories appear to agree that failure of timber under longitudinal tensile stressing is basically by shear unless density is low, in which case trans-wall failure occurs. However, under certain conditions the pattern of tensile failure may be abnormal. Firstly, at temperatures in excess of 100°C, the lignin component is softened and its shear strength is reduced. Consequently, on stressing, failure will occur within the cementing material rather than within the cell wall.

Secondly, trans-wall failure has been recorded in weathering studies where the mode of failure changed from shear to brittle as degradation progressed; this was interpreted as being caused by a breakdown of the lignin and degradation of the cellulose, both of which processes would be reflected in a marked reduction in density (Turkulin and Sell, 1997).

Finally, in timber that has been highly stressed in compression before being pulled in tension, it will be found that tensile rupture will occur along the line of compression damage which, as will be explained below, runs transversely. Consequently, failure in tension is horizontal, giving rise to a brittle type of fracture (see *Fig. 55.1*).

In the literature a wide range of tensile failure criteria is recorded, the most commonly applied being some critical strain parameter, an approach that is supported by a considerable volume of evidence, though its lack of universal application has been pointed out by several workers.

Compression strength parallel to the grain

Few attempts have been made to derive a mathematical model for the compressive strength of timber. One of the few, and one of the most successful, is that by Easterling *et al.* (1982). In modelling the axial and transverse compressive strength of balsa, these authors found that their theory – which related the axial strength linearly to the ratio of the density of the wood to the density of the dry cell wall material, and the transverse strength to the square of this ratio – was well supported by experimental evidence. It also appears that their simple theory for balsa may be applicable to timber of higher density.

Compression failure is a slow-yielding process in which there is a progressive development of structural change. The initial stage of this sequence appears to occur at a stress of about 25% of the ultimate failing stress (Dinwoodie, 1968), though Keith (1971) considers that these early stages do not develop until about 60% of the ultimate. There

showed that the direction and level of shear stress in the various wall layers are such as to initiate failure between the S_1 and S_2 layers. Mark's treatise received a certain amount of criticism on the grounds that he treated one cell in isolation, opening it up longitudinally in his model to treat it as a two-dimensional structure; nevertheless, the work marked the beginning of a new phase of investigation into elasticity and fracture and the approach has been modified and subsequently developed. The extension of the work has explained the initiation of failure at the S_1–S_2 boundary, or within the S_1 layer, in terms of either buckling instability of the microfibrils, or the formation of ruptures in the matrix or framework giving rise to a redistribution of stress.

is certainly a very marked increase in the amount of structural change above 60%, which is reflected by the marked departure from linearity of the stress–strain diagram illustrated in *Fig. 53.3*. The former author maintains that linearity here is an artefact resulting from insensitive testing equipment and that some plastic flow has occurred at levels well below 60% of the ultimate stress.

Compression deformation assumes the form of a small *kink* in the microfibrillar structure, and because of the presence of crystalline regions in the cell wall, it is possible to observe this feature using polarisation microscopy (*Fig. 54.9*). The sequence of irreversible anatomical changes leading to failure originates in the tracheid or fibre wall at that point where the longitudinal cell is displaced vertically to accommodate the horizontally running ray. As stress and strain increase, these kinks become more prominent and increase numerically, generally in a preferred lateral direction, horizontally on the longitudinal–radial plane (*Fig. 54.10*) and at an angle to the vertical axis of from 45° to 60° on the longitudinal–tangential plane. These lines of deformation, generally called a *crease* and comprising numerous kinks, continue to develop in width and length; at failure, defined in terms of maximum stress, these creases can be observed by eye on the face of the block of timber (Dinwoodie, 1968). At this stage there is considerable buckling of the cell wall and delamination within it, usually between the S_1 and S_2 layers. Models have been produced to simulate buckling behaviour, and calculated crease angles for instability agree well with observed angles (Grossman and Wold, 1971).

Dinwoodie (1974) has shown that, the angle at which the kink traverses the cell wall (*Fig. 54.9*) varies systematically between earlywood and latewood, between different species, and with temperature. Almost 72% of the variation in the kink angle could be accounted for by a combination of

Fig. 54.9 *Formation of kinks in the cell walls of spruce timber (Picea abies) during longitudinal compression stressing. The angle θ lying between the plane of shear and the middle lamella varies systematically between timbers and is influenced by temperature (magnification × 1600, polarized light) (© Building Research Establishment).*

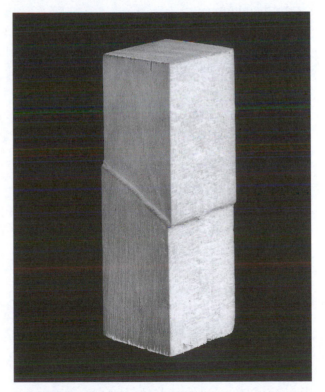

Fig. 54.10 *Failure under longitudinal compression at the macroscopic level. On the longitudinal radial plane the crease (shear line) runs horizontally, while on the longitudinal tangential plane the crease is inclined at 65° to the vertical axis (© Building Research Establishment).*

the angle of the microfibrils in the S_2 layer and the ratio of cell-wall modulus of elasticity in longitudinal and horizontal planes.

Attempts have been made to relate the size and number of kinks to the amount of elastic strain or the degree of viscous deformation. Under conditions of prolonged loading, total strain and the ratio of creep strain to elastic strain (relative creep) appear to provide the most sensitive guide to the occurrence of cell-wall deformation; the gross creases appear to be associated with strains of 0.33% (Keith, 1972).

The number and distribution of kinks depend on temperature and moisture content. Increasing moisture content, though resulting in a lower strain to failure, results in the production of more kinks, although each is smaller in size than its 'dry' counterpart; these are to be found in a more even distribution than they are in dry timber. Increasing temperature results in a similarly wider distribution of the kinks.

Static bending

In the bending mode timber is subjected to compression stresses on the upper part of the beam and tensile on the lower part. Since the strength of clear timber in compression is only about one third that in tension, failure will occur on the compression side of the beam long before it will do so on the tension side. In knotty timber, however, the compressive strength is often equal to and can actually exceed the tensile strength. As recorded in the previous section, failure in compression is progressive and starts at low levels of stressing. Consequently, the first stages of failure in bending in clear straight-grained timber will frequently be associated with compression failure, and as both the bending stress and consequently the degree of compression failure increase, so the neutral axis will move progressively downwards from its original central position in the beam (assuming uniform cross-section), thereby allowing the increased compression load to be carried over a greater cross-sectional area. Fracture occurs when the stress on the tensile surface reaches the ultimate strength in bending.

Toughness

Timber is a tough material, and in possessing moderate to high stiffness and strength in addition to its toughness, it is favoured with a unique combination of mechanical properties emulated only by bone which, like timber, is a natural composite. As we discussed in Chapter 4, toughness is generally defined as the resistance of a material to the propagation of cracks. In the comparison of materials it is usual to express toughness in terms of *work of fracture*, which is a measure of the energy necessary to propagate a crack, thereby producing new surfaces.

In timber the work of fracture, a measure of the energy involved in the production of cracks at right angles to the grain, is about 10^4 J/m^2; this value is an order of magnitude less than that for ductile metals, but is comparable with that for the manmade composites. Now the energy required to break all the chemical bonds in a plane cross section is of the order of 1–2 J/m^2, that is, four orders of magnitude lower than the experimental values. Since pull-out of the microfibrils does not appear to happen to any great extent, it is not possible to account for the high work of fracture in this way (Gordon and Jeronimidis, 1974; Jeronimidis, 1980).

One of the earlier theories to account for the high toughness of timber was based on the work of Cook and Gordon (1964), who demonstrated that toughness in fibre-reinforced materials is associated with the arrest of cracks made possible by the presence of numerous weak interfaces. As these interfaces open, so secondary cracks are initiated at right angles to the primary, thereby dissipating the energy of the original crack. This theory is applicable to timber, as *Fig. 54.11* illustrates, but it is doubtful whether the total discrepancy in energy between experiment and theory can be explained in this way.

Subsequent investigations have contributed to a better understanding of toughness in timber (Gordon and Jeronimidis, 1974; Jeronimidis, 1980). Prior to fracture it would appear that the cells separate in the fracture area, and on further stressing these individual and unrestrained cells buckle inwards, generally assuming a triangular shape. In this form they are capable of extending up to 20% before final rupture thereby absorbing a large quantity of energy. Inward buckling of helically wound cells under tensile stresses is possible only because the microfibrils of the S_2 layer are wound in a single direction. Observations and calculations on timber have been supported by glass-fibre models, and it is considered that the high work of fracture can be accounted for by this unusual mode of failure. It appears that increased toughness is possibly achieved at the expense of some stiffness, since increased stiffness would have resulted from contrawinding of the microfibrils in the S_2 layer.

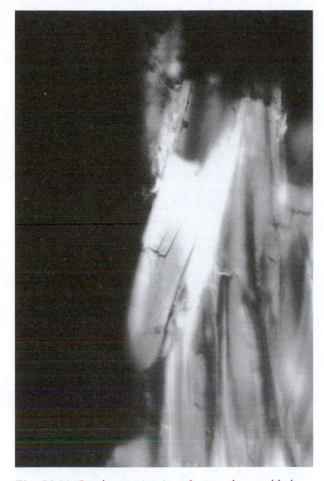

Fig. 54.11 Crack-stopping in a fractured rotor blade. The orientation of the secondary cracks corresponds to the microfibrillar orientation of the middle layer of the secondary cell wall (magnification × 500, polarized light) (© Building Research Establishment).

So far, we have discussed toughness in terms of only clear timber. Should knots or defects be present, timber will no longer be tough and the comments made earlier as to viewpoint are particularly relevant here. The material scientist sees timber as a tough material, but the structural engineer will view it as a brittle material because of its inherent defects, a theme that is developed in section 54.7.2.

Loss of toughness, however, can arise not only on account of the presence of defects and knots, but also through the effects of acid, prolonged elevated temperatures, fungal attack, or the presence of compression damage with its associated development of cell-wall deformations; these result from overstressing within the living tree, or in the handling

or utilisation of timber after conversion (section 55.1.4 and *Fig. 55.1*) (Dinwoodie, 1971; Wilkins and Ghali, 1987). Under these abnormal conditions the timber is said to be *brash* and failure occurs in a brittle mode.

Fatigue

Fatigue, which we discussed in Chapter 2, section 2.8, is usually defined as the progressive damage and failure that occur when a material is subjected to repeated loads of a magnitude smaller than the static load to failure; it is, perhaps, the repetition of the loads that is the significant and distinguishing feature of fatigue.

In fatigue testing the load is generally applied in the form of a sinusoidal or a square wave. Minimum and maximum stress levels are usually held constant throughout the test, though other wave forms, and block or variable stress levels, may be applied. The three most important criteria in determining the character of the wave form are:

- the stress range, $\Delta\sigma$, where $\Delta\sigma = \sigma_{max} - \sigma_{min}$
- the R-ratio, where $R = \sigma_{min}/\sigma_{max}$, which is the position of minimum stress (σ_{min}) and maximum stress (σ_{max}) relative to zero stress. This will determine whether or not reversed loading will occur. It is quantified in terms of the R-ratio, e.g. a wave form lying symmetrically about zero load will result in reversed loading and have an R-ratio of −1
- the frequency of loading.

The usual method of presenting fatigue data is by way of the S–N curve, where log N (the number of cycles to failure) is plotted against the mean stress S; a linear regression is usually fitted.

Using test pieces of Sitka spruce, laminated Khaya and compressed beech, Tsai and Ansell (1990) carried out fatigue tests under load control in four-point bending. The tests were conducted in repeated and reversed loading over a range of five R-ratios at three moisture contents (*Fig. 54.12*). Fatigue life was found to be largely independent of species when normalised by static strength, but was reduced with increasing moisture content and under reversed loading. The accumulation of fatigue damage was followed microscopically in test pieces fatigued at $R = 0.1$ and was found to be associated with the formation of kinks in the cell walls and compression creases in the wood.

In related work Bonfield and Ansell (1991) investigated the axial fatigue in constant-amplitude tests in tension, compression and shear in both Khaya and Douglas fir using a wide range of R-ratios, and

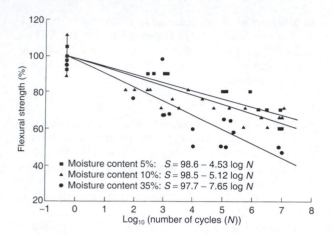

Fig. 54.12 *The effect of moisture content on sliced Khaya laminates fatigued at R = 0. The maximum peak stresses are expressed as a percentage of static flexural (bending) strength (from Tsai and Ansell (1990), by permission of Kluwer Academic Publishers).*

confirmed that reversed loading is the most severe loading regime. Fatigue lives measured in all-tensile tests (R = 0.1) were considerably longer than those in all-compression tests (R = 10), a result that they related to the lower static strength in compression relative to tension. S–N data at different R-ratios yielded a set of constant lifelines when alternating stress was plotted against mean stress; these lifelines possessed a point of inflection when loading became all compressive. More information on fatigue in timber is to be found in Dinwoodie (2000).

54.6.2 ENGINEERING APPROACH TO STRENGTH AND FRACTURE

Fracture mechanics, which we introduced in Chapter 4, provides a second approach to the concept of strength and failure. This is a more practical one and is based on the premise that all materials contain flaws or minute cracks, and that performance is determined solely by the propagation of cracks arising from these defects. The largest flaw will become self-propagating when the rate of release of strain energy exceeds the rate of increase of surface energy of the propagating crack. The application of fracture mechanics to timber did not take place until as late as 1961. Part of the reason is due to the modelling of wood as an orthotropic material, and consequently there are six values of the fracture toughness (see Chapter 4) for each of the three principal modes of crack propagation. In timber, however, macroscopic crack extension almost always occurs parallel to the grain even though it is

initiated in a different plane, thereby giving rise to a mixed-mode type of failure.

The value of fracture toughness (K_c) depends not only on orientation (as implied above), but also on the opening mode, orientation, timber density, moisture content, specimen thickness, and crack speed (see e.g. Dinwoodie, 2000). Thus, the value of K_{Ic} (K_c in mode I) in the four weak parallel-to-the-grain systems is about one-tenth that in the two tough across-the-grain systems. K_c increases with increasing density and with increasing specimen thickness.

Fracture mechanics has been applied to various aspects of timber behaviour and failure, e.g. the effect of knots, splits and joints, and good agreement has been found between predicted values using fracture mechanics and actual strength values. Examples are to be found in Dinwoodie (2000).

54.7 Structural design in timber

Timber, like many other materials, is graded according to its anticipated performance in service, but because of its inherent variability distinct grades of material must be recognised. The grading of timber for structural use may be carried out visually or mechanically.

54.7.1 VISUAL GRADING

Visual grading, as the title implies, is a visual assessment of the quality of a piece of structural timber, and is carried out against the permissible defects limits given in BS EN 14081-1. However, visual grading is a laborious process since all four faces of the timber should be examined. Furthermore, it does not separate naturally weak from naturally strong timber and hence it has to be assumed that pieces of the same size and species containing identical defects have the same strength. Such an assumption is invalid and leads to a most conservative estimate of strength.

The permissible defects limits are set out in BS 4978 and BS 5756 if using the national (British) system of grading and the various visual grade and species combinations are attributed to strength classes in BS EN 1912. If adopting a European approach, the permissible defects limits are set out in BS EN 14081-1, and the various visual grade and species combinations are again attributed to strength classes in BS EN 1912.

54.7.2 MACHINE GRADING

Many of the disadvantages of visual grading can be removed by machine grading, a process that was introduced commercially in the 1970s with the use of bending machines that either placed the timber under a constant load and measured deflection or subjected the timber to a constant deflection and measured the load that had to be applied. The principle underlying this process is the high correlation that has been found to exist between the moduli of elasticity and rupture, which was described in section 54.5.1. Grading machines based on this relationship usually provide higher yields of the higher grades than are achieved by visual grading. Since the above relationship varies among different species, it is necessary to set the grading machine for each species or species group and its geographical location (see below).

Over the last decade a number of different types of machine have been developed to assess the measurable parameters that can be related to timber strength. Some of these are based on X-rays or stress waves, either acoustical or vibrational in origin. For each of the machine types mentioned above, includindg the original bending machine, BS EN 14081 requires that the relationship between timber strength and the machine-indicated parameters (which is used to derive the settings to operate the machine in grading) is appropriate for the species of timber and the geographical region in which the tree was growing (BS EN 14081-1 and BS EN 14081-2).

Europe is now divided into four geographical regions for this purpose. Each region has been responsible for producing machine settings for the timber species that grow in that region and are used structurally in that region. The machine settings from the four regions are set out in BS EN 14081-4 while additional requirements for factory production control are presented in BS EN14081-3. We should note in passing that the moisture content of the timber has a marked influence on the selection of grading machine. Thus while dry timber can be graded by all machine types, X-ray and stress-wave machines are normally limited to a timber moisture content of around 20%. Grading of green timber can currently only be done on bending machines, as the larger sectional size of green timber conveniently offsets the lower modulus.

Under the new European system the grade mark will include the specification number used in the grading (BS EN 14081):

- the species or species group
- the timber condition (green or dry)
- the strength class (see below)
- the grader or company name

- the company registration number
- the certification body (logo).

54.7.3 STRENGTH CLASSES
Graded timber is assigned to the strength classes contained in BS EN 338, a standard that provides the characteristic values for each of the strength properties and density for each of the eighteen strength classes of timber. A reduced version of Table 1 of BS EN 338, containing only six of the strength classes, is given in *Table 54.6* to illustrate the type of data presented.

54.7.4 STRUCTURAL DESIGN
The formation of the European Union as a free trade area and the production of the Construction Products Directive (CPD) led automatically to the introduction and implementation of new European standards and the withdrawal of conflicting national standards. Such an approach has much merit, but as far as the UK is concerned, it has led to changes in both the derivation and use of working stresses. First, test results are now expressed in terms of a characteristic value expressed in terms of the lower 5th percentile, in contrast to the former use in the UK of a mean value and its standard deviation. Second, in the design of timber structures the new Eurocode 5 is written in terms of limit state design in contrast to the former UK use of permissible stress design. Third, the number of strength classes in the new European system is greater than in the UK system, thereby giving rise to mismatching of certain timbers.

As noted above, the design of timber structures must now be in accordance with Eurocode 5 (BS EN 1995-1-1:2004). The characteristic values for timber given in BS EN 338 must be reduced according to the period of loading and service class (defined in terms of level of humidity). Values of K_{mod} (duration of load and service class) and K_{def} (creep and service class) are set out in Eurocode 5 for each of the three service classes. It is understood that although BS 5268-2:2007 is likely to be withdrawn officially at the end of October 2012, it is most likely that permissible stress design using the grade stresses for the various strength classes set out in BS 5268-2:2007 will continue for at least another decade in certain quarters of the industry employing where relevant Parts 1 and 2 of BS 5268-6. Certain grades of the various wood-based panels described in section 56.1.2 may be used for structural purposes. The characteristic values are calculated according to BS EN 1058 and are listed in BS EN12369-1 and BS EN 12369-2.

Table 54.6 An extract from Table 1 of BS EN 338:2003 illustrating the characteristic values for certain selected strength classes at a moisture content of 12% for each of the strength and stiffness parameters

| | | Strength classes and characteristic values | | | | | |
| | | Softwood | | | | Hardwood | |
Property		C16	C18	C22	C24	D30	D40
Strength properties (MPa)							
Bending	$f_{m,k}$	16	18	22	24	30	40
Tension parallel	$f_{t,0,k}$	10	11	13	14	18	24
Tension perpendicular	$f_{t,90,k}$	0.5	0.5	0.5	0.5	0.6	0.6
Compression parallel	$f_{c,0,k}$	17	18	20	21	23	26
Compression perpendicular	$f_{c,90,k}$	2.2	2.2	2.4	2.5	8.0	8.8
Shear	$f_{v,k}$	1.8	2.0	2.4	2.5	3.0	3.8
Stiffness properties (GPa)							
Mean modulus of elasticity, parallel	$E_{o,mean}$	8	9	10	11	10	11
5% modulus of elasticity, parallel	$E_{0,05}$	5.4	6.0	6.7	7.4	8.0	9.4
Mean modulus of elasticity, perpendicular	$E_{90,mean}$	0.27	0.30	0.33	0.37	0.64	0.75
Mean shear modulus	G_{mean}	0.50	0.56	0.63	0.69	0.60	0.70
Density (kg/m^3)							
Density	ρ_k	310	320	340	350	530	590
Average density	ρ_{mean}	370	380	410	420	640	700

References

1. Standards and specifications

ASTM Standard D143-52:1972. Standard methods of testing small clear specimens of timber, *American Society for Testing Materials*.

BS 373:1957. Methods of testing small clear specimens of timber. *BSI, London*

BS 4978:2007. Visual strength grading of softwood. Specification. *BSI, London*.

BS 5268-2:2007. Structural use of timber. Code of practice for permissible stress design, materials and workmanship, *BSI, London*.

BS 5268-6.1:1996. Structural use of timber. Code of practice for timber frame walls.

BS 5268-6.2:2001. Structural use of timber. Code of practice for timber frame walls. Buildings other than dwellings not exceeding four storeys.

BS 5756:2007. Visual strength grading of hardwoods. Specification. *BSI, London*.

BS 5820:1979. Methods of test for determination of certain physical and mechanical properties of timber in structural sizes, *BSI, London*. (Now withdrawn).

BS EN 338:2003. Structural timber. Strength classes.

BS EN 384:2004. Structural timber. Determination of characteristic values on mechanical properties and density.

BS EN 408:2003. Timber structures. Structural timber and glued laminated timber. Determination of some physical and mechanical properties.

BS EN 1058:1996. Wood based panels. Determination of chacteristic values of mechanical properties and density.

BS EN 1912:2004. Structural timber. Strength classes. Assignment of visual grades and species.

BS EN 12369-1:2001. Wood-based panels. Characteristic values for use in structural design. Part 1: Particleboards, OSB and Fibreboards.

BS EN 12369-2:2004. Wood-based panels. Characteristic values for use in structural design. Part 2: Plywood.

BS EN 14081-1:2005. Timber structures. Strength graded structural timber with rectangular cross section. General requirements.

BS EN 14081-2:2005. Timber structures. Strength graded structural timber with rectangular cross section. Machine grading. Additional requirements for initial type testing.

BS EN 14081-3:2005. Timber structures. Strength graded structural timber with rectangular cross section. Machine grading. Additional requirements for factory production control.

BS EN 14081-4:2005. Timber structures. Strength graded structural timber with rectangular cross section. Machine grading. Grading machine settings for machine controlled systems.

BS EN 1995-1-1:2004. Eurocode 5: Design of timber structures. General. Common rules and rules for buildings.

2 Literature

Barrett JD (1996). *Duration of load – the past, present and future*. In Proceedings of the International Conference on Wood Mechanics, Stuttgart, Germany (ed. Aicher S), EC COST 508 Action, pp. 121–137.

Bonfield PW and Ansell MP (1991). Fatigue properties of wood in tension, compression and shear. *Journal of Material Science*, **26**, 4765–4773.

Cave ID (1969). The longitudinal Young's modulus of *Pinus radiata*. *Wood Science and Technology*, **3**, 40–48.

Cook J and Gordon JE (1964). A mechanism for the control of crack propagation in all brittle systems. *Proceedings of the Royal Society of London, A*, **282**, 508.

Dinwoodie JM (1968). Failure in timber, Part I: Microscopic changes in cell wall structure associated with compression failure. *Journal of the Institute of Wood Science*, **21**, 37–53.

Dinwoodie JM (1971). Brashness in timber and its significance. *Journal of the Institute of Wood Science*, **28**, 3–11.

Dinwoodie JM (1974). Failure in timber, Part II: The angle of shear through the cell wall during longitudinal compression stressing. *Wood Science and Technology*, **8**, 56–67.

Dinwoodie JM (2000). *Timber: its nature and behaviour*, E & F N Spon, London, 256 pages.

Easterling KE, Harrysson R, Gibson LJ and Ashby MF (1982). The structure and mechanics of balsa wood. *Proceedings of the Royal Society of London*, **383**, 31–41.

Fridley KJ, Tang RC and Soltis LA (1991). Moisture effects on the load-duration behaviour of lumber. Part I: Effect of constant relative humidity. *Wood and Fiber Science*, **23** (No. 1), 114–127.

Fridley KC, Tang RC and Soltis LA (1992a). Load-duration effects in structural lumber: strain energy approach. *Journal of Structural Engineering, Structural Division of the ASCE*, **118** (No. 9), 2351–2369.

Fridley KJ, Tang RC and Soltis LA (1992b). Moisture effects on the load-duration behaviour of lumber. Part II: Effect of cyclic relative humidity. *Wood and Fiber Science*, **24** (No. 1), 89–98.

Gerhards CC (1982). Effect of moisture content and temperature on the mechanical properties of wood: an analysis of immediate effects. *Wood and Fiber*, **14** (No. 1), 4–26.

Gordon JE and Jeronimidis G (1974). Work of fracture of natural cellulose. *Nature (London)*, **252**, 116.

Grossman PUA and Wold MB (1971). Compression fracture of wood parallel to the grain. *Wood Science and Technology*, **5**, 147–156.

Hoffmeyer P (1990). *Failure of wood as influenced by moisture and duration of load*. State University of New York, Syracuse, New York, USA, 123 pages.

Jeronimidis G (1980). The fracture behaviour of wood and the relations between toughness and morphology. *Proceedings of the Royal Society of London, B*, **208**, 447–460.

Keith CT (1971). The anatomy of compression failure in relation to creep-inducing stress. *Wood Science*, **4** (No. 2), 71–82.

Keith CT (1972). The mechanical behaviour of wood in longitudinal compression. *Wood Science*, **4** (No. 4), 234–244.

Lavers GM (1969). *The strength properties of timber*. Bulletin 50, Forest Products Research Laboratory, UK, 2nd edition, HMSO (2nd edition revised by Moore G, 1983).

Mark RE (1967). *Cell Wall Mechanics of Tracheids*, Yale University Press, New Haven.

Morlier P, Valentin G and Toratti T (1994). *Review of the theories on long term strength and time to failure*. Proceedings of the Workshop on Service Life Assessment of Wooden Structures, Espoo, Finland (ed. Gowda SS), Management Committee of EC COST 508 Action, pp. 3–27.

Pearson RG (1972). The effect of duration of load on the bending strength of wood. *Holzforschung*, **26** (No. 4), 153–158.

Tang RC (1994). *Overview of duration-of-load research on lumber and wood composite panels in North America*. Proceedings of the Workshop on Service Life Assessment of Wooden Structures, Espoo, Finland (ed. Gowda SS), Management Committee of EC COST 508 Action, pp. 171–205.

Tsai KT and Ansell MP (1990). The fatigue properties of wood in flexure. *Journal of Material Science*, **25**, 865–878.

Turkulin J and Sell J (1997). *Structural and fractographic study on weathered wood*. Fotschungs-und Arbeitsbericht 115/36 Abteilung Holz, EMPA, Switzerland, 4 pages.

Wilkins AP and Ghali M (1987). Relationship between toughness, cell wall deformations and density in *Eucaluptus pilularis*. *Wood Science and Technology*, **21**, 219–226.

Wood LW (1951). *Relation of strength of wood to duration of load*. Forest Products Laboratory, Madison, Report No 1916.

Chapter 55

Durability of timber

Durability is a term that has different connotations for different people; it is defined here in the broadest possible sense to embrace the resistance of timber to attack from a whole series of agencies whether physical, chemical or biological in origin.

By far the most important are the biological agencies, fungi and insects, both of which can cause tremendous havoc given the right conditions. In the absence of fire, fungal or insect attack, timber is really remarkably resistant, and timber structures will survive, indeed have survived, incredibly long periods of time, especially when it is appreciated that it is a natural organic material with which we are dealing. Examples of well-preserved timber items now over 2000 years old are to be seen in the Egyptian tombs.

Attack by both fungi and insects is described in section 55.2, together with the importance of the natural durability of timber. Another important aspect of durability of timber is its reaction to fire, which is discussed in section 55.3. The effects of photochemical, chemical, thermal and mechanical action are usually of secondary importance in determining durability. These will be briefly considered first in section 55.1.

55.1 Chemical, physical and mechanical agencies affecting durability and causing degradation

55.1.1 PHOTOCHEMICAL DEGRADATION

On exposure to sunlight the coloration of the heartwood of most timbers – e.g. mahogany, afrormosia and oak – will lighten, though a few timbers will actually darken, e.g. Rhodesian teak. Indoors the action of sunlight will be slow and the process will take several years, but outdoors the change in colour is very rapid, taking place in a matter of months, and is generally regarded as an initial and very transient stage in the whole process of *weathering*.

In weathering the action of light energy (photochemical degradation), rain and wind results in a complex degrading mechanism that renders the timber silvery-grey in appearance. More important is the loss of surface integrity, a process that has been quantified in terms of the residual tensile strength of thin strips of wood (Derbyshire and Miller, 1981; Derbyshire *et al.*, 1995). The loss in integrity embraces the degradation of both lignin, primarily by the action of ultraviolet light, and cellulose, by shortening of the chain length, mainly by the action of energy from the visible part of the spectrum. Degradation results in erosion of the cell wall and in particular the pit aperture and torus. Fractography using scanning electron microscopy has revealed that the progression of degradation involves initially the development of brittleness and the reduction in stress transfer capabilities through lignin degradation, followed by reductions in microfibril strength resulting from cellulose degradation (Derbyshire *et al.*, 1995).

However, the same cell walls that are attacked act as an efficient filter for those cells below, and here the rate of erosion from the combined effects of UV, light and rain is very slow indeed; in the absence of fungi and insects the rate of removal of the surface by weathering is of the order of only 1 mm every 20 years. Nevertheless, because of the continual threat of biological attack, it is unwise to leave most timbers completely unprotected from the weather, and it should be appreciated that during weathering, the integrity of the surface layers is markedly reduced, thereby adversely affecting the performance of an applied surface coating. In order to effect good adhesion the weathered layers must first be removed (see section 56.4).

55.1.2 CHEMICAL DEGRADATION

As a general rule, timber is highly resistant to a large number of chemicals and its continued use for various types of tanks and containers, even in the face of competition from stainless steel, indicates that its resistance is very good; it is certainly cost-effective. Timber is far superior to cast iron and ordinary steel in its resistance to mild acids and for very many years timber was used as separators in lead-acid batteries. However, in its resistance to alkalis timber is inferior to iron and steel. Dissolution of both the lignin and the hemicelluloses occurs under the action of even mild alkalis.

Iron salts are frequently very acidic and in the presence of moisture result in hydrolytic degradation of timber; the softening and darkish-blue discoloration of timber in the vicinity of iron nails and bolts is due to this effect.

Timber used in boats is often subjected to the effects of chemical decay associated with the corrosion of metallic fastenings, a condition frequently referred to as *nail sickness*. This is basically an electrochemical effect, the rate of activity being controlled by the availability of oxygen.

55.1.3 THERMAL DEGRADATION

Prolonged exposure of timber to elevated temperatures results in a reduction in strength and a very marked loss in toughness (impact resistance). Thus, timber heated at 120°C for one month loses 10% of its strength, while at 140°C the same loss in strength occurs after only one week (Shafizadeh and Chin, 1977). Tests on three softwood timbers subjected to daily cycles of 20 to 90°C for a period of three years resulted in a reduction in toughness to only 44% of the value of samples exposed for only one day (Moore, 1984). It has been suggested that degradation can occur at temperatures as low as 65°C when exposed for many years.

Thermal degradation results in a characteristic browning of timber with associated caramel-like odour, indicative of burnt sugar. Initially this is the result of degradation of the hemicelluloses, but with time the cellulose is also, affected with a reduction in chain length through scission of the β-1–4 linkage. Commensurate degradation occurs in the lignin, but usually at a slower rate (see section 55.3).

55.1.4 MECHANICAL DEGRADATION

The most common type of mechanical degradation is that which occurs in timber when stressed under load for long periods of time. The concepts of duration of load (DOL) and creep have been explained fully in sections 54.5.12 and 53.3.2, respectively. Thus, it was illustrated in *Figs 54.7* and *54.8* how there is a loss in strength with time under load, such that after being loaded for 50 years the strength of timber is reduced by approximately 50%. Similarly, there is a marked reduction in elasticity that manifests itself as an increase in extension or deformation with time under load, as is illustrated in *Figs 53.10* and *53.13*. The structural engineer, in designing his timber structure, has to take into account the loss with time of both strength and modulus of elasticity by applying two time-modification factors.

A second and less common form of mechanical degradation is the induction of compression failure within the cell walls of timber, which can arise either in the standing tree first in the form of a *natural compression failure* due to high localised compressive stress, or second, as *brittleheart* due to the occurrence of high growth stresses in the centre of the trunk, or under service conditions where the timber is over-stressed in longitudinal compression with the production of kinks in the cell wall, as described in section 54.6.1. Loss in tensile strength due to the induction of compression damage is about 10–15%, but the loss in toughness can be as high as 50%. An example of failure in a scaffold board in bending resulting from the prior induction of compression damage due to malpractice on site is illustrated in *Fig. 55.1*.

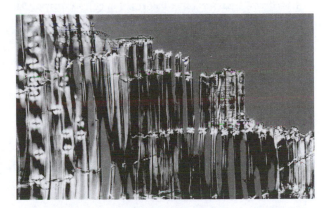

Fig. 55.1 Kink bands and compression creases in the tension face of a scaffold board that failed in bending on site. Note how the crack pathway has followed the line of the top compression crease, which had been induced some time previously (magnification × 100 under polarised light) (© Building Research Establishment).

55.2 Natural durability and attack by fungi and insects

55.2.1 NATURAL DURABILITY

Generally when the durability of timber is discussed reference is being made explicitly to the resistance of the timber to both fungal and insect attack; this resistance is termed *natural durability*.

Recalling that timber is an organic product it is surprising at first to find that it can withstand attack from fungi and insects for very long periods of time, certainly much longer than its herbaceous counterparts. This resistance can be explained in part by the basic constituents of the cell wall, and in part by the deposition of extractives (sections 52.1 and 52.3.1; *Table 52.2*).

The presence of lignin – which surrounds and protects the crystalline cellulose – appears to offer a slight degree of resistance to fungal attack. Certainly the resistance of sapwood is higher than that of herbaceous plants. Fungal attack can commence only in the presence of moisture, and the threshold value of 20% for timber is about twice as high as the corresponding value for non-lignified plants.

Timber has a low nitrogen content, of the order of 0.03–0.1% by mass and, since this element is a prerequisite for fungal growth, its presence in only such small quantities contributes to the natural resistance of timber.

However, the principal factor conferring resistance to biological attack is undoubtedly the presence of extractives in the heartwood. The far higher durability of the heartwood of certain species compared with the sapwood is attributable primarily to the presence in the former of toxic substances, many of which are phenolic in origin. Other factors such as a decreased moisture content, reduced rate of diffusion, moderate density and deposition of gums and resins also play a role in determining the higher durability of the heartwood.

Considerable variation in durability can occur within the heartwood zone. In a number of timbers the outer band of the heartwood has a higher resistance than the inner region, owing, it is thought, to the progressive degradation of toxic substances by enzymatic or microbial action.

Durability of the heartwood varies considerably among different species, and is related to the type and quantity of extractives present; the heartwood of timbers devoid of extractives has a very low durability. The sapwood of all timbers is susceptible to attack owing not only to the absence of extractives, but also to the presence in the ray cells of stored starch, which constitutes a ready source of food for the invading fungus.

BS EN 350-1 defines natural durability as 'the inherent resistance of wood to attack by wood destroying organisms' and specifies the techniques to be used in assessing the durability of wood against wood-destroying fungi, beetles, termites and marine borers. The normal test method, using heartwood stakes in ground contact, is described in EN 252 and BS 7282. However, provision is made in BS EN 350-1 for the use of laboratory tests to provide an initial indicator of the potential durability of a new species of wood.

The results of these tests lead to the placement of a species of wood in a five-band classification of natural durability against each of four biological agents, namely wood-destroying fungi, *Hylotrupes bajulus*, *Anobium punctatum* and termites. The durability classes for each agent are:

- *Class 1* – very durable.
- *Class 2* – durable.
- *Class 3* – moderately durable.
- *Class 4* – slightly durable.
- *Class 5* – not durable.

This classification is presented in BS EN 350-2 and illustrated for a selection of British-grown timbers in *Table 55.1*.

The relationship between service environment and risk of attack by wood-destroying organisms is defined in BS EN 335-1 by employing a set of five '*use classes*' (formerly '*hazard classes*') as shown in *Table 55.2*, while BS EN 335-2 relates to the application of these use classes to solid wood and BS EN 335-3 relates to their application to wood-based panels. Lastly, BS EN 460 sets out the durability requirement of the wood to be used in each use class. It should be appreciated that to meet the requirements of some of the use classes the wood may have to have its inherent durability enhanced by the introduction of preservatives (as described

Table 55.1 Natural durability of the heartwood of certain British-grown species in ground contact

Wood species	Natural durability of heartwood
Oak	Class 2
Douglas fir	Class 3–4
Larch	Class 3–4
Scots pine	Class 3–4
Beech	Class 5

Table 55.2 'Use classes' as defined in BS EN 335-1

Use class	Description	Principal biological agency	Example of service situation
1	Above ground, covered, permanently dry	Insects	Internal joinery, intermediate floor joists
2	Above ground, covered, occasional risk of wetting	Fungi and insects	Tiling battens, timbers in flat roofs and timber-frame structures and where there is a risk of condensation
3	Above ground, not covered, exposed to frequent wetting	Fungi	External joinery, fence rails, gates
4	In contact with ground or with fresh water	Fungi	Fence posts, deck supports, poles, sleepers
5	In contact with sea water	Marine borers	Marine piling

in section 56.2.1), by thermally modifying the wood (as discussed in section 56.3), or by changing the chemistry of the wood (as described in section 56.2.1).

55.2.2 THE NATURE OF FUNGAL DECAY

Some fungi, e.g. the moulds, are present only on the surface of timer and although they may cause staining they have no effect on the strength properties. A second group of fungi, the sapstain fungi, live on the sugars present in the ray cells, and the presence of their hyphae in the sapwood imparts a distinctive coloration to that region of the timber, which is often referred to as 'blue-stain'. One of the best examples of sapstain is that found in recently felled Scots pine logs. In temperate countries the presence of this type of fungus results in only inappreciable losses in bending strength, though several staining fungi in the tropical countries cause considerable reductions in strength.

By far the most important fungi are those that cause decay of timber by chemical decomposition; this is achieved by the digesting action of enzymes secreted by the fungal hyphae. Two main groups of timber-rotting fungi can be distinguished:

1. *The brown rots*, which consume the cellulose and hemicelluloses, but attack the lignin only slightly. During attack the wood usually darkens and in an advanced stage of attack tends to break up into cubes and crumble under pressure. One of the best known fungi of this group is *Serpula lacrymans*, which causes *dry rot*. Contrary to what its name suggests, the fungus requires an adequate supply of moisture for its development.
2. *The white rots*, which attack all the chemical constituents of the cell wall. Although the timber

may darken initially, it becomes very much lighter than normal at advanced stages of attack. Unlike timber under attack from brown rot, timber with white rot does not crumble under pressure, but separates into a fibrous mass.

In very general terms, the brown rots are usually to be found in constructional timbers, whereas the white rots are frequently responsible for the decay of exterior joinery.

Decay, of course, results in a loss of strength, but it is important to note that considerable strength reductions may arise in the very early stages of attack, toughness being particularly sensitive to the presence of fungal attack. Loss in mass of the timber is also characteristic of attack, and decayed timber can lose up to 80% of its air-dry mass. The principal types of fungal attack of wood in the standing tree, of timber in felled logs and of timber in service are set out in *Table 55.3*. More information on fungal attack of timber is to be found in Desch and Dinwoodie (1996) and Bravery *et al.* (1987).

55.2.3 THE NATURE OF INSECT ATTACK

Although all timbers are susceptible to attack by at least one species of insect, in practice only a small proportion of the timber in service actually becomes infested. Some timbers are more susceptible to attack than others and generally the heartwood is much more resistant than the sapwood; nevertheless, the heartwood can be attacked by certain species particularly when decay is also present.

Timber is consumed by the adult form of certain insects, the best-known example being termites whose adult, but sexually immature, workers cause most damage. Few timbers are immune to attack by these voracious eaters and it is indeed fortunate

Table 55.3 The principal types of fungal attack

| Fungus | Location of attack | | | Effect on the timber | |
	Tree	Logs	Timber	Gross	Micro
Brown rot	✓*	✓	✓	Darkening of timber with cuboidal cracking	Attacks cellulose and hemicellulose
White rot	✓*	✓	✓	Bleaching of timber, which turns fibrous	Attacks cellulose, hemicellulose and lignin
Soft rot	–	–	✓	Superficial; small cross–cracking; mostly occurring in ground contact	Attacks cellulose of S_2 layer
Sapstain (blue-stain)	–	✓	✓	Stains the sapwood of timber in depth	Stain due to colour of hyphae
Moulds	–	✓	✓	Superficial staining due to spores	Live on cell contents; may increase permeability of timber
Bacteria	✓	✓	✓	Subtle changes in texture and colour	Increases permeability; get significant decay in ground contact

*Commonly as pocket rots.

that these insects generally cannot survive the cooler weather of the UK. They are to be found principally in the tropics, but certain species are present in the Mediterranean region, including southern France.

In the UK insect attack is mainly by the grub or larval stage of certain beetles. The adult beetle lays its eggs on the surface of the timber, frequently in surface cracks, or in the cut ends of cells; these eggs hatch to produce grubs, which tunnel their way into the timber, remaining there for periods of up to three years or longer. The size and shape of the tunnel, the type of detritus left behind in the tunnel (frass) and the exit holes made by the emerging adults are all characteristic of the species of beetle.

The principal types of insect attack of wood in the standing tree, of timber in the form of felled logs and of timber in service that may be free from decay or partially decayed are set out in *Table 55.4*. More information on insect attack of timber is to be found in Desch and Dinwoodie (1996) and Bravery *et al.* (1987).

55.2.4 MARINE BORERS
Timber used in salt water is subjected to attack by marine-boring animals such as the shipworm (*Teredo* spp.) and the gribble (*Limnoria* spp.). Marine borers are particularly active in tropical waters, nevertheless, around the coast of Great Britain *Limnoria* is fairly active and *Teredo*, though spasmodic, has still to be considered a potential hazard. The degree of hazard will vary considerably with local conditions, and relatively few timbers are recognised as having heartwood resistant under all

conditions. The list of resistant timbers includes ekki, greenheart, okan, opepe and pyinkado.

55.3 Performance of timber in fire

The performance of materials in fire is an aspect of durability that has always attracted much attention, not so much from the research scientist, but rather from the material user who has to conform with the legislation on safety and who is influenced by the weight of public opinion on the use of only 'safe' materials. While various tests have been devised to assess the performance of materials in fire there is a fair degree of agreement on the unsatisfactory nature of many of these tests, and an awareness that certain materials can perform better in practice than is indicated by these tests. Thus, while no one would doubt that timber is a combustible material showing up rather poorly in both the 'spread of flame' and 'heat release' tests, nevertheless in at least one aspect of performance, namely the maintenance of strength with increasing temperature and time, wood performs better than steel.

There is a critical surface temperature below which timber will not ignite. As the surface temperature increases above 100°C, volatile gases begin to be emitted as thermal degradation slowly commences. However, it is not until the temperature is in excess of 250°C that there is sufficient build-up of these

Table 55.4 The principal types of insect attack of timber

| Type of insect | Location of attack | | | | Comments |
| | Tree | Logs | Timber in service | | |
			Sound	Decayed	
Pin-hole borers (ambrosia beetle)	✓	✓	–	–	Produce galleries 1–2 mm diameter, which are devoid of bore dust and are usually darkly stained; attack is frequently present in tropical hardwoods
Forest longhorn	✓	(✓)	–	–	Galleries oval in cross-section; no bore dust but galleries may be plugged with coarse fibres; exit holes oval, 6–10 mm diameter
Wood wasp	(✓)	✓	–	–	Galleries circular in cross section and packed with bore dust; attacks softwoods; exit holes circular, 4–7 mm diameter
Bark borer beetle (Ernobius mollis)	(✓)	✓	(✓)	–	Requires presence of bark on timber; galleries empty and mainly in bark, but will also penetrate sapwood; exit holes circular, 1–2 mm diameter
Powder-post beetle (Lyctus brunneus and L. lineraris)	–	✓	✓	–	Attack confined to sapwood of hardwoods having large-diameter vessels; bore dust fine, talc-like; exit holes circular, 1–2 mm diameter
Common furniture beetle (Anobium punctatum)	–	–	✓	–	Attacks mainly the sapwood of both softwoods and European hardwoods; bore dust lemon-shaped pellets; exit holes circular, 1–2 mm diameter
House longhorn beetle (Hylotrupes bajalus)	–	–	✓	–	Attacks sapwood of softwoods mainly in the roof space of houses in certain parts of Surrey; bore dust sausage-shaped; exit holes few, oval, often ragged, 6–10 mm diameter
Death-watch beetle (Xestobium rufovillosum)	–	–	(✓)	✓	Attacks both sap- and heartwood of partially-decayed hardwoods, principally oak; bore dust bun-shaped; exit holes circular, 3 mm diameter
Weevils (e.g. Euophryum confine)	–	–	–	✓	Attacks decayed softwoods and hardwoods in damp conditions; exit holes small, ragged, about 1 mm diameter
Wharf borer (Narcerdes melanura)	–	–	–	✓	Attacks partially decayed timber to produce large galleries with 6 mm diameter oval exit hole

gases to cause ignition of the timber in the presence of a pilot flame. Where this is absent, the surface temperature can rise to about 500°C before the gases become self-igniting. Ignition, however, is related not only to the actual temperature, but also to the time of exposure at that temperature, since ignition is primarily a function of heat flux.

Generally chemical bonds begin to break down at about 175°C. It is known that the first constituent of timber to degrade is the lignin, and this degradation continues slowly up to 500°C. The hemicelluloses degrade much more quickly between 200 and 260°C, as does the cellulose within the temperature range 260–350°C. Degradation of the cellulose results in the production of flammable volatile gases and a marked reduction in its degree of polymerisation (chain length) (le Van and Winandy, 1990).

The performance of timber at temperatures above ignition level is very similar to that of certain reinforced thermosetting resins that have been used as sacrificial protective coatings on space-return capsules. Both timber and these ablative polymers undergo thermal decomposition with subsequent removal of mass, leaving behind sufficient residue to protect the residual material. The onset of pyrolysis in timber is marked by a darkening of the timber and the commencement of emission

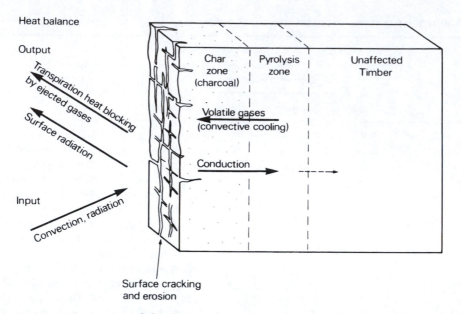

Fig. 55.2 Diagrammatic representation of the thermal decomposition of timber (© Building Research Establishment).

of volatile gases; the reaction becomes exothermic and the timber changes to a carbonised char popularly known as charcoal (*Fig. 55.2*). The volatiles, in moving to the surface, cool the char and are subsequently ejected into the boundary layer, where they block the incoming convective heat, a very important phenomenon known as transpirational cooling. High surface temperatures are reached and some heat is emitted by thermal radiation; the heat balance is indicated in *Fig. 55.2*. The surface layers crack badly both along and across the grain and surface material is continually, but slowly, being lost.

A quasi-steady state is reached, therefore, with a balance between the rate of loss of surface and the rate of recession of the undamaged wood. The rate at which the front recedes is a product of:

• the heat flux
• the temperature of combustion
• the oxygen level.

In the UK, evidence has been produced (Hall and Jackman, 1975) that demonstrates that under exposure to a fully developed fire as represented by the fire resistance test (BS 476: Part 8: now withdrawn) the surface of large softwood and medium-density hardwood timber sections recedes at a rate of approximately 0.64 mm/min. For high-density hardwoods of an equivalent size, the value is given as 0.5 mm/min. In mainland Europe, where furnaces are fuelled by oil, which provides a greater radiant

component to the flames from the burners that are used to heat the furnace, they have used higher softwood charring rates, e.g. 0.8 mm/min. The values set out in Eurocode 5: Part 1–2 (BS EN 1995-1-2): *General Rules – Structural Fire Design* for timber of a stated minimum dimension reflect mainland Europe's experience and should be used in all calculations. It should be appreciated that small sections of timber will always char at an enhanced rate, and caution must be exercised when applying the Eurocode rates to smaller sections.

The formation of the char protects the unburnt timber, which may be only a few millimetres from the surface. Failure of the beam or strut will occur only when the cross-sectional area of the unburnt core becomes too small to support any static or dynamic load. By increasing the dimensions of the timber above those required for structural considerations, it is possible to guarantee structural stability and/or load-bearing capacity in a fire for a given period of time. This is a much more desirable situation than that presented by steel, where total collapse of the beam or strut occurs at a critical temperature that is proportional to the stress in the member.

55.3.1 METHODS OF ASSESSING REACTION TO FIRE OF CONSTRUCTIONAL MATERIALS

The formation of the European Community (EC) as a free trade area and the production of the

Construction Products Directive (CPD) led automatically to the introduction and implementation of new European standards that would allow products manufactured in one country in the EC to be used in all the other countries of the EC. With reference to the reaction to fire of constructional materials, EN 13501-1 classifies all constructional products into one of six Euroclasses (A–F) according to their reaction-to-fire performance in fire tests. Two of these tests classify the least combustible materials (Euroclasses A_1 and A_2); these two new tests are a small furnace test for non-combustibility (EN ISO 1182) and an oxygen bomb calorimeter test, which measures the heat of combustion of the material (EN ISO 1716).

At the lower end of the range of Euroclasses (classes E and F), construction products of appreciable combustibility will be assessed using a simple ignitability test (EN ISO 11925-2). Products that fall into Classes A_2, B_1 C and D (and D contains timber and wood-based panels) will be tested using the single burning item test (EN ISO13823), except where the products are used as floor coverings, for which the critical flux (radiant panel) test (EN ISO 9239-1) will be used to determine performance in Euroclasses B–E. For both floor and non-floor applications, generally only two of the above tests will be required to characterise the performance of any one product.

The UK position in late 2008

Appendix A to both volumes of Approved Document B (Fire Safety) of the Building Regulations (dwelling houses and other buildings) sets out the recommendations in England and Wales for the fire performance of materials, products and structures using two alternative systems, one of which utilises EN standards, while the second continues to permit the use of national (British) standards. Similar dual systems apply to Scotland and Northern Ireland.

In the section relating to the reaction to fire of non-combustible materials, performance is defined in terms of *National classes* (clause 8a), defined by testing to BS 476-4:1970 (non-combustible test for materials) or BS 476-11:1982 (method for assessing the heat emission from building products), or *European classes* (clause 8b) as class A1 in accordance with BS EN 13501-1 and when tested to BS EN ISO 1182 and BS EN ISO 1716.

For materials of limited combustibility (which includes timber and panel products), performance is again defined in terms of *National classes* (clause 9a) using the method specified in BS 476-11:1982 or in terms of *European classes* when classified as class A2-s3,d2 in accordance with BS EN 13501-1 when tested to BS EN ISO1182 or BS EN ISO 1716.

A similar picture relates to wall linings, for which where national or European standards may be used.

The relevant national standards are BS 476-7:1997 and BS 476-6:1989. The most commonly recommended rating, class 1, used for the majority of rooms and circulation spaces, is proven by testing the lining (including the substrate) to BS 476-7:1997. Should the circulation space form part of the dedicated escape route from the building, then the lining must also be tested by the method described in BS 476-6: 1989, 'the fire propagation test', and satisfy the criteria given in the approved document or other guidance documents. Materials meeting the dual requirement are known as class '0' materials.

Small rooms in both dwellings and other buildings (as defined in the guidance) are permitted to be lined with class 3 materials as determined by BS 476-7. Timber in its undecorated state is assumed to be class 3, but by treatment can be made to meet both class 1 and class 0 criteria (see section 56.3.1, Flame retardants).

The relevant European standard is BS EN 13501-1. For the majority of rooms and circulation areas, lining materials must have a rating of C-s3,d2, while for circulation areas that form part of the dedicated escape route, the lining must have a B-s3,d2 classification, and for small residential rooms of less than 4 m^2, a lining having a lower classification of D-s3,d2 may be used. When the classification includes 's3,d2' this means that there is no limit set for smoke production and/or flaming droplets or particles. Further information and guidance on the use of the dual system is given in BS 476-10, published 2008.

The use of national and CEN standards

The inclusion of certain parts of BS 476 in approved documents B of the UK Building Regulations means that products manufactured abroad (excluding the other CEN countries) and tested to the relevant parts of BS 476 according to Appendix A of approved document B may be used in construction within the UK. A further consequence of this inclusion is that a UK manufacturer can have his product certified to BS 476 for its use in construction solely in the UK rather than employing the more costly route through EN certification. Products manufactured in Europe, or elsewhere in the world when tested in accordance with the relevant reaction to fire ENs (as of 2008) have free access to the UK market in building construction, owing to the dual system in the guidance documents for the UK building regulations.

The above reaction-to-fire tests relate to the material or product. When that product is incorporated into a building element, the fire resistance of that element will be determined by a whole series of other tests that are outside the scope of this textbook.

References

1. Standards and specifications

BS 476-4:1970. Fire tests on building materials and structures. Non-combustibility tests for materials.

BS 476-6:1989. Fire tests on building materials and structures. Method of test for fire propagation for products.

BS 476-7:1997. Fire tests on building materials and structures. Method of test to determine the classification of the surface spread of flame of products.

BS 476-10:2009. Fire tests on building materials and structures. Part 10: Guide to the principles, selection, role and application of fire testing and their outputs.

BS 476-11:1982. Fire tests on building materials and structures. Method for assessing the heat emission from building materials.

BS 7282:1990 and EN 252:1989. Field test method for determining the relative protective effectiveness of a wood preservative in ground contact.

BS EN 335-1:2006. Durability of wood and wood-based products. Definitions of use classes. General.

BS EN 335-2:2006. Durability of wood and wood-based products. Definitions of use classes. Application to solid wood.

BS EN 335-3:1996. Hazard classes of wood and wood-based products against biological attack. Application to wood-based panels.

BS EN 350-1:1994. Durability of wood and wood-based products. Natural durability of solid wood. Guide to the principles of testing and classification of natural durability of wood.

BS EN 350-2:1994. Durability of wood and wood-based products. Natural durability of solid wood. Guide to natural durability and treatability of selected wood species of importance in Europe.

BS EN 460:1994. Durability of wood and wood-based products. Natural durability of solid wood. Guide to the durability requirements for wood to be used in hazard classes.

BS EN 1995-1-2:1994. Design of timber structures. General rules. Structural fire design.

BS EN 13501-1:2007. Fire classification of construction products and building elements. Classification using data from reaction to fire tests.

BS EN ISO 1182:2002. Reaction to fire tests for building products. Non-combustibility test.

BS EN ISO 1716:2002. Reaction to fire tests for building products. Determination of the heat of combustion.

BS EN ISO 9239-1:2002. Reaction to fire tests. Horizontal surface spread of flame on floor-covering systems. Determination of the burning behaviour using a radiant heat source.

BS EN ISO 11925-2:2002. Reaction to fire tests. Ignitability of building products subjected to direct impingement of flame.

BS EN ISO 13823:2002. Reaction to fire tests for building products. Building products excluding floorings exposed to thermal attack by a single burning item test.

2. Literature

Bravery AF, Berry RW, Carey JK and Cooper DE (1987). *Recognising wood rot and insect damage in Buildings*. Building Research Establishment Report (Ring-bound pocket handbook).

Derbyshire H and Miller ER (1981). The photodegradation of wood during solar radiation. *Holz als Roh-und Werkstoff*, **39**, 341–350.

Derbyshire H, Miller ER, Sell J and Turkulin H (1995). Assessment of wood photodegradation by microtensile testing. *Drvna Industrija*, **46** (No.3), 123–132.

Desch HE and Dinwoodie JM (1996). *Timber – structure, properties, conversion and use*, 7th edition, Macmillan, Basingstoke, 306 pages.

Hall GS and Jackman PE (1975). Performance of timber in fire. *Timber Trades Journal*, 15 Nov 1975, 38–40.

le Van SL and Winandy JE (1990). Effects of fire retardant treatments on wood strength: a review. *Wood and Fiber Science*, **22** (No. 1), 113–131.

Moore GL (1984). The effect of long-term temperature cycling on the strength of wood. *Journal of the Institute of Wood Science*, **9** (No. 6), 264–267.

Shafizadeh F and Chin PPS (1977). Thermal degradation of wood. In *Wood Technology: Chemical Aspects* (ed. Goldstein IS), ACS Symposium Series 4, American Chemical Society, Washington, DC, pp. 57–81.

Chapter 56

Processing and recycling of timber

After felling, a tree has to be processed in order to render the timber suitable for man's use. Such processing may be basically mechanical or chemical in nature or even a combination of both. On the one hand timber may be sawn or chipped, while on the other it can be treated with chemicals that markedly affect its structure and properties. In some of these processing operations the timber has to be dried, a technique that has already been discussed in section 52.7.4 and that will not be referred to again in this chapter.

The many diverse mechanical and chemical processes for timber have been described in great detail in previous publications and it is certainly not the intention to repeat such description here; readers desirous of such information are referred to the excellent and authoritative texts listed under *Further reading*. In looking at processing in this chapter, the emphasis is placed on the properties of the timber as they influence or restrict the type of processing. For convenience, the processes are subdivided below into mechanical and chemical, but frequently their boundaries overlap.

56.1 Mechanical processing and recycling

56.1.1 SOLID TIMBER

Sawing and planing

The basic requirement of these processes is quite simply to produce as efficiently as possible timber of the required dimensions having a quality of surface commensurate with the intended use. Such a requirement depends not only on the basic properties of the timber, but also on the design and condition of the cutting tool. Many of the variables are inter-related and it is frequently necessary to compromise in the selection of processing variables.

In section 52.6 the density of timber was shown to vary by a factor of ten from about 120 to 1200 kg/m^3. As density increases, so the time taken for the cutting edge to become blunt decreases; whereas it is possible to cut over 10 000 feet of Scots pine before it is necessary to resharpen, only one or two thousand feet of a dense hardwood such as jarrah can be cut. Density will also have a marked effect on the amount of power consumed in processing. When all the other factors affecting power consumption are held constant, this variable is highly correlated with the density of the timber, as illustrated in *Fig. 56.1*.

Timber of high moisture content never machines as well as that at lower moisture levels. There is a tendency for the thin-walled cells to be deformed rather than cut because of their increased elasticity when wet. After the cutters have passed over, these deformed areas slowly resume their previous shape,

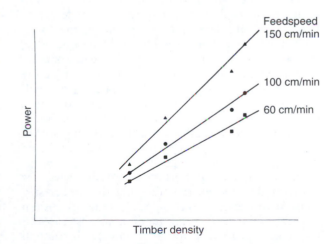

Fig. 56.1 *Effect of timber density and feedspeed on the consumption of power using a circular saw to cut along the grain (rip-sawing) (© Building Research Establishment).*

resulting in an irregular appearance to the surface that is very noticeable when the timber is dried and painted; this induced defect is known as *raised grain*.

The cost of timber processing is determined primarily by the cost of tool maintenance, which in turn is related not only to properties of the timber, but also to the type and design of the saw or planer blade. In addition to the effect of timber density on tool life, the presence in certain timbers of gums and resins has an adverse effect because of the tendency for the gum to adhere to the tool thereby causing overheating. In saw blades this in turn leads to loss in tension, resulting in saw instability and a reduction in sawing accuracy.

A certain number of tropical hardwood timbers contain mineral inclusions that develop during the growth of the tree. The most common is silica, which is present usually in the form of small grains within the ray cells. The abrasive action of these inclusions is considerable, and the life of the edge of the cutting tool is frequently reduced to almost one-hundredth of that obtained when cutting timber of the same density, but free of silica. Timbers containing silica are frequently avoided unless they possess special features that more than offset the difficulties resulting from its presence.

The moisture content of the timber also plays a significant role in determining the life of cutting tools. As moisture content decreases, so there is a marked reduction in the time interval between resharpening both saw and planer blades. The fibrous nature of tension wood (section 52.4) will also increase the wear on tools.

Service life will also depend on the type and design of the tool. Although tungsten carbide-tipped blades are considerably more expensive than steel ones, their use extends the life of the cutting edge, especially where timbers are either dense or abrasive. Increasing the number of teeth on the saw or the number of planer blades on the rotating stock will increase the quality of the surface, provided that the feedspeed is sufficient to provide a minimum bite per revolution; this ensures a cutting rather than a rubbing action, which would accelerate blunting of the tool's edge.

One of the most important tool design variables is the angle between the edge and the timber surface. As discussed in section 52.2, timber is seldom straight grained, the grain tending in most cases to be in the form of a spiral of low pitch, though occasionally it is interlocked or wavy. In these circumstances, there is a strong tendency for those cells that are inclined towards the direction of the rotating cutter to be pulled out rather than cut

cleanly, a phenomenon known as *pick-up* or *tearing*. This defect can be eliminated almost completely by reducing the cutting angle (rake angle) of the rotating blades, though this will result in increased power consumption.

The cost of processing – though determined primarily by tool life – will be influenced also by the amount of power consumed. In addition to the effect of the density of the timber, as previously discussed, the amount of energy required will depend on the feedspeed (*Fig. 56.1*), tool design and, above all, on tool sharpness.

Steam bending

Steam bending of certain timbers is a long-established process that was used extensively when it was fashionable to have furniture with rounded lines. The backs of chairs and wooden hat stands are two common examples from the past, but the process is still employed at the present time, albeit on a much reduced scale. The handles of certain garden implements, walking sticks and some sports goods are all produced by steam bending.

The mechanics of bending involves a pre-steaming operation to soften the lignin, swell the timber, and render the timber less stiff. With the ends restrained, the timber is usually bent round a former, and after bending the timber must be held in the restrained mode until it dries out and the bend is *set*. In broad terms the deformation is irreversible, but over a long period of time, especially with marked alternations in humidity of the atmosphere, a certain degree of recovery will arise, especially where the curve is unrestrained by some fixing. Although most timbers can be bent slightly, only certain species, principally the hardwood timbers of the temperate region, can be bent to sharp radii without cracking. When the timber is bent over a supporting, but removable, strap, the limiting radius of curvature is reduced appreciably. Thus, it is possible to bend 25 mm thick ash to a radius of 64 mm and walnut to a radius of only 25 mm.

56.1.2 BOARD MATERIALS

As a material, timber has a number of deficiencies:

- it possesses a high degree of variability
- it is strongly anisotropic in both strength and moisture movement
- it is dimensionally unstable in the presence of changing humidity
- it is available in only limited widths.

Such deficiencies can be improved appreciably by reducing the timber to small units and subsequently reconstituting it, usually in the form of large, flat sheets, though moulded items are also produced, e.g. trays, bowls, coffins, chair backs. The degree to which these boards assume a higher dimensional stability and a lower level of anisotropy than is the case with solid timber depends on the size and orientation of the component pieces of timber and the method by which they are bonded together.

By comparison with timber, board materials possess a lower degree of variability, lower anisotropy, and higher dimensional stability, and they are also available in very large sizes. The reduction in variability is due quite simply to the random repositioning of variable components, the degree of reduction increasing as the size of the components decreases.

The area of board materials is regarded as the fastest growing area within the timber industry over the last three decades. Not only does this represent a greater volume of construction (particularly in the domestic area) and of consumer goods (e.g. furniture), but it also reflects a large degree of substitution of board materials for solid timber.

There is a vast range of board types though there are only four principal ones – plywood, particleboard (chipboard), oriented strand board (OSB) and medium density fibreboard (MDF).

World production of timber and wood-based panels in 2005 (the last year for which complete data are available) is set out in *Table 56.1*. From this table it will be appreciated that the world production of wood-based panels has increased to a point where in 2005 it was equal to almost 50% of the volume of solid timber used. The volume of particleboard and plywood each comprised a third of the total volume of wood-based panels.

In the UK the consumption of wood-based panels in 2006 was 6.44×10^6 m^3, of which 49% was particleboard, 21% was plywood, 17% was MDF, 7% was OSB and 6% was other fibreboards, the total value of which was £1.2 billion. Home production accounted for 54% of total consumption of panels (*Table 56.2*). The value of wood-based panels (without secondary processing) consumed by the UK in 2006 was £1.18 billion (TTF, 2008). The individual board materials are discussed in turn below.

Plywood

World consumption of plywood in 2005 was 68.9×10^6 m^3, of which about 1.4×10^6 m^3 was consumed in the UK. This is, in fact, approximately the UK's annual consumption, most of which is made from softwood and is imported from the United States, Canada and Finland. Plywoods made from temperate hardwoods are imported mainly from Germany (beech) and Finland (birch – or a birch/spruce combination) while plywoods made from tropical hardwoods come predominately from South-East Asia (mainly Indonesia and Malaysia), South America and to a lesser, but increasing extent, from Africa.

Logs, the denser of which are softened by boiling in water, may be sliced into thin veneer for surface decoration by repeated horizontal or vertical cuts, or, for plywood, peeled by rotation against a slowly advancing knife to give a continuous strip. After drying, sheets of veneer for plywood manufacture are coated with adhesive, laid up with the grain direction at right angles in alternate layers and then pressed. Plywood frequently contains an odd

Table 56.1 World consumption of timber and wood-based panels in 2005 (from Wadsworth, 2007)

Material	World production (10^6 m^3)	Production by sub-total (10^6 m^3)	Percentage of sub-total	Percentage of total production
Timber				
Softwood timber	323.1		76	51
Hardwood timber	102.7		24	16
		(425.8)		(67)
Wood-based panel				
Particleboard	72.4		35	11
Plywood	68.9		33	11
MDF	41.4		20	7
OSB	26.7		13	4
		(209.4)		(33)
Total	635.2	635.2		100

MDF, medium density fibreboard; OSB, oriented strand board.

Table 56.2 UK consumption and production of wood-based panels in 2006 (from TTF, 2008)

Panel type	UK consumption (10^6 m^3)	Percentage of total consumption of all panels	UK production as a percentage of consumption
Particleboard	3.17	49	74
Plywood	1.36	21	0
MDF	1.11	17	79
OSB	0.43	7	68
Other fibreboards	0.37	6	0
Total	6.44	100	54

MDF, medium density fibreboard; OSB, oriented strand board.

number of plies so the system is balanced around the central veneer; some plywoods, however, contain an even number of plies, but with the two central plies having the same orientation, thereby ensuring that the plywood is balanced on each side of the central glue line.

As the number of plies increases, so the degree of anisotropy in both strength and movement drops quickly from the value of 40:1 for timber in the solid state. With three-ply construction and using veneers of equal thickness, the degree of anisotropy is reduced to 5:1, while for nine-ply this drops to 1.5:1. However, cost increases markedly with number of plies and for most applications a three-ply construction is regarded as a good compromise between isotropy and cost.

The common multilayered plywood is technically known as a *veneer plywood*, in contrast to the range of *core plywoods*, in which the surface veneers overlay a core of blocks or strips of wood.

Plywood (veneer type) for use in construction in Europe must comply with the requirements of one part of the European specification BS EN 636, of which the most important requirement is that of bond performance.

The mechanical and physical properties of plywood, therefore, depend not only on the type of adhesive used, but also on the species of timber selected. Both softwoods and hardwoods within a density range of 400–700 kg/m³ are normally utilised. Plywood for internal use is produced from the non-durable species and urea–formaldehyde adhesive (UF), while plywood for external use is generally manufactured using phenol–formaldehyde (PF) resins. However, with the exception of marine-grade plywood, in the UK durable timbers, or permeable non-durable timbers that have been treated with preservative, are seldom used.

It is not possible to talk about strength properties of plywood in general terms since not only are there different strength properties in different grain directions, but these are also affected by configuration of the plywood in terms of number, thickness, orientation and quality of the veneers and by the type of adhesive used. The factors that affect the strength of plywood are the same as those set out in Chapter 54 for the strength of timber, though the effects are not necessarily the same. Thus intrinsic factors, such as knots and density, play a less significant part than they do in the case of timber, but the effects of extrinsic variables such as moisture content, temperature and time are very similar to those for timber.

Plywood is the oldest of the timber sheet materials and for very many years has enjoyed a high reputation as a structural sheet material. Its use in the Mosquito aircraft and gliders in the 1940s, and its subsequent performance in the construction of small boats, in sheathing in timber-frame housing, and in the construction of web and hollow-box beams all bears testament to its suitability as a structural material.

When materials are compared in terms of their specific stiffness (modulus of elasticity per unit mass), plywood is stiffer than many other materials, including mild steel sheet; generally, plywood also has a high specific strength. Another important property of plywood is its resistance to splitting, which permits nailing and screwing relatively close to the edges of the boards. This is a reflection of the removal of a line of cleavage along the grain, which is a drawback of solid timber. The impact resistance (toughness) of plywood is very high, and tests have shown that to initiate failure a force greater than the tensile strength of the timber species is required.

Plywoods tend to fall into three distinct groups. The first comprises those that are capable of being used structurally. Large quantities of softwood structural plywood are imported into the UK from North America, supplemented by smaller volumes from Sweden and Finland. The latter country also produces a birch/spruce structural plywood. The use of this group of structural plywoods in Europe is controlled in that they must first comply with BS EN 636 and second, where limit state design is used with Eurocode 5, the characteristic values for use in design must have been derived from semi-sized test pieces according to European test methods. These values are presented in BS EN 12369-2.

The second group of plywoods comprises those that are used for decorative purposes, while the third group comprises those for general-purpose use. The latter are usually of very varied performance in terms of both bond quality and strength and are frequently used indoors for infill panels and certain types of furniture.

Particleboard (chipboard)

In the UK boards made from wood chips and resin were originally known as chipboards, but with the advent of European standards, the product is now referred to as *particleboard*. The particleboard industry dates from the mid-forties and originated with the purpose of utilising waste timber. After a long, slow start, when the quality of the board left much to be desired, the industry has grown tremendously over the last four decades, far exceeding the supplies of waste timber available and now relying to a very large measure on the use of small trees for its raw material. Such a marked expansion is due in no small part to the much tighter control in processing and the ability to produce boards with a known and reproducible performance, frequently tailor-made for a specific end use. In 2006 the UK consumption of particleboard was 3.09×10^6 m^3, 74% of which was home-produced (TTF, 2008). Particleboard consumption for the whole of Europe in the same year was 33.48×10^6 m^3.

In the manufacture of particleboard the timber, which is principally softwood, is cut by a series of rotating knives to produce thin chips, which are dried and then sprayed with adhesive. Usually the chips are blown onto flat platens in such a way that the smaller chips end up on the surfaces of the board and the coarse chips in the centre. The mat is usually first cut to length before passing into a single or multi-daylight press where it is held for 0.10–0.20 min per mm of board thickness at temperatures of up to 200°C. The density of the boards produced ranges from 450 to 750 kg/m^3, depending on end-use classification, while the resin content varies from about 9–11% in the outer layers to 5–7% in the centre layer, averaging out for the board at about 7–8% on a dry mass basis. Over the last two decades most of the new particleboard plants have installed large continuous presses. As the name implies, the mat is fed in at one end to reappear at the other end as a fully-cured board. This type of press has the advantage of being quick to respond to production changes in board thickness, adhesive type or board density.

Particleboard can also be made continuously using an extrusion process in which the mat is forced out through a heated die. Though a simple process, this results in the orientation of the chips at right angles to the plane of the board, which reduces both the strength and stiffness of the material. Extruded board is used primarily as a core in the manufacture of doors and composite panels.

The performance of particleboard, like that of plywood, is very dependent on the type of adhesive used. Much of the particleboard produced in Europe is made using urea–formaldehyde (UF) resin which, because of its sensitivity to moisture, renders this type of particleboard unsuitable for use where there is a risk of the material becoming wet, or even being subjected to marked alternations in relative humidity over a long period of time. More expensive boards possessing some resistance to the presence of moisture are manufactured using melamine fortified urea–formaldehyde (MUF), or phenol–formaldehyde (PF) or isocyanate (IS) adhesives; however, a true external-grade board has not yet been produced commercially.

Particleboard, like timber, is a viscoelastic material and an example of its deformation over an extended period of time has already been presented (*Fig. 53.10*). However, the rate of creep in particleboard is considerably higher than that in timber, though it is possible to reduce it by increasing the amount of adhesive or by modifying the chemical composition of the adhesive.

Within the new framework of European specifications, six grades of particleboard are specified in BS EN 312, of which four are rated as load bearing (i.e. they can be used in structural design) and two are non-load bearing. The characteristic values for the load-bearing grades for use in structural design are given in BS EN 12369-1.

Particleboards are also produced from a wide variety of plant material and synthetic resin, of which flaxboard and bagasse board are the best known examples.

491

MDF (dry-process fibreboard)

There has been a phenomenal increase in the production of MDF worldwide, with a forty-fold increase from 1980 to 2005. Production in 2005 was over 41×10^6 m^3 and with planned expansion, production capacity would increase to nearly 57×10^6 m^3 by the end of 2008 (Wadsworth, 2007). European production rose from 0.58×10^6 m^3 in 1986 to 14.2×10^6 m^3 in 2005, of which 0.8×10^6 m^3 was produced in the UK. In 2006 UK production was 0.84×10^6 m^3, a volume that represented 68% of consumption.

MDF is manufactured by a dry-process in contrast to other types of fibreboard. The fibre bundles are first dried to a low moisture content before being sprayed with an adhesive and formed into a mat, which is hot-pressed to produce a board with two smooth faces similar to the production of particleboard. Both multi-daylight and continuous presses are employed. Various adhesive systems are employed; where the board will be used in dry conditions a UF resin is employed, while a board with improved resistance to moisture for use in humid conditions is usually manufactured using an MUF resin, though PF or IS resins are sometimes used.

The European specification for MDF (BS EN 622-5) includes both load-bearing and non load-bearing grades for both dry and humid end uses. Characteristic values for structural use of the former are given in BS EN 12369-1, however it should be noted that the use of load-bearing panels under humid conditions is restricted to only short periods of loading.

A very large part of MDF production is taken up in the manufacture of furniture, where non-load bearing grades for dry use are appropriate. Up until 2005 MDF was produced in thicknesses greater than 8 mm, but recently (2006–2008) a number of new plants have come on-stream producing 'thin MDF' (1.5–4.5 mm) thereby extending the use of MDF into a large number of new applications.

Wet-process fibreboard

Fibreboard can also be produced using a wet process, which was the original method of fibreboard production before the advent of MDF. Production levels of wet-process boards have fallen over the years, but it is still used in certain applications, such as insulation and the linings of doors and backs of furniture in the UK, and as a cladding and roofing material in Scandinavia.

The process of manufacture is quite different from that of the other board materials in that the timber is first reduced to chips, which are then steamed under very high pressure in order to soften the lignin, which is thermoplastic in behaviour. The softened chips then pass to a defibrator, which separates them into individual fibres or fibre bundles without inducing too much damage.

The fibrous mass is usually mixed with hot water and formed into a mat on a wire mesh; the mat is then cut into lengths and, like particleboard, pressed in a multi-platen hot press at a temperature of from 180 to 210°C. The board produced is smooth on one side only, the underside bearing the imprint of the wire mesh. By modifying the pressure applied in the final pressing, boards of a wide range of density are produced, ranging from *softboard* (with a density of less than 400 kg/m^3) through *medium-board* (with a density range of 400–900 kg/m^3) to *hardboard* (with a density exceeding 900 kg/m^3). Fibreboard, like the other board products, is moisture sensitive, but in the case of hardboard a certain degree of resistance can be obtained by passing the material through a hot oil bath, thereby imparting a high degree of water repellency to the material, which is referred to as *tempered hardboard*.

The European specifications for the various types of wet-process fibreboards are BS EN 622-2 for hardboard, BS EN 622-3 for medium board and BS EN 622-4 for softboard. UK consumption of wet-process fibreboard in 2006 was 0.37×10^6 m^3, all of which was imported.

OSB (oriented strand board)

Like MDF, OSB production capacity has grown and continues to grow at a very fast rate. From a worldwide capacity in 1997 of 16×10^6 m^3 this has grown to 30×10^6 m^3 in 2006, 85% of which was produced in North America. European production in 2006 was 3.9×10^6 m^3, and UK production from one mill was 0.265×10^6 m^3.

Strands up to 75 mm in length with a maximum width of half its length are generally sprayed with an adhesive at a rate corresponding to about 2–3% of the dry mass of the strands. It is possible to work with much lower resin concentrations than with particleboard manufacture owing to the removal of dust and 'fines' from the OSB line prior to resin application. In a few mills powdered resins are used, though most manufacturers use a liquid resin. In the majority of mills a PF resin is used, but in one or two mills a MUF or IS resin is employed.

In the formation of the mat the strands are aligned either in each of three layers, or only in the outer two layers of the board. The extent of orientation varies among manufacturers with property level ratios in the machine to cross direction of 1.25/1 to 2.5/1, thereby emulating plywood. Indeed, the success of OSB has been as a cheaper replacement for plywood, but it must be appreciated that its strength and stiffness are considerably lower than those of high-quality structural-grade plywood, though only marginally lower than those of many of the current structural softwood plywoods. It is widely used for suspended flooring, sheathing in timber-frame construction and flat roof decking.

The European specification for OSB (EN 300) sets out the requirements for four grades, three of which are load bearing, covering both dry and humid applications. Characteristic values for structural use of the load-bearing grades are given in EN 12369-1.

CBPB (cement bonded particleboard)

This is very much a special end-use product manufactured in relatively small quantities. It comprises by mass 70–75% Portland cement and 25–30% wood chips similar to those used in particleboard manufacture. The board is heavy, with a density of about $1200–kg/m^3$, but it is very durable (owing to its high pH of 11), is more dimensionally stable under changing relative humidity (owing to the high cement content), has a very good performance in reaction to fire tests (again because of the high cement content) and has poor sound transmission (owing to its high density). The board is therefore used in high-hazard situations with respect to moisture, fire or sound. The European specification for CBPB (BS EN 634-2) sets out the requirements for a single grade, while characteristic values have to be obtained from the manufacturer.

Comparative performance of the wood-based boards

With such a diverse range of board types, each manufactured in several grades, it is exceedingly difficult to select examples in order to make some form of comparative assessment. In general terms, not only are the strength properties of good quality structural softwood plywood considerably higher than those of all the other board materials, but they are usually similar to, or slightly higher than, those of softwood timber. Next to a good quality structural plywood in strength are the hardboards, followed by MDF and OSB. Particleboard is of lower strength, but still stronger than the mediumboards and CBPB. *Table 56.3* provides the 5th percentile strength values included in the EN product specifications, with the exception of plywood, for which actual test data for Douglas fir plywood have been used. In passing, it is interesting to note the reduc-

tion in anisotropy in bending strength from 4.5 for 3-ply construction to 1.8 for 7-ply lay-up. Other structural softwood plywoods can have strength values lower than those of Douglas fir, being similar to, or only slightly above, those of OSB of high quality. Actual strength values of individual manufacturer's products of non-plywood panels may be higher than these minimum specification values. It should be realised that these specification values are only for the purpose of quality control and must never be used in design calculations.

Comparison of the behaviour of these products with the effect on their behaviour of 24 hours' cold water soaking is also included in *Table 56.3*. CBPB is far superior to all other boards. Even higher swell values than those recorded in the table, of 25% for 15 mm OSB/1 (general purpose board) and 35% for 3.2 mm HB.LA hardboard (load-bearing, dry), can be found. For those boards listed in the specifications for use under humid conditions that are included in *Table 56.3*, the table provides information on their moisture resistance in terms of their retention of internal bond strength following either the cyclic exposure test (BS EN 321), the boil test (BS EN1087-1) or both.

56.1.3 LAMINATED TIMBER

The process of cutting up timber into strips and gluing them together again has three main attractions. Defects in the original piece of timber such as knots, splits, reaction wood, or sloping grain are redistributed randomly throughout the composite member, making it more uniform in quality than the original piece of timber, where the defects often result in stress raisers when load is applied. Consequently, the strength and modulus of elasticity of the laminated product will usually be higher than those of the timber from which it was made. The second attraction is the ability to create curved beams or complex shapes, while the third is the ability to use shorter lengths of timber, which can be end-jointed.

Glulam

Glulam – the popular term for laminated timber – has been around for many years, and can be found in the form of large curved beams in public buildings and sports halls. In manufacture, strips of timber about 20–30 mm in depth are coated with adhesive on their faces and laid up parallel to one another in a jig, the whole assembly being clamped until the adhesive has set. Generally, cold-setting adhesives are used because of the size of these beams. For dry end use a UF resin is employed,

Table 56.3 Five percentile strength and stiffness and 95 percentile swell values for timber and board materials: quality control values in the EN product specifications

Material	Thickness (mm)	Density (kg/m³)	EN 310 Bending strength (MPa)		Elastic modulus (MPa)		EN 319 Internal bond (MPa)	EN 317 24-hr swelling thickness (%)	Moisture resistance: Internal bond (MPa) after: Cyclic test EN 321	Boil test EN 1087-1
			Parallel	Perpen	Parallel	Perpen				
Solid timber – Douglas fir										
Small clear test pieces	20	590	80*	2.2*	16 400*	1100*	–	–	–	–
Structural timber	100	580	22	–	8110	–	–	–	–	–
Plywood (EN 636-3)										
Douglas fir (three ply)	4.8	520†	51†	11†	8462†	624†	–	–	–	–
Douglas fir (seven ply)	19	600†	42†	23†	7524†	2496†	–	–	–	–
OSB/3 (EN 300)	18	670†	20	10	3500	1400	0.32	15	0.15	0.13
OSB/4	18	670†	28	15	4800	1900	0.45	12	0.17	0.15
Particleboard (EN 312)										
Type P4 (Load-bearing dry)	15	720†	15		2150		0.35	15	–	–
Type P7 (Heavy load-bearing humid)	15	740†	20		3100		0.70	8	0.36	0.23
CBPB (EN 634)	18	1000	9		4000		0.50	1.5	0.3	–
Fibreboard (EN 622)										
Part 2 hardboard (load-bearing humid) HB.HLA1	3.2	900†	38		3800		0.80	15	–	0.5
Part 3 mediumboard (load-bearing dry) MBH.LA1	10	500†	18		1800		0.10	15	–	–
Part 5 MDF (load-bearing humid) MDF.HLS	12	790†	32		2800		0.80	10	0.25	0.15

*Mean value. †Not in the specifications. CBPB, cement bonded particleboard; MDF, medium density fibreboard; OSB, oriented strand board; Perpen, perpendicular.

while for humid conditions a resorcinol–formaldehyde (RF) resin is employed. The individual laminae are end-jointed using either a scarf (sloping) or finger (interlocked) joint. Structural characteristic values for glulam are determined by the srength class of the timber(s) from which it is made, factored for the number and type(s) of laminates used.

Vertical studs and structural beams

The need to increase yields of medium-density structural timber has focused attention in the last decade on the upgrading of lower-grade timber. This has been achieved in a manner similar to that used for glulam in that battens are cross-cut to remove knots and other defects, dried to 10–12% moisture content, their ends finger-jointed and coated with a durable adhesive (usually PF) prior to assembly into a long batten which, because it is only a single member in thickness, can be heat-cured (unlike glulam). These composite beams and studs are again much stronger and stiffer than the original component parts.

A fairly recent extension of this concept has been the gluing of green timber, thereby eliminating the time and cost of kilning. This process has been made possible by the development of adhesives with low penetration of the timber and high rates of curing. The best known of these adhesives resulted from research initiated in 1988 by the New Zealand Forest Research Institute, which led to the creation of the *Greenweld* process in 1990. In 1993 the first Greenweld mill in New Zealand was commissioned, and subsequent mills have been built in New Zealand, Australia, Canada and the USA. The process uses a specific phenol–resorcinol–formaldehyde resin and an accelerator to give a 5-minute closed-press time using conventional finger-jointing machines on timber with a moisture content of up to 180%, and at temperatures down to 0°C. Greenweld-bonded timber may be used for structural applications in the USA provided it has been certified as passing a number of specific tests (Stephens, 1995; Garver, 1998). So far (2008), Greenweld products have not been included in any European standard for structural timber.

56.1.4 ENGINEERED STRUCTURAL LUMBER

The following three products are similar in concept to glulam, but are formed from much smaller wooden components.

Laminated veneer lumber (LVL)

LVL is produced from softwood logs that are rotary peeled to produce veneers 3 mm in thickness which, after kiln drying, are coated with a PF adhesive and bonded together under pressure to produce a large panel 24 m in length. This is then sawn into structural battens. Characteristic values for design use, which are from 50% to 100% higher than corresponding structural softwood timber, are available from the two European and one Canadian manufacturer. The European specification for LVL is BS EN 14279.

Parallel strand lumber (PSL)

PSL is a North American product in which the 2.5 mm thick rotary peeled veneer of Douglas fir or Southern pine is cut into strands 2.4 m in length and 3 mm in width, which are then coated with a resin, pressed together and microwave-cured to produce battens up to 20 m in length.

Laminated strand lumber (LSL)

LSL is another North American product in which aspen veneer is cut into strands 300 mm in length and 10 mm in width and coated with IS resin before being aligned parallel to each other and pressed into thick sheets, which are cut up to produce battens.

56.1.5 MECHANICAL PULPING

Pulp may be produced by either mechanical or chemical processes and it is the intention to postpone discussion on the latter until later in this chapter. In the original process for producing mechanical pulp, logs with a high moisture content are fed against a grinding wheel, which is continuously sprayed with water in order to keep it cool and free it of the fibrous mass produced. The pulp so formed, known as stone groundwood, is coarse in texture, comprising bundles of cells rather than individual cells, and is mainly used as newsprint. To avoid the necessity to adopt a costly bleaching process only light-coloured timbers are accepted. Furthermore, because the power consumed on grinding is a linear function of the density of the timber, only low-density timbers with no or only small quantities of resin are used.

Much of the mechanical pulp now used is produced by disc-refining. Wood chips, softened in hot water, by steaming, or by chemical pretreatment, are fed into the centre of two high-speed counter-rotating, ridged, metal plates; on passing from the centre of the plates to the periphery the chips are reduced to fine bundles of cells or even individual cells. This process is capable of accepting a wider range of timbers than the traditional stone groundwood method.

56.1.6 RECYCLING OF TIMBER WASTE

In 2003 (the latest year for which complete data are available) the total wood waste in the UK was 10.5×10^6 tonnes, comprising municipal waste (1.0×10^6 tonnes), commercial and industrial waste (4.5×10^6 tonnes) and waste from construction and demolition (5.0×10^6 tonnes) (WRAP, 2005). Owing to the lack of good quality data, WRAP stress that these figures should be treated as indicative rather than definitive. In 2003 the manufacture of wood-based panels, principally particleboard, consumed about 1.0×10^6 tonnes of this waste, while horticulture and animal bedding together used 0.2×10^6 tonnes and a similar amount was burned as fuel (WRAP, 2005). Much of the remaining 9×10^6 tonnes would have been incinerated or dumped in landfill.

Since 2003 there has been a significant increase in the amount of wood waste that is now recycled owing in part to significant increases in the cost of timber and in part to the greater public awareness of the need to conserve resources by recycling. Additionally, recent changes to the ROC regulations now mean that waste in the form of wood-based panels can be chipped and burnt as a fuel in the generation of electricity. A direct result of this has been the commissioning in 2007 of two biomass power-stations, which are discussed below as case studies. More of these stations are in construction or at the planning stage in the UK.

The current (2008) use of recycled wood and panels in the UK is set out in *Fig. 56.2*. Although it is not possible to apportion amounts to each end use, it can be safely assumed that particleboard manufacture (see case study 1 below) will be the largest consumer of waste, with energy generation lying in second place (see case studies 2 and 3). A further good example of recycling in Europe is presented as Case study 4.

Case study 1

Many particleboard manufacturers in the UK use a small percentage of chips from waste in the core layer of the board. However, the Sonae plant at Knowsley near Liverpool, with an annual capacity of 450 000 m³, now produces particleboard that comprises 97% recycled wood in the form of chips, 60% of which originated as pallets. The chips are inspected and contaminants (ferrous and non-ferrous metals, silica, plastic and grit) are removed prior to board formation (WBPI, 2007).

Case study 2

The Sembcorp power station (Wilton 10) was officially opened on 19 November 2007. It cost £60 million to build, is located in the Tees valley and is designed to generate 30 MW of electricity from an annual input of 150 000 tonnes of bone-dry timber (about 300 000 tonnes of wet wood). Forty per cent of this is recycled timber, much of which was previously sent to landfill, and includes wood-based panels, demolition timber, sawmill waste and some sawdust. A further 40% will come from small roundwood and 20% from short-rotation willow coppice (Sembcorp, 2008).

Case study 3

The 'e-on' power station at Steven's Croft, near Lockerbie, Scotland, was commissioned in December 2007. It cost £90 million to construct and was designed to produce 44 MW of electricity from 480 000 tonnes of oven-dry timber per year, 20% of which will be recycled timber and boards, 20% from short-rotation willow coppice and 60% as small roundwood and sawmill co-products (slabs, edgings and sawdust) (e-on, 2008).

Case study 4

Another good example of recycling in the manufacture of boards is provided by the Italian Group

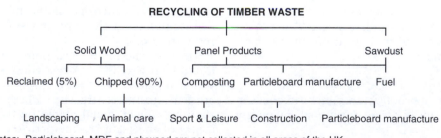

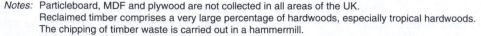

Notes: Particleboard, MDF and plywood are not collected in all areas of the UK.
Reclaimed timber comprises a very large percentage of hardwoods, especially tropical hardwoods.
The chipping of timber waste is carried out in a hammermill.

Fig. 56.2 The recycling of wood and panel waste in the UK.

Mauro Saviola at their factory in Viadana in northern Italy. Since 1997, 100% of its particleboard, amounting to over a million m^3 in 2007 (but due to increase in 2008), has been produced from post-consumer waste. This is collected by 200 trucks from the urban areas of northern Italy and over 1000 trains gathering material from further afield in Europe; the collected material is first cleaned on site prior to its processing. It is claimed that the use of recycled wood saves the felling of 10 000 trees per day (WBPI, 2008).

56.2 Chemical processing

56.2.1 TREATABILITY

The ease with which a timber can be impregnated with liquids, especially wood preservatives, is generally referred to as its *treatability*. Treatability is related directly to the permeability of timber, which was discussed in some detail in section 52.8, where the pathways of flow were described. It will be recalled that permeability was shown to be a function not only of moisture content and temperature, but also of grain direction, sapwood/heartwood, earlywood/latewood and species.

Longitudinal permeability is usually about $10^4 \times$ transverse permeability owing principally to the orientation of the cells in the longitudinal direction. Heartwood, owing to the deposition of both gums and encrusting materials, is generally much less permeable than sapwood, while earlywood of the sapwood in the dry condition has a much lower permeability than the same tissue in the green state, owing to aspiration of the bordered pits in the dry state.

Perhaps the greatest variability in ease of impregnation occurs between species. Within the softwoods this can be related to the number and distribution of the bordered pits and to the efficiency of the *residual* flow paths, which utilise both the latewood bordered pits and the semi-bordered ray pits. Within the hardwoods variability in impregnation is related to the size and distribution of the vessels and to the degree of dissolution of the end walls of the vessel members.

Four arbitrary classes of treatability are recognised in BS EN 350-2. These are:

1. Easy to treat.
2. Moderately easy to treat.
3. Difficult to treat.
4. Extremely difficult to treat.

Different timbers are assigned to these classes according to the depth and pattern of penetration.

Table 56.4 Treatability of the heartwood and sapwood of certain British-grown species (BS EN 350-2)

Species	Treatability of heartwood	Treatability of sapwood
Oak	Class 4	Class 1
Douglas fir	Class 4	Class 2–3
Larch	Class 4	Class 1
Scots pine	Class 3–4	Class 1
Beech	Class 1	Class 1

The treatability classification for selected hardwoods and softwoods of importance in Europe are also given in BS EN 350-2 and illustrated for a selection of British-grown timbers in *Table 56.4*. This classification is derived primarily for preservatives, but is equally applicable to impregnation by flame retardants or dimensional stabilisers, for although differences in viscosity will influence the degree of penetration, the treatability of the different species will remain in the same relative order.

Preservatives and preservation

Except where the heartwood of a naturally durable timber (see section 55.2.1) is being used, timber should always be treated with a wood preservative if there is any significant risk that its moisture content will rise above 20% during its service life. At and above this moisture content, wood-destroying fungi can attack. The relationship between service environment and risk of attack by wood-destroying organisms is defined in BS EN 335-1, 335-2 and 335-3 using the '*use*' (formerly '*hazard*') classification of biological attack (see *Table 55.2*), while BS EN 460 sets out the durability requirements for wood to be used in these use classes. The natural durability (see section 55.2.1) and treatability of certain timbers is given in BS EN 350-2, and clearly those timbers of greater permeability will take up preservatives more easily and are to be preferred over those that are more difficult to treat.

It is normally not necessary to protect internal woodwork, which should remain dry (use class 1). However, where the risk of water spillage, leakage from pipes or from the roof, or where condensation is seen as likely or significant (use class 2), application of wood preservative becomes necessary for most timbers.

A variety of methods for the application of wood preservatives are available. Short-term dipping and surface treatments by brush or spray are the least effective ways of applying a preservative because of the small loading and poor penetration achieved. In these treatments only the surface layers are penetrated and there is a risk of splits occurring during service that will expose untreated timber to the risk of attack by wood-destroying organisms. Such treatments are usually confined to do-it-yourself treatments, or treatments carried out during remediation or maintenance of existing woodwork.

The most effective methods of timber impregnation are industrial methods in which changes in applied pressure ensure controlled, more uniform penetration and retention of preservative. The magnitude of the pressure difference depends on the type of preservative being used. Essentially, the timber to be treated is sealed in a pressure vessel and a vacuum drawn. While under vacuum, the vessel is filled with the preservative and then returned to atmospheric pressure, during which some preservative enters the wood. At this point, an over-pressure of between zero and 13 bar is applied, depending mainly on the preservative being used, but also on the treatability of the timber. This can be held for between several minutes and many hours, after which the vessel is drained of preservative. A final vacuum is often applied to recover some of the preservative and to ensure that the treated timber is free of excess fluid. The degree of penetration and retention of the preservative in solid wood is classified according to BS EN 351-1 in one of nine penetration classes; these should be used as a basis for specifying preservative treatments for particular products. The procedures to be used in the preparation of samples for the determination of the penetration and retention of preservative are described in BS EN 351-2. The performance of different preservatives according to use class is determined by biological testing as set out in BS EN 599-1, while 599-2 specifies for each of the five use classes (defined in BS EN 335-1) requirements for classifying and marking wood preservative products according to their performance and suitability for use. The UK has an interpretive document (BS 8417) which smoothes the transition from the previous 'process-based' British specifications to these new 'results-based' European specifications described above.

There are three main types of preservative. The first group comprises the tar oils, of which coal tar creosote is the most important. Its specification is given in BS 144. Its efficacy as a preservative lies not only in its natural toxicity, but also in its water repellency properties. It has a very distinctive and heavy odour, and treated timber cannot be painted unless first coated with a metallic primer.

Creosote, however, was withdrawn from use in the DIY market by EU member states on 30 November 2003, though it can still be used for industrial applications such as telegraph poles, sleepers, bridges and piles.

The second group comprises the water-borne preservatives. In the past, the most common formulation was that containing copper, chromium and arsenic compounds (CCA preservative). However, since September 2006 approval for the use of CCA preservative has been withdrawn in Europe, though for an interim period it is possible to import into Europe timber treated with CCA for use only in accordance with a restricted list of applications. Combinations of copper/chromium and copper/chromium/boron have also been withdrawn. A number of new water-borne preservatives have appeared on the market, most of them based on copper/organic compounds such as the copper azoles, copper HDO and ammoniacal quaternary compounds either with or without boric acid (Reynolds *et al.*, 2007). All these preservatives are usually applied by a vacuum–high-pressure treatment. The chemicals react once in the wood and become *fixed*, i.e. they are not leached out in service and can be used for ground-contact conditions.

Inorganic boron compounds are also used as water-borne preservatives, but have the disadvantage that they do not become fixed within the wood and therefore can be leached from the wood during service. Their use is therefore confined to environments where leaching cannot take place.

The third group comprises the solvent-type preservatives, which currently represent about five percent of preservative-treated wood. These preservatives tend to be more expensive than those of the first two groups, but they have the advantage that machined timber can be treated without the grain being raised, as would be the case with aqueous solutions. The current formulations of the solvent type are based on a variety of compounds including copper and zinc naphthanates, copper and zinc versatate, zinc ocoate, and the metallic soaps known as acypetacs zinc and acypetacs copper. It should be noted that pentachlorophenol has now been withdrawn from use and that tri-n-butyltin is now rarely used. Some organic solvent preservatives include insecticides and water repellents. These preservatives find uses in the DIY and industrial sectors. Industrial treatment processes include double vacuum and immersion techniques, while DIY and on-site treatment includes dipping and brushing.

In looking at the application of these different types of preservative, creosote and the new water-borne preservatives based on copper/organic compounds are able to protect timber in high-hazard situations such as ground contact, while organic solvent preservatives are used in timber out of ground contact and preferably protected with a coat of paint. It should be noted that the use of creosote is restricted to only industrial applications. Guidance on the selection of appropriate preservatives for use in external timber structures is given by Reynolds *et al.* (2007).

Although it is not an impregnation process as defined above, it is convenient to examine here the diffusion process of preservation. The timber must be in the green state and the preservative must be water-soluble. Timber is immersed for a short period in a concentrated (sometimes hot) solution of a boron compound, usually disodium octoborate tetrahydrate, and then close-stacked under cover for several weeks to allow the preservative to diffuse into the timber. Although colourless, odourless and low in cost, boron preservatives can be leached out of the timber if it is wetted in service and their use must be restricted to use classes where there is no or little risk of wetting.

For many years there was some difference of opinion as to whether preservatives merely lined the walls of the cell cavity or actually entered the cell wall. However, it has been demonstrated by electron microanalysis that whereas creosote only coats the cell walls, the water-borne preservatives do impregnate the cell wall. It is doubtful whether solvent-type preservatives *in general* penetrate the wall owing to their large molecular size and by being carried in a non-polar solvent.

There is considerable variation in preservative distribution in treated dry timber. In softwoods only the latewood tends to be treated owing to aspiration of the earlywood bordered pits (as described in section 52.8.1). In hardwoods, treatment is usually restricted to the vessels and tissue in close proximity to the vessels, again as described in section 52.8.1.

In those timbers that can be impregnated it is likely that the durability of the sapwood after pressure impregnation will be greater than the natural durability of the heartwood, and it is not unknown to find telegraph and transmission poles the heartwood of which is decayed while the treated sapwood is perfectly sound.

Mention has been made already of the difficulty of painting timber that has been treated with creosote. This disadvantage is not shared by the other preservatives and not only is it possible to paint the treated timber, but it is also possible to glue together treated components.

A considerable amount of attention over the last two decades has been focused on the application of artificial preservatives to wood-based panels, but with very limited success. The main difficulty lies in achieving efficacy of preservative treatment without loss in performance of the panel. As a general rule, all types of preservative treatment of the manufactured panel result in considerable losses in its mechanical and physical properties together with a marked increase in its price. Some success has been achieved in reducing the degree of loss in panel properties by treating the chips or strands prior to resin application (e.g. Goroyias and Hale, 2004), the addition of powdered preservatives either in the resin or after resin application or using gaseous diffusion of trimethyl borate on actual panels (e.g. Turner and Murphy, 1998).

Flame retardants

Flame-retardant chemicals may be applied as surface coatings or by vacuum-pressure impregnation, thereby rendering the timber less easily ignitable and reducing the rate of flame spread. Intumescent coatings will be discussed later and this section is devoted to the application of flame retardants by impregnation. There is little evidence to suggest that when flame retardant-treated timber is subjected to fully developed fire conditions it makes the timber burn any slower. Indeed there is evidence to show that some of the treatments can lead to enhanced char rates owing to the nature of the action of the impregnated salts. However, in all cases the treatment will suppress the tendency to ignite. If and when it does ignite, the combustion process will be accompanied by less flaming.

The salts most commonly employed in the UK for the vacuum pressure impregnation process are mono-, di- and poly-ammonium phosphate and ammonium sulphate. These chemicals vary considerably in solubility, hygroscopicity and effectiveness against fire. Most proprietary flame retardants are mixtures of such chemicals formulated to give the best performance at reasonable cost. The fact that these chemicals are applied as an aqueous solution means that a combined water-borne preservative and fire retardant solution can be used, which has distinct economic implications. Quite frequently, corrosion inhibitors are incorporated when the timber is to be joined by metal connectors. Treatment of the timber involves high pressure/vacuum processes; as aqueous solutions are involved redrying of the timber is required.

Considerable caution has to be exercised in determining the level of heating to be used in drying the timber following impregnation. The ammonium phosphates and sulphate tend to break down on heating, giving off ammonia and leaving an acidic residue that can result in degradation of the wood substance. Thus, it has been found that drying at 65°C following impregnation by solutions of these salts results in a loss of bending strength of from 10 to 30%. Drying at 90°C, which is adopted in certain kiln schedules, results in a loss of 50% of the strength and even higher losses are recorded for the impact resistance or toughness of the timber. It is essential, therefore, to dry the timber at as low a temperature as possible and also to ensure that the timber in service is never subjected to elevated temperatures that would initiate or continue the process of acidic degradation. Most certainly, timber that has to withstand suddenly applied loads should not be treated with this type of fire retardant, and care must also be exercised in the selection of glues for construction. The best overall performance from timber treated with these flame retardants is obtained when the component is installed and maintained under cool, dry conditions.

Conscious of the limitations of flame retardants based on ammonium salts, a number of companies have developed effective retardants of very different chemical composition, many of which are polymer-based formulations. These newer products are less likely to produce significant strength losses, are more leach resistant and are usually non-corrosive to metal fixings. However, they are considerably more expensive than those products based on ammonium salts.

Dimensional stabilisers and durability enhancers

In section 53.1 timber, because of its hygroscopic nature, was shown to change in dimensions as its moisture content varied in order to come into equilibrium with the vapour pressure of the atmosphere. Because of the composite nature of timber such movement will differ in extent in the three principal axes.

Movement is the result of water adsorption or desorption by the hydroxyl groups present in all the matrix constituents. Thus it should be possible to reduce movement (i.e. increase the dimensional stability) by eliminating or at least reducing the accessibility of these groups to water. This can be achieved either by chemical changes or by the introduction of physical bulking agents. Dimensional stability can be imparted to wood by swelling of the substrate by means of chemical modification, since the bonded groups occupy space within the cell wall. At high levels of modification, wood is

swollen to near its green volume, and anti-shrink efficiencies close to 100% are achieved. After extended reaction, swelling in excess of the green volume can occur, which is accompanied by cell-wall splitting. Enhacement of durability appears to be an important side-effect.

Various attempts have been made to substitute the hydroxyl groups chemically by less polar groups, the most successful of which has been by acetylation (Rowell, 1984). In this process acetic anhydride is used as a source of acetyl groups. A very marked improvement in dimensional stability is achieved with only a marginal loss in strength. Using carboxylic acid anhydrides of varying chain length, Hill and Jones (1996) obtained good dimensional stabilisation that was attributed solely to the bulking effect, a conlusion that has been supported by further investigation (Papadopoulos and Hill, 2003).

One commercial product based on acetylation came on the European market in 2007. It is claimed that the swelling and shrinkage of this product is reduced by at least 75% compared to untreated timber, while the durability is increased to Class 1 (BS EN 350-2). The product can be obtained in the UK and an example of its use in the UK is given in Suttie (2007).

Another chemical modification reagent is furfuryl alcohol. In a commercial operation that started in 2003, this reagent was prepared from plant waste and reacted with the timber. Not only are the dimensional stability and durability markedly improved, but the hardess of the timber is also increased. The product is also available in the UK (Suttie, 2007).

Good stabilisation can also be achieved by reacting the wood with formaldehyde, which forms methylene bridges between adjacent hydroxyl groups. However, the acid catalyst necessary for the process causes acidic degradation of the timber.

You can find more information on chemical modification in the review by Rowell (1984) and the textbook edited by Hon (1996).

In contrast to the above means of chemical modification, a variety of chemicals have been used to physically stabilise the cell wall. These impregnants act as bulking agents and hold the timber in a swollen condition even after water is removed, thus minimising dimensional movement. Starting in the mid-forties and continuing on a modest scale to the present time, some solid timber, but more usually wood veneers, are impregnated with solutions of phenol–formaldehyde. The veneers are stacked, heated and compressed to form a high-density material with good dimensional stability, which still finds wide usage as a heavy-duty insulant in the electrical distribution industry.

Considerable success has also been achieved using polyethylene glycol (PEG), a wax-like solid that is soluble in water. Under controlled conditions, it is possible to replace all the water in timber by PEG by a diffusion process, thereby maintaining it in a swollen condition. The technique has found application, among other things, in the preservation of waterlogged objects of archaeological interest, the best-known examples of which are the Swedish warship *Wasa* and Henry VIII's *Mary Rose*. The *Wasa* was raised from the depths of Stockholm Harbour in 1961, having foundered in 1628. From 1961 the timber was sprayed continuously for over a decade with an aqueous solution of PEG, which diffused into the wet timber, gradually replacing the bound water in the cell wall without causing any dimensional changes. The *Mary Rose* was launched in 1511 and sank in 1545; its exact location was discovered in 1971 and a large section of the hull was recovered in 1982. It was sprayed continuously with fresh water until 1994 when the spray was changed to an aqueous solution of PEG, a process that will continue for up to 25 years.

PEG may also be applied to dry timber by standard vacuum impregnation using solution strengths of from 5 to 30%. Frequently, preservative and/or fire-retardant chemicals are also incorporated in the impregnating solution. It will be noted from *Fig. 56.3* that following impregnation with PEG, the amount of swelling has been reduced to one-third that of the untreated timber.

Developments in the production and use of water-repellent preservatives based on resins dissolved in low-viscosity organic solvents have resulted in the ability to confer on timber a low, but none the less important, level of dimensional stability. Their application is of considerable proven practical significance in the protection of joinery out-of-doors, which is discussed further in section 56.4.

56.2.2 CHEMICAL PULPING

The size of the pulping industry has already been discussed, as has the production of mechanical pulp. Where paper of a higher quality than newsprint or corrugated paper is required, a pulp must be produced consisting of individual cells rather than fibre bundles. To obtain this type of pulp the middle lamella has to be removed, which can be achieved only by chemical means.

A number of chemical processes are described in detail in the literature. All are concerned with the removal of lignin, which is the principal constituent of the middle lamella. However, during the pulping process lignin will also be removed from within the cell wall as well as from between the cells; this is both acceptable and desirable, since lignin imparts a greyish coloration to the pulp, which is unacceptable for the production of white paper.

However, it is not possible to remove all the lignin without also dissolving most of the hemicelluloses that not only add to the mass of pulp produced, but also impart a measure of adhesion between the fibres. Thus, a compromise has to be reached in determining how far to progress with the chemical reaction, and the decision depends on the requirements of the end product. Frequently, though not always, the initial pulping process is terminated when a quarter to a half of the lignin still remains and this is then removed in a subsequent chemical operation known as bleaching, which, though

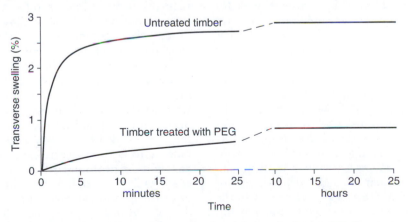

Fig. 56.3 *The comparative rates of swelling in water of untreated pine timber and timber impregnated with a 50% (by mass) solution of polyethylene-glycol (PEG). This is equivalent to a 22% loading on a dry-wood basis (adapted from R.E. Morén (1964) with permission).*

expensive, has relatively little effect on the hemi-celluloses. The yield of chemical pulp will vary considerably depending on the conditions employed, but it will usually be in the range of 40–50% of the dry mass of the original timber.

The yield of pulp can be increased to 55–80% by semi-chemical pulping. Only part of the lignin is removed in an initial chemical treatment that is designed to soften the wood chips. Subsequent mechanical treatment separates the fibres without undue damage. These high-yield pulps usually find their way into card and board-liner, which are extensively used for packaging where ultimate whiteness is not a prerequisite.

56.2.3 OTHER CHEMICAL PROCESSES

Brief mention must be made of the *destructive distillation* of timber, a process that is carried out either for the production of charcoal alone or for the additional recovery of the volatile by-products such as methanol, acetic acid, acetone and wood-tar. The timber is heated initially to 250°C, after which the process is exothermic. Distillation must be carried out either in the complete absence of air, or with small, controlled amounts of air.

Timber can be softened in the presence of ammonia vapour as a result of plasticisation of the lignin. The timber can then be bent or moulded using this process, but, because of the harmful effects of the vapour, the process has never been adopted commercially.

56.3 Thermal processing

Good dimensional stabilisation and an improvement in durability can be obtained by heating timber for short periods of time to very high temperatures (250–350°C). A reduction of 40% in movement has been recorded after heating timber to 350°C for short periods of time (Rowell and Youngs, 1981). It is possible to achieve a reduction in swelling of from 50 to 80% at lower temperatures (180–200°C), again in short periods of time, by heating the sample in an inert gas at 8–10 bar (Giebeler, 1983). The use of lower temperatures (120–160°C) in the presence of air necessitates exposure for several months in order to achieve a similar reduction in swelling. It appears that it is the degradation of the hemicelluloses that reduces the propensity of the timber to swell (Stamm, 1977). Unfortunately, all these treatments result in thermal degradation of timber with considerable loss in strength and especially toughness, unless oxygen is either removed or

reduced appreciably, in which case the magnitude of the loss is also reduced considerably.

There has been renewed interest in themal modification of wood over the last decade with the commercialisation of the process by four European companies to produce slightly different products. In the best known and best documented of these products, timber is first pre-heated to 150°C for 48 hours before the temperature is raised to 240°C for up to 4 hours in an atmosphere low in oxygen. The timber is then allowed to cool and stabilise for 24 hours. The product has been used for cladding and solar shading for over 10 years and it is claimed that it is 50% more stable than untreated softwood, and also that durability is enhanced. Thus, one grade has a durability class to EN350-2 of 3, while a more superior grade has a rating of 2; this performance is in excess of Scots pine heartwood, which has a rating of only 4. Loss in strength varies from 10 to 30% depending on property, time at elevated temperature, exposure temperature and the degree of reduction of oxygen in the atmosphere. Examples of the use of this product in the UK are given by Suttie (2007).

56.4 Finishes

Finishes have a combined decorative and protective function. Indoors they are employed primarily for aesthetic reasons, though their role in resisting soiling and abrasion is also important. Outdoors, however, their protective function is vital. In section 55.1.1, the natural weathering process of timber was described in terms of the attack on the cell-wall constituents by ultraviolet light and the subsequent removal of breakdown products by rain. The function of finishes is to slow down this weathering process to an acceptable level, the degree of success varying considerably among the wide range of finishes commercially available.

In sections 52.1–52.3 the complex chemical and morphological structure of timber was described, while in section 52.7 the hygroscopic nature of this fibre composite and its significance in determining the movement of timber was discussed. The combined effects of structure and moisture movement have a most profound effect on the performance of coatings. For example in the softwoods the presence of distinct bands of early- and latewood with their differential degree of permeability results not only in a difference in sheen or reflectance of the coating between these zones, but also in marked differences in adhesion. In Douglas fir, where the latewood is most conspicuous, flaking of paint from the late-wood is a common occurrence. In addition, the radial movement of the latewood has been shown to be as high as six times that of the earlywood, consequently the ingress of water to the surface layers results in differential movement and considerable stressing of the coatings. In those hardwoods characterised by the presence of large vessels, the coating tends to sag across the vessel and it is therefore essential to apply a paste filler to the surface prior to painting; even with this, the life of a paint film on a timber such as oak (see *Fig. 52.6*) is very short. For this reason, the use of exterior wood stains (see later) is common, as this type of finish tends not to exhibit the same degree of flaking. The presence of extractives in certain timbers (see section 52.3.1 and *Table 52.2*) results in an inhibition of drying of most finishes; with iroko and Rhodesian teak, many types of finish may never dry.

Contrary to general belief, deep penetration of the timber is not necessary for good adhesion, but it is absolutely essential that the weathered cells on the surface are removed prior to repainting. Good adhesion appears to be achieved by molecular attraction rather than by mechanical keying into the cell structure.

Although aesthetically most pleasing, fully exposed varnish, irrespective of chemical composition, has a life of only a very few years, principally because of the tendency of most types to become brittle on exposure to ultraviolet radiation, thereby cracking and disintegrating because of the stresses imposed by the movement of the timber under changes in moisture content. Ultraviolet light can readily pass through the majority of varnish films, degrading the timber at the interface and causing adhesion failure of the coating.

A second type of natural finish that overcomes some of the drawbacks of clear varnish is the *water-repellent preservative stain* or *exterior wood stain*. There are many types available, but all consist of resin solutions of low viscosity and low solids content. These solutions are readily absorbed into the surface layers of the timber. Their protective action is due in part to the effectiveness of water-repellent resins in preventing the ingress of water, and in part to the presence of finely dispersed pigments, which protect against photochemical attack. The higher the concentration of pigments the greater the protection, but this is achieved at the expense of loss of transparency of the finish. Though easy to apply and maintain these thin films, however, offer little resistance to the transmission of water vapour into and out of the timber. Compared

with a paint or varnish the water-repellent finish will allow timber to wet up and dry out at a much faster rate, thereby eliminating problems of water accumulation that can occur behind impermeable paint systems. The presence of a preservative constituent reduces the possibility of fungal development during periods of high moisture uptake. The films do, however, require more frequent maintenance, but nevertheless have become well established for the treatment of cladding and hardwood joinery.

By far the most widely used finish, especially for external softwood joinery, is the traditional opaque alkyd gloss or flat paint system embracing appropriate undercoats and primers; a three- or four-coat system is usually recommended. Multiple coats of oil-based paint are effective barriers to the movement of liquid and vapour water, however, breaks in the continuity of the film after relatively short exposure constitute a ready means of entry of moisture, after which the surrounding, intact film will act as a barrier to the escape of moisture, thereby increasing the likelihood of fungal attack. The effectiveness of the paint system is determined to a considerable extent by the quality of the primer. Quite frequently window and door joinery with only a priming coat is left exposed on building sites for long periods of time. Most primers are permeable to water, are low in elasticity and rapidly disintegrate owing to stresses set up in the wet timber; it is therefore essential that only a high-quality primer is used. Emulsion-based primer/undercoats applied in two consecutive coats are more flexible and potentially more durable than the traditional resin-based primers and undercoats.

A new range of exterior quality paints – that are either solvent-borne or water-borne formulations – has been produced in the last two decades. Some of the formulations have a higher level of moisture permeability than conventional paint systems and have been described as *microporous*. These are claimed to resist the passage of liquid water, but to allow the passage of water vapour, thereby allowing the timber to dry out. However, there appears to be no conclusive proof for such claims.

Solvent-borne exterior paints come in many forms, for example, as a three-layer system based on flexible alkyd resins, which produce a gloss finish, or a one-can system that is applied in two coats, and which produces a low-sheen finish.

Water-borne exterior paints are based on acrylic or alkyd-acrylic emulsions applied in either two- or three-coat systems. Water-borne systems have a higher level of permeability than solvent-borne systems. Even more important is the high level of film extensibility of water-borne systems, which is retained on ageing (Miller and Boxall, 1994) and which contributes to their better performance on site than solvent-borne exterior paints.

Test work has indicated that the pretreatment of surfaces to be coated with a water-repellent preservative solution has a most beneficial effect in extending the life of the complete system, first by increasing the stability of the wood surface and thereby reducing the stresses set up on exposure, and second by increasing adhesion between the timber surface and the coating. This concept of an integrated system of protection employing preservation and coating, though new for timber, has long been established for certain other materials. Thus it is common practice prior to the coating of metal to degrease the surface to improve adhesion.

One specialised group of finishes for timber and timber products is that of the *flame-retardant coatings*. These coatings are designed to be applied on-site, unlike impregnation treatments, and yet allow timber and wood-based panels to comply with the relevant surface-spread-of-flame regulations. Almost exclusively, they are intumescent in action, i.e. the film expands on heating thereby coating the substrate with an insulated char. Some of the finishes are pigmented, but the vast majority are clear so that the natural beauty of the timber shows through. The modern versions achieve the necessary performance rating with much thinner coatings than the first-generation products and, more importantly, have a harder surface, thereby reducing their tendency to pick up finger marks. However they are not suitable for external environments. They do not modify the charring rate of timber, but under fire resistance conditions they will delay ignition for approaching fifteen minutes, i.e. during the critical 'escape' stage of a building on fire.

References

1. Standards and specifications

BS 144:1997. Specification for coal tar Creosote for wood preservation.

BS 8417:2003. Preservation of timber. Recommendations.

BS EN 300:2006. Oriented strand boards (OSB). Definitions, classification and specification.

BS EN 312:2003. Particleboards. Specifications.

BS EN 321:2002. Wood-based panels. Determination of the moisture resistance under cyclic test conditions.

BS EN 335-1:2006. Durability of wood and wood-based products. Definitions of use classes. General.

BS EN 335-2:2006. Durability of wood and wood-based products. Definitions of use classes. Application to solid wood.

BS EN 335-3:1996. Hazard classes of wood and wood-based products against biological attack. Application to wood-based panels.

BS EN 350-2:1994. Durability of wood and wood-based products. Natural durability of solid wood. Guide to natural durability and treatability of selected wood species of importance in Europe.

BS EN 351-1:2007. Durability of wood and wood-based products. Preservative-treated solid wood. Classification of preservative penetration and retention.

BS EN 351-2:2007. Durability of wood and wood-based products. Preservative-treated solid wood. Guidance on sampling for the analysis of preservative-treated wood.

BS EN 460:1994. Durability of wood and wood-based products. Natural durability of solid wood. Guide to the durability requirements for wood to be used in hazard classes.

BS EN 599-1:1997. Durability of wood and wood-based products. Performance of preservatives as determined by biological tests. Specification according to hazard class.

BS EN 599-2:1997. Durability of wood and wood-based products. Performance of preservatives as determined by biological tests. Classification and labelling.

BS EN 622–2 :1997. Fibreboards. Specifications. Requirements for hardboards.

BS EN 622-3:2004. Fibreboards. Specifications. Requirements for medium boards.

BS EN 622-4 :2004. Fibreboards. Specifications. Requirements for softboards.

BS EN 622-5:2006. Fibreboards. Specifications. Requirements for dry process boards (MDF).

BS EN 634-2:2007. Cement-bonded particleboards. Specifications. Requirements for OPC bonded particleboards for use in dry, humid and external conditions.

BS EN 636-1:2003. Plywood. Specifications.

BS EN 1087-1:1995. Determination of moisture resistance. Particleboards. Determination of moisture resistance. Boil test.

BS EN 1995-1-1:2004. Eurocode 5. Design of timber structures. Part 1.1 General. Common rules and rules for buildings.

BS EN 12369-1:2001. Wood-based panels. Characteristic values for use in structural design-Part 1: Particleboards, OSB and Fibreboards.

BS EN 12369-2:2004. Wood-based panels. Characteristic values for use in structural design-Part 2: Plywood.

BS EN 14279:2004. Laminated veneer lumber. Definition, classification and specification.

2. Literature

e.on (2008). Steven's Croft biomass power station. www. eon-uk.com/generation/stevenscroft.aspx.

Garver JW (1998). *The Adhesive system for finger-jointing green lumber*. Proceedings of the 49th meeting of the Western Dry Kiln Association, Reno, Nevada, pp. 24–27.

Giebeler E (1983). Dimensionsstabilisierung von holzduruch eine feuchte/warme/druck-behandlung. *Holz als Roh-und Werkstoff*, **41**, 87–94.

Goroyias GJ and Hale MD (2004). The mechanical and physical properties of strand boards treated with preservatives at different stages of manufacture. *Wood Science and Technology*, **38**, 93–107.

Hill CAS and Jones D (1996). The dimensional stabilisation of Corsican pine sapwood by reaction with carboxylic acid anhydrides. *Holzforschung*, **50** (No. 5), 457–462.

Hon DNS (ed.) (1996) *Chemical modification of lignocellulosic materials*, Marcel Dekker Inc., New York.

Miller ER and Boxall J (1994). *Water-borne coatings for exterior wood*, Building Research Establishment Information Paper IP 4/94.

Morén RE (1964). *Some practical applications of polyethylene glycol for the stabilisation and preservation of wood*. Paper presented to the British Wood Preserving Association annual convention.

Papadopoulos AN and Hill CAS (2003). The sorption of water vapour by anhydride modified softwood. *Wood Science and Technology*, **37**, 221–231.

Reynolds T, Suttie E and Coggins C (2007). *External Timber Structures. Preservative treatments and durability*. Building Research Establishment Digest 503, 8 pages.

Rowell RM (1984). Chemical modification of wood. *Forest Product Abstracts*, **6**, 75–78.

Rowell RM and Youngs RL (1981). *Dimensional stabilization of wood in use*. US Dept. of Agriculture, Forest Products Laboratory Research Note FPL – 0243.

Sembcorp (2008). Looking to a greener future. www.sembutilities.co.uk/wilton10/index.html

Stamm AJ (1977). Dimensional changes of wood and their control. In *Wood Technology: Chemical Aspects* (ed. Goldstein IS). ACS Symposium Series 4, American Chemical Society, Washington, DC, pp. 115–140.

Stephens PL (1995). *Improved recovery through green finger jointing – the greenweld process*. Proceedings of the 46th meeting of the Western Dry Kiln Association, Reno, Nevada, pp. 17–22.

Suttie E (2007). *Modified wood: an introduction to products in UK construction*. BRE Digest 508.

TTF (2008). *The Timber Industry Statistical Review 2006*, The Timber Trade Federation, London.

Turner P and Murphy RJ (1998). Treatments of timber products with gaseous borate esters. Part 2: Process improvement. *Wood Science and Technology*, **32**, 25–31.

Wadsworth J (2007). *Panel market developments – a personal view*. Proceedings of the International Panel Products Symposium, Cardiff.

WBPI (2007). Revenue from 'waste'. *Wood Based Panels International*, June/July 2007, 54–56.

WBPI (2008). Even Greener. *Wood Based Panels International*, August/September 2008, 46–47.

WRAP (2005). *Reference document on the status of wood waste arisings and management* – June 2005. www.wrap.org.uk/applications/publications. 2670kb.

Further reading for Part 8 Timber

Textbooks etc.

Barnett JR and Jeronimidis G (eds) (2003). *Wood quality and its biological basis*, Blackwell Publishing, Oxford, 30–52.

Barrett JD, Foschi RO, Vokey HP and Varoglu E (eds) (1986). Proceedings of the International Workshop on Duration of Load in Lumber and Wood Products, held at Richmond, BC, Canada, 1985. Special publication No. SP-27, Forintek Canada Corp., 115 pages.

Barrett JD and Foschi RO (eds) (1979). Proceedings of the First International Conference on Wood Fracture, Banff, Alberta, 1978, Forintek Canada Corp., 304 pages.

Bonfield PW, Dinwoodie JM and Mundy JS (eds) (1995). *Workshop on mechanical properties of panel products.* Proceedings of a workshop held at Watford, 1995, and published by the management committee of EC COST Action 508, 317 pages.

Bodig J and Jayne BA (1982). *Mechanics of wood and wood composites*, Van Nostrand Reinhold, New York, 712 pages.

Bravery AF, Berry RW, Carey J K and Cooper DE (1987). *Recognising wood rot and insect damage in buildings.* Building Research Establishment Report, BRE, Watford, England, 120 pages.

Desch HE and Dinwoodie JM (1996). *Timber – structure, properties, conversion and use*, 7th edition, Macmillan, Basingstoke, England, 306 pages.

Dinwoodie JM (2000). *Timber – Its nature and behaviour*, E&FN Spon, London, 256 pages.

Gordon JE (1976). *The new science of strong materials*, 2nd edition, Penguin, 269 pages.

Hearmon RFS (1961). *An introduction to applied anisotropic elasticity*, Oxford University Press, Oxford, 136 pages.

Hillis WE (1987). *Heartwood and tree exudates*, Springer-Verlag, Berlin, 268 pages.

Hoffmeyer P (ed.) (1997). *International Conference on Wood–Water Relations.* Proceedings of a conference held in Copenhagen in 1997, published by the management committee of EC COST Action E8, 469 pages.

Jane FW (1970). *The structure of wood*, 2nd edition, Adam & Charles Black, London, 478 pages.

Kettunen PO (2006). *Wood Structure and Properties*, Trans Tech Publications Ltd, Switzerland, 401 pages.

Kollmann FFP and Côté WA (1968). *Principles of wood science and technology. I: Solid wood*, Springer-Verlag, Berlin, 592 pages.

Kollmann FFP, Kuenzi EW and Stamm AJ (1975). *Principles of wood science and technology. II: Wood-based materials*, Springer-Verlag, Berlin, 703 pages.

Morlier P (ed.) (1994). *Creep in timber structures*, RILEM report 8, E. & F. N. Spon, London, 149 pages.

Morlier P, Valentin G and Seoane I (eds) (1992). *Workshop on fracture mechanics in wooden materials.* Proceedings of a workshop held at Bordeaux, France, April 1992, and published by the management committee of EC COST Action 508, 203 pages.

Pizzi A (1983). *Wood adhesives: chemistry and technology*, Marcel Dekker, New York, 364 pages.

Preston RD (1974). *The physical biology of plant cell walls*, Chapman and Hall, London, 491 pages.

Siau JF (1988). *Transport processes in wood*, Springer-Verlag, Berlin, 245 pages.

Skaar C (1988). *Wood–water relations*, Springer-Verlag, Berlin, 283 pages.

Wise LE and Jahn EC (1952). *Wood chemistry*, 2nd edition, Rheinhold Publishing Corp., New York, 688 pages.

PART 9

GLASS

Graham Dodd

Introduction

Glass is a prominent material in modern construction but it is still commonly misunderstood. On one hand it is very simple, being mostly of a single composition, but on the other hand the complexity of multiple processing stages, the variability of its strength and the unforgiving nature of its brittleness make it a challenge to select the most appropriate type for a particular application and to detail its interfaces with other materials.

When glass is used structurally a number of functions are integrated into the glass elements. This requires the simultaneous solution of, for example, strength, robustness, vision, heat and light requirements. Load and impact tests are often performed before breaking the glass but it is not until the glass has been broken that a real understanding of the safety and robustness of glass can be gathered.

The glass processing industry is continually increasing its capability to deliver glass of consistent quality in a wider range of sizes, thicknesses and heat treatment conditions. Heat treatment and lamination are the two most important techniques for controlling the structural properties of glass and the state of processing capability has a direct influence on the state of the art in design and construction. We therefore concentrate on these issues in the first chapter in this section.

The challenge of designing structural applications of glass is to take full advantage of its desirable qualities while compensating efficiently for its shortcomings as a structural material. We discuss the properties in the second chapter and then introduce some of the consequences for design and application in the third chapter. We finish with some comments on durability and end-of life disposal.

Chapter 57

Manufacture and processing

57.1 Manufacturing of flat glass

Flat glass manufacturing is a key industrial process and has been subject to continuous development to achieve lower cost and higher quality since its transition from the craft of mouth-blown glass. Float glass is the dominant process for flat glass, while rolled plate remains in use for patterned (textured) glass and wired glass.

57.1.1 GLASSMAKING MATERIALS

The key constituent is sand, which provides the silica (SiO_2) matrix, and glass making historically evolved in locations with a source of pure silica sand. Pure silica melts sharply at about 1600°C and forms a dense glass with high refractive index on cooling. On slow cooling in nature, silica forms crystals of quartz or coloured gem stones like amethyst, ruby and sapphire according to the presence of small amounts of impurities (elements other than silicon). The high melting temperature and narrow temperature range over which the material can be formed make pure silica glasses impractical for most purposes.

The process of making glass in craft or industry involves the addition of 'fluxes', which are other minerals that lower the melting point and widen the range of workability. Adding soda (Na_2O) in the form of soda ash (Na_2CO_3) lowers the melting point by about 500°C but leaves the glass soluble in water.

Adding lime (CaO) to the soda glass, in the form of limestone, calcium carbonate ($CaCO_3$) makes the glass insoluble and widens its working range. Cullet (broken glass) forms a key part of the batch, improving heat transfer during melting and acting as a flux. Soda lime glass produced industrially also contains dolomite, which adds some magnesium oxide (MgO), and a number of other metal oxides in small quantities to control the melting point, working range and colour. The range of float glass composition is shown in *Table 57.1*, but each float

Table 57.1 Range of float glass composition

Oxide	Range (% by wt)
Silicon dioxide (SiO_2)	69–74
Calcium oxide (CaO)	5–12
Sodium oxide (Na_2O)	12–16
Magnesium oxide (MgO)	0–6
Aluminium oxide (Al_2O_3)	0–3

plant tends to be optimised to a different mix to take advantage of the locally available minerals and the design of the equipment, so chemical analysis can sometimes be used to trace a sample of float glass to a particular plant or manufacturer.

57.1.2 COMPOSITION

Iron oxide in the raw materials gives the glass a light green colour, which has become standard in Europe, but commonly produced glass in other parts of the world has a variety of tones. There are several 'low iron' glasses available that have a much whiter appearance.

Glass used in buildings is almost exclusively soda lime silicate glass, which is defined by a series of European standards in its basic form and when processed in a variety of ways (BS EN 572 and BS 952).[1] Borosilicate glass is also produced industrially by the sheet and float processes, and is used principally for its fire resistance because it has a lower coefficient of thermal expansion, which means it is less likely to crack when heated rapidly in a fire. It is also very widely used in tube form for handling chemicals, for instance as pipe work in laboratories and chemical plant. The chemical composition of borosilicate glass is approximately 70% silica,

---

[1] A list of standards mentioned in the text is included in 'Further reading' at the end this part of the book.

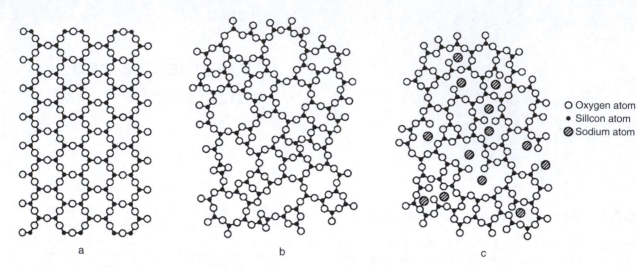

Fig. 57.1 Microstructure of (a) crystalline silica, (b) silica glass, (c) soda glass.

10% boric oxide, 8% sodium oxide, 8% potassium oxide, and 1% calcium oxide.

57.1.3 CONSTITUENTS AND MICROSTRUCTURE OF GLASS

The microstructure of glass is dominated by the nature and arrangement of the bonds between the silicon and oxygen atoms that form the silica matrix (see Chapter 3, section 3.4). The fundamental characteristic of a glass is that it has an amorphous microstructure, meaning no shape or order, so there are no crystalline regions, as illustrated in *Fig. 57.1*.

The significant thing about an amorphous structure is that there are no slip planes, dislocations or grain boundaries to enable plastic flow and impart toughness. Once a crack has been created in a glass, it will encounter no change in properties as it progresses through the material, so it can grow deeper and become a more intense stress raiser until it causes fracture or the load is relieved.

57.1.4 HISTORICAL PROCESSES

Flat glass making started as a craft process using simple tools and developed into the highly mechanised industry of today through several steps. Crown glass was produced by glass blowing techniques and, with considerable skill, a 'crown' of up to 1.1 m diameter could be spun. The surfaces were smooth and fire polished but the thickness varied and circumferential lines were common. Each crown would be cut into rectangular sizes and the 'bull's eye' at the centre was the least valuable, only used where necessary in cottage windows.

Cylinder glass also started by blowing but the bulb was swung on the end of the blowpipe to elongate it as far as possible. The resulting cylinder had its ends cut off and was split lengthwise to open out and anneal on an iron slab. Larger sizes could be made by this process once it became mechanically assisted, but the surface quality was not so good because the glass lay in contact with the iron slab while hot.

Sheet glass is drawn up from a tank through a slot in a ceramic die and gripped by toothed rollers at the edges of the ribbon. Drawn sheet glass has very smooth, fire polished surfaces, tends to have continuous 'draw lines' and is still commonly seen in houses built in the 1930s.

57.1.5 ROLLED GLASS (INCLUDING WIRED AND POLISHED WIRED)

Patterned glass and low optical quality glass for horticultural use is produced by the rolled plate process, in which the molten glass is poured between a pair of temperature-controlled iron rollers. The ribbon is transported horizontally on ceramic rollers as it stiffens and cools until annealed and cut automatically. The lower iron roller is engraved with a pattern to create the texture in the glass, which may be for decorative or obscuring purposes (*Fig. 57.2*).

Wired glass is produced by the rolling process, but with the flow of molten glass divided into two streams, one above the other, two ribbons of hot glass are brought together continuously with wire mesh fed between the two. Further rolling joins the ribbons of glass around the wire and forms textured

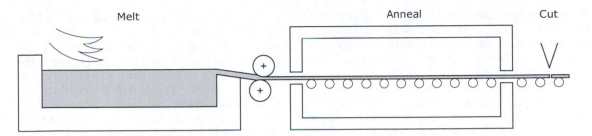

Fig. 57.2 Rolled plate glass-making process.

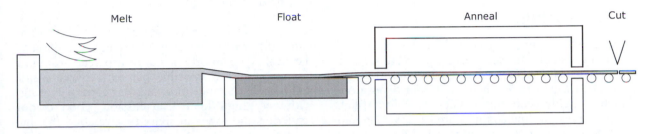

Fig. 57.3 Float glass process.

surfaces to the finished glass. If wired glass with smooth surfaces is required it is mechanically polished to remove the surface texture. Wired glass was the first kind of glass to have any structural integrity after breakage, so it was often used for overhead applications before the introduction of laminated glass, but it does not have the toughness to be an effective safety glass and it is weaker than float glass because of the internal defects caused by the encapsulated wire. Wired glass can provide 30 minutes integrity against fire when properly framed, and is often used for this property in doors and small windows.

57.1.6 FLOAT GLASS

The float process provides over 90% of flat glass production, and has been dominant since its rapid licensing to all the major glass manufacturers by Pilkington during the 1970s, following its development in the early 1960s. In a continuous process, glass is melted from the mineral batch, refined in a tank where the molten glass moves slowly to allow bubbles to escape, then flows onto a bath of molten tin, where it floats and naturally forms a 'ribbon' about 6 mm thick. The interface between the glass and the molten tin is perfectly flat and the top surface is smooth and 'fire polished'. As the ribbon is drawn across the tin bath it is allowed to cool from about 1100°C to around 600°C, at which temperature it is rigid, and it progresses onto a series of ceramic rollers. The rollers carry the continuous ribbon through the 'lehr', a continuous oven in which the temperature of the glass is lowered gradually and evenly so that it emerges in the 'annealed' condition, i.e. with very low residual stress (*Fig. 57.3*).

The cutting of float glass is fully automated and the glass is stacked by robots into packs of typically 2 tonnes for distribution in loads of 20 t. The standard maximum size in Europe is 6000 mm by 3210 mm, and is known as a 'jumbo sheet'. Longer 'super jumbos' up to 8 or 9 m long are available from some float lines to special order, and a very few lines can produce extremely long sheets up to around 12 m, but elaborate handling equipment has to be arranged to off-load and transport such glass.

A typical float line produces around 500 t of glass per day and can be adjusted to produce a range of thickness. The thinner 'substances' have to be stretched out as they cross the float tank, and to make heavy glass the ribbon has to be constrained to build up the thickness. Changing from one thickness to another takes time as the glass flows continuously, and the process is allowed to settle within the tolerances of the next thickness. Float lines operate a system of 'campaigns', in which the substance is stepped up progressively to the maximum and then down to the minimum, with a planned volume of production at each thickness. Float glass for buildings is available in standard thicknesses (defined in BS EN 572-2) of 2, 3, 4, 5, 6, 8, 10, 12, 15, 19 and 25 mm.

Some float lines specialise in body tinted (coloured) solar control glass, where additional minerals are added to pigment the glass to absorb more infrared energy and visible light. Body tints in green, grey, bronze, blue and pink are available in a limited range of thicknesses. It can take a matter of days for a float line to change from one colour to another, as the old composition flows out of the tank and the new composition flows through consistently, and the thickness and quality settle down.

57.2 Coatings

Float glass is commonly coated to alter its transmission of energy or other surface properties. Some coatings can be applied on the float line while the glass is hot, by the process of chemical vapour deposition, which is highly economical. These 'pyrolitic' (high temperature), on-line coatings are hard, strong and durable because they are intimately joined to the glass surface.

Other coatings, known as 'off line' or 'soft' are applied by the separate process of *magnetron sputtering* at room temperature but under high vacuum. The glass enters a continuous series of vacuum chambers, in which the top surface of the glass is exposed to an electric plasma that bounces metal atoms off a solid metal 'target' to deposit as a layer on the glass. Traces of gas can be introduced to the plasma so that the deposited layer is a pure metal, an oxide or a nitride. Pure silver layers are used as part of a 'stack' of up to 12 coatings designed to work in combination to achieve the desired spectral properties.

57.2.1 LOW-EMISSIVITY GLASS

Glass naturally has an emissivity of about 0.9 to the broad spectrum of infrared energy emitted around room temperature. Since emissivity and absorbtivity are two expressions of the same physical property, this means that glass will absorb 90% of heat energy that is radiated to it from the people and interior of a room, and will radiate it equally readily across the insulating cavity of a plain double-glazed unit. To improve the insulating performance of double glazing, low emissivity – 'low-e' – coatings are applied very widely by on-line (pyrolitic) and off-line (sputtering) processes.

57.2.2 SOLAR-CONTROL GLASS

Direct solar energy consists of a few percent ultraviolet, about 47% visible light and about 50% short wave infrared radiation, which passes easily through the glass to the interior. In cold climates and greenhouses this 'solar gain' is beneficial and can be retained by the use of 'low-e' coatings (see above) but in hot climates, or buildings with an excess of internal heat gains and large glass areas, the solar gain can place a large load on cooling systems.

Solar-control coatings are designed to reflect the short wave 'near infrared' radiation back into the environment and allow visible light through to the interior. The best modern coatings have very little effect on the balance of visible wavelengths and so the light retains it natural colour, which is measured as the 'colour rendering index'.

57.2.3 SELECTIVE, HIGH-PERFORMANCE GLASS

Late 20th, early 21st century architecture pursued the ideal of a comfortable but transparent clear glass building with an invisible barrier between inside and outside. A range of clear solar-control coatings with very high light transmission, good colour rendering and low emissivity to give good thermal insulation, were developed. They are known as 'high performance' or 'highly selective' because they select between the visible light to be transmitted and the infrared to be reflected outwards or inwards.

57.2.4 SELF-CLEANING GLASS

There are two competing strategies for making glass 'self cleaning' or 'reduced cleaning'. The first products were liquid applied coatings offered as after-market treatments, which make the glass water repellent or 'hydrophobic'. The objective is that water does not wet the glass surface but forms beads that run off easily, carrying away dust particles. The effect is similar to that of the surface of a well polished car, a lotus flower or a fresh cabbage leaf.

The latest generation of self-cleaning coatings are based on the 'super-hydrophilic strategy', in which the glass surface is made highly attractive to water, so that surface tension is overcome and the water wets the surface thoroughly. The coating of titanium dioxide is applied on the float line, so it is an integral part of the surface and highly durable. The coating acts as a catalyst under ultraviolet light, which breaks down organic deposits assisting the self cleaning action.

57.3 Strengthening processes

57.3.1 TOUGHENING (TEMPERING) WITH HEAT SOAK TEST

Thermal toughening, known as tempering in North America, is based on the phenomenon that glass contracts further at high temperatures if it is cooled

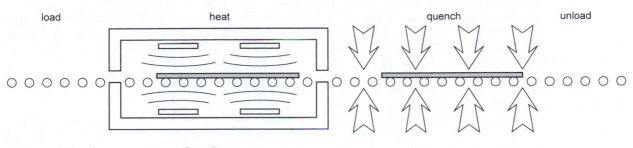

Fig. 57.4 *Toughening process for glass.*

more slowly. The technique is to raise the temperature of the glass evenly above its transition point, say to about 620°C, when it is starting to become soft, and then cool it evenly on both surfaces (*Fig. 57.4*). The surfaces harden quickly and the core of the glass sheet contracts as it cools more slowly, producing a balanced distribution of tension in the centre of the sheet and compression at the surfaces. Typically the surface compression would be around 100 MPa and the central tension about half this value, with the distribution being parabolic through the thickness (*Fig. 57.5*).

With the surface of the glass in a state of permanent compression, any applied stresses must overcome the residual stress before opening surface flaws that could cause fracture. The core of the glass is generally free from defects and so is able to resist the residual tension permanently. Toughened glass can be broken by penetrating the compressive layer at its surface so that a crack enters a zone where the residual stress is tensile. This can be caused by hard impact or by sustained contact with hard materials, especially where sharp features concentrate the contact stress.

The residual compression stress effectively increases the strength by a factor of about four and provides a proportion of strength that is unaffected by the duration of loading. Secondly, when broken, the stored strain energy drives a process of crack branching that spreads throughout the pane in a fraction of a second and divides it into roughly cubic fragments. This characteristic fracture pattern is much less prone to cause cutting and piercing injuries and for this reason properly toughened glass can be classified as a 'safety glass'. Its first widespread use was in car windscreens, and it is still used for side windows and rear screens but has been replaced in windscreens by laminated glass.

Toughened glass is only able to sustain the high residual tensile stress that exists at its core because there are no flaws or cracks in that region to weaken it. However, certain types of inclusion, notably those of nickel sulphide (NiS), undergo a solid-state phase transformation that can cause cracks that lead to what is known as 'spontaneous breakage'. Nickel-bearing ores and metallic nickel, such as is found in stainless steel, are eliminated from the whole process of mining, transporting and preparing the batch for glass making as far as possible, but occasional contamination is inevitable in processes consuming hundreds of tonnes of raw material per day. Nickel in the melt reacts with sulphur, which is a common contaminant, and tiny globules of nickel sulphide form in the glass.

At the high temperature of the float tank, nickel sulphide adopts a close-packed crystal structure know as the 'alpha phase' but at room temperature its stable 'beta' phase is a less dense structure, so nickel sulphide inclusions tend to expand over time as they transform. In annealed glass this is not a problem but when this happens in toughened glass it can cause a tiny crack that initiates fracture. Nickel sulphide inclusions are only critical if they exist in the tensile region of the toughened glass, and are large enough to generate cracks. Critical inclusions tend to be between 40 and 250 μm, which is too small to be obvious on visual inspection, although large enough to see with the naked eye once located.

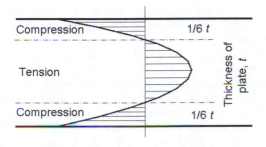

Fig. 57.5 *Distribution of residual stress in toughened glass.*

The unexpected breakage of toughened glass can be inconvenient and costly, alarming, and sometimes dangerous, so a great deal of effort has been dedicated to preventing it. The most effective method has been found to be 'heat soaking', in which the toughened glass is re-heated to a temperature at which critical nickel sulphide inclusions will undergo the transition to the beta phase and cause breakage then and there. Heat soaking is defined in the European standard for heat-soaked thermally toughened soda lime silicate glass BS EN 14179, and involves raising the glass to a temperature of between 280°C and 300°C for two hours.

57.3.2 HEAT STRENGTHENING

The additional strength of toughened glass is valuable for providing resistance to wind pressure and thermal stress, but its fracture pattern makes it prone to collapse once broken and this can be a safety issue if it falls from a tall building. Heat-strengthened glass has about twice the strength of annealed (float) glass and enough thermal shock resistance to suit any practical building example, but its residual compressive stress is low enough not to break the pane into multiple fragments. The intention is that heat-strengthened glass breaks more or less like annealed glass, with a radiating pattern of cracks and very little branching, so that all the fragments extend to the perimeter and can be retained by the glazing system. The European standard for heat-strengthened glass (EN1863) defines the required fracture pattern and sets a minimum characteristic bending strength.

Heat strengthening is carried out with the same machinery as toughening, and the process steps are the same, with the exception of the quenching stage, and the glass is cooled more slowly so that the temperature difference between the core and the surfaces is carefully controlled.

Heat-strengthened glass is not a safety glass because when it breaks it does not break in a safe manner that would avoid injury to someone impacting it. The combination of higher strength, thermal shock resistance and radial breakage pattern makes heat-strengthened glass very useful when laminated with other panes for structural applications.

57.3.3 CHEMICAL STRENGTHENING

Chemical strengthening is an alternative way of establishing compressive stress in the surface of the glass without introducing distortion, and which can be applied to any thickness of glass. The process is, in essence, one of replacing sodium ions (from the soda flux) at the surface of the glass with potassium ions, which are bigger. The wedging action caused by the substitution generates a compressive stress in a very shallow layer of the surface, which is balanced by a very low tensile stress in the core. It is carried out by placing the glass in a bath of hot potassium chloride, the temperature, concentration and duration of immersion being proprietary parameters.

Despite its cost and other limitations, chemical strengthening is the only option for some shaped glasses, such as conics, and is widely used for the layers of laminated aircraft windows, where high-strength thin glass is required for weight saving. Very little strain energy is stored in the panel by the chemical strengthening process, so the treated glass does not fragment like toughened glass, and it is not a safety glazing material unless laminated.

57.4 Forming processes

57.4.1 BENDING

Flat glass can be bent by heating to around 700°C and allowing to slump into or over a mould, often made of steel sheet or tubes draped with refractory fabric. Such moulds can be relatively inexpensive and are suitable for ruled surfaces, especially cylindrical or sinusoidal forms. If a laminated panel is required, a pair of blanks or up to four to make an insulating unit, are stacked together, separated with a mineral powder to prevent adhesion when the glass is hot. The blank is then placed on the mould and the kiln closed and fired. The glass softens and sags into or slumps over the mould and the temperature is lowered to the annealing range, through which it is lowered slowly to allow the stresses of bending to relax and to prevent the creation of residual stress.

Double curves such as segments of a sphere or ellipsoid, which require stretching, are often approximated by sag bending on a 'skeleton mould'. This consists of a shaped steel rail defining the required perimeter profile of the glass, supported by a welded lightweight steel framework, with a reference point indicating the maximum depth of curvature of the required form (*Fig. 57.6*).

An over-sized flat blank of glass is placed on the high points of the skeleton mould, and the kiln heated to the bending temperature. While the glass temperature rises, it has to be observed as it softens and deflects, and the kiln temperature quickly dropped to arrest its flow when the glass reaches the desired depth of curvature. Great skill and experience are required in the design of the mould and the application of heat to the glass, to influence

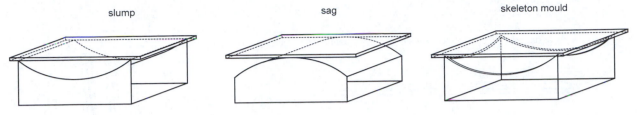

| slump | sag | skeleton mould |

Fig. 57.6 Glass-bending processes.

how and where it flows to get the best approximation of the intended form.

The kiln bending process is adapted to create a wide variety of unique textures and patterns in a process known as 'kiln forming' or 'casting'. Here the flat glass blank is placed on a full surface mould of ceramic or steel and heated until it conforms to the surface texture of the mould.

57.4.2 BENDING AND TEMPERING

Curved or bent toughened glass cannot be made by the processes described above, because both surfaces have to be exposed to cooling air in order to achieve the required residual stress. Special tempering machines are used, in which the rollers of the quenching section are mounted on numerically-controlled mechanisms, and in some cases are segmented and flexible. When the hot glass enters the quenching section, the rollers are moved to the programmed shape and continue to oscillate the glass as it bends under its own weight, before air is blown from above and below to temper the glass. If laminated bent toughened glass is required, each sheet of glass has to be processed separately and with sufficient consistency that pairs can be laminated together. There are no standards for the quality of curved glass at present, so designers have to specify the requirements for individual projects in considerable detail.

57.4.3 CHANNEL GLASS

In an adaptation of the rolled plate process, a rigid glass channel can be created, which has much greater strength and stiffness in one direction than a sheet of glass. Longitudinal wires can also be introduced into the glass to provide a very limited degree of stabilisation after breakage. Glass channels have been widely used as an inexpensive means of providing daylight, especially for industrial buildings, but more recently for architectural applications, because they can span between floors without the need for framing.

57.5 Decoration processes

57.5.1 SAND BLASTING

A simple way to modify the transparency of glass is to sand blast one surface to create a texture that scatters the light and diffuses the image seen through the pane. A range of textures can be produced and the surface can be sculpted to achieve surface relief. Automated sand blasting is used to achieve consistent texture, and can be combined with masking to apply patterns and graphic designs, while manual blasting allows more creative effects. Abrasive granules are fired at the surface of the glass by a stream of compressed air and create a mass of pits and tiny cracks, which act to reflect and refract light in random directions. The surface tends to absorb grease and oils readily, so it shows finger marks and is difficult to clean. Therefore a sand-blasted surface is commonly sealed with a proprietary dirt-repellent treatment before delivery. Sand-blasted surfaces provide higher friction than smooth glass surfaces when wet, so are sometimes used to reduce the risk of slipping on glass flooring.

57.5.2 ACID ETCHING

Glass is resistant to most chemicals but can be dissolved by hydrofluoric acid. The areas not to be etched are masked and the glass set up so that the surface can be flooded with hydrofluoric acid. Variations of the solution and the etching time are used to create a subtle range of textures and, being a skilled craft, acid etching is a costly process. Acid-etched glass diffuses light rather less than sand-blasted because the surface is more undulating and granular, lacking the numerous tiny fracture faces that characterise the sand-blasted finish. However, acid-etched glass is easier to clean and does not mark so readily.

57.5.3 FRITTING

Ceramic 'frit' is a mixture of low melting-point clear glass and mineral pigments in the form of a water-based paste, which can be applied to glass,

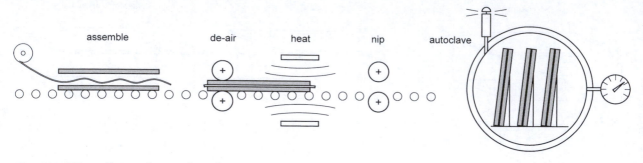

Fig. 57.7 Nip roller and autoclave laminating process.

dried and then fired to create permanent enamel. A wide range of colours is available to special order but most processors offer a limited range of standards, usually including black and white, a few muted colours and a translucent white 'mist frit' that very effectively mimics sand-blasted or acid-etched glass. Abrasive granules can be added to frit to provide a non-slip surface. The frit is fired during the heating phase of the toughening or bending process, and the vast majority of fritted glass is toughened or heat strengthened.

Patterns such as dots, meshes and lines, are usually applied by 'silk-screen printing', in which the frit is forced through a masked screen by a squeegee. Recently, a digital process has been devised by means of which masking of the screen can be carried out, which allows images to be incorporated easily, but the cost of screens remains significant. Some processors are able to apply several layers of different colour frit by successive printing steps.

Frit can also be applied directly to the glass by the 'ink-jet' process on a scale of up to 3 m by 4 m. Direct digital printing in this way can reproduce digital artwork in colour but is currently a slow process, and therefore more suited to individual designs rather than repetitive patterns.

57.5.4 STAINED GLASS

The ancient technique of 'stained glass', in which pigments were fired directly into the surface of clear glass, has been updated recently. Instead of ink-jet printing with a liquid frit, this process deposits dry powder pigments onto the glass under digital control. When the glass is fired, the pigments flow and merge, which results in less definition than a digitally printed frit but much higher translucency and blending of colour.

57.5.5 PRINTING

When glass is laminated, the plastic interlayer material can be digitally printed with patterns or images in full colour. The resolution of printing onto plastics using organic inks is currently better and more flexible than directly printing frit onto the glass.

57.6 Laminating

The brittleness of glass itself cannot be modified, but lamination provides a means of producing a glass pane with a form of ductility. Lamination involves sandwiching a layer of ductile plastic between layers of glass, and can be used to make products with a range of performance, from a modest level of protection in the case of human impact, to the ability to stop high velocity ballistics.

Most glass is laminated by the 'nip roller and autoclave' process, using a polyvinyl-butyral (PVB) clear interlayer. The glass is usually in the annealed form, but can be toughened or heat strengthened and may be coated or decorated before lamination. A sheet of PVB is laid onto one pane and another placed on top, in a 'clean room' section of the factory where the atmosphere is controlled and the potential for contamination minimised (*Fig. 57.7*).

The pane passes through a pair of rubber-coated rollers, which squeeze out most of the air trapped between the glass and the textured surface of the PVB. Once the sandwich has been heated and passed through another pair of 'nip rollers', the glass is stacked on edge upon a 'stillage' and loaded into an autoclave, where the temperature is raised to around 130°C and the pressure is raised to around 16 bar (16 atmospheres or around 1600 kPa). The small amount of trapped air dissolves in the hot PVB interlayer, which wets the surface completely. The temperature and pressure are reduced in a controlled way and the glass is removed from the autoclave.

Glass that is curved, deeply textured or very thick, is not suitable for the nip roller process and a vacuum bag is used in its place. This is a flat bag, formed around the assembled glass from sheets of polyester film and sealed at the perimeter, from which the air is evacuated through a valve.

PVB is the most widely used interlayer because it is reasonably economical, extremely ductile and energy absorbing and resistant to ultraviolet light. One of the alternatives to PVB is ethylene-vinyl acetate (EVA), which is also supplied in sheet form on a roll, and is widely used in the lamination of photo-voltaic cells into modules encapsulated between panes of glass. Panes laminated with EVA tend to be more resistant to high temperatures and less affected by moisture ingress at the edges, although EVA is not as ductile and tough as PVB.

Polyurethane (PU) sheet interlayers have good heat resistance, remaining rubbery and elastic over a wide range of temperatures, and bond well to sheet plastics like polycarbonate. This makes them a preferred choice for some security glazing applications against ballistics and manual attack. The clarity of polyurethane interlayers is good but the cost is higher than for other interlayers.

The latest sheet interlayer to have a significant impact on architectural glazing is only available from one supplier, but has been adopted by a number of processors to offer different mechanical properties from PVB. DuPont Sentry Glass has an 'ionomer' interlayer; developed from a class of plastics originally used for the skin of golf balls, it has high strength and stiffness, which it retains up to around 50°C. Within the polymer structure are ionic bonds that are mobile at processing temperatures but prevent the polymer chains from sliding over each other at service temperatures. Ionomer interlayer bonds strongly to glass and very strongly to metals. The principal advantage of ionomer interlayer is its high shear modulus and resistance to stress relaxation, which allows designers to take advantage of composite action between panes of glass when laminated together, providing the service temperature is moderate.

A number of alternative laminating materials are also available, collectively known as 'cast in place' (CIP), 'cold pour' or 'resin' laminates. These all have the advantage for the processor of being usable with very little investment in equipment. Cast in place interlayers may be chemically cured, such as two-part acrylic (PMMA) or two-part acrylate, polymetyl methacrylate, or single-part resins cured by ultra-violet light (UV), which are for the most part acrylic/acrylate.

Additional layers of material can be incorporated into the laminating process, to enhance technical performance or provide decoration. Multi-layer drawn polyester films provide a dimensionally stable substrate for coating and printing, which can be laminated between glass using two layers of PVB. Decorative materials like fine fabrics, thin wood veneers, metal mesh, expanded metal and even leaves, have also been laminated between glasses, generally by the vacuum bag process.

57.7 Insulating unit manufacture

Solid glass is a poor conductor of heat compared to metals, but not an adequate insulator for the level of thermal performance required of modern buildings. The addition of layers of plastics by laminating does not enhance the thermal performance to any useful degree, so the majority of glass currently used in building façades is in the form of insulating glazing units (IGU), either double- or triple-glazed. The idea of an IGU is to trap a layer of dry air or gas between panes of glass and permanently seal the perimeter to prevent the gas escaping and to prevent moisture penetrating into the dry cavity. It is important to keep the gas in the cavity dry because the outer pane of glass will reach low temperatures during cold weather and condensation would form on its inside face unless the moisture had been removed. The width of the cavity should be sufficient to provide an insulating layer of still air or gas, but not so wide that convection currents can become established, which would reduce the insulation effect by transferring heat between the panes.

The spacer between the panes contains a desiccant material to absorb moisture from the air trapped in the cavity when the unit is assembled. Desiccants were originally silica gel but are now 'molecular sieve' materials that can absorb large quantities of moisture and lower the dew point inside the unit to around −80°C when first sealed. The low humidity within a unit creates a large vapour pressure difference between outside and inside the cavity, which drives moisture past or through the seal materials over time. The primary vapour seal between the metal spacer and the glass is a bead of poly-isobutylene (PIB) in a good quality unit because PIB has very low vapour permeability and remains sticky and flexible over a wide range of temperatures. The bead of PIB, also known simply as 'butyl' has to be continuous and to close any junctions in the spacer to create a complete seal (*Fig. 57.8*).

Changes in the relative pressure between the inside and outside of the unit tend to force the glass away from the butyl primary seal on the spacer, so a 'secondary sealant' is used to hold the two panes together and locate the spacer. Polysulphide

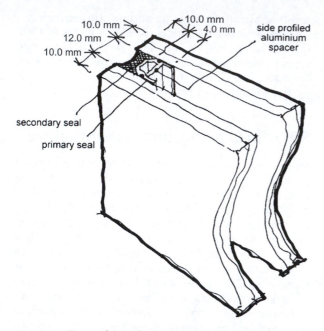

10.0 mm
12.0 mm
10.0 mm

10.0 mm
4.0 mm

side profiled
aluminium
spacer

secondary seal

primary seal

*Fig. 57.8 Typical insulating glass unit edge seal
(courtesy D Richards, Arup).*

secondary sealants are popular for units that will be framed so that the polysulphide will not be exposed to UV radiation, which breaks down its adhesion to glass, and some polyurethane sealants are used for the same purpose. Units whose edges will be exposed to sunlight, for frameless glazing or structural silicone glazing, for example, are sealed with silicone secondary sealants, which are resistant to UV but often tend to be more flexible.

Extruded aluminium spacers are increasingly being replaced with folded stainless steel, pultruded fibre-reinforced plastics or polymer foam spacers, which have reduced thermal conductivity. These types are collectively known as 'warm-edge' spacers because they reduce the cold-edge effect that results from using a conductive metal spacer, and they have a range of thermal performance and a variety of individual advantages.

57.8 Fire–resisting glasses

The process of making wired glass was described above, and it is still commonly used for smaller panes where integrity against fire is required and appearance is not paramount. Non-wired fire-resisting glasses fall broadly into three performance groups and two different technology groups.

The performance of fire-resisting glass is classified according to BS EN 357 in terms of 'integrity only' (E), 'integrity with radiation protection' (EW) and 'integrity with insulation' (EI). If a glass prevents the passage of smoke and flame for a specified period under test conditions then it can be classified as providing 'integrity' for the given period – 30 minutes, 60 minutes or 120 minutes, for example. Higher-performing 'integrity and insulation' glasses have an insulating effect and prevent the non-fire side of the glass rising to dangerous temperatures that would prevent people passing by the outside during a fire.

Bororsilicate glass is successful at providing 'integrity only' protection, because it has a lower coefficient of thermal expansion than soda lime glass, so it is able to resist more severe thermal shock and it is usually toughened, which further increases its thermal shock resistance. Toughened borosilicate has been successfully fire tested in certain sizes up to 60 minutes when framed on top and bottom edges only, with special sealant in the butt joint between panes.

Some laminated soda lime glass products incorporating mineral based interlayers are able to provide 60 or 120 minutes integrity when the interlayer foams to provide the second layer of glass with enough protection to control thermal shock. There are also some fire-resistant products that are essentially highly toughened soda lime glass with smooth edge work in narrow frames that maximise the strength of the glass and minimise the thermal shock it experiences during the heating phase of a fire test.

Multiple layers of annealed low-iron glass, laminated with a transparent layer of hydrated salts, can provide a combination of integrity and insulation. When exposed to fire, the hydrated salt interlayer turns into a foam (intumesces) and expands as the glass softens. The foamed interlayer insulates subsequent layers of glass and blocks the transfer of radiant heat, so that the non-fire side of the glass does not present a burn hazard. This kind of glass can be used to protect an escape route past a window or glass screen because the foamed interlayer prevents the passage of heat and masks the fire from view. Other products consisting of two layers of glass enclosing a thick intumescent gel provide similar performance.

Hardwood frames with deep rebates or steel frames covered in thermal insulation are the preferred forms of fire framing, and the combination of glass and framing is crucial at any particular size of panel and duration of fire resistance.

Chapter 58

Properties and performance

58.1 Physical properties

58.1.1 DENSITY, THERMAL EXPANSION COEFFICIENT, THERMAL CONDUCTIVITY, EMISSIVITY AND SELECTIVITY OF COATED GLASSES

The physical properties of float glass produced in Europe do not vary significantly between manufacturers and can be taken as standard values according to BS EN 572-2. Float glass produced in other parts of the world may vary in colour but the physical properties are very similar. The properties are summarised in *Table 58.1*.

Rolled plate glass and drawn sheet glass have slightly higher densities because the viscosity required for those processes is higher, but this is unlikely to be significant in design.

58.2 Mechanical properties

58.2.1 PATTERNS OF BREAKAGE OF GLASS

The three commonly used heat-treatment conditions of soda lime glass are most clearly distinguished by the manner in which they break.

Annealed glass, which is the standard condition in which it is manufactured, stocked and cut, has

Table 58.1 Physical properties of float glass

Property	Value
Density	2500 kg/m^3
Young's modulus (e)	70 GPa
Poisson's ratio	0.2
Specific heat capacity	720 J/kgK
Thermal expansion coefficient	9×10^{-6} K^{-1}
Thermal conductivity	1 W/mK
Refractive index (average for visible wavelengths)	1.5

a characteristic bending strength of about 45 MPa (not a design value). When broken, cracks run as far as they are driven by the applied force, which may be low, such as a thermal stress, or high such as from impact or wind pressure, in which case the cracks branch and propagate to the edges of the pane (*Fig. 58.1*).

Heat-strengthened glass, produced to BS EN 1863, has a finely balanced level of residual stress, such that its characteristic bending strength is at

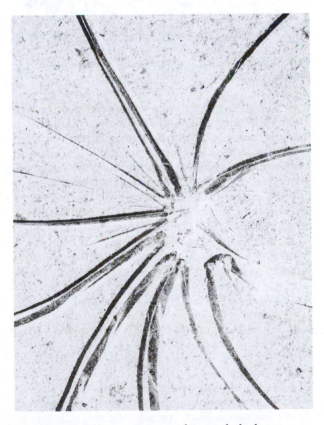

Fig. 58.1 Breakage pattern of annealed glass.

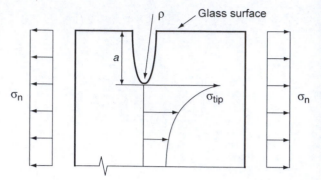

Fig. 58.3 Stress concentration at crack tip.

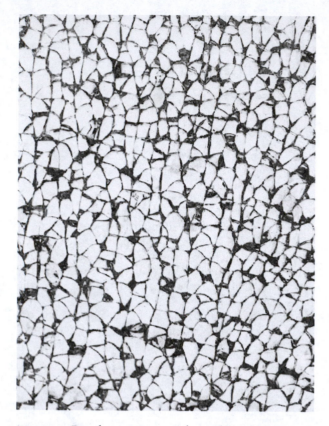

Fig. 58.2 Breakage pattern of toughened glass.

least 70 MPa (not a design value), but propagating cracks do not branch so that often fragments or 'islands' are produced, which could become displaced from the broken pane. The residual stress of heat-strengthened glass ensures that any cracks will propagate to the edges, where compressive stress along the edge often causes the crack to branch by 180° and run parallel for a short distance before breaking out.

Toughened glass, also known as 'fully tempered', whether heat soaked or not, should break into a large number of roughly cubic fragments (*Fig. 58.2*). The size of the fragments is related to the thickness by the standards BS EN 12150 and BS EN 14179. For example, 10 mm glass should break into not less than 40 fragments in a 50 mm square within 5 minutes of breaking.

58.2.2 STRENGTH OF GLASS

Glass has a high theoretical strength (over 30 GPa) because of strong bonds between its molecules but, as we discussed when introducing the concepts of fracture mechanics in Chapter 4, the practical strength

is determined by brittle fracture originating at surface defects. The absence of crystalline structure prevents plastic flow on a macro scale and so glass exhibits virtually perfect linear elastic behaviour until brittle fracture occurs.

When glass is tested to destruction it is common to obtain results considerably higher than the design stress, or even the characteristic stress, because the surface condition of the test sample is in a better condition than we can assume it will be after many years in service.

On the surface of a glass plate there will be a range of flaws such as scratches or pits, and for the purposes of fracture mechanics, flaws are idealised as semi-elliptical cracks normal to the surface, of depth a, with a radius at the tip ρ (*Fig. 58.3*). The stress at the tip of the crack is represented by the stress concentration expression (we first came across this as equation 4.5 in Chapter 4):

$$\sigma_{tip} = 2\sigma_n \sqrt{(a/\rho)} \qquad (58.1)$$

The radius is usually taken to be $\sim 10^{-9}$ mm, and critical depths of flaws in annealed glass are much less than a millimetre, depending on the applied stress.

It is useful to combine the severity of a surface flaw with the applied stress when considering the conditions for brittle fracture, and Griffith introduced the stress intensity factor, K_I:

$$K_I = Y\sigma_n \sqrt{(\pi a)} \qquad (58.2)$$

(where Y is a geometrical factor ranging from 0.56 to 1.12 according to the shape of the crack). The fracture toughness of a material can then be represented by a critical stress intensity factor, K_{Ic}, and any anticipated service condition compared with that limit (i.e. the Griffith failure criterion):

$$K_I \geq K_{Ic} \qquad (58.3)$$

The critical stress intensity factor for soda lime glass is around 0.75 MPa m$^{1/2}$. By way of comparison, the fracture toughness K_{Ic}, for mild steel is of the order of 100 MPa m$^{1/2}$.

58.2.3 STATIC FATIGUE

Soda lime glass is particularly prone to a type of stress corrosion cracking known as 'static fatigue' that makes it weaker under continuous loading than under short-term load. Although glass is brittle, it is actually more resistant to a short-term load like the impact of a football than a long-term load like the pressure of water in a fish tank. Water molecules from the environment can diffuse down a crack in the glass. At the very tip of a crack, if the individual bonds between atoms that are resisting its progress are under enough tension, water molecules can attach themselves and break the bonds, allowing the crack to grow minutely. This process of slow crack growth can start and stop with variations in loading, and can go undetected for long periods.

The strength of glass is found to be highest when measured rapidly because surface flaws under stress will grow, so the strength is usually expressed as the 'short-term strength' or 'sixty-second strength'. Any value of glass strength that is not qualified with the duration of loading should be treated with suspicion.

When the pre-existing flaws grow slowly by static fatigue, their stress intensity increases at an accelerating rate until $K_I \geq K_{Ic}$ and the glass cracks visibly, and usually audibly. There is a threshold stress intensity, K_{I0}, below which a flaw will not grow, which is around 0.25 MPa m$^{1/2}$. Some glass design methods use a factor of between 2.6 and 3.0 to reduce the short-term strength when considering long-term loads.

The relationship between strengths of glass measured over different time periods of steadily increasing load until failure was represented by Charles (1958), in his classic work on why glass is weak when loaded for long duration, with the following relationship:

$$\sigma_{f2} = \sigma_{f1}(t_1/t_2)^{1/n}$$

where n is a material factor, found to be 16 for float glass in air, and t_1 and t_2 are the times to failure in seconds.

58.2.4 POST-BREAKAGE CHARACTERISTICS OF LAMINATED GLASS COMBINATIONS

Laminated glass can consist of any combination of processed glass types with a choice of interlayers with different properties, as described in Chapter 57.

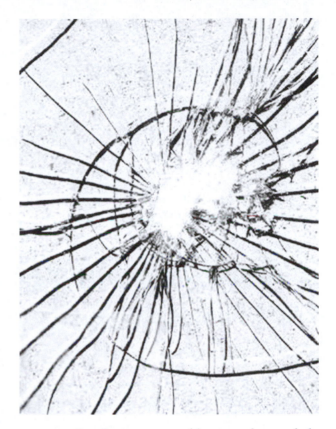

Fig. 58.4 Breakage pattern of laminated annealed glass.

The following combinations are all described assuming a typical PVB interlayer.

Annealed/annealed

This is by far the most common combination and is usually described just by the generic term 'laminated glass', and it is the standard material for vehicle front windscreens. If it is broken by a soft-body impact, a pattern of cracks like a spider's web is formed (*Fig. 58.4*). Radial cracks caused by bending stress and membrane stress in the glass panes are crossed by circumferential cracks where the triangular shards are subjected to bending stress. Hard-body impact may create a small star of cracks or a 'Hertzian cone' if the projectile is fast moving. The cracks in each layer of the laminate tend to follow similar paths if the breaking force is high, but can deviate when the applied load is less. Thermal fracture from edge damage to laminated glass will often break both plies from the same location, with the individual cracks following different paths. If the edge is undamaged, thermal stress may generate cracks from different places in the two plies, or only in one ply.

Heat strengthened/heat strengthened

This combination tends to behave similarly to annealed/annealed because of the similar breakage characteristics of the glass types. This gives the laminated combination a good degree of stability and the capacity to carry small loads once both layers of glass are broken. It is commonly used for glass floors, particularly for outside applications where thermal shock resistance is required and the breakage pattern similar to that of annealed glass would be preferred to that of toughened glass.

Toughened/toughened

This combination offers high ultimate strength but little residual strength after both leaves are broken. When toughened glass fragments it is able to resist compressive loads, but the small particles do nothing to transfer tensile forces. Therefore, a broken panel can only resist bending by virtue of the tensile capacity of the interlayer and tends to fold easily, especially when warm. The tear resistance of a normal PVB interlayer is rarely adequate to support a broken panel on point fixings.

Toughened/heat strengthened

This combination is popular in bolted, or 'point fixed' glazing systems, where toughened/toughened would be at risk of tearing away from the attachment points, particularly in inclined applications. The toughened glass provides high bending strength when the panel is intact, and the heat-strengthened glass provides large chunks and unbroken zones to lock onto the fittings after the panel is broken.

Reference

Charles RJ (1958). Static fatigue of glass, II. *Journal of Applied Physics*, **29** (No. 11), 1554–1560.

Chapter 59

Design and applications

59.1 Design of glazing and selection of type of glass

The selection of type of glass involves consideration of a range of factors and combination of a number of processes to achieve the required set of properties. In some cases the design will be constrained by the limits of one or more processing steps, particularly when it comes to panel size, and the inner and outer panes of insulating units are commonly different types. Only after the glass type has been selected can a suitable thickness be calculated. In the UK, the current standard for this is BS 6262.

In many structural applications of glass there is no agreed method or precedent for design and in these instances an approach based on first principles is necessary. The fundamental approach is to anticipate scenarios in which the glass will get broken and provide levels of redundancy by duplication or by an alternative load-path that does not rely upon the integrity of the glass. Design guidance, references to standards and built details for a wide range of structural glass applications is contained in the Institution of Structural Engineers guide *Structural use of glass in buildings* (I Struct E, 1999).

59.2 Deflection limits for glazing

There are no mandatory requirements for deflection limits on window glass specified in codes of practice commonly used in the UK, US, Australia and mainland Europe.

59.2.1 DEFLECTION CRITERIA
Insulating glass unit (IGU) edge seals can be over-strained if the edges are allowed to deflect significantly, which leads to the ingress of water vapour into the cavity, premature internal 'misting' and early unit breakdown. Some manufacturers have robust edge-seal constructions that will accommodate deflections of up to span/50 at full design wind load. The response of glazing to hand pressure seems to be acceptable if its natural frequency is above 4 Hz, whereas glazing with a natural frequency of around 2 Hz feels flimsy because it wobbles readily if you push it.

59.2.2 GUIDANCE FROM STANDARDS
Most glass design codes allude to the need to limit deflection, but avoid direct specification. The UK codes for patent glazing (BS 5516) and the structural design of aluminium (BS 8118) set the following limits for mullion deflection:

- span/125 for single glazing
- span/175 for insulated glazing.

This deflection limit for single glazing limits the stress in the supported edges, which is important for annealed and wired glass. The limit for double glazing is intended to protect the edge seals of IGUs. The deflection limits also tend to be applied to the unsupported edges in one-way spanning glass.

59.3 Design stresses and load factors

There has historically been a lack of consensus regarding glass design stresses, with varying views held between industry and consulting engineers, and also across the continents, where the basis of design codes varies widely. There is no current UK code of practice providing a general basis for glass design.

The permissible stresses in *Table 59.1* have been used without load factors as general rules of thumb, but would not be appropriate for precise design of critical glass elements.

Progress has been made on limit state design of glass, although the combination of loads of different duration combined with the time-dependent strength of glass makes it important to account for the stress

Table 59.1 Approximate permissible stresses for glass according to duration of load

Type of glass	Typical allowable stresses (MPa)		
	Short (secs)	Medium (hours)	Long (years)
Toughened glass	50	50	50
Heat-strengthened glass	25	25	25
Annealed glass	17	8.4	7

history that is expected for a structural glass element, and so the combination of load factors becomes complex. Haldiman *et al.* (2008) have published on behalf of The International Association for Bridge and Structural Engineering a structural engineering document on the structural use of glass, in which this subject is covered in great depth. This is an effective state-of-the-art work for practising engineers.

59.3.1 STRENGTH OF LAMINATED GLASS

The strength and stiffness of laminated glass depend on the extent to which the interlayer couples the two sheets of glass in shear. If the shear transfer between two sheets is fully effective, they will behave as one, with thickness equal to the two combined. However, plastic interlayer materials are viscoelastic to varying degrees, so they tend to relax and flow when stressed. At low temperatures or when loaded for a very short time, a viscoelastic interlayer may appear rigid and a laminated pane may be almost as stiff and strong as a monolithic pane of the combined thickness. This is known as 'composite' action. At higher temperatures, or when loaded for a longer period, the individual plies slide over each other, allowing the deflection and bending stress in the glass to increase, which is known as 'layered' action.

59.4 Windows

Standards for the design of glazing for windows exist in most developed countries, although there is wide variation in the methods of calculation and the values for the strength of glass used when calculating the required thickness. Large window panes tend to deflect by two or three times their own thickness under full design wind load, so membrane stresses are generated, being tensile in the centre of the pane and compressive around the edge. These stresses act in combination with the bending stiffness of the glass and can significantly reduce the deflection and maximum principle stress compared with what is predicted by simple, first-order, bending theory.

Large-deflection theory solutions to the bending of plates are more complex to calculate but some design codes, such as ASTM E-1300, provide reference charts to save the designer lengthy calculations. Some other codes use simple bending calculations but factor up the design stress to allow for the fact that the calculated stress will be overestimated. This is one of the reasons why design stresses from window codes cannot be taken to be safe values for more general application to glass engineering.

59.4.1 DESIGN OF INSULATING UNITS

When two panes of glass are sealed around their perimeter to create an insulating unit, the fixed mass of air between them obeys the General Gas Law (pressure × volume/temperature = constant). Consequently, the pressure inside the unit will rise if one pane is displaced because it changes the volume of the cavity (*Fig. 59.1*). The wider the cavity, the larger the initial volume contained and the less significant are small changes in volume caused by displacement of the panes, so units with thinner air spaces will share load between panes to a greater degree.

When the temperature of the enclosed gas is raised, for instance in hot weather, the internal pressure rises and the panes of glass flex outwards, and in cold weather the effect is reversed. The pressure changes caused by temperature, atmospheric pressure and a change in altitude are considered together as the 'climatic load' on an insulating unit by some window-design methods. Climatic loads are likely to be insignificant for large flat units with modest cavity width, because the glass deflects easily and the sealed volume changes rapidly, so the differential is equalised. However, climatic loads can become important to consider along with other loads if the units are very long and narrow, or curved.

59.5 Glass walls and structural glass assemblies

The use of glass as a structural element in a wall is invariably linked to the desire for transparency or translucency. Glazed areas can function as collectors or dissipaters of energy under different weather conditions. The gain or loss of energy may not be beneficial at all times and glazed walls are increasingly required to be double glazed and incorporate selective coatings to control the transmission of

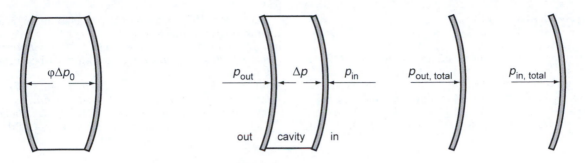

Fig. 59.1 *Pressures acting on the panes of a double-glazed unit.*

incoming solar radiation and the loss of heat from the building. The use of double-glazed panels and coated glass adds to the challenge of using the glass safely and reliably in a structural role.

Glass in a wall could present a safety hazard in two ways if it should break and fall from the opening; it could injure someone below and it could allow someone inside the building to fall out. Given the availability of heat-treated and laminated glass, it is reasonable to apply the criterion that if falling glass would present a danger to people then glazing should retain a reasonable level of stability after breakage, at least sufficient to allow further measures to be taken before any glass should fall. In regions subject to hurricanes the most onerous requirements are placed upon glass, such as impact damage from a baulk of timber followed by several thousand reversing load cycles.

59.6 Skylights

The criteria for glass roofs fall between the requirements for walls and floors; in some cases the roof of a building has full public access and is effectively a floor. However, most glass roofs will need pedestrian access for cleaning and maintenance and, even if safe routes are provided to avoid walking on the glass, it is essential to avoid the hazards associated with fragile roofing. The Centre for Window and Cladding Technology in Bath has published a technical note, TN42, which gives guidance on glazed-roof design and sets out a test regime to verify that the glazing will prevent someone falling through, even if the glass is broken (CWCT, 2004).

59.7 Floors and stairs

The use of glass as a floor introduces the need for slip resistance and increases the level of robustness required. It also puts glass into a location that invites rough treatment. In a floor application the loading may remain applied for long durations, which means reducing the allowable stress according to the heat-treatment condition of the glass being used and taking account of the viscoelastic properties of the polymer interlayers that will affect the degree of composite action between panes (Smith and Dodd, 2003).

59.8 Glazing for security

Security against manual attack, 'anti-bandit' or 'anti-break-in' glass is intended to delay or prevent someone breaking through the glass with a variety of tools. The design objective is to absorb energy from the impacts of something like a hammer or an axe and avoid the interlayer being cut, torn or penetrated such that it would allow someone to gain access through the window.

Glazing to resist ballistics is classified according to tests in which weapons of different calibre are fired at glass samples, usually in a group of three shots in a triangular pattern. The kinetic energy imparted by some weapons requires many layers of glass and interlayer, which results in a thick and heavy construction. Polycarbonate sheet is often laminated in place of some glass layers, using polyurethane interlayer because of its potential to absorb energy.

When glazing is exposed to a blast, the primary design objective is to protect people inside the building from injuries caused by flying glass. The broken panel must be kept attached to the frame if it is to prevent fragments flying across the room, so the glazing details have to be deep and robust. The frame construction and its attachment to the building structure also have to be designed to carry the reactions from blast loading, which are sometimes calculated using non-linear analysis of the dynamic response of the whole system.

References

Centre for Window and Cladding Technology (2004). *Safety and fragility of glazed roofing – guidance on specification and testing*. Technical Note 42, Centre for Window and Cladding Technology, Bath.

Haldiman M, Luible A and Overend M (2008). *Structural Use of Glass – Structural Engineering Document 10*, International Association for Bridge and Structural Engineering.

Institution of Structural Engineers (1999). *Structural use of glass in buildings*, Institution of Structural Engineers, London.

Smith A and Dodd G (2003). *Performance Criteria and Tests for Novel Glass Construction*. Proceedings, Glass Performance Days (GPD), Tampere, Finland, 2003. Papers are available at GLASSFILES.COM

Chapter 60

Service and end of life

60.1 Durability

Float glass is highly durable in most environments encountered in construction applications, and it does not degrade or change over time, other than the accumulation of microscopic surface damage, which reduces the strength of new glass but tends to stabilise after several years. Therefore monolithic single glazing can usually be regarded as maintenance free for the lifetime of a glazing system, aside from routine cleaning and replacement in the event of breakage.

Some environments can have a detrimental effect on glass. Most notably, alkaline conditions created by contact with cementitious materials can lead to etching of the surface. Even a puddle of water on a glass roof that is too flat to drain properly can become highly alkaline around the perimeter as it dries because some of the alkali fluxes dissolve from the glass. This results in permanent etched marks around the puddle.

When glass is assembled into a double-glazed unit, the edge seals are expected to offer a limited service life of up to 25 years or so. Similarly, interlayers in laminated glazing may not be as durable as the glass itself owing to mechanisms such as yellowing when exposed to UV radiation, loss of plasticisers or delamination.

60.1.1 CLEANING

Glass is usually cleaned with a solution of mild detergent in water, using a mop and squeegee. The mop wets the glass and loosens the dirt, aided by the detergent, and the squeegee is used to sweep the dirty water off the glass, leaving it dry. The process does not use a great quantity of water, and because it is not left on the glass, it does not matter that the water becomes dirty. If droplets of water are left to dry on glass, they leave faint white rings where salts are leached from the surface and deposited as the water dries.

Self-cleaning treatments on glass (section 57.2.4), either hydrophobic or hydrophilic, rely on regular wetting by rain to carry away dust and dirt. The drying out of droplets is avoided, either by forcing them to run off as beads or by drawing them out over the surface of the glass until dry, according to the technology employed.

60.1.2 PROTECTION ON SITE

Glass can easily be damaged on a construction site by impact, particularly on the edges and corners, and minor damage before glazing can result in premature failure in service when thermal and other stresses start to act on the glass. Welding and grinding works pose a less obvious risk to unprotected glass because sparks, spatter and dust can fuse to the surface of glass. Hot metal particles cause pits and can initiate vents (cracks) into the surface, which may substantially weaken glass, cause scratching when dislodged by window cleaning, or create rust stains when exposed to the weather.

Stacked glass also has to be kept dry because it can be permanently marked if wetted when in contact with other glass or packing materials, especially in the presence of cementitious dust.

60.1.3 FAILURE OF DOUBLE GLAZED UNITS

Insulating glass units eventually absorb enough moisture that condensation occurs within the cavity during cold weather, which damages or negates the effect of any coating and spoils the view out. The service life is affected by the quality of the original manufacturing and the conditions in which the unit is used. If the edge seals are exposed to liquid water for long periods, the life can be dramatically shortened.

60.1.4 DELAMINATION OF LAMINATED GLASS

Laminated glass is generally resistant to occasional wetting of its edges, if they are allowed to dry out but, like insulating units, can be rapidly damaged by standing in water. Early effects may be seen as a white 'fogging' of the interlayer, followed by progressive loss of adhesion between the interlayer and the glass.

60.2 What to do if glass breaks

Broken glass is often viewed as dangerous rubbish to be disposed of as swiftly as possible, but in many cases someone will want to know why the glass broke, and that can only be determined if the evidence is preserved.

If glass presents a safety risk because it might fall and cause injury then the first priority should be to exclude people from the area where it might fall. If it has to be removed for safety, it is usually helpful to photograph the breakage pattern as clearly as possible and collect all the glass fragments for later examination. Cracks in all materials have markings that relate to the manner of breaking and the properties of the material. The study of those markings is known as fractography, and it is especially useful in the understanding of brittle materials like glass.

Whenever glass is tested to destruction or glass breaks in service there are lessons to be learnt by the designer from a study of the fracture markings. There are several texts on the subject of fractography, which should be required reading for any engineer wishing to work with glass or other brittle materials (e.g. Quinn, 2007).

60.3 Disposal and recycling

The raw materials required to produce glass are available in abundant supply. However, the energy cost of actually producing glass from the raw materials is high owing to the temperatures involved. Further high-energy procedures such as toughening and heat-soak testing may be carried out once the float process has been completed. This should be set against the fact that glass is a durable material, and therefore offers the benefit of prolonged, low-maintenance service superior to many of the possible alternatives. Such materials, for example historic stained-glass windows, can be seen in many old buildings, where material hundreds of years old has survived since construction and continues to function as intended.

The stability of glass makes disposal difficult, as it will not readily break down. Recycling is a viable option, by crushing, re-melting and reforming waste glass into a new product. This process is not difficult to carry out, however it tends to lead to contamination and it is difficult to produce recycled glass of the highest optical quality. Therefore recycled material is mostly used in non-architectural applications, such as coloured drinks bottles, where visual quality is less critical. Reusing crushed glass directly, for example as a secondary aggregate for concrete or screed, is also possible (see Chapter 62, section 62.5.3) though the level of demand is limited.

Reference

Quinn G (2007). *Fractography of Ceramics and Glasses*, Special Publication 960-16, NIST, USA. (Required reading for anyone wanting to diagnose glass breakage. Free download.)

PART 10

SELECTION AND SUSTAINABLE USE OF CONSTRUCTION MATERIALS

Peter Domone

Introduction

The preceding parts of this book have dealt with each of the major construction materials in turn. However, in the early stages of designing a structure, the engineer will need to choose the most appropriate material for the task. In some cases the choice will be obvious, for example high strength steel for the cables of a suspension bridge, but in many cases two or more materials may be appropriate, e.g. timber, reinforced concrete or steel (or a combination of all three) for a bridge with a modest span.

The choice will depend on many factors. Uppermost must be the ability of the material to fulfil the structural requirements, and in this respect many of the properties already described in this book will be important, e.g. strength, stiffness, toughness, durability, etc. etc. A number of other factors are also critical, including:

- cost (both initial and during the lifetime of the structure)
- availability (often linked to cost)
- fabrication, either on-site or, increasingly, off-site for rapid assembly on-site
- the energy required to produce the structure and consumed during its lifetime

- the maintenance required to ensure that the structure remains fit-for-purpose
- the end-of-life properties, particularly re-use and recycling.

In this part of the book we bring together some of the more important properties of construction materials that need to be considered when making a selection. In the first chapter we make some comparisons between the mechanical properties of the materials that have been discussed in earlier chapters, and see how this can contribute to their selection for any given application. The second chapter is concerned with the increasingly important subject of sustainability. We first discuss some of the background issues involved and the initiatives that the construction industry is taking as part of the drive to increasing the sustainability of society as a whole. We then go on to consider the uses of and developments in each of the major construction materials in relation to these initiatives. We will not discuss costs, important though these are. These can vary and fluctuate widely in different regions and at different times, and so any figures that we give are likely to be immediately out of date.

Chapter 61

Mechanical properties of materials

In this chapter we bring together the values of some of the important mechanical properties of materials that have been discussed in the preceding sections. We will consider the ranges of these properties, and some others derived from them, and make some comparisons; this will help in understanding how this contributes to their selection and effective use.

61.1 Ranges of properties

Table 61.1 shows typical values or ranges of values of density, elastic modulus (stiffness), strength (or ultimate stress) and toughness for some of the individual and groups of materials that we have described in this book, together with those for one or two other materials for comparison (we are not suggesting using diamond as a load-bearing material, but it is included as example of an 'extreme' material).

It is immediately apparent that there is a wide range for each property, which gives engineers substantial scope either for selecting the most appropriate material for the job in hand, or for assessing the benefits of an alternative material. However, we normally do not get something for nothing and in many cases an advantageous property can be offset by some disadvantage, e.g. the reduced toughness of higher-strength steel and the low tensile strength of concrete. Strategies are therefore required for coping with this, e.g. modifications to design rules or by forming composites of two or more materials.

It is also interesting to compare the overall ranges of each property, which are more clearly apparent in *Figs 61.1* to *61.4*, in which the materials have been grouped into the broad classes of metals, ceramics, polymers and composites. Density (*Fig. 61.1*) varies by about two orders of magnitude from the

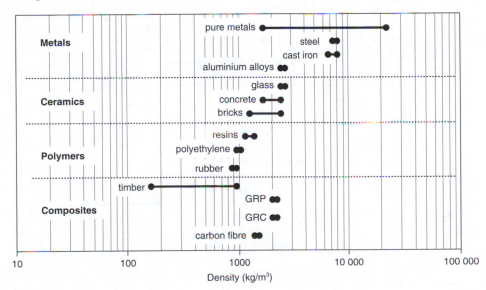

Fig. 61.1 *Densities of structural materials.*
(GRC = glass reinforced cement; GRP = glass fibre reinforced plastic (resin).)

Table 61.1 Selected properties of a range of materials (Sources: Ashby and Jones (2005) and earlier chapters in this book)

Material	Density (kg/m³)	Stiffness (E) (GPa)	Poisson's ratio	Strength or limiting stress (MPa)*	Toughness (G_c) (kJ/m²)	Fracture toughness (MN/m³ᐟ²)
Diamond	3500	700–1200	0.1–0.29	8600–16 500		
Pure metals	1750–21 000	15–290	0.25–0.35	10–80	100–1000	150–300
Structural steel	7850	195–205	0.3	235–960	100–50	150–100
High-strength steel	7850	205	0.3	500–1900	75–15	125–50
Cast iron (SG)	6900–7800	170	0.26	220–1000	3–0.2	5–20
Aluminium alloys	2700	70	0.33	80–505	30–8	50–25
Soda glass	2500	70	0.23	50–70†	0.01	0.8
Concrete	1800–2500	20–50	0.15–0.22	2–12 (tens) 20–150 (comp)	0.03	0.8–1.2
Clay bricks	1480–2400	14–18		5–110 (comp)	0.02	0.6
Timber‡	170–980 (dry)	3–21	0.25–0.49	10–80 (tens) 15–90 (comp)	20–8	7–13
Epoxy resin	1100–1400	2.6–3		30–100	0.1–0.3	0.5–1
Glass fibre comp – GRP	1400–2000	6–50		40–1250	10–100	20–50
Glass fibre comp – GRC	1900–2100	10–20		10–30**	25–10	10–20
Carbon fibre composite	1500–1600	70–200		600–700	30–5	30–45
Polyethyelene (high-density)	960	1.1	0.4–0.45	20–30	6–7	2.5
Rubber	830–910	0.1–1	0.5	15–30		

*In tension unless stated, yield or proof strength for metals, ultimate strength for other materials. †Modulus of rupture.
‡From tests on small clear specimens, load parallel to grain. tens, in tension; comp, under compression.

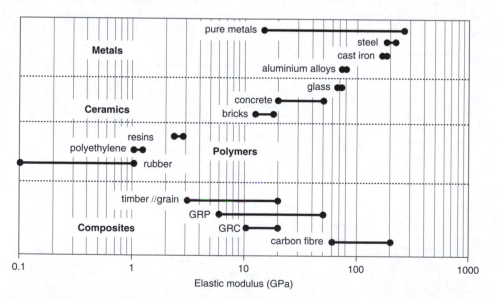

Fig. 61.2 Elastic moduli (stiffness) of structural materials.
(GRC = glass reinforced cement; GRP = glass fibre reinforced plastic (resin).)

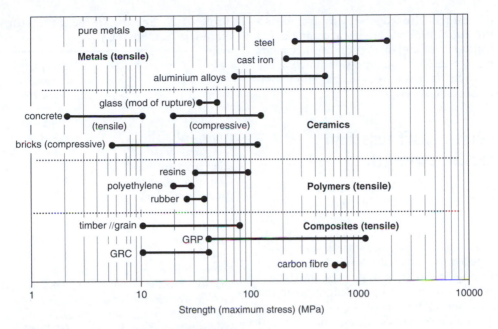

Fig. 61.3 *Strength (ultimate stress) of structural materials.*
(GRC = glass reinforced cement; GRP = glass fibre reinforced plastic (resin).)

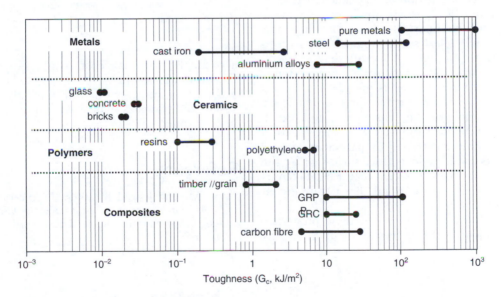

Fig. 61.4 *Toughness of structural materials.*
(GRC = glass reinforced cement; GRP = glass fibre reinforced plastic (resin).)

least to the most dense (timber to metals). Stiffness (*Fig. 61.2*) and strength (*Fig. 61.3*) both vary by a little more than three orders of magnitude, but toughness (*Fig. 61.4*) has the greatest range of all – five orders of magnitudes (from glass to the toughest metals).

The great range of the last property is perhaps the most significant of all. As we explained in Chapter 4, it is a measure of how easy it is to break a material, particularly under impact loading, and how well it copes with minor flaws, cracks etc.; it should not be confused with strength. Low values are extremely difficult for engineers to deal with – low strength and stiffness can be accommodated by bigger section sizes and structural arrangements (within limits), but low toughness is much more difficult to handle. It is one reason why fibre composites have become so popular.

61.2 Specific stiffness and specific strength

It is also interesting to make comparison between materials on a weight-for-weight basis or, in other words, to see what we get for a given weight. *Table 61.2* shows values of stiffness/density and strength/density, known as *specific stiffness* and *specific strength*, respectively. (If we use a minor subterfuge and use unit weight rather than density, the dimensions of both quantities end up as length, as in the table.) The specific stiffnesses of the most common structural materials – metal alloys, concrete and timber – all turn out to be remarkably similar, somewhere between 1 and 3×10^6 m. This means that stiffness is largely a function of how much material is present, and not how it is organised or structured.

These specific strength values show much greater variation, which is not surprising as it is governed by a far greater number of factors than stiffness, from nano- to macrostructural scales, such as atomic and molecular bond strengths and defects, cracks and flaws. What is most interesting is that if we rank the materials on this basis then the order is significantly different from that which we get from ranking by strength alone. In particular, timber shows up very well, and is only surpassed by the highest-strength steels. It seems that it has taken us a long time to get to the point that nature got to long before we were around.

We must, however, restate that these properties are not sufficient in themselves to justify the use of

Table 61.2 Specific stiffness and stiffness of the materials in table 61.1

Material	Specific stiffness $(m \times 10^6)$*	Specific strength† $(m \times 10^3)$†
Diamond	20–35	250–480
Pure metals	0.9–1.4	0.4–0.6
Structural steel	2.6	3–12.5
High-strength steel	2.7	6.5–25
Cast iron (SG)	2.4	3–14
Aluminium alloys	2.6	3–19
Soda glass	2.9	2.5
Concrete	1.1–2	1.1–6.1
Clay bricks	0.6–1.25	0.3–4.7
Timber	1.8–2.2	6–8
Epoxy resin	0.23	2.87–7.3
Glass fibre comp – GRP	0.4–2.6	3–63
Glass fibre comp – GRC	0.5–1.0	0.5–1.5
Carbon fibre composite	4.7–13	40–45
Polyethyelene (high-density)	0.01–0.1	2.1
Rubber	0.01–0.1	3.2

*Modulus of elasticity/unit weight. †Strength or limiting stress/unit weight.

a particular material; cost, availability, durability, fabrication procedures etc. must also be taken into account, along with the increasingly important sustainability issues, which we will consider in the next chapter.

Reference

Ashby MF and Jones DR (2005). *Engineering Materials 1: An introduction to properties, applications and design*, 3rd edition, Elsevier Butterworth Heinemann, Oxford.

Chapter 62

Sustainability and construction materials

62.1 Global considerations

Since the start of the industrial revolution in the late 18th century, there has been an exponential increase in our exploitation of materials for use in the technologies that have driven economic growth and increased the prosperity and living standards in much of the world. These advances have not, of course, been uniformly experienced owing to the wide variation in political, economic and social conditions in different regions and countries. Much of the growth has only been possible with the associated development and fabrication of infrastructure, which has required enormous quantities of construction materials with controlled and reliable properties. Furthermore, throughout the last two hundred years developments in materials science and technology have led to innovative forms of construction of increasingly spectacular scope and size, e.g. large-span suspension bridges in the first half and pre-stressed concrete in the second half of the 20th century were made possible by the development of high-strength steel.

Although such developments have had enormously beneficial consequences for both society as a whole and the people that it comprises, they were not without their drawbacks. In the materials context, there were problems associated with both production – e.g. harmful emissions from cement factories – and use – e.g. asbestos fibres for insulation. Such issues were generally addressed and eliminated (or at least controlled) on what could be considered as a relatively local level, e.g. dust-extraction systems for emissions and banning the use of asbestos fibres once their harmful carcinogenic properties had been established.

However, before the latter part of the 20th century, there was little consideration of possible concerns at wider, more global levels. Developments proceeded assuming limitless or near limitless supplies of materials and resources, albeit at increasing cost

if, say, local aggregate or cement supplies were exhausted and imported materials were required. Of course, the construction industry was not unique in this respect – much of the developed world thought that advances in all aspects of technology would lead to increasingly higher living standards, levels of consumption, etc., etc. more or less indefinitely.

By the mid-20th century the exponential nature of much of the growth – economic, population, use of resources, pollution etc. – was clear. A report published in 1972 by a team from the Massachusetts Institute of Technology (Meadows *et al.*, 1972) considered the implications of this for the future; they concluded that, without a concerted attack on the problems, the consequences were potentially catastrophic.

Environmental issues subsequently received increasing attention, and in 1983 the United Nations set up its World Commission on Environment and Development. In 1987 this produced the Brundtland Report, which defined sustainable development as 'development that meets the needs of the present without compromising the needs of future generations to meet their own needs' (United Nations, 1987). Extensive work on monitoring, modelling and forecasting global resources, climate and environmental changes has followed, much indicating the alarming consequences of inaction, particularly in relation to energy production, carbon emissions and global warming. The amount of publicity that this has received at the popular as well as the scientific level means that few people cannot be aware of this. It is beyond the scope and purpose of this book to discuss these issues in any detail, but suffice it to say that:

- Atmospheric carbon dioxide and other gases cause the 'greenhouse effect', whereby some of the energy received from the sun is trapped in the atmosphere. This in itself is a good thing, and

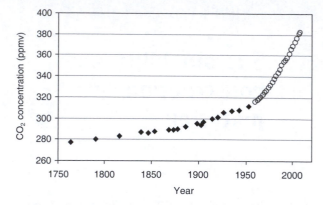

Fig. 62.1 *Atmospheric carbon dioxide levels since 1750 (Carbon Dioxide Information Analysis Center, 2009).* ◆ *Siple Station ice cores,* ○ *Manua Loa.*

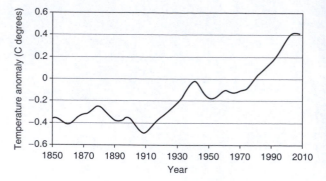

Fig. 62.3 *Global surface air temperatures since 1850 (Temperature anomaly = difference from 1961–90 mean) (Climatic Research Unit, University of East Anglia, 2009).*

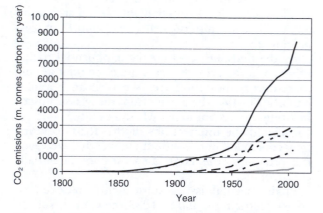

Fig. 62.2 *Global carbon dioxide emissions from fossil-fuel burning and cement production since 1800 (Carbon Dioxide Information Analysis Center, 2009).* — *Total,* ···· *Coal,* –– *Oil,* --- *Gas,* — *Cement.*

carbon dioxide levels with a period of about 100 000 years.

- There is considerable debate and significant disagreement over:

  - The cause and effect of global warming, since there are many other factors to be considered, such as the role of water vapour in the atmosphere and the release of carbon dioxide as the oceans warm. However, the Intergovernmental Panel on Climate Change in its report in 2007 stated that 'most of the observed increase in globally average temperatures since the mid-20th century is *very likely* due to the observed increase in anthropogenic greenhouse gas concentrations' (IPCC, 2007).

  - The likely magnitude of the warming, the resulting changes in climate, rainfall, sea levels etc. and their consequences.

The levels of concern are such that the general consensus is that all our activities should be modified to at least reduce the risk of uncontrolled disaster. Not surprisingly, many initiatives have arisen, at local, governmental and intergovernmental level, of varying effectiveness, and targets of varying, but generally increasing, severity have been set. Again, it is not a purpose of this book to review or discuss these, but we will see in this chapter that the construction industry has a large part to play. We will first outline the general approaches of the construction industry to the sustainable use of materials, with some specific examples mostly from the UK, and then discuss and compare the sustainability issues and initiatives relating to the materials whose production and properties we have already described in this book.

leads to temperatures high enough for life to exist, but increased levels of greenhouse gases will lead to higher temperatures, i.e. global warming.

- Carbon dioxide levels in the atmosphere have increased at an increasing rate since the early 1800s (*Fig. 62.1*), which coincides with the rise in emissions from the burning of increasing quantities of fossil fuels for energy production (*Fig. 62.2*).

- Global temperatures are increasing rapidly at a seemingly unprecedented rate (*Fig. 62.3*), but it is worth noting that changes in global temperature and atmospheric carbon dioxide levels are nothing new. Over at least the past 400 000 years there have been cycles of global warming and cooling coincident with cycles of atmospheric

Table 62.1 Annual use of main structural materials, 2006 (from: Iron and Steel Statistics Bureau, European Convention for Structural Steelwork, International Iron and Steel Institute, British Geological Society, Timber Trade Federation, Construction Statistics Annual (Department for Business, Enterprise and Regulatory Reform), Brick Development Association)

Material	World		UK	
	Million tonnes	Tonnes/person	Million tonnes	Tonnes/person
Structural steel	1244	0.19	3.6	0.06
Cement	2000	0.30	12	0.19
Concrete	12 000	1.8	84 (35 m. m^3)	1.4
Aggregates			275	4.6
Timber			9.6* (12 m. m^3)	0.19 m^3
Bricks			5.8	0.10

*Assuming an average density of 800 kg/m^3 – the volume figure is more exact. UK population, 60.5 million; world population, 6670 million.

62.2 Sustainability and the construction industry

62.2.1 USE OF MATERIALS

Table 62.1 shows figures compiled from a variety of sources of the annual use of the main construction materials, both worldwide and in the UK, for 2006. Aggregates are included as a separate category; as well as forming the major proportion of concrete and bituminous composites, as discussed in earlier sections, they are also used extensively for fill, pavement sub-base etc. The average amounts per person help to show the magnitude of the quantities involved and hence the potential for benefits from improvements in practice. Indeed, in 2002 the UK total for all construction materials of about 420 million tonnes was approximately two-thirds of the consumption of all materials (Lazarus, 2002).

Global rates of use of all the materials have increased significantly in the last 10 years, for example the annual production of cement and steel have both increased by about 25%, but this is mainly due to activities in developing countries, with rates of use in, for example, the UK being approximately constant. Much depends of course on the prevailing economic climate at any time, so production and use can fluctuate significantly from year to year, but it is likely that global consumption rates will continue to increase. The resources of the raw materials that are required for production of these materials are, of course, finite but in overall terms they are all abundant and present in sufficient quantities for many centuries of exploitation at current rates

(unlike, perhaps, oil and gas). This does not therefore place any restriction on increasing use, but extraction may be increasingly costly both in economics and environmental terms – mining and quarrying are often noisy, dusty and unsightly, and require restoration, and transport can cause local disruption.

Much of the output of the construction industry comprises buildings and structures that subsequently consume vast quantities of energy for heating, lighting, maintenance etc. while in service. Over their entire lifespan, structures are responsible for (Toyne, 2007):

- 40% of the world's energy use
- 40% of the world's solid waste generation
- 40% of the world's greenhouse gas emissions
- 33% of resource use
- 12% of water use.

For most structures, particularly occupied buildings, the majority of each of these will occur during the service life, but the materials' production, the construction procedures employed and eventual demolition will all make a significant contribution. The choice of materials and structural form will also influence the performance throughout a structure's life, e.g. high-durability materials will need less maintenance and thus reduce the whole-life costs.

62.2.2 LIFE-CYCLE ASSESSMENT

An overall analysis of the environmental impact of any product, process or construction system is therefore required alongside consideration of its technical and functional performance. A key task in this is life-cycle assessment (also known as life-cycle analysis or, more conveniently, LCA). The assessment can be

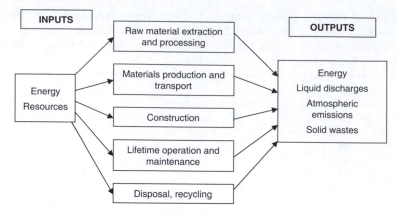

Fig. 62.4 *Summary of life-cycle assessment procedure (adapted from Royal Society of Chemistry, 2005).*

carried out over the entire life cycle from conception to disposal – the so-called 'cradle-to-grave' approach, or for only a part of this time, e.g. up to delivery at a construction site – 'cradle to site'. We do not have space to consider this in any detail here (you can find some general guides to this in 'Further reading' at the end of the chapter), but a few brief comments will be useful.

The general assessment procedure is summarised in *Fig. 62.4*. The input and output of material, energy and waste to and from all stages of a structure's life are quantified and their sum is a measure of the overall environmental impact. Different structural forms and systems can then be compared on this basis and decisions taken on the most appropriate one(s). The value of implementing measures such as reducing waste or using recycled materials can also be demonstrated.

The general framework, principles and requirements for carrying out and reporting LCAs have been specified in international standards (see standards list in 'Further reading' at the end of Part 10), and a number of methods and tools of varying complexity for the LCA itself have been produced. LCAs have also been included in methods for rating the overall environmental performance of structural elements, such as the BRE *Green Guide to Specification* (Anderson *et al.*, 2009) and provide information that can be used to evaluate the performance of a whole building or structure, for example the *Building Research Establishment Environmental Assessment Method* (BRE, 2008).

Carrying out an LCA is not necessarily straightforward. It is necessary at an early stage to define the system boundaries (for example, should the environmental impact of manufacturing the construction equipment be taken into account?), and the collection of data about the inputs and outputs to produce the so-called life-cycle inventory for the analysis is often difficult.

Information on many of the properties of the materials already discussed in this book will be important in an LCA, particularly those relating to durability and maintenance during service, but further essential information is the *embodied energy* and the *embodied carbon* content of materials, defined as:

- embodied energy (EE): the amount of energy required to mine, collect, crush, refine, extract, synthesise and process the materials into the form that we can use
- embodied carbon (EC): the amount of carbon dioxide emitted during the above processes, but taking into account the source of the energy and its impact on the environment.

These relate to a material's production but not to operations during or after construction. They are normally given either as values immediately as produced (called 'cradle-to-gate') or after delivery for use (called 'cradle-to-site'); these clearly contribute to the overall cradle-to-grave analysis of the complete LCA.

Some typical values are given in *Table 62.2*, compiled from Hammond and Jones (2008); these are thought to be the 'best' single figures that are representative of the type of material, but in many cases they are the average of a number of published values, which often have a wide variation. The values reflect both the amount of raw material that

Table 62.2 Embodied energy (EE) and embodied carbon content EC (expressed as CO_2) of construction materials (cradle-to-site)

Material	EE (MJ/kg)*	EC (kg CO_2/kg)*	EE/unit strength (MJ/m^3/MPa)	EC/unit strength (kg CO_2/m^3/MPa)
Aggregates	0.11	0.005		
Aluminium alloys	218	11.46	8000	400
Asphalt	2.4	0.14		
Bitumen	47	0.48		
Bricks (clay)	3	0.22	120	10
Cast iron	25	1.91	300	25
Cement – Portland	4.6	0.83		
Concrete – medium strength	1.1	0.16	45	6
Concrete – reinforced (3% steel)	3.5	0.32		
Copper	70	3.83		
Glass	15	0.85	650	40
Glass fibres	28	1.53	50	3
Grc	10.9	2.1	3000	570
Grp	100	8.10	1050	85
Mild steel	35.3	2.75	700	55
Polyethelene	83	1.94	3200	75
Resins	140	5.91	2585	110
Rubber	101	3.18	4040	125
Stainless steel	56.7	6.15	1500	160
Timber	8.5	0.46	102	6

*Source: Hammond and Jones (2008). Values for metals are for 'virgin' material without any recycled content. Values for timber exclude its calorific value.

needs to be obtained, transported and processed, and the energy intensity of the extraction and preparation processes. In general, resins and metals have the highest values of both properties. Although, as might be anticipated, there seems to be some broad correlation between the two properties they are not directly proportional; much depends on the specific energy sources used for a material's processing. For example, processes that require electrical energy produced by methods that have a production efficiency of around 30% will result in higher CO_2 emissions that those that run on low-grade heat energy that can be generated with efficiencies of around 80%. Also, the heat requirements can sometimes be met from waste heat from other parts of the process, further reducing the embodied CO_2. Embodied carbon is therefore the preferred measure for climate change analysis.

As we have stressed, when selecting materials for structural use we have to consider the combination of a whole range of properties. In the case of EE and EC we might, for example, be interested in comparing not only the EE and EC values per unit weight, but also how much load-carrying capacity we get for the energy input or carbon emission. We can do this by dividing the EE or EC per unit volume with the strength, to give the EE or EC per unit strength. Combining the density and mid-range strength data for the materials given in *Table 61.1* and the EE and EC, the values in columns 2 and 3 of *Table 62.2* give columns 4 and 5 in the table. Within the limitation that these are average or mid-range values, they do show some interesting behaviour. Metals are now joined by composites and polymers as high-energy-cost materials, with masonry, timber and concrete all low-cost.

62.2.3 THE GREEN HIERARCHY

It is clear that society will benefit both in economic and sustainability terms from the implementation of the so-called green hierarchy for materials:

reduce:
- use of materials
- energy for production and construction
- energy during use.

reuse:
- components
- adapt structures for change of use.

recycle:
- materials after demolition
- waste.

recover:
- energy from materials with few recycling options.

dispose:
- only if no other alternative.

This includes consideration of the production and selection of materials, ensuring that they are used efficiently with minimum waste, and promoting reuse and recycling wherever possible. It has been incorporated in various forms in recommendations and regulations in many countries. For example in the UK, a Sustainable Construction Task Group set up by the government identified the following 'Themes for action' in 2000 (DTI, 2000):

- re-use existing built assets
- design for minimum waste – before, during and after the structure's life
- aim for lean construction
- minimise energy in construction and in use
- do not pollute
- preserve and enhance biodiversity
- conserve water resources
- respect people and their local environment
- set targets, monitor and report.

These have subsequently been developed into more specific aims, and in 2008 a joint industry and Government initiative produced a *Strategy for Sustainable Construction* (UK Government and Strategic Forum for Construction, 2008), which included the aims of:

- Reducing the total UK carbon dioxide (CO_2) emissions by at least 60% on 1990 levels by 2050 and by at least 26% by 2020.
- A 50% reduction of construction, demolition and excavation waste to landfill by 2012 compared to 2008 levels.
- The materials used in construction to have the least environmental and social impact as feasible both socially and economically.
- 25% of products used in construction projects to be from schemes recognised for responsible sourcing by 2012.

Most construction is of course carried out in a competitive economic climate, so incentives for the industry to change practice must go beyond the long-term benefits to society. Although a company's reputation and general perception of its corporate responsibility can benefit considerably from the implementation of green policies (and indeed this is highlighted in publicity information, annual reports etc.), in reality the incentive must largely be either economic or legislative or some combination of the two. In the UK, recycling of waste rather than disposal to landfill is a good example of the two working in combination. A landfill tax, introduced in 1996 (Aggregain, 2008), imposed a levy on waste disposed to landfill; this has been gradually increased thereafter (in 2008 it was £2.50 per tonne for inert or inactive waste and £32 per tonne for all other wastes) and has led to increased rates of recycling, reductions of over-ordering and more efficient site practices.

The materials discussed in this book have a wide range of production methods, properties and uses and so, not surprisingly, can make different contributions to improvements in sustainability and environmental impact and to achieving overall industry targets such as those listed above. There is a large and burgeoning amount of information available in many forms on these issues, most of which is, understandably, specific to individual materials. For most material groups, consortiums of associations, trade organisations and producers have formed 'task forces' or 'action committees' to document and improve the sustainability properties of their material. An internet search readily leads to a whole host of the resulting websites and publications – several examples from the UK are given in the reference list at the end of this chapter. Making comparisons is not immediately straightforward since each consortium has been formed to promote the use of its own material, and therefore makes claims that, for the uninitiated, would seem to indicate that all materials can make an equally good contribution to sustainability.

A selection of the titles of various publications bears testament to this:

- *Steel – the sustainable facts* (BCSA and Corus)
- *Civil Engineering: Sustainable solutions using concrete* (The Concrete Centre, 2005)
- *Asphalt – The Environmental Choice* (The Asphalt Institute)
- *Brick – The Case for Sustainability* (Brick Development Association).

The remainder of this chapter considers the sustainability issues relating to the main construction materials discussed earlier in this book, with the addition of a section on aggregates since they are

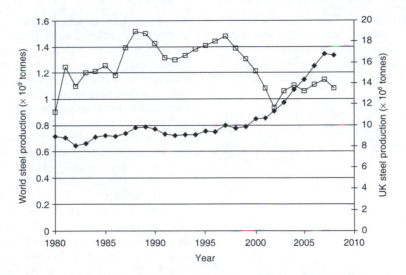

Fig. 62.5 *World and UK steel production (World Steel Association, 2009).* ◆ *World,* ⊟ *UK.*

used in large quantities in both an unbound state and when bound in concrete, bituminous materials and masonry construction. Our discussion cannot be exhaustive – that would take a whole book by itself – but it is hoped that it will help you to understand the various issues, claims and counterclaims and put them in some sort of perspective. Much of the information and data discussed relates to UK practice, but similar arguments apply in many countries.

62.3 Steel

World steel production nearly doubled between the mid-1990s and 2008 (*Fig. 62.5*), and is expected to double again by 2050, with some fluctuations due to the global economic conditions (World Steel Association, 2009). Production in the UK declined by about 40% in the same period. Steel of course has many uses, and in the UK about 29% of the total produced is used in construction, half of which is structural steel (Iron and Steel Statistics Bureau, 2008).

As shown in *Table 62.2*, the embodied energy (EE) and embodied carbon (EC) for steel are at the higher end of those for construction materials that are used in large quantities. However, the figures given are for primary steel production from iron ore, which involves the two stages of reduction to pig iron in a blast furnace followed by conversion to steel, both of which require high temperatures. The alternative, production of recycled steel from scrap, only involves the second stage and results in steel with much lower EE and EC, as shown in

Table 62.3 Embodied energy and carbon content for various types of steel (cradle-to-site) (from Hammond and Jones (2008))

Type of steel	Embodied energy (MJ/kg)	Embodied carbon (kg CO_2/kg)
General		
Virgin	35.3	2.75
Recycled	9.5	0.43
Bar and rod		
Virgin	36.4	2.68
Recycled	8.8	0.42
Plate	48.4	3.19
Galvanised	39.0	2.82
Stainless	56.7	6.15

Table 62.3. The electric arc furnace is particularly suited to processing scrap, which can be up to 100% of its charge (see Chapter 11). Currently, primary steel accounts for about 75% of world steel production, and recycled steel about 25% (World Steel Association, 2008). In the UK, structural steel has a recycled content of about 60% (BCSA and Corus, 2008), and reinforcement for concrete is 100% recycled steel (The Concrete Centre, 2007). *Table 62.3* also shows that adding other processes to steel production (e.g. galvanising) or alloying with other elements to produce special steels (e.g. stainless steel) also require energy.

Other major sustainability benefits claimed for steel construction relating to design considerations, construction procedures, performance in service and end-of-life are (Steel Construction Sector Sustainability Committee, 2002; British Constructional Steelwork Association, 2008):

- factory production of structural elements minimises site activity, leading to higher quality, rapid construction, fewer site operatives and less waste
- steel structures can be designed to have long-span flexible spaces that are adaptable for reuse
- considerations at the design stage can produce structures of which the elements and components are capable of being demounted and reused
- even without this, it may be possible to reclaim structural elements and buildings from demolition if there is no deterioration in properties from the as-built state, particularly if they have bolted connections
- the magnetic properties of steel make separation of scrap from other metals and materials straightforward for recycling into the steel production process; this can, in principle, be carried out an unlimited number of times. From demolition in the UK, 10% of the steel is reused and 84% is recycled.

62.4 Aggregates

Aggregates are used throughout construction in quantities significantly greater than any other material. We have seen in earlier sections of this book that they form the bulk of the volume of concrete and asphalts, but they are also used in mortars for masonry and brickwork and in an unbound state in sub-bases for roadways and pavements, as railway ballast, as fill for embankments, as drainage channels in gravity dams and as facings for sea defences etc. etc.

In the UK, the annual sale of aggregate in 2007 was about 270 million tonnes, with 36% of this used in concrete, 22% in asphalt and roadstone and 35% elsewhere in construction (Quarry Products Association, 2008).

Aggregates can be divided into three main types depending on their source:

- *Primary*: from natural sources, which can be either crushed rocks from bulk rock or sands and gravels from river or sea-bed deposits.
- *Secondary*: by-products of other industrial processes not previously used in construction.
- *Recycled*: from previously used construction materials.

The EE and EC of aggregates are very low compared with those of other construction materials (*Table 62.2*) and primarily result from the energy required for extraction, processing (including crushing where necessary) and transport. The figures for secondary and recycled aggregates may be a little higher than those for primary aggregates, so there is no incentive to use them to reduce the carbon footprint.

There are however considerable pressures to reduce the consumption of primary aggregates for other environmental reasons. Extracting large quantities of material from quarries or gravel pits can cause loss of valuable or scenic land, dust and noise emissions and extra traffic on unsuitable rural roads. The resulting large holes eventually require landscaping or conversion to other uses such as water parks or other leisure facilities. Near-shore dredging can disturb wave and current flow, causing unwanted seabed movement.

In the UK, the aggregates levy – a tax on aggregates from natural sources – was introduced in 2002 as an economic incentive to reduce the use of primary aggregates. In 2008 it stood at £1.95 per tonne of sand and gravel rock extracted. This and the landfill tax, which we discussed earlier in the chapter, therefore act as a double incentive to use less primary aggregate and more secondary or recycled aggregate, which would otherwise need disposal as waste. In 2005, secondary and recycled aggregates comprised 25% of the total used; the target for 2010 is 30% (WRAP-Aggregain, 2009).

We have discussed the use of secondary and recycled aggregates in new concrete in Chapters 17 and 26, respectively, and for bituminous mixtures for roads, pavements etc. in Chapter 30. There is considerable scope for the use of a wide variety of secondary and recycled aggregates elsewhere in construction. A major example is the use of recycled concrete aggregate and recycled aggregate as sub-base for pavements, particularly during demolition and redevelopment, as the processing of the demolition waste can be carried out on site, minimising the need for transport. Other recent examples include the use of glass as a fill material and the use of tyres for embankments. A web-site with number of case studies is maintained by WRAP-Aggregain (2009)

62.5 Cement and concrete

62.5.1 CEMENT

The quantity of cement used worldwide has risen markedly in the past few years (*Fig. 62.6*), a trend that looks set to continue for the foreseeable future,

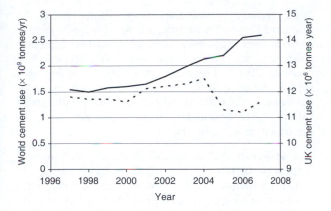

Fig. 62.6 World and UK cement use (United States Geological Survey, 2009; Office for National Statistics, 2008). — World, ---- UK.

although global economic conditions may cause some short-term fluctuations. Its use in the UK has been approximately constant. The consequent levels of carbon emissions from cement production are significant enough to be included in overall global emissions data such as those used to compile *Fig. 62.2*. Some simple arithmetic using the data on cement production, embodied carbon and total global emissions in *Tables 62.1* and *62.2* shows that in 2006 cement production accounted for about 5.5% of the global CO_2 emissions, as shown in *Fig. 62.2*.

The production of Portland cement, which we described in Chapter 13, requires high temperatures firstly to decompose the calcium carbonate to calcium oxide and carbon dioxide (a process called *calcining*), and then to fuse the calcium oxide with the silicates, aluminates and ferrites to form the cement compounds. Carbon dioxide emissions therefore occur as a result of the burning of the fuels to produce the high temperatures, the breakdown of the calcium carbonate and the production of the energy required for raw material extraction, clinker grinding and transport of the finished cement.

A simple stoichiometric analysis of the calcining reaction to give the typical calcium oxide content of Portland cement (about 66%) shows that about 0.52 kg of CO_2 is given off for each kg of cement produced. This is about 60% of the embodied carbon content of cement (0.83 kg CO_2/kg, *Table 62.2*) and cannot be reduced – it is a necessary consequence of the process. It is, however, possible to reduce the other 40% by more energy-efficient processes, and indeed there has been significant progress in this respect in the last decade or so. For example, in the UK between 1998 and 2007 CO_2 emission from

cement plants was reduced from 0.92 to 0.82 kg/kg cement, with dust and other gaseous emissions such as NO_X also significantly reduced (British Cement Association, 2008).

An indirect reduction in CO_2 emissions has been obtained by the use of alternative fuels in the cement kiln. Waste products with significant calorific value, including bone meal, contaminated meat, waste oils, paper and paper sludge, used vehicle tyres, plastics, textiles and sewage sludge can be used in combination with the conventional oil or coal, generally after some pre-treatment, e.g. drying, shredding, blending or grinding. In 2006 such fuels accounted for 18% of the total fuel use (Cembureau, 2009). Without such use, disposal of the wastes by incineration or landfill may cause increased greenhouse gas emissions, and in some cases the burning process is an effective way of dealing with any hazardous contaminants, although care has to be taken to ensure that these are either destroyed or bound within the cement. Clearly, the quantity of fossil fuel required is also reduced.

Some waste materials that contain alumina, silica or iron oxides in suitable quantities can also be used to replace some of the raw materials extracted from quarries (chalk, limestone, clay etc.). Examples include contaminated soil, waste from road cleaning, coal ash, blast furnace slag and municipal solid waste incinerator ash (MSWIA). In Europe in 2006, these constituted about 5% of the total raw materials used (Cembureau, 2009).

The importance of the issues involved led to the formation in 1999 of the Cement Sustainability Initiative by ten of the world's leading cement companies. Membership was increased to 18 companies in 2007. As well as gathering data and hence identifying trends in cement production, emissions and energy use, this sets targets and promotes improvement by its member companies in all of the aspects described above (World Business Council for Sustainable Development, 2008).

The unavoidable carbon dioxide emissions from calcining limit the contribution that the manufacture of Portland cement can make to the sustainability agenda. However, combining Portland cement with one or more of the additions described in Chapter 15 can have significant benefits both for the embodied energy/carbon emissions of the resulting concrete and for many other properties. *Table 62.4* gives values for three of the most widely used additions:

- limestone powder, which is available in large quantities, and requires energy for grinding from the parent rock

Table 62.4 Embodied energy and carbon content for concrete constituents and classes (cradle-to-site) (from Hammond and Jones (2008))

Material	Embodied energy (MJ/kg)	Embodied carbon (kg CO_2/kg)
Constituent		
Cement (typical CEM I)	4.6	0.83
GGBS	1.33	0.07
Fly ash	0.1	0.01
Limestone powder		0.03
Concrete (binder = CEM I)		
Class C16/30	0.85	0.11
Class C30/37	1.08	0.15
Class C50/60	1.41	0.21
Reinforced concrete (mean)	2.1	0.25

- fly ash (pulverised fuel ash, PFA), which requires little energy input other than perhaps some mechanical processing to remove unwanted particles
- ground granulated blast furnace slag (GGBS), which requires energy for grinding after rapid cooling.

Fly ash and GGBS have the added advantage of being by-products of other processes. All three will have similar energy requirements to Portland cement for handling, transport etc.

The embodied energy/carbon of any particular blend will be a weighted average, and the advantages of using blends are clear from *Table 62.4*. However, as we explained in Chapter 15, the limiting proportions for use in structural concrete of acceptable properties for most purposes are different for each addition: typically 20% for limestone powder, 40–50% for PFA and 80–85% for GGBS. Blended cements are becoming more popular and, in the UK at least, low percentages of additions (i.e. CEM II cements) are now routinely used.

The potential benefits of additions are such that considerable work is being carried out either on modifying their characteristics, e.g. by removing deleterious particles and/or refining their particle size distribution, or on developing products from other waste materials or by-products with similar compositions. An example of the latter is the 'waste-derived cements', which we discussed in Chapter 16. In the next few years it is likely that a range of technically proven and commercially viable materials with performance levels better than hitherto will become available.

To entirely avoid the sustainability issues of Portland cement, a number of novel 'non-Portland' cements (termed low-energy or low-carbon cements) either have been or are being developed (Taylor, 2009; Gartner, 2004). These are generally based on non-traditional processes (involving lower temperatures than required for Portland cement) or raw materials (often waste products). Some examples that we discussed in Chapter 16 are:

- Alkali-activated cements, in which the cementitious properties of a pozzolanic material (such as fly ash, municipal solid waste incinerator ash, meta-kaolin or slags) are activated by an alkali silicate such as sodium silicate and/or sodium hydroxide.
- Geopolymer cements, which are produced by low-temperature firing of alumina- and silica-containing raw materials and therefore require less energy and produce less CO_2 during manufacture.
- Magnesium oxide-based cements derived from:
  - Magnesium carbonate. Considerable quantities of carbon dioxide are released when this is converted to magnesium oxide during the cement's production, but at least some will be subsequently reabsorbed by carbonation of the magnesium hydroxide. The net CO_2 balance therefore needs to be assessed over the whole service life.
  - Magnesium silicate. Less CO_2 than with magnesium carbonate is emitted when converting this to magnesium oxide, and theoretically at least a greater amount can be absorbed during the concrete's life. It is claimed that this could therefore be a 'carbon-negative' cement (Novacem, 2009).

Although such cements have some significant attractions and potential benefits, considerable work is required to ensure that they are 'fit for purpose' for use as an alternative to Portland cement, including life-cycle assessments as well as laboratory testing. In some cases this work is well advanced but nevertheless it is probable that Portland cement, albeit blended with additions, will continue to be the dominant cement used in concrete construction for the foreseeable future (Price, 2009).

62.5.2 AGGREGATES FOR CONCRETE

The EE and EC values for primary aggregates (*Table 62.2*) result mainly from extraction and processing (crushing, if required, and grading) and transport.

In Chapter 17 we described aggregates for concrete from natural sources (i.e. primary aggregates) and in Chapter 25 manufactured lightweight aggregates, some of which are produced from by-products of other industries, particularly fly ash and blast furnace slag (i.e. secondary aggregates). As we have discussed above, there are considerable pressures to reduce the use of primary aggregates (not just in concrete) and therefore to increase the use of secondary aggregates (of any density) and recycled aggregates, defined as those obtained from previously used construction materials.

As mentioned in Chapter 17, secondary aggregates that are suitable for use in concrete can, in principle, be any inert material that has a particle size distributions, strength, cleanliness etc. similar to those of primary aggregates. Many by-products and waste materials have been exploited for this purpose. Examples include crushed glass, ferro-silicate ash from zinc production and shredded vehicle tyres. Some case studies from the UK can be found at WRAP-Aggregain (2009).

We discussed the use of aggregates produced by sorting and crushing demolition waste in Chapter 26, making a distinction between recycled concrete aggregate (RCA), which is all or nearly all crushed concrete and can be used in new structural-strength-grade concrete, and recycled aggregates (RAs), which contain significant quantities of other demolition waste, such as plaster, masonry etc. and can only be used for lower-strength concrete.

62.5.3 CONCRETE

The EE and EC values for concrete result from those of its constituents – with the aggregates that make up the bulk of the volume having much lower values than the cement or additions – and its mixing, production and transport. Typical figures for mixes with various strength grades and with 100% Portland cement as the binder are given in *Table 62.4*; clearly the higher cement contents required for higher-strength concrete result in higher EE and EC, but a doubling in strength results in only about a 40% increase in either value. The figures for concrete in which part of the cement is replaced with an addition will be proportionately lower, an important factor in favour of the use of greater quantities of additions if the properties of the concrete are not to be compromised.

As we have seen in Chapter 24, provided sufficient attention is given to the selection of the constituent materials, mix design and production and placing procedures, structural concrete can have excellent durability in nearly all environments, thus being suitable for long service lives with low maintenance costs. It also has significant fire resistance.

The thermal mass and other thermal properties of concrete have been exploited to reduce the costs of heating and cooling buildings throughout their life (de Saulles, 2009). In common with other heavyweight building materials such as brick and stone, concrete has an ideal combination of a high specific heat capacity (1000 J/kg.°C), high density (2400 kg/m^3) and moderate thermal conductivity (1.75 W/m.°C). This means that, if there is sufficient exposed concrete within a building, during the summer months much of the heat gain during the warm days is absorbed by the concrete, preventing an excessive temperature rise, and is then released during the cooler nights. If warm air is not allowed to enter the building during the day and the windows are shaded, and the building is then vented at night, the daily minimum and maximum internal peak temperatures are both delayed, by up to six hours, and reduced, by up to 6–8°C, thus reducing air-conditioning costs. During the winter months, provided there are sufficient south facing windows, sufficient heat is gained during the daytime to be stored by the concrete and then released at night, thus reducing the need for supplementary heating. The process requires a significant diurnal temperature variation (an average of 10°C or more) to draw the heat out of the concrete – climates such as those in the UK provide this. As well as reducing the energy requirements for heating and cooling, the working or living conditions within the building are improved.

We discussed the possibilities for the recycling of fresh and hardened concrete in Chapter 26, and considerable efforts have been made to reduce the amount of waste from both sources sent to landfill in recent years.

62.6 Asphalt and bituminous materials

Eighty per cent of the world's annual bitumen production of 100 million tonnes is used for pavement construction and maintenance, generally combined with aggregates to give asphalt. As with cement and concrete, the embodied energy and carbon of the composite are significantly less than those of the binder, owing to the low embodied energy and embodied carbon of the aggregate filler that forms the bulk of the volume (*Table 62.2*).

Maintenance and life-cycle costs of pavements have been reduced by the developments of so-called 'perpetual pavements' (Washington Asphalt Pavement Association, 2008). These consist of a hot-mixed asphalt base layer designed specifically to resist fatigue cracking, a stable and durable intermediate layer designed to carry most of the traffic load, and a top layer or wearing surface to resist surface-initiated distresses such as top-down cracking and rutting. These are claimed to last for more than 50 years without maintenance, other than needing periodic renewal of the wearing surface.

The need for controlled drainage of surface water in urban areas to avoid storm-water overflow causing flooding has led to the development of porous asphalt paving systems for areas such as car parks, foot and cycle paths, minor roads and brown-field sites. These form part of sustainable drainage systems.

New pavement can successfully incorporate secondary aggregates from a variety of sources, and as we have seen in Chapter 30, at the end of their working life pavement materials can be either recycled as sub-base or, after processing, mixed with further bitumen in new surface layers.

62.7 Masonry

The output of the brick production industry in the UK varied between about 6.5 and 5.8 million tonnes in the period from 2001 to 2007 (Brick Development Association, 2008). The figure depends on the prevailing economic climate, which can rapidly affect the building industry but, as with most construction materials, there is no reason to suppose that over the long term production and use will not increase, both in the UK and worldwide.

Clay bricks have higher EE and EC contents than concrete (*Table 62.2*) and, as with other high-volume materials, quarrying of the raw materials creates aesthetic and other environmental problems. The brick-production industry in many countries has sought to minimise the environmental impact by making production plants more efficient, e.g. by optimising the recycling of heat during production, and by moving to more highly perforated bricks, which minimise resource usage.

Reducing waste and energy for brick production has perhaps been more difficult than in other industries owing to the large number of brick factories, generally sited near to sources of clay, but many of which are relatively small, a pattern repeated throughout the world (Brick Development Association, 2001). Nevertheless in the decade to 2007 the energy consumption per unit of output was reduced by about 10% (Brick Development Association, 2008), which followed a reduction of 20% in the preceding 20 years (Beardmore, 1998).

A major factor contributing to the sustainability of masonry is that, if correctly specified and constructed, it has a very long life with very low maintenance costs. The average life, to date, of all the standing stock of masonry dwellings in the UK is around 58 years, but a significant proportion is over 125 years old. The significance of this long, low-maintenance life is shown by the fact that, if the production energy of the brickwork in a typical dwelling is spread over an average life of, say, 100 years this gives a figure of about 0.5 GJ/year, which is between 0.5 and 1% of the annual heating requirement (Shorrock *et al.*, 1992, 1993).

For structures with shorter lives modern masonry runs into problems with recycling. Portland cement-based mortars adhere strongly to bricks and therefore make recovery and recycling of whole bricks after demolition difficult. The current UK code of practice does not provide a basis for the design of slender walls constructed with a lime mortar, which would facilitate recycling. However, this is not a significant problem because the replacement rate of housing in the UK is only of the order of 0.25% per annum, and the overall construction rate is only just over 1%. The current replacement rate implies that the UK expects its dwellings to last several hundred years, which is, perhaps, the reason why long-lived materials are popular.

When demolition is necessary, crushed brickwork can be used as a low-grade fill material for pavement construction and trenches, or as aggregate for low-strength concrete, thus reducing the need for primary aggregates.

Masonry buildings can give similar thermal-mass advantages to those described above for concrete.

62.8 Glass

We have seen in Part 9 that glass is being increasingly used as a structural material, and that it has an essential role when used in the building envelope in controlling energy gain and loss.

The UK manufactures 750 000 tonnes of flat glass each year, three quarters of which goes into glazing products for buildings. Removal of windows produced 90 000 tonnes of glass in 2008, much of which goes to landfill; this is expected to double in the next ten years and so there is considerable potential for recycling more flat glass waste (WRAP, 2007). This and waste container glass can be re-melted and added to the mix of raw materials to produce new glass; in 2008 the recycled content of flat glass produced in the UK was between

20 and 30% (WRAP, 2008). This clearly saves energy (reducing the EE and EC values given in *Table 62.2*) and reduces waste, but the glass used must be clear and free of contaminants, as structural flat glass for use in buildings has strict quality requirements.

Apart from being able to be incorporated in the feedstock for new glass production, waste glass cullet can be used as alternative to primary aggregates for concrete (see section 62.5) and bitumen-bound pavements (see section 62.6) and as fill for pavement sub-base, embankments, etc.

62.9 Polymers and fibre composites

The production of polymers is an energy-intensive process and therefore their EE and EC are in the upper part of the range for construction materials, along with metals (*Table 62.2*). Composites of polymers with glass, carbon or synthetic fibres all have similar or higher values owing to the EE and EC of the fibres themselves and the production process of the composites. Similar arguments apply to fibre cements and concrete composite, in which the fibres generally have much higher EE and EC than the cement or concrete matrix.

As we have seen in Chapters 42 and 48 a wide range of cellulose-based natural fibres have been used for many years to form composites with polymers, cement and concrete. Fibres can be obtained from plant stems (e.g. jute and flax), leaves (e.g. sisal), or woody parts (waste structural wood or bamboo) by processing to extract them from their organic matrices. These offer significant advantages for sustainability, as they are generally locally available – often in large quantities – and renewable and have much lower EE and EC contents than manufactured fibres such as glass and steel. However, as we also saw, the properties of the fibres depend heavily on the extraction process, and the performance of the resulting composite, particularly in terms of moisture sensitivity and durability, is not as high as with manufactured fibres.

The sustainability advantages of polymers and fibre composites therefore result from their effect on the construction procedures and their performance during service. It will be apparent from Parts 6 and 7 that these include:

- *polymers*: a wide range of adhesives are available that can assist rapid and durable construction, and form the basis for effective repairs

- *polymer and fibre cement composites*: light, strong and durable structural elements can be produced in controlled factory conditions, which results in reduced transport costs, lower foundation loads, ease of construction and low maintenance
- *fibre concretes*: conventional reinforcement can be reduced and there can be increased long-term structural performance, particularly in relation to cracking and, again, increased durability and lower maintenance costs.

As we noted in earlier sections, reuse of the composite elements may be possible in some cases, but recovery and recycling of the fibres is, in effect, impossible. Most recycling therefore consists of the production of aggregates from demolition within the constraints of allowable composition, as discussed earlier in the chapter.

62.10 Timber

Table 62.2 gives EE and EC values for 'general timber'; specific values for timber and timber products are given in *Table 62.5*. The figures exclude the calorific value of the wood, and in several cases are averages of a wide range of data. The values for sawn hardwood and softwood result from the logging, transport and processing operations, and the higher values for manufactured timber products depend on the intensity of the processes required for their production.

The sustainability issues regarding timber are somewhat different from those of the other materials discussed in this book. The initial source of timber is a natural, growing material and so it can be considered as being harvested rather than quarried

Table 62.5 Embodied energy and carbon content for timber and timber products (cradle-to-site) (from Hammond and Jones (2008))

Material	Embodied energy (MJ/kg)	Embodied carbon (kg CO_2/kg)
Sawn softwood	7.4	0.45
Sawn hardwood	7.8	0.47
Glued laminated timber	12	0.65
Laminated veneer	9.5	0.51
Particle board	9.5	0.51
MDF	11	0.59
Plywood	15	0.81
Hardboard	16	0.86

or mined. Clearly supplies of new timber will only be sustainable if the rate of harvesting does not exceed the rate of growth, but other issues, such as the forest maintaining its ecological functions of biodiversity, climate and water cycles, and local people being involved in the benefits from the forest are equally important.

There have been, and still are, many examples of unsustainable deforestation both for the supply of timber for construction and other markets and to release land for the growth of other more lucrative crops. However, there have also been significant moves towards ensuring that timber is obtained from sustainably managed forests. A major example is the work of the Forest Stewardship Council (FSC), an international, independent non-governmental organisation set up in 1992. They have introduced a certification scheme, recognised in most countries, which accredits supplies as coming from well-managed and sustainable sources. The certification process considers such aspects as management planning, harvesting, conservation of biodiversity, pest and disease management, and the social impact of the forestry operations. This often takes the form of recognising or approving schemes set up nationally or regionally, such as the UK Woodland Assurance Scheme, which was established in 1999. In 2008 the FSC certified 100 million hectares of forest in 79 countries (Forest Stewardship Council, 2009). The construction industry in many countries therefore now has the opportunity of ensuring that timber suppliers are only obtained from certified sustainable sources.

The recovery of sound timber sections after demolition may be possible, with reuse after planing and sawing. Also, we saw in Chapter 60 that there is considerable potential for the recycling of timber waste in particleboard, horticulture, animal bedding and as a fuel for power stations. The proportion of waste that goes into these uses, rather than ending up as landfill or domestic fuel (although the latter could be considered as recycling) has been steadily increasing in recent years.

References

Aggregain (2008). *Landfill Tax*. http://www.aggregain.org.uk/waste_management_regulations/waste_management_regulations_ni/background/landfilltax.html (accessed 2/10/08).

Anderson J, Shiers D and Steele K (2009). *The Green Guide to Specification: an environmental profiling system for building materials and components*, 4th edition, BRE Press, Watford, England.

British Constructional Steelwork Association and Corus (2008). *Steel – The sustainable facts*. http://www.sustainablesteel.co.uk/facts.html (accessed 22/9/08).

Beardmore C (1998). Fuel usage in the manufacture of clay building bricks (16), *Ceram Research RP812*, CERAM, Staffordshire.

Brick Development Association (2001). *A Sustainability Strategy for the Brick Industry*. http://www.brick.org.uk/industry-sustainability.html (accessed 4/2/09).

Brick Development Association (2008). *Sustainability Strategy for the Brick Industry – An update*. http://www.brick.org.uk/industry-sustainability.html (accessed 4/2/09).

British Cement Association (2008). *Performance*, November, BCA, Surrey. www.cementindustry.co.uk (accessed 3/4/09).

British Constructional Steelwork Association (2008). *Sustainable steel construction: Steel Industry Guidance Note SN29*. http://www.corusconstruction.com/ (accessed 4/3/09).

Building Research Establishment (2008). *BREEAM: BRE Environmental Assessment Method*. http://www.breeam.org/ (accessed 2/3/09).

Cembureau (2009). *Sustainable cement production – Co-processing of alternative fuels and raw materials in the European cement industry*. Cembureau, Brussels. www.cembureau.eu (accessed 8/7/09).

Carbon Dioxide Information Analysis Center (2009). http://cdiac.esd.ornl.gov/ftp/trends/co2 (accessed 5/3/09).

Climatic Research Unit, University of East Anglia (2009). Information sheet 1. http://www.cru.uea.ac.uk/cru/info/warming/ (accessed 5/3/09).

de Saulles T (2009). *Thermal mass explained*, The Concrete Centre, Camberley, Surrey. www.concretecentre.com/publications (accessed 24/5/09).

Dept of Trade and Industry (2000). *Building a better quality of life – a strategy for more sustainable construction*, April. http://www.berr.gov.uk/files/file13547.pdf (accessed 22/10/08).

Forest Stewardship Council (2009). *Global FSC certificates: type and distribution*, FSC, March. http://www.fsc.org (accessed 5/5/09).

Gartner G (2004). Industrially interesting approaches to 'low CO2' cements. *Cement and Concrete Research*, **34**, 1489–1498.

Hammond G and Jones C (2008). *Inventory of Carbon and Energy version 1.6a*. www.bath.ac.uk/mech-eng/sert/embodied (accessed 22/9/08).

IPCC (2007). *Climate Change 2007: Synthesis Report. Contribution of Working Groups I, II and III to the Fourth Assessment Report of the Intergovernmental Panel on Climate Change* (eds Pachauri RK and Reisinger A), IPCC, Geneva, Switzerland, 104 pages.

Iron and Steel Statistics Bureau (2008). *Global Trade Statistics* (accessed 22/9/08).

Lazarus N (2002). *BedZED: Construction Materials Report. Toolkit of Carbon Neutral Developments. Part 1*. www.bioregional.com (accessed 2/10/08).

Meadows DH, Meadows DL, Randers J and Behrens WW (1972). *The limits to growth*, Earth Island Ltd, London.

Novacem (2009). *What we do*. http://www.novacem.com/ (accessed 23/5/09).

Office for National Statistics (2008). *Construction Statistics Annual, 2008*. http://www.statistics.gov.uk/ (accessed 6/4/09).

Price WF (2009). Cementitious materials for the 21st century. *Proceedings of the Institute of Civil Engineers – Civil Engineering 162*, May, Paper 08-00043, 64–69.

Quarry Products Association (2008). *Annual Report*. http://www.qpa.org/sus_report01.htm (accessed 4/3/09).

Royal Society of Chemistry – Environmental, Health and Safety Committee (2005). *Life Cycle Assessment*, RSC London.

Shorrock LD, Henderson G and Bown JHF (1992 and 1993). *Domestic energy fact file. BR 220 and BR 251 (update report)*, Construction Research Communications Ltd, Garston.

Steel Construction Sector Sustainability Committee (2002). *Sustainable steel construction – building a better future*. http://www.steel-sci.org/NR/rdonlyres/12A3E33E-C13D-40FB-9B7D-57B93BC7BAAE/2896/SSC.pdf (accessed 29/4/09).

Taylor MG (2009). *Novel cements: Low energy, low carbon cements*. Fact Sheet 12, British Cement Association, Camberley, January.

The Concrete Centre (2007). *Sustainable concrete: the environmental, social and economic sustainability credentials of concrete*. www.concretecentre.com/pdf/MB_Sustainable_Concrete_Feb07.pdf (accessed 12/11/09).

Toyne P (2007). *New Civil Engineer*, 29 November, p17.

UK Government and Strategic Forum for Construction (2008). *Strategy for Sustainable Construction*. http://www.berr.gov.uk/whatwedo/sectors/construction/sustainability/page13691.html (accessed 23/8/09).

United Nations (1987). *Our common future (the Brundtland report)*, UN World Commission on Environment and Development, Oxford University Press, Oxford.

United States Geological Survey (2009). *Minerals information; cement*. http://www.minerals.usgs.gov/ (accessed 6/4/09).

Washington Asphalt Pavement Association (2008). *Perpetual Pavements*. http://www.asphaltwa.com/wapa_web/modules/06_structural_design/06_perpetual.htm (accessed 26/5/09).

World Business Council for Sustainable Development (2008). *The cement sustainability initiative – Climate actions*, WBCSD, Geneva, November. www.wbcsdcement.org (accessed 4/3/09).

World Steel Association (2008). *Fact sheet: Steel and energy*. http://www.worldsteel.org (accessed 5/4/09).

World Steel Association (2009). *Statistics archive*. http://www.worldsteel.org (accessed 5/4/09).

WRAP (2007). Can you recycle your flat glass waste? http://www.wrap.org.uk/ (accessed 4/2/09).

WRAP (2008). *Collection of flat glass for use in flat glass manufacture – A Good Practice Guide*. http://www.wrap.org.uk/ (accessed 4/2/09).

WRAP-Aggregain (2009). *Sustainable Aggregates*. http://www.aggregain.org.uk (accessed 6/6/09).

Further reading for Part 10 Selection and sustainable use of construction materials

TEXTBOOKS ETC.

Khatib J (ed.) (2009). *Sustainability of construction materials*, Woodhead Publishing Limited, Abingdon UK

> *Contains separate chapters by individual authors on all the major groups of construction materials. Comprehensive but expensive. Read it in your library.*

Uno de Haes HA and Heijungs R (2007). Life-cycle assessment for energy analysis and management. *Applied Energy*, **84** (No. 7–8), 817–827.

> *A good place to start with LCA. Gives a not-too-technical account of the development and background to LCA over the past 30 years, some explanation and comments on current methods and some thoughts on future trends.*

Anderson J, Shiers D and Steele K (2009). *The Green Guide to Specification*, 4th edition, 2009, BRE Press, Watford, UK (on-line version available http://www.thegreenguide.org.uk/).

> *Underpinned by LCA, with much detailed information useful at the design stage; ideal for reference, but not for general reading!*

de Vekey RC (1999). *Clay bricks and clay brick masonry*, BRE Digest 441, Part 2.

de Vekey RC (2001). *Bricks blocks and masonry made from aggregate concrete: Part 2 – Appearance and environmental aspects*, BRE Digest 460, Part 2.

These two parts of these digests have valuable contributions on the sustainability of masonry.

BRITISH/EUROPEAN STANDARDS FOR LIFE CYCLE ASSESSMENT

BS EN ISO 14040:2006 Environmental management. Life cycle assessment. Principles and framework.

> *This specifies the general framework, principles and requirements for conducting and reporting LCA studies. However, it does not describe the life-cycle assessment technique in detail.*

BS EN ISO 14044:2006 Environmental management. Life cycle assessment. Requirements and guidelines.

> *This specifies the requirements and the procedures necessary for the compilation and preparation of a life-cycle assessment (LCA), and for performing, interpreting and reporting a life-cycle inventory analysis and a life-cycle impact assessment.*

These standards do not prescribe a methodology, but are intended to provide information on how to conduct, review, present and use an LCA. The documents cover LCA terminology.

Index